Neutron Stars: Theory and Observation

NATO ASI Series

Advanced Science Institutes Series

A Series presenting the results of activities sponsored by the NATO Science Committee, which aims at the dissemination of advanced scientific and technological knowledge, with a view to strengthening links between scientific communities.

The Series is published by an international board of publishers in conjunction with the NATO Scientific Affairs Division

A Life Sciences **B Physics**	Plenum Publishing Corporation London and New York
C Mathematical and Physical Sciences **D Behavioural and Social Sciences** **E Applied Sciences**	Kluwer Academic Publishers Dordrecht, Boston and London
F Computer and Systems Sciences **G Ecological Sciences** **H Cell Biology** **I Global Environmental Change**	Springer-Verlag Berlin, Heidelberg, New York, London, Paris and Tokyo

NATO-PCO-DATA BASE

The electronic index to the NATO ASI Series provides full bibliographical references (with keywords and/or abstracts) to more than 30000 contributions from international scientists published in all sections of the NATO ASI Series.
Access to the NATO-PCO-DATA BASE is possible in two ways:

– via online FILE 128 (NATO-PCO-DATA BASE) hosted by ESRIN, Via Galileo Galilei, I-00044 Frascati, Italy.

– via CD-ROM "NATO-PCO-DATA BASE" with user-friendly retrieval software in English, French and German (© WTV GmbH and DATAWARE Technologies Inc. 1989).

The CD-ROM can be ordered through any member of the Board of Publishers or through NATO-PCO, Overijse, Belgium.

Series C: Mathematical and Physical Sciences - Vol. 344

Neutron Stars: Theory and Observation

edited by

Joseph Ventura
Department of Physics,
University of Crete
and IESL, Research Center of Crete,
Heraklion, Crete, Greece

and

David Pines
Department of Physics,
University of Illinois,
Urbana-Champaign, IL, U.S.A.

Kluwer Academic Publishers

Dordrecht / Boston / London

Published in cooperation with NATO Scientific Affairs Division

Proceedings of the NATO Advanced Study Institute on
Neutron Stars: An Interdisciplinary Field
Agia Pelagia, Crete, Greece
September 3–14, 1990

Library of Congress Cataloging-in-Publication Data

Neutron stars : theory and observation / edited by Joseph Ventura and David Pines.
p. cm. -- (NATO ASI series. Series C, Mathematical and physical sciences ; vol. 344)
Proceedings of a NATO Advanced Study Institute.
Includes bibliographical references and indexes.
ISBN 0-7923-1397-6 (HB : acid free paper)
1. Neutron stars--Congresses. I. Ventura, Joseph. II. Pines, David, 1924- . III. Series: NATO ASI series. Series C, Mathematical and physical sciences ; no. 344.
QB843.N4N49 1991
523.8'874--dc20 91-25032

ISBN 0–7923–1397–6

Published by Kluwer Academic Publishers,
P.O. Box 17, 3300 AA Dordrecht, The Netherlands.

Kluwer Academic Publishers incorporates the publishing programmes of
D. Reidel, Martinus Nijhoff, Dr W. Junk and MTP Press.

Sold and distributed in the U.S.A. and Canada
by Kluwer Academic Publishers,
101 Philip Drive, Norwell, MA 02061, U.S.A.

In all other countries, sold and distributed
by Kluwer Academic Publishers Group,
P.O. Box 322, 3300 AH Dordrecht, The Netherlands.

Printed on acid-free paper

Printed in the Netherlands

TABLE OF CONTENTS

4. PULSAR AND X-RAY EMISSION

5. BURSTS AND TeV EMISSION

PREFACE

Some twenty-three years after the discovery of pulsars and their identification as rotating neutron stars, neutron star physics may be regarded as coming of age. Pulsars and accreting neutron stars have now been studied at every wavelength, from the initial radio observations, through optical, X-, and γ-ray, up to the very recent observations in the TeV region, while theorists have studied in some detail relevant physical processes both outside and inside neutron stars. As a result, comparison of theory with observation provides a test of our theoretical ideas in fields as diverse as neutron and nuclear matter, superfluidity and superconductivity, the acceleration of high energy particles, and the generation and maintenance of intense magnetic fields. For example, through observations of glitches and post glitch behavior of pulsars, it has become possible to establish the presence of superfluid neutron matter in the inner crust of neutron stars, and to determine some of its properties, while neutron stars in compact binary systems offer one of the most efficient energy generation mechanisms known. It is in fact the interactive interpretation of these diverse pieces of information that can lead to major advances in our understanding of the physics of these exotic objects, and justifies the characterization of neutron stars as hadron physics laboratories.

The NATO Advanced Study Institute, whose proceedings are presented in this volume, was therefore designed to bring together theorists and observers in an effort to confront theory with observation, to determine where either further theoretical work is called for to understand observations, or where future observations might resolve conflicting theoretical proposals. In so doing, we were fortunate to have among the participants a number of the pioneers of neutron star physics whose papers written in the late sixties and early seventies have served as benchmarks in the field.

During the meeting, a number of recent observational findings, including millisecond pulsars, the presence of QPO oscillations in the temporal variation of several low mass X-ray binaries, and their proposed evolutionary link to millisecond pulsars, as well as the recent observations of very high energy gamma-rays from binaries, were reviewed and discussed. The recent successful deployment of *ROSAT* with its demonstrated improved capability for observing faint sources in the soft X-ray region, raises the expectation that some of the outstanding problems in neutron star physics may come to their resolution. Gamma ray bursts have long been thought to originate at the surface of neutron stars, a hypothesis strengthened by recent *GINGA* cyclotron line observations. Theoretical ideas concerning these violent events are extensively discussed in the ensuing pages.

Particular attention was paid to the secular evolution of neutron star magnetic fields; from analyses of recent observations, both speakers and participants concluded that the commonly held view, that solitary pulsar fields decay spontaneously within time-scales of the order of several million years, can no longer be supported. Recent evidence appears, in fact, to link field decay to mass accretion onto the neutron star during its evolution in a binary. Millisecond pulsars would then result after the final detachment or "evaporation" of the neutron star's companion. Recent theoretical work reviewed further indicates that the core magnetic field—which exists in the form of fluxoids imbedded in the neutron superfluid—might decrease as a result of the pulsar spin-down, due to the expulsion of the pinned fluxoids as the density of the rotational vortices decreases with the spin-down.

In addition to recent advances in understanding the nuclear constitution of neutron stars, new work on the core conductivity, and its role in determining the evolution of the magnetic field, was reported. Recent calculations of proton superconductivity, which may also influence the secular timescale of the magnetic field, are likewise reviewed.

We have thus attempted to have a good mix of theory and observation, as well as between the tutorial and current research aspects in the lectures, and thanks to the close cooperation of the lecturers, we feel that we have come close to achieving this goal.

In the first part of these proceedings, our present understanding of the formation, structure, and dynamics of neutron stars is reviewed. The second part is devoted to the formation and evolution of neutron stars, while the third part focuses on observations. In the fourth chapter, the plasma physics of accreting neutron stars and theories of pulsar and X-ray emission are reviewed. The last chapter is devoted to theoretical and observational work on X- and gamma-ray bursts, and TeV emission.

The Institute would not have been possible without the grant of the NATO Advanced Study Institute Program; we thank the Committee for their generous support. We are also indebted to the Research Center of Crete and the Greek Secretariat for Research and Technology for their support. In selecting the lecturers, and in running this Institute, we benefitted greatly from the advice and assistance of our fellow members of the scientific organizing committee, J. P. Lasota, H. Ögelman, C. J. Pethick, and J. Trümper. We are further indebted to Joachim Trümper and the Max Planck Institut für extraterrestrische Physik for helping support the Institute by enabling a sizeable number of its scientists to participate. It is finally a special pleasure to thank Nick Kylafis for his continuing assistance and Lia Papadopoulou for her devoted services.

J. Ventura, Heraklion, Crete
D. Pines, Urbana-Champaign
June 1991

LIST OF PARTICIPANTS

Aysun Akyûz	Cukurova University, Turkey
Ali Alpar	Middle East Technical University, Turkey
E. Asseo	Ecole Polytechnique, France
Zsolt Bagoly	Roland Eotvos University, Hungary
Matthew Bailes	NRAL, UK
Menashe Banit	Columbia University, USA
Altan Baykal	Middle East Technical University, Turkey
Gordon Baym	University of Illinois-Urbana, USA
Werner Becker	MPI für Extraterrestrische Physik, Germany
Tomaso Belloni	MPI für Extraterrestrische Physik, Germany
D. Bhattacharya	University of Amsterdam, The Netherlands
Javier Bonatti	Universität Tubingen, Germany
C. Marcella Carolo	MPE, Germany
Vittorio M. Canuto	NASA, USA
K.S. Cheng	University of Hong Kong, Hong Kong
Jacqueline Cote	University of Amsterdam, The Netherlands
Giancarlo Cusumano	I.F.C.A.I./C.N.R., Italy
Bhaskar Datta	Indian Institute of Astrophysics, India
O.C. de Jager	Potchesfstroom University, South Africa
Stefano del Sordo	I.F.C.A.I./C.N.R., Italy
Roger S. Foster	University of California, USA
Andrew S. Fruchter	Carnegie Institution of Washington, USA
Ulrich Geppert	Institut für Kosmosforchung, Germany
Rolf Geprägs	Universität Tübingen, Germany
Pranab Ghosh	University of Illinois-Urbana, USA
Janusz A. Gil	Pedagogical University, Poland
Fatma Gök	Middle East Technical University, Turkey
Theofanis Grammenos	University of Athens, Greece
Jochen Greiner	Institut für Kosmosforschung, Germany
Isabelle Grenier	CEN Saclay, France
M. Donata Guarnieri	Osservatorio Astronomico-Torino, Italy
Jean-Marie Hameury	Observatoire Meudon, France

Wolfgang Hillebrandt	MPI für Astrophysik, Germany
Naoki Itoh	Sophia University, Japan
Simon Johnston	NRAL, UK
Martin H. Kaiser	Universität Tübingen, Germany
Vassiliki Kalogera	13 Leoforos Stratou, Thessaloniki, Greece
Eckhard Kendziorra	Universität Tübingen, Germany
Alan D. Kerrick	University of California, USA
Andrew King	University of Leicester, UK
Umit Kiziloglü	Middle East Technical University, Turkey
Wlodzimierz Kluzniak	Columbia University, USA
Chryssa Kouveliotou	University of Athens, Greece
Wolfgang Kundt	Universität Bonn, Germany
Mathias Kunz	Universität Tübingen, Germany
Nick Kylafis	University of Crete, Greece
Don Q. Lamb	University of Chicago, USA
Frederick K. Lamb	University of Illinois-Urbana, USA
Jean-Pierre Lasota	Observatoire Meudon, France
Bennett Link	Los Alamos National Lab., USA
Andrew G. Lyne	University of Manchester, UK
Apostolos Mastichiadis	MPI für Kernphysik, Germany
Fotis Mavromatakis	University of Crete, Greece
Peter Mészáros	Pennsylvania State University, USA
Bruno Mony	Universität Tübingen, Germany
H.I. Nel	Potchefstroom University, South Africa
Hans-Peter Nollert	Universität Tübingen, Germany
Ken'ichi Nomoto	University of Tokyo, Japan
Laura Norci	CNR-Istituto di Astrofisica Spaziale, Italy
Hakki Ögelman	MPI für Extraterrestrische Physik, Germany
Marina Orio	Osservatorio Astronomico Torino, Italy
Dany P. Page	Columbia University, USA
Hara Papathanasiou	Evzonon 42, Greece
Costas Pavlakis	University of Crete, Greece
George Pavlov	Ioffe Institute, USSR

C.J. Pethick	NORDITA, Denmark
David Pines	University of Illinois-Urbana, USA
Phillipp Podsiadlowski	University of Cambridge, UK
O.R. Pols	University of Amsterdam, The Netherlands
Denis Priou	Observatoire Meudon, France
Demetris Psaltis	13 Leoforos Stratou, Thessaloniki, Greece
Joanna Rankin	University of Vermont, USA
Andreas Reisenegger	CALTECH, USA
Harald Riffert	Universität Tübingen, Germany
Malvin Ruderman	Columbia University, USA
Christophe Sauty	FORTH, Greece
Jannis Seiradakis	University of Thessaloniki, Greece
Jacob Shaham	Columbia University, USA
Noriaki Shibazaki	Rikkyo University, Japan
Patrick Slane	Harvard-Smithsonian Center for Astrophysics, USA
Ian A. Smith	Washington University, USA
Nikos Spyrou	University of Thessaloniki, Greece
Rudiger Staubert	Universität Tübingen, Germany
Marco Tavani	University of California, USA
Joachim Trümper	MPI für Extraterrestrische Physik, Germany
Jochen Wambach	Institut für Kernphysik, Germany
Hans-Joachim Wiebicke	Institut für Kosmosforchung, Germany
Ralf A.M.J. Wijers	University of Amsterdam, The Netherlands
Xuejun Wu	Universität Tübingen, Germany
E.P.J. van den Heuvel	University of Amsterdam, The Netherlands
M. van der Klis	University of Amsterdam, The Netherlands
M.H. van Kerkwijk	University of Amsterdam, The Netherlands
Jan van Paradijs	University of Amsterdam, The Netherlands
Joseph Ventura	University of Crete, Greece
Kyriaki Xylouri	University of Crete, Greece
Sophia Yancopoulos	Columbia University, USA
D.G. Yakovlev	Ioffe Institute, USSR
G.A.A. Zwarthoed	University of Amsterdam, The Netherlands

1. INSIDE NEUTRON STARS

AN INTRODUCTION TO MATTER AT SUBNUCLEAR DENSITIES

C. J. PETHICK and D. G. RAVENHALL
Nordita, Blegdamsvej 17, DK-2100 Copenhagen Ø, Denmark, and
Department of Physics, University of Illinois, 1110 West Green Street, Urbana, Illinois 61801, U.S.A.

ABSTRACT. An overview is given of important aspects of the physics of dense matter below nuclear matter density. Among topics discussed are electronic properties, the surface tension of neutron-rich and hot nuclei, the reduction of the Coulomb energy as a consequence of the close packing of nuclei, and the possibility of non-spherical nuclei. For the most part, the discussion of nuclear properties is based on generalizations of the liquid-drop model, but results of calculations based on more microscopic approaches are also given.

1. Introduction

In a neutron star one encounters matter at densities ranging from laboratory densities or below close to the stellar surface, to ones well above nuclear matter density at the center. While the bulk of matter in neutron stars is at densities well above that of nuclear matter, properties of the regions of the star at lower densities are important for a number of reasons. First of all, heat transported from the stellar core by conduction and radiative transport must make its way through the surface regions of the star before being radiated from the surface, and therefore the surface radiation is sensitive to the properties of relatively low-density matter. Second, the outer parts of the star are likely to be solid over a large range of densities, and they play a large part in models to account for the rotational and magnetic evolution of neutron stars (see, e.g., Ruderman, 1991). Third, the physics of the surface layers is important in accreting neutron stars, because matter arriving at the surface is compressed to higher and higher densities as further matter arrives. In this article, we give an introduction to a number of important themes in the physics of matter at densities sufficiently low that nuclei have not merged to form a uniform nuclear liquid.

Temperatures of neutron stars are of order 10 MeV ($\sim 10^{11} K$) immediately after formation, but fall rapidly thereafter. For most of the life of a neutron star, it is a good approximation in all but the outermost centimetres to assume that matter is cold. We therefore consider first the properties of cold matter. In section 2

J. Ventura and D. Pines (eds.), Neutron Stars: Theory and Observation, 3–20.

we consider electronic properties, and in section 3 we consider matter at higher densities, at which the electron fraction of matter is reduced. In section 4 we discuss neutron drip, the density at which all bound neutron states in nuclei are occupied, and neutrons begin to populate continuum states. Section 5 is devoted to a discussion of properties of matter above the neutron-drip density, and section 6 considers the properties of hot matter, and matter out of beta equilibrium.

2. Electronic Properties

At the lowest densities the most stable nucleus is ^{56}Fe, and matter in its ground state consists of iron nuclei immersed in an electron liquid. At terrestrial densities, electrons are a strongly interacting liquid, and the most difficult task in calculating the properties of cold matter is the evaluation of electronic energies. However, effects of electron interaction become less important with increasing density, as one can see from the following argument.

2.1 ELECTRON-ELECTRON INTERACTION

We suppose that matter is composed of nuclei with charge Z and of density n_N. It is convenient to introduce a length r_c, which is defined by the condition that the volume per nucleus is equal to the volume of a sphere of radius r_c. We therefore have

$$n_N = \frac{1}{\frac{4}{3}\pi r_c^3}. \tag{1}$$

In order of magnitude, r_c is the inter-nuclear spacing. Likewise we define a corresponding length, r_e, for the electrons which is given in terms of their density by

$$n_e = \frac{1}{\frac{4}{3}\pi r_e^3}. \tag{2}$$

Because bulk matter must be electrically neutral, we have

$$n_e = Z n_N, \tag{3}$$

or, from Eqs. (1) and (2)

$$r_c = Z^{1/3} r_e. \tag{4}$$

To estimate how important Coulomb interactions are for the electrons, we observe that the typical distance of an electron from an ion is of order r_c, and therefore the Coulomb energy per electron is of order $Ze^2/r_c \sim Z^{2/3}e^2/r_e$. On the other hand, the kinetic energy of an electron is $\sim p_e^2/2m_e$, where p_e is the electron Fermi momentum and m_e is the electron mass. The Fermi momentum is of order $\hbar/r_e$, and therefore the electron kinetic energy is $\sim \hbar^2/(2m_e r_e^2)$. Electrons will be essentially unaffected by interaction with ions, and will behave as free electrons, provided

$$\frac{\hbar^2}{2m_e r_e^2} >> Z^{2/3}\frac{e^2}{r_e} \tag{5}$$

or

$$r_e << \frac{a_o}{Z^{2/3}}, \tag{6}$$

where $a_o = \hbar^2/m_e e^2$ is the Bohr radius. The corresponding density of matter will be

$$\rho = (Am)n_N = (Am)\frac{n_e}{Z} = \frac{Am}{Z}\frac{1}{\frac{4}{3}\pi r_e^3} >> \frac{m}{a_o^3}\ AZ \ \sim\ 10\ AZ\ g\ cm^{-3} \tag{7}$$

where m is the nucleon mass. For hydrogen the density above which electrons are free is a few g cm^{-3}, and for iron it is $\sim 10^4 g\ cm^{-3}$. In the discussion above we have neglected the electron-electron interaction. This is justified because it gives an energy of order e^2/r_e per electron, which is less than the electron-ion interaction by a factor $Z^{2/3}$.

Another way to arrive at the criterion (7) is to observe that in Thomas-Fermi theory the radius of an atom is

$$r_{atom} \sim \frac{a_o}{Z^{1/3}}. \tag{8}$$

When the spacing between nuclei is much less than the atomic radius, electrons will not be localized in the vicinity of one nucleus, but will move freely between nuclei. This will occur when

$$r_c << r_{atom} \tag{9}$$

which, with the help of Eqs. (1) and (8), is easily seen to be equivalent to condition (6). When the electrons are essentially free, the number density and pressure are those of a free Fermi gas:

$$n_e = \frac{p_e^3}{3\pi^2\hbar^3}, \tag{10}$$

$$p = \frac{(3\pi^2)^{2/3}\hbar^2}{5m_e} n_e^{5/3} = \frac{(3\pi^2)^{2/3}\hbar^2}{5m_e\ m^{5/3}} \left(\frac{Z}{A}\right)^{5/3} \rho^{5/3}$$

$$= 9.95 \times 10^{12}\ \rho^{5/3} \left(\frac{Z}{A}\right)^{5/3} dyne\ cm^{-2}, \tag{11}$$

where in the last formula ρ is in $g\ cm^{-3}$. Recall that $10^{13}\ dyne\ cm^{-2} = 10^7\ bar$, and therefore one can easily understand why for typical terrestrial solids megabar pressures are required to produce significant changes in volume.

2.2 RELATIVISTIC ELECTRONS

As the spacing between electrons becomes smaller, their velocities increase and they become relativistic: $p_e >> m_e c$, or

$$r_e << \frac{\hbar}{m_e c} = \lambda_C \tag{12}$$

where λ_C is the Compton wavelength of the electron divided by 2π. From the expression (12) for the density we deduce that electrons are relativistic if

$$n_e >> \frac{1}{3\pi^2}\frac{1}{\lambda_C^3} \sim \frac{1}{\alpha^3}\frac{1}{3\pi^2}\frac{1}{a_o^3} \tag{13}$$

where $\alpha = e^2/\hbar c \approx 1/137$ is the fine structure constant. The corresponding condition on the density is given by

$$\rho >> \frac{A}{Z}\frac{1}{3\pi^2}\frac{m}{a_o^3}\frac{1}{\alpha^3} \sim 10^6\ g\ cm^{-3}. \tag{14}$$

(The density $m/a_o^3 \approx 11.3\ g\ cm^{-3}$ is typical of terrestrial matter, which contains of order one electron and one nucleon per volume a_0^3.) The pressure at such densities is given by

$$\begin{aligned} p &= \frac{(3\pi^2)^{1/3}}{4}\frac{\hbar c}{m^{4/3}}\left(\frac{Z}{A}\right)^{4/3}\rho^{4/3} \\ &= 1.2 \times 10^{15}\left(\frac{Z}{A}\right)^{4/3}\rho^{4/3}\ dyne\ cm^{-2}, \end{aligned} \tag{15}$$

where in the second formula ρ is in $g\ cm^{-3}$. At a density of order $10^6\ g\ cm^{-3}$ the pressure is $\sim 10^{17} bar$.

3. Equilibrium Nuclei at Higher Densities.

When electrons are relativistic, their energies exceed $m_e c^2$ ($\sim 0.511 MeV$), which is significant on the scale of energies important in nuclear physics. When the electron energy becomes high enough, it becomes favourable to capture electrons on protons:

$$e^- + p \rightarrow n + \nu_e.$$

Thus the electron fraction of matter decreases with increasing density, and this in turn affects the equilibrium nuclei to be expected at any density.

3.1 THE SEMI-EMPIRICAL MASS FORMULA

A simple calculation of the most favourable nucleus may be made using the liquid drop model of the nucleus. According to the simplest version of this model, the energy per nucleus is

$$\begin{aligned} E(A,Z) \ = \ (A-Z)m_n c^2 \ + \ Z m_p c^2 \ - \ E_b A \ + \ E_s(1-2x)^2 A \\ + \ E_\sigma A^{2/3} \ + \ E_C x^2 A^{5/3}, \end{aligned} \tag{16}$$

where m_n is the neutron rest mass, m_p is the proton rest mass, $x = Z/A$ is the proton fraction, E_b is the bulk binding energy per nucleon, E_s is the symmetry energy coefficient, E_σ is the surface energy coefficient, and E_C is the Coulomb energy coefficient. Consider matter with n_N nuclei per unit volume, and $n_e = Z n_N$ electrons per unit volume. The total energy per unit volume is

$$E = E_e(n_e = Z n_N) + n_N E(A,Z), \tag{17}$$

where E_e is the electronic energy per unit volume as a function of electron density. To determine the equilibrium state, one must minimize the energy subject to the condition that the total nucleon density $n = An_N$ be fixed. Neutrinos do not have to be taken into account, since they are assumed to be able to escape freely from the matter, and, since we neglect their mass, the number of leptons per unit volume may be changed at no cost in neutrino energy.

It is convenient to minimize the energy first with respect to A, and then with respect to x. The first condition is

$$\left(\frac{\partial E}{\partial A}\right)_{x,n} = \frac{n_N}{A}\left(\frac{2}{3}E_C x^2 A^{5/3} - \frac{1}{3}E_\sigma A^{2/3}\right) = 0, \tag{18}$$

or

$$E_\sigma A^{2/3} = 2\ E_C x^2 A^{5/3}, \tag{19}$$

which states that the surface energy per nucleus is twice the Coulomb energy. Because the variation we have considered preserves the total number of protons and the total number of neutrons, the rest mass, bulk, and symmetry terms do not enter the equilibrium condition, which is determined solely by finite size effects. Equation (19) may be written

$$A = \frac{E_\sigma}{2E_C x^2}, \tag{20}$$

which shows that the equilibrium nuclear size increases as the proton fraction decreases. We remark in passing that the equilibrium nuclei are very stable against spontaneous fission. A liquid drop is unstable to a quadrupolar distortion if the surface energy is less than *one half* the Coulomb energy (Bohr and Wheeler, 1939), while for the equilibrium nuclei it is *twice* the Coulomb energy.

To find the equilibrium proton fraction, we substitute Eq. (20) for A into Eqs. (16) and (17), and differentiate with respect to x. The sum of Coulomb and surface energies may be written as $(3A/2)(2E_\sigma^2 E_C x^2)^{1/3}$ and the equilibrium condition is

$$\mu_e + m_p c^2 - m_n c^2 - 4E_s(1-2x) + (2E_C E_\sigma^2)^{1/3} x^{-1/3} = 0, \tag{21}$$

where $\mu_e = \partial E_e/\partial n_e$ is the electron chemical potential, or Fermi energy. Equation (21) is an expression of the thermodynamic equilibrium condition

$$\mu_e + \mu_p = \mu_n, \tag{22}$$

where μ_p and μ_n are the proton and neutron chemical potentials, including rest mass terms. This states that in equilibrium it costs no energy to convert an electron and a proton into a neutron, i.e. electron capture leads to no energy reduction.

Equation (21) relates the electron chemical potential to the proton fraction. If we assume the electrons to be an ideal extremely relativistic Fermi gas, the chemical potential is $p_e c$, and the density, Eq.(10), is therefore given by

$$n_e = \frac{1}{3\pi^2}\left(\frac{\mu_e}{\hbar c}\right)^3. \tag{23}$$

Thus Eq. (21) gives x as a function of n_e, or alternatively of the mass density ρ, and one can then determine the equilibrium nuclear size from Eq. (20). With rising density, μ_e increases and therefore, from Eq. (21), one can see that the proton fraction x decreases.

3.2 SHELL EFFECTS

A more realistic determination of equilibrium nuclei must incorporate shell effects. In early work this was done by using a liquid drop model which contained such effects, whose magnitude was determined by fits to laboratory nuclei. Shell effects have little influence on the proton fraction as a function of density, since that is most sensitive to bulk nuclear properties, but they do affect the nuclear size, which is independent of bulk properties of nuclear matter and depends on finite-size effects. The early calculations indicated that neutron shell effects dominated the energetics, and many of the equilibrium nuclei had closed shells with 82 neutrons.

Recently masses of neutron-rich nuclei have been estimated theoretically using the Hartree-Fock-Bogoliubov method (Haensel, Dobaczewski and Zdunik, 1989). The parameters of the model for the nucleon-nucleon interaction are chosen to fit properties of laboratory nuclei, but no assumption is made about the dependence of shell effects on neutron richness - rather this is an output from the calculation. In these calculations it is found that nuclei with proton, rather than neutron, closed shells dominate, with isotopes of Ni ($Z = 28$) being favoured for densities below $3.3 \times 10^{11}\ g\ cm^{-3}$ (a little less than the density for neutron drip) and isotopes of Zr ($Z = 40$) for densities above. These results match nicely those of Negele and Vautherin (1973), for densities above neutron drip. They found isotopes of Zr to be the equilibrium nuclides for densities below $\sim 3 \times 10^{12} g\ cm^{-3}$.

The reason for neutron shell effects being less pronounced is presumably that as nuclei become more neutron rich, the most energetic bound neutron levels occupied approach the continuum of extended states. The most energetic proton states, on the other hand lie deep below the proton continuum.

The other interesting feature of the above results is that the prominent closed shell has $Z = 40$, rather than $Z = 50$ as in terrestrial nuclei. This must reflect the fact that the spin-orbit interaction is weaker in neutron-rich nuclei, since 40 nucleons corresponds to filling the $1f, 2p$ shell without the 10 $1g_{9/2}$ levels that contribute to the 50 nucleon closed shell in terrestrial nuclei.

4. Neutron Drip

As the density increases, nuclei become increasingly neutron rich until eventually the last neutron is no longer bound to a nucleus. The physical condition for this is that the chemical potential (the energy to add a neutron) be equal to its rest mass:

$$\mu_n = m_n c^2. \tag{24}$$

This is referred to picturesquely as the point of "neutron drip". The neutron chemical potential is found by differentiating the energy expression (17) with respect to the total neutron number density $n_n = n_N N$, where $N = A - Z$ is the neutron number of the nucleus. Only the nuclear energy is of importance, and one finds

$$\mu_n = \left(\frac{\partial E(A,Z)}{\partial N}\right)_Z = m_n c^2 - E_b + E_s(1 - 4x^2) + \frac{1}{2}(2E_C E_\sigma^2)^{1/3} x^{2/3}, \quad (25)$$

and therefore the condition (24) for neutron drip is

$$E_b = E_s(1 - 4x_{drip}^2) + \frac{1}{2}(2E_C E_\sigma^2)^{1/3} x_{drip}^{2/3}. \quad (26)$$

The surface and Coulomb terms are less important than the others, and if one neglects them one finds

$$x_{drip}^2 \approx \frac{1}{4}(1 - \frac{E_b}{E_s}) \quad (27)$$

which for standard liquid drop model parameters, $E_b \approx 16MeV$ and $E_s \approx 24MeV$, gives $x_{drip} \approx 0.29$. If the surface and Coloumb terms are retained, with $E_C \approx 0.72MeV$ and $E_\sigma \approx 18MeV$, one finds $x_{drip} \approx 0.32$, and the corresponding equilibrium nucleus has $Z \approx 39$ and $A \approx 122$. The electron chemical potential is about $25MeV$ and the density is $\approx 3.5 \times 10^{11} g\ cm^{-3}$, of order one thousandth of nuclear matter density.

5. Matter Above Neutron Drip Density

At densities above that for neutron drip, nuclei will be immersed in a sea of neutrons. To describe nuclei under such conditions a number of refinements must be incorporated in the liquid drop model. We now discuss some of these.

As nuclei become more neutron rich, the surface tension decreases. Above neutron drip density, the density of neutrons increases with increasing density, and the proton fraction of matter in nuclei decreases. As the matter inside nuclei becomes increasingly similar to the neutron sea outside, the surface tension decreases, and falls to values far below those for laboratory nuclei. We shall discuss surface effects in more detail in Sec.5.2.

A further effect of increasing neutron richness is that the saturation density of nuclear matter decreases. For sufficiently small proton fractions, it is in fact impossible for nuclear matter to be in equilibrium at zero pressure. However, in dense matter nuclei can still exist because the pressure of the neutrons outside prevents a nucleus from expanding.

To describe matter above neutron drip density it is necessary to generalize the model we used to describe properties at lower densities. i) The surface (or interface) energy must be modified to take into account the fact that nuclei are neutron rich, and that there are neutrons outside nuclei. ii) The fact that matter inside nuclei can be very neutron rich means that it is inadequate to calculate the energy of a nucleus to second order in the neutron excess parameter, as one does in the simplest semiempirical mass formula (16). iii) The energy of neutrons outside nuclei must be included. iv) When the neutrons outside nuclei become dense, they exert a significant pressure on nuclei, as we mentioned above. v) At neutron drip density and above, the Coulomb energy among nuclei is comparable with the Coulomb energy of the isolated drop model.

All these features are taken into account in the compressible liquid drop model (Baym, Bethe and Pethick, 1971) which is a development of work on nuclear mass formulae by Myers and Swiatecki (1966, 1969). (For more recent work on mass formulae, see Myers (1977).) In the model, one imagines nuclei to be spheres of radius r_N containing nuclear matter with some density n_i, and proton concentration, x_i. These are surrounded by a neutron gas of density n_n. The total energy is made up of the bulk energy of the nuclear matter inside and the neutrons outside, a surface energy, a Coulomb energy, and the energy of the electrons. The energy density of the neutrons outside nuclei and the nuclear matter inside are described by the same function of the density and proton concentration of the matter. This automatically ensures that the two sorts of matter, which are quite similar when the nuclei are very neutron rich, are treated in a consistent fashion. We shall describe the Coulomb and surface energies in more detail below. To determine the equilibrium properties, one minimizes the energy per unit volume, keeping the total density of nucleons fixed. If one fixes the proton density, this gives three conditions. First, the neutron chemical potentials of matter inside and outside nuclei must be the same, which ensures that no reduction in energy is possible by transferring neutrons from the nuclei to the neutron gas outside, or vice versa. Second, the interface between the nuclei and the neutron gas must be in pressure equilibrium. Third, the size of the nuclei must be such as to maximize the binding energy per nucleon. In the calculations with the semiempirical mass formula, only the third condition came in; the density of nuclear matter in nuclei was fixed, and consequently there was no pressure condition, and there were no neutrons outside nuclei, and therefore there was no condition on the neutron chemical potential. The three conditions above give the equilibrium conditions if the proton fraction is held fixed. However, in the case of neutron stars, we wish to find the proton fraction that leads to the lowest energy. This corresponds to allowing β-decay processes to take place. The minimization of the energy with respect to proton fraction gives the chemical equilibrium condition (22), just as it did in the simpler case, but for the more general problem considered here, the neutron and proton chemical potentials are given by different expressions. When the density of neutrons outside nuclei is comparable to that of nuclear matter, nuclei merge to form a uniform liquid of neutrons, protons and electrons.

5.1 COULOMB ENERGY

In estimating Coulomb energies we have up to now assumed this to be the Coulomb energy of isolated nuclei, of order $\left(\frac{3}{5}\right)Z^2e^2/r_N$, where r_N is the nuclear radius. However, at the density for neutron drip and above, the Coulomb interaction *between* nuclei can be important. Even though the nuclei fill only about one thousandth of space at neutron drip, the typical nuclear separation, r_c, is only of order 10 times r_N, and consequently the internuclear Coulomb energy $\sim Z^2e^2/r_c$, is $\sim 10\%$ of the Coulomb energy of an isolated nucleus. To make a quantitative estimate of the total Coulomb energy we shall use the Wigner-Seitz approximation first used in band-structure calculations. One regards matter as being made up of spherical cells centered on the individual nuclei. The radius of the cells is taken to be r_c, Eq. (1), which ensures that they are electrically neutral. We assume the Coulomb energy to be the sum of the Coulomb energies of individual cells, and we shall neglect overlap of cells. The charge density in a cell has nuclear and electron contributions, and therefore we may write the Coulomb energy as

$$E_{coul} = E_{nuc-nuc} + E_{e-e} + E_{e-nuc}. \tag{28}$$

We assume the nucleus to be a uniformly charged sphere, and the electrons to be uniformly distributed, and therefore

$$E_{coul} = \frac{3}{5}\frac{Z^2e^2}{r_N} + \frac{3}{5}\frac{Z^2e^2}{r_c} + \int V_e(r)\tilde{\rho}_{nuc}(r)d^3r, \tag{29}$$

where $\tilde{\rho}_{nuc}$ is the charge density of the nucleus and $V_e(r)$ is the electrostatic potential due to the electrons, which is given by

$$V_e(r) = -\frac{3}{2}\frac{Ze}{r_c} + \frac{1}{2}\frac{Ze}{r_c}\left(\frac{r}{r_c}\right)^2, \quad r < r_c, \tag{30}$$

as one can show by solving Poisson's equation. Thus we find

$$\begin{aligned}\int V_e(r)\tilde{\rho}_{nuc}(r)d^3r &= -\frac{3}{2}\frac{Z^2e^2}{r_c} + \frac{1}{2}\frac{Z^2e}{r_c^3}\langle r^2\rangle_{nuc} \\ &= -\frac{3}{2}\frac{Z^2e^2}{r_c} + \frac{3}{10}\frac{Z^2e^2}{r_c}\left(\frac{r_N}{r_c}\right)^2\end{aligned} \tag{31}$$

where $\langle r^2\rangle_{nuc}$ is the r.m.s. nuclear charge radius, equal to $\frac{3}{5}r_N^2$ for a uniform sphere. Hence the total Coulomb energy is

$$E_{Coul} = \frac{3}{5}\frac{Z^2e^2}{r_N} - \frac{9}{10}\frac{Z^2e^2}{r_c} + \frac{3}{10}\frac{Z^2e^2r_N^2}{r_c^3}. \tag{32}$$

The first term is the Coulomb energy of an isolated nucleus, and the second is the so-called *lattice energy*. The name is not altogether a felicitous one, since the magnitude of the Coulomb energy does not depend crucially on whether or not there is a lattice present. The "lattice energy" is close to the value in Eq. (32) even if the ions form a liquid, provided the thermal energy is small compared with the lattice energy. The Wigner-Seitz result for the lattice energy is remarkably close to the results obtained from detailed calculations for real lattices, the differences being less than 1% for the body-centered, face-centered and simple cubic lattices. The third term in Eq. (32) represents the contribution from the finite size of the nucleus. This is the term responsible for the isotope shift in atomic spectra, but it is much larger in dense matter because of the high electron density. The Coulomb energy (32) may be rewritten as

$$E_{Coul} = \frac{3}{5}\frac{Z^2e^2}{r_N}\left(1 - \frac{3}{2}\frac{r_N}{r_c} + \frac{1}{2}\left(\frac{r_N}{r_c}\right)^3\right), \tag{33}$$

which clearly exhibits the fact that as nuclei fill an increasing fraction of space, the Coulomb energy is reduced, and it eventually vanishes for $r_N \to r_c$, at which point the net charge density vanishes since the nuclear and electronic charge densities cancel exactly.

5.2 SURFACE TENSION OF NEUTRON-RICH MATTER

The energy of an interface between nuclear matter and pure neutrons is reduced as the nuclear matter becomes increasingly rich. This is physically plausible because the phases of matter on the two sides of the interface become increasingly similar as the neutron density of the pure neutron phase increases and the proton concentration of the nuclear matter decreases. An important point is that for a nuclear matter phase which does not contain equal densities of protons and neutrons, variations of the neutron and proton densities in the interface do not follow each other: in the case of nuclear matter with a neutron excess there is a neutron skin. This skin must be allowed for in calculating energies of the interface, and, as discussed by e.g. Landau and Lifshitz (1980), it is most natural to work in terms of the surface tension, which is the contribution per unit area of the surface to the thermodynamic potential, Ω, rather than to the energy. This choice simplifies evaluation of equilibrium conditions. We use the same symbol as above, because even for nuclei with no neutrons outside, there is a neutron skin and if this is allowed for, the equilibrium nuclear size, for example, is determined from the results given above, with σ defined in the same way as for the case when there are neutrons outside nuclei. Detailed calculations of interfaces have been made in the Hartree-Fock approximation with a nucleon-nucleon interaction of the Skyrme type (Ravenhall, Bennett and Pethick, 1972). As the proton concentration of nuclear matter decreases, the surface tension decreases, and for small values of x, it lies well below the value for symmetric nuclear matter.

5.3 THE SIZE OF NUCLEI ABOVE NEUTRON DRIP

We now use the estimate of the Coulomb energy to determine the size of nuclei above neutron drip. The minimization of the total energy with respect to nuclear size is carried out in a manner quite analogous to that at lower density, and one finds that the optimal nuclear radius is determined by the competition between Coulomb and surface effects. Mathematically, the condition is

$$4\pi r_N^2\sigma(x_i) = 2\,\frac{3}{5}\frac{Z^2e^2}{r_N}\left(1 - \frac{3}{2}u^{1/3} + \frac{u}{2}\right) \tag{34}$$

where $Z = \frac{4}{3}\pi r_N^3 n_i x_i$ is the atomic number of the nucleus, and $u = (r_N/r_c)^3$ is the fraction of space filled by nuclei. Equation (34) is a generalization of Eq. (19), which states that the surface energy is twice the Coulomb energy. Note that the total Coulomb energy, including the lattice contribution, appears here. As the density increases, the coefficient of the Coulomb energy decreases because the lattice contribution becomes more important. This tends to decrease the nuclear size. On the other hand, nuclei become increasingly neutron rich as the density increases, and this leads to a smaller surface tension, and therefore to a larger nuclear size. Detailed calculations using this model show that above neutron drip density the atomic number, Z, of nuclei remains rather constant and close to its value at neutron drip ~ 40 (Ravenhall, Bennett and Pethick, 1972).

As the density of matter increases, nuclei fill an increasing fraction of space, and eventually, when the density is roughly nuclear matter density, they merge to form a uniform liquid. At densities just below that, nuclear matter may not be in the form of spherical nuclei, but may have other shapes. The range of densities over which this occurs for cold matter in its ground state is small, and may be vanishingly small, and we shall defer discussion of these effects until Sec.6, since the effects are likely to be more important for matter in stellar collapse, which has a higher electron fraction.

5.4 MICROSCOPIC CALCULATIONS OF MATTER ABOVE NEUTRON DRIP

In the semi-empirical mass formula described above, shell effects are not included. Negele and Vautherin (1973) carried out Hartree-Fock calculations, which implicitly include shell effects, and found that the equilibrium nuclei were isotopes of Zr ($Z = 40$) for $\rho \leq 4 \times 10^{12} g\ cm^{-3}$. For densities above this value, but $\leq 4 \times 10^{13} g\ cm^{-3}$, they found that the equilibrium nuclei were isotopes of Sn ($Z = 50$). At the next higher density point, with $\rho \approx 8 \times 10^{13} g\ cm^{-3}$, they found $Z = 40$, while at the highest density point calculated, with ($\rho \approx 1.3 \times 10^{14} g\ cm^{-3}$) just below the density at which nuclei merge to form a uniform liquid, they found Ni ($Z = 28$). These results are dominated by shell effects. The Z=50 and Z=28 shell closings are well known for terrestrial nuclei, while Z=40 is not. However, as described in Sec.3.2, calculations of equilibrium below neutron drip indicate that the $Z = 40$ shell closing is also important there. The microscopic reasons for the importance of the $Z = 40$ closed shell for nuclei remain to be elucidated. It is important to note that shell effects do not affect the general scale of nuclear sizes, which are well-described by a generalized liquid drop model.

5.5 MATTER AT NUCLEAR DENSITIES AND ABOVE

In order to calculate the properties of matter at nuclear matter density and below, the necessary input physics can be obtained from low energy nuclear physics experiments, and the properties of matter can be predicted with a high degree of confidence. On the other hand, it is much more difficult to calculate properties of matter at higher densities, because of uncertainties in the input physics, and the difficulty of making reliable calculations for strongly interacting systems. However, the properties of matter at higher densities are of crucial importance for the structure of neutron stars, because the bulk of the matter in neutron stars is at such densities. Properties of such matter are discussed in detail in G. Baym's article (Baym, 1991) in this volume.

6. Matter out of beta equilibrium and at finite temperatures

For application to stellar collapse, the calculations above must be generalized in a number of respects. First, effects of finite temperature must be allowed for. A second effect is that in stellar collapse, neutrino mean free paths can be small compared with the radius of the collapsing star, and typical times for a neutrino to diffuse out of the stellar core exceed the collapse time for matter densities in excess of about 10^{12} g cm^{-3}. As a consequence, neutrinos produced by electron capture on protons are trapped inside the star, and the total lepton number in the star remains fixed for the duration of the collapse. In equilibrium at fixed lepton number, it must cost no energy to convert an electron or a proton into a neutron or a neutrino, or

$$\mu_e + \mu_p = \mu_n + \mu_{\nu e}, \tag{35}$$

which differs from (22) only by the presence of the neutrino chemical potential, $\mu_{\nu e}$. In some circumstances the characteristic times for weak interaction processes may be long compared with the time scales of interest, and in that case it is of interest to calculate the properties of matter for a given proton fraction, whose magnitude is determined by other considerations. In stellar collapse, trapping of neutrinos prevents the proton fraction falling below about 0.3, whereas at nuclear densities the proton fraction of matter in its ground state is of order 0.05.

6.1 NUCLEAR SIZES

The first question we wish to address is, when does one expect nuclei to be present in hot dense matter. To do this, we generalize our earlier discussion to finite temperatures. Instead of working with energy as a function of the baryon density, for a system at a fixed temperature T one must work with the Helmholtz free energy. The free energy may then be expressed in terms of bulk, surface, Coulomb and other contributions, just as the energy was at zero temperature. The simplest approximation is to neglect all contributions to the free energy except bulk ones. The problem of determining when nuclei are present then reduces to finding the conditions under which hot nuclear matter can be in equilibrium with a nucleon vapour. At a sufficiently low (but finite) temperature, matter forms a vapour at low densities. As the density is increased, a point will be reached at which drops of nuclear matter begin to condense. With further increase in density, the fraction of matter in the vapour decreases, and eventually nuclear matter fills all of space. Condensation will not occur if the temperature is higher than some temperature which is usually close to the critical temperature. The conditions for coexistence of two bulk phases is that the neutron chemical potentials in the two phases be equal, that the proton chemical potentials be equal, and that the pressures be equal. A review of such calculations is given by Lattimer (1981), and in Fig. 1, we show, as the line marked "bulk equilibrium", the boundary of the two-phase region calculated for the Skyrme 1′ interaction by Lattimer and Ravenhall (1978). Notice that at densities approaching nuclear density, the two-phase region extends to temperatures of order 20 MeV. In nuclear physics language, the persistence of nuclei to such high temperatures is due to the importance of nuclear excited states, which increase the statistical weight of the droplets of nuclear matter (see, e.g., Mazurek, Brown and Lattimer, 1978; Bethe, Brown, Applegate and Lattimer, 1979). In the bulk equilibrium calculation, such effects are automatically included

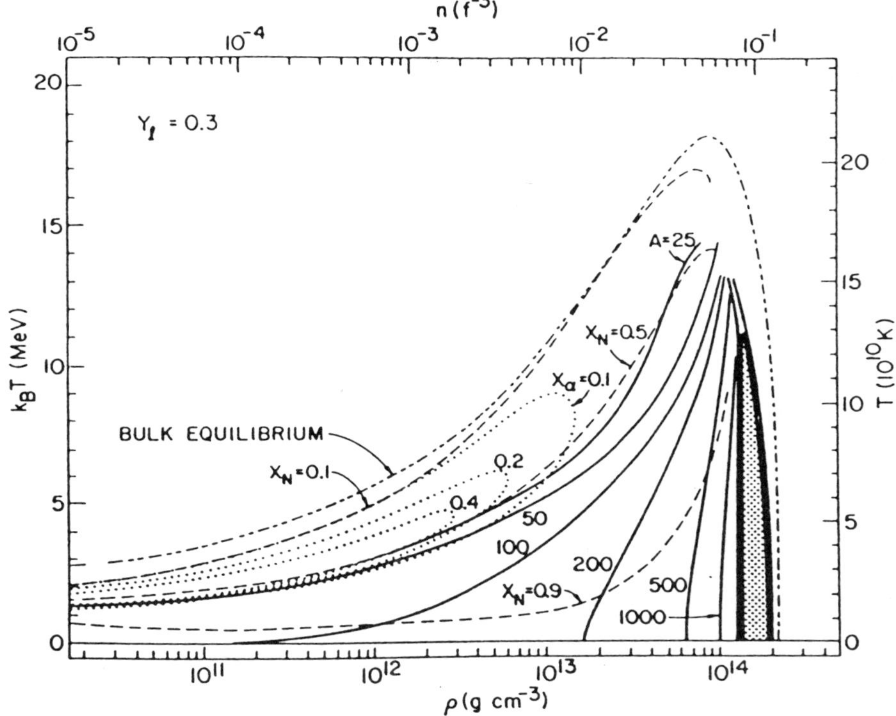

Fig. 1: Properties of hot dense matter at subnuclear densities for a lepton fraction of 0.3. The line labelled "Bulk equilibrium" denotes the boundary of the two-phase region when surface and Coulomb effects are ignored. Finite nucleus effects were calculated using the methods of Lamb, Lattimer, Pethick and Ravenhall (1978) and Lattimer, Pethick, Ravenhall and Lamb (1985). Contours of constant nuclear mass number A are shown, as well as the mass fractions of nucleons in nuclei. X_α is the fraction of baryons in the form of alpha particles, and X_N the fraction in heavy nuclei (including alpha particles). Bubbles are present in the dotted region, and mixtures (nuclei-bubbles, bubbles-uniform matter, or nuclei-uniform matter) in the surrounding black region.

through the finite-temperature contributions to the nucleon chemical potentials of the denser phase.

The bulk equilibrium calculations are a good guide to the conditions under which nuclei are likely to be present, but do not give any indication of the nuclear size, which, as in the zero temperature case, is determined by finite nucleus effects. There are quite a number of these, as is discussed in detail by Lamb, Lattimer, Pethick and Ravenhall (1978) and Lattimer, Pethick, Ravenhall and Lamb (1985), but at low temperatures, the most important finite nuclear size contributions to the free energy are those from the Coulomb interaction and nuclear surface. As long as the nuclei are highly correlated, that is, the Coulomb lattice energy $\sim Z^2e^2/r_c$ is large compared with the thermal energy k_BT, or, equivalently, the dimensionless coupling constant $\Gamma = Z^2e^2/(r_c k_B T)$ is large compared with unity, one may neglect finite temperature contributions to the Coulomb free energy, which is therefore given by (33). The surface thermodynamic potential is given by

$$\Omega_{surf} = 4\pi r_N^2 \sigma, \tag{36}$$

where the surface tension now depends on temperature. With increasing temperature, the nuclear surface becomes increasingly diffuse, which leads to a decrease in surface tension. Calculations of the dependence of the surface tension on proton concentration and temperature have been carried out in the Thomas-Fermi approximation by Ravenhall, Pethick and Lattimer (1983), using the Skyrme 1′ nucleon-nucleon interaction.

The optimal nuclear size is given by the condition

$$\Omega_{surf} = 2E_{Coul}, \tag{37}$$

which is a generalization of Eq. (19), or, with the values of the coefficients given before,

$$A = \frac{12.5}{x^2}\left(\frac{\sigma(x,T)}{\sigma(0.5,T=0)}\right)\left(\frac{n_i(x=0.5)}{n_i(x)}\right)\frac{1}{1-\frac{3}{2}u^{1/3}+\frac{1}{2}u}. \tag{38}$$

Here $n_i(x)$ is the saturation density of nuclear matter as a function of proton concentration. This result enables us to understand the variation of A with density and temperature obtained in the detailed calculations of Lamb, Lattimer, Pethick and Ravenhall (1978) and Lattimer, Pethick, Ravenhall and Lamb (1985), which are displayed in Fig.1. With increasing density, u increases, which leads to a decrease in the Coulomb energy and an increase in A. As the temperature increases, the surface tension decreases, and therefore A decreases.

The detailed calculations included a number of physical effects not discussed above, such as the translational degrees of freedom of the nuclei, and the presence of alpha particles as well as heavy nuclei, but space does not allow us to go into them here. We conclude this section by remarking that finite temperature Hartree-Fock calculations have been performed for hot dense matter by Bonche and Vautherin (1981, 1982). Their results show that at finite temperature, just as in the zero-temperature case considered earlier, the generalized liquid drop model gives a good account of the properties of matter at subnuclear densities.

6.2 NUCLEAR SHAPES

So far we have assumed nuclei to be spherical, but that ceases to be the case at densities approaching nuclear saturation density, since the Coulomb potential due to neighbouring ions has the symmetry of the crystal lattice and is therefore aspherical. This leads to a number of interesting consequences for nuclear shapes. For simplicity, we consider the case of zero temperature, and take into account only bulk, surface and Coulomb contributions to the energy. We also consider proton concentrations so high that there are no neutrons outside the dense nuclear matter phase. When nuclei fill half of space, $u = 1/2$, the Coulomb and surface energies are both unchanged if nuclei are turned inside out, so that matter consists of a

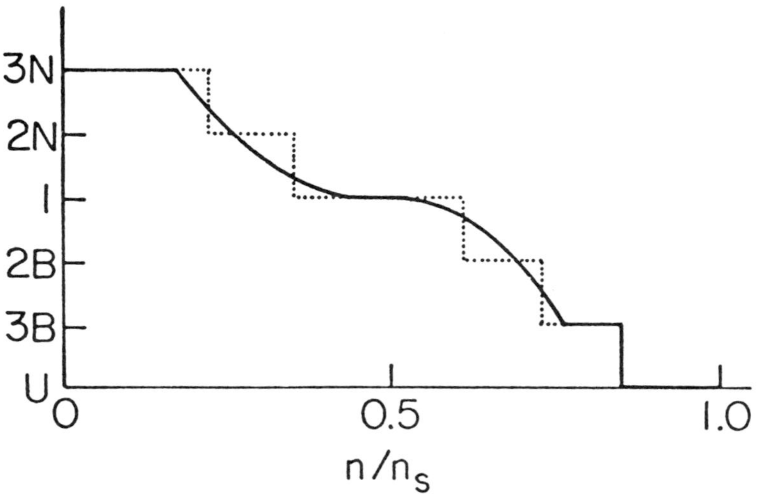

Fig. 2: Optimal dimensionality of nuclei as a function of density. The continuous curve is what is obtained if d is treated as a continuous variable. If d is allowed to take on only the discrete values 1 (plates), 2 (rods) and 3(spheres) one obtains the dotted line. U denotes uniform matter.

lattice of spherical voids or bubbles in otherwise uniform nuclear matter (Baym, Bethe and Pethick, 1971; Lamb, Lattimer, Pethick and Ravenhall, 1978, 1983). When $u \leq 1/2$, the state with nuclei has the lower energy, while when $u \geq 1/2$, the bubble phase is energetically favourable. The transition between the nuclei and bubble states occurs when $u \approx 1/2$, or $n \approx n_s/2$. In Fig. 2, the region where the bubble state is favoured is also shown. Detailed calculations of the nuclei-bubble transition have been carried out by Lamb, Lattimer, Pethick and Ravenhall (1983).

It is interesting that shapes other than spherical nuclei or spherical bubbles are favoured at certain densities. The first possibilities considered (Ravenhall, Pethick and Wilson, 1983) were rod-like and plate-like nuclei, which have been dubbed spaghetti and lasagna, respectively, together with the bubble analogue of spaghetti. (Since the bubble analogue of lasagna is still lasagna, it does not represent a new possibility.) With increasing density, the sequence of most favourable nuclear states is spherical nuclei, spaghetti, lasagna, the bubble analogue of spaghetti, spherical bubbles, and, finally, uniform nuclear matter. The densities at which the various states are energetically favoured are shown by the dotted curve in Fig. 2, where N stands for nuclei and B for bubbles. 3 denotes spheres, 2 denotes rods, and 1 denotes plates. U stands for uniform nuclear matter. The curve is symmetrical about $u = 1/2$ if the nuclear compressibility is neglected.

One of the surprising conclusions of these calculations is that very aspherical nuclei are favourable at densities as low as $\sim 0.1 n_s$. The basic physical reason for this is that with increasing density, the nuclear size increases, due to the reduction of the nuclear Coulomb energy coefficient. Eventually, the nuclear size becomes so large that the nuclei are unstable to spontaneous elongation, as was considered in the classic fission calculation of Bohr and Wheeler (1939). The essential point is that the lattice energy has a large effect on the equilibrium nuclear size, but has much less influence on the fission condition (Pethick and Ravenhall, 1983).

The simple shapes considered above are clearly over-simplifications of the actual nuclear shapes. For example, the cross section of "spaghetti" will not be circular, due to the Coulomb interactions with other strands. To attempt to model these more complicated situations, we regard matter as being made up of objects whose shape is specified by a parameter, d, which gives the number of spatial dimensions in which the nuclear matter density varies. The states with nuclei, spaghetti and lasagna therefore correspond to $d = 3$, 2 and 1, respectively. To allow for intermediate shapes, we treated d as a continuous variable, whose value is to be determined by minimizing the energy with respect to d. The optimal value of d is shown in the continuous curve in Fig. 2.

A natural question which arises is whether the conclusions reached on the basis of the simplified calculations hold if one allows for more general configurations. Williams and Koonin (1985) have investigated this by considering cubic cells of matter with a proton fraction of 1/2, with periodic boundary conditions, which they studied using differential Thomas-Fermi theory. The gross features of the arrangements of nucleons they find to be most favourable are similar to the ones in the simplified calculation, except that they find in addition a phase intermediate between plates and rods, which may be regarded either as plates pierced by a lattice of holes, or as rods which have partially fused together to form planar structures. Again a number of first-order phase transitions are found, as one expects on general grounds.

While progress has been made in elucidating the properties of the novel phases, more work remains to be done. For example, it is still not known what the optimal configuration is for an incompressible charged fluid, with a surface tension, in the presence of a rigid charged background to ensure electrical neutrality. Also one does not know how many phase transitions occur as one varies the fractions of space filled by the charged fluid. Williams and Koonin's calculations were performed for a very specific cell, a cube, and for $x = 1/2$, and one may ask to what extent the results will be altered if other symmetries of the unit cell were considered, and more realistic proton fractions used. Another open question is what happens at finite temperature. The energy differences among the phases are due solely to Coulomb and surface effects, and are therefore rather small. Consequently, thermal fluctuations may be important even at rather modest temperatures.

When there are neutrons outside nuclei, as there are in neutron star matter, the density range where one expects aspherical nuclei would be expected to extend from about $0.1n_s + 0.9n_d$ to n_s, where n_d is the density of "dripped" neutrons. Since n_d is only slightly less than n_s, this range is much less that n_s, and when the finite compressibility of matter is taken into account, it may be the case that in this density range the favoured state of matter is a two-phase region with one phase consisting of roughly spherical nuclei immersed in dripped neutrons, and the other of uniform phase of neutrons, protons, and electrons.

7. Concluding remarks

At subnuclear densities and at temperatures low enough that subnuclear degrees of freedom are not excited, the properties of matter can be predicted with a fair degree of certainty, because many of the important physical parameters needed can be measured directly. A good quantitative description of the equation of state may be obtained by using a generalized liquid-drop model of the nucleus, and microscopic calculations are essential only for predicting detailed effects, such as ones due to shell structure.

Among important uncertainties are the extrapolation to very neutron-rich matter, which requires that the neutron-neutron interaction be known, and that the energy of neutron-rich matter be calculated by many-body theory. Another uncertainty, of a statistical-mechanical nature, concerns the description of states with aspherical nuclei. At present studies have been made only for simplified models, and at low temperatures, and there is room for further work.

This work was supported in part by U. S. National Science Foundation grants PHY86-00377 and PHY89-21025, and NASA grant NAGW-1583.

8. References

Baym, G. 1991, these proceedings.
Baym, G., Bethe, H. A., and Pethick, C. J. 1971, *Nucl. Phys.*, **A175**, 225.
Bohr, N. and Wheeler, J. A. 1939, *Phys. Rev.*, **56**, 426.
Bonche, P. and Vautherin, D. 1981, *Nucl. Phys.*, **A372**, 496.
Bonche, P. and Vautherin, D. 1982, *Astr. Ap.*, **112**, 168.
Haensel, P., Zdunik, J. L., and Dobaczewski, J. 1989, *Astr. Ap.*, **222**, 353
Lamb, D. Q., Lattimer, J. M., Pethick, C. J., and Ravenhall, D. G. 1978, *Phys. Rev Lett.*, **41**, 1623.
Lamb, D. Q., Lattimer, J. M., Pethick, C. J., and Ravenhall, D. G. 1983, *Nucl. Phys.*, **A411**, 449.
Landau, L. D. and Lifshitz, E. M. 1980, *Statistical Physics, Part I*, Pergamon, Oxford, 524.
Lattimer, J. M. 1981, *Ann. Rev. Nucl. Part. Sci.*, **31**, 337.
Lattimer, J. M., Pethick, C. J., Ravenhall, D. G., and Lamb, D. Q. 1985, *Nucl. Phys.*, **A432**, 646.
Mazurek, T. J., Lattimer, J. M., and Brown, G. E. 1979, *Ap. J.*, **229**, 713.
Myers, W. D. 1977, *Droplet Model of Atomic Nuclei*, Plenum, New York.
Myers, W. D. and Swiatecki, W. J. 1966, *Nucl. Phys.*, **81**, 1.
Myers, W. D. and Swiatecki, W. J. 1969, *Ann. Phys.*, **55**, 395.
Negele, J. W. and Vautherin, D. 1973, *Nucl. Phys.*, **A207**, 298.
Pethick, C. J., and Ravenhall, D. G. 1985, in *Numerical Astrophysics*, J. Centrella, R. Bowers, and J. LeBlanc (eds.), Jones and Bartlett, Boston, 461.
Ravenhall, D. G., Bennett, C. D., and Pethick, C. J. 1972, *Phys. Rev. Lett.*, **28**, 978.
Ravenhall, D. G., Pethick, C. J., and Lattimer, J. M. 1983, *Nucl. Phys.*, **A407**, 571.

Ravenhall, D. G., Pethick, C. J., and Wilson, J. R., 1983, *Phys. Rev. Lett.*, **50**, 2066.
Ruderman, M.A. 1991, this volume.
Williams, R. D. and Koonin, S. E. 1985, *Nucl.Phys.*, **A435**, 844.

THE HIGH DENSITY INTERIORS OF NEUTRON STARS*

GORDON BAYM
Loomis Laboratory of Physics
University of Illinois at Urbana-Champaign
1110 W. Green Street, Urbana, Illinois 61801
U. S. A.

ABSTRACT. This talk reviews current understanding of the states of matter in the liquid interior of neutron stars – including nuclear matter, pion and kaon condensation, and normal and strange quark matter – and their effect on neutron star structure and dynamics. Learning about high density matter in the laboratory by means of ultrarelativistic heavy-ion collisions is also discussed.

1. Introduction

The outer lower density regions of neutron stars are solidified and form a crust. At around nuclear matter density, $\rho_0 \simeq 2.8 \times 10^{14}$ g/cm^3 $\simeq 0.16$ baryons/fm^3 (1 fm $= 10^{-13}$cm), the nuclei of the crust become so close that they merge, to form a nuclear matter liquid. To a major extent, the properties of matter in this liquid interior determine the gross structure of neutron stars, such as the mass density profile, $\epsilon(r)/c^2$ – and hence the radius, R, and moment of inertia, I, of a star of given mass, M – and the maximum possible neutron star mass, M_{max}. The early cooling of neutron stars is determined by the neutrino emissivity of the matter in the liquid interior.

Substantial progress has been made in determining the properties of matter in the liquid near ρ_0 by extrapolation from laboratory nuclear physics, the topic I will first discuss. Considerable uncertainty remains however in pinning down the state of matter in the deep interiors. Various exotic states, such as pion condensed and strangeness condensed matter, quark matter, and strange quark matter, have been proposed at higher densities; their description will be the topic of the second part of this talk. (See also early reviews in Baym and Pethick, 1975, 1979). Finally I will close with a few words on how one can attempt to determine the properties of matter under extreme conditions from future ultrarelativistic heavy-ion collisions in the laboratory.

* Supported in part by NSF Grant DMR88-18713.

J. Ventura and D. Pines (eds.), Neutron Stars: Theory and Observation, 21–36.

2. Nuclear Matter in the Interior

In the liquid interior at densities near ρ_0, matter in neutron stars consists primarily of degenerate neutrons, likely superfluid, with a small admixture of degenerate protons and electrons. The protons are also likely superfluid (i.e., superconducting), while the electrons are too weakly interacting themselves to be superfluid. [In fact the transition temperature for electron superconductivity is $\sim T_{fe}e^{-\hbar c/e^2}$, where T_{fe} is the electron Fermi temperature, of order hundreds of MeV, and $e^2/\hbar c = 1/137$ is the fine structure constant.] The number density of electrons must, by charge neutrality, equal the number density of protons. The relative number of neutrons and protons is determined by condition of equilibrium under the beta processes,

$$n \to p + e^- + \bar{\nu}_e, \quad e^- + p \to n + \nu_e, \tag{1}$$

which implies that the chemical potentials of the particles obey

$$\mu_n = \mu_p + \mu_e. \tag{2}$$

Evaluation of (2) for non-interacting neutrons, protons and electrons (a simple and instructive exercise) implies that the proton concentration is of order 5%.

To calculate the equation of state of the nuclear matter in neutron stars is a several step procedure. First, one must extract the nucleon-nucleon interaction from pp and pn scattering experiments, at energies below $\sim$ 300 MeV, constrained by fitting the properties of the deuteron. Examples of modern fits to the interactions are the Paris, Urbana-v_{14} (UV14), and Argonne-v_{14} (AV14) two-body potentials, described in Wiringa et al. (1988 – referred to as WFF hereafter); the 14's refer to the number of different components, such as central, spin-orbit, etc., included in the interactions. The second step is to solve the many-body Schrödinger equation to find the energy density as a function of baryon number. Powerful variational techniques (reviewed in WFF, and Pandharipande and Wiringa 1989) have been developed recently that enable one to solve the Schrödinger equation accurately by computer with the rather complex modern potentials.

Two important checks on the calculation are to use the two-body potentials to compute the saturation energy and density, ρ_0, of symmetric nuclear matter, and the binding of light nuclei ($A \leq 4$). One finds that while the two-body potentials give a reasonable binding energy of nuclear matter, the equilibrium density is too high. Similarly two-body potentials fail to produce sufficient binding of light nuclei, e.g., for ^{3}He, the theoretical binding energy is 7.5 MeV compared with 8.5 MeV experimentally, and for ^{4}He, 24.5 MeV theoretically vs. the measured 28.3 MeV. To cure the binding problems one must take into account intrinsic three-body forces acting between nucleons, processes that cannot be described as a sequence of two-nucleon scatterings, such as two of the nucleons scattering and becoming internally excited to intermediate isobar states (Δ) while the third nucleon scatters from one of the isobars. The three-body forces must increase the binding in the neighborhood of ρ_0, but, to avoid overbinding nuclear matter, they must become repulsive at higher densities. This repulsion leads to a stiffening of the equation of state of neutron star matter at higher densities over that computed from two-body forces alone. Figure 1a illustrates the stiffening effects on the energy per nucleon, $E = \epsilon/\rho$, of neutron matter from inclusion of three-body forces, for the phenomenological

(a)

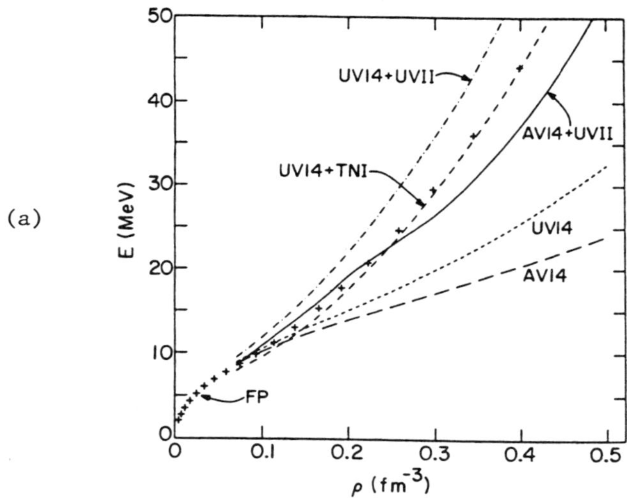

(b)

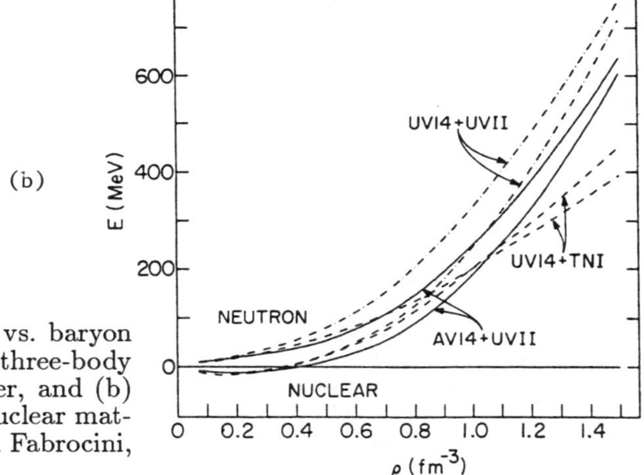

Figure 1. Energy per nucleon vs. baryon density (a) with and without three-body forces in pure neutron matter, and (b) for neutron and symmetric nuclear matter. From Wiringa, Fiks, and Fabrocini, 1988.

Urbana "Three-Nucleon" (TNI) and three-body Urbana VII (UVII) interactions, while fig. 1b shows E vs. baryon density ρ in both symmetric nuclear and neutron matter. The pressure P is directly calculable from E by the relation $P = \rho^2 \partial E/\partial\rho$. Note the uncertainties in the equation of state arising from uncertainty in the three-body interaction, particularly at higher densities.

The energy of neutron star matter in beta equilibrium at given ρ is $\sim 15\%$ smaller than that of pure neutron matter. Slightly above ρ_0 the electron chemical potential begins to exceed the mass of the muon, and a small fraction of electrons

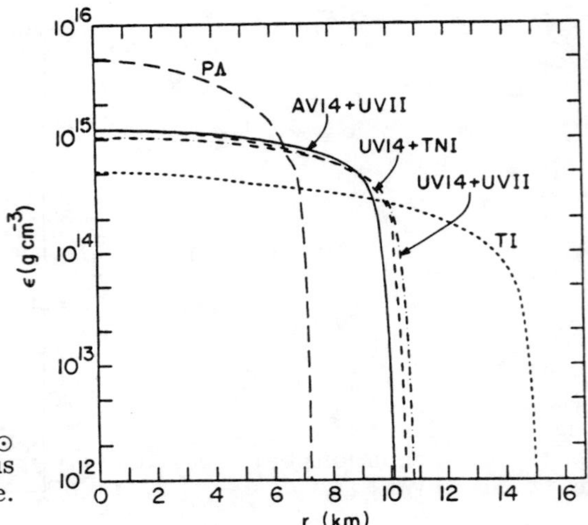

Figure 2. Density profiles of a $1.4M_\odot$ neutron star calculated for various neutron matter equations of state. From WFF.

become converted to muons (via the process $e^- \rightarrow \mu^- + \nu_e + \bar{\nu}_\mu$ taking place in the many-particle medium).

Despite the uncertainties in the equation of state of neutron matter, the modern potentials supplemented by reasonable three-body interactions yield very similar models of neutron stars. To calculate a neutron star model for a given equation of state, one integrates the Tolman-Oppenheimer-Volkoff (TOV) equation,

$$\frac{\partial P(r)}{\partial r} = -G\frac{(m(r) + 4\pi P(r)r^3/c^2)(\epsilon(r) + P(r))}{r^2c^2(1 - 2m(r)G/rc^2)}, \tag{3}$$

from the center outwards; here ϵ is the energy density, and $m(r)$, the mass interior to radius r, equals $\int_0^r 4\pi r'^2 dr' \epsilon(r')/c^2$. The (gravitational) mass of the star is $M = m(R)$, where the stellar radius R is the point where the pressure falls to zero.

The mass density profile of a $1.4M_\odot$ star, calculated by WFF with the Urbana and Argonne two-body, and Urbana three-body potentials, is shown in fig. 2; this figure also shows, for comparison, the density profile for a very soft equation of state (PΛ) – that of Friedman and Pandharipande (1981) – and for a very stiff equation of state, the tensor interaction model (TI) of Pandharipande and Smith (1975). The $1.4M_\odot$ models calculated with the modern potentials have a radius $\simeq$ 10.4-11.2 km, and a central density $\sim 6\rho_0$. The crusts of these stars are only $\sim$ 1 km thick, very thin compared with that of the stiff TI model. Figure 3 shows the gravitational mass vs. central density for families of stars calculated with the same equations of state as in fig. 2. The maximum mass for the modern equations of state is $\sim 2M_\odot$. The horizontal line in fig. 3 is the lower mass limit, $1.55M_\odot$, inferred for the X-ray binary 4U0900-40, a mass that should be accommodated by any family of acceptable models.

The corresponding mass vs. radius of the families of models is shown in fig. 4; as one sees, the radii of the modern models vary little with mass, except at the extremes. Finally, fig. 5 shows the binding energy $E_b = (Nm - M)c^2$ of the various neutron star models (where N is the number of nucleons in the star, and m is the

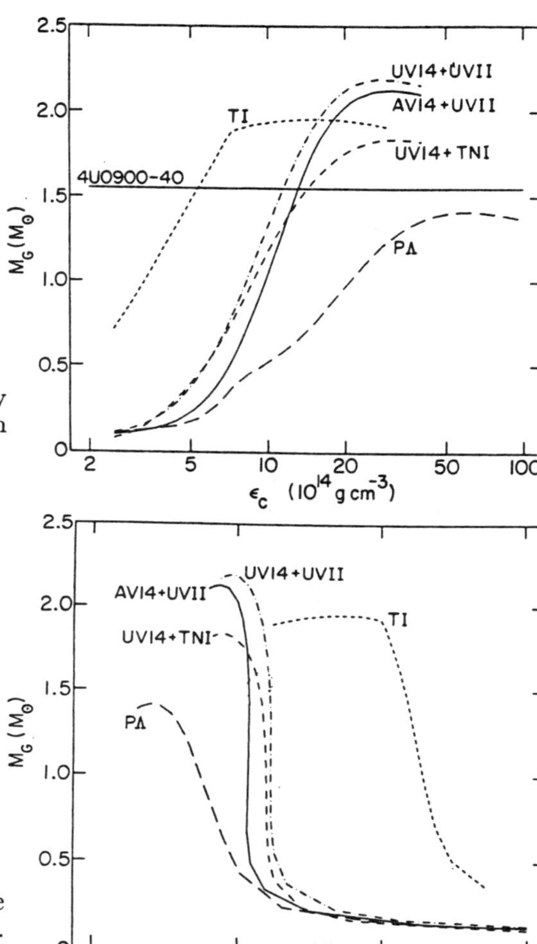

Figure 3. Mass vs. central density for the same equations of state as in fig. 2. From WFF.

Figure 4. Mass vs. radius for the same equations of state as in fig. 2. From WFF.

nucleon mass) as a function of the observed radius, R_{obs}, which is the stellar radius R multiplied by the redshift, $1 + z = (1 - 2MG/Rc^2)^{-1/2}$. This figure also shows the regions, with level of confidence, in the E_b, R_{obs} plane in which the neutron star in the supernova SN1987a is expected, as deduced by Loredo and Lamb (1989) from the energy distribution of the neutrinos observed from SN1987a. As we see, the neutrino observations are consistent with the modern nuclear models, but they fail to rule out any of the equations of state presently considered.

While the recent nuclear equations of state undoubtedly give an accurate description of neutron star matter in the neighborhood of ρ_0, they have several fundamental limitations. The first is that they fail to take into account the rich variety of hadronic and quark degrees of freedom in the nuclear system that become important with increasing density. More generally, one no longer expects, beyond a few times ρ_0, that the forces between particles can be described in terms of static

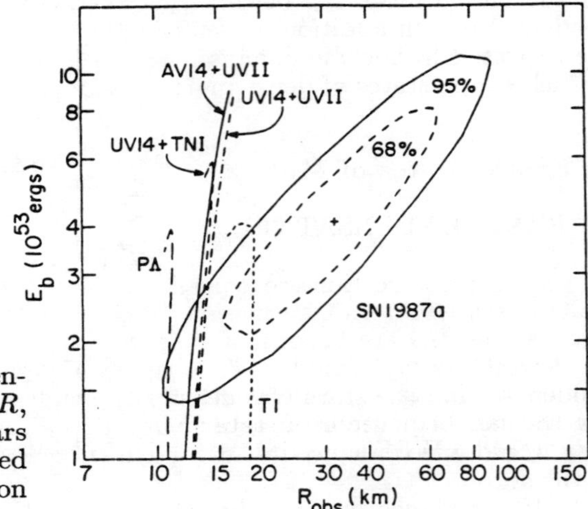

Figure 5. Gravitational binding energy vs. observed radius, $(1+z)R$, for the same families of neutron stars in fig. 4, compared with the expected location in the plane of the neutron star in SN1987a. From WFF.

few-body potentials; nor can one continue to assume at higher densities that the system can even be described in terms of well-defined "asymptotic" laboratory particles. As one sees in fig. 2, the central core of a typical neutron star is well above ρ_0; equations of state and neutron star models based on consideration of nuclear matter alone should not be regarded as definitive.

Further hadronic degrees of freedom include Δ's, hyperons, and higher baryon resonances, as well as mesonic degrees of freedom. The Δ's can both be excited as real states in the matter – as would occur when $\mu_n + \mu_e > m_\Delta$ in a picture in which the interactions of the Δ's are neglected – and contribute indirectly through the three-body forces among nucleons. The major uncertainties in including them in the equation of state are their interactions with nucleons and with other Δ's, interactions that become increasing important with increasing density.

Strange baryons, e.g., Λ and Σ, can similarly be excited via processes such as weak conversion of a neutron into a Λ, or electron capture, $e^- + n \rightarrow \Sigma^-$, with accompanying neutrino escaping from the star (in fact, the latter process should occur at lower density than the former as long as the electron Fermi energy exceeds the Σ - Λ mass difference). Again lack of detailed knowledge of the interactions of hyperons prevents an accurate description of their effect on the equation of state, and to infer consequences of their presence on dense neutron star matter one must turn to particular models of their interactions. Models based on SU(3) flavor invariance (e. g., Pandharipande and Garde 1972; reviewed in Baym and Pethick 1975) indicate that hyperons make only a brief appearance, if at all, and have only small effect on the equation of state. Very recently, the group of Kapusta (1990), Olive, and Ellis (1990) have discussed, on the basis of a parameter exploration of an equation of state calculated within the framework of a model of the lowest spin-1/2 baryon octet (no Δ's) interacting via mean meson fields, the sensitivity of the maximum neutron star mass to the effective hyperon-meson field couplings.

The meson fields that mediate the low-energy interactions between baryons also become dynamical degrees of freedom with increasing density. In dense matter the meson fields can become macroscopically excited, leading to pion or kaon condensation. In addition, at sufficiently high densities, quark degrees of freedom are expected to become dominant. Let us turn now to describing the possibilities of such exotic states of dense matter.

3. Exotic States of Matter

3.1 PION CONDENSATION

In the previous section we considered only the presence of fermion particles in neutron star matter. Can mesons, such as pions or kaons, exist in dense matter? Because mesons are bosons, if they do appear, they will macroscopically occupy the lowest available mode, i.e., form a condensate, as in ordinary Bose-Einstein condensation. The state of matter with condensed charged pions would be superconducting. Such states in neutron stars would be astrophysically important since they would have enhanced neutrino emission and hence lead to more rapid cooling of the stars.

If one naively ignores the interactions of pions with the medium, one would conclude that charged pions make their appearance once $\mu_n - \mu_p = \mu_e$ exceeds m_π = 139.6 MeV, the charged pion mass, for it is then energetically favorable for a neutron at the top of the neutron Fermi sea to turn into a proton and a π^-:

$$n \rightarrow p + \pi^- . \tag{4}$$

In neutron star matter in beta equilibrium, $\mu_n - \mu_p$ reaches $\sim$ 110 MeV at nuclear matter density, and so one might expect π^- to appear at slightly higher densities.

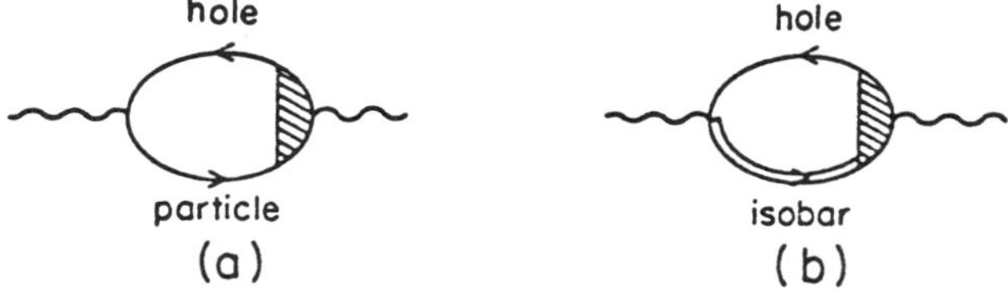

Figure 6. Coupling of pions to (a) nucleon particle-hole and (b) Δ-nucleon hole states leading to collective spin-isospin modes in matter.

However, the coupling of the nucleon particle-hole states and Δ-nucleon hole states to the pionic degrees of freedom, illustrated in fig. 6, where the wavy line is a pion, produces an important mixing of these states, and leads to a somewhat different physical picture of the onset of condensation. [The theory of condensation and numerical calculations are reviewed in Brown and Weise 1976, Migdal 1978, Baym 1978, Baym and Campbell 1979.] Essentially, nuclear matter has a collective mode with the quantum numbers of the charged pion – an oscillation of the matter with spatially-varying nucleon spin (S=1) and isospin (I=1) – which at a certain critical density, ρ_π, goes to zero frequency at a critical wavevector, k_c. This "softening" of the collective mode causes the nucleon eigenstates to become rotated in isospin space; instead of the states being pure neutron $|n\rangle$ and proton $|p\rangle$, the eigenstates spontaneously undergo a [chiral $SU(2) \otimes SU(2)$] rotation to become linear superpositions of neutron and proton states of the form:

$$\begin{aligned} |N'\rangle &= \cos\theta|n\rangle + \sin\theta|p\rangle \\ |P'\rangle &= \cos\theta|p\rangle - \sin\theta|n\rangle, \end{aligned} \tag{5}$$

where the condensation angle θ grows from zero as the density increases above ρ_π. To conserve charge as the nucleon eigenstates are rotated, the system develops a macroscopic spatially-varying (p-wave) pion field, $\langle\pi\rangle \sim e^{i\mathbf{k}\cdot\mathbf{r}}$ – the condensate – of net negative charge. The magnitude of the condensation wavevector k begins as k_c at ρ_π. An analogous neutral pion condensed state can also be formed through softening of the neutron particle-hole collective mode; the neutral condensed state is characterized by a spatially-varying finite expectation value of the neutral pion field.

Early estimates predicted the onset of charged pion condensation at $\rho_\pi \sim 2\rho_0$. However, these estimates are very sensitive to the strength of the effective neutron-hole proton repulsion in the I=1, S=1 channel (described by the Landau Fermi-liquid parameter g', see Bäckman and Weise 1979), which tends to suppress the condensation mechanism. Recent measurements in nuclei tend to indicate that the repulsion is too strong to permit condensation (Brown et al. 1988).

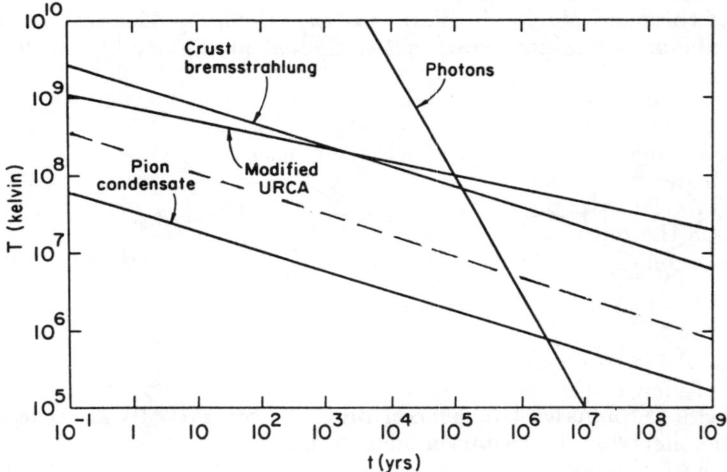

Figure 7. Schematic cooling over time of neutron star cores produced by various processes, were each acting alone. The enhanced cooling from a quark core is qualitatively the same as that from a pion-condensed core. The dashed line shows the cooling of a strangeness condensed core.

Pion condensation would enhance the cooling rates of neutron stars by allowing a much larger phase space for the URCA process (1) than the Pauli principle permits in ordinary degenerate neutron star matter. The process (1) cannot occur at low temperatures while conserving energy and momentum, since the neutron Fermi momentum is much larger than that of the protons and electrons, and also much larger than that available to the emitted low energy neutrino. Only by transferring the excess momentum to a bystander nucleon can the normal process work. However, the density of states available to the bystander in degenerate matter is suppressed by a factor $\sim (T/T_f)^2$, where T is the temperature, and T_f is the Fermi momentum of the bystander. The resulting neutrino luminosity for this "modified URCA process" is $L_\nu \sim (6 \times 10^{39}\text{erg/s})(M/M_\odot)(\rho/\rho_0)T_9^8$, where T_9 is the temperature in units of 10^9K, $M/M_\odot$ is the stellar mass in solar masses, and ρ is a characteristic interior density. In charged-pion condensed matter, however, the rotated nucleon states, N' and P', can directly participate in the URCA process, effectively β-decaying into themselves, e.g., $N' \to N' + e^- + \bar{\nu}_e$. The spatially-varying condensate serves to absorb the excess momentum without the need for a bystander nucleon; effectively, the nucleons are Bragg-scattered by the condensate. The resulting neutrino luminosity, for small condensation angle θ, is increased to $L_\nu^\pi \sim (1.5 \times 10^{46}\text{erg/s})\theta^2(M/M_\odot)(\rho_0/\rho)T_9^6$. The dramatic effect of this increase in luminosity on the interior temperature of a neutron star as a function of time is shown schematically in fig. 7.

3.2 STRANGENESS CONDENSATION

An alternative, and perhaps more likely, form of condensation involves spontaneous formation of K mesons (kaons). The underlying chiral SU(3)⊗SU(3) symmetry of strong interactions, exact in the limit that the up (u), down (d) and strange (s) quark masses vanish, implies that K mesons have an effective attractive interaction with nucleons of the form $H_{eff} \sim -\rho\bar{K}K$, where ρ is the baryon density, and K is the K meson field. This interaction acts as a density dependent term in the kaon effective mass which lowers its energy in the matter. As noted by Kaplan and Nelson (1986), the energy of a K^+ falls below $\mu_p - \mu_n$, the chemical potential for K^+, at a critical density $\rho_K \sim 2.5\text{-}3\ \rho_0$; above this density the system should form a kaon or strangeness condensate, with a macroscopic expectation value of the charged K field.

Analogous to the rotation of the nucleon eigenstates in pion condensation, in kaon condensation the nucleons undergo a chiral $SU(3) \otimes SU(3)$ rotation in which a neutron state becomes a linear superposition of a neutron and a Σ^-,

$$|N'\rangle = \cos\theta_K|n\rangle + \sin\theta_K|\Sigma^-\rangle, \tag{6}$$

while a proton state is rotated into a linear superposition of a proton, Σ^0 and Λ. From the point of view of the underlying quark structure of the baryons (n = udd, p = uud, Σ^- = dds, Σ^0 = uds, Λ = uds), the u and s components are rotated into each other by the angle θ_K. The rotation leads to a non-zero field expectation value in matter, $\langle\bar{s}u\rangle$, with the quantum numbers of the K^+ (where u and s are here the quark fields); the condensed state spontaneously breaks the chiral $SU(3) \otimes SU(3)$ symmetry. The matter can also form an η meson condensate, i.e., a state with a non-vanishing expectation value, $\langle\bar{s}s\rangle$, with the quantum numbers of the η meson.

One major consequence of kaon condensation, should it occur in neutron star matter, would also be enhanced neutrino luminosity. Even though the condensate is expected to be spatially uniform (s-wave), the rotated nucleons have phase space to β-decay into themselves and thus undergo an URCA cycle. As estimated by Brown et al. (1988), the neutrino luminosity for small θ_K is $L_\nu^K \sim$

$(5 \times 10^{43}\text{erg/s})\theta_K^2(M/M_\odot)(\rho_0/\rho)^{1/3}T_9^6$, not as large as the enhanced cooling for pion-condensed matter, but still significantly larger than the normal neutrino luminosity. See the dashed line in fig. 7. Detailed calculations of neutron star cooling in the presence of a kaon condensate are given by Page and Baron (1990). The uncertainties in the parameters of the effective interaction of kaons and the effective strange-baryon nucleon-hole couplings in neutron star matter lead, however, to large uncertainties in the condensation angle expected at given density.

3.3 QUARK MATTER

At lower densities in the star when nuclei are squeezed together the matter merges into a continuous fluid of neutrons and protons; one similarly expects that a fluid of nucleons, when squeezed to sufficiently high density (or sufficiently heated), turns into a liquid of uniform quark matter, composed of quarks, and at a finite temperature, antiquarks and gluons as well – a *quark-gluon plasma*– in which the quarks and gluons are no longer confined in individual hadrons but are free to roam over macroscopic distances in the matter. Quark matter at low temperatures consists of Fermi seas of u and d quarks. The regions where this new state is expected to occur can be seen in the phase diagram of nuclear matter in the temperature-baryon density plane, fig. 8. In the low temperature - low baryon density region the basic degrees of freedom are hadronic, those of nucleons, mesons and isobars, while in the high temperature or high baryon-density regions the basic degrees of freedom should become those of quarks and gluons.

The only reliable approach at present to determining the transition from hadronic to quark matter is through Monte-Carlo calculations of lattice gauge theory (reviewed by Ukawa 1990, Gottlieb 1991). At zero baryon density these calculations indicate that the transition to a quark-gluon plasma is likely first order, at a critical temperature $\sim$ 200 MeV (or 10^{12}K), with a latent heat of order a few GeV/fm^3. However, lattice gauge calculations at non-zero baryon density are beset by technical problems; to date one does not have a reliable estimate of the transition density at zero temperature from nuclear to quark matter or even compelling evidence that there is a sharp phase transition. For a rough estimate one can ask whether nuclear matter or quark matter has a lower energy per baryon, as a function of baryon density; such comparisons (which are based on inequivalent physical descriptions of the two phases, and thus cannot be expected to give an accurate description of the transition) imply a first-order transition with a discontinuity in the baryon density, at $\rho \sim 5 - 10\rho_0$. Although this density range is possibly above the central density found in models of neutron stars with $M \sim 1.4M_\odot$ based on nuclear equations of state, the question of whether neutron stars can have quark matter cores remains open. In the absence of information about the equation of state at very high densities, the issue of whether a distinct family of quark stars with higher central densities than neutron stars can exist also remains open.

Quark matter, as well as pion and kaon condensed matter, should have a much higher neutrino luminosity than ordinary neutron star matter. The fundamental URCA processes are

$$d \to u + e^- + \bar{\nu}_e, \;\; e^- + u \to d + \nu_e. \tag{7}$$

Massless non-interacting u and d quarks have essentially no phase space for these reactions. However, the strong interactions increase the energy of the individual u and d quarks [in lowest order perturbation theory, $E_p = (1 + (2/3\pi)\alpha_c)cp$, where

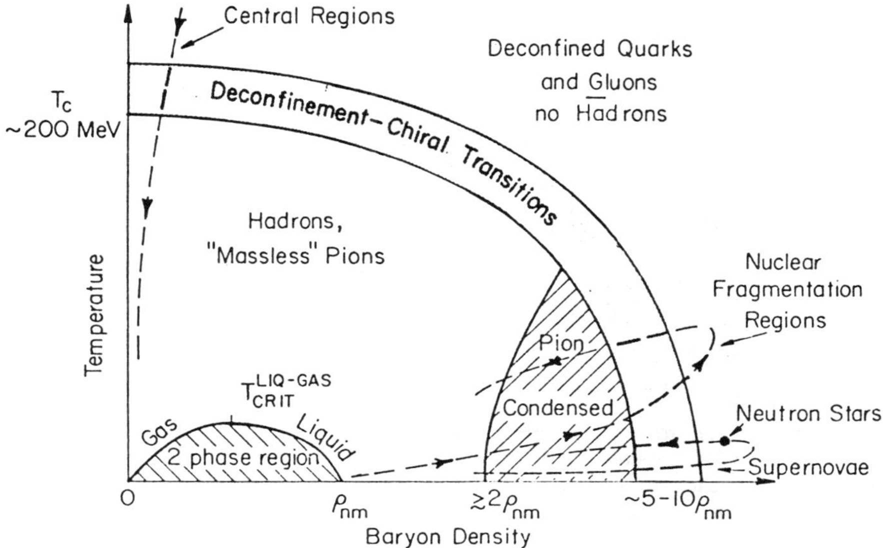

Figure 8. Phase diagram of nuclear matter in the baryon density (here ρ_{nm}) – temperature plane. The dashed curve, "supernovae," indicates the trajectory of the central density of the core, were it to rise above the deconfinement transition; the point labelled "neutron stars" indicates the central density if they have quark matter cores.

p is the quark momentum, and α_c is the quantum chromodynamic fine structure constant], thus allowing phase space for the URCA process (Iwamoto 1980). The resulting luminosity is $L_\nu^Q \sim (10^{45}\text{erg/s})\alpha_c(M/M_\odot)T_9^6$. Within the uncertainites in parameters the effect of a quark core on the cooling is roughly that of a pion-condensed core; see fig. 7.

3.4 STRANGE QUARK MATTER

The absolute ground state of matter at given baryon number, when all available energy has been extracted, is commonly believed to be in the form of ^{56}Fe (and indeed solid and ferromagnetic). To convert a nucleon in this state to a strange baryon would clearly require energy. However Witten (1984) in discussing the hadronization transition in the early universe, raised anew the question of whether strangeness-rich quark matter can have lower energy than the usual nuclear state. If so the absolute ground state would be self-bound strange quark matter, with zero pressure, at a certain finite baryon density ρ_s. (See reviews in Alcock and Olinto 1988, Baym 1988.)

The argument is based on the observation that at sufficiently high densities matter is in the form of a non-strange quark plasma, as described above; however by converting a u or d quark in the quark matter to a strange quark s, via the interactions

$$d + u \rightarrow s + u, \quad e^- + u \rightarrow s\,(+\nu_e), \tag{8}$$

one creates further Fermi seas for the same number of fermions, thus lowering the Fermi energy of the system. The cost of the conversion, neglecting interaction effects for the moment, is the increased mass of the s quark, $m_s \sim 150$ MeV. If the gain in Fermi energy exceeds the mass cost, such conversions can lower the energy of the quark matter, possibly, one cannot rule out, to below that of usual nuclear matter – in which case strange quark matter would be the absolute ground state. [Since ordinary nuclear matter is stable against turning into such quark matter, the minimum energy of the quark matter must be greater than that of nuclear matter at its minimum.] Strange quark matter is a state distinct from strangeness-condensed hadronic matter discussed above.

Whether strange quark matter can be the absolute ground state depends critically on the interactions among the constituents of the matter. Within the framework of nucleonic bag models and perturbation theory, Farhi and Jaffe (1984) found that over a reasonable range of interaction strengths, bag constants, and strange quark masses one can have strange quark matter of lower energy per baryon, E, than nuclear matter, and at the same time have non-strange quark matter unbound with respect to nuclear matter. However, such a perturbative approach is not expected to be valid at lower densities, and one must regard the issue as unsettled theoretically.

If strange quark matter is the most favorable state of matter, then one would expect the existence of a class of neutron stars, ***strange stars***, made of such matter. A strange star would be made in a supernova explosion through the initial formation of a normal neutron star. At the very high central densities in the neutron star, a seed of strange quark matter can form either spontaneously or through a large density fluctuation. If strange matter is more favorable than normal hadronic matter, then once formed the seed would begin to "burn" the matter around it into strange matter. The burning front would first convert the liquid core of the neutron star to exotic matter; the heat ahead of the front would melt the crust of the neutron star, as well as melt the nuclei in the crust into normal fluid nuclear matter, and within an hour or so, the entire star would be converted into a strange neutron star (Baym et al. 1985, Olinto 1987). If strange stars can be made, then one would expect many, if not all, neutron stars to be strange.

The gross structure of stars made of strange quark matter is not very different from that of normal neutron stars. For a first estimate of their structure one can take a simple bag model equation of state $P = (\epsilon - 4B)/3$, relating the pressure and energy density, where $B \sim 100$ MeV/fm^3 is the bag constant. Integrating the TOV equation (3) with this toy equation of state, from a central density ρ_{cent} ($> 4B/c^2$) to the surface, where $P = 0$, one finds (Alcock and Olinto 1988, and references therein) that the maximum mass of a strange star scales with the assumed B as $M_{max} = 2.03(B_0/B)^{1/2} M_\odot$, where B_0=56 MeV/fm^3; the central density for the maximum mass is $\rho_{cent}(M_{max}) = 1.92 \times 10^{15}(B/B_0)$ g/cm^3. While the radius of a normal neutron star generally decreases with increasing mass, that of a strange quark star – made of self-bound matter – increases with increasing mass, reaching a value for the bag model equation of state, $R(M_{max}) = 11.1(B_0/B)^{1/2}$ km. These parameters are well within the accepted range for normal neutron stars.

In common with other exotic states of matter, strange matter should also have a higher neutrino luminosity than normal nuclear matter, via processes (7), as well as $s \rightarrow u + e^- + \bar{\nu}_e$, and $e^- + u \rightarrow s + \nu_e$. Thus strange stars, if they exist, would cool faster than normal neutron stars.

To what extent do astrophysical observations provide evidence for or against the existence of strange stars? One the one hand, they provide the basis of an amusing explanation for the reported deep underground muons with source in the direction of, and having the 4.79 hour period of Cygnus X-3, a compact source composed of a neutron star in close orbit with an ordinary star (Marshak et al. 1985, Battistoni et al. 1985). Such muons could be understood as the decay product of a new long-lived neutral hadron originating in a strange rather than a normal neutron star in Cygnus X-3 (Baym et al. 1975). Strange stars have also been proposed as an exotic model for the transient γ-ray source of 5 March 1979 (Alcock, Farhi, and Olinto 1986).

A crucial question, however, is whether strange neutron stars can operate as pulsars. First, the surface of the neutron star must not be so tightly bound that matter there cannot be removed to supply particles for the magnetosphere. Second, the remarkable timing stability of pulsars requires a reasonably thick crust to anchor the magnetic field. The crust is also the most likely site for the mechanism of sudden pulsar speedups or glitches. Unless strange stars can have substantial crusts, they cannot be pulsars as observed (Alpar 1987).

One can imagine several possibilities: The surface of the strange star could be bare quark matter, at the saturation density of strange quark matter; such a star would have no crust. The electrons at the surface would extend out slightly further (on the scale of a screening length) than the quarks, forming a dipole charge layer that gives rise to an electric field adequate to suspend a thin crust of normal nuclei above the strange matter (Alcock et al. 1986). The normal nuclei would be prevented by the field from touching the strange matter and being burnt immediately to strange matter itself. Such a crust can reach at most the nuclear neutron drip density, $\rho_{drip} = 4.3 \times 10^{11}$ g/cm^3, since free neutrons present above this density would be unimpeded by Coulomb barriers from falling into the strange matter. The burning of a neutron star to form a strange star should proceed all the way through to the surface, and thus a normal crust on a strange star would only be expected to form by gentle accretion of matter in a binary system after the formation of the strange star. Since pulsars are found primarily in isolated neutron stars, it is difficult to see how they could be strange stars with normal crusts. [On the other hand, the Cygnus X-3 model of Baym et al. (1985) requires the surface of the strange star, which is in an accreting binary, to be strange.] A way out of the dilemma could be found if lower density strange quark matter is unstable against breaking into droplets, as normal nuclear matter at low pressure breaks up into nuclei; in this case a strange star could have a crust made of strange quark droplets arranged in a lattice.

3.5 RAPIDLY ROTATING STARS AND THE EQUATION OF STATE

Understanding of the equation of state of neutron star matter was put to a severe test by the reported discovery in early 1989 of a pulsar in the remnant of SN1987a with a half-millisecond period, since it is not at all clear that a neutron star can rotate that fast. (See Alpar et al., 1989.) Although the discovery turned out to be spurious, it lead to a valuable examination of the maximum rotation rate that a neutron star constructed from a given equation of state can have – a constraint that observations of very rapid pulsar place on equations of state.

The point is that a neutron star cannot rotate so fast that centrifugal forces begin to remove matter from the equator; a rough estimate of the maximum rotation frequency $\Omega_m = 2\pi/P_m$ is to equate the gravitational force per unit mass at the surface, GM/R^2, with the centrifugal force, $\Omega^2 R$, there, yielding a minimum period $P_m \sim (5 \times 10^4 \mathrm{s}) R_{10}^{3/2} (M_\odot/M)^{1/2}$, where R_{10} is the radius in units of 10 km. Note that P_m scales as $1/\sqrt{\bar{\rho}}$, where $\bar{\rho}$ is the mean mass density in the star. The actual minimum period is larger, $\sim (8 \times 10^4 \mathrm{s}) R_{10}^{3/2} (M_\odot/M)^{1/2}$ (Haensel and Zdunik 1989), since rotational deformation of the star decreases the gravity at the equator, while increasing centrifugal forces. A stiff equation of state allows a larger mass, but it holds the matter out at larger radius with a smaller mean density, increasing the minimum period. Decrease of the minimum period requires a softer equation of state; however, the softer the equation of state, the smaller the maximum neutron star mass. The maximum mass neutron star must be above $1.44 M_\odot$, that observed in PSR1913+16, and likely above the lower limit $1.55 M_\odot$ inferred for the X-ray binary 4U0900-40.

As Friedman, Ipser, and Parker (1986, 1989) showed, none of the standard nuclear matter equations of state that yield neutron stars with large enough masses permits the models to rotate with a 0.5 ms period. For example, models derived from the rather soft equation of state of Friedman and Pandharipande (1981) only marginally reach this period. On the other hand, as many authors pointed out, neutron stars constructed from schematic equations of state based on exotic states of matter, such as stars with pion condensed cores or strange quark stars, could accommodate such small periods and the mass constraints.

4. Ultrarelativistic Heavy-Ion Collisions

Our knowledge of neutron stars is limited by the lack of experimentally-based information on the equation of state of matter in their interiors beyond a few times nuclear matter density. Over the past decade, nuclear and high-energy experimentalists have begun to create nuclear matter under extremely conditions in the laboratory by colliding large atomic nuclei together at ultrarelativistic energies. Among the hopes of these studies is to discover and examine the quark-gluon plasma, as well as to learn more generally about the properties of matter at high energy and baryon density. (Useful comprehensive sources on ultrarelativistic nucleus-nucleus collisions are the recent proceedings of the ongoing conferences on quark matter; Baym, Braun-Munzinger, and Nagamiya 1990, Blaizot, Gerschel, and Romana 1991.)

An initial program of fixed-target experiments with lighter nuclear projectiles has been running at CERN in the SPS accelerator and at Brookhaven in the AGS accelerator since the autumn of 1986. CERN experiments have been carried out at lab energies of 60 and 200 GeV per nucleon (GeV/A) with ^{16}O and ^{32}S beams, as well as proton beams. The Brookhaven program has been carried out at lab energies of 10-14.5 GeV per nucleon with ^{16}O, ^{28}Si, and p beams. The second generation of these experiments at both laboratories is currently being implemented and will be running over the next few years. The next step is to employ heavier beams. A new booster ring adjacent to the AGS will allow injection of fully stripped ions of arbitrarily large mass into the AGS; similarly, CERN is designing a "Pb Injector" for the SPS. Experiments with heavy beams, at ~12 GeV/A at the AGS, and ~170 GeV/A at the SPS, should begin within the next several years.

Looking further into the future, the Brookhaven Relativistic Heavy Ion Collider (RHIC), on which construction has just begun, with beams expected by 1997, will provide the capability of colliding nuclei as heavy as Au on Au at 100 GeV/A in the center-of-mass (equivalent to 20 TeV/A in a fixed target experiment). CERN is also discussing building the Large Hadron Collider (LHC) with the possibility of injecting heavy-ion beams from the SPS and having one intersection region for heavy-ion detectors, thus enabling heavy-ion collisions at ~ 4 TeV/A in the center-of-mass. If built, the LHC could have heavy-ion beams as early as 1998.

The physics of ultrarelativistic heavy-ion collisions changes with increasing energy. In the lower energy AGS regime, one can picture the two colliding nuclei as effectively stopping each other (in their center of mass), forming, to a crude first approximation, a high-density fireball. [In reality, of course, parts of the nuclei pass through the collision volume rather than remaining in a fireball. With light projectiles, as employed so far in the AGS and SPS experiments, matter in the collision volume, which is small transverse to the beam direction, ought not to achieve thermal equilibration necessary for a fireball description.] Collisions between heavy nuclei in this regime may reach energy densities of order a few Gev/fm^3 – possibly high enough to reach a quark-gluon plasma – and baryon densities several times normal nuclear matter density, ρ_0. The trajectory of the matter in the collision is shown schematically in fig. 8 as the curve, "nuclear fragmentation regions." The high density matter produced in such collisions is relatively baryon rich, and should be a good testing ground for neutron star matter. An important problem is that the matter is produced in a highly excited state, and disperses over times at most tens of fm/c, some 10^{-22}s, rather than cooling to the ground state; one must extrapolate its properties to the low temperatures of matter in neutron stars.

As the beam energy increases, the colliding nuclei begin to pass through each other, becoming highly excited internally, and at the same time, leaving the vacuum between them in a highly excited state, containing matter of low baryon concentration. Such "nuclear transparency" begins to become important in the regime of the CERN fixed target experiments, and should be completely manifested at RHIC and LHC collider energies. The highly-excited fragments of the original nuclei, which recede from each other at the speed of light, contain essentially all the baryons of the original nuclei, and can be used for studying matter at high baryon density. The central region, to a first approximation, has no baryon excess, and would resemble the hot vacuum of the early universe. After being produced and thermalized, the matter follows the trajectory labelled "central regions" in fig. 8.

5. References

Alcock, C. and Olinto, A. 1988 *Ann. Rev. Nucl. Part. Sci.* **38**, 161.

Alcock, C., Farhi, E., and Olinto, A. 1986, *Phys. Rev. Lett.* **57**, 2088.

Alpar, M. A. 1987, *Phys. Rev. Lett.* **58**, 2152.

Alpar, M. A., Fushiki, I., Lamb, F. K., Miller, G. S., Park, M.-G., and Pines, D. 1989, *Nature* **338**, 295.

Bäckman, S.-O. and Weise, W. 1979 in *Mesons in Nuclei*, v. 3, M. Rho and D. Wilkinson (eds.) North-Holland Publ. Co., Amsterdam, p. 1095.

Battistoni, G. et al. 1985, *Phys. Lett.* **155B** 465.

Baym, G. 1978, in *Nuclear Physics with Heavy Ions and Mesons, Les Houches session XXX*, R. Balian, M. Rho, and G. Ripka (eds.), North-Holland Publ. Co., Amsterdam, p. 745.

Baym, G. 1988, *Nucl. Phys.* **A479**, 27c.
Baym, G., Braun-Munzinger, P., and Nagamiya, S. (eds.) 1989, *Quark Matter '88, Proc. 7th Intl. Conf. on Ultra-relativistic Nucleus-Nucleus Collisions, Nucl. Phys.* **A498**.
Baym, G. and Campbell, D. K. 1979, in *Mesons in Nuclei*, v. 3, M. Rho and D. Wilkinson (eds.) North-Holland Publ. Co., Amsterdam, p. 1031.
Baym, G., Jaffe, R., Kolb, E. W., McLerran, L., and Walker, T. P. 1985, *Phys. Lett.* **160B**, 181.
Baym, G. and Pethick, C. 1975, *Ann. Rev. Nucl. Sci.* **25**, 27.
Baym, G. and Pethick, C. 1979, *Ann. Rev. Astron. Astrophys.* **17**, 415.
Blaizot, J.-P., Gerschel, C., and Romana, A. (eds.) 1991, *Quark Matter '90, Proc. 8th Intl. Conf. on Ultra-relativistic Nucleus-Nucleus Collisions, Nucl. Phys.* **A525**.
Brown, G.E., Kubodera, K., Page, D. and Pizzochero, P. 1988, *Phys. Rev. D37*, 2042.
Brown, G. E. and Weise, W. 1976, *Phys. Rept.* **27**, 1.
Ellis, J., Kapusta, J. I. and Olive, K. A. 1990, preprint CERN-TH.5741/90.
Farhi, E. and Jaffe, R. L. 1984, *Phys. Rev.* **D30**, 2379.
Friedman, B. and Pandharipande, V. R. 1981, *Nucl. Phys.* **A361**, 502.
Friedman, J. L., Ipser, J. R., and Parker, L. 1986, *Ap. J.* **304**, 115.
Friedman, J. L., Ipser, J. R., and Parker, L. 1989, *Phys. Rev. Lett.* **62**, 3015.
Gottlieb, S. 1991, in *Intl. Symp. LATTICE '90, Nucl. Phys.* **B** (Proc. Suppl.), in press.
Haensel, P. and Zdunik, J. L. 1989, *Nature* **340**, 617.
Iwamoto, N. 1980, *Phys. Rev. Lett.* **44**, 1637.
Kaplan, D. B. and Nelson, A. E. 1986, *Phys. Lett.* **B175**, 57.
Kapusta, J. I. and Olive, K. A. 1990, *Phys. Rev. Lett.* **64**, 13.
Loredo, T. J. and Lamb, D. Q. 1989, *Ann. N. Y. Acad. Sci.* **571**, 601.
Marshak, M. L. et al. 1985, *Phys. Rev. Lett.* **54**, 2079, **55**, 1965.
Migdal, A. B. 1978, *Rev. Mod. Phys* **50**, 107.
Olinto, A. V. 1987 *Phys. Lett.* **192B**, 71.
Page, D., and Baron, E. 1990 *Ap. J.* **354**, L17.
Pandharipande, V. R. and Garde, V. K. 1972, *Phys. Lett. B39*, 608.
Pandharipande, V. R. and Smith, R. A. 1975, *Nucl. Phys.* **A237**, 507.
Pandharipande, V. R. and Wiringa, R. B. 1989, in *The Nuclear Equation of State*, W. Greiner and H. Stöcker (eds.), Plenum Press, New York, p. 585.
Ukawa, A. 1990, *Nucl. Phys.* **B** (Proc. Suppl.) **17**, 118.
Wiringa, R. B., Fiks, V., and Fabrocini, A. 1988, *Phys. Rev.* **C38**, 1010.
Witten, E. 1984, *Phys. Rev.* **D30**, 272.

PAIRING INTERACTIONS IN NEUTRON STARS

J. WAMBACH†, T. L. AINSWORTH* and D. PINES
Department of Physics
University of Illinois at Urbana-Champaign
1110 West Green Street, Urbana, Il 61801

ABSTRACT. A microscopic theory for calculating $^1S_0$-pairing gaps for neutrons and protons in neutron star matter is presented. While short-range correlations are treated in a semi-empirical way, long-range fluctuations are incorporated consistently by solving coupled two-body equations for the particle-hole interaction and the scattering amplitude at the Fermi surface. The solutions determine the quasiparticle effective mass and the pairing matrix element from which the gaps are found in weak coupling BCS-theory. Implications for neutron star models are briefly discussed.

1. INTRODUCTION

Neutron superfluidity and proton superconductivity are believed to play an important role in the evolution of neutron stars. At early times the cooling of the star is dominated by neutrino emission involving the modified Urca process as well as neutron-neutron and neutron-proton bremsstrahlung in the liquid interior [1]. It is well established that the existence of pairing gaps suppresses the neutrino processes by reducing the available phase space in the final state. At the same time the total heat capacity of the star may be drastically lowered. Neutron superfluidity in the crust, on the other hand, plays an important role in the modeling of glitch phenomena. The vortex-pinning model [2] which interprets the glitch as a sudden unpinning of vortex lines in the neutron superfluid from lattice nuclei in the inner crust has two key parameters. These are the pinning energy and the coherence length which depend sensitively the $^1S_0$ neutron gap parameter. Finally the time evolution of the magnetic field may be influenced by the interaction between neutron vortex lines and superfluid protons.

2. A MICROSCOPIC THEORY FOR S-WAVE PAIRING

The evaluation of pairing properties in nucleonic matter from microscopic theory has been notoriously difficult as can be seen from Fig. 1. Here we display a representative

J. Ventura and D. Pines (eds.), Neutron Stars: Theory and Observation, 37–48.

set of calculations for the density-dependence of the ${}^1S_0$ neutron gap.

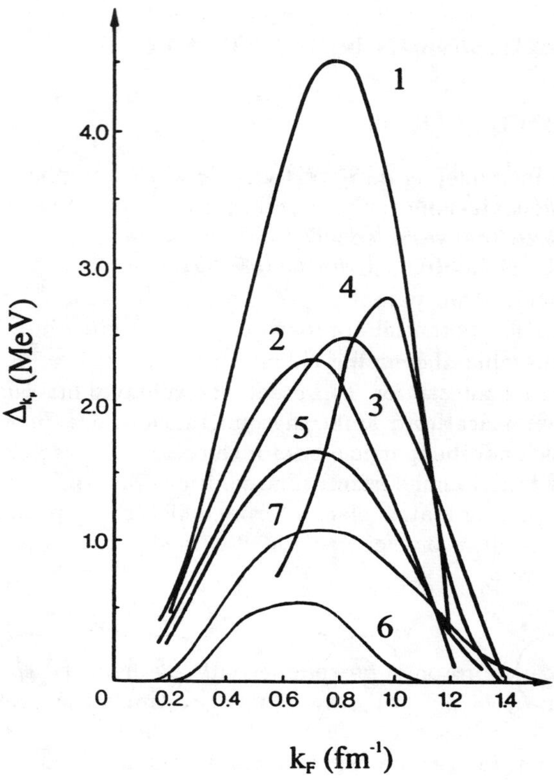

Fig. 1 Density-dependence of the ${}^1S_0$-neutron pairing gap from various calculations: 1 ref.[4], 2 ref.[5], 3 ref.[6], 4 ref.[7], 5 ref.[8], 6 ref.[9] and 7 ref.[10].

The predictions for the maximum gap vary by almost one order of magnitude (the situation is similar for the corresponding proton gaps). To appreciate the difficulties consider the s-wave gap equation

$$\Delta_k = -\frac{1}{2}\sum_{k'} P_{kk'} \frac{\Delta_{k'}}{((e_{k'}-\mu)^2+\Delta_{k'})^{1/2}} \tag{1}$$

where $P_{kk'} \equiv \langle k\uparrow -k\downarrow|\mathcal{A}|k\uparrow -k\downarrow\rangle$ is the pairing matrix element involving the quasi-particle scattering amplitude $\mathcal{A}$, e_k are the single-particle energies of the normal state and μ is the chemical potential. Given the Fermi energy $e_F = k_F^2/2m^*$ and the pairing matrix element $P_{k_F k_F}$ at the Fermi surface the BCS weak coupling approximation [3]

$$\Delta_{k_F} = 2e_F \exp[1/N(0)P_{k_F k_F}] \tag{2}$$

where $N(0) = m^* k_F/\pi^2$ is widely used to estimate the gap (at k_F) which is accurate to within factors of 1.5-2.0. Since the dependence on the pairing matrix element and the effective mass is exponential the extreme sensitivity of the results to the microscopic model is not surprising.

2.1 THE MANY-BODY EQUATIONS

Part of the difficulties in modeling gaps arises from a strong medium dependence of the pairing interaction via long-range correlations, as was first pointed out by Niskanen and Sauls some ten years ago [7]. To treat such correlations carefully we have proposed a Fermi liquid model [10] which retains the complexity of the interaction and yet provides transparent physical insight into the relevant processes. The basis is the polarization potential approach of Aldrich and Pines [11] which has been extended by Ainsworth and Bedell [12].

To compute the pairing interaction $P_{k_F k_F}$ and the effective mass m^* which enter the BCS formula (2) we start from a set of coupled non-linear equations for the scattering amplitude $\mathcal{A}$ and the particle-hole interaction $\mathcal{F}$ (Fig. 2) in which all momenta are restricted to lie on the Fermi surface.

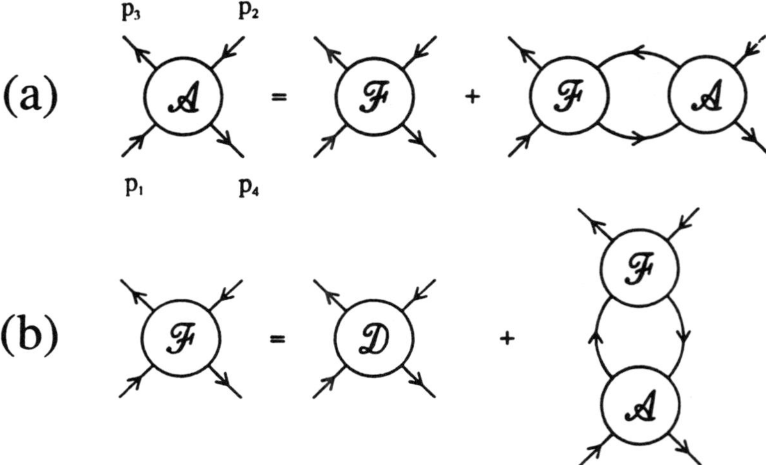

Fig. 2 Diagrammatic representation of the coupled equations for the scattering amplitude $\mathcal{A}$ (a) and the particle-hole interaction $\mathcal{F}$ (b).

For $\mathcal{A}$ we have the well known RPA equation (Fig. 2a) which includes collective excitations such as density- or spin-density fluctuations. If the scattering amplitude is to be antisymmetric then the particle-hole interaction must also contain the exchange of these excitations (Fig. 2b). This part is usually called the 'induced interaction' [13]. In addition there is a driving term, the 'direct interaction' $\mathcal{D}$, which is particle-hole irreducible in all momentum channels and specifies $\mathcal{F}$ and $\mathcal{A}$ completely. $\mathcal{D}$ mostly contains short-range correlations and is to be modeled on the basis of the bare nucleon-nucleon interaction (see below).

In some more detail, given the structure of the bare interaction

$$V = V_0 + V_\tau \boldsymbol{\tau}\cdot\boldsymbol{\tau} + V_\sigma \boldsymbol{\sigma}\cdot\boldsymbol{\sigma} + V_{\sigma\tau}\boldsymbol{\sigma}\cdot\boldsymbol{\sigma}\boldsymbol{\tau}\cdot\boldsymbol{\tau} \tag{3}$$

(for s-wave pairing only central components need to be considered) $\mathcal{F}$ may be decomposed as

$$\mathcal{F} = N(0)^{-1}[F + F'\boldsymbol{\tau}\cdot\boldsymbol{\tau} + G\boldsymbol{\sigma}\cdot\boldsymbol{\sigma} + G'\boldsymbol{\sigma}\cdot\boldsymbol{\sigma}\boldsymbol{\tau}\cdot\boldsymbol{\tau}] \tag{4}$$

and similarly for $\mathcal{A}$. At the Fermi surface the magnitude of the quasiparticle momenta is restricted to k_F and the components $\mathcal{F}$ and $\mathcal{A}$ can be expanded in two angles Θ_L and Θ_q as

$$\begin{aligned} F^\alpha &= \sum_{lm} F^\alpha_{lm} P_l(\cos\Theta_L) P_m(\cos\Theta_q) \\ A^\alpha &= \sum_{lm} A^\alpha_{lm} P_l(\cos\Theta_L) P_m(\cos\Theta_q) \end{aligned} \tag{5}$$

$(F^\alpha = F, F', G, G')$. Here Θ_L denotes the angle between $p_1 + p_3/2$ and $p_2 + p_4/2$ and Θ_q is the angle between p_1 and p_3. The RPA-channel momentum transfer is given by $q = 2k_f \sin\Theta_q/2$. In the limit $q \to 0$, Θ_L reduces to the usual 'Landau angle', i.e. the q-dependent coefficients $F^\alpha_l(q) = \sum_m F^\alpha_{lm} P_m(\cos\Theta_q)$ become the Landau parameters of nuclear matter.

According to the diagrams given in Fig. 2 the components $F^\alpha_l(q)$ and $A^\alpha_l(q)$ obey coupled integral equations given by [10]

$$F^\alpha_l(q) = D^\alpha_l(q) + \mathcal{P}_l(q,q')\Big\{\sum_{l',m,n,\beta} c^F_{\alpha\beta} F^\beta_m(q')\chi^{mn}_{l'}(q')A^\beta_n(q')\Big\} \tag{6}$$

and

$$A^\alpha_l(q) = F^\alpha_l(q) - \sum_{m,n,\beta} c^A_{\alpha\beta} F^\beta_m(q)\chi^{mn}_l(q)A^\beta_n(q) \tag{7}$$

where q' denotes the magnitude of the momentum in the exchange particle-hole channel and $\mathcal{P}_l$ projects onto momentum q of the direct particle-hole channel (RPA-channel). $D^\alpha_l(q)$ denotes the components of the 'direct' interaction $\mathcal{D}$, $\chi^{mn}_l(q)$ are particle-hole Lindhard functions, and $c^F_{\alpha\beta}$ and $c^A_{\alpha\beta}$ are spin- and isospin recoupling coefficients. The incorporation of the momentum dependence of the quasiparticle interaction and scattering amplitude guarantees the latter will be antisymmetric for all scattering angles. Eqs. (6) and (7) for the coefficients F^α_{lm} and A^α_{lm} are non-linear and have to be solved self-consistently for a truncated set in (l,m), chosen large enough to ensure convergence. Note again that $\mathcal{D}$ completely specifies $\mathcal{F}$ and $\mathcal{A}$.

2.2 THE DIRECT INTERACTION

To model the direct interaction $\mathcal{D}$ which drives the many-body equations (6) and (7) we start from the central components of the bare nucleon-nucleon interaction

$$V(r) = V_0(r) + V_\tau(r)\boldsymbol{\tau}\cdot\boldsymbol{\tau} + V_\sigma(r)\boldsymbol{\sigma}\cdot\boldsymbol{\sigma} + V_{\sigma\tau}(r)\boldsymbol{\sigma}\cdot\boldsymbol{\sigma}\boldsymbol{\tau}\cdot\boldsymbol{\tau}. \tag{8}$$

In particular we consider the Argonne potential [14] which gives excellent fits to the scattering shifts up to 400 MeV and describes the low-energy properties of the two-nucleon system very well.

For convenience and physical insight the bare interaction may be decomposed into spin-parallel and spin-antiparallel parts for protons and neutrons as

$$\begin{aligned}
V_{nn}^{\uparrow\uparrow}(r) &= V_0(r) + V_\sigma(r) + (V_\tau(r) + V_{\sigma\tau}(r)) \\
V_{nn}^{\uparrow\downarrow}(r) &= V_0(r) - V_\sigma(r) + (V_\tau(r) - V_{\sigma\tau}(r)) \\
V_{pn}^{\uparrow\uparrow}(r) &= V_0(r) + V_\sigma(r) - (V_\tau(r) + V_{\sigma\tau}(r)) \\
V_{pn}^{\uparrow\downarrow}(r) &= V_0(r) - V_\sigma(r) - (V_\tau(r) - V_{\sigma\tau}(r))
\end{aligned} \tag{9}$$

The radial dependencies are displayed in Fig. 3.

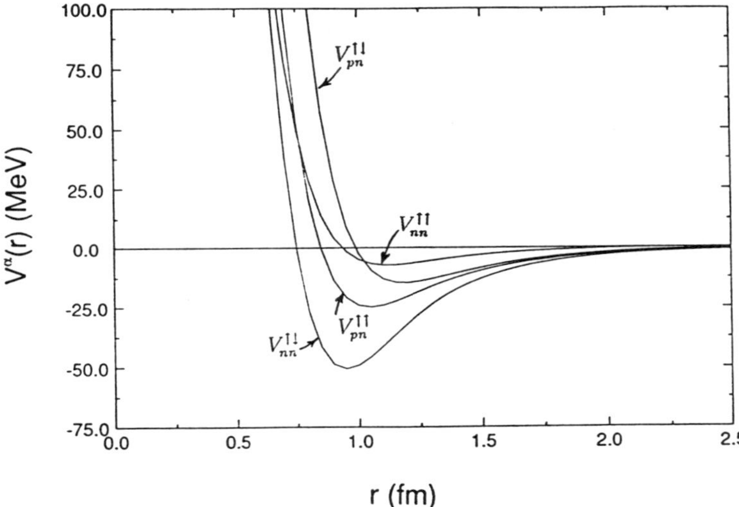

Fig. 3 The central components of the Argonne bare nucleon-nucleon interaction in r-space [14].

All interactions have strong repulsive cores. Correlations which inhibit particles from sampling this strong repulsion are incorporated phenomenologically by fixing the maximum value the cores can attain. We use a simple cut-off prescription by defining 'pseudo-potentials' $\tilde{V}$ as

$$\tilde{V}_{nn}^{\uparrow\uparrow}(r) = \min\{a_{nn}^{\uparrow\uparrow}, V_{nn}^{\uparrow\uparrow}(r)\}; \qquad etc. \tag{10}$$

which, below a certain distance, sets the potentials equal to a constant 'core-height' $a_{nn}^{\uparrow\uparrow}$ etc. These core-heights are parameters which generally depend on the matter density and are determined from the equation of state [10]. Given the r-space pseudo-potentials $\tilde{V}^{\alpha}(r)$, the components D_l^{α} of the direct interaction in eq. (6) are obtained by Fourier transformation and proper antisymmetrization

$$D_l^{\alpha}(q) = \mathcal{P}_l^{asym}\left(\int_0^{\infty} drr^2\, j_0(qr)\tilde{V}^{\alpha}(r)\right) \tag{11}$$

where $\mathcal{P}_l^{asym}$ is the antisymmetrizer for given Landau moment l.

By definition the pseudo-potentials $\tilde{V}$ are intended to incorporate the same physical processes as are typically contained in a G-matrix interaction. We have verified that, by suitable choice of core-heights, microscopic calculations can, indeed, be reproduced. Alternatively one may use G-matrix interactions directly. A simple parameterization which has proven very successful in nuclear structure calculations and low-energy nucleon-nucleus scattering [15] is given by

$$\tilde{V}(r) = [-36 + 18\boldsymbol{\tau}\cdot\boldsymbol{\tau} + 6\boldsymbol{\sigma}\cdot\boldsymbol{\sigma} + 12\boldsymbol{\sigma}\cdot\boldsymbol{\sigma}\boldsymbol{\tau}\cdot\boldsymbol{\tau}]\frac{e^{-r}}{r}. \tag{12}$$

It only acts in states even under spatial exchange. By including a density-dependence in the spin-isospin independent strength parameter of this interaction, adjusted to reproduce the equation of state, we have obtained very similar results for the $^1S_0$ neutron gaps compared to the pseudo-potentials derived from the Argonne potential with core-heights.

In the case of proton pairing the direct interaction also has to account for the coupling of the protons to the neutron matter [16]. This gives rise to polarizations effects which are indicated in Fig. 4. In this case the momentum space interaction is given by

$$\bar{V}_{pp}^{\alpha}(q) = \tilde{V}_{pp}^{\alpha}(q) + \sum_{\alpha} |\tilde{V}_{pn}^{\alpha}(q)|^2 R_{nn}^{\alpha}(q) \tag{13}$$

where $\tilde{V}_{pp}^{\alpha}(q)$ and $\tilde{V}_{pn}^{\alpha}(q)$ are Fourier transforms of the r-space pseudo-potentials and

$$R_{nn}^{\alpha}(q) = \chi_0(q)[1 + A_0^{\alpha}(q)\chi_0(q)] \tag{14}$$

are the density- and spin-density response functions of the neutron background.

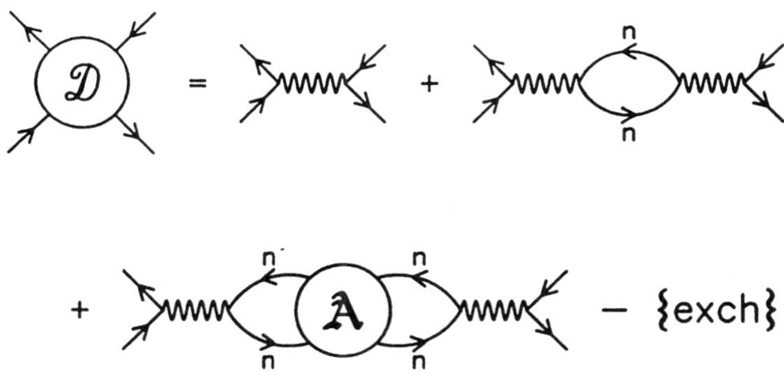

Fig. 4 Diagrammatic representation of the direct interaction $\bar{V}^{\alpha}_{pp}(q)$ (eq. (13)). The first term is the pp pseudo-potential, while the second and third term describe the coupling to the neutron background.

2.3 PAIRING INTERACTION AND S-WAVE GAPS

Once $\mathcal{D}$ is determined, it is straightforward to calculate the effective interactions $\mathcal{F}$ and $\mathcal{A}$ by using the coupled equations (6) and (7). The quasiparticle effective mass at the Fermi surface m^*/m which enters the Fermi energy e_F and the density of states $N(0)$ in the BCS formula (2) is then obtained from the spin-isospin independent $l = 1$-moment of $\mathcal{F}$ in the limit $q \to 0$ as $m^*/m = 1 + F_1(0)/3$. While the neutron effective mass is always close to unity in the relevant density range, the proton effective mass m_p^*/m_p is strongly modified by its coupling to the neutron background. Results are shown in Fig. 5 as a function of the proton Fermi momentum $k_F(p)$. One of the observational parameters, directly related to m_p^*/m_p, is the time scale τ_d for crust-core coupling via e^--scattering from the inhomogeneous magnetization surrounding a neutron vortex line. This time scale can be estimated as [17]

$$\tau_d \approx 100\left(\frac{m_p}{m_p - m_p^*}\right)^2 P \tag{15}$$

where P is the rotation period in seconds. With the values for m_p^*/m_p given in Fig. 5 we obtain $\tau_d \approx 4500P$, $2100P$, and $1500P$ at densities $\rho_{14} = 2.6$, 4.0 and

5.2 respectively ($\rho_{14} = 10^{14}$ g/cm^3). Such values exclude crust-core coupling as an explanation for post-glitch relaxation.

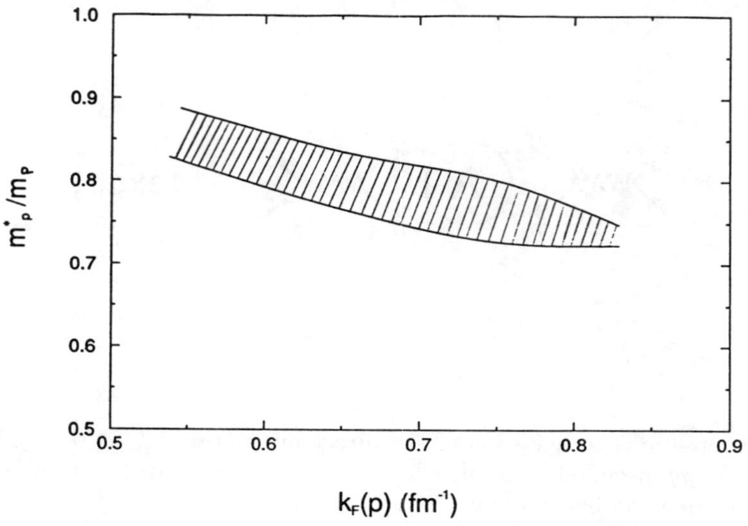

Fig. 5 Density-dependence of the proton effective mass m_p^*/m_p. The hatched area indicates the uncertainty in the pn pseudo-potential (see text).

To obtain the pairing matrix element $P_{k_F k_F}$ at the Fermi surface the scattering amplitude $\mathcal{A}$ has to be transformed to the particle-particle channel by going from Θ_L and Θ_q defined in sect. 2.1 to the angle θ between incoming and outgoing relative momenta of the scattered quasiparticles and the angle ϕ between the scattering planes. This transformation is determined by the relations

$$\begin{aligned} \cos\theta &= -1 + \frac{1}{2}(1+\cos\Theta_L)(1+\cos\Theta_q) \\ \cos\phi &= \frac{2\cos\Theta_q - \frac{1}{2}(1+\cos\Theta_L)(1+\cos\Theta_q)}{2 - \frac{1}{2}(1+\cos\Theta_L)(1+\cos\Theta_q)}. \end{aligned} \tag{16}$$

From the transformed scattering amplitude the s-wave pairing interaction $P_{k_F k_F}$ can now be obtained by the method of Patton and Zaringhalam [18]. For two quasiparticles of zero total momentum at the Fermi surface one has

$$P_{k_F k_F} = \frac{1}{8}\int_{-1}^{1} d\cos\phi \Big[A^0(\theta=\pi,\phi) - 3A^1(\theta=\pi,\phi)\Big] \tag{17}$$

where $A^0(\theta,\phi)$ and $A^1(\theta,\phi)$ are the transformed spin-symmetric and spin-antisymmetric scattering amplitudes. Together with the effective mass this is then used to compute the gaps according to the BCS-formula (2).

Results for the $^1S_0$ neutron gaps Δ_{k_F} are shown in Fig. 6 as a function of neutron k_F.

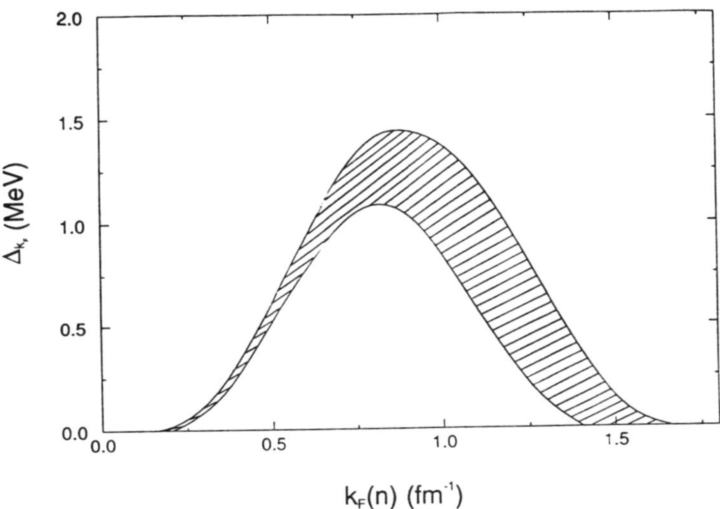

Fig. 6 Density-dependence of the $^1S_0$ neutron gap [10].

The hatched area indicates the uncertainty in predicting those gaps from the present theory. The boundaries were obtained by allowing for physically reasonable values of the core-heights in the pseudo-potentials, which specify the direct interaction $\mathcal{D}$ [10]. We have recently verified, that the G-matrix parameterization (12) gives similar results [16]. This is not surprising, since $\mathcal{F}$ and $\mathcal{A}$ are relatively insensitive to the details of the direct interaction due to screening effects via the induced interaction. These screening effects are crucial in obtaining relatively small gaps . Without such corrections one would obtain values which are larger by factors 4-6. It should be mentioned that a recent variational calculation which perturbative corrections [19] gives results which are consistent with ours.

The calculated gaps are of about the right magnitude to explain the post-glitch observations of pulsars [20]. Application of vortex creep theory to observations of all glitching pulsars shows that in the stellar crust one finds primarily regimes of weak and superweak pinning which require gaps of the order of 1.0-1.5 MeV.

For the proton $^1S_0$-gaps the sensitivity to the direct interaction is more pronounced. This arises from the coupling to the background neutrons via a strong spin-independent proton-neutron pseudo-potential (eq. (13) and Fig. 4). Without this coupling the results would be the same as for neutrons at the same k_F as indicated by the dashed line in Fig. 7.

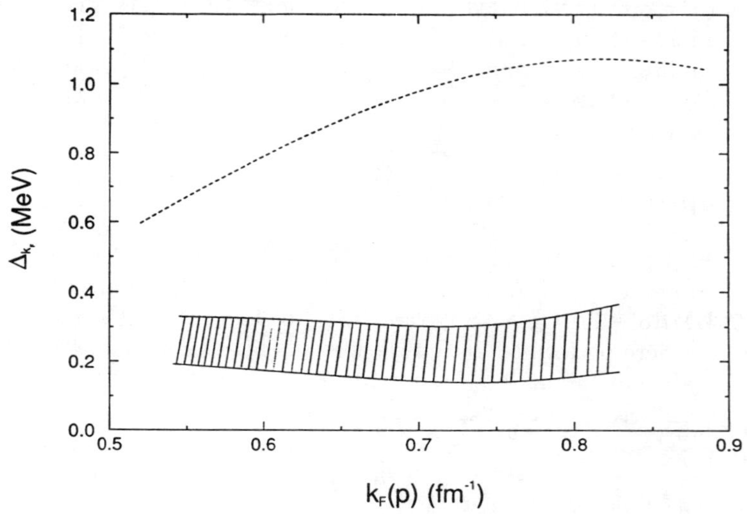

Fig. 7 Density-dependence of the $^1S_0$ proton gap [16]. The dashed line gives the result in the absence of coupling to the neutron background.

The neutron background, however, quenches these gaps to about 0.2-0.3 MeV over a wide range of proton densities which corresponds to transition temperatures T_c of 100-200 keV. To determine the proton fraction which specifies $k_F(p)$ we have used results from the equation of state of Wiringa et al. [21]. The hatched area in Fig. 7 indicates the uncertainty in $\tilde{V}^{\alpha}_{pn}(q)$ which chiefly arises from the density-dependence of the spin-isospin independent part of $\tilde{V}$.

3. SUMMARY

A new approach for calculating quasiparticle interactions at the Fermi surface has been discussed. It combines ideas from polarization potential theory with an induced interaction model. While short-range correlations, induced by the core of the bare NN-interaction, are treated phenomenologically via short-distance regularization or in terms of simple parameterizations of the Brueckner G-matrix, long-range correlations are incorporated explicitly. While the details of the short-range correlations are not important at the densities of astrophysical interest, a proper treatment of long-range fluctuations is crucial. For neutron $^1S_0$ pairing in the inner crust we find values compatible with the vortex-pinning model for postglitch relaxation. The proton s-wave gaps in the liquid core are strongly influenced by the coupling to the dense neutron background, which leads to further reduction. We obtain transition temperatures of the order of 100-200 keV which is somewhat lower than earlier estimates. Furthermore $m_p^*/m_p \sim 0.8$ in the relevant density range implies time scales for crust-core coupling via superfluid drag of several minutes. This time scale is much too short to either account for postglitch relaxation or affect vortex-pinning models of postglitch relaxation.

ACKNOWLEDGEMENT

We thank M. A. Alpar, J. W. Clark, V. A. Madsen and C. J. Pethick for helpful discussions and suggestions. This work was supported by NSF grants PHY86-00377, PHY89-21025, DMR85-211041, NASA grant NSG-7653 and NATO grant RG 85/093. Most calculations were performed at the National Center for Supercomputer Applications at Urbana-Champaign.

REFERENCES

1. O. V. Maxwell, *Ap.J.*, **231** (1979) 201.
2. D. Pines and M. A. Alpar, *Nature*, **316** (1985) 27; and references therein.
3. J. Bardeen, L. N. Cooper and J. R. Schrieffer, *Phys.Rev.*, **108** (1957) 1175; J. Schrieffer, Theory of superconductivity (Benjamin, New York, 1964).
4. E. Østgaard, *Zeit.Phys.*, **243** (1971) 79.
5. C.-H. Yang and J. W. Clark, *Nucl.Phys.*, **A174** (1971) 49.
6. T. Takatsuka, *Prog.Theor.Phys.*, **48** (1972) 1517.
7. J. A. Niskanen and J. A. Sauls, preprint (1981, unpublished).
8. L. Amundsen and E. Østgaard, *Nucl.Phys.*, **A437** (1985) 487.
9. J. M. C. Chen, J. W. Clark, E. Krotscheck and R. A. Smith, *Nucl.Phys.*, **A451** (1986) 509.
10. T. L. Ainsworth, J. Wambach and D. Pines, *Phys.Lett.*, **B228** (1989) 173.
11. C. H. Aldrich III and D. Pines, *J.Low Temp.Phys.*, **32** (1978) 689.
12. T. L. Ainsworth and K. S. Bedell, *Phys.Rev.*, **B35** (1987) 8425.
13. S. Babu and G. E. Brown, *Ann.Phys.*, **78** (1973) 1.
14. R. B. Wiringa, R. A. Smith and T. L. Ainsworth, *Phys.Rev.*, **C29** (1984) 1207.

15. F. Petrovich, H. McManus, V. A. Madsen and J. Atkinson, *Phys.Rev.Lett.*, **22** (1969) 895.

16. T. L. Ainsworth, J. Wambach and D. Pines, to be published.

17. M. A. Alpar and J. A. Sauls, *Ap.J.*, **327** (1988) 723.

18. B. R. Patton and A. Zaringhalam, *Phys.Lett.*, **55A** (1975) 95.

19. J. W. Clark, private communication.

20. M. A. Alpar and D. Pines, in 'Timing Neutron Stars', Proc. NATO ASI, Geşume, Turkey (1988), ed. H. Ögelmann and E. J. van den Heuvel, (Kluwer, 1989).

21. R. B. Wiringa, V. Fiks and A. Fabrocini, *Phy.Rev.*, **C38** (1988) 1010.

---

† present address: Institut für Kernphysik, Forschungszentrum Jülich, D-5170 Jülich, Fed. Rep. Germany

∗ present address: Department of Physics, Texas A&M University, College Station, TX 77843

SUPERFLUID DYNAMICS IN NEUTRON STARS

M. Ali ALPAR
Physics Department
Middle East Technical University
Ankara 06531, Turkey

ABSTRACT. This lecture reviews some basic points of superfluid dynamics in neutron stars. Recent developments and open problems are pointed out.

1. Introduction

Superfluidity is a consequence of the occupation of a single ground state by a macroscopic number of particles. This results in coherent flow properties involving all the particles in the condensate. Dissipative processes are hindered by energy gaps between the ground state and the lowest nontrivial excitations. For bosons superfluidity results from symmetry under particle exchange while interacting fermion systems at low enough temperatures undergo a phase transition through the formation of Cooper pairs. The ground state consists of correlated Cooper pairs, with all particles described by a single wave function and currents y the phase of this wave function. In the laboratory three classes of terrestrial superfluid systems have been studied extensively. These are the two isotopes of Helium, He^4 which becomes superfluid at 2.2 °K and He^3 which has several interesting superfluid phases at millidegrees, and the electron gas in many solids, which become superconducting at low temperatures. The best general introductions to superfluidity and superconductivity are to be found in the texts by Feynman (1965, 1972).

In neutron stars, the neutrons and protons are both expected to be superfluid, because at the densities in neutron stars which are of the order of nuclear matter density, attractive interparticle interactions lead to the formation of Cooper pairs. Calculations of the energy gap and the transition temperature of the neutron star superfluids and the constraints on the superfluid parameters obtained from astrophysical observations are discussed in the contributions of Pines and of Wambach to this volume. Typical transition temperatures are in the MeV (10^{10} °K) range. Neutron stars are at these temperatures only at their birth and rapidly cool to much lower temperatures. They are thus the ultimate high T_c systems, with their actual temperatures effectively zero compared to T_c. The observational evidence for superfluidity in neutron stars comes from pulsar timing observations. Slow dynamical relaxation following sudden disturbances (glitches) in the rotation and spindown rates of pulsars is ascribed to superfluid interior components of the neutron star. An excellent review of neutron star superfluidity was given by Sauls (1989). The specific dynamical

J. Ventura and D. Pines (eds.), Neutron Stars: Theory and Observation, 49–56.

model of vortex creep was reviewed by Alpar and Pines (1989). This lecture will outline some aspects of the dynamics of the neutron star superfluids and point out some of the open problems.

2. Some Comments on the Observations

Pulsar glitches provide clues for the interior structure and dynamics of neutron stars. These are sudden increases of the rotation and spindown rates that do not seem to involve any signs of change in the external (electromagnetic) torque. This is a very important point as it implies that glitches must reflect some change in the pulsar interior. The rotation and spindown rates of the pulsar at least partially relax back to an extrapolation of the preglitch situation on timescales of days, weeks, months, years. The argument for the existence of superfluidity in neutron stars stems from these long timescales: although the viscous diffusion time for a normal fluid of neutrons could be comparably long, a normal fluid in rotation does not wait for laminar viscous diffusion to equilibrate, there are other processes like Ekman pumps which always operate in normal fluids, and through these the normal fluid would relax in a matter of minutes (Easson, 1979). Hence the observed very long relaxation times are taken to indicate the presence of superfluid inside neutron stars.

Glitches seem to be experienced by all pulsars, though less frequently as a pulsar ages (Alpar and Ho, 1983). The most recent (8th) glitch of the Vela pulsar is the first glitch to have occurred during an observing session (McCulloch *et al.*, 1990) and has indicated the presence of prompt postglitch relaxation down to the earliest resolved timescales (Flanagan,1990). McCulloch *et al.* have also made the first claim of a discontinuous change in a pulsar's electromagnetic signature at the time of a glitch; a sudden change in the dispersion measure has been reported. If confirmed this would be very important as all dynamical models of the neutron star interior use glitch data with the assumption that glitches are changes in the interior of the neutron star and not in the torques on the star. More recent glitches from other pulsars (McKenna,1989) have shown that the distribution of glitch sizes $\Delta\Omega/\Omega$ is not bimodal as the earlier situation seemed to indicate, but is roughly uniform in logarithmic bins of glitch size. However, the sizes of glitches from an individual pulsar is not similarly distributed, as the Crab and Vela pulsars demonstrate. All events down to size $\Delta\Omega/\Omega \sim 10^{-9}$ involve an increase in the rotation rate and also in the absolute value of the spindown rate. The smallest resolved events, with $\Delta\Omega/\Omega \sim 10^{-10}$, display both signs of change in either variable (Cordes, Downs and Krause-Polstorff, 1988). Events smaller than this cannot be resolved, and the timing noise they form collectively is not a series of small scale glitches (Alpar, Nandkumar, and Pines, 1985). All these observations may give clues for the mechanism of glitches. Glitches are expected to originate in some instability of the superfluid (Packard, 1972), probably having to do with the sudden unpinning of vortex lines (Anderson and Itoh, 1975). The sudden motion of the unpinned vortices transfers angular momentum from superfluid components of moments of inertia I_i, whose rotation rates decrease by the amounts $\delta\Omega_i$, to the pulsar crust to give the observed jump $\Delta\Omega_c$ in the rotation rate of the crust

$$I_c \, \Delta\Omega_c = \Sigma \, I_i \; \delta\Omega_i \, . \tag{1}$$

A detailed physical model has not yet been proposed. Differences between the Crab and Vela pulsar glitches suggest the existence of more than one mechanism for glitches. For the frequent large glitches of Vela and of pulsar 1737-30, and the large glitches from old pulsars whose glitch occurrence frequencies are expected to scale down with their spindown rates, the alternative model for glitches, starquakes through the storage and release of elastic energy in a solid part of the star, is not viable, and some superfluid instability remains as the only energetically viable alternative. The starquake model is viable for the small Crab pulsar glitches and may operate in conjunction with vortex unpinning.

Unlike the origin of the glitches, more detailed models exist for the postglich relaxation in terms of superfluid dynamics in the interior of the star. The postglitch relaxation is observed to consist of several components with distinct timescales. These components are either in the form of prompt exponential decay ("linear response") or as long term or very slowly recovering offsets in the spindown rate ("nonlinear response"). For either type of response the fractional moment of inertia involved in the response can be extracted from the postglitch data. Postglitch response of both categories from all glitching pulsars observed so far has always yielded moment of inertia fractions of order 10^{-3} - 10^{-2} . This coincides with the fraction of a neutron star's moment of inertia carried by the superfluid in the star's crust, and comprises the most important and telling regularity in the postglitch relaxation data.

3. Superfluid Neutrons and Protons: Rotation and Magnetic Field

A superfluid rotates by forming an array of quantized vortices (Feynman, 1972). The minimum free energy state of a multicomponent system carrying angular momentum is the state of rigid rotation. A superfluid cannot exactly achieve rigid body rotation because its velocity is given by the gradient of the phase of a quantum mechanical wavefunction,

$$v_n = h/(4\pi m) \text{ grad } \phi_n \tag{2}$$

and a gradient field cannot carry rotation unless it contains singular structures. These are the quantized vortices. Each vortex contains a cylindrical core region of radius $\xi \sim E_F/(k_F \Delta)$ where E_F and k_F are the neutron Fermi energy and wavenumber respectively, and Δ the energy gap of the superfluid. Within the core superfluid neutron pairs are replaced with normal neutrons. Around this core the neutron superfluid flows with a velocity that decreases as 1/r with distance r from the axis of the vortex, and a circulation that is constant and quantized. For the neutron superfluid the quantum of circulation is $\kappa = h/(2m)$, where h is Planck's constant and m the neutron mass. The vortex lines are aligned parallel with the rotation axis. Collectively they set up a superfluid velocity field that is very close to rigid body rotation. If the equilibrium rotation rate is Ω, a uniform density of vortex lines n is set up so that the vorticity 2Ω of classical rigid rotation is achieved, with $n\kappa = 2\Omega$. The energy for the creation of vortices is available from the free energy gain achieved by bringing the superfluid to a macroscopic state that is very close to rigid body rotation.

The superconducting protons in the neutron star rotate without forming any vortices. This is because they are charged and gauge invariance gives a contribution $-eA/m_pc$ to the proton velocity in the presence of a vector potential A; so that

$$v_p = h/(4\pi m_p) \text{ grad } \phi_p - eA/m_pc. \tag{3}$$

The curl of the vector potential, which is the magnetic field, then gives the curl of the velocity field. Rigid body rotation of the protons at rate Ω is achieved without any contribution from the phase gradient term, by setting up a "London" field $b_L = -(2m_pc/e)\Omega$. This is a tiny field whose cyclotron frequency is 2Ω : for a pulsar rotation rate of 100 rad/s, b_L is only $2 * 10^{-4}$ Gauss. This field is easily set up by a small fraction of the free energy gain from getting the protons to rotate. The London field is not prohibited by a Meissner effect because the kinetic energy it entails is actually required for achieving the free energy minimum in the presence of angular momentum.

By contrast bringing the proton superfluid to a minimum free energy state in the presence of the huge magnetic flux on the pulsar requires avoiding the exorbitant particle velocities the penetration of a uniform field would require. This is achieved by introducing singularities to the phase of the proton superconductor wavefunction such that the phase gradient can cancel the vector potential contribution in the expression for the particle velocities. This leads to quantized magnetic flux lines each carrying a flux quantum hc/(2e).

To summarize, the neutron superfluid rotates by forming quantized vortices. The charged proton superconductor rotates without vortices, through the formation of a tiny London field. The independent constraint of the huge magnetic flux on a pulsar requires the presence of quantized flux lines in the proton superconductor.

4. Dynamics of the Rotating Superfluid

Pulsars spin down as a result of the electromagnetic torques on them. The interior of the neutron star must follow this spindown to remain in dynamical equilibrium with the crust as the total angular momentum of the star decreases. As the rotation rate of a superfluid is determined by the density of vortex lines in it, its spindown rate is determined by the rate of radially outward motion of the vortex lines. To be more precise, the rotation rate along a circular contour at radius r from the rotation axis is given by the total number of vortex lines going through the area enclosed by the contour. The rate of change of the superfluid rotation rate on this contour is therefore given by the change in the number of vortex lines enclosed, that is, by the current of vortex lines crossing the contour. A radially outward vortex current gives spindown of the superfluid:

$$\dot{\Omega} = - n\, k\, V_r / r \tag{4}$$

where V_r, the radially outward flow rate of the vortex lines, is determined by the physical coupling mechanism between the superfluid and the crust. The observed crust of the neutron star responds to the internal torques from the various superfluid components i in the interior,

$$I_c \dot{\Omega}_c = N_{ext} - \Sigma\, I_i \dot{\Omega}_i \tag{5}$$

along with the external torque N_{ext}. Equations of the form of Eq.(4) for each superfluid component, together with Eq. (5), specify the dynamics of the neutron star. The radial flow of vortices transfers angular momentum from the superfluid to the "crust"; microscopically this vortex motion must be initiated by the spindown of the "crust", defined to include all normal (nonsuperfluid) matter coupled tightly to the crust and spinning down with it. The equation of motion of vortex lines indeed provides for a radially outward motion in response for an azimuthal drag force exerted on the vortex line by normal matter moving at a slower rotation rate than the superfluid:

$$\mathbf{f} = r\, \kappa\, X\, (\mathbf{V}_s - \mathbf{V}_L)\,. \tag{6}$$

In this Magnus equation $\mathbf{f}$ is the force on unit length of vortex line, r the superfluid density, κ the vorticity vector directed along the vortex line, $\mathbf{V}_L$ the velocity of the line and $\mathbf{V}_s$ that of the ambient superfluid. We label the direction of the rotation axis and vortex lines as the z direction. As the normal matter acquires an azimuthal velocity lag with respect to the vortex lines, and a drag force in the $-\phi$ direction is exerted on the vortex line, $\mathbf{v}_L$ acquires a radially outward component.

5. The Core Superfluid

There are two distinct types of environment for vortex lines in the neutron star superfluid. In the inner crust of the neutron star, the superfluid coexists with a crystal lattice, and the motion of the vortex lines is constrained by pinning to the lattice. In the core of the neutron star, at densities above about $2 * 10^{14}$ gm/cm^3 there is a homogeneous mixture of superfluid neutrons and superconducting protons. The number of protons is a few percent of the neutron number. This core region contains most of the mass and moment of inertia of the neutron star. It is expected to couple to charged particles (electrons) and the crust of the star very tightly: The two superfluids are coupled by neutron-proton interactions which drag protons along with the neutron supercurrent around a neutron vortex line, and the resulting magnetization of the vortex line leads to a drag force on it when electrons move with the spinning down crust, at a relative velocity with the vortex lines (Alpar, Langer and Sauls, 1984; Alpar and Sauls,1988). The resulting dynamical coupling time τ is given by the approximate expression

$$\tau = 100\,(m/\delta m)^2\,P \text{ sec} \tag{7}$$

where m and δm are the bare mass and the difference of effective from bare mass for protons or neutrons; P is the rotation period of the pulsar. The meaning of t is that an equilibrium lag $\omega = \Omega - \Omega_c = |\dot{\Omega}|\tau$ between the neutron superfluid and crust provides the drag force necessary to set up the radial vortex flow V_r at exactly the rate for steady state coupling of the core superfluid so that it spins down at the same rate $\dot{\Omega}$ as the crust. After a perturbation to the crust, the core superfluids come into equilibrium with the crust within this timescale τ. This means that on the observational timescales after a glitch the core superfluid has already equilibrated with the crust, and the effective "crust" moment of inertia I_c in Eq. (5) actually includes the core superfluid; I_c is therefore almost the entire moment of inertia of the star and only the crust superfluid with a few percent of the star's total moment of inertia is left to account for the couplings in Eq. (5) that can be applied to postglitch relaxation phenomena on the observed timescales. This is in agreement with the moment of inertia fractions deduced from the postglitch relaxation. It is disappointing that the core superfluid is so tightly coupled to the crust as to be practically unobservable. The recent 8th glitch of the Vela pulsar, which was resolved down to 120 s after the event, implies that τ is even shorter and hence that (δm/m) must be at least 0.3. Timing data from the accretion powered X-ray pulsar Vela X-1 has been tested for evidence of the core superfluid coupling time, as τ could fall in the observable range for this pulsar by virtue of the long period P = 283 s. Analysis of timing data shows that at the 99% confidence level at most 80% of the star's moment of inertia can have a coupling time longer than 1d (Boynton et al, 1984, Baykal, 1991). If a superfluid core is present than it must couple to the crust on a timescale shorter than 1 d, which implies that (δm/m) is at least 0.6. An opposite line of constraints can come from the 35 d cycle of Her X-1. If this neutron star is indeed precessing (Trümper *et al.*, 1986), then the core superfluid applies a torque N= 1.5 $*10^{37}$ I_{45} $(\delta m/m)^2$ gm cm^2 s^{-2} to damp the precession, and external torques applied by the accretion disk on the neutron star must also be of at least this magnitude to keep the precession going (Alpar and Ögelman, 1987). An upper limit on the external torques will translate into an upper limit on ($\delta m/m$) and a lower limit on τ that may be in conflict with the above limits. It will be very interesting to observe the coupling time τ and to use it to obtain checks on astrophysical scenarios like the precession of Her X-1, or to measure the effective masses of nucleons in the neutron star core. At present it should be kept in mind that there is no observational evidence of superfluidity for the neutron star core.

The interactions of neutron vortex lines and proton flux lines may provide an alternative coupling mechanism that may turn out to be important both for the dynamical coupling of the core superfluids (Sauls,1989) and for linking the magnetic field decay in neutron stars to their spin histories (Srinivasan *et al.*.1990).

6. The Crust Superfluid

The motion of vortex lines in the crust superfluid is constrained by the possibility of pinning to the lattice. There are many unknowns in the physics of pinning and of vortex motion in the presence of pinning. Intrinsic pinning of vortices by nuclei in the lattice has been estimated and several different regimes of pinning strength have been delineated (Alpar,1977, Anderson *et al.*,1982, Alpar, Cheng and Pines,1989). Given a pinning energy E_p (the gain in energy when a vortex line is pinned), one can calculate the rate of vortex flow by thermal activation over the energy barriers. V_r is related to the lag $\omega = \Omega - \Omega_c$ between the crust and the pinned superfluid in the crust as:

$$V_r = 2V_o \exp(-E_p/kT) \sinh(E_p/kT.\omega/\omega_{cr}) \quad (8)$$

where V_o is a microscopic vortex velocity and ω_{cr} is a critical lag at which vortices unpin. The vortex creep model obtained from Eqs. (5) and (6) using Eq.(8) to model V_r has been discussed extensively and applied successfully to postglitch relaxation data from several pulsars (Pines and Alpar, 1989, Alpar, Cheng and Pines, 1989, and references therein). It has two distinct dynamical regimes: in young and hot pulsars and for weak pinning conditions a steady state can be achieved at a small lag ω and the theory can be linearized. In the opposite regime large lags build up to bring the system into steady state and the full nonlinear dependence of V_r on ω comes into play. In the linear regime the response to a perturbation in ω, as introduced by a glitch, is linear in the initial perturbation and decays exponentially in time. Indeed, in this linear regime the model is formally identical to the two component theory (Baym, Pethick, Pines and Ruderman, 1969). The spindown rate undergoes a sudden change

$$\Delta\dot{\Omega}/\dot{\Omega} = I_i/I.\delta\omega_i/\tau_i \quad (9)$$

in response to a perturbation $\delta\omega_i$ to the lag in some region i of the crust superfluid that is in the linear regime which promptly starts to decay exponentially with a relaxation time τ_i. The total jump in the observed $\dot{\Omega}$ of the crust consists of contributions from many regions. For regions in the nonlinear regime, the contribution to the jump in $\dot{\Omega}$ either persists for a long time or decays very slowly and nonexponentially. According to vortex creep theory for such terms the initial jump directly equals the fractional moment of inertia of the nonlinear region involved,

$$\Delta\dot{\Omega}/\dot{\Omega} = I_i/I . \quad (10)$$

In either case one can identify the kind of response from its time dependence (whether an exponential decay or slow or no recovery) and apply the appropriate formula, Eq.(9) or Eq.(10) to extract the moment of inertia involved. Remarkably, the amplitudes of both kinds of response have always given I_i/I of order 10^{-3}-10^{-2} in agreement with the moment of inertia fraction in the crust superfluid. In the case of linear response Eq. (9) together with Eq. (1) can be used to test whether the perturbation $\delta\omega$ in the lag is simply the observed jump $\Delta\Omega_c$ in the crust rotation rate or whether it involves a larger $\delta\omega$ indicating that the rotation rate of the superfluid also underwent a reduction $\delta\Omega$. The latter is the case for exponential relaxation following the Crab pulsar glitches. In terms of a generic two component model the Crab postglitch relaxation indicates that some internal component of the star has experienced a change in the fluid rotation rate. For the specific example of the vortex creep model this is evidence of vortex motion at the time of the glitch. The interpretation of the observed relaxation times in terms of the pinning energy and temperature can give some clues to the properties of the crust superfluid but is subject to some uncertainty in view of several possible pinning regimes. One prediction of the model is that old pulsars will display nonlinear response rather than simple exponential relaxation following glitches. This evolution in the nature of the response is because the vortex creep model is based on thermal activation. As a pulsar cools loss of activation moves it to larger lags and to the nonlinear response regime. An alternative possibility for exponential relaxation comes from conditions of continuous scattering of vortex lines from the phonons of the lattice without going through a series of thermally activated pinnings and unpinnings (Jones,1990). This alternative for linear response is basically temperature independent. The evolution of dynamical response with pulsar age can provide a valuable test to distinguish between the alternatives.

ACKNOWLEDGEMENTS

I thank D. Pines, K.S.Cheng and the other participants of the ASI for useful discussions, and H.F. Chau for many communications.

REFERENCES

Alpar,M.A. (1977) *Ap.J.* **213**, 527.

Alpar,M.A. and Pines,D. (1989), in Timing Neutron Stars, H.Ögelman and E.P.J. van den Heuvel, eds., pp. 441-454, Kluwer, Dordrecht.

Alpar,M.A. and Ho,C. (1983) *Mon.Not.Royal astr.Soc.* **204**, 655.

Alpar,M.A.,Langer,S.A. and Sauls,J.A. (1984) *Ap.J.* **282**, 533.

Alpar,M.A.,Nandkumar,R. and Pines,D. (1985) *Ap.J.* **288**, 191.

Alpar,M.A. and Ögelman,H. (1987) *Astron. and Astrophys.* **185**, 196.

Alpar,M.A. and Sauls,J.A. (1988) *Ap.J.* **327**, 723.

Alpar,M.A., Cheng,K.S. and Pines,D. (1989) *Ap.J.* **346**, 823.

Anderson,P.W.,Alpar,M.A.,Pines,D. and Shaham,J. (1982) *Phil.Mag.A* **45**, 227.

Anderson,P.W. and Itoh, N. (1975) *Nature* **256**, 25.

Baykal,A. (1991) unpublished Ph.D. thesis, Middle East Technical University.

Baym,G., Pethick,C.J., Pines,D. and Ruderman,M. (1969) *Nature* **224**, 872.

Boynton,P.E.,Deeter,J.E.,Lamb,F.K.,Zylstra,G.,Pravdo,S.H.,White,N.E.,Wood,K.S., & Yentis,D.J. (1984) *Ap.J.Lett.* **283**, L53.

Cordes,J.M.,Downs,G.S.,and Krause-Polstorff,J. (1988) *Ap.J.* **330**, 847.

Easson, I. (1979) *Ap.J.* **228**, 257.

Feynman,R.P. (1965), Lectures on Physics, Vol.III, Addison Wesley.

Feynman,R.P. (1972), Statistical Mechanics, Benjamin.
Flanagan,C. (1990) *Nature* **345**, 416.
Jones,P.B. (1990) *Mon.Not.Royal astr.Soc.* **243**, 257.
Lyne,A.G., and Pritchard,R.S. (1987) *Mon.Not.Royal astr.Soc.* **229**, 223.
McCulloch,P.M., Hamilton,P.A.,Mc Connell,D.& King,E.A.(1990) *Nature* **346**, 822.
McKenna,J. (1989) in Timing Neutron Stars, H.Ögelman and E.P.J. van den Heuvel, eds., pp. 143-152, Kluwer Academic, Dordrecht.
Packard,R.E. (1972) *Phys.Rev. Lett.* **28**, 1080.
Sauls,J.A. (1989), in Timing Neutron Stars, H.Ögelman and E.P.J. van den Heuvel, eds., pp. 457-490 Kluwer Academic, Dordrecht.
Srinivasan,G.,Bhattacharya,D.,Muslimov,A.G.& Tsygan,A.I. (1990) *Current Science* **59**,31.
Trümper,J.,Kahabka,P.,Ögelman,H.,Pietsch,W.& Voges,W. (1986) *Ap.J. Lett.* **300**, L63.

NEUTRON STARS AS COSMIC HADRON PHYSICS LABORATORIES: WHAT GLITCHES TEACH US

DAVID PINES
Physics Department
University of Illinois at Urbana-Champaign
1110 West Green Street
Urbana, IL 61801, USA

ABSTRACT. Our present theoretical understanding of the crustal neutron superfluid is reviewed, with particular attention to the information concerning its existence and behavior which can be derived from observations of pulsar glitches and postglitch behavior.

1. Introduction

In late February, 1969, the Vela pulsar, which had been discovered some months earlier, and, in common with all previously studied pulsars, had appeared to be spinning down in extremely regular fashion, exhibited a remarkable glitch (Radhakrishnan and Manchester, 1969; Reichley and Downs, 1969). The jump, $\Delta\Omega_c$ in its rotation rate, Ω_c was enormous; the jump $\Delta \dot{\Omega}_c$ in its spindown rate, $\dot{\Omega}_c$ larger yet:

$$\left|\left(\frac{\Delta\Omega_c}{\Omega_c}\right)\right| = 2.34 \times 10^{-6}$$

$$\left|\left(\frac{\Delta\dot{\Omega}_c}{\dot{\Omega}_c}\right)\right| = 6.8 \times 10^{-3}$$

Equally remarkable was its post-glitch spin behavior, in that much of the glitch appeared to decay away in a time of the order of months. That summer, a group of theorists (all of whom are present at this workshop) working together at the Aspen Center for Physics suggested that the physical origin of the glitch was a starquake, induced by pulsar spindown and accompanied by a sudden shrinking of the stellar crust, and that the postglitch behavior represented the response of the core neutron superfluid to the sudden change in the stellar crust rotation (Baym et al., 1969). Taken together, theory and observation provided the first observational evidence for the presence of neutron superfluid inside neutron stars, and opened up the possibility that future postglitch timing observations would provide further information on that superfluid.

Astrophysics is not like terrestrial physics. We cannot propose a theory and test it by experiment; rather we must proceed by constructing *scenarios* which link theory with existing observations; the test of a given body of theory is then whether the scenarios to which it gives rise are both consistent with existing observations and borne out by subsequent observations. The question which always confronts the astrophysicist is whether a given scenario is not simply plausible, but is possibly unique. In other words, have the links between theory and a sufficient body of

J. Ventura and D. Pines (eds.), Neutron Stars: Theory and Observation, 57–70.

observation matured to the point that existing and subsequent observations can be regarded as the functional equivalent of a terrestrial laboratory?

We celebrate at this workshop the twenty-first anniversary of observed celestial superfluid behavior, and of the first proposed use of pulsars as a cosmic hadron physics laboratory. During this period the Vela pulsar has been observed to glitch seven more times, the Crab pulsar four times, and some ten glitches have been observed in six other, older, pulsars. The approximate two-year recurrence period for the Vela pulsar glitches rules out their explanation in terms of oblateness-changing starquarkes, and attention has shifted to the role played by the crustal neutron superfluid, whose vortices are pinned to the crustal nuclei with which it coexists. Because of this pinning, the crustal superfluid, which contains ~ 1% of the pulsar's moment of inertia, can be out of equilibrium with the crust on timescales ranging from hours to years; it is now believed to be responsible for both pulsar glitches and the observed post-glitch behavior (Pines and Alpar, 1985). Thus the observed postglitch relaxation times tell us about microscopic pinning processes and the creep of pinned vortex lines in portions of the stellar crust which lie a kilometer or more beneath the pulsar surface.The quantum liquid core, whose superfluid properties continue to be difficult to calculate (Wambach, Ainsworth and Pines, 1991), is now believed to rotate rigidly with the crust on timescales $\gtrsim$ a few minutes, and hence plays no observable dynamic role (Alpar, 1991).

In the first part of this lecture I review briefly our present theoretical understanding of neutron star structure and the behavior of the crustal neutron superfluid. I then consider the observation of glitches and postglitch behavior and review the arguments which demonstrate that it is the *crustal* neutron superfluid which is involved in the giant glitches, and that the observed postglitch behavior represents the response of that superfluid to a glitch. I conclude with a discussion of what glitches have taught us about the crustal neutron superfluid and the behavior of hadron matter in neutron stars.

2. An Overview, with Emphasis on Crustal Behavior

The mass radius relation, crustal extent, distribution of moment of inertia, and crustal density of a neutron star are all determined by the equation of state of neutron matter at densities $\rho \gtrsim \rho_o$, the density of nuclear matter (Pandharipande, Pines, and Smith, 1976). Present observations of the masses of neutron stars in binary pulsars and pulsating X-ray sources yield masses which are all close to 1.4 $M_\odot$, while, as discussed elsewhere in these proceedings (Baym 1991), perhaps the best current calculation of the equation of state of neutron matter at densities greater than ρ_o is that of Wiringa et al. (1989). A representative neutron star of 1.4$M_\odot$ may reasonably be expected to have a radius somewhat greater than 10 km. Beneath a solid outer crust made up of nuclei and relativistic electrons, one finds an inner crust of somewhat less than a kilometer in extent, in which superfluid neutrons co-exist with neutron-rich nuclei. At densities of the order of ρ_o, these nuclei become unstable, and one gets a transition to a quantum liquid regime, containing, it is presently believed, superfluid neutrons and superconducting protons. For densities greater than $2\rho_o$ it is possible that this quantum liquid core in turn gives way to a pion condensate.

THE CRUSTAL NEUTRON SUPERFLUID

At distances $\gtrsim$ 1 fm, the interaction between neutrons in neutron matter is attractive, while it becomes strongly repulsive at distances $\lesssim$ 1 fm. It is therefore physically reasonable that at densities such that the neutrons do not sample too much of the strong short-range repulsion, the net attraction between the neutrons will prevail, and the neutrons will become superfluid at a temperature given by

$$T_c \cong E_F \exp - 1/N(0)V \tag{2.1}$$

where E_F is their Fermi energy, N(0) the density of states per unit energy, and V their effective interaction in the vicinity of the Fermi surface. An accurate calculation of the transition temperature is difficult, since it depends exponentially on an effective interaction which is determined by the influence of the background liquid in which the neutrons move. Thus short-range correlations brought about by the strong repulsive interaction will tend to keep the neutrons from sampling too much of that interaction, while the particle-hole correlation which describes the polarization of the background liquid will tend to screen out the long-range interaction.

Early variational calculations of the transition temperature and energy gap, Δ, took the short-range correlations into account, but did not allow for long-range polarization effects. The latter have been calculated in two ways: using a Fermi-liquid approach (Ainsworth, Wambach, and Pines, 1989; Wambach, Ainsworth, and Pines, 1991) and using correlated basis functions in a variational approach (for a recent review, see Clark, 1991). What is encouraging is that the two rather different approaches now yield quite similar answers. Ainsworth, Wambach, and Pines find that the shape of the density dependence of the energy gap (or transition temperature) is just what might be expected on the basis of quite general arguments; at low densities ($\rho << \rho_o/8$) the pairing interaction is ineffective because the neutron matter density of states is low, while at densities of the order of ρ_o, the neutrons sample so much of the short-range repulsion that the transition temperature will be quite low; the energy gap therefore becomes maximum with $\Delta_{m\tilde{a}x}$ 1.4 MeV, at densities ~ $\rho_o/8 \cong 3.5 \times 10^{13}$ g cm^{-3}. When one compares the energy gap calculated by Ainsworth et al. with the pioneering variation calculation of Hoffberg et al. (1970), the influence of the liquid correlations is seen to be substantial; while the density variation of the superfluid gaps is quite similar, the Fermi liquid energy gaps are seen to be substantially smaller.

CRUSTAL NEUTRON SUPERFLUID VORTEX PINNING

What makes the crustal neutron superfluid key to understanding glitches and postglitch behavior is the fact that over much of the inner crust it is *pinned* to the lattice of neutron-rich nuclei with which it coexists. Over a wide range of angular velocities (10^{-14} s$^{-1} \lesssim \Omega \lesssim 10^{20}$ s^{-1}) it is energetically favorable for quantized vortices to form in a rotating superfluid. As Onsager and Feynman first showed, the vortex structure enables the superfluid to mimic solid body rotation; thus for a superfluid rotating at an angular velocity Ω, one has an array of quantized vortex lines, each of which has a circulation $\kappa = nh/2m$, whose area density $n = \kappa/2\Omega$. The rotational dynamics of the superfluid is determined by the distribution and motion of its vortex lines. Since the nuclei in the lattice have dimensions and spacing comparable to the size, ξ, of the normal fluid core of a vortex line, it is natural to expect that these nuclei will act as *intrinsic* pinning centers for the superfluid vortices (Pines, 1971).

As discussed by Alpar (1991) elsewhere in this volume, since the superfluid velocity can only change through the motion of vortices, vortex pinning means that as a pulsar spins down the superfluid angular velocity will necessarily lag the crustal angular velocity, Ω_c, by an amount

$$\omega = \Omega - \Omega_c \tag{2.2}$$

Unpinning of a vortex line will occur when the Magnus force on the line produced by the lag

$$f_M = \rho\kappa \times \omega \times \mathbf{r} \tag{2.3}$$

exceeds the pinning force, f_p. The latter is determined by the pinning energy, E_p, which measures the strength of the pinning of the vortex core to a given nucleus, by the core size, ξ, and by the distance, b, between successive pinning centers along a vortex line,

$$f_p = \frac{E_p}{\xi b} \ . \tag{2.4}$$

Thus one has a natural limit,

$$\omega_{cr} = \frac{E_p}{\rho\kappa r b\xi} \tag{2.5}$$

on the amount by which the superfluid angular velocity can lag the crustal angular velocity.

As the pulsar spins down, this intrinsic pinning of the crustal neutron superfluid provides a natural reservoir of both angular momentum and rotational energy. Released catastrophically, and suddenly, it can give rise to a glitch in the pulsar rotation, as first suggested by Anderson and Itoh (1975); released gradually, through the creep of vortex lines in response to the lag ω (Alpar et al., 1984a, b), it provides a significant source of energy dissipation, and may be expected to determine the temperature of a neutron star once its initial heat content has been radiated away (Alpar, Nandkumar, and Pines, 1985; Shibazaki and Lamb, 1989). Moreover, as we shall see, the creep of pinned vortex lines following a glitch provides a natural explanation for the postglitch behavior of pulsars. Vortex pinning may also be responsible for giant glitches, by stressing the lattice sufficiently to induce starquakes (Ruderman, 1991).

VORTEX CREEP THEORY

Vortex creep theory is a phenomenological description of the dynamic behavior of the neutron superfluid in terms of the thermal motion of vortex lines in a inhomogeneous medium of pinning centers. As a result of the Magnus force, this random potential is biased in favor of motions of vortex lines radially outward from the rotation axis so that the superfluid can spin down (for reviews, see Pines and Alpar, 1985; Alpar and Pines, 1989; Alpar, 1991). The superfluid lag at a distance r from the rotation axis is given by:

$$\dot{\omega}(r) \cong \dot{\Omega}_\infty \left\{ 1 - \sinh\left(\frac{\omega}{\omega_{cr}(r)} \frac{E_p(r)}{kT}\right) / [\eta(r)] \right\} \tag{2.6}$$

where η, the vortex creep response parameter is given by

$$\eta(r) = \frac{|\dot{\Omega}_\infty| r}{4\Omega_c v_o} \exp \frac{E_p(r)}{kT} \tag{2.7}$$

and v_o is a microscopic velocity typically taken to be ~ 10^7 cms^{-1}. Where conditions are such that $\eta \ll 1$, $\omega \ll \omega_{cr}$, the system responds linearly to departures of ω from steady state conditions. Under these circumstances, Eq. (2.6) reduces to

$$\dot{\omega} = |\dot{\Omega}_c| - \omega/\tau_\ell \ . \tag{2.8}$$

When combined with the equation of motion for the crust rotation rate,

$$I_c\dot{\Omega} = N_{ext} - I_p\dot{\Omega} \tag{2.9}$$

these coupled equations are formally identical to the two-component model of Baym et al. (1969), albeit in a different context, since now the superfluid component is the "creeping" pinned crustal neutron superfluid (not the core neutron superfluid) and the crust-superfluid coupling time,

$$\tau_l = \frac{kT}{E_p} \frac{\omega_{cr} r}{4\dot{\Omega}_c v_o} \exp \frac{E_p}{kT} \tag{2.10}$$

depends exponentially on the pinning energy and stellar temperature.

Where $\eta \gg 1$, the steady state value of ω is close to ω_{cr}, and the approach to steady state depends *non-linearly* on the initial conditions. Eq. (2.6) reduces to

$$\dot{\omega} = |\dot{\Omega}_c| - (1/2\eta) \exp \frac{\omega}{\omega_{cr}} \frac{E_p}{kT} \tag{2.11}$$

and the observed behavior of the crust is given by

$$\dot{\Omega}_c = \dot{\Omega}_\infty - \frac{I_p}{I} \dot{\Omega}_\infty \left[\frac{1}{1 + \left[\exp \frac{t_o}{\tau_{nl}} - 1 \right] \exp - t/\tau_{nl}} - 1 \right] \tag{2.12}$$

where the coupling time is now given by

$$\tau_{nl} = \frac{kT}{E_p} \frac{\omega_{cr}}{|\dot{\Omega}|_\infty} \tag{2.13}$$

and the non-linear dependence on initial conditions is reflected in the "offset" time,

$$t_o = \frac{(\omega_\infty)_{nl} - \omega(0)}{|\dot{\Omega}_\infty|} \tag{2.14}$$

which measures how far away one is from the non-linear steady state conditions.

At a given temperature, the response of a given region in the star will be linear or non-linear depending on the strength of the pinning there. The transition value of E_p corresponds to a region in which $\eta = 1$, or what is equivalent, $\tau_l = \tau_{nl}$, and hence

$$\left(\frac{E_p}{kT}\right)_{tr} = \ln \frac{4\Omega_c}{|\dot{\Omega}_\infty|} \frac{v_o}{r} \cong 35 + \ln t_{s,6} \tag{2.15}$$

where $t_{s,6}$ is the spin-down time, $t_s \equiv \Omega_c/2|\dot{\Omega}_\infty|$, in units of 10^6 y, and typical values, $v_o = 10^7$ cms^{-1} and $r = 10^6$ cm are used (Alpar, Cheng, and Pines 1989). Where the pinning is such that $E_p > (E_p)_{tr}$, the vortex creep response will be non-linear, while for more weakly pinned regions it will be linear. For a given range of pinning energies it is evident that as a pulsar grows older its temperature and spin down rate will decrease, so that a region which may exhibit linear response in a young pulsar will shift to non-linear response as the pulsar ages.

PINNING ENERGIES AND CRITICAL ANGULAR VELOCITIES

The pinning energy per nuclear cluster, E_p, is given by

$$E_p(\Delta,\rho) = \frac{3}{8}\gamma\frac{\Delta^2(\rho)}{E_F(\rho)} \bullet \frac{4\pi R_N^3}{3} \quad ; \tag{2.16}$$

γ is a factor of order unity which takes into account the overlap of the vortex core and the cluster, and the differences in condensation energies within and without the nuclear cluster. The pinning parameters are quite sensitive to the superfluid energy gap, $\Delta(\rho)$, for the density region of interest, $1/8 \lesssim (\rho/\rho_o) \lesssim (1/2)$. The physical quantities of interest in this region are the following:

Nuclear spacing	$30 \text{ fm} \lesssim b_z \lesssim 50 \text{ fm}$
Neutron Fermi wave vector	$0.82 \text{ fm}^{-1} \lesssim k_f \lesssim 1.2 \text{ fm}^{-1}$
Nuclear cluster size	$R_N \sim 7$ fm
Superfluid coherence length:	ξ (fm) ~ 13 k_f (fm^{-1})/Δ (MeV)
Pinning energy:	E_p (MeV) ~ k_f (fm^{-1})Δ^2 (MeV)
Energy to displace a nucleus a distance ξ from equilibrium site	E_L (MeV) ~ $(Z^2e^2\xi^2)/b_z^3$ ~ bk_f^2 (fm^{-1})/Δ^2 (MeV) $(b_z/50)^3$.

Depending then on $\Delta(r)$, pinning comes in one of three strengths.

1. *Strong pinning*. It is energetically favorable to displace a nuclear cluster so that it pins more effectively to the vortex line. The condition for this is $E_p \gtrsim E_L$. In this regime $\xi \lesssim R_N$, while the distance between pinning centers is $b^s \sim b_z$, from which it follows that the criterion for being in the strong pinning regime is

$$\Delta^S(\rho) \gtrsim 1.5\, k_f^{1/4}(\text{fm})\,(b_z/50)^{-3/4} \text{ MeV} \tag{2.17}$$

and

$$\omega_{cr}^S(\rho) \gtrsim 6.5 k_f\,(\text{fm}^{-1})/\rho_{14} r_6 \Delta\,(\text{MeV})\,(b_z/50)^4 \tag{2.18}$$

where ρ_{14} is the density in units of 10^{14} g cm^{-3}, and r_6 the radial distance in units of 10^6 cm. Present microscopic calculations suggest that there are no appreciable regions of strong pinning in pulsars; Alpar et al. (1987) have derived from observation an upper limit, $\langle I_{p,43}\, \omega_{cr} \rangle \lesssim 0.7$ rad s^{-1}, which means that $(\omega_{cr})_{max} \lesssim 4$ for regions which possess inertial moments in excess of 0.1% of that of the star.

2. *Weak pinning*. Vortices pin to whatever nuclear clusters are encountered. The spacing between pinning centers is then

$$b^w = b_z^3/\pi\xi^2 \cong 230\,\frac{(b_z/50)^3\,\Delta^2\,(\text{MeV})}{k_f^2}\,\text{fm} \tag{2.19}$$

and

$$\omega_{cr}^{w} \cong 0.4 \frac{\Delta\,(\mathrm{MeV})}{r_6 k_f\,(\mathrm{fm}^{-1})\,(b_z/50\ \mathrm{fm})^3}\ \mathrm{rad\ s^{-1}} \tag{2.20}$$

is typically an order of magnitude smaller than $\omega_{cr}^{s}(\rho)$. Weak pinning exists for coherence lengths ξ such that

$$\xi^{w} \lesssim b_z/2; \tag{2.21}$$

in other words, a vortex pins to only one nuclear cluster at a time.

3. *Superweak pinning*. When the coherence length becomes comparable to, or larger than, the average spacing between nuclear clusters,

$$\xi^{sw} \gtrsim b_z/2 \tag{2.22}$$

vortices pin weakly to two or more adjacent nuclear clusters. Under these circumstances it is obviously easier for a vortex line to move; there is no simple way to estimate either the effective pinning energy, per cluster, or the corresponding critical lag; we may, however, expect that

$$(E_p)_{eff}^{sw} << E_{\rho}^{w} \tag{2.23a}$$

$$\omega_{cr}^{sw} << \omega_{cr}^{w} \tag{2.23b}$$

3. Glitches and Postglitch Behavior

The two glitching pulsars whose timing behavior has been most thoroughly studied are the Crab and Vela pulsars, and it is the postglitch behavior of these pulsars I consider here. I shall follow Alpar et al. (1984a) in assuming that in a glitch, vortices unpin in a given part of the star and move outwards through the stellar crust until they repin or reach the boundary of the inner crust. The region through which the vortices move then responds to the glitch; that response may be linear or non-linear depending on the local pinning energies and associated relaxation times. Here I show how observations of post-glitch behavior can be interpreted using a highly simplified model of the pinned crustal superfluid response in which distinct regions with inertial moments, I_p^i, through which vortices have not moved, respond linearly with relaxation times τ_i to the initial jump, $\Delta\Omega_c(0)$, in the crustal angular velocity, while the region through which vortices have moved responds both linearly and non-linearly to the glitch (See Fig. 1). On making use of the results of the two-component theory for the linear response and the vortex creep theory of Alpar et al. (1984a) for the non-linear response, the change in the crust rotation rate at a time t following the glitch may be written in the following form [compare Eq. (2.8)],

$$I_c\Delta\dot{\Omega}_c = \sum_i I_p^i \frac{\Delta\Omega_c(0)}{\tau_i} e^{-t/\tau_i} + \Delta N_\ell(t) + \Delta N_{n\ell}(t) \tag{3.1}$$

where

$$\Delta N_\ell(t) = I_c|\dot{\Omega}_\infty| \left\{\frac{I_\ell}{I}\ \frac{\delta\Omega}{\Omega_c}\ \frac{2t_{sd}}{\tau_\ell}\right\} e^{-t/\tau_\ell} \tag{3.2}$$

$$\Delta N_{n\ell}(t) = I_c|\dot{\Omega}_\infty|\left\{\frac{I_{n\ell}}{I_c} - \frac{I_A}{I_c}\frac{t}{t_A}\right\}. \tag{3.3}$$

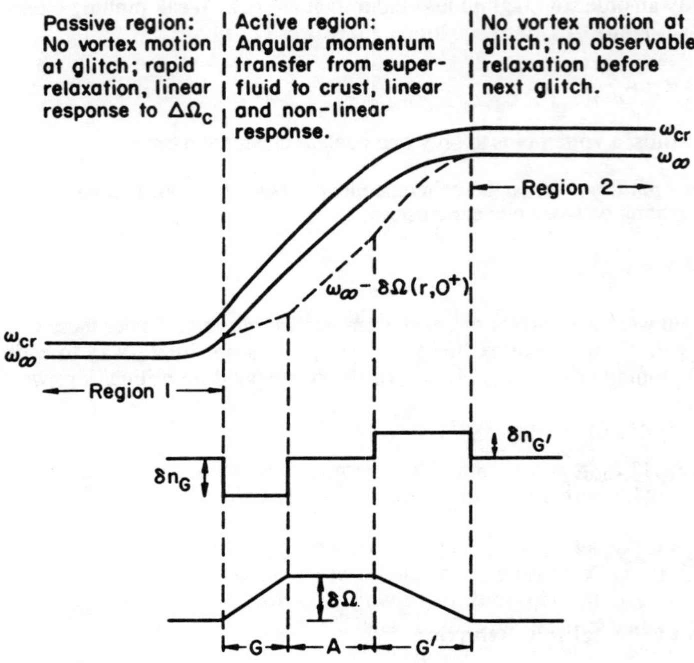

Figure 1. A toy model of a glitch, in which vortices unpin at the boundary of region 1 and repin at the boundary of region 2. The distance from the rotation axis increases to the right and density to the left; regions 1 and 2 are "passive" regions, through which vortices do not move and the response of the superfluid to a glitch is set by the change in $\Delta\Omega_c$. Based on the microscopic calculations of $\Delta(r)$, pinning energies increase to the right: in region 1, and possibly parts of the "active" region, GAG', the pinning of vortices is expected to be superweak; elsewhere it is expected to be weak. The corresponding relaxation times for vortex creep response are assumed to increase to the right, in such a way that the time scale for the response of region 2, if it contains superfluid neutrons, is longer than the time between glitches. The upper part of the figure gives the possible variation of ω_{cr} and the steady-state lag, ω_∞; the mid-part shows the change in vortex density at a glitch; the bottom part shows the change in superfluid velocity at the glitch.

In writing (3.2) and (3.3) I have assumed that the average change in the relative velocity of the pinned superfluid and the crust in the region through which vortices have moved is $\delta\Omega$ ($>>\Delta\Omega_c$); $\Delta N_\ell(t)$ is the torque on the crust produced by the part of the region which responds linearly to the change $\delta\Omega$ ($>>\Delta\Omega_c$) of the superfluid velocity with a characteristic relaxation time τ_ℓ [cf. Eq. (2.8)], while $\Delta N_{n\ell}(t)$ represents the torque on the crust produced by those parts of the region which respond non-linearly to the glitch. Part of the latter response is associated with regions in which the offset time, t_o (Eq. 2.14) is greater than the time between glitches; these contribute a constant offset, $I_{n\ell}\ |\dot{\Omega}_\infty|$, in the spin-down rate; the other part of $N_{n\ell}$ (t) represents the gradual non-linear recoupling of vortices in the boundary regions in which vortices have unpinned or repinned, which Alpar et al. (1984a) have shown gives rise to a torque which varies linearly with time. In these equations, I have set $\dot{\Omega}_c$ where appropriate to its long-term value, set by the external torque,

$$\dot{\Omega}_c = -|\dot{\Omega}_\infty| = -(\Omega_c/2t_{sd}) \tag{3.4}$$

and introduced

$$t_A = \frac{\delta\Omega}{|\dot{\Omega}_c|_\infty} \equiv \left(\frac{\delta\Omega}{\Omega_c}\right) 2t_{sd} \tag{3.5}$$

as the required time for the normal stellar spin down to give rise to a net angular velocity shift, $\delta\Omega$.

Since angular momentum is conserved in a glitch, we may write, following Alpar et al. (1984a);

$$I_c\, \Delta\Omega_c(o) = (I_\ell + I_{n\ell} + \frac{I_A}{2})\, \delta\Omega \tag{3.6}$$

To the extent that I_ℓ, $I_{n\ell}$, and I_A may be determined from observations following a given glitch, Eq. (3.6) enables one to determine the average net change in superfluid velocity in the glitch, $\delta\Omega$.

Finally, in what follows I shall assume that unpinning occurs in the vicinity of the transition from superweak to weak pinning in the stellar crust, where a rapid change in ω_{cr} is to be expected; in that case, the region through which vortices have not moved is primarily the "superweakly" pinned crustal superfluid, and the relaxation times which characterize its response may be expected to be short compared to τ_ℓ.

3.1 THE CRAB PULSAR

Let us begin by applying this simple model of pinned crustal superfluid response to the observations following the 1975 and 1989 glitches of the Crab pulsar for which Lyne and Pritchard (Lyne, private communication) have shown that the post-glitch behavior takes a particularly simple form. A portion, $[\Delta\dot{\Omega}_c/\dot{\Omega}_\infty]_\ell$, of the observed initial jump in the spin-down rate, $|\Delta\dot{\Omega}_c(o)/\dot{\Omega}_\infty|$, decays exponentially in a characteristic time, 14^d, following the glitch; the remaining part, $[\Delta\dot{\Omega}_c/\dot{\Omega}_\infty]_{\ell t} = \{[\Delta\dot{\Omega}_c(o) - (\Delta\dot{\Omega}_c)_\ell]/\dot{\Omega}_\infty\}$ represents a long-term off-set in the spin-down rate. Thus they find the following expression:

$$\left|\frac{\Delta\dot{\Omega}_c(t)}{\dot{\Omega}_c}\right| = \left|\frac{\Delta\dot{\Omega}(0)}{\dot{\Omega}}\right|_\ell \exp{-t/\tau_\ell} + \left(\frac{\Delta\dot{\Omega}_c}{\dot{\Omega}_\infty}\right)_{\ell t} \tag{3.7}$$

with $\tau_\ell = 14d$ provides an excellent fit to the post-glitch behavior of the Crab pulsar. The relevant parameters for each glitch obtained by Lyne and Pritchard are given in Table 1.

TABLE 1. Crab pulsar glitch and post-glitch parameters. The observational quantities, $|\Delta\Omega_c(o)/\Omega_c|$, $|\Delta \dot{\Omega}_c(o)/\dot{\Omega}_c|$, and $(I_{n\ell} / I) \equiv (\Delta \dot{\Omega}_c/\dot{\Omega}_c)_{\ell t}$ are those determined by Lyne and Pritchard (Lyne, private communication); the remaining quantities are derived using the simple model described here. $[\Delta\Omega_c(o)/\Omega_c]_{-8} = [\Delta\Omega_c(o)/\Omega_c] \times 10^8$, etc.

Glitch	$[\frac{\Delta\Omega_c(o)}{\Omega_c}]_{-8}$	$[\frac{\Delta\dot{\Omega}_c(o)}{\dot{\Omega}_c}]_{-3}$	$[\frac{\Delta\dot{\Omega}_c(o)}{\dot{\Omega}_c}]^{\ell}_{-4}$	$(\frac{I_{n\ell}}{I})_{-4}$	$(\frac{\delta\Omega}{\Omega})_{-5}$	$(\frac{I_\ell}{I})_{-4}$
03 Feb, 1975	3.7	2.1	19	2	3.3	9.2
29 Aug, 1989	7.4	4.5	41	4	2.1	31

Three aspects of their results deserve special mention. First, the extraordinary similarity between the 1975 and 1989 post-glitch behavior provides perhaps the strongest argument to date that we can use pulsar observations to probe a particular crustal region, in this case one for which the superfluid response is 14 days. Second, the 1989 glitch was observed at the time of the glitch, and followed initially, for an hour, before the Crab pulsar was no longer observable during that session. Analysis of the immediate post-glitch behavior may thus provide a constraint on the crust-core coupling time. Third, a straightforward application of the model described above enables us to deduce $I_{n\ell}$, I_ℓ, and $\delta\Omega$ for each glitch.

On comparing the fit, Eq. (3.7) with Eqs. (3.1) - (3.3), I conclude that what has been observed is the linear response, $N_\ell(t)$, with t_ℓ= 14d, of part of the region through which the vortices have moved, and that the long-term offset in the spin-down rate is a non-linear response from the remaining parts of the region, with relaxation times $\tau_{n\ell}$ in excess of years, and a net inertial moment $I_{n\ell}$. Neither gradual recoupling of the boundary region or a linear response from regions through which vortices have not moved is seen. Since $N_\ell(t)$, $I_{n\ell}$ (and I_A) are known from observation, it is straightforward to obtain the remaining unknowns, I_ℓ and $\delta\Omega$ by making use of Eq. (3.2) and Eq. (3.4). The results of that analysis are likewise given in Table 1.

We see that the relative glitch-induced change in superfluid angular velocity for the two glitches is comparable, and is $\sim 3 \times 10^{-5}$, corresponding to the unpinning and subsequent motion of $\sim 3 \times 10^{11}$ vortices in each glitch. The analysis suggests that in the 1989 glitch, which was approximately twice the magnitude of that observed in 1975, both the non-linear and linear contributions to the post-glitch spin-down torque likewise approximately doubled, with the latter increase coming from a substantially larger region of 14^d linear creep response to the glitch.

3.2 THE VELA PULSAR

Eight "superglitches" of the Vela pulsar have now been observed, with the last, Christmas 1988, glitch occurring during the observing sessions in both Tasmania (McCulloch et al., 1990) and South Africa (Flanagan, 1990). Alpar, Chau, Cheng, and I (Alpar et al., 1991) have recently been able to fit the post-glitch behavior following each of these glitches using the model described above. For the first seven glitches it sufficed to combine a linear response, with $\tau \sim 3$d, from a superweak region through which vortices did not move, with two contributions from regions through which vortices had moved: a linear torque of the form Eq. (3.2), with $\tau_\ell \sim 33$d, and the full non-linear torque, Eq. (3.3). In the eighth glitch, Flanagan found a new rapid linear response, with $\tau_\ell \sim 0.4$d, which Alpar et al. (1990) have shown can come from a "second" superweak region through which no vortices have moved; because the relaxation time of this region is quite rapid, its contribution to the post-glitch response is only seen unambiguously if observations begin at times quite close to the time of the glitch.

In making the fits, Alpar et al. assume that the linear response from regions through which vortices do not move is the same after every glitch, and is given by the two-component results (see Eq. 3.1) for the torque exerted by the superfluid on the crust

$$I_1 \frac{\Delta\Omega_c(o)}{\tau_1} e^{-t/\tau_1} + I_2 \frac{\Delta\Omega_c(o)}{\tau_2} e^{-t/\tau_2}; \tag{3.8}$$

thus I_1 and τ_1, and I_2 and τ_2, are not assumed to vary from glitch to glitch. They find that a good fit to the rapid post-glitch response is obtained with the following parameters:

$$\tau_1 = 0.4\text{d} \qquad (I_1/I) = 5.7 \times 10^{-3} \tag{3.9}$$

$$\tau_2 = 3.2\text{d} \qquad (I_2/I) = 1.5 \times 10^{-3} \tag{3.10}$$

On the other hand, they take the torques associated with regions through which vortices move to be given by N_ℓ (t) and $N_{n\ell}$ (t), Eqs. (3.2) and (3.3), with only $\tau_\ell \cong 33\text{d}$ held constant from glitch to glitch. The results they obtain are given in Table 2, and I discuss these briefly.

Table 2. Vela pulsar glitches: post-glitch response parameters for regions through which vortices have moved.

Glitch date	2/28/69	8/29/71	9/28/75	7/3/78	11/11/81	8/10/82	7/12/85	12/24/88
$[\Delta\Omega_c(o)/(\Omega_c]_{-6}$	2.35	2.05	1.99	3.06	1.14	2.05	1.30	1.81
$(\Delta\dot{\Omega}_c(o)/\dot{\Omega}_\infty)_{-2}$	1.3	1.8	1.1	1.8	0.90	2.0	1.5	16
$(I_{n\ell}/I)_{-4}$	7.1	7.2	7.2	6.6	6.3	6.0	6.5	4.7
$(\delta\Omega/\Omega)_{-4}$	9.5	8.1	7.1	11	4.4	9.3	5.7	9.8
$(I_A/I)_{-3}$	3.5	3.6	4.2	4.4	3.8	3.2	3.2	2.7
$[(I_{n\ell} + I_A)]_{-3}$	4.2	4.3	4.9	5.1	4.4	3.8	3.9	3.2
t_A (d)	7900	6700	5900	8900	3700	7800	4800	8200

Inspection of Table 2 shows that it is $\delta\Omega$, and thus the net number of vortices unpinned, which varies most significantly from glitch to glitch (compare, for example, the the 1978 and 1981 glitches); the sum of the inertial moments, $I_{n\ell} + I_A$, of the pinned crustal superfluid in the region through which vortices have moved is nearly the same for all eight glitches. The "anomalously" large value of $|\Delta\dot{\Omega}_c(o)/\dot{\Omega}_\infty|$ observed for the eighth glitch simply reflects the fact that the immediate post-glitch observations of Flanagan picked up the very large internal torque, $[I_p^{(1)} \Delta\Omega_c(o)/\tau_1]$ which can only be seen for observations at times $\lesssim 1^d$ of the glitch. On doing the bookkeeping for the net moment of inertia for the pinned crustal superfluid,

$$I_p = I_1 + I_2 + I_\ell + I_{n\ell} + I_A \tag{3.11}$$

I find, on making use of Eqs. (3.9) and (3.10), and Table 2, that

$$\frac{I_p}{I} \gtrsim 0.013. \tag{3.12}$$

This result provides perhaps the strongest constraint to date on the hadron matter equation of state at densities greater than nuclear matter density, since for very soft equations of state, such as that calculated using the Reid potential, the stellar crust of a $1.4M_\odot$ neutron star is so thin that its total inertial moment is less than 0.02 I.

It is instructive to compare the glitch parameters for the Crab and Vela pulsars. The values obtained for $I_{n\ell}$ from observation are not markedly dissimilar; what is responsible for the much larger values of the Vela pulsar glitches is primarily the larger number of vortices released in a glitch. This result is perhaps to be expected on evolutionary grounds; as a neutron star cools and spins down more slowly, a given set of pinning sites becomes more effective in holding vortices back; at a glitch, then, more vortices will be released. The overall difference in the Vela and Crab post-glitch behavior can also be qualitatively understood in terms of their different ages and temperatures; those "superweak" regions in the Vela pulsar described by (I_1, τ_1) and (I_2, τ_2) would not give an observable response in a pulsar as young and hot as the Crab pulsar.

4. The Pinned Crustal Superfluid: What Glitches Teach Us

The fit to postglitch behavior provided by vortex creep theory is strong evidence for vortex unpinning and subsequent repinning in pulsar glitches, and hence for the presence of pinned neutron superfluid in the stellar crust. Not only does the response of the pinned crustal superfluid provide a natural explanation for the simultaneous presence of regions which respond linearly to sudden changes in the crustal and superfluid angular velocity, and of regions which respond in non-linear fashion (observed as a long-term off-set in $\dot{\Omega}_c$ and an internal torque which increases linearly in time), but the fractional moments of inertia of the distinct regions whose response is observed are all $\lesssim 10^{-2}$, which is what microscopic theory gives for regions located in the inner crust. Moreover the changes in glitch size and postglitch behavior from the Crab to the Vela pulsar can be understood as a result of pulsar evolution, and do not require the assumption that the pulsars possess different masses.

Thus the response of that region of the Crab pulsar with (I_ℓ/I) $\lesssim 3 \times 10^{-3}$ which responds linearly (with $\tau_\ell \sim 14^d$) to a glitch may be expected to evolve in such a way that when the pulsar possesses a temperature and spindown rate comparable to the Vela pulsar, it will respond non-linearly in a fashion similar to that seen for region "A" of the latter pulsar; a torque which increases linearly in time, with (I_A/I) $\lesssim 4 \times 10^{-3}$. Similarly, the regions of the Vela pulsar which at present respond linearly would, at an epoch in which the Vela pulsar resembled the Crab pulsar, have responded so rapidly to a glitch that their response could not have been observed.

As Alpar, Cheng, and Pines (1989) have shown, one can take the further step of combining the results of the phenomenological analysis with microscopic theory to obtain useful limits on pinning energies and crust superfluid angular velocities for both pulsars. Thus the simultaneous presence of linear and non-linear response in the Crab pulsar means, according to Eq.(2.15), that the pinning energies in that part of the star which responded non-linearly must exceed ~1 Mev (for a Crab pulsar temperature of ~35 kev). Application of Eq.(2.10) to the region of the star which responded linearly with $\tau \sim 14^d$ gives a "characteristic" pinning energy only slightly below this value. (Since τ_ℓ depends exponentially on E_p and the latter quanity may be expected to vary over a region which has an inertial moment $\sim 10^{-3}$ I, it is to be expected that what has been observed is the integrated response of many smaller regions with a range of pinning energies, such that one arrives at $\tau \sim 14^d$ on computing the integrated response.) On applying similar arguments to the Vela pulsar, for an assumed Vela temperature of 11 kev, I conclude that the characteristic pinning energy for the region whose linear response time is $\sim 33^d$ is some 330 kev, and that for regions with pinning energies slightly larger than this the response must necessarily be non-linear.

Finally, I call attention to the fact that this deduced range of pinning energies is consistent with those calculated from Eq.(2.16), using the energy gaps calculated by Ainsworth, Wambach, and Pines (1989). Were the calculated energy gaps either a factor of two larger or smaller, even qualitative agreement between theory and these deduced observational constraints would be difficult, if not impossible, to achieve.

Given the many simplications made in developing and applying vortex creep theory, and the difficulty of carrying out microscopic calculations of energy gaps, pinning energies and the neutron matter equation of state, the extent to which the present microscopic calculations of the consequences of hadron interaction are consistent with glitch observations is both pleasing and surprising. Indeed it would seem that the study of neutron stars may now have progressed to the point that during the next decade, by combining theory and observation, we can use glitching pulsars as a cosmic hadron physics laboratory, not simply to verify the presence of pinned crustal superfluid, but to deduce the properties of that superfluid deep within the stellar crust.

Acknowledgements

I should like to thank Ali Alpar, H. F. Chau, and K. S. Cheng for stimulating discussions and correspondence on these and related topics, Andrew Lyne for communicating to me in advance of publication the results of his analysis with Pritchard of the timing behavior of the Crab pulsar, and Ali Alpar for a critical reading of this manuscript. The research described in this lecture has been supported by the National Science Foundation through Grants NSF PHY 86-00377 and NSF INT 89-18665, and by NATO.

References

Ainsworth, T., Wambach, J., and Pines, D. (1989), *Phys. Rev. Lett.* **B**, 228 **222**, 173.

Alpar, M. A. (1991), These proceedings.

Alpar, M. A., Chau, H. F., Cheng, K. S., and Pines, D., in preparation.

Alpar, M. A., Cheng, K. S., and Pines, D., (1990) *Nature*, **348**, 707.

Alpar, M. A. and Pines, D. (1989), in Timing Neutron Stars, ed. H. Ögelman and E.P.J. van den Heuvel (Kluwer:Dordrecht), p. 441-456.

Alpar, M. A., Anderson, P. W., Pines, D., and Shaham, J. (1984a), *Ap. J.*, **276**, 325.

Alpar, M. A., Anderson, P. W., Pines, D., and Shaham, J. (1984b), *Ap. J.*, **278**, 791.

Alpar, M. A., Cheng, K. S., and Pines, D. (1989), *Ap. J.*, **346**, 823.

Alpar, M. A., Nandkumar, R., and Pines, D. (1985), *Ap. J.*, **288**, 191.

Anderson, P. W., and Itoh, N. (1975), *Nature*, **256**, 25.

Baym, G. (1991), These proceedings.

Baym, G., Pethick, C., Pines, D., and Ruderman, M. (1969), *Nature*, **224**, 872.

Clark, J. (1991), in Structure and Evolution of Neutron Stars, ed. D. Pines, R. Tamagaki, and S. Tsuruta, (Addison-Wesley), in the press.

Flanagan, C. (1990) *Nature*, **345**, 416.

Hoffberg, M., Glassgold, A. E. Richardson, R. W. and Ruderman, M. (1970), *Phys. Rev. Letters*, **24**, 175.

McCulloch, P. M., Hamilton, P. A., McConnell, D. and King, E. A. (1990), *Nature*, **346**, 822.

Pandharipande, V. R., Pines, D. and Smith, R. A. (1976), *Ap. J.*, **208**, 550.

Pines, D. (1971) in Proc. 12th International Conference on Low Temperature Physics, ed. E. Konda (Tokyo:Kaigeku), p. 7.

Pines, D., and Alpar, M. A. (1985), *Nature*, **316**, 27.

Radhakrishnan, V., and Manchester, R. N. (1969), *Nature*, **222**, 228.

Reichley, P .E., and Downs, G. S. (1969), *Nature*, **222**, 229.

Ruderman, M. (1991), These proceedings.

Shibazaki, N., and Lamb, F. K. (1989), *Ap. J.*, **346**, 808.

Wambach, J., Ainsworth, T., and Pines, D. (1991), These proceedings.

Wiringa, R., Fiks, V. and Fabrocini, A. (1988), *Phys. Rev.* **C 38**, 1010.

NEUTRON STAR PLATE TECTONICS

M. RUDERMAN
Physics Department and Columbia Astrophysics Laboratory†
Columbia University
538 West 120th Street
New York, NY 10027 and
University of Oxford
Department of Theoretical Physics
1 Keble Road, Oxford OX1 3NP

ABSTRACT. Spinning down (or up) neutron star crusts can be stressed beyond their yield strengths by neutron superfluid vortex line pinning or magnetic field stresses. The spin history of a fast pulsar determines radio emission features which can be compared to those observed in millisecond pulsars. In young warm spinning-down neutron stars such as the Crab pulsar, crust flow should be largely plastic. Crusts of cooler stars such as the Vela pulsar may crack and give period glitches. In more slowly rotating (cool) neutron stars the glitches should be associated with intense bursts of gamma-rays.

1. Introduction

The superfluid superconducting core of a neutron star is surrounded by a relatively thin crustal lattice of nuclei. The crust thickness (ℓ) is of order 10^5cm. The deep interior of the crust has an extremely high electrical conductivity (of order 10^{11} that of copper) [1]. In so far as the crust is rigid any magnetic field through it would be frozen on timescales up to the eddy current diffusion time

$$t_e \sim \frac{3\ell_5^2}{T_8^2} 10^8 \text{ yrs.} \tag{1}$$

Since the inner crust temperature $T < 10^{8}\,^{\circ}$K for all neutron stars except the very youngest (and $T_8 < 4$ even for the Crab pulsar), a rigid crust would freeze any magnetic field through it on timescales of interest in neutron star evolution. However, stresses which build up within the crust as the star's spin rate changes can cause the crust to move and carry with it embedded magnetic field. In millisecond radiopulsars crust motion should be the critical feature determining magnetic field geometry, especially the orthogonality or alignment of magnetic dipoles and spin [2]. In canonical young radiopulsars it would give a period (P) dependent dipole moment which is reflected in the spin-down index relating $\dot{P}$ and P. In cooler pulsars the deep crust is brittle. If brittle crust motion entails large scale crust cracking rather than very small scale crumbling, such events would give period glitches similar to those observed in radiopulsars much older than 10^3 years. The sudden twisting of surface magnetic field accompanying such cracking should give bursts of hard

† Permanent Address

J. Ventura and D. Pines (eds.), Neutron Stars: Theory and Observation, 71–86.

X-rays and γ-rays in neutron stars with sufficiently low pre-glitch magnetosphere charge density. The expected number and energy of such events closely resembles those needed for the observed Gamma-Ray Burst sources. The stresses in the crust and the crust's response to them can, therefore, play a crucial role in a variety of pulsar phenomena. We turn to these crust properties in the next section.

2. Stress and Strain in the Stellar Crust

2.1. Crust Properties

The electron Fermi energy in the deep crust is so large that the relativistic degenerate electron sea is only slightly perturbed by the Coulomb fields of the nuclei imbedded in it. If electron screeening is neglected the lattice nuclei has a bcc structure with melting temperature

$$T_m = \frac{(Ze)^2}{180 b_Z k_b} \sim 4 \cdot 10^9 \left(\frac{Z}{40}\right)^{5/3} \quad {}^\circ\mathrm{K} \tag{2}$$

and shear modulus

$$\mu = 0.3 \frac{(Ze)^2}{b_Z^4} \sim 2 \cdot 10^{29} \left(\frac{Z}{40}\right)^{2/3} \quad \mathrm{dyne\ cm}^{-2}, \tag{3}$$

with b_Z the internuclear spacing, k_B the Boltzman constant and Z the atomic number of the nuclei. The numerical estimates here (and below) assume $b_Z = 5 \cdot 10^{-12}$cm, $Z = 40$, and total crust density $\rho \sim 5 \cdot 10^{13}\mathrm{gcm}^{-3}$ (near where neutron vortex line pinning is strongest) [3]. The BCS correlation length of the internuclear neutron superfluid (ξ) around this density is expected to be greater than the nuclear radius ($R_N \sim 7 \cdot 10^{-13}$cm) but less than b_Z. In this region, the superfluid transition temperature $T_c \sim 10^{10}{}^\circ$K $\sim T_m$, all much greater than the lower crust temperature of neutron stars more than a few decades old.

In the relevant classical regime where lattice quantum effects are unimportant the dimensionless strain θ in a pure Coulomb lattice such as that above ($\theta \equiv$ change in length per unit length under tension or compression) can depend upon stress (σ) only in the form

$$\sigma = \mu\theta f(\theta, T/T_m, x), \tag{4}$$

where x is a dimensionless measure of the crystal's dislocation complex. Stress experiments on ionic crystals at ordinary densities in the laboratory may be extrapolated to the superdense crystalline matter of a neutron star crust. Stress measurements on very slowly stretched and compressed tiny (Mg stabilized) Li crystals are shown in Figs. 1 and 2 [4]. The crystal melting temperatures (T_m) are around 700°K. The following extrapolatable features are significant for neutron star crusts:

a) $\sigma = \theta\mu$ until an elastic strain limit $\theta_{\max}$ is reached;
b) $\theta_{\max} \sim 5 \cdot 10^{-3}$ at $T \sim 10^{-1} T_m$;
c) $\theta_{\max} \sim 10^{-2}$ for $T \ll 10^{-1} T_m$;
d) $\theta_{\max} \ll 10^{-2}$ for $10^{-1} T_m \ll T < T_m$.

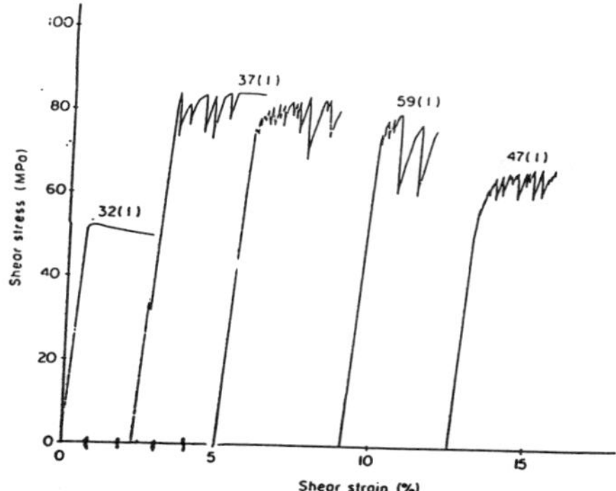

Fig. 1. Shear stress vs. strain for Li alloy crystals (stabilized with 57% Mg) under tension. The 32 (1) curve is for $T = 10°$K; the others are different samples at $T = 77°$K [4].

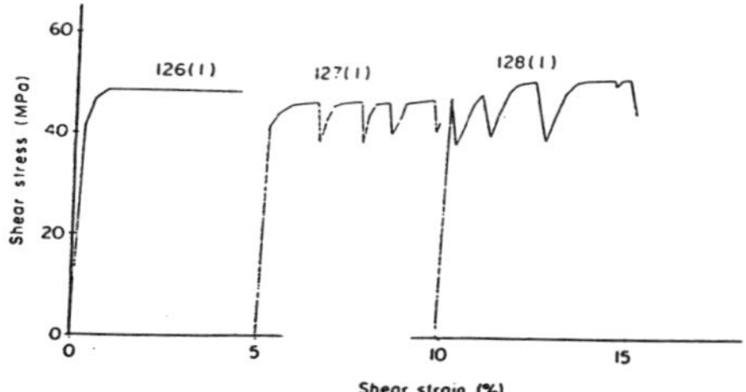

Fig. 2. Shear stress vs. strain for Li alloy crystals (stabilized with 33% Mg) under compression. $T = 77°$K for sample 126 (1), 65°K for 128 (1), [4].

When $\theta > \theta_{\rm max}$, σ remains near $\mu\theta_{\rm max}$. The crystal lattice yields, and continues to deform without further increase in stress. When $T \lesssim 10^{-1}T_m$ the lattice yields discontinuously: it continually but eratically "breaks" on a scale $\Delta\theta < \theta_{\rm max}$. Then we expect in a neutron star crust

e) a transition in yeilding by plastic flow to a brittle response over a small temperature range (of order $10^{-2}T_m$) when $T \sim 10^{-1}T_m$.

A crucial and unresolved problem is not so much the extrapolation through almost 14 orders of magnitude from the density of small laboratory crystals to that within a neutron star crust, but rather extrapolating from the behavior of small single crystals or grains to that of an entire crust, a 10^{17} lattice constants in dimension. How does a brittle neutron star crust break when stress slowly grows to that at which such a crust must yield? Does it yield only by very many microscopic breaks so that its large scale response still resembles that of plastic flow or does it yield by rarer but more substantial cracks? In Sections 6 and 7 we shall see the observational consequences when the large scale crust response to a stress beyond its yield strength resembles that of the small crystals of which the crust is composed.

In numerical extimates below we shall conservatively assume $\theta_{\mathbf{max}} \sim 10^{-2}$. It was long ago proposed [6,7] that for the crust as a whole an effective $\theta_{\mathbf{max}}$ as small as 10^{-3} to 10^{-5} may be plausible. Smaller values of the strain elastic limit would enlarge the domain of applicability of the conclusions below.

2.2. Shear Stresses on a Neutron Star Crust

Two major kinds of shearing stresses act on a neutron star crust.

1) The neutron superfluid which fills the space between lattice nuclei rotates (average superfluid angular velocity $\equiv \mathbf{\Omega}_n$) only through a quasiparallel array of quantized vortex lines each of which has a circulation $\pi\hbar/m_n$ ($m_n \equiv$ neutron mass). The area density of these vortex lines is

$$n_V = \frac{2m_n\Omega_n}{\pi\hbar}. \tag{5}$$

In most of the crust, where $\rho \sim 10^{13} - 10^{14}\mathrm{gcm}^{-3}$, these vortex lines are expected to be pinned to crust nuclei [7,8]. When the crust spin rate and crustal neutron superfluid rotation rates are equal there is no shear stress on the lattice to which the vortex lines are pinned. Otherwise there is a force density on the lattice within the pinning region of the crust

$$\mathcal{F}_0 = 2\boldsymbol{\omega} \times (\mathbf{\Omega}_n \times \boldsymbol{r})\rho_n \tag{6}$$

for a relative rotation velocity

$$\boldsymbol{\omega} \equiv \mathbf{\Omega}_n - \mathbf{\Omega}\,, \tag{7}$$

where ρ_n is the internuclear superfluid neutron density, and $\mathbf{\Omega}$ the angular velocity of the lattice and of the vortices pinned to them. Only part of the lattice body force of Eq. (6) can be balanced by pressure from the deformed lattice and its degenerate electrons together with the gravitation force. The rest is opposed by the lattice shear strength [2]. The lattice elastic strength will be exceeded when ω exceeds

$$\omega_B = 4\cdot 10^{-2}\left(\frac{P}{10^{-3}\mathrm{s}}\right)\left(\frac{\theta_{\mathrm{max}}}{10^{-2}}\right)\left(\frac{5\cdot 10^{13}\mathrm{gcm}^{-3}}{\rho}\right)\left(\frac{Z}{40}\right)^{2/3}\ \mathrm{s}^{-1}. \tag{8}$$

If ω reaches ω_B the lattice yields to plastic flow, crumbling, or cracking, and will not support any increase in stress beyond that needed to achieve $\omega = \omega_B$.

2) A second kind of stress on the crust originates in the pull of core magnetic field which passes through it. If the core magnetic field were homogeneous this stress could be at most

$$S_1(B) \sim \frac{B^2}{8\pi} \sim 4\left(\frac{B}{3\cdot 10^{12}G}\right)^2 10^{23}\ \mathrm{dyne\ cm}^{-2}, \tag{9}$$

which is expected to be much smaller than the maximum possible stress the crust could bear before yielding,

$$S_{\mathrm{max}} \sim \frac{\ell}{R}\mu\theta_{\mathrm{max}} \sim 3\cdot 10^{26}\left(\frac{\theta_{\mathrm{max}}}{10^{-2}}\right)\ \mathrm{dyne\ cm}^{-2}. \tag{10}$$

However, a superconducting core will organize any magnetic field which threads it into structures such that $\langle B^2 \rangle \gg \langle B \rangle^2$. If the core is a type II superconductor (proton plasma frequency $\omega_p < c\xi^{-1}$) the core magnetic field forms an array of quantized magnetic flux tubes in the core's superconducting proton sea which terminate near the base of the crust. There they open out and traverse the crust as a normal magnetic field. These core flux tubes move in response to changes in the positions of core neutron superfluid vortex lines as a neutron star spins up or down [9]. Each quantized magnetic flux tube in the subcrust core contains an average magnetic field $B_c \sim 10^{15} G$. (It is mainly because the neutron vortex line cores also contain a similarly magnetized core [10] that flux and vortex cores do not easily pass through each other.) If the crust were to remain rigid and immobile the shear stress on the base of the highly conducting crust from core magnetic flux tube motion could grow to reach

$$S_2(B) \sim \frac{BB_c}{8\pi} \sim \left(\frac{B}{3 \cdot 10^{12} G}\right) 10^{26} \text{ dyne cm}^{-2}, \tag{11}$$

where B is the average magnetic field through the crust. If the core protons form a type I superconductor B_c in Eq. (11) is replaced by the magnetic field needed to quench proton superconductivity. This is still expected to be of order $10^{15} G$.

We shall consider first the family of "millisecond pulsars" which have small magnetic dipole moments. For these the pinned vortex stresses in the crust are expected to be much more important than those from pull of the core magnetic field. To describe consequences of the pinning stresses it is necessary to compare the magnitude of the pinning stresses which would deform the crust beyond its yield limit to those which would cause vortices to become unpinned.

2.3. When and Where does Vortex Unpinning Preempt Crust Breaking in Limiting Stresses on the Lattice?

There is not yet a definitive calculation of the strength of the pinning of neutron superfluid vortex cores to lattice nuclei in the deep crust. In the approximation that superfluid energy gaps and correlation lengths can be treated as local functions even over distances smaller than the nuclear radii ($\xi > R_N$) the pinning force is directly related to calculated BCS energy gaps for homogeneous neutron matter. The pinning is strongest in the density regime $2 - 7 \times 10^{13}$gcm^{-3}. Some calculations for the critical angular velocity difference (ω_{cr}) which would cause unpinning in this density regime are given in Table 1 for two different assumptions about lattice crystal axis alignment at a vortex core. If the lattice forms before the neutron superfluid transition temperature T_c no special correlation between lattice axes and neutron star spin (the vortex direction) are expected. The ω_{cr} for pinning by a randomly oriented lattice are given in the "unaligned lattice" column. the last column is for the aligned case. When the lattice forms after the superfluid vortices the lowered energy of a nucleus situated within a vortex line's core favors the local crystallization patterns which minimize the spearation between nuclei within such cores. The neutron superfluid transition temperature T_c is related to the BCS energy gap Δ by

$$T_c = 5 \left(\frac{\Delta}{\text{MeV}}\right) \cdot 10^9 \text{ °K}. \tag{12}$$

If Δ in the $2-7\times10^{13}$gcm^{-3} regime exceeds 1 MeV, $T_c > T_m$ of Eq. (2), and considerable initial lattice alignment should exist in the strongest pinning region. This inequality is satisfied for the gap energy calculations used in Table 1 (e.g. $\Delta > 1.3$ MeV in refs. [11] and [12]). A stronger reason for assuming alignment is based upon the evolution of the new hot crust in a very young neutron star. As long as the crust temperature $T \gtrsim 10^{-1}T_m$ the crust's elastic yield strength is much smaller than the value it ultimately achieves when $T < 10^{-1}T_m$. A "hot" crust has an effective $\theta_{\mathrm{max}} \ll 10^{-2}$ and will flow in response to spin-up or spin-down induced stresses as indicated in the next section. The moving crustal matter is continually subducted into the core and replaced by core matter flowing up from below. This core matter will, of course, form nuclei and a lattice after it has moved into the crust but this will be at the existing crust temperature rather than that at which the original crust solidified. One consequence is that continual formation of new crust will favor micro-crystal alignment along the already existing crustal neutron superfluid cores.

Table 1. Maximum angular frequency differences ω_{cr} between crustal neutron superfluid and pinned vortex lines before unpinning. Assumed lattice spacing $b_Z = 5\cdot10^{-12}$cm and superfluid density $\rho_n = 5\cdot10^{13}$gcm^{-3}.

MODEL	$\omega_{cr}(\mathrm{s}^{-1})$	
	Aligned Lattice	Unaligned Lattice
AAPS	18	2
ACP	4	0.5
R	5	0.7
EB	10^2	10

The AAPS [11], ACP [12] and R [2] estimates assume ξ and Δ are local functions. EB [13] introduce a correction for non-locality. The ACP ratio between aligned and unaligned lattices is assumed for all models. The EB ω_{cr} is inferred from the maximum of the calculated derivative of pinning energy with respect to separation of vortex axis from nucleus center.

From Table 1 and ω_B of Eq. (8) it seems plausible to answer the title question of this section as follows.

1) Even if θ_{max} for the large scale structure of the crust is 10^{-2} and the crustal lattice is always randomly aligned at a vortex core, lattice yield strengths are exceeded before unpinning when $P \lesssim 10$ ms.
2) Arguments for alignment at vortex cores suggest extending the lattice breaking before unpinning regime to periods $P \lesssim 10^2$ ms.
3) If the lattice has large scale "fault planes" it may be much weaker than the micro-crystals of which it is formed. The above range of pulsar spin periods may be an underestimate but would hardly be expected to be an overestimate. An effective $\theta_{\mathrm{max}} \ll 10^{-2}$ and a range of periods $P \gg 10^2$ ms has not been excluded.

It should be noted that as long as all vortex lines in the crust are nearly parallel to the crustal spin $\mathbf{\Omega}$ a considerable fraction of the crustal neutron superfluid vortex array would always unpin in response to growing ω or may never be pinned at all. Vortex lines near the stellar equator do not reach to the base of the crust (cf. Fig. 3) and may not even pass through regions of crustal density above $10^{13}\mathrm{gcm}^{-3}$ below which pinning is ineffective. Thus post-glitch analyses [14,15] of crustal superfluid response can focus on a different region of the neutron vortex array than the more tightly bound part discussed above.

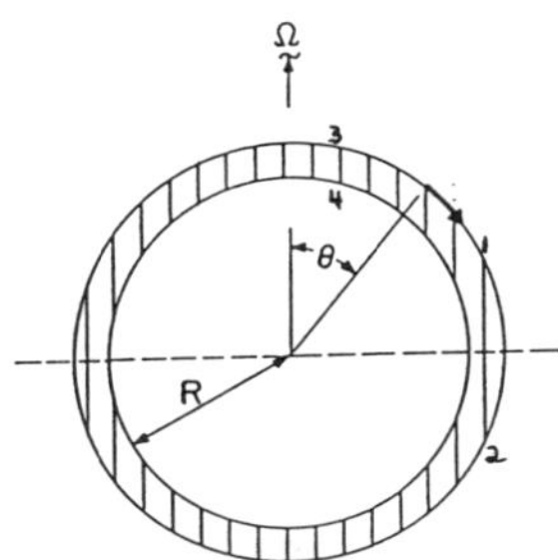

Fig. 3. Vortex array in the crust of a spinning neutron star. The vortex line $\overline{34}$ is strongly pinned; the line $\overline{12}$ can unpin before the local lattice yields or may not be pinned at all.

3. Plate Motion

We have seen in Sect. 2.3 that the crusts of very rapidly spinning pulsars with P in the range given in 2.3 1), 2), or 3) must continually yield by creep or cracking as the pulsar spins-down or is spun-up. On long timescales $t > t_B = \omega_B/\dot{\Omega}$ the crustal "plate" flows like a very viscous fluid. The plate consists of the entire crustal lattice including the parts where pinning is ineffective, as well as the electrons which move with it. On these scales crust motion keeps the local average superfluid angular speed very near that of the spinning-up or -down crust: $|\Omega - \Omega_n| \lesssim \omega_B$. Then on the average crustal plates for a spinning-down millisecond pulsar must move as indicated in Figs. 3 and 4. Plates move with a velocity $\boldsymbol{v}$ whose perpendicular component to the spin axis $r_\perp$ away is

$$\boldsymbol{v}_\perp = -\frac{\boldsymbol{r}\,\dot{\Omega}}{2\,\Omega} \,. \tag{13}$$

The tangential component of plate velocity

$$\boldsymbol{v}_t = \frac{R\dot{\Omega}\sin\theta}{2\Omega\cos\theta}\hat{\boldsymbol{\theta}}. \tag{14}$$

It reverses sign at the equator where the downward flowing northern hemisphere plate meets the upward moving sourthern hemisphere one. (This plate motion is also the same as that

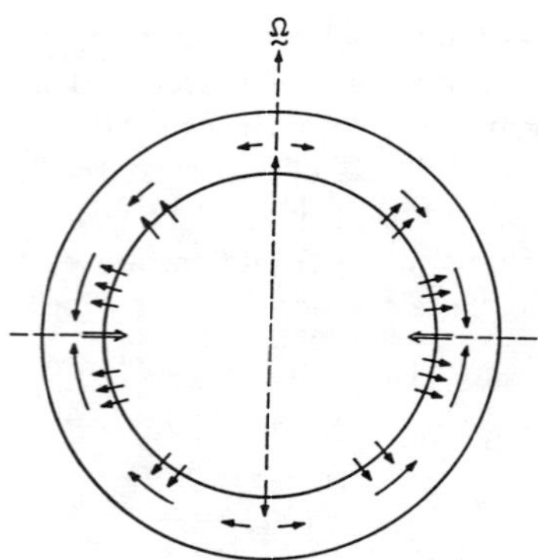

Fig. 4. Plate flow and core matter motion into the plate of a spinning down millisecond pulsar [2]. Some possible crust backflow close to the surface is not shown.

which would exist if plates were coupled to core superfluid vortices either by the strong magnetic field which connects them in canonical pulsars or by core vortices extending into the crust lattice.) The equatorial region is mainly a plate "subduction zone" where crustal plate matter from both hemispheres is pushed into the core. There nuclei dissolve into the core's neutron-proton sea. Elsewhere the moving plates are fed matter from core over which they move. Some crustal backflow near the surface may accompany the subduction. New plate nuclei from the newly injected nucleons continually form to replace those ultimately lost from the moving plates in the subduction zone. They will form crystals with an aligned lattice axis within a neutron vortex core.

For a plate mass per unit area Σ ($\sim 10^{18}\mathrm{gcm}^{-2}$) the average rate at which new mass enters a moving plate from the core is [2]

$$\dot{\Sigma} = \Sigma \nabla_2 \cdot \boldsymbol{v}_t = \frac{\Sigma}{2}\left(1 + \frac{1}{\cos^2\theta}\right)\frac{\dot{\Omega}}{\Omega} \, . \tag{15}$$

If a millisecond pulsar is spinning-*up* the stress of Eq. (6) is a push *inward.* This results in a compression or bucking around the poles (rather than a stretching) and an equatorial tearing of the plate lattice when yield strengths are ultimately exceeded by the pinned vortex forces. Plate matter penertrates into the core except near the equator where core matter now enters into the plate. On the average a time reversal of the situation indicated in Fig. 4 is expected.

4. Magnetic Dipole Evolution in Very Rapidly Spinning Pulsars

Because the base of a crustal plate is an extraordinarily good electrical conductor any traversing magnetic field moves with the crust, or more specifically with its electron-lattice component. When the neutron star field is weak, as in millisecond pulsars and LMXB's, magnetic stresses on crustal plates are negligible relative ot the crust breaking stresses of Sect. 2.2 (2) ($B^2/8\pi\mu\theta_{max} \sim 10^{-9} B_9^2$ for $\theta_{max} = 3 \cdot 10^{-3}$).

The movements of magnetic field emerging from and reentering the surface of a spinning-up and a spinning-down millisecond pulsar are shown in Figure 5 and 6. In a spinning-down star all poles move into the equatorial plate subduction zone. There they can disappear only if North and South poles meet or can be pulled together through the crust. This is suppressed in very weak field neutron stars. Until magnetic poles carried into the thin subduction zone find partners with which to recombine, the magnetic field strength there builds up. Unless the initial magnetic field is very weak the subduction zone is then subject to large growing stresses which should ultimately cause it continually to crack, crumble, or creep – all of which give it fluid-like properties over long timescales. North-South pole attraction in the subduction zone can then increase the recombination rate above that sustained merely by initial overlap when they entered the zone. The regime of fields sufficienntly weak to avoid such destruction is roughly [2]

$$B < 5 \cdot 10^9 \left(\frac{P}{10^{-3}\mathrm{s}}\right)\left(\frac{\theta_{max}}{10^{-2}}\right) G \ . \tag{16}$$

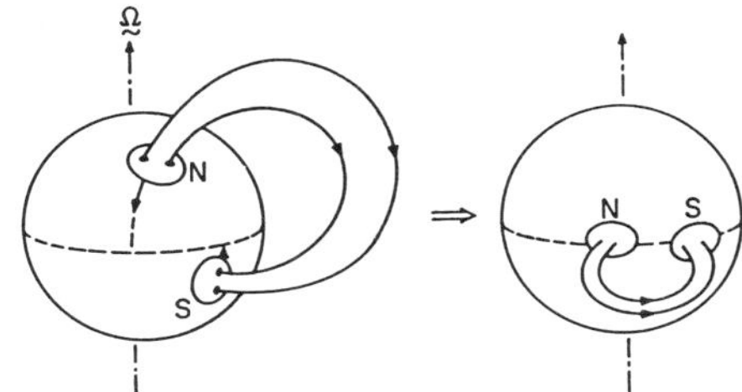

Fig. 5. Evolution of the surface magnetic field of a short period spinning down neutron star when equatorial zone magnetic field reconnections are ignored [2].

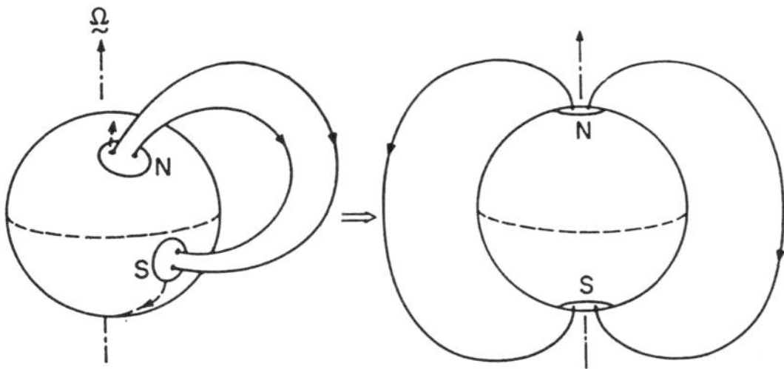

Fig. 6. Evolution of the surface magnetic field of a short period spinning-up neutron star when flux lines connect the two hemispheres [2].

We are led to expect that unless initial field geometries are very special

a) spun-up millisecond pulsars have almost spin aligned magnetic dipoles. If spin-up has increased Ω by a large factor the pulsar magnetic field should be almost axially symmetric unless the initial field is in the category discussed in the next paragraph.
b) spun-down very weak field ($B \lesssim 10^9 G$) millisecond pulsars have perpendicular magnetic dipoles. After large spin-down almost all poles lie in the equatorial plate subduction zone (a zonal strip of angular width $\lesssim \ell/R \sim 10^{-1}$).

It should be noted, however, and perhaps emphasized, that the predicted magnetic field alignment for a strongly spun-up pulsar depends crucially upon the assumption that the magnetic flux *out* of a hemisphere is not almost exactly balanced by the magnetic flux *into* that same hemisphere. When this is not the case there is no net flux out of each of the two shrinking polar caps into which the initial crust hemispheres of a spun-up pulsar evolve. Since the radii of these polar caps is approximately $(\Omega_0/\Omega)^{1/2}R$ as the pulsar angular spin rate grows from Ω_0 to Ω, one consequence is a greatly reduced stellar magnetic dipole moment whose orientation is typically orthogonal to the pulsar spin to within an angle $(\Omega_0/\Omega)^{1/2}$. In this very special case, spin-up might also lead to orthogonal dipole moments (cf. Fig. 7). If a fraction exceeding $(\Omega_0/\Omega)^{1/2}$ of flux from one hemisphere reenters the star at the other then the contribution to the dipole from field lines between the polar caps exceeds that from each polar cap separately and Fig. 6 geometry is dominant. For spin-up from, say, $P_0 \sim 1$ sec to $P \sim 10^{-3}$ sec total flux cancellation in each hemisphere separately would then have to be good to within three percent for Fig. 7 to be appropriate.

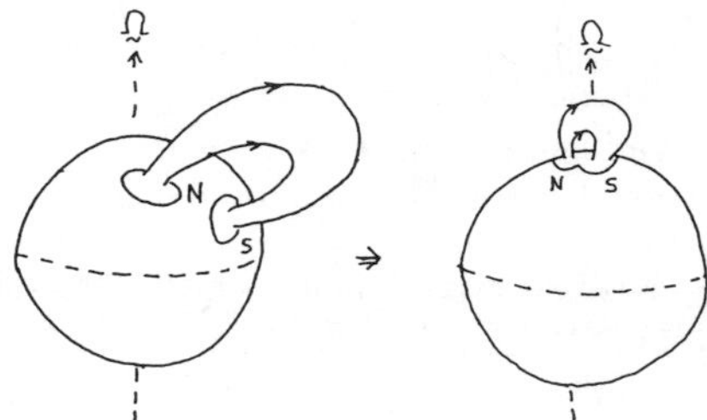

Fig. 7. Evolution of the magnetic field of a spinning-up short period neutron star when all flux leaving a hemisphere reenters the same hemisphere.

Another assumption that needs further investigation is that any crustal backflow *above* the strongly stressed $10^{13} - 10^{14}$gcm^{-3} region of a moving crustal plate does not change the long timescale evolution of the surface magnetic field.

5. Implications for Millisecond Pulsar and LMXB Observations

1) There is rather general agreement that LMXB's are rapidly spinning weak field neutron stars which are being spun-up (or have already been spun-up) by a companion-fed accretion disk. Usually these neutron stars should then be members of the a) class of Sect. 4. Almost all should have nearly axisymmetric magnetic fields so that direct

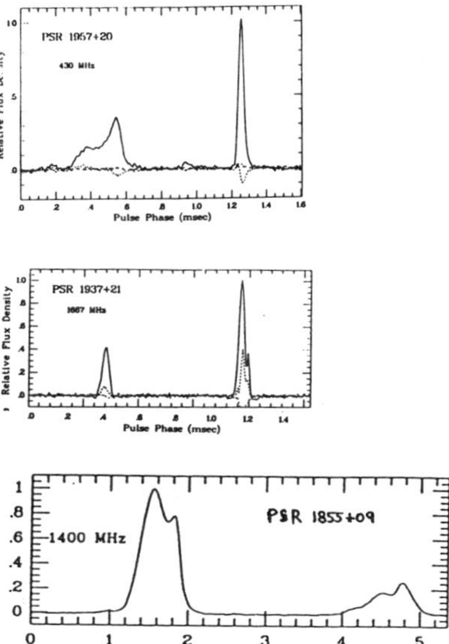

Fig. 8. Pulse shapes of the three fastest millisecond pulsars [33].

observation of the stellar spin should be very difficult. (See, however, the *caveats* in the previous paragraphs.) This would support the absence, so far, of such observations.

2) Of the 4 so far identified millisecond pulsars with magnetic fields less that $10^9 G$ at least 3 seem to be orthogonal rotators: PSR 1855+09 [16], PSR 1937+21 [17], and PSR 1957+20 [18]. They have double pulses of similar intensities (to within a factor of e) with an almost 180° frequency independent separation (cf. Fig. 8). (A 180° separation of open field line bundles far from the star is a consequence of any orthogonal rotator, even when the N and S poles are far from the equator). If three of the fastest millisecond lowest magnetic field pulsars are orthogonal rotators, the magnetic field evolution patterns in Sect. 4, Figs. 5 and 6 admit only the one in which these millisecond pulsars have been mainly spun-down rather than spun-up as has been widely suggested. One consequence is that, unless the initial magnetic field distribution is very restricted so that Fig. 7 is appropriate, the accretion spun-up LMXB's are not the ancestors of these millisecond pulsars. The apparent problem of too few observed LMXB's to account for the observed number of millisecond pulsars [19,20] would then disappear for them. The even more rapidly spinning neutron stars which are the ancestors of the present orthogonally rotating millisecond pulsars would, presumably, have been formed in the very high spin state. Accretion induced collapse of a white dwarf would then seem to be the most plausible way of forming that original neutron star population [21,22,23]. Because of the accretion induced spin-up of the white dwarf the resulting neutron star would have an initial period $P_0 \sim 10^{-3}$s if of order $10^{-1} M_{\odot}$

has been accreted [24]. These neutron stars spin-*down* to become the low B millisecond pulsars. The low B neutron star is a consequence of the "low" magnetic field of conventional white dwarfs ($B < 10^5 G$). This B is amplified by a factor of order 10^5 in the collapse from a near Chandrasekhar mass dwarf to a neutron star. In this scenario the canonical accretion "spin-up line" plays no special role: millisecond pulsars have not yet been observed considerably above it (i.e., with shorter periods) only because the time to spin-down to it from the initial P_0 period ($\sim 10^7 P_{-3}^2 B_9^{-2}$ yrs $\sim 10^8$ yrs) is so much less than millisecond radiopulsar lifetimes. The early spin-down power from a newborn millisecond pulsar $\sim 10^{38} B_9^2$ erg s^{-1}, much larger than presently observed ones. Evaporation of some companions to leave isolated millisecond pulsars may be accomplished most easily in that early phase.

While the above three of the five fastest millisecond pulsars appear to be orthogonal rotators, the other two, PSR 1953 and PSR 1821 have been identified as almost aligned by Lyne and Manchester [25] (It is probably not yet possible to exclude the possibility that these two are also nearly orthogonal. PSR 1821 is also double pulsed but with 120° subpulse sseparation and the PSR 1953 pulse structure is especially broad with a very small subpulse 180° from the main pulse structure.) If putative ancestral white dwarfs for these two pulsars were initially much closer to their Chandrasekhar limits they would have acquired correspondingly less angular momentum before collapsing to neutron stars. These neutron stars would then have been formed spinning more slowly than those which spun-down to become orthogonal rotators. Instead these two could have been spun-up by subsequent accretion to the canonical "spin-up line" from which they later spun-down to their present periods. If the spin history of these pulsars after neutron star formation contained more early spin-up than later spin-down they should now be almost aligned. Low magnetic field very fast millisecond pulsars should end as either orthogonal or aligned depending only upon the amount of mass accretion before collapse. Pulsars with the same present B and P could have had either history. (Alternatively present pulsar alignment could have been achieved by long LMXB spin-up regardless of the neutron star ancestry.) However, the more conventional spin-up genesis [34] cannot be excluded. Orthogonality would then result from the spin-up of neutron stars with those special initial field geometries which lead to Fig. 7. (Alternatively there may be a possibility of near surface crustal backflow opposite to the deep crust motion which, at least temporarily, controls surface field.)

6. Crust Motion in Canonical Pulsars: Glitches

Canonical radiopulsars such as the Crab and Vela may have been born with an initial short period $P_0 \lesssim 10$ ms (shorter certainly than the Crab's present $P - 33$ ms). The sum of the vortex pinning and magnetic stresses of Sect. 2.2 would then be expected to move the initially warm crust easily and continuously. If the stress of Eq. (11) is sufficiently strong even vortex unpinning would be irrelevant in determining crust motion. A recent estimate for the internal Crab pulsar temperature gives $T \sim 4 \cdot 10^8$°K $\sim 10^{-1} T_m$ [12]. According to Sect 2.1 the Crab would then be at the transition temperature where further cooling would make the crust brittle. Even before that transition is reached the crustal flow related evolution of pulsar magnetic field should have observable consequences. The "spin-down index" $n \equiv \Omega\ddot{\Omega}/\dot{\Omega}^2$ should differ from 3, its value for a fixed dipole. (Values in the range $n < 2$ obtain for spin-*down* with the geometry of Fig. 7 but are not excluded for various other kinds of field distributions.) The temperature at the base of the Vela crust has been estimated to be $T = 1.3 \times 10^8$°K [12] for which $T/T_m \sim 3 \cdot 10^{-2}$, well into the brittle crust response phase of Sect. 2.1. If we assume crust response to spin-down induced stress and strain to be that of Fig. 1, Vela's crust would always be stressed with $\sigma \sim \mu\theta_{\mathbf{max}} \sim 10^{-2}\mu$ and have repeated sudden small jumps in θ of $\Delta\theta$ (small next to $\theta_{\mathbf{max}}$) accompanied by small relaxations in stress. The sudden outward crust motions would carry crust pinned vortex lines outward a distance $\sim R\Delta\theta$. This spins-down the crustal neutron superfluid by an amount $\Delta\Omega_n \sim -\Delta\theta\Omega_n$. The response to this by the rest of the star is a spin frequency jump (a "glitch") [26] by an amount $\Delta\Omega = \Delta\theta\Omega I_n/I$, where I_n is the involved crust superfluid's moment of inertia and I is that of the star. It is difficult to predict $\Delta\theta$. Moreover, a large part of the crust may partially relax with a very small $\Delta\theta$ or small parts may relax less often but with larger $\Delta\theta$. However, the interval between glitches (t_g) and $\Delta\theta$ may be combined to give a $\Delta\theta$ indpendent "glitch activity" rate

$$\frac{\Delta\Omega}{\Omega t_g} = \frac{\dot{\Omega}}{\Omega}\frac{I_n}{I} \sim \frac{10^{-5}\mathrm{yr}^{-1}}{(\mathrm{Age}/10^3\mathrm{yrs})} . \tag{17}$$

The numerical estimate assumes a canonical value $I_n/I = 10^{-2}$. It is compared to observations of young pulsars in Table 2.

Table 2. Glitch activity in young radiopulsars [27]

PULSAR	Age	Glitch Activity (10^{-7} yr^{-1})	
	(10^3 yrs)	Observed	Eq. (17)
0531	1.2	0.1	...
1509	1.5	~ 0	...
0540	1.7	~ 0	...
0833	11	8	9
1800	16	?	6
1737	20	4	5
1823	21	5	5

The value of $\Delta\theta$, and hence of $\Delta\Omega$ and t_g, is not predicted: the observed glitch repetition rates give an effective $\Delta\theta \sim 2 \cdot 10^{-4}$ for Vela and $4 \cdot 10^{-5}$ for PSR 1737. The very low glitch activity in the Crab family is here attributed soley to the fact that unlike Vela their lower crusts are warm enough to respond to growing stress mainly by plastic flow [26,27]. The radiopulsars much older than Vela should continue to have glitches, but with much longer inter-glitch intervals τ_g so that few have been observed so far.

If a post-Crab pulsar crust relaxes with large scale cracking the crust will break up into "platelets" [2]. New matter from the core would flow up into the cracks between them as indicated in Fig. 9. The platelets move apart as stellar spin-down continues but maintain their integrity: they will keep the pinned vortices and the magnetic field they had when they were formed. The average magnetic field in a hemisphere drops with stellar spin-down as platelets move apart and collide in the equatorial zone where they cause field line recombination. However, where magnetic field exists, i.e., in platelets, it retains the value it had when the platelets began to separate. Because the vortex line density in each

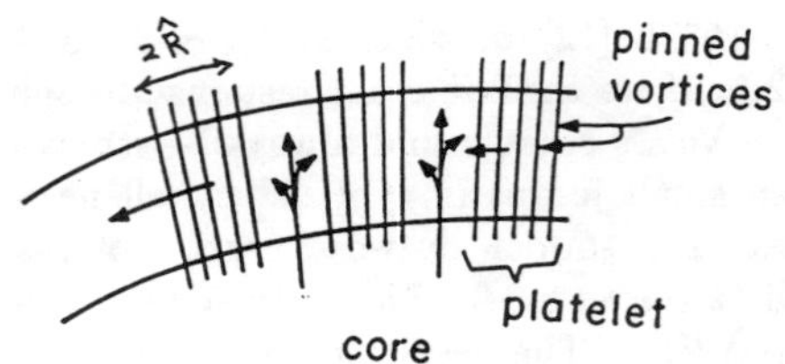

Fig. 9. Motion of platelets, each of which carries its originally pinned vortices and frozen in magnetic field. Core matter continually flows up to fill the growing space between separating platelets [2].

platelet is also fixed even though Ω drops as the star spins-down, platelets will continue to feel increasing vortex line pinning stresses until they again reach the stress which caused the initial fragmentation and cause another glitch. Thus, as long as spin-down continues so will glitches, always with similar $\Delta\Omega$ but longer τ_g.

Because of the frozen B in each platelet measurements of B in older spun-down neutron stars should give conflicting results. Those based upon the magnitude of the magnetic dipole moment (spin-down in radio pulsars, accretion torques in X-ray pulsars) should give smaller values of B than those inferred from surface cyclotron resonance features in the X-ray spectrum (some X-ray pulsars and Gamma-Ray Burst sources).

It should be noted that as long as the crustal vortex array remains approximately parallel to $\mathbf{\Omega}$, vortex lines whch pass through the equatorial plane are not as strongly pinned to the crust as those that penetrate to the crust base and pass through the strongest pinning region, $\rho \sim (2-8)\cdot 10^{13}\mathrm{gcm}^{-3}$. Many may not be pinned at all [15]. It is the readjustment of these to the glitch which determines the post-glitch period response [14,15]. For a warmer pulsar such as the Crab and possibly even for Vela, crust lattice creep may also be significant [26].

7. Gamma-Ray Bursts

The total number of expected glitches during the spin-down of a pulsar from Ω_1 (the angular spin frequency at which crust cracking begins) to Ω_f (the spin frequency reached after 10^{10} yrs.) is

$$N_g = \frac{1}{\Delta\theta}\ln\left(\frac{\Omega_1}{\Omega_f}\right) \sim 10^5 \ . \tag{18}$$

The glitching will continue after pulsar turnoff. The sudden release of elastic strain energy from the crust in a glitch is

$$\mathcal{E}_g \sim \ell \hat{R}^2 \mu \theta_{\max} \Delta\theta \sim 10^{40} \hat{R}_6^2 \text{ ergs}, \tag{19}$$

with $\hat{R}$ the radius of a platelet. The total number of large glitches and the amount of the associated sudden energy release are both suggestive of those needed to account for

Gamma Ray Bursts from the Galactic population of isolated spinning-down neutron stars [28]. Other features which support this association [30,31] include submillisecond GRB rise times. A crack (or slip) is expected to propagate with the deep crust shear velocity $v_s = (\mu/\rho_L)^{1/2} \sim 2 \cdot 10^8$ cm s^{-1} for a lattice density ρ_L (which does not include the internuclear superfluid neutrons). Cracking timescales ℓ/v_s or $\hat{R}/v_s$ of less than 10^{-3}s should be characteristic of glitches together with post-crack crust shear svibrations with frequencies $\gtrsim 1$ KHz. Blaes *et al.* [29] have shown how such high frequency elastic vibrations will reach the surface as Alfven waves with amplitudes greatly amplified from those of the initial deep crust oscillation. The total time for repeatedly reflected crust oscillation energy to pass through the crust surface is $0.1-1$ s. (The refocussing time for shear waves travelling around the star, $2\pi R/v_s \sim 30$ ms, is suggestive of the observed 23 ms quasi-periodicity in the famous March 5, 1979 GRB.)

The surface potential drop at the surface above a platelet oscillating with frequency $\omega(\sim v_s/\ell)$ is

$$\Delta V = \left[\frac{8\pi\mathcal{E}_g}{(\text{burst duration})\, c}\right]^{1/2} \cos\omega t \sim 10^{17} \cos\omega t \; V, \tag{20}$$

enough, in most neutron stars, to allow large near surface pair production and hard X-ray and γ-ray synchrotron radiation from near the stellar surface. (Even in the older neutron stars platelet $B \sim 3{\cdot}10^{12}G$.) A necessary condition for a GRB is that the available potential drop and power be injected into a region with very few ambient particles to absorb the injected energy.

No GRB was observed during a Vela pulsar glitch, which limits the hard X-ray emission during Vela glitches to not much above 10^{38} ergs. As emphasized by Blaes *et al.* [29] only the older pulsars whose charge-separated magnetospheres have sufficiently low charge density should have strong GRB's. Ultrarelativistic particle acceleration as a response of the magnetosphere to the oscillating surface electric field of Eg. (20) may be suppressed because of the current that flows along B from movement of the ambient initial charge-separated density $(\vec{\Omega} \cdot \vec{B}/2\pi c)$ already in magnetosphere [32]. Simply moving that charge and the associated magnetic field energy from its current (would use most of the glitch ΔV unless $\Delta V \gg \dot{J} \times \text{Inductance} \sim \omega(\Omega B/2\pi)(\pi\hat{R}^2)(\hat{R}/c^2)$, an inequality which would exclude Vela if $\hat{R} \sim 10^6$ cm. Alternatively GRB's may result only in a magnetosphere starved of charge because there is no longer the steady magnetospheric $e^{\pm}$ production of radiopulsars whch have not yet reached the canonical "death line".

It is a pleasure to thank J. Groves and G. Taylor for information about the yield strengths of alkali crystals and P. Jones for enlightening discussions on this and other topics. This research has been supported in part by NSF AST89-01681. This work is contribution number 451 of the Columbia Astrophysics Laboratory.

8. References

[1] Sang, Y. and Cahnmugam, G. 1987, *Ap.J. (Letters)*, **323**, L61.
[2] Ruderman, M. 1991, *Ap.J.*, in press.
[3] Negele, J. and Vautherin, D., 1975, *Nucl. Phys.* , **A207**, 298.
[4] Siedersleben, M. and Taylor, G. 1989, *Phil. Mag. A.*, **60**, 631.
[5] Ruderman, M. 1969, *Nature*, **223**, 597.

[6] Smoluchowski, R. and Welch, D. 1970, *Phys. Rev. Letters*, **24**, 1191.
[7] Alpar, A. 1991. These proceedings.
[8] Pines, D. 1991. These proceedings.
[9] Srinivasan, G., Bhattacharya, D., Muslimov, A., and Tsygan, A. 1990, *Current Science*, **59**, 31.
[10] Alpar, M., Langer, S., and Sauls, J. 1989, *Ap.J.*, **282**, 533.
[11] Alpar, M., Anderson, P., Pines, D., and Shaham, J. 1984, *Ap.J.*, **278**, 791.
[12] Alpar, M., Cheng, K., and Pines, D. 1990, *Vortex Creep and the Internal Temperature of Neutron Stars*, Univ. of Ill. preprint.
[13] Epstein, R. and Baym, G. 1988, *Ap.J.*, **328**, 680.
[14] Alpar, M., Anderson, P., Pines, D., and Shaham, J. 1984, *Ap.J.*, **276**, 325.
[15] Jones, P. 1990, *M.N.R.A.S.*, in press.
[16] Segelstein, D., Rawley, L., Stinebring, D., Fruchter, A., and Taylor, J. 1986,. Nature323714
[17] Cordes, J. and Stinebring, D. 1984, *Ap.J. (Letters)*, **277**, 53.
[18] Fruchter, A., Stinebring, D., and Taylor, J. 1988, *Nature*, **333**, 237.
[19] Kulkarni, S., and Narayan, R. 1989, *Ap.J.*, **355**, 755.
[20] Romani, R. 1990 in *Supernovae and Stellar Evolution*, eds. A. Ray and T. Velusamy (World Scientific: Singapore), in press.
[21] Michel, F. 1987, *Nature*, **329**, 310.
[22] Grindlay, J. and Baily, C. 1988, *Nature*, **336**, 48; and Baily, C. and Grindlay, J. 1990, *Ap.J.*, **353**, 159.
[23] Ray, A. and Kluźniak, W. 1990, *Nature*, **344**, 415.
[24] Narayan, R. and Popham, R. 1989, *Ap.J. (Letters)*, **346**, L25.
[25] Lyne, M. and Manchester, R. 1988, *M.N.R.A.S.*, **234**, 477.
[26] Ruderman, M. 1976, *Ap.J.*, **203**, 213.
[27] McKenna, J. and Lyne, A. 1990, Nature, **343**, 349.
[28] Liang, E. and Petrosian, V. 1986, *Gamma Ray Bursts*, AIP Conf. Proc. **141**.
[29] Blaes, O., Blandford, R., Goldreich, P. and Madau, P. 1989, *Ap.J.*, **343**, 839.
[30] Pacini, F. and Ruderman, M. 1974 *Nature*, **251**, 399.
[31] Epstein, R. 1988, *Phys. Rep.*, **163**, 155.
[32] Goldreich, P. and Julian, W. 1959, *Ap.J.*, **157**, 869.
[33] Thorsett, S. and Stinebring, D. 1990, *Ap.J.*, **361**, 644.
[34] Srinivasan, G. 1989, *Astron. Astrophys. Rev.*, **1**, 209.

THERMAL EMISSION OF PULSARS

HAKKI ÖGELMAN
Department of Physics, University of Wisconsin
Madison, Wisconsin 53706, U.S.A.
and
Max Planck Institut für Extraterrestrische Physik
D-8046, Garching bei München, Germany

ABSTRACT. Theoretical models of cooling and re-heating of single neutron stars is reviewed and compared with the observations of *EINSTEIN* and *EXOSAT* satellites. The expected contribution of the *ROSAT* satellite is discussed.

1. Introduction

Back during the infancy of X-ray astronomy, some three decades ago, the first schemes for discovering neutron stars were based on the detection of thermal radiation from a cooling neutron star in the X-ray band. Considering that neutron stars have no internal energy sources, one expects that they should be cooling from the moment they were created. Following the advent of X-ray astronomy in the early 1960s with the discovery of the brightest X-ray source Sco X-1 (Giacconi *et al.* 1962), detailed calculations of cooling rates and cooling curves started to appear in the literature. Early calculations (Chiu 1964; Chiu and Salpeter 1964; Tsuruta 1964) indicated that the neutron star are born very hot ($T \geq 10^{10}$ K) and cool rapidly by neutrino emission from the interior. The surface temperatures are expected to come down to $\sim 2 \times 10^6$ K in few hundred years, slowly decrease to $\sim 5 \times 10^5$ K by about 10^5 years and rapidly cool thereafter via photon emission from the surface. Bachall and Wolf (1965) subsequently showed that if Sco X-1 was a cooling neutron star, it had to be $\lesssim 100$ yr old even if it was as close as 100 parsec; at a distance of few kpc it had to be no more than few hours old! As no supernova, the birth signal of neutron stars, so close and so recent would have gone unnoticed the cooling neutron star explanation was discarded. The discovery of neutron stars had to wait till 1968 when Hewish *et al.* (1968) observed the pulsed electromagnetic radiation from a rotating magnetized neutron star. Eventually, discovery of pulsed Sco X-1 like sources by the *UHURU* satellite confirmed Shklovskii's (1967) explanation that these sources were also neutron stars, but accreting mass from close binary companions and converting the gravitational energy of this mass to X-rays emitted from the surface.

J. Ventura and D. Pines (eds.), Neutron Stars: Theory and Observation, 87–100.

Through recent major advances in the astronomical technology, namely the construction and deployment in orbit of imaging X-ray telescopes, we have attained the possibility of detecting thermal radiation from isolated neutron stars. The first such satellite payload was the *EINSTEIN* observatory (1978-1981), followed by the European mission *EXOSAT* (1983-1986). The third member of this family *ROSAT* has very recently been placed in orbit (1 June 1990) and has already started giving high quality results. With the exception of an upper limit of 3×10^6 K for the surface temperature of the Crab pulsar through lunar occultation experiments (Wolff *et al.* 1975; Toor and Seward 1977), no experimental information on neutron star surface temperatures existed prior to the *EINSTEIN -EXOSAT - ROSAT* era.

Despite the fact that the radiation emitted from cooling neutron stars has not been the easiest way to discover them, this information should contribute significantly to our understanding of the internal structure of the neutron star. The structure is of topical interest not only for astrophysics, it also provides a unique opportunity to study the poorly understood hadron equation of state at nuclear ($\rho_o \simeq 3 \times 10^{14}$ g cm^{-3}) and [Higher densities (for a recent review see Pines 1989).

The outline of this lecture note is as follows: In section 2 the thermal features of a neutron star is outlined; section 3 summarizes the neutrino processes in a cooling neutron star; in section 4 the possible re-heating processes are considered; section 5 examines the various cooling scenarios; section 6 compares the observational results with the theoretical calculations; section 7 is a look at the near future prospects with the *ROSAT* satellite.

2. Thermal features of a neutron star

The most simple model of a cooling neutron star consists of an isothermal interior which cools via neutrino processes where the neutrinos leave the interior directly without any further interactions, and an external crust which conducts the heat from interior to the surface which subsequently radiates it via photon emission (Ögelman 1987). In this section we review the general structure of a neutron star with a stiff equation of state and the relevant features concerning its thermodynamic evolution. These constituents are summarized in Table 1.

Going from the surface towards the center we encounter the following regions:

Atmosphere consists of a thin, ~1 cm deep gas of non-degenerate electrons and ions in the density range 1 to 10 gm cm^{-3} corresponding to an optical depth of unity in soft X-ray peak of $\sim 10^6$ K surface temperature. Its chemical composition will modify the emergent photon spectrum causing departures from blackbody and giving rise to absorbtion edges (Romani 1987). It is also possible that the presence of strong magnetic fields may enhance the atomic binding energies and modify the opacities which determine the photon spectrum (Ventura 1989). In general, a blackbody assumption for the photon flux appears acceptable within the limits of of other theoretical and experimental uncertainties.

Outer crust consists of a solid lattice of nuclei and degenerate electrons extending to about 300 meters from the surface where the density reaches $\sim 4 \times 10^{11}$ gm cm^{-3} .

Table 1: The profile of various regions of a 1.4 $M_{\odot}$ neutron star with stiff equation of state depicting the constituents of these regions and their importance in the thermodynamic evolution of neutron stars.

R (km)	ρ (g cm^{-3})	constituents	thermodynamic significance
12.0	0	*SURFACE*	
atmosphere		atoms, ions and electrons	Determines the spectrum of the emerging photon flux and its departure from blackbody
12.0 - 10^{-5}	10		
outer crust		solid crust of nuclei and degenerate electrons	The thermal conductivity determines the heat conduction from the interior to the surface as well as the dependence of the surface temparature to that of the interior. Below this layer the temperature is uniform.
11.7	4.10^{11}		
inner crust		lattice of neutron-rich nuclei, relativistic electrons and superfluid neutrons	Pinning site of the superfluid neutron vortices to the nuclear lattice; generates heat when pulsar is spinning up or down. Cools mainly via neutrino bremsstrahlung.
10.3	2.10^{14}		
core		electrons, superfluid neutrons and superconducting protons	Cools mainly via URCA processes
8	6.10^{14}		
inner core		pion and kaon condensates ? quark matter?	If exotic matter exists, it enhances the neutrino luminosities by large amounts
0	1.10^{15}	*CENTER*	

The conductive properties of this layer determines the temperature gradient of the neutron star from the isothermal interior T_i to the surface T_s . Calculations by Gudmundsson, Pethick and Epstein (1982) have shown that this layer can be considered independent of the internal temperature and the two temperatures can be related by the simple relationship $T_s \simeq 0.93 \times 10^6\ (T_i/10^8)^{0.55} K$. A more recent calculation by Van Riper (1988), included the effects of the magnetic field on the heat transport; it is concluded that this effect is small compared to the other uncertainties.

Inner crust consists of a lattice of neutron rich nuclei, relativistic electrons and superfluid neutrons(see Pines and Alpar 1985, and references therein). It extends from densities of 4×10^{11} gm cm^{-3} about 1 km downwards to densities of 2×10^{14} gm cm^{-3} . This region can emit neutrinos via electron-ion (crust) bremsstrahlung and cool directly. A very important feature of this region is the coexistence of the neutron superfluid together with the normal lattice of neutron-rich nuclei. Since neutron stars rotate, the neutron superfluid will contain an array of vortex lines parallel to the rotation axis. These vortices are expected to be pinned to the normal lattice. Subsequently, as the neutron star spins down (or up) the vortices have to move in order to adjust to the changing angular velocity. Due to the pinning forces, the movement of the vortices are expected to be in a creep like fashion, thus generating frictional heating. Consequently, this region of the neutron star can be a heat source.

Core consists of mainly superfluid neutrons with a few percent of superconducting protons and normal electrons. It contains the bulk of the neutron stars heat capacity. It cools by emitting neutrinos via the modified URCA processes. It has been suggested that the magnetic field decay in the core of neutron stars can lead to heating in this region (Haensel, Urpin and Yakovlev 1990).

Inner core consists of the regions where the densities are in in excess of the nuclear densities ($\rho_o \gtrsim 3 \times 10^{14}$ gm cm^{-3}). The constituents of the core at densities $\gtrsim 2\rho_o$ are not well known. Appearance of various Bose-Einstein condensations of pions (see Brown and Weise 1976 for a review) and kaons (Kaplan and Nelson 1986) have been discussed in the literature. If the central densities exceed $5\rho_o$ quark matter may also exist (Baym and Pethick 1979). If such exotic matter exist, it would greatly enhance the neutrino cooling rates from the interior.

3. Neutrino processes in cooling neutron stars

Although it is believed that the neutron stars are formed with internal temperatures in excess of 10^{11} K, they cool very fast; in time-scales of few days the internal temperatures come down to $\sim 10^9$ K and surface temperatures $\sim 5 \times 10^6$ K. Around these temperatures, the dominant mechanism of cooling is neutrino emission; the photon emission from the surface becomes noticeable after the interior temperatures falls down to 10^8 K (for a review see Baym and Pethick 1979). In the interior, one of the most important cooling mechanisms is the modified URCA process:

$$n + n \rightarrow n + p + e^- + \bar{\nu}_e \text{ and } n + p + e^- \rightarrow n + n + \nu_e$$

where the estimated luminosity is (Friman and Maxwell 1979):

$$L_\nu^{URCA} \simeq 6 \times 10^{39} \Big(\frac{M}{M_\odot}\Big) \Big(\frac{\rho_o}{\rho}\Big)^{1/3} \Big(\frac{T_i}{10^9}\Big)^8 \text{ erg s}^{-1}$$

This process is in principle just the neutron decay $n \to p + e^- + \bar{\nu}_e$, with rearrangement and addition of a bystander neutron in order to conserve energy and momentum in the degenerate environment. The strong dependence on T_i is due to the available phase space for the involved fermions. If the nucleons are in a superfluid state, the above rate is reduced further by factors like $exp(-T_{crit}/T_i)$ due to the reduction in thermal excitations (T_{crit} is the condensation temperature of the superfluid).

When neutral currents are considered, nucleon pair bremsstrahlung

$$n + n \to n + n + \nu + \bar{\nu} \text{ and } n + p \to n + p + \nu\bar{\nu}$$

can contribute to the cooling. The luminosity due to this process also goes as T_i^9 but it is about a factor 30 less than the URCA process (Friman and Maxwell 1979).

Another important cooling process that can operate in the crust where nuclei are present, is neutrino pair bremsstrahlung (also called crust bremsstrahlung):

$$e^- + (Z, A) \to e^- + (Z, A) + \nu + \bar{\nu}$$

The magnitude of this process is estimated as (Maxwell and Friman 1979):

$$L_\nu^{crust} \simeq 5 \times 10^{39} \Big(\frac{M_{crust}}{M_\odot}\Big) \Big(\frac{T_i}{10^9}\Big)^6 \text{ erg s}^{-1}$$

where the mass of the crust depends on the assumed equation of state (EOS). Soyeur and Brown (1979) estimate that the crust mass to total mass ratio can range from about 1% for soft EOS to 30% for a stiff one.

In the inner core of the neutron star where densities reach $\geq 2\rho_o$, pion condensates and other exotic matter may appear and significantly increase the luminosity since the restriction of bystander fermions disappear. The neutrino luminosity due to pion condensates is estimated to be (Maxwell *et al.* 1977):

$$L_\nu^\pi \simeq 1.5 \times 10^{46} \theta^2 \Big(\frac{M}{M_\odot}\Big) \Big(\frac{\rho_o}{\rho}\Big) \Big(\frac{T_i}{10^9}\Big)^6 \text{ erg s}^{-1}$$

where $\theta \sim 0.3$ is the degree of pion condensation.

Additional exotic cooling agents such as kaons, quark matter and axions have also been discussed in the literature; they all lead to very high luminosities in comparison with the URCA process.

4. Re-heating processes in neutron stars

The most likely location for the re-heating processes in the neutron star is the inner crust where the neutron superfluid coexist with the lattice of neutron-rich nuclei.

The neutron superfluid will contain an array of vortex lines mimicking the rotation of the solid crust. As the neutron star crust spins down (or up), these vortices are also forced to move radially outward (inward) to keep up with the changing angular velocity. However, since it is energetically favorable for vortex lines to have their centers coincident with the nuclei of the lattice, their movement in response to angular velocity changes is not a smooth process but a creep (Alpar *et al.* 1984). Internal energy dissipation in the neutron star is expected due to this frictional force of vortex creep. This process can be characterized by the parameters I_i and ω_i where an interior layer i of the star with superfluid moment of inertia I_i rotates with a lag $\omega_i \equiv \Omega_i - \Omega_c$ which is the difference between the rotation rate Ω_i of that layer and the rotation rate Ω_c of the crust. The magnitude of ω_i depends on the efficiency of the coupling between layer i and the crust. The associated energy dissipation rate is $\dot{E}_{diss} = |\dot{\Omega}|\Sigma\ I_i\omega_i$. The term in the summation can be considered the angular momentum in the differential rotation of the frictionally coupled superfluid J. The theoretical values of J come from microscopic physics and are sensitively dependent on the uncertain details of vortex-lattice interactions; best estimates are in the 10^{41} to 10^{43} gm cm^2 rad s^{-1} (see Shibazaki and Lamb 1989 and references therein).

In case the crust cracks due to the stresses of vortex pinning or magnetic field, the internal heat dissipation can be characterized by similar parameters (Ruderman 1990)

Another heating source is the bombardment of the polar caps by the relativistic particles from the magnetosphere (polar cap heating). Cheng and Ruderman (1980) give the relation $T \simeq 3 \times 10^5\ E_B(keV)$K between the polar cap temperature T and the neutron star surface binding energy E_B. For $R = 10^5$ cm (the polar cap area $A_c = \pi R^2$), $T \lesssim 3.6\times10^5$K, and $E_B \lesssim 1.2$ keV, while for $R = 4\times10^4$ cm, $T \lesssim 5\times10^5$K and $E_B \lesssim 1.7$ keV. These limits are compatible with recent approximate calculations of the binding energy, which yield $E_B \lesssim 1$ keV (Jones 1985). For the Arons (1981) model of electron flow from a polar cap, the polar cap area is given as $A_c = \frac{2\pi R_*^3}{cP}$ where R_* is the neutron star radius and P is the pulsar period.

It has also been pointed out that the electrical resistivity of the interior of neutron stars can significantly increase in the presence of magnetic fields and cause field decay in a time-scale of 10^7 years thus producing heat release which can delay the cooling process (Haensel, Urpin and Yakovlev 1990).

The possibility of older neutron stars accreting from the interstellar medium has been considered in the literature (Paczynski 1990; Hartmann, Epstein and Woosley 1990) and it has been predicted that slow old ($v \lesssim 10$ km/s) neutron stars maybe re-heated on the surface to temperatures in excess of 10^6 K (Treves and Colpi 1990).

5. Cooling scenarios

For interior temperatures below 10^{11} K, fulfilled within few seconds after birth, the neutrinos which are emitted in the cooling processes listed above will leave the neutron star without further interactions. Hence the simplest form of the cooling equation will be:

$$C_v \frac{dT}{dt} = -L_\nu - L_\gamma + H$$

where C_v is the heat capacity, L_ν the internal neutrino luminosity, L_γ the surface photon luminosity and H is the re-heating due to internal (or external) processes. In the case of no superfluidity, the heat capacity is given by (Maxwell 1979):

$$C_v \simeq 1.9 \times 10^{39} \left(\frac{T_i}{10^9}\right) \text{ erg K}^{-1}$$

In the case of superfluid interior below the critical temperature, the heat capacity is expected to decrease by about a factor of 10 (Shibazaki and Lamb 1989).

The surface photon luminosity can be assumed to be blackbody-like giving $L_\gamma = 4\pi\sigma T_s^4$. Using the relationship $T_s \simeq 0.93 \times 10^6\ (T_i/10^8)^{0.55} K$. (Gudmundsson, Pethick and Epstein 1982) the photon luminosity can also be expressed in terms of the internal temperature as $L_\gamma = 2 \times 10^{15} T_i^{2.2}$ erg s^{-1} .

For the simple case of no exotic cooling agents and no superfluidity, the solution of the cooling equation yields the following general features: For about 10^5 years following its formation, the neutron star cools predominantly by neutrino emission and $T_i \propto t^{-1/6}$ ($T_s \propto t^{-1/11}$). After this time, the photon luminosity becomes the dominant cooling process where $T_i \propto t^{-5}$ ($T_s \propto t^{-11/4}$). At the change over point, $T_i \sim 2 \times 10^8$ K and $T_s \sim 10^6$ K.

A number of authors have added more thought and physics to the simple calculation above and come up with various alternatives and corrections to the cooling curve. The parameters considered in these models have been: 1) the mass of the neutron star; 2) the equation of state; 3) the existence of superfluidity and its critical temperature; 4) finite time-scale of thermal relaxation of the crust; 5) existence of exotic matter; 6) re-heating. In figure 1 I plot a number of published cooling scenarios depicting the effect of the above parameters. As can be seen from the figure, these scenarios cover a large range of possible surface temperatures which can only be weeded out via experimental observations. Table 2 contains the tabulated form of some of the curves plotted in figure 1.

We can summarize the influence of these effects on the cooling curve as follows:

a) The mass and the equation of state introduces some minor modifications on the standard cooling curve; these factors may also influence the shape of the cooling curve during the thermal relaxation stage or determine the state of the inner core and hence the existence of exotic cooling agents (see Nomoto and Tsuruta 1987).

b) Superfluidity has the effect of decreasing the heat capacity of the interior as well as reducing the crust bremsstrahlung rate. Compared to the standard cooling curve, the reduced crust bremsstrahlung rate leads to higher internal and surface temperatures during the earlier stages of the cooling curve up to $t \lesssim 10^5$ years. Subsequently, the reduced heat capacity causes earlier turnover to the photon cooling era.

c) Finite time-scale of the thermal relaxation of the crust affects the early part of the cooling curve ($t \lesssim 10^2$ years) where the surface temperatures remain above the standard cooling scenario until the crust comes to equilibrium with the heat flow from the interior (see Nomoto and Tsuruta 1987).

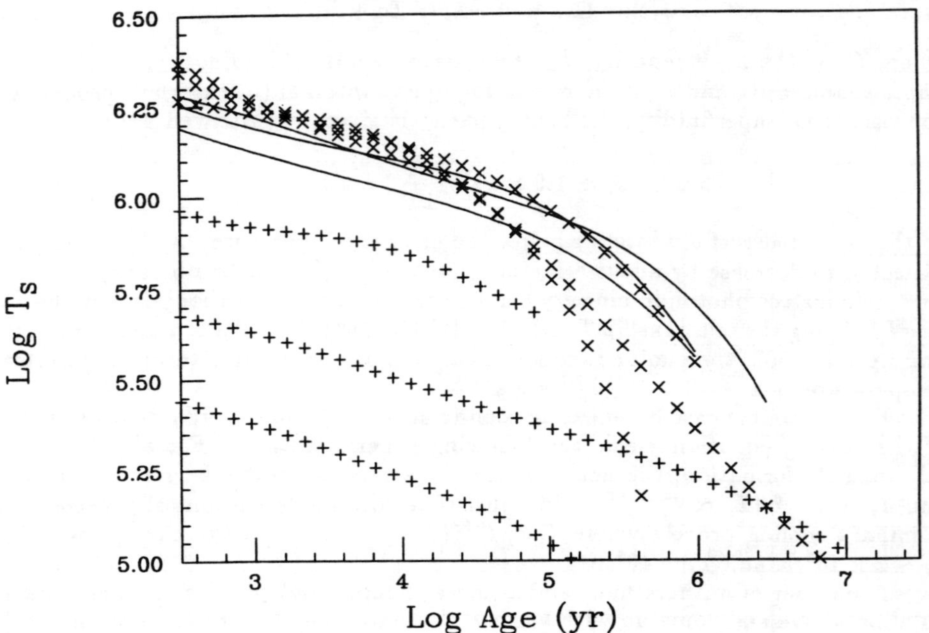

Figure 1: Various neutron star cooling curves published in the literature. (Solid lines): no superfluid and no exotic matter; (×): superfluid and no exotic matter; (+): superfluid and exotic matter.

d) Existence of exotic matter **like pion/kaon condensates or quark matter has the most dramatic influence on the standard cooling curve in the sense that the highly enhanced cooling rates can reduce the interior and surface temperatures by an order of magnitude.**

e) Re-heating **modifies the standard cooling curve most noticeably at the later stages by decreasing the slope of the cooling curve during the photon cooling era. It also keeps a higher temperature during the neutrino cooling era (see Shibazaki and Lamb 1889).**

6. Comparison to experimental results

The recent progress in construction and deployment in orbit of imaging X-ray telescopes, has brought the detection of thermal radiation from isolated neutron stars

Table 2: Tabulation of various neutron star cooling curves published in the recent literature. In most cases the data is presented in graphical form thus reducing the accuracy with which it can be read out.

Log Age(yr)	Log $T_{surface}$ for different models								
	1	2	3	4	5	6	7	8	9
0.00	6.38	6.38	6.55	6.43	-	-	-	-	-
0.50	-	-	-	6.42	-	-	-	-	-
1.00	6.35	6.35	6.51	6.41	6.40	-	-	-	-
1.50	-	-	-	6.39	6.27	-	-	-	-
2.00	6.32	6.30	6.33	6.25	6.03	-	-	-	-
2.50	-	-	-	6.17	5.98	6.37	5.44	6.34	5.68
3.00	6.23	6.23	6.19	6.13	5.94	6.26	5.38	6.27	5.62
3.50	-	-	-	6.07	5.88	6.19	5.30	6.20	5.57
4.00	6.08	6.14	6.09	6.03	5.83	6.10	5.23	6.12	5.50
4.50	-	-	-	5.94	5.78	6.00	5.15	6.02	5.43
5.00	5.95	5.96	5.96	5.87	5.66	5.79	5.04	5.85	5.36
5.50	-	-	-	5.74	-	5.33	4.86	5.56	5.29
6.00	5.70	5.53	5.56	5.55	-	4.50	4.45	5.34	5.22
6.50	-	-	-	-	-	-	-	5.14	5.14
7.00	-	-	-	-	-	-	-	4.95	5.02
7.50	-	-	-	-	-	-	-	4.76	4.85
8.00	-	-	-	-	-	-	-	4.58	4.67

1-3) Calculations by Nomoto and Tsuruta (1987) for a $1.4 M_{\odot}$ neutron star including effects of the thermal inertia of crust. (1) EOS=PS, no superfluid, no exotic matter; (2) EOS=PS, superfluid, no exotic matter; (3) EOS=BPS, no superfluid, no exotic matter. 4-5) Calculations by Page and Baron (1990) for $1.4 M_{\odot}$ star with EOS=AV14. (4) no superfluid, no exotic matter; (5) superfluid and kaon condensates ($\rho_c = 4\rho_o$). 6-8) Calculations by Shibazaki and Lamb (1989) of a $1.4 M_{\odot}$ star with superfluidity. (6) no exotic matter, no internal re-heating; (7) pion condensates and no internal re-heating (8) same as 6 but with re-heating ($J = 3.1 \times 10^{43}$); (9) same as 7 but with re-heating ($J = 1.7 \times 10^{43}$).

within observational realm. The first such satellite payload was the *EINSTEIN* observatory (1978-1981), followed by the European mission *EXOSAT* (1983-1986). The third member of this family *ROSAT* has very recently been placed in orbit (1 June 1990) and has already started giving high quality results. With the exception of an upper limit of 3×10^6 K for the surface temperature of the Crab pulsar through lunar occultation experiments (Wolff *et al.* 1975; Toor and Seward 1977), no experimental information on neutron star surface temperatures existed prior to the *EINSTEIN* -*EXOSAT* - *ROSAT* era.

The early results of the *EINSTEIN* survey of thermal emission from radio pulsars yielded about 8 positive detections out of nearly two dozen observed candidates (Helfand 1983). A more complete analysis by Seward and Wang (1988) lists the detected count rates for 10 pulsars and upper limits to another 14. The bright soft X-ray source PSR 0656+14 was discovered in a survey of ultrasoft sources in the *EINSTEIN* data base (Cordova et. al. 1989). *EXOSAT* observed 10 pulsars of which all but one (PSR 0031-07) are included in the *EINSTEIN* list. The published *EXOSAT* results include PSR 0540-69 (Ögelman and Hasinger 1990), PSR 0833-45 (Ögelman and Zimmermann 1989), PSR 1055-52 (Brinkmann and Ögelman 1987), PSR 1929+10 (Alpar *et al.* 1987) and PSR 1951+32 (Ögelman and Buccheri 1987).

The measured X-ray fluxes from the young pulsars PSR 0531+21 (Crab), PSR 0540-69 (50 ms pulsar in LMC), PSR 1509-58, and PSR 1951+32 are due to magnetospheric and nebular emission of synchrotron radiation. In the case of the Crab pulsar there has been a detailed investigation of the possible contribution of the thermal emission to the light curve (see Toor and Seward 1977); for the others, no detailed investigation has been published in the literature. As a rough calculation, one can assume that the observed count rates are upper limits to the thermal emission. Similar problems of distinguishing the thermal emission from magnetospheric or nebular emission arises for pulsars PSR 0950+08, PSR 1929+10 and PSR 1642-03 where the very low count rates do not allow reliable spectral and spatial analysis.

In conclusion, there are only 3 pulsars for which we have some spectral evidence for thermal emission from the surface, namely PSR 0656+14 (Cordova *et al.* 1989), PSR 0833-45 (Ögelman and Zimmermann 1989) and PSR 1055-52 (Brinkmann and Ögelman 1987).

The experimental results are summarized in Figure 2 where we plot the surface temperatures or upper limits as a function of the apparent age $(P/2\dot{P})$ together with a typical no superfluid no exotic matter and no re-heating cooling curve. For the *EINSTEIN* data these temperatures and upper limits to have been calculated by simply folding the blackbody spectrum of a 8 and 15 km radius neutron star at the given distance with interstellar absorbtion (N_H) and *EINSTEIN* IPC response and matching it to the count rates listed in Seward and Wang (1988). The N_H values and distances are also taken from the reference above. Similar analysis technique was applied to the unpublished *EXOSAT* upper limits for pulsars 0031-07, 0355+54 0950+08, 1937+21.For the case of pulsars with published temperature fits these results have been used instead.

From figure 2, it is apparent that about a dozen measurements are in a position to restrict the theoretical models. In particular the temperatures of the Vela pulsar

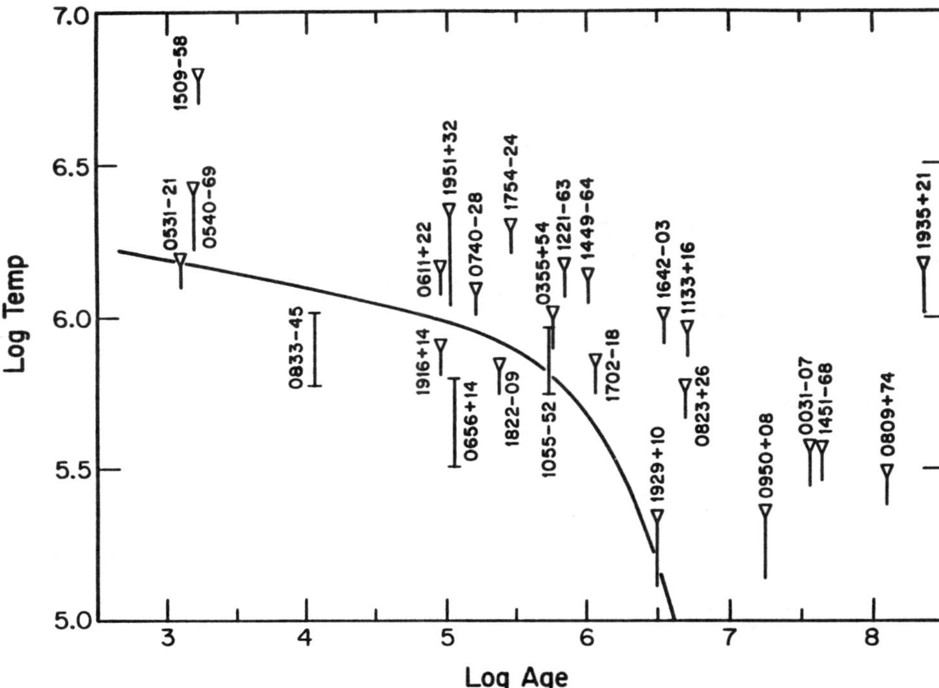

Figure 2: Summary of the existing X-ray measurements of pulsar surface temperatures versus their apparent age $(P/2\dot{P})$. For the *EINSTEIN* data the temperature upper limits have been calculated by simply folding the blackbody spectrum of a 8 to 15 km radius neutron star at the given distance with interstellar absorption (N_H) and *EINSTEIN* IPC response and matching it to the count rate limits listed in Seward and Wang (1988). The N_H values and distances are also taken from the above reference. The range of upper limit bar are just calculated by assuming two different radii for the neutron star, 8 and 15 km. Similar simple analysis has been applied to the unpublished *EXOSAT* data for pulsars 0031-07, 0355+54, 0950+08 and 1937+21. All but three of these pulsars (PSR's 0833-45, 1055-52 and 0656+14) can be considered as upper limits to the surface temperature. For comparison, the curve displays a typical no superfluid, no exotic matter cooling curve.

PSR 0833-45 and PSR 0656+14 are sufficiently below the standard cooling scenarios to make us consider exotic cooling agents, and the upper limit to the temperature of PSR 1929 (Alpar *et al.* 1987) is sufficiently low to exclude strong pinning of vortex lines in the inner crust.

I would like to close this section with a word of caution concerning the temperature upper limits shown on figure 2. The part of the black-body like spectrum of the neutron stars transmitted through interstellar space and sampled with the X-ray detectors is very sensitive to the value of the interstellar absorbtion by cold gas. If no spectral information or reliable determination of N_H from other channels is not available, as in the case of the upper limits drawn on figure 2, we may have sizable errors. The range of upper limit bar on figure 2 is just calculated by assuming two different radii for the neutron star, 8 and 15 km.

7. Future prospects with the *ROSAT* satellite

Following its successfull launch on 1 June 1990, *ROSAT* satellite has already started an all-sky survey on 31 July 1990 which will last till 1 Feb 1991. During this survey the sky will be scanned with effective exposures of about 600 seconds at the ecliptic equator; the exposure should increase as 600sec θ where θ is the ecliptic latitude. The angular resolution of the survey is about an arc-minute. With a background rate of about 4×10^{-4} counts/sec per resolution element, the detection sensitivity is basically source count limited implying that the *ROSAT* survey should be able to detect sources with count rates of 8×10^{-3} cps even at the lowest exposure areas. This rate corresponds to an average energy flux of $\sim 5 \times 10^{-14}$ erg cm^{-2} s^{-1} within the 0.1 to 2.2 keV energy bandpass of ROSAT. The total expected number of sources that will be detected in this survey is about 10^5, about an order of magnitude larger than the X-ray sources detected by *EINSTEIN* . The preliminary examination of the *ROSAT* survey data confirms these expectations. There are some 450 known radio pulsars of which only 10 are known to be X-ray sources. We anticipate that the *ROSAT* survey will at least double this list.

The longer pointed observations will allow a more detailed study of these objects. For example, in the case of PSR 0833-45 (the Vela pulsar) the .089 s pulsations of the point-like source can be examined down to a sensitivity of about 1% of the total point-like emission, about an order of magnitude better than the previous measurements. These limits in turn imply that we will be able to observe temperature variations of the surface to a sensitivity of about 0.3%. Using the PSPC data we can also investigate the energy spectrum of the 2' diameter compact nebula and the emission from the central object by using the differential radial counts as a function of radius. An estimate shows that i) there are considerable number of counts per differential radius step to perform a good spectral fit, and ii) the point-like source and the compact nebula counts are well resolved to perform a spectral fit to both components separately as a function of radius.

As a listing of newly discovered survey sources start becoming available, the tedious task of optical identification program will start. Almost all of the $\sim 10^5$

sources will either be extragalactic or nearby stellar sources. However, it is possible that a small fraction of the unidentified sources could be potential neutron stars which may have gone unnoticed or which may be nearby very old neutron stars accreting material from the interstellar medium. Supplementary information in other energy channels such as deeper optical searches, correlations with GRO results may yield the necessary information that will make their physical identification possible.

The *AXAF* satellite is expected to be in orbit by late 90's. With the improved area of the mirrors and angular resolution of the High-resolution camera (HRC) over *ROSAT* , *AXAF* should be an excellent detector to further our understanding of neutron star interiors.

8. References

Alpar, A. M., Anderson, P. W., Pines, D., and Shaham, J.: 1984, *Astrophys. J.* **278**, 791

Alpar, A., Brinkmann, W., Kızıloğlu, Ü., Ögelman, H. and Pines D.: 1987, *Astron. Astrophys.* **177**, 101

Arons, J.: 1981, *Astrophys. J.* **248**, 1099

Bachall, J.N. and Wolf, R.A.: 1965, *Phys. Rev. B*, **140**, 1452

Baym, G. and Pethick, C.: 1979, *Ann. Rev. Astr. Astrophys.* **17**, 415

Brinkmann, W., and Ögelman, H.: 1987, *Astron. Astrophys.* **182**, 71

Brown, G.E. and Weise, W.: 1976, *Phys. Rep.*, **27 C**, 1

Cheng, A.F., and Ruderman, M.A.: 1980, *Astrophys. J.* **235**, 576

Chiu, H-Y.: 1964, *Ann. Phys.*, **26**, 364

Chiu, H-Y. and Salpeter, E.E.: 1964, *Phys. Rev. Lett.* **12**, 413

Cordova, F.A., Hjellming, R.M., Mason, K.D. and Middleditch, J.: 1989, *Astrophys. J.* **345**, 451

Eichler, D. and Cheng, A.F.: 1989, *Astrophys. J.* **336**, 360

Friman, B.L. and Maxwell, O.V.: 1979, *Astrophys. J.* **232**, 541

Giacconi, R., Gursky, H., Paolini, F.R. and Rossi, B.B: 1962, *Phys. Rev. Lett.* **9**, 439

Greenstein, G., and Hartke, G. J.: 1983, *Astrophys. J.* **271**, 283

Gudmundsson, E.H., Pethick, C.J. and Epstein, R.I.: 1982, *Astrophys. J. (Letters)* **259**, L19

Haensel, P., Urpin, V.A. and Yakovlev, D.G.: 1990, *Astron. Astrophys.* **229**, 133

Hartmann, D., Epstein, R.I. and Woosley, S.E.: 1990, *Astrophys. J.* , **348**, 625

Helfand, D.J.: 1983, in *IAU Symposium 101, Supernova Remnants and Their X-ray Emission*, ed. J. Danziger and P. Gorenstein (Dodrecht: Reidel), p. 471

Hewish, A., Bell, S.J., Pilkington, J.D.H., Scott, P.F., and Collins, R.A.: 1968, *Nature,* **217**, 709

Jones, P.B.: 1986, *Mon. Not. R. astr. Soc.* **218**, 477

Kaplan, D.B. and Nelson, A.E.: 1986, *Phys. Lett. B*, **175**, 57

Maxwell, O.V.: 1979, *Astrophys. J.* **231**, 201

Maxwell, O., Brown, G.E., Campbell, D.K., Dashen, R.F. and Manassah, J.T.: 1977,

Astrophys. J. **216**, 77
Nomoto, K. and Tsuruta, S.: 1987, *Astrophys. J.* **312**, 711
Ögelman, H.: 1987, *Europhys. News*, **18**, 98
Ögelman, H., and Buccheri, R.: 1987, *Astron. Astrophys.* **186**, L17
Ögelman, H. and Zimmermann, H.-U.: 1989, *Astron. Astrophys.* **214**, 179
Ögelman, H. and Hasinger, G.: 1990, *Astrophys. J. (Letters)* **353**, L21
Paczynski, B.: 1990, *Astrophys. J.* **348**, 485
Pines, D. and Alpar, M.A.: 1985, *Nature,* **316**, 27
Pines,D.: 1987, in *High Energy Phenomena Around Collapsed Stars*, ed. F. Pacini (Reidel, Dordrecht), 193
Romani, R.W.: 1987, *Astrophys. J.* **313**, 718
Ruderman, M.: 1990, preprint, Columbia Astrophy. Lab. Contribution No:421
Seward, F.D. and Wang, Z-R.: 1988, *Astrophys. J.* **332**, 199
Shibazaki, N. and Lamb, F.K.: 1989, *Astrophys. J.* **346**, 808
Shklovskii, I.S.: 1967, *Astron. Zhur.*, **44**, 930
Toor, A. and Seward, F.D.: 1977, *Astrophys. J.* **216**, 560
Treves, A. and Colpi, M.: 1990, *Astron. Astrophys.* submitted
Tsuruta, S.: 1964, Ph.D. Thesis, Columbia University
Van Riper, K.A.: 1988, *Astrophys. J.* **329**, 339
Ventura, J.: 1989, in *Timing Neutron Stars*, ed. H. Ögelman and E.P.J. van den Heuvel, (Kluwer, Dordrecht), p. 491
Wolff, R.S., Kestenbaum, H.L., Ku, W. and Novick, R.: 1975, *Astrophys. J. (Letters)* **202**, L77

2. FORMATION AND EVOLUTION

THE FORMATION OF NEUTRON STARS IN THE GALAXY

DIPANKAR BHATTACHARYA*
Astronomical Institute 'Anton Pannekoek'
University of Amsterdam
Roetersstraat 15, 1018 WB Amsterdam
The Netherlands

and

Centre for High Energy Astrophysics
NIKHEF-H, Kruislaan 409, 1098 SJ Amsterdam
The Netherlands

ABSTRACT. Two different ways for the formation of neutron stars have been suggested in the literature: gravitational collapse of the cores of massive stars, and accretion-induced collapse of heavy white dwarfs. This article discusses constraints on both scenarios based on observations of supernovae, supernova remnants, binary radio pulsars and radio pulsars in globular clusters.

1 Introduction

Neutron stars are among the densest objects known in the universe. The formation of such highly dense objects must involve gravitational collapse of ordinary matter. According to our present understanding, such a collapse may take place under two possible circumstances:

1. At the end of its nuclear evolution a star produces a inert, fuel-exhausted core of matter which can no longer be supported against gravitational collapse by thermal pressure. If the mass of this core is below the Chandrasekhar limit $M_{\rm lim}$ ($\sim 1.4\,M_\odot$), electron degeneracy pressure is able to support the core and a white dwarf remnant is left. If the mass of the core exceeds this limit, it collapses to form a neutron star (and perhaps in some cases a black hole). Cores massive enough to undergo such a collapse occur in stars with original main-sequence mass larger than a certain limit $M_{\rm U}$. According to the standard picture of stellar evolution, $M_{\rm U}$ has a value $\sim 8\,M_\odot$ (see e.g. Woosley and Weaver 1986) .

2. Accretion of matter on a white dwarf in a binary system may allow its mass to grow beyond the Chandrasekhar limit, upon which it may collapse to form a neutron star (Miyaji et al. 1980) . Several conditions are to be met for this to actually occur. First, the mass of the white dwarf and the accretion rate must be large

*On leave from Raman Research Institute, Bangalore, India

J. Ventura and D. Pines (eds.), Neutron Stars: Theory and Observation, 103–123.

enough for the mass to grow beyond $M_{\rm lim}$ during the mass transfer phase. Second, the accretion rate and the composition of the infalling matter must be such that the nuclear burning of the accumulated matter proceeds steadily; otherwise nova explosions would *erode* the white dwarf rather than let it grow. Third, the thermal state of the white dwarf, its composition, and its initial mass should be right for a non-explosive transition to $M_{\rm lim}$; failing which the entire white dwarf may explode leaving no compact remnant. The exact conditions under which accretion-induced collapse may take place are still somewhat uncertain; most recent models indicate accretion rates $\dot{M} \gtrsim 10^{-8}\, M_\odot\ {\rm yr}^{-1}$ and/or initial white dwarf masses above $1.2\, M_\odot$ may be necessary (Nomoto and Kondo, 1990; Canal, Isern and Labay, 1990; Nomoto, these proceedings).

The formation of a neutron star is accompanied by the release of a huge amount ($\sim 10^{53}$ erg) of gravitational binding energy. Most of this energy is carried away by neutrinos, but even a small fraction ($\lesssim$ 1%) of this is enough to eject the overlying material at a very high speed. In case of a core collapse, this would produce a supernova (SN) outburst by expelling the stellar envelope.

In an accretion-induced collapse, however, the amount of material ejected will be very small. This is likely to make the event too faint to detect by optical means. A question that has often been asked is whether a significant fraction of neutron stars may originate in such "silent supernovae". In what follows we shall examine this question, in the context of neutron stars observed in different forms and settings, namely, as isolated radio pulsars in the galatic disk, as neutron stars in binary systems, and as radio pulsars in globular clusters.

2 Neutron stars and supernovae

The idea that supernova outbursts may be associated with neutron star formation goes back to Baade and Zwicky (1934). Their conjecture was dramatically confirmed with the discovery of a neutron star (a radio pulsar) in the Crab nebula (Staelin and Reifenstein 1968), the remnant of the supernova seen in 1054 AD. However, careful searches have been unable to find neutron stars in the remnants of several other historical supernovae, e.g. those recorded in AD1006, AD1572 (Tycho's SN) and AD1604 (Kepler's SN). This most probably indicates that not all supernovae are associated with neutron star formation. The question we shall now address is whether the number of neutron star producing supernovae can account for the observed population of neutron stars in the galaxy.

To do so, one must first enumerate what kinds of supernovae are likely to leave neutron stars. An excellent recent discussion of this issue can be found in the review by Branch (1990). Supernovae are classified into two major categories based on the presence or absence of hydrogen lines in their spectra. The latter category is called "type I" and the former "type II". More recently, finer divisions of these classes have been made. Among the type I supernovae, some show a very strong silicon line feature at 6150 Å near maximum light. These have been named supernovae of "type Ia". In the rest of the type I supernovae, namely those of "type Ib", this silicon feature is absent but distinct Helium lines are seen. Type II supernovae have also been divided into two subclasses, based on the shape of their light curves: Those which display a post-maximum "plateau" are called type II-plateau or "type II-P" supernovae, and the others are known as type II-linear or "type II-L" supernovae.

Which of these classes of supernovae produce neutron stars? The observed characteristics of SNIa appear to be well reproduced by accreting white dwarfs that explode due to carbon deflagaration, leaving no compact remnant (Nomoto, Thielemann and Yokoi 1984; Woosley, Axelrod and Weaver 1984; Sutherland and Wheeler 1984; Branch et al. 1985; Wheeler et al. 1986; Graham 1987; see also Nomoto, this volume). The limited information available from historical records suggests that SN1572 and SN1604 were probably of this category (Clark and Stephenson 1977), which would be consistent with the non-detection of neutron stars in their remnants. The statistics of occurrence sites also suggest that SNIa involve rather low-mass stars, possibly less massive than M_{U}, and thus neutron star formation is not expected from them: Unlike SNII, which occur only near star forming regions in spiral galaxies, SNIa are also found to occur in ellipticals, and SNIa in spirals are not associated with either spiral arms or giant HII regions (Branch 1990 and references therein).

The indication from occurrence sites that SNII involve massive stars also make them promising candidates for neutron star production. About 25% of SNII are of type II-L; it is possible that some of these are SNIa-like explosions occurring within hydrogen envelopes, but there does not appear to be any statistically significant difference between the occurrence sites of SNII-P and SNII-L (Branch 1990). Thus the assumption that most SNII leave neutron stars does not seem to be a bad one. A small fraction, coming from the very massive stars (probably $M \gg 20\,M_\odot$), may produce black holes or (for $M \gtrsim 100\,M_\odot$) complete disruption due to pair creation instability (see Trimble, 1982 for a review), but this would hardly make any difference in total numbers.

The nature of SNIb progenitors has been a subject of debate in recent literature. It was pointed out by Van den Bergh (1988) that SNIb tend to be strongly associated with star forming regions. This, along with some very high estimates of the mass of ejected oxygen from late time spectra, led to the conjecture that SNIb come from very massive, probably Wolf-Rayet, stars that have lost their hydrogen envelopes due to strong stellar wind (Van den Bergh 1988). However, more recent work (Panagia and Laidler 1988) shows that the association of SNIb with star forming regions appears to be looser than that of SNII. Further, recent spectral modelling has considerably reduced the estimated oxygen mass, which now stands at $\lesssim 1\,M_\odot$ (Fransson and Chevalier 1989; Axelrod 1988). This favours a mass range for SNIb progenitors somewhere in between that of SNII and SNIa (Branch 1990).

Whether or not SNIb leave neutron stars is very much an open question. If the masses of the progenitors of SNIb are indeed in between that of SNII and SNIa, then the outcome would depend on what these masses really are. If SNIb progenitors are massive enough to undergo core collapse, the same stars should then produce SNIIs if they do not lose their hydrogen envelope. There seems to be some evidence to this effect from SN 1987K, which had hydrogen lines near maximum light but a few weeks later began to display a SNIb spectrum (Filippenko 1988). A possible hypothesis seems to be that SNIb arise from stars initially more massive than M_{U} which have lost their envelopes to their binary companions. However, pending detailed calculations of the expected spectrum, models involving complete explosion of accreting white dwarfs cannot be ruled out (Branch 1990).

What then, is the galactic formation rate of neutron stars from supernovae? The relative frequency of SNIa, SNIb and SNII has been estimated by Van den Bergh, McClure and Evans (1987) to be ~3:4:11 in a typical Shapley-Ames galaxy. Assuming these relative rates to be valid also for our own galaxy, and taking into account that ~25% of all SNII are "linear", one obtains the ratio SNIa:SNIb:SNII-P:SNII-L ~ 3:4:8:3. The fraction f_{NS} of supernovae producing neutron stars then lies between 0.4 and 0.8, the lower value assumes that only SNII-P makes neutron stars, while the higher value counts also SNIb and SNII-L among the neutron star progenitors.

The total galactic supernova rate is also quite uncertain. This number can be estimated in two ways: (*a*) by counting supernovae in external galaxies similar to ours in morphology and size, and (*b*) from the historical supernovae recorded in our own galaxy. Both methods are subject to considerable uncertainties owing to the unknown amount of incompleteness in the observed samples. Recent estimates of the interval τ_{SN} between galactic supernovae using these two methods stand at about 50 yr (Van den Bergh, McClure and Evans 1987) and 20–45 yr (Strom 1990) respectively.

Putting the above estimates together, one obtains the neutron star formation rate from supernovae: $f_{NS} \times \tau_{SN}^{-1}$ ~1 in 25–125 yr, a rather wide range which probably still does not reflect the full uncertainty in the estimates. This makes it hard to decide, based on this number alone, whether or not one needs an alternative channel to supply a significant fraction of neutron stars. Different estimates of the birthrate of radio pulsars in the galaxy lie in the range ~1 in 30–120 yr (see, e.g. Narayan 1987; Lyne, Manchester and Taylor 1985). Within uncertainties, this appears to be quite consistent with the rate that can be supplied by supernovae.

3 The association between pulsars and supernova remnants

There is, of course, another direction from which this problem can be approached. Supernovae produce visible remnants due to the interaction of the ejected matter with the surrounding interstellar medium. Young neutron stars would be expected to be associated with such supernova remnants (SNRs). Since young neutron stars are expected to function as radio pulsars, the origin of most neutron stars in supernovae would imply that radio pulsars should be associated with a fraction f_{NS} of supernova remnants. This naive expectation is, however, in stark contrast with available observations. Although about 200 SNRs and over 500 radio pulsars are now known, the number of associations between them stand at a paltry seven, two of which are uncertain. How is this discrepancy to be understood? It is to be noted that radio pulsars have much longer active lifetimes than SNRs, so the absence of SNRs around many pulsars is not a real problem, but the converse is!

The major reason for such a poor association between pulsars and supernova remnants lies in the numerous selection effects that plague the detection of radio pulsars. Most of the known radio pulsars lie very close to the sun, detection of distant ones being very difficult on account of their low radio fluxes. Further, broadening of the pulses due to scattering and dispersion in the interstellar medium markedly reduces the degree of modulation of the received signal, thus making it difficult to recognize its pulsed nature.

Supernova remnants, on the other hand, do not suffer from the above selection effects. Their radio luminosities are, in general, much larger than those of pulsars, and dispersion and scattering do not affect their detectability. A SNR usually disappears at an age $\gtrsim$ a few thousand years when its dropping radio luminosity and increasing size result in too low a surface brightness for it to be recognizable against the diffuse galactic radio background.

The known supernova remnants cover a much larger region of the galaxy than the radio pulsars: $\gtrsim$60% of known radio pulsars are situated within a projected distance on the galactic plane of 3 kpc from the sun, an area which contains $\lesssim$15% of known SNRs. This greatly reduces the number of possible associations one could observe. In addition, even if a pulsar is present within a supernova remnant, we can only see it if its emission cone points towards us. The probability of an observer being within the radiation cone of a pulsar (commonly known as the "beaming factor") is estimated to be ~0.2 for a typical spin period of ~1 s, rising to about unity at much shorter spin periods (Narayan and Vivekanand 1983; Lyne and Manchester 1988). This reduces the probability of observing a pulsar-SNR association even further. Given all these selection effects, the expected number of observable pulsar-SNR associations may not exceed about 10, not inconsistent with observations. It is to be noted in this context that all known radio pulsars with spindown age $\tau(\equiv P/2\dot{P}$; P=spin period, $\dot{P}$=spindown rate) less than $\sim 10^4$ yr do have associated SNRs.

4 Synchrotron nebulae produced by pulsars

While the non-detection of pulsars within supernova remnants may be attributed to the above selection effects, it was pointed out by Radhakrishnan and Srinivasan (1980) that one should be able to detect the Crab-like synchrotron nebulae (plerions) such pulsars would produce by pumping relativistic particles and magnetic field into the SNR cavity. Thus a supernova remnant harboring a young pulsar should have a plerionic (centre-filled) or a composite (plerion within a shell) morphology. However, the vast majority of supernova remnants have the morphology of a hollow shell, with hardly any central emission. The number of plerions and composites together amount to less than 10% that of all known SNRs.

A straightforward interpretation of this might be that most SNRs do *not* contain functioning pulsars. But before one jumps to this conclusion, one must make a careful evaluation of the factors that may work against the detection of a plerionic component of a supernova remnant, even if there is one. The recognition of a plerionic component embedded in a supernova remnant depends on at least three factors: (*a*) the relative surface brightness of the plerion and the shell, (*b*) the dynamic range of the available radio maps, and (*c*) the amount of spectral information available, since the plerionic components usually have a flatter spectrum than the shell.

The expected relative brightness of the two components can be estimated by modelling the evolution of the composite remnant. The evolution of the luminosity of the plerionic component would depend on the total rotational energy content of the pulsar (i.e. the initial spin period P_0), the rate at which this energy is released into the SNR (determined by the magnetic field strength B_* of the pulsar), and the speed of expansion of the nebula, which determines the amount of energy the particles lose in expansion. Models of spectral evolution of such expanding bubbles of relativistic fluid were constructed first by Pacini and Salvati (1973) , which have been extended to include non-uniform expansion by several authors (Reynolds and Chevalier 1984; Bandiera, Pacini and Salvati 1984; Bhattacharya 1987). These models can be readily used, in conjunction with models for the evolution of the shell emission, to follow the evolution of the morphology of composite SNRs (Bhattacharya 1990).

Unfortunately the available models for the evolution of radio emission from the shell are at present much less developed than those for the plerionic component. Nevertheless, a crude estimate of the luminosity can be obtained by using the results of the numerical models by Gull (1973). The composite models may then be constructed for different values for the ambient explosion energy, interstellar density, ejected mass, P_0 and B_*, assuming that the plerionic component fills a fraction of the interior volume of the SNR (Bhattacharya 1990).

This result of such an exercise suggests that at the presently available sensitivity and dynamic range radio plerions should be detectable within shell SNRs if the initial spin periods of the embedded pulsars were $P_0 \lesssim$ 30–50 ms, and their magnetic fields were $B_* > 10^{12}$ G (Srinivasan, Bhattacharya and Dwarakanath 1984; Bhattacharya 1990). It has also been argued that X-ray synchrotron nebulae would be detectable if $P_0 \lesssim 100$ ms and $B_* > 10^{12}$ G (Chevalier and Emmering 1986).

The paucity of such detectable nebulae therefore indicates that most pulsars are born "weak" (spinning slowly and/or with low magnetic fields), or in events that do not produce observable supernova remnants. Of these, long initial spin periods seems the most attractive possibility, since some evidence for this has also emerged from statistical analyses of pulsar properties (Vivekanand and Narayan 1981; Chevalier and Emmering 1986; Narayan 1987; Emmering and Chevalier 1989). The failure to detect short-period ($P \lesssim 100$ ms) pulsars in the Princeton-Arecibo survey, despite a good sensitivity, also indicates that many pulsars are born with relatively long spin periods (Stokes et al. 1986).

It is to be noted that the studies of pulsar statistics similar to those of Vivekanand and Narayan (1981), Narayan (1987) and Emmering and Chevalier (1989) find that the majority of pulsars are born in the period range 0.5–0.7 s. While this is consistent with the absence of detectable plerions in most SNRs, the present level of detectability of radio plerions imposes a much less severe constraint on initial spin periods (see above). The above conclusion from pulsar statistics depends on the assumed model of the dependence of the radio luminosity of a pulsar on its period and magnetic field. The same result is indicated, however, even in a model-independent approach (Narayan 1987), but with a marginal statistical significance. It will be fair to say that though there seems to be a good amount of evidence that many pulsars are born spinning slow, the form of the birthrate as a function of initial spin period is yet to be determined with any certainty. More sensitive pulsar surveys, and better mapping of radio SNRs will be required to improve our understanding in this regard.

To summarize, according to the present estimates, the galactic birthrate of neutron stars (radio pulsars) appears to be consistent with the estimated galactic rate of neutron-star-producing supernovae; the poor association between observed pulsars and supernova remnants can be attributed entirely to selection effects; and the paucity of bright plerionic nebulae can be understood in terms of most pulsars being born with spin periods in excess of a few tens of milliseconds. Thus it appears that conventional supernovae can account for the origin of most, if not all, neutron stars in the galaxy.

5 Masses of neutron star progenitors

So far we have discussed the rate of formation of neutron stars, and the rate of occurrence of supernovae as estimated from observations. These rates can be used, in conjunction with the stellar mass function and the stellar lifetimes, to infer the lower mass limit of the required progenitors. A galactic pulsar birthrate of 1 in ~40 yr would need all stars with initial masses above ~8 $M_\odot$ to form neutron stars. A supernova rate of 1 in ~20 yr will need stars above ~6 $M_\odot$ to form supernovae, but, as discussed above, this may just be a formal number, since many supernovae may come from white dwarfs exploding due to accretion.

There are, of course, large uncertainties in these numbers because the rates cannot be accurately determined. To avoid some of these problems, Blaauw (1985,1987) carried out a comparison of the *local* pulsar population with *local* star counts. Restricting himself to pulsars within a distance projected on the galactic plane of 0.5 kpc from the sun, he found that the maintenance of this population would require a *local* birthrate of $\sim 33/\bar{f}_b$ pulsars kpc^{-2} per 4.6 Myr, where $\bar{f}_b$ is the average "beaming factor".

From the star counts in the solar neighbourhood, on the other hand, he estimated that local O-B associations can provide only $\lesssim$15 stellar deaths kpc^{-2} per 4.6 Myr; which implies that the large majority of pulsars must be contributed by field stars. From the observed local density of field stars, the estimated lower mass limit for neutron star formation then followed: 6 $M_\odot$ for $\bar{f}_b = 0.4$, 7 $M_\odot$ for $\bar{f}_b = 0.6$, rising to 8 $M_\odot$ only for a very large value of the average beaming factor, $\bar{f}_b = 0.8$. Since the value of the average beaming factor for the local pulsar population estimated from observations does not exceed about 0.4, this would lead one to the surprising conclusion that stars with initial main-sequence masses as low as 6 $M_\odot$ must form neutron stars, contrary to what is expected from conventional stellar evolution theory.

The initial mass boundary M_{U} below which stellar cores do not become heavy enough to undergo gravitational collapse is estimated to be about 8 $M_\odot$ in the conventional stellar evolution schemes. The location of this mass boundary, however, is quite sensitive to the treatment of convection in the computation of stellar evolution. It has been pointed out by several authors (Bertelli, Bressan and Chiosi 1985; Castellani et al. 1985; Maeder and Meynet 1989) that the inclusion of somewhat more than usual convective overshooting can lead to a reduction of M_{U} to about 6.5 $M_\odot$.

There is another interesting situation where stars with initial masses below M_{U} may be persuaded to leave neutron star remnants. This happens if the star is in a close binary system with a more massive companion. As the more massive component evolves, it grows in size, and eventually transfers several solar masses of envelope matter onto its companion before exploding as a supernova. The total mass of the originally less massive component can now exceed M_{U}, and if so, a neutron star would be left as its final evolutionary product. For the primary (originally more massive) stars, however, the evolution in close binary systems may *increase* the lower mass limit for neutron star formation somewhat above that for single stars (see e.g. Habets 1985; Van den Heuvel 1983; Van den Heuvel and Habets 1985)

6 Formation of neutron stars in binary systems: origin of X-ray binaries and binary radio pulsars in the galactic disk

Apart from modifying the value of the limiting mass M_U, binary systems also provide an additional route for neutron star formation: namely, the accretion-induced collapse of a white dwarf (see above). In the rest of this article we shall discuss in which class of systems the latter is likely to take place, and whether there is any observational evidence for this actually happening.

6.1 X-ray binaries

If the neutron star in a binary system is formed due to a supernova explosion, then the accompanying violent ejection of matter may unbind the binary system if the amount of ejected matter exceeds the remaining total mass of the system (Blaauw 1961). Further, if the matter ejection is asymmetric, then the binary may be disrupted even if a smaller amount of matter is ejected, though in favourable circumstances the asymmetry may also *prevent* the disruption of a binary system.

The observed interacting binaries containing neutron stars may be classified into two main categories (see Van Paradijs, this volume): (*i*) the Massive X-ray binaries, where the mass of the companion object is much larger than that of the neutron star. This class contains the so-called "standard massive X-ray binaries" and the B-emission X-ray binaries; and (*ii*) the Low-Mass X-ray Binaries (LMXBs), where the mass of the companion is $\lesssim$ the neutron star mass.

The present mass-donating components in the massive X-ray binaries have masses typically above 6 $M_\odot$. It is most probable that the originally more massive components in these systems had masses above M_U, and underwent core collapse to form neutron stars. Many of the orbits of the massive X-ray binaries have substantial eccentricities, a characteristic one would expect to result from a supernova explosion in the binary.

The formation of the second class of systems, namely, the low-mass X-ray binaries, presents an interesting problem. If the neutron star in such a system was formed in a supernova, its progenitor must have been a fairly massive ($\gtrsim$8 $M_\odot$) star. If most of the envelope matter of this star was to be expelled in the supernova explosion, the binary system would very likely have become unbound, even accounting for asymmetries.

There are thus two possible explanations for the formation of LMXBs. Either the primary component of the original binary had a mass below M_U, and left a white dwarf which eventually formed a neutron star via accretion-induced collapse; or the primary, starting out with a mass larger than M_U, lost most of its mass *before* the supernova explosion occurred. The latter is, in fact, *expected* to happen in binaries with mass ratio far from unity (see Van den Heuvel (this volume) for details). In such binaries, when the primary evolves and begins to transfer matter, the mass transfer proceeds at the thermal timescale (the timescale in which a star, once disturbed, can restore its thermal equilibrium: see, e.g., Cox and Giuli 1968) of the donor. The low-mass secondary, on the other hand, cannot accept this matter and remain in thermal equilibrium, since *its* thermal timescale is much longer ($\tau_{th} \propto M^{-2}$). The secondary swells as a result, and a "common envelope" around both stars is formed. The orbital separation shrinks within this envelope, the secondary "spirals in" towards the core of the primary, and a large quantity of mass is lost from the system. This spiral-in may eventually end when the separation of the two stars becomes close to the size of the helium core of the primary. Due to the short duration of this phase, and the low rate of mass settlement on the secondary (due to its long τ_{th}), the net enhancement of the mass of the low-mass star will be very little; but the high-mass primary would have lost most of its envelope. The survival of the binary system after the eventual supernova explosion will thus be determined by the mass of the post-spiral-in core of the primary. If helium cores as small as $\sim 2.2\,M_\odot$ can produce neutron stars (Habets 1985; Van den Heuvel and Habets 1985), then the formation of low-mass X-ray binaries in this manner would not pose much of a problem.

One must remember, however, that the details of the spiral-in process are so far very poorly understood. The evidence that spiral-in does occur at least in some cases comes from the existence of the close double neutron star systems like PSR 1913+16, PSR 2303+46 and PSR 1534+12, a close binary pulsar with a white dwarf companion such as PSR 0655+64 and double cores of planetary nebulae (see Bhattacharya and Van den Heuvel 1990; De Kool 1990; and Van den Heuvel (this volume) for detailed discussions). But given many uncertain factors and the unknown number of binaries with very large ratio of primary-to-secondary mass, it is not possible to estimate at present the relative contribution of spiral-in and accretion-induced collapse to the formation of low-mass X-ray binaries. Kinematics of LMXBs may provide a clue in this regard: the *systemic* velocities of LMXBs originating from AIC are likely to be much smaller than those for systems in which the neutron stars were born in supernovae. Reliable determination of the velocities of a significant number of LMXBs will be required before a firm conclusion can be reached in this respect.

6.2 Radio pulsars in low-mass binaries

The accretion of matter by a neutron star in an interacting binary allows angular momentum to be transferred from the orbit to the spin of the neutron star, causing the neutron star to spin-up. For a given accretion rate, this spin-up can proceed up to an "equilibrium" value at which the spin frequency of the neutron star roughly matches the keplerian frequency at the magnetospheric boundary (see F. Lamb (this volume) and Ghosh (this volume) for detailed discussions). For the maximum possible accretion rate, namely, the Eddington rate ($\sim 10^{-8}\,M_\odot\ \mathrm{yr}^{-1}$), the equilibrium spin period is determined chiefly by the magnetic field strength of the neutron star (see, e.g. Srinivasan and Van den Heuvel 1982): $P_{\mathrm{min}} \sim 1.9$ ms $(B_*/10^9\mathrm{G})^{6/7}$. Once the accretion ceases, these neutron stars are expected to make their appearance as rapidly spinning (especially so if the magnetic field of the neutron star is small, e.g. $\lesssim 10^9$ G for the millisecond pulsars) radio pulsars, often referred to in the literature as "recycled" radio pulsars.

The low-mass X-ray binaries discussed above have been widely recognized as the ideal progenitors of recycled radio pulsars in binaries with low-mass ($\lesssim 0.4\,M_\odot$) white dwarf companions (Webbink, Rappaport and Savonije 1983; Savonije 1983; Joss and Rappaport 1983; Paczyński 1983; see Verbunt 1990 and Bhattacharya and Van den Heuvel 1990 for recent reviews). Several such systems are now known in the galactic disk.

The evolution of low-mass X-ray binaries proceeds along two different tracks depending on the initial orbital period of the system. If the orbit is initially wide, with $P_{\rm orb} > 10^{\rm h}$, the mass transfer is driven by the nuclear evolution, and the consequent radius expansion of the donor. The orbit widens as the mass transfer proceeds, and when the binary finally detaches, the degenerate core of the donor is left as a white dwarf in a wide orbit around the neutron star. The widest orbit of such a binary found so far is that of PSR 0820+02, the present orbital period being 1232.5 days. It can be estimated that the initial orbital period of its progenitor LMXB must have been $\gtrsim$300 days. Such an wide-orbit LMXB is, however, difficult to form from a spiral-in evolution. Ejection of matter from the neutron star progenitor during the common envelope phase occurs at the cost of the total orbital energy. The price of reducing the envelope mass enough so as to keep the orbit bound in a subsequent supernova would have reduced the orbital period to $\lesssim$15 days at the end of spiral-in, contrary to what is required to explain the present wide orbit. It seems likely, therefore, that in this system the neutron star originated from the accretion-induced collapse of a white dwarf. One cannot, however, be entirely certain about this, since even with a large amount of ejected matter the orbit could still have remained bound in the case of a very favourable asymmetry in mass ejection. This issue can, in principle, be settled by measuring the centre-of-mass velocity of this system.

The velocity transverse to the line of sight has recently been measured (Ryba and Taylor 1990) for one of the low-mass binary pulsar systems, namely, PSR 1855+09. The measured velocity turns out to be rather small, only $\sim$10 km s^{-1}, favouring an accretion-induced collapse origin also for this neutron star.

Low-mass X-ray binaries that start out with $P_{\rm orb} < 10^{\rm h}$ evolve with substantial angular momentum loss due to gravitational radiation and magnetic braking. Mass transfer starts while the neutron star companion is still on the main sequence. The orbit shrinks and in absence of any interruption would pass through a minimum orbital period of $\sim$80 minutes, at which point the donor star becomes degenerate, its radius begins to expand on loss of mass, and the orbit begins to widen (see Van den Heuvel (this volume); Verbunt 1990; Bhattacharya and Van den Heuvel 1990).

In reality, the evolution perhaps seldom proceeds this far. Once the neutron star has been spun up to a few millisecond period due to mass accretion, it may begin to function as a pulsar, and the pulsar wind may halt further mass accretion onto the neutron star, curtailing the active X-ray lifetime of the system (Ruderman, Shaham and Tavani 1989; see Shaham, this volume). Eventually the pulsar wind may erode the companion star and vaporize it away (as seems to be happening in the PSR 1957+20 system, and also in the PSR 1744−24A system in the globular cluster Terzan 5). If the pulsar does manage to completely evaporate its companion, a solitary millisecond pulsar would be the final evolutionary product (Ruderman, Shaham and Tavani 1990; Krolik and Sincell 1990). This appears to be the most promising explanation for the formation of the single millisecond pulsars PSR 1937+21 and PSR 1257+12 in the galactic disk.

The curtailment of low-mass X-ray binary lifetimes may also help remove another outstanding problem: that of the discrepancy between the birthrates of low-mass binary pulsars and low-mass X-ray binaries in the galactic disk. From the observed population of low-mass binary pulsars it has been estimated (Kulkarni and Narayan 1988) that the birthrate of low mass binary pulsars (short $P_{\rm orb}$ systems) in the galactic disk exceeds that of low-mass X-ray binaries by about an order of magnitude, if the 'conventional' lifetimes of $\sim 10^8 - 10^9$ yr for low-mass X-ray binaries are correct. This has led Kulkarni and Narayan (1988) to argue that the progenitors of low-mass binary pulsars should be sought among systems other than low-mass X-ray binaries. However, one must note that it requires only $\lesssim 10^7$ years of mass accretion to spin a neutron star up to a few millisecond period (provided its magnetic field is low enough). If the neutron star is able to stop the accretion soon after this, the LMXB lifetimes could be shortened by at least an order of magnitude, increasing the estimated birthrate of LMXBs to match that of low-mass binary pulsars (Coté and Pylyser 1989; Tavani 1989).

7 Origin of radio pulsars in globular clusters

The origin of neutron stars in globular clusters has generated a considerable amount of debate in the current literature; the formation of a large majority of them in accretion-induced collapse being seriously contemplated by several authors. In what follows I shall attempt an assessment of this question based on recent developments.

Neutron stars in globular clusters were first discovered as low-mass X-ray binaries, of which some ten are known at present. It was noticed quite early that the globular clusters seem to be extremely rich in LMXBs: all globular clusters put together contain $\lesssim 10^{-3}$ of the total mass in the galactic disk, but ~10 of the ~50 known LMXBs. Following the realization that the evolutionary products of LMXBs may be rapidly spinning recycled radio pulsars, pulsar searches in globular clusters have been undertaken. These have resulted in an avalanche of discoveries, $\gtrsim 25$ have already been found, more than half of them within the past one year.

The stars in globular clusters are $\gtrsim 10^{10}$ yr old, and the only surviving ones are of low mass ($\lesssim 0.8\,M_\odot$). Neutron stars formed in core collapse of massive stars must therefore be nearly 10^{10} yr old. Those formed in accretion-induced collapse could be younger, but given the total number of potential accretion-induced collapse candidates, most neutron stars formed this way must also be very old. If the neutron stars formed in globular clusters had magnetic field strengths similar to the young neutron stars in the galactic disk, they would have spun down to long spin periods and ceased to function as pulsars long ago. It seems most likely, therefore, that the radio pulsars we see in globular clusters have all been "recycled" in binary systems. The overabundance of neutron star binaries in globular clusters certainly helps to accomplish this task.

Of some importance in addressing the question of the origin of neutron stars in the globular clusters is an estimate of the total population. This has to be done, starting from the observed number of radio pulsars, and assuming a certain distribution of the number over different clusters, by accounting for the possible selection effects in pulsar surveys. The first estimate based on this procedure yielded a total number of $\sim 10^4$ active radio pulsars in the whole globular cluster system (Kulkarni, Narayan and Romani 1990). This estimate was, however, subject to large uncertainties (see Bhattacharya and Van den Heuvel 1990 for a discussion). Following this, other authors have tried to improve upon this estimate by taking into account the observed total continuum radio flux from a few globular clusters, and comparing them with that expected from the population of possibly undetected pulsars in each of these clusters. The total number of pulsars in the entire globular cluster system estimated this way turns out to be $\lesssim 1000$ (Fruchter and Goss 1990; Wijers and Van Paradijs

1990). Any scenario proposed for the formation of radio pulsars in globular clusters must, apart from explaining their characteristics, also account for this estimated total number.

7.1 Formation of neutron star binaries

The overabundance of neutron star binaries in globular clusters has been traditionally attributed to the unique processes of binary formation through stellar encounters that operates due to the very high stellar density found in these clusters. Two objects, initially unbound, may form a binary system in the presence of a sink of orbital energy. In globular clusters, this can be achieved in at least three different ways:

1. In a collision involving three stars, one of the objects may take up the excess energy and escape, leaving the other two in a bound system. Since this process requires a simultaneous convergence of three independent trajectories, the probability of this happening is rather small.

2. If a neutron star interacts with an already existing binary, then an *exchange* may take place, in which the neutron star replaces one of the components of the original binary system.

3. During the close passage of a compact star a normal star would undergo substantial tidal deformation at the cost of a part of the relative kinetic energy of the orbit. Most of this energy will eventually be dissipated through oscillations and heating. If the amount of energy thus lost exceeds the total positive energy of the initial unbound orbit a bound system will result. This mechanism, known as "tidal capture", is thought to be one of the most important binary formation routes in globular clusters.

Tidal capture The conditions necessary for tidal capture to occur, and the cross section of this process were first computed by Fabian, Pringle and Rees (1975) . Since then several more refined calculations have been made (Press and Teukolsky 1977; Giersz 1986; Lee and Ostriker 1986; McMillan, Mcdermott and Taam 1987), which yield very similar results. The following are the main features:

1. For a tidal capture to occur, the distance of closest approach between the two stars must be $\lesssim 3$ times the radius R of the extended star.

2. Due to gravitational focussing, the cross section for an encounter with a closest approach d is proportional to d. Since $d \lesssim 3R$ for capture to occur at all, this means that about one-third of all tidal captures would lead to a direct collision between the two stars. The outcome of a direct collision would depend on the evolutionary state of the extended star. If the star is a giant or a subgiant with an already formed degenerate core, a spiral-in may result in the formation of a close binary containing this core and the compact star. If the star is on the main sequence, however, the star is likely to be completely disrupted, leaving a massive disk around the compact object.

3. By taking into account the relative number of stars in different evolutionary phases, and their appropriate cross sections, it has been estimated (Verbunt and Hut 1987;

Verbunt 1988; Verbunt and Meylan 1988) that about 60–90% of tidal captures occur on main-sequence stars, and the rest on giants and subgiants.

Exchange encounters A close encounter between a compact star and an existing binary can lead to the substitution of one of the components of the original binary by the incoming compact star. The compact star should approach the original binary to within a few times the orbital separation a for this to take place (Hut 1983). The exchange may occur either directly, or via a resonance encounter in which a temporary triple system is formed at first, from which one member, usually the least massive one, is eventually ejected (Hut 1983). For the typical conditions in a globular cluster, a resonant exchange is roughly twice as probable as a direct one (Hut 1984).

Exchange encounters gain over tidal capture by having a larger cross section (the relevant scale being a instead of $3R$), but lose because of the smaller number density of binaries than of single stars. The number of primordial binaries in globular clusters is not well known: results of searches indicate that ordinary binaries are much less common in the globular clusters than in the galactic disk. Significant contribution from exchange encounters is, however, expected if the number of stars in primordial binaries exceed $\sim 10\%$ of total, and recent work (Pryor et al. 1987; Hut, Murphy and Verbunt 1990; Phinney and Kulkarni 1990) shows that this may well be the case. The distribution of orbital separations and companion types in the binaries formed by exchange encounters are expected to be different from those resulting from tidal capture. In particular, a larger number of binaries with more massive giant and subgiant stars are expected to form than with less massive main-sequence stars (this is a consequence of the high probability of losing the least massive member of a three-body system). However, without a clear knowledge of the abundance of primordial binaries and the distribution of their orbital periods, it is difficult at present to estimate the real contribution of exchange encounters.

7.2 Evolution of neutron star binaries in globular clusters

Binaries formed in globular clusters will follow two distinct evolutionary tracks exactly as LMXBs in the galactic disk do. Giant secondaries will drive mass transfer via radius expansion, and the orbit will widen. In binaries with main-sequence companions, gravitational radiation and/or magnetic braking will be responsible for bringing the system into contact and sustaining mass transfer. These systems will shrink in course of evolution. But if the mass of the main-sequence secondary is close to the "turn-off mass" (the minimum mass for which a star can evolve away from the main-sequence within the lifetime of the cluster) nuclear evolution may also play an important role and some of these orbits will widen.

As the final evolutionary product a giant star binary is expected to leave a rapidly spinning pulsar in a wide orbit with a low mass, degenerate companion (see above). Since the binding energy of such an orbit will be low, it would also become feasible for the neutron star to be released from the binary system in stellar encounters, especially in dense clusters. Close passage of a third star may either ionize the binary (Romani, Kulkarni and Blandford 1987) or the neutron star may change places with the third star (Verbunt et al. 1987).

Mass transfer in close binaries may, at some point, be interrupted by the wind from the spun-up pulsar (which may eventually evaporate the companion away to leave a solitary

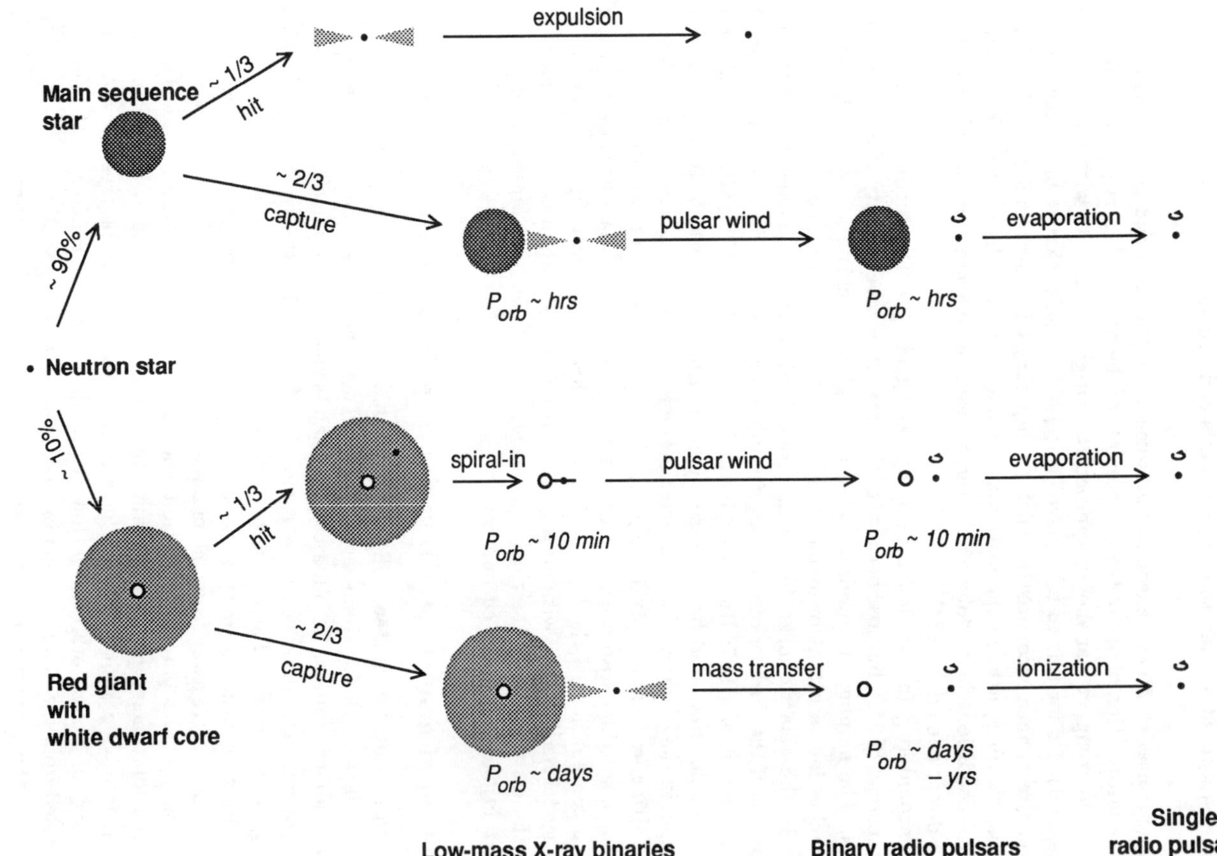

Figure 1. Evolutionary routes for the binaries formed by tidal capture in globular clusters

radio pulsar). If this does not happen, then these binaries are unlikely to produce functioning, let alone rapidly spinning, radio pulsars—first, because detachment may not occur in a Hubble time for many of these systems and second, the neutron star, even if spinning rapidly to begin with, will probably slow down to a very long period during the prolonged, slow mass transfer in the final phases of the evolution (Jeffrey 1986).

Very close white dwarf plus neutron star binaries that formed due to direct collisions with giants will evolve via gravitational radiation. The orbit will shrink till the system is brought into contact and will subsequently widen. The orbital period in the contact phase will be a few minutes. As in the case of original main-sequence donors these, too, will end their evolution when the secondary either becomes planet-sized or gets evaporated by the pulsar wind.

A massive disk formed around a neutron star in the direct collision with a main-sequence star is likely to empty itself quite soon by a highly super-Eddington mass flow. Only a small fraction of this would actually accrete onto the neutron star and the rest would be blown away. It has been proposed that under certain circumstances the neutron star may be spun up enough to leave a single millisecond pulsar after the disk disappears (Krolik 1984). Whether this can actually occur in real systems, however, remains debatable.

A summary of the different evolutionary paths described above is presented in figure 1.

7.3 The observed radio pulsars

Figure 2 summarizes the observed spin periods and orbital periods of the pulsars detected so far in globular clusters. As can be seen, more than 50% of these are solitary. As discussed above, spun up pulsars in globular clusters can become single in several ways:

1. Wide systems may release the pulsar in an 'ionization' or 'exchange' encounter.
2. Companions of rapidly spinning pulsars may be evaporated by the pulsar wind.
3. Pulsars spun up from massive disks will appear single after the disk disappears.

For a neutron star to have been liberated from a binary due to a stellar encounter, the original binary must have been wide enough to have a significant cross section for this to take place within the cluster lifetime. Calculations show (Rappaport, Putney and Verbunt 1989; Bhattacharya and Van den Heuvel 1990) that in all cases fairly wide orbits are needed, ones that can result only from the evolution of binaries formed with giant stars. If tidal capture is the dominant process of binary formation, then only a small fraction of all binaries can achieve such long periods. If, however, exchange encounters with primordial binaries make an appreciable contribution, then it would be possible to form a larger number of isolated pulsars this way.

It is more likely, however, that many of the observed radio pulsars have become single by vaporizing their companions. At least two basic conditions should be met to accomplish this: (*a*) the pulsar should have a sufficiently large spindown luminosity $L_{\rm sd}$ and (*b*) the total rotational energy $E_{\rm rot}$ of the pulsar, multiplied by the efficiency of utilization of the spindown luminosity in the evaporation process, must exceed the gravitational binding energy of the companion.

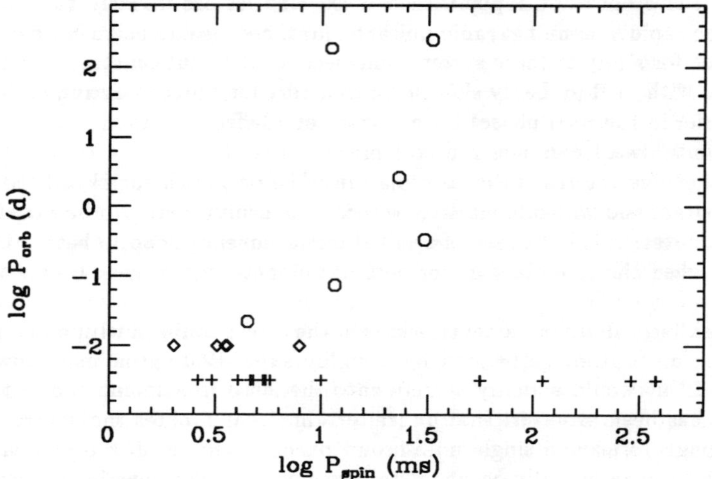

Figure 2. The spin periods and the orbital periods of radio pulsars discovered in globular clusters. Solitary pulsars are shown as crosses. The diamonds represent pulsars in binaries whose orbital periods are yet to be determined.

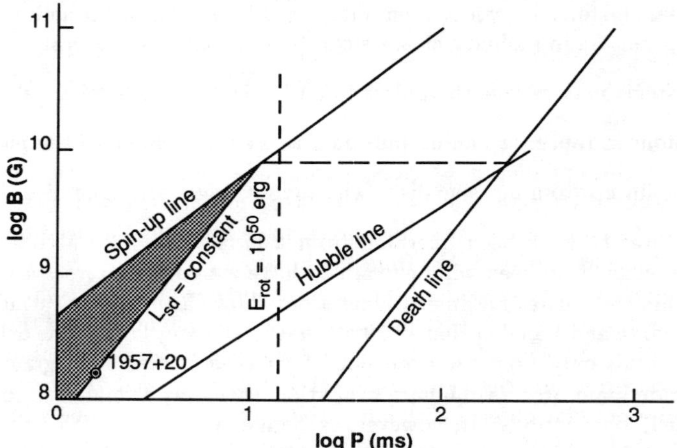

Figure 3. The expected location in the $B - P$ diagram of recycled radio pulsars that eventually vaporize their companions. Such a pulsar should probably be spun-up to within the shaded wedge, bounded by the "spin-up line" (corresponding to maximum possible spin-up for a neutron star accreting at Eddington rate) and a spindown luminosity equal to that of PSR 1957+20 (see text). Given $\lesssim 10^{10}$ yr to spin down, however, such a pulsar should be found anywhere between this region and the "Hubble line". As can be seen from the diagram, if the pulsar started out near the top of the "wedge", its present period could be as long as $\sim$200 ms.

The binding energy of a main-sequence star is $\sim 10^{48}(M/M_{\odot})$ erg. If $\sim$ 10% of the spindown luminosity is intercepted (as it would be for the typical size of a Roche-lobe filling companion) and $\sim$ 10% of the intercepted energy is used in evaporation (as estimated in the case of PSR1957+20: see e.g. Krolik and Sincell 1990), then the energy requirement (b) becomes $E_{rot} \gtrsim 10^{50}$ erg, which means that the pulsar should have been spun up to a spin period $P \lesssim 15$ ms for evaporation to take place.

The luminosity criterion (a) is less certain, but as a safe limit one may assume that the pulsar should be at least as luminous as PSR1957+20, which is seen to be evaporating its companion. This restricts the parameters of the pulsar after spin up to within the shaded wedge in field-period diagram shown in figure 3. If all pulsars spun up to within this region eventually become single due to evaporation, then one may ask with what characteristics these pulsars are likely to be found at present, allowing for possible evolution in a $\lesssim 10^{10}$ yr period since spin-up. If the magnetic field strength of the pulsar remains constant during this period, as all available evidence seems to indicate (see Bhattacharya and Srinivasan, this volume), then they would move horizontally in the $B - P$ diagram, but remain to the left of the "Hubble line", on which the spindown age equals the age of the galaxy. For them to be observable they should also be above the "death line", beyond which pulsar activity stops.

As can be seen from figure 3, the present spin periods of these pulsars could be as long as $\sim$ 200 ms for those starting out near the top of the wedge, but $\lesssim 10$ ms at the other end. This means that almost all the single pulsars in globular clusters qualify as candidates for this evolutionary route, contrary to suggestions recently made in the literature (see e.g. Ray and Kluźniak 1990). The measurement of period derivatives (and hence the magnetic fields) of these pulsars will provide an additional criterion to judge whether the relatively slow ($P \gtrsim 30$ ms) single pulsars could have had such a history (Bhattacharya and Van den Heuvel 1990).

The third alternative, namely, spin-up from a massive disk, is the least certain of all. Since there is no clear prediction of the spin periods expected after spin up, it is difficult to identify possible candidates. This may be a promising mechanism to produce single pulsars too slow to evaporate their companions, since it avoids the need for a second close encounter to release the neutron star from the binary.

That the evaporation of companion stars in close binaries may be a common occurrence is also indicated by the paucity of short-period ($P_{orb} < 1^{d}$) systems among the observed radio pulsars. PSR 1744−24A with $P_{orb} = 1.8^{h}$ is, in fact, seen to be evaporating its companion. Two other short-period binaries (PSR 2127+11C: $P_{orb} = 8^{h}$, and *PSR 0021−72A: $P_{orb} =$ 32 min) are quite peculiar objects with high orbital eccentricities (0.68 and 0.32) and large companion masses ($> 0.9\,M_{\odot}$ and $\sim 0.8\,M_{\odot}$). These were most probably formed in 3-body resonance encounters in which the former companions of the neutron stars were expelled (see Bhattacharya and Van den Heuvel 1990 for a detailed discussion).

7.4 Primordial neutron stars versus accretion-induced collapse

Finally, one has to estimate whether the inferred total number of radio pulsars in globular clusters can be contributed by primordial neutron stars alone, or whether a significant

*confirmation awaited

contribution from accretion-induced collapse is needed. Some authors (see e.g. Bailyn and Grindlay 1990 and references therein) have preferred the latter, based on the following arguments:

1. The neutron stars formed in supernova explosions would have had large space velocities (radio pulsars in the galactic disk have velocities $\sim$200 km s^{-1}) and most of them would have escaped from the globular clusters.

2. The ratio of the number of radio pulsars to that of LMXBs in globular clusters is so large that an alternative origin of radio pulsars, avoiding the LMXB phase, is needed. (This problem is no longer so serious with the more recent estimates revising the number of radio pulsars, and the lifetimes of LMXBs, downwards by about an order of magnitude.)

3. The initial mass function of globular clusters must have been quite steep, with a much smaller fraction of massive stars than in the galactic disk. Otherwise the mass ejected from the massive stars in supernova explosions would have unbound the clusters (Chernoff and Weinberg 1990). Steep mass functions would produce a larger number of heavy white dwarfs (capable of undergoing accretion-induced collapse) relative to neutron stars.

4. The preponderance of single radio pulsars and pulsars in wide binaries indicates a preference for wide systems (this assumes that complete evaporation of companions in close binaries rarely occurs). The mass accretion rates required for accretion-induced collapse to take place (Nomoto 1987; Nomoto and Kondo 1990; Nomoto, this volume) also seem to be achievable only in wide binaries. Thus accretion-induced collapse origin of most neutron stars would naturally explain the observed orbital characteristics of the radio pulsars (Romani 1990; Ray and Kluźniak 1990).

These arguments, however, fail to withstand close scrutiny. First, the limit on the initial mass function from cluster disruption turns out to be an artifact of the assumed lower cutoff of the mass distribution at $0.4\,M_\odot$; extension of this limit to $\sim 0.1\,M_\odot$ would allow much flatter mass functions (B. Murphy and E. S. Phinney, in preparation, 1990). Second, even with a mass function as steep as $dN/dM \propto M^{-3}$, and with 90% of neutron stars escaping from the clusters at birth, one can still comfortably produce several thousand recycled pulsars (Bhattacharya and Van den Heuvel 1990); recent work (e.g. Phinney 1990) indicates that the real mass functions may be even flatter than this. Finally, halting of mass accretion due to pulsar activity followed by the evaporation of the companion stars and/or the contribution of 3-body encounters in the formation of neutron star binaries can satisfactorily account for the pulsar/LMXB ratio, and the orbital characteristics of globular cluster pulsars. Thus at present there appears to be no need to invoke accretion-induced collapse for the formation of a major fraction of neutron stars in globular clusters.

8 Summary

The three main points I have tried to make in this article are as follows:

1. The rate of neutron-star-producing supernovae, the supernova remnant birth rate and the pulsar birth rate are, within uncertainties, in reasonable agreement in the galactic disk, and imply a lower mass limit of 6–8 $M_\odot$ for neutron star progenitors. The majority of neutron stars in the disk are born in standard core collapse supernovae.

2. The observed association between radio pulsars and supernova remnants is very poor at present. This can, however, be attributed entirely to selection effects.

3. So far no convincing need for the formation of neutron stars in globular clusters by accretion-induced collapse of white dwarfs has been demonstrated. In the galactic disk, PSR 0820+02 and PSR 1855+09 are the only objects which show some signs of having formed in accretion-induced collapse.

References

Axelrod, T. S.: 1988, *Lecture Notes in Physics* **305**, 3754

Baade, W. and Zwicky, F.: 1934, *Proc. Nat. Acad. Sci.* **20**, 254

Bailyn, C. D. and Grindlay, J. E.: 1990, *Astrophys. J.* **353**, 159

Bandiera, R., Pacini, F., and Salvati, M.: 1984, *Astrophys. J.* **285**, 134

Bertelli, G., Bressan, A. G., and Chiosi, C.: 1985, *Astron. Astrophys.* **150**, 33

Bhattacharya, D.: PhD thesis, Indian Institute of Science, Bangalore, 1987

Bhattacharya, D.: 1990, *J. Astrophys. Astron.* **11**, 125

Bhattacharya, D. and van den Heuvel, E. P. J.: 1990, *Physics Reports*, in press

Blaauw, A.: 1961, *Bull. Astron. Inst. Netherlands* **15**, 265

Blaauw, A.: 1985, in Boland, W. and van Woerden, H., eds., *Birth and Evolution of Massive Stars and Stellar Groups*, page 211, D. Reidel, Dordrecht

Blaauw, A.: 1987, in Helfand, D. J. and Huang, J. H., eds., *IAU Symp. No. 125: Origin and Evolution of Neutron Stars*, page 48, D. Reidel, Dordrecht

Branch, D.: 1990, in Kundt, W., ed., *Neutron Stars and Their Birth Events*, page 281, Kluwer Academic Publishers, Dordrecht

Branch, D., Doggett, J. B., Nomoto, K., and Thielemann, F.-K.: 1985, *Astrophys. J.* **294**, 619

Canal, R., Isern, J., and Labay, J.: 1990, *Ann. Rev. Astron. Astrophys.* **28**, 183

Castellani, V., Chieffi, A., Pulone, L., and Tornambe, A.: 1985, *Astrophys. J., Lett.* **294**, L31

Chernoff, D. and Weinberg, M.: 1990, *Astrophys. J.* **351**, 121

Chevalier, R. A. and Emmering, R. T.: 1986, *Astrophys. J.* **304**, 140

Clark, D. H. and Stephenson, F. R.: 1977, *Historical Supernovae*, Pergamon Press, Oxford

Coté, J. and Pylyser, E. H. P.: 1989, *Astron. Astrophys.* **218**, 131

Cox, J. P. and Giuli, R. T.: 1968, *Stellar Structure, vol. I and II*, Gordon and Breach, New York

Emmering, R. T. and Chevalier, R. A.: 1989, *Astrophys. J.* **345**, 931

Fabian, A. C., Pringle, J. E., and Rees, M. J.: 1975, *Mon. Not. R. astr. Soc.* **172**, 15P

Filippenko, A. V.: 1988, *Astron. J.* **96**, 1941

Fransson, C. and Chevalier, R. A.: 1989, *Astrophys. J.* **343**, 323

Fruchter, A. S. and Goss, W. M.: 1990, *Astrophys. J. Lett.*, in press

Giersz, M.: 1986, *Acta Astron.* **36**, 181

Graham, J. R.: 1987, *Astrophys. J.* **315**, 588

Gull, S. F.: 1973, *Mon. Not. R. astr. Soc.* **161**, 47

Habets, G. M. H. J.: PhD thesis, Univ. of Amsterdam, 1985

Hut, P.: 1983, *Astron. J.* **88**, 1549

Hut, P.: 1984, *Astrophys. J. Suppl. Ser.* **55**, 301

Hut, P., Murphy, B., and Verbunt, F.: 1990, submitted to *Astron. Astrophys.*

Jeffrey, L. C.: 1986, *Nature* **319**, 384
Joss, P. C. and Rappaport, S. A.: 1983, *Nature* **304**, 419
Kool, M. D.: 1990, *Astrophys. J.* **358**, 189
Krolik, J. H.: 1984, *Astrophys. J.* **282**, 452
Krolik, J. H. and Sincell, M. W.: 1990, *Astrophys. J.* **357**, 208
Kulkarni, S. R. and Narayan, R.: 1988, *Astrophys. J.* **335**, 755
Kulkarni, S. R., Narayan, R., and Romani, R. W.: 1990, *Astrophys. J.* **356**, 174
Lee, H. M. and Ostriker, J. P.: 1986, *Astrophys. J.* **310**, 176
Lyne, A. G. and Manchester, R. N.: 1988, *Mon. Not. R. astr. Soc.* **234**, 477
Lyne, A. G., Manchester, R. N., and Taylor, J. H.: 1985, *Mon. Not. R. astr. Soc.* **213**, 613
Maeder, A. and Meynet, G.: 1989, *Astron. Astrophys.* **210**, 155
McMillan, S. L. W., Mcdermott, P. N., and Taam, R. E.: 1987, *Astrophys. J.* **318**, 261
Miyaji, S., Nomoto, K., Yokoi, K., and Sugimoto, D.: 1980, *Publ. Astron. Soc. Jpn.* **32**, 303
Narayan, R.: 1987, *Astrophys. J.* **319**, 162
Narayan, R. and Vivekanand, M.: 1983, *Astron. Astrophys.* **122**, L45
Nomoto, K.: 1987, in Helfand, D. J. and Huang, J. H., eds., *IAU Symp. No. 125: The Origin and Evolution of Neutron Stars*, page 281, D. Reidel, Dordrecht
Nomoto, K. and Kondo, Y.: 1990, submitted to *Astrophys. J.*
Nomoto, K., Thieleman, F. K., and Yokoi, K.: 1984, *Astrophys. J.* **286**, 644
Pacini, F. and Salvati, M.: 1973, *Astrophys. J.* **186**, 249
Paczyński, B.: 1983, *Nature* **304**, 421
Panagia, N. and Laidler, V. G.: 1988, in Kundt, W., ed., *Supernova Shells and their Birth Events*, page 187, Springer-Verlag, Berlin
Phinney, E. S.: 1990, submitted to *Mon. Not. R. Astron. Soc.*
Phinney, E. S. and Kulkarni, S. R.: 1990, *Nature*, in press
Press, W. H. and Teukolsky, S. A.: 1977, *Astrophys. J.* **213**, 183
Pryor, C. P., McClure, R. D., Hesser, J. E., and Fletcher, J. M.: 1987, *Bull. Am. Astron. Soc.* **19**, 676
Radhakrishnan, V. and Srinivasan, G.: 1980, *J. Astrophys. Astron.* **1**, 25
Rappaport, S. A., Putney, A., and Verbunt, F.: 1989, *Astrophys. J.* **345**, 210
Ray, A. and Kluźniak, W.: 1990, *Nature* **304**, 415
Reynolds, S. P. and Chevalier, R. A.: 1984, *Astrophys. J.* **278**, 630
Romani, R. W.: 1990, *Astrophys. J.* **357**, 493
Romani, R. W., Kulkarni, S. R., and Blandford, R. D.: 1987, *Nature* **329**, 309
Ruderman, M., Shaham, J., and Tavani, M.: 1989, *Astrophys. J.* **336**, 507
Ryba, M. F. and Taylor, J. H.: 1990, *Astrophys. J.*, in press
Savonije, G. J.: 1983, *Nature* **304**, 422
Srinivasan, G., Bhattacharya, D., and Dwarakanath, K. S.: 1984, *J. Astrophys. Astron.* **5**, 403
Srinivasan, G. and van den Heuvel, E. P. J.: 1982, *Astron. Astrophys.* **108**, 143
Staelin, D. H. and Reifenstein, E. C.: 1968, *IAU Circ.* 2110
Stokes, G. H., Segelstein, D. J., Taylor, J. H., and Dewey, R. J.: 1986, *Astrophys. J.* **311**, 694
Strom, R. G.: 1990, in Kundt, W., ed., *Neutron Stars and their Birth Events*, page 253, Kluwer Academic Publishers, Dordrecht
Sutherland, P. G. and Wheeler, J. C.: 1984, *Astrophys. J.* **280**, 282
Tavani, M.: 1989, in J. Hunt and B. Battrick, ed., *ESA SP-296: Proceedings of the 23rd ESLAB Symposium on X-ray Astronomy, vol. 1*, page 241, European Space Agency, Paris
Trimble, V.: 1982, *Rev. Mod. Phys.* **54**, 1183
van den Bergh, S.: 1988, *Astrophys. J.* **327**, 156
van den Bergh, S., McClure, R. D., and Evans, R.: 1987, *Astrophys. J.* **323**, 844
van den Heuvel, E. P. J.: 1983, in Lewin, W. H. G. and van den Heuvel, E. P. J., eds., *Accretion Driven Stellar X-ray Sources*, page 303, Cambridge University Press, Cambridge, U. K.

van den Heuvel, E. P. J. and Habets, G. M. H. J.: 1985, in Srinivasan, G. and Radhakrishnan, V., eds., *Supernovae, Their Progenitors and Remnants*, page 129, Indian Acad. Sciences, Bangalore

Verbunt, F.: 1988, in White, N. E. and Filipov, L., eds., *The physics of compact objects, theory vs. observation*, page 529, Pergamon Press, Oxford

Verbunt, F.: 1990, in Kundt, W., ed., *Neutron Stars and Their Birth Events*, page 179, Kluwer Academic Publishers, Dordrecht

Verbunt, F. and Hut, P.: 1987, in Helfand, D. J. and Huang, J. H., eds., *The origin and evolution of neutron stars*, page 187, D. Reidel, Dordrecht

Verbunt, F. and Meylan, G.: 1988, *Astron. Astrophys.* **203**, 297

Verbunt, F., van den Heuvel, E. P. J., van Paradijs, J., and Rappaport, S. A.: 1987, *Nature* **329**, 312

Vivekanand, M. and Narayan, R.: 1981, *J. Astrophys. Astron.* **2**, 315

Webbink, R. F., Rappaport, S. A., and Savonije, G. J.: 1983, *Astrophys. J.* **270**, 678

Wheeler, J. C., Harkness, R. P., Barkat, Z., and Schwartz, D.: 1986, *Publ. Astron. Soc. Pac.* **98**, 1018

Wijers, R. A. M. J. and van Paradijs, J. A.: 1990, submitted to *Astron. Astrophys. Lett*

Woosley, S. E., Axelrod, T. S., and Weaver, T. A.: 1984, in Chiosi, C. and Renzini, A., eds., *Stellar Nucleosynthesis*, page 263, D. Reidel, Dordrecht

Woosley, S. E. and Weaver, T. A.: 1986, *Ann. Rev. Astron. Astrophys.* **24**, 205

CORE COLLAPSE SUPERNOVAE AND NEUTRON STAR FORMATION

WOLFGANG HILLEBRANDT

Max-Planck-Institut für Physik und Astrophysik
Institut für Astrophysik
D–8046 Garching b. München, Fed. Rep. Germany

1. Introduction

Supernovae of type II, i.e. explosions of stars which show strong Balmer lines of hydrogen in their spectra, and possibly also type Ib and Ic events, are likely to originate from massive stars, $M \gtrsim 8M_\odot$, at the end of their quiet hydrostatic nuclear burning and the final outcome is generally believed to be the birth event of most neutron stars. If this picture is correct the energy observed in the explosion must ultimately come from the gain in gravitational binding of the central core of the star and/or from the binding energy of the newly born neutron star. However, although various mechanism have been proposed which are potentially able to transform a small fraction of the gravitational energy into outward momentum of the stellar envelope, no satisfactory explanation has been given so far (see, e.g. Hillebrandt (1987) and Müller (1990) for recent reviews).

In this article several open questions directly related to this problem will be discussed. We will start in section 2 by describing the micro- and macro-physics ingredients which enter into all numerical simulations of stellar collapse and supernova explosions. Here, in particular, the nuclear equation of state and certain weak interaction rates seem to be of key importance, and we are far from knowing them well enough to draw any unambiguous conclusions. A second class of problems arises from the fact that certain approximative neutrino transport schemes are commonly used in numerical models which are far from being satisfactory. Finally, we may not even know the stellar properties at the onset of collapse with sufficient accuracy to make any firm statements concerning the final outcome of the collapse

J. Ventura and D. Pines (eds.), Neutron Stars: Theory and Observation, 125–141.

of the core of a massive star.

In section 3 present day's favorite supernova models will be discussed. It will be argued that given the obvious difficulties of prompt explosion models neutrino heating possibly supported by some convective energy transport seems to be the only promising scenario. But again it appears impossible at present to draw any firm conclusions, mainly because detailed models will require multi-dimensional simulations including up to date micro-physics input, realistic stellar models, and very accurate neutrino transport schemes, and such models are still not feasible.

In section 4 possibilities and difficulties to use supernova observations for the diagnostics of core collapse will be discussed. It will be demonstrated that only for galactic events neutrino astronomy may provide information on details of the dynamics of core collapse and that the same is probably true for the next generation of gravitational wave antennas. Therefore, the prospects for probing our theoretical ideas by direct observations do not look very promising. Finally, a summary and outlook is presented in section 5.

2. Basic input physics

In constructing a supernova model one has to solve the hydrodynamic equations, some field equations describing gravity, rate equations for composition changes, and transport equations for particle numbers and energy for a given set of initial conditions (densities, entropies, velocities, composition variables, etc.) and material functions (equation of state, reaction rates, interaction cross-sections, etc.). It is obvious that this set of equations cannot be solved in full generality and that many approximations are necessary in order to make the problem tractable.

Concerning numerical methods for solving the hydro-equations great progress has been made during the last several years (see, e.g. Müller (1991) for a review) provided Newtonian mechanics and gravity is a valid prescription of the system. However, in stellar collapse peak velocities approach several tenths of the velocity of light and general relativistic effects are not negligibly small, in particular at core bounce and during the early cooling and deleptonization phase of the newly born neutron star. The use of the Newtonian approximation, therefore, is rather questionable. For general relativistic hydrodynamics, on the other hand, the numerical techniques are much less advanced, and if one wants to compute stellar collapse for rather general micro-physics input data one is restricted so far to spherical symmetry, which again not always may be a good approximation. Moreover, even if one argues that relativity may not change the general dynamics of stellar collapse and that, therefore, Newtonian physics is a fair approximation, multidimensional simulations are still only affordable if severe simplifications concerning the equation of state, reaction rates, and neutrino transport are adopted. So, up to now, not even that part of the problem is treated with convincing accuracy which does not involve poorly known physics, and one has to keep this situation in mind when discussing uncertainties in and implications from micro-physics input data.

One of the most important ingredients for both neutron star and supernova models is the equation of state (EOS) and, consequently, also a major fraction of the uncertainties result from our incomplete knowledge of the EOS at densities from about 10^{-2} to 10 times nuclear matter density ($\varrho_0 \simeq 3 \times 10^{14}$g cm^{-3}). The EOSs of supernova and early neutron star matter have in common that the entropy per nucleon is low ($S/k_B \lesssim$ a few), and that the lepton fraction is significantly lower than in normal matter. Therefore, there is little hope that we can learn much about the EOS from laboratory experiments such as heavy ion collisions, but we have to rely on theoretical models and, possibly, on interpretations of astrophysical observations.

At rather low densities, $\varrho \lesssim 10^{-2}\varrho_0$, in principle the EOS can easily be calculated from a Boltzmann gas approximation for nucleons and nuclei, provided nuclear binding energies and partition functions are known. This, however, is not the case, because most of the nuclei present in the interior of a collapsing stellar core would be very short-lived under laboratory conditions, and most of them have not even be synthesized yet by experiments. Therefore one has to rely on extrapolations from the properties of stable and "mildly" unstable nuclei. Moreover, collective effects only present at high matter density may be crucial.

The nuclear partition function may serve as an example. It is well known that the concentration of free protons in a collapsing star more or less determines the effective electron capture rate and thus the time scale of collapse, the mass of the homologous core, etc. In nuclear statistical equilibrium, in turn, the proton fraction depends, among others, strongly on the nuclear partition function, because nuclei will have high excitation energies and, therefore, the question arises at which excitation energy free nucleons are leaking out of the nuclei. Moreover, it is a non-trivial problem to decide which states should be counted as "bound" states still belonging to a "nucleus" (see, e.g., El Eid and Hillebrandt (1980), Tubbs and Koonin (1979), and Engelbrecht and Engelbrecht (1990)), and various approaches mainly based on the so-called back-shifted Fermi gas model (Gilbert and Cameron,1965) give significantly different results. Similar problems and questions arise as far as the binding energies of neutron-rich nuclei are concerned.

At very high densities the Boltzmann-gas approach to the nuclear part of the EOS is no longer valid. This happens if the nuclear radius R_N becomes comparable to the Coulomb interaction radius R_c. Phenomenologically one finds $R_N \sim A^{1/3}$ and R_c is roughly proportional to $(A/\varrho)^{1/3}$ leading to $\frac{R_N}{R_c} \simeq 10^{-5}\varrho^{1/3}$. Thus we can conclude that if we require $R_N/R_c \ll 1$, ϱ has to be smaller than 10^{12}g cm^{-3}. Therefore, at high densities a self-consistent model has to be invented, and at even higher densities, $\varrho \gtrsim 0.1\varrho_0$, similar arguments show that also nucleon-nucleon interactions have to be included in such a self-consistent model.

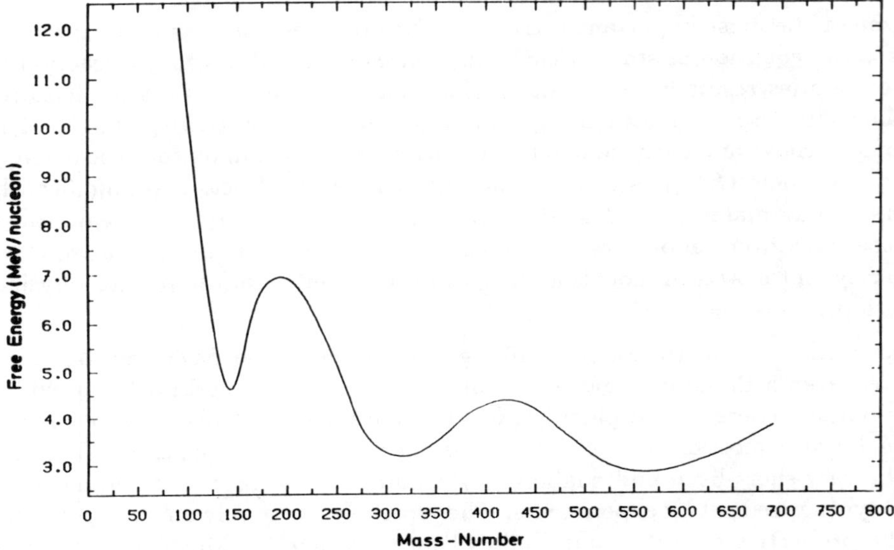

Figure 1: Free energy per nucleon in Hartree-Fock approximation as a function of nucleon number in a Wigner-Seitz cell (from Hillebrandt *et al.* (1984)). A density of $6 \times 10^{13} g/cm^3$ and a temperature of 2.5 MeV were chosen. The electron concentration was 0.35.

The most advanced method which has been applied to the supernova problem so far is the temperature dependent Hartree-Fock method (Bonche and Vautherin, 1981; Hillebrandt and Wolff, 1985) but very little progress has been made in the last several years. In the Hartree-Fock method one starts from a model Hamiltonian incorporating the nucleon-nucleon interactions by an effective potential which is tailored in a way to reproduce certain properties of finite nuclei and cold nuclear matter. Temperature effects are included in a statistical way. Although the conception of this approach is clear and simple, it leads to several practical and principle problems which should be noted. Firstly, in the deep interior of a collapsing star nuclei will form a Coulomb lattice and, therefore, the Hartree-Fock equations have to be solved for a unit cell with periodic boundary conditions. This immediately leads to a 3-dimensional problem which cannot be solved for the whole range of temperatures, densities, and lepton concentrations present in a supernova core. One therefore in general assumes spherical symmetry and applies the Wigner-Seitz method to include Coulomb-lattice effects. This then means that one cannot impose periodic boundary conditions on the single-particle wave functions leading to some arbitrariness in the boundary conditions and consequently in the solutions.

It is likely that some of the ongoing discussion concerning the geometrical structure of matter near the transition to homogeneous nuclear matter is a result of these ambiguities.

Next one has to admit that in reality matter will be a mixture of various types of "nuclei" and free nucleons in nuclear statistical equilibrium, but that self-consistent models always assume that only the configuration with the lowest free energy (or free enthalpy) is present. Fig. 1 shows an example of a free energy curve obtained from a Hartree-Fock calculation with three distinct local minima. The lowest state would be a "nucleus" with about 560 nucleons, but the barrier between the various minima is less than kT and therefore a mixture of very many "nuclei" with rather similar abundances will be present in equilibrium. It is obvious that this effect will have some impact on the EOS.

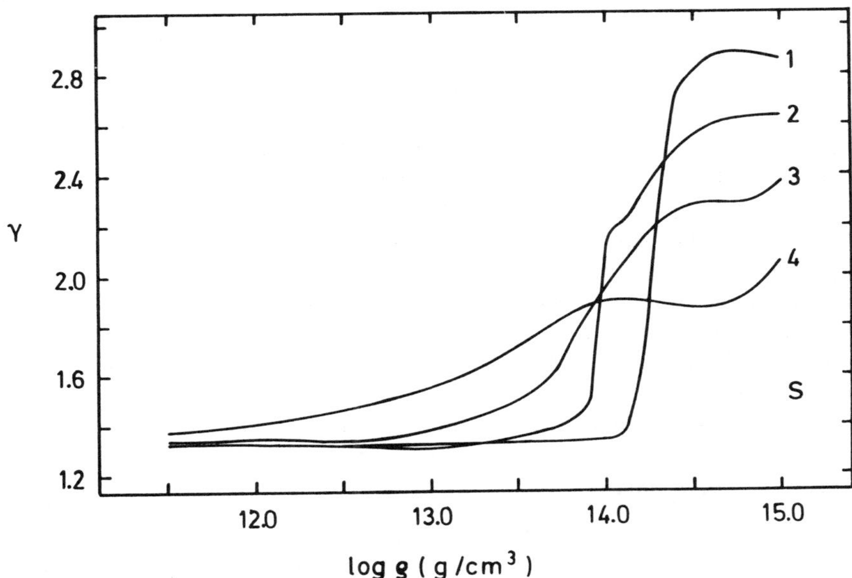

Figure 2: Adiabatic index for constant electron fraction $y_e = 0.35$ and various entropies per nucleon as a function of density (from Hillebrandt *et al.* (1989)).

Finally, the Hartree-Fock method is a single-particle approach and does not treat correlations in a self-consistent way. It therefore will gradually loose reliability with encreasing density. Moreover, the results will be sensitive to the choice of the effective interactions (Hillebrandt and Wolff, 1985). Moreover, it should be noted that in most supernova simulations even further simplifications of the method outlined so far are used or the EOS is parametrized in a rather arbitrary way. Numerical results, therefore, should be considered with some care.

One might ask why core collapse should be so sensitive to the nuclear part of the EOS because during most of the collapse phase the pressure is dominated by leptons. The main reason is that for relativistic leptons the adiabatic index is equal to the critical value of 4/3 and that deviations from this value which are caused by nuclear effects are therefore crucial (see also Fig.2). In addition, the density at which the core bounces is completely determined by the nuclear part of the EOS and therefore also the energy available to the shock is in part fixed. We will come back to this equation later in section 3.

Next we want to discuss briefly some of the uncertainties entering through our incomplete knowledge of weak interaction rates. It is well known that during most of core collapse and during the early cooling phase of newly born neutron stars typical weak interaction timescales are of the same order as the dynamical or evolutional time scale of the stellar core or star, respectively. So in contrast to strong and electromagnetic interactions weak rates have to be known explicitly. Moreover, because in some cases neutrino energy distributions are not in equilibrium, it is not even sufficient to calculate energy-averaged rates. Unfortunately, in this field very little progress has been made during the last several years, mainly because for weak interaction rates involving very neutron-rich nuclei transition matrix elements cannot be determined by experiments and thermally excited states have to be considered. Generally speaking, the so-called β-strength function defined as the energetical average of the squared nuclear matrix element times the final nuclear level density has to be determined, and it is well known from cases where experimental information is available that sometimes this strength function is dominated by one or a few single transitions. Best models available to date are based on shell model wave functions (Fuller *et al.*, 1982; Mathews *et al.*, 1983) or use the quasiparticle random phase approximation (Möller*et al.*, 1990), but from the sensitivity of those results to details of the nuclear model one can safely conclude that the calculated rates are uncertain to within a factor of five on the average, and possibly by orders of magnitude in some particular cases.

We conclude this section with some remarks on neutrino interactions and transport and on stellar evolution models. In principle, important processes such as neutrino-electron scattering, neutrino absorption by free nucleons, etc., can be computed numerically exactly (e.g., Tubbs and Schramm (1975), Yueh and Buchler (1976), Tubbs (1979), Bruenn (1985)), but in most numerical simulations of stellar collapse and supernova explosions approximations to the neutrino distribution function are used which may be sufficiently accurate in almost opaque regions of the star but which will fail in transparent regions where, on the other hand, neutrinos may have the largest impact on the stellar dynamics. Recently it has been demonstrated that due to the variation of the opacity with roughly the square of the energy, the neutrino spectra do have a non-thermal shape (Janka and Hillebrandt, 1989a). It is possible, however, to approximate the actual distribution function by normalized Fermi-Dirac functions with an "effective degeneracy" parameter η_ν and a neutrino temperature T_ν which in general is different from the matter temperature. Moreover, both T_ν and η_ν in general do not correspond to the thermodynamic conditions at a given position in the star (Janka and Hillebrandt, 1989b) (see also Fig.3). It will be demonstrated later in section 3 that these uncertainties have a large impact on the reliability of those supernova models which make use of energy and/or momentum transfer by neutrinos to trigger the explosion.

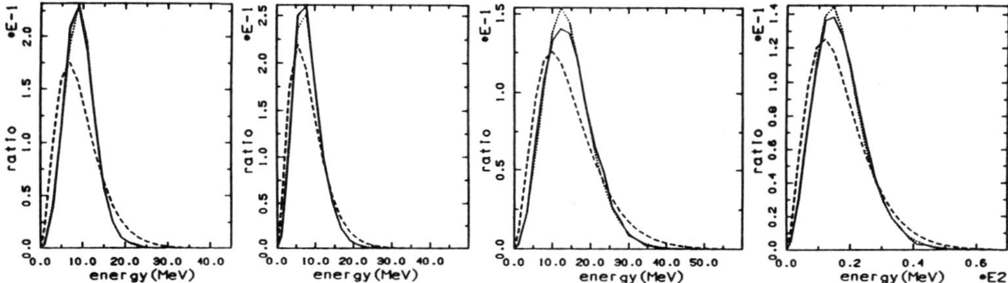

Figure 3: Typical Monte Carlo spectra (solid lines) for early and late time electron neutrinos, electron antineutrinos and muon neutrinos (from left to right) in comparison with blackbody spectra repesenting the mean neutrino energy correctly (dashed line) and Fermi-Dirac fit distributions with an effective chemical potential (dotted lines). The discrepancies indicate the non-thermal nature of the true spectra, but those fits which reproduce the first three moments of the Monte-Carlo energy distribution are very accurate. (From Janka and Hillebrandt (1989a)

An even more complicated problem is the stellar structure at the onset of collapse which effects the final outcome in a variety of different ways (see, e.g., Baron and Cooperstein (1990) for a summary). In particular the mass of the iron core at the onset of collapse is crucial, but not well determined because it depends sensitively upon the model of convective energy transport and mixing one is willing to choose. In general it seems that prompt explosions, if possible at all, require very small core masses, whereas neutrino-driven supernovae are less sensitive to this quantity in particular. We also want to call attention to the fact that all binary neutron stars of which there exist accurate mass determinations have gravitating masses surprisingly close to $1.4 M_\odot$, corresponding to a baryonic mass of about $1.6 M_\odot$, whereas stellar evolution models predict a spread in core masses of at least $0.5 M_\odot$, depending on the main sequence mass. Whether this discrepancy reflects a short-coming of the stellar evolution models or whether (more likely!) it is a consequence of the true explosion mechanism is not known, but the question certainly deserves some attention.

3. Today's supernova models (and their problems)

Over the years a variety of ideas have been investigated none of which, however, can explain supernova explosions beyond doubt. Generally speaking they can be put in three classes depending on the mechanism responsible for energy and momentum transfer from the stellar interior to the envelope, namely a hydrodynamical shock wave, a thermonuclear burning front, or neutrino flux, or combinations of the three.

At present it seems plausible that only under very special circumstances a shockwave created by the rebounding inner core can lead to the prompt ejection of the stellar envelope. Roughly speaking, the argument is as follows. We can obtain a rough estimate of the energy put into the shockfront from

$$E_s \simeq \left(\frac{GM_{\rm CH}}{r_s} - \frac{\bar{\varepsilon}}{\bar{\varrho}} \right) M_{\rm CH} \quad , \tag{1}$$

where $\bar{\varrho}$ and $\bar{\varepsilon}$ are average values of energy density and density inside the sonic radius r_s at bounce and $M_{\rm CH} = M_{\rm CH}(y_e)$ is the Chandrasekhar mass. The first term in eq. (1) gives the gravitational binding of the inner core per gramme whereas the second term is the internal energy per gramme. So the energy available to the shock is, roughly speaking, given by the kinetic energy of the inner core (see, e.g. Yahil and Lattimer (1982)). For a polytropic structure and an appropriate equation of state, eq. (1) can be written as

$$E_S \simeq (K_1 \, M_{\rm CH}^{2/3} \, \bar{\varrho}^{1/3} - K_2 \bar{\varrho}^{\gamma-1}) \, M_{\rm CH} \quad , \tag{2}$$

From eq. (2) it is obvious that the shock energy increases with increasing $\bar{\varrho}$. Therefore "soft" EOS at densities beyond nuclear matter density will favour prompt hydrodynamically driven explosions because the collapse will be stopped at higher densities. In addition, because $M_{\rm CH}$ is proportional to y_e^2, a "low" initial entropy leading to a small concentration of free protons and thus to fewer electron captures also works in favour of an explosion. Finally, a "stiff" EOS at $\varrho \lesssim \varrho_0$, i.e. an adiabatic index γ very close to the critical value of 4/3 will ensure homology up to ϱ_0 and, therefore, again will give high shock energies. But even in the most favourable cases the shock energy will not exceed 8×10^{51}erg, and more typical values are around $(4-5) \times 10^{51}$erg (see, e.g., Hillebrandt (1987)). It then follows that the most energetic shocks at best can dissociate $0.4 M_\odot$ of heavy elements into free neutrons and protons and, because $M_{\rm CH} < 0.8 M_\odot$ in all recent simulations, the iron core mass has to be less than $1.2 M_\odot$. The latter condition is (if at all) fulfilled in stars with main sequence masses of less than 10 or $12 M_\odot$, but certainly not for the progenitor of SN 1987A (see Fig.4).

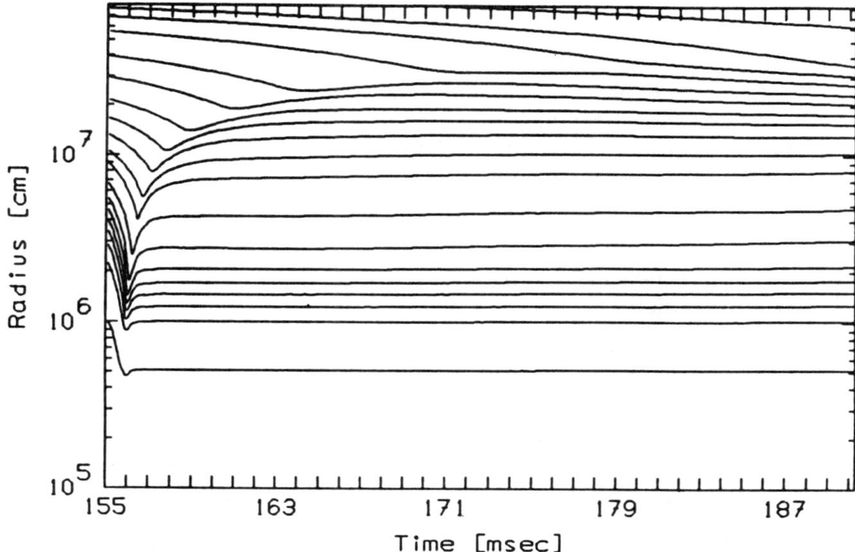

Figure 4: Radius versus time for various mass zones of a collapsing stellar model of $20M_{\odot}$. It is obvious that the shock stalls at a radius of about 300 km (from Hillebrandt (1987)).

Therefore, it seems more likely that neutrinos in one way or another are needed in order to drive a shock wave out into the stellar envelope. A possibility which has been discussed extensively is the so-called delayed explosion mechanism (Wilson, 1985; Bethe and Wilson, 1985), the idea being that a few hundred milliseconds after core-bounce energy transfer by neutrinos may revive a stalled shock. In its original version this explosion mechanism was based on the reactions

$$\begin{aligned} e^- + p &\to n + \nu_e \\ e^+ + n &\to p + \bar{\nu}_e \\ e^+ + e^- &\to \nu_x + \bar{\nu}_x \end{aligned} \tag{3}$$

in the hot shocked and dissociated gas. These neutrinos and those diffusing out of the unshocked core can interact with the outer layers via the inverse reactions

$$\begin{aligned} \nu_e + n &\to e^- + p \\ \bar{\nu}_e + p &\to e^+ + n \end{aligned} \tag{4}$$

and neutrino-electron scattering. The net effect is some heating, the amount being sensitive to the neutrino energy distribution and the density structure near the neutrino sphere. Moreover, the process is likely to be a self-regulating one: Neutrino heating proceeds on a timescale much longer than the hydrodynamical timescale. Therefore, once enough heat has been added to lift the layers in the gravitational potential of the protoneutron star the density will drop and thus the heating will be turned off. Accordingly, one would expect that at most roughly the binding energy of the stellar mantle can be released which is of the order of a few 10^{50}erg, in agreement with some numerical simulations (Wilson *et al.* (1986) see Fig.5), but in contrast to the much higher explosion energy observed in SN 1987A (see, e.g., Hillebrandt and Höflich (1989)).

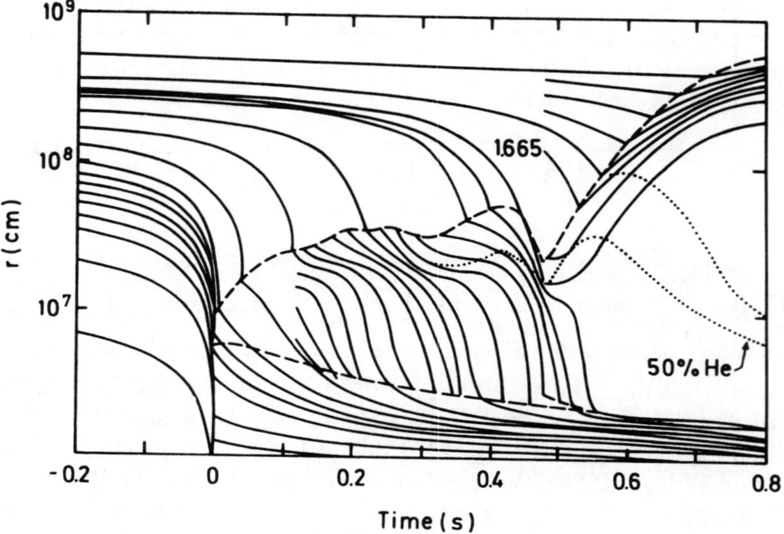

Figure 5: Same as Fig.4, but for a $25M_\odot$ model (from Wilson (1985)). Time is measured fro core-bounce. The position of the shock front (upper broken curve) and that of the neutrino spher are also shown. The stalled shock is revived by neutrino heating 0.5s after bounce.

A way out of these problems might be the inclusion of neutrino-antineutrino annihilation radiation in the numerical models, as was first suggested by Goodman, Dar, and Nussinov (1987). Basis of this explosion mechanism is the observation that most of the binding energy of the neutron star is radiated away in all three neutrino flavors on timescales of several seconds. According to numerical simulations by Wilson and Mayle (1989) this energy may lead to the formation of a hot (high entropy, low density) bubble from the reactions $\nu_x \bar{\nu}_x \leftrightarrows e^+ e \leftrightarrows 2\gamma$ and $(\nu_e, \bar{\nu}_e)$ (e^+, e^-)-scattering. It is obvious that the $\nu_x \bar{\nu}_x$-annihilation cross-section is only large for head-on collisions and therefore, the heating will be limited to a region very near to the proto-neutron star. Moreover, in the mass-zones that are to be heated the density should be low, $\varrho \lesssim 10^5 \mathrm{g\ cm}^{-3}$, in order to raise the entropy to values above $100 k_B/nucleon$, which are necessary to push the matter outwards. So, generally speaking, this explosion mechanism requires a very steep density gradient in the outer layers of the proto-neutron star in order to be efficient, and in the "delayed explosion" models Wilson *et al.*(1986) this seems to be in fact the case.

However, there are several problems which make the results obtained by Wilson and Mayle (1989) uncertain. First of all, the neutrino spectra obtained from their multi-group flux-limited diffusion scheme seem to overestimate the high-energy part as was first critizised by Janka (1987) and by Janka and Hillebrandt (1989a) (see also Fig. 3). Secondly, their simulations assumed spherical symmetry and therefore, a certain averaging procedure had to be applied in order to get rid of the angle distribution of the neutrinos. This averaging, however, is crucial because the annihilation rate is proportional to

$$Q \sim \int d\tau (\varepsilon_\nu + \varepsilon_{\bar{\nu}}) F_\nu F_{\bar{\nu}} (1 - \hat{\Omega}_\nu \hat{\Omega}_{\bar{\nu}})^2 \tag{5}$$

where $d\tau = d\varepsilon_\nu d\varepsilon_\nu d\Omega_\nu d\Omega^*_{\bar{\nu}}/(4\pi)$ is the volume element in energy space and ε_ν, $\varepsilon_{\bar{\nu}}$ are neutrino energies, F_ν and $F_{\bar{\nu}}$ are the neutrino fluxes, and $\hat{\Omega}_\nu$ and $\hat{\Omega}_{\bar{\nu}}$ are unit vectors in the direction of the neutrino and antineutrino, respectively, and therefore only head-on collisions contribute significantly to the rate. Janka (1990) has argued that neutrino back-scattering in the semitransparent layers may encrease the effective annihilation rate considerably very near to the proto-neutron star, as compared to the vacuum approximation applied by Goodman *et al.* (1987). However, limb darkening should then also be included which will, in contrast, reduce the heating by a similar factor of 2 to 3, but in layers further away from the core. Both effects are hard to treat by flux-limited diffusion schemes and require more accurate neutrino transport models (Janka, 1990). Given these uncertainties and shortcomings of the Wilson and Mayle (1989) simulations it seems precipitate to conclude that the "hot bubble" explosion mechanism is the answer to the question how supernovae explode (Colgate, 1989, 1991). It may in fact be that the explosion mechanism is even more complicated than anticipated so far. In particular, in the models of Wilson and Mayle (1989) the layers that finally form the hot bubble are Rayleigh-Taylor unstable already a few tenths of a second after core-bounce, and the growth-time of the instability is of the order of or even less than the hydrodynamic timescale. It is, therefore, likely that there will be a considerable transport of energy from the surface of the proto-neutron star to the matter in the forming bubble by macroscopic motions of the matter. Certainly, this effect deserves further attention. One may even speculate, that the rather high velocities of layers containing heavy elements observed in SN 1987A could have resulted from Rayleigh-Taylor instabilities in the innermost layers of the ejecta. But a realistic treatment of these effects will require 3-dimensional rather than spherically symmetric simulations.

4. Supernova observations as diagnostic tools of core collapse

With the exception of the explosion energy, which can be obtained from models for the light curve and from synthetic spectra, provided a sufficiently accurate set of observational data is available (see, e.g., Höflich (1990)), information on the mass-cut, which can be inferred from a comparison of predicted and observed abundances of heavy elements (Thielemann *et al.*, 1990), and possibly information on large-scale deviations from spherical symmetry in polarization measurements (Höflich, 1990), only neutrinos and gravity waves can be used to trace the dynamics of stellar collapse. For example, the first detection of neutrinos from SN 1978A by the KAMIOKANDE (Hirata *et al.*, 1987) and IMB (Bionta *et al.*, 1987) experiments were the first proof that the cores of massive stars do indeed collapse to neutron star densities. However, even for a supernova as near as SN 1987A the information available is not sufficient to draw any firm conclusions, neither on the total energy nor on the spectrum, and also the time evolution of the neutrino emission is uncertain. Concerning the prospects of gravity waves, the situation is also not very promising because recent numerical models predict that the energy emmitted from a rotating stellar core should be of the order of $10^{-7} M_{\odot} c^2$ only and, therefore, can hardly be detected from outside the Local Group, even future detectors (Mönchmeyer *et al.*, 1990). In this section we will discuss neutrino detections in some detail, but also briefly mention a few other possibilities.

As was mentioned earlier, the mass of the newly born neutron star is crucial for the explosion mechanism. Therefore it is important to ask whether there are observational constraints from neutrino detections. Of course, the neutrino burst associated with SN 1987A has clearly shown that a compact object, presumably a neutron star, has formed during the collapse of the central core of its progenitor. Because the total energy emitted in form of neutrinos cannot exceed the binding energy of the final cold neutron star, a lower limit on its mass can be obtained. Unfortunately, the total $\bar{\nu}_e$-energies estimated from the KAMIOKANDE and IMB data are only weakly consistent, leaving uncertainties of about a factor of two (see fig.6). Moreover, we do not know whether some of the events were caused by (ν, e)-scattering rather than ($\bar{\nu}_e$, p)-reactions and whether some of the events were due to noise in the detectors. Finally, if the progenitor was a rapidly spinning star anisotropies of the neutrino emission would add further uncertainties (Janka and Mönchmeyer, 1988). So in conclusion, we can only state that the neutrino data are consistent with the assumption that a neutron star of about 1.5 $M_{\odot}$ was born in SN 1987A, and has radiated away a significant fraction of its binding energy in form of thermal neutrinos during the first few seconds of its life, but neutrino observations do not prove this assumption.

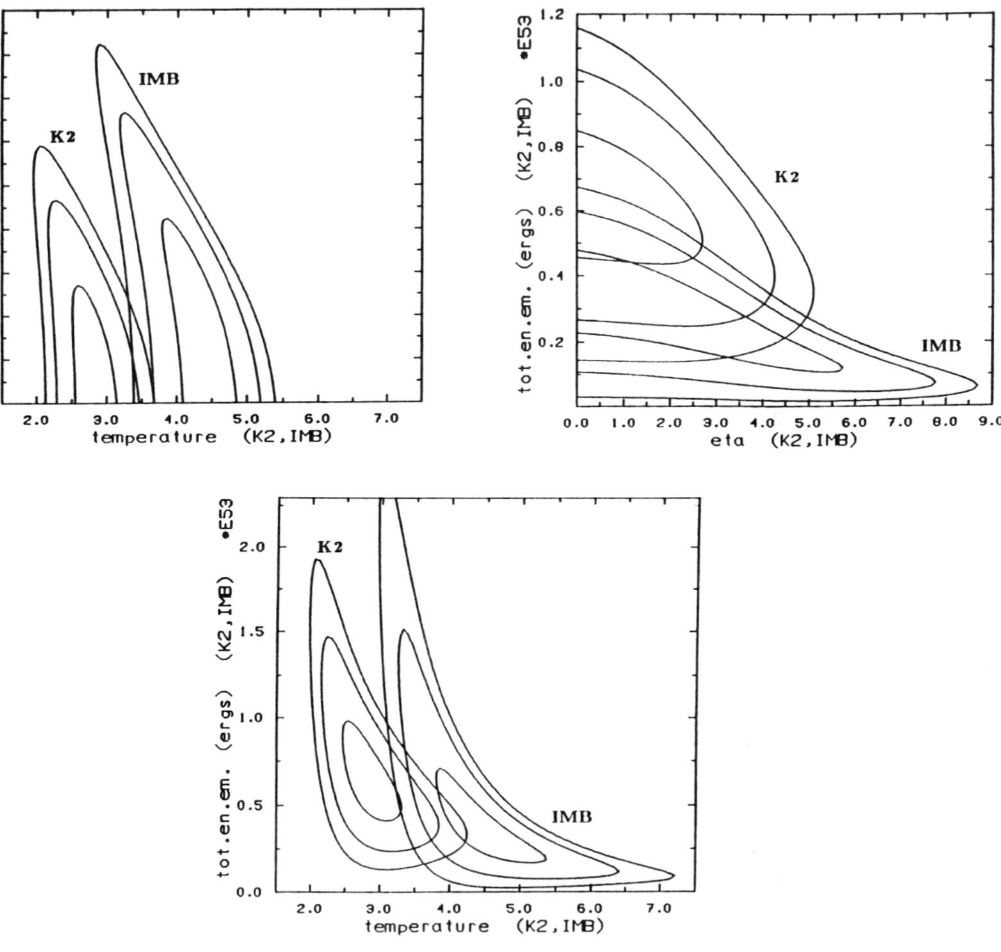

Figure 6: Maximum likelihood regions in the space of chemical potential, temperature and total energy for neutrinos observed by the KAMIOKANDE (K2) and IMB detectors for assumed Fermi-Dirac distributions. The cut planes are located at the best fit values for the degeneracy parameter eta, total energy emitted and neutrino temperature in MeV for both detections. It is obvious that both detections are only weakly consistent (the contours mark 68%, 95%, and 99% confidence levels). (From Janka and Hillebrandt (1989b))

There is infact additional evidence for significant large scale anisotropies in SN 1987A. During the first few months after the initial outburst frequency dependent linear polarizations have been observed by several groups (Barrett, 1987; Schwarz, 1987; Mendez et al., 1988). It is not clear yet whether these polarizations were caused by prolate or oblate deformations, but deviations from spherical symmetry seem to be of the order of 5 to 20% (Höflich et al., 1989). From their recent Speckle observations at optical wavelengths Papaliolios et al. (1989) find that the expanding envelope is rather asymmetric and that the ratio between the minor and major axes is about 2 to 3. It is difficult to imagine a mechanism for such large asymmetries other than a large deformation of the progenitor's envelope. Therefore it is tempting to conclude that Sk–69°202 was in fact a fast rotator, in which case we would have to revise several of the ideas outlined in the previous section.

The delayed explosion models, on the other hand, seem to fit well both the energy spectrum and the duration of the neutrino pulse from SN 1987A (Mayle et al., 1988). This success indicates that the newly born neutron star is indeed massive, $M > 1.5 M_\odot$, which in turn would argue for a stiff equation of state. Again, this conclusion should be regarded with some care because the neutrino fluxes and spectra have been computed by means of the flux limited diffusion approximation and are not very reliable (Janka and Hillebrandt, 1989a). Moreover, as was discussed above, the neutron star binding energies estimated from both the KAMIOKANDE and the IMB experiments are only weakly consistent and do not provide enough information to determine accurately the mass of the neutron star.

5. Concluding remarks

We have demonstrated that the theory of core collapse supernovae still suffers from severe short-comings both in our knowledge of the fundamental physics and in the technical aspects of the problem. Although a lot of effort has gone into the development of very detailed numerical models our present knowledge is not much better than it was about ten years ago. In particular, we still do not know by what mechanism the stellar envelope is ejected, leaving still room for many speculations.

Of course, it would help if the various classes of models could be constraint by observations. Unfortunately until now there is no even a direct proof for the existence of a neutron star in SN 1987A, the only good test case we have for our theoretical models. Optical pulses have been searched for but have not been found (Ögelman *et al.*, 1990). Moreover, there is no indication for a X-ray point source (Trümper, private communication) nor for pulsed radio emmission. However, the light curve shows extra luminosity of the order of 10^{38}erg/sec (Bouchet *et al.*, 1991), which seems to be approximately constant since fall 1989 (although variations of a factor of two cannot be excluded) and which cannot easily be accounted for by radioactive decay models. A possible explanation would be that the compact object in the center of SN 1987A is still surrounded by an accretion disk and that what we observe is accretion luminosity. In fact, this interpretation would, in a natural way, explain why the luminosity is so close to the Eddington value. It could account for the apparent variations in the luminosity by means of disk instabilities similar to what is found in dwarf novae (Meyer and Meyer-Hofmeister, 1984, 1991). Such a disk should be the normal case in core collapse provided the progenitor star possessed some angular momentum and some matter fell back onto the newly born compact object. Therefore, roughly constant bolometric light curves with Eddington luminosities over several years should be a common phenomenon for core collapse supernovae. But for the time being these speculations do not help to answer the questions raised in this article.

References

Baron E., Cooperstein J., 1990 *Astrophys.J.* **353** 597

Barrett P., 1987 in *SN 1987A*, ed. by I.J. Danziger, ESO, Garching, p. 173

Bethe H.A., Wilson J.R., 1985 *Astrophys.J.* **295** 11

Bionta R.M. *et al.*, 1987 *Phys.Rev.Lett* **58** 1494

Bonche P., Vautherin D., 1981 *Nucl.Phys.* **A372** 496

Bouchet P., Danziger I.J., Lucy L.B., 1991 *Astron.J.*, in press, and this volume

Bruenn S.W., 1985 *Astrophys.J.Suppl.* **58** 771

Bruenn S.W., 1988 *Astrophys.Space Sci.* **143** 15

Colgate S.A., 1989 *Nature* **341** 489

Colgate S.A., 1991 preprint, and this volume

El Eid M.F., and Hillebrandt W., 1986 *Astron.Astrophys.Suppl.* **42** 215

Engelbrecht C.A., Engelbrecht J.D., 1990 *Annals of Phys.*, in press

Fuller G., Fowler W.A., Newman M., 1982 *Astrophys.J.* **252** 447

Gilbert A., Cameron A.G.W., 1965 *Can.J.Phys.* **43** 1248

Goodman J.,Dar A.,Nussinov S., 1987 *Astrophys.J.* **314** L7

Hillebrandt W., 1987 in *High Energy Phenomena around Collapsed Stars*, ed. by F. Pacini, NATO-ASI **C195**, Reidel, Dordrecht, p. 73

Hillebrandt W., Höflich P., 1989 *Rep.Prog.Phys.* **52** 1421

Hillebrandt W., Wolff R.G., 1985 in Nucleosynthesis: Challenges and New Developments, ed. by W.D. Arnett, and J.W. Truran, Univ. Chicago Press, Chicago, p. 131

Hillebrandt W., Müller E., Mönchmeyer R., 1989 in *The Nuclear Equation of State, Part A*, ed. by W. Greiner and H. Stöcker, Plenum Press, N.Y., p.689

Hillebrandt W., Nomoto K., Wolff R.G., 1984 *Astron.Astrophys.* **133** 175

Hirata K. *et al.*, 1987 *Phys.Rev.Lett.* **58** 1490

Höflich P., 1990 , Habilitationsschrift, LMU München

Höflich P., Sharp Ch.M., Zorec J., 1989, in *Particle Astrophysics* ed. by E.B.Norman, World Scientific, Singapore, p.186

Janka H.-T., 1990 , preprint MPA546, *Astron.Astrophys.*, in press

Janka H.-T., Hillebrandt W., 1989a *Astron.Astrophys.Suppl.* **78** 375

Janka H.-T., Hillebrandt W., 1989b *Astron.Astrophys.* **224** 49

Janka H.-T., Mönchmeyer R., 1988 *Astron.Astrophys.* **209** L5

Mathews G.J., Bloom S.D., Hausman R.F., 1983 *Phys.Rev.* **C28** 1367

Mayle R., Wilson J.R., Ellis J., Olive K., Schramm D.N., Steigman G., 1988 *Phys.Lett.* **B203** 188

Möller P., Nix R.J., Kratz K.-L., Howard W.M., 1990 in *Nuclear Astrophysics*, ed. by H. Oberhummer and W. Hillebrandt, Proceedings MPA/P4, Max-Planck-Institut, Garching

Müller E., 1990 *J.Phys.G* **16** 1571

Müller E., 1991, preprint MPA573, Proc. Les Houches, in press

Mendez M., Clocchaiatti A., Benvenuto G., Feinstein C., Marraco U.G., 1988 *Astrophys.J.* **334** 295

Meyer F., Meyer-Hofmeister E., 1984 *Astron.Astrophys.* **132** 143

Meyer F., Meyer-Hofmeister E., 1991, in preparation

Mönchmeyer R., Schäfer G., Müller E., Kates R.E., 1990, preprint MPA508, *Astron.Astrophys.*, in press

Ögelman H., *et al.*, 1990 , preprint MPE188, *Astron. Astrophys.Lett.*, in press

Papaliolios C.,*et al.*, 1989 *Nature* **338** 565

Thielemann F.-K., Nomoto K., Hashimoto M., 1990 *Astrophys.J.* **349** 222

Tubbs D.L., 1979 *Astrophys.J.* **231** 846

Tubbs D.L., Koonin S.E., 1979 *Astrophys.J.* **232** L59

Tubbs D.L., Schramm D.N., 1975 *Astrophys.J.* **201** 467

Schwarz H.E., 1987 in *SN 1987A*, ed. by I.J. Danziger, ESO, Garching, p. 167

Wilson J.R., 1985 in *Numerical Astrophysics*, ed. by J. Centrella, J. Le Blanc, and R. Bowers, Jones and Bartlett, p. 422

Wilson J.R., Mayle R.W., 1989 in *The Nuclear Equation of State, Part A*, ed. by W. Greiner and H. Stöcker, Plenum Press, N.Y., p.731

Wilson J.R. Mayle R.W., Woosley S.E., Weaver T.A., 1986 *Ann. NY Acad.Sci.* **479** 267

Yahil A., Lattimmer J., 1982 in *Supernovae: A Survey of Current Research*, ed. by M.J. Rees and R.J. Stoneham, Reidel, Dortrecht, p.53

Yueh W.R., Buchler J.R., 1976 *Astrophys.Space Sci.* **41** 221

NEUTRON STAR FORMATION IN CLOSE BINARY SYSTEMS

K. NOMOTO, H. YAMAOKA, T. SHIGEYAMA, S. KUMAGAI
T. TSUJIMOTO
Department of Astronomy, Faculty of Science, University of Tokyo, Tokyo

ABSTRACT. Possible evolutionary scenarios for the formation of neutron stars in interacting binaries are summarized. (1) Firstly conditions for the occurrence of accretion-induced collapse of white dwarfs are examined in relation to the origin of low-mass X-ray binaries and binary millisecond pulsars. The outcome of the evolution of accreting white dwarfs is summarized as functions of accretion rate and the initial mass of the white dwarf. (2) Secondly discussed are the neutron star formation in Type Ib/Ic supernovae (SNe Ib/Ic), which are likely to be the core collapse of helium stars in binary systems. The light curves of SNe Ib/Ic (maximum brightness and decline rate) are used to infer the progenitors' masses and the neutron star masses. (3) Thirdly the evolutionary origin of double neutron stars are discussed in relation to SNe Ib/Ic.

1. Introduction

The final forms of the stars in interacting binary systems would be either white dwarfs or helium stars of mass $\gtrsim 2.5\ M_{\odot}$ after the loss of their hydrogen-rich envelope by Roche lobe overflow. The helium stars evolve through the Fe core collapse and would undergo supernova explosions. White dwarfs, if accreting matter from companion stars, would either explode or collapse depending on the conditions of binary systems. These scenarios of neutron star formation in binary stars have attracted growing attention because these are related to the origin of subtypes of Type I supernovae and binary millisecond pulsars. Relevant observations and motivations for the current studies are summarized below.

1.1. *Type I Supernovae*

Supernovae have been spectroscopically classified into Type I and Type II according to the absence and presence of hydrogen in their optical spectra. Type I supernovae (SNe I) are further subclassified into Ia, Ib, and Ic. As shown in Figure 1 (Branch *et al.* 1991), the early-time ($t \sim 1$ month past maximum) photospheric spectra of SNe I define such subtypes. SNe Ia are characterized by the presence of a deep absorption trough near 6150 Å produced by blueshifted Si II λ6355. SNe Ib and Ic, by contrast, do not show this line. Moderately strong He I lines, especially He I

J. Ventura and D. Pines (eds.), Neutron Stars: Theory and Observation, 143–170.

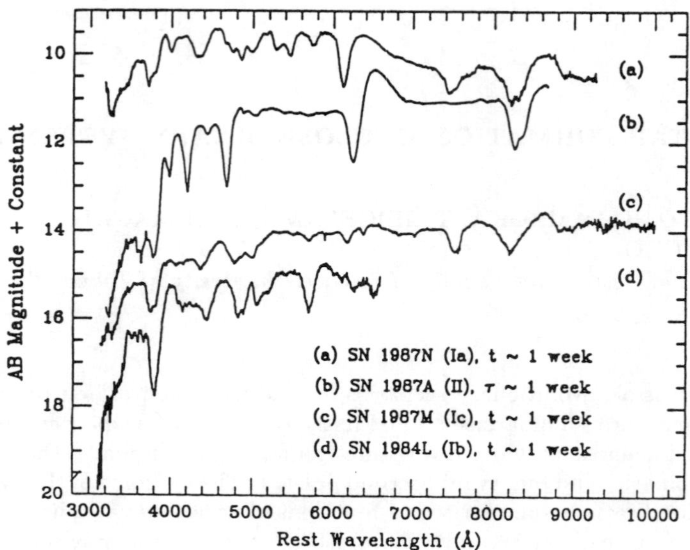

Fig. 1: Spectra of supernovae, showing the early–time distinctions between the four major types and subtypes (Branch *et al.* 1991). The variable t is taken to indicate time after observed visual maximum, whereas τ represents time after core collapse. AB magnitude $= -2.5 \log f_\nu - 48.6$, where the units of f_ν are ergs s^{-1} cm^{-2} Hz^{-1}.

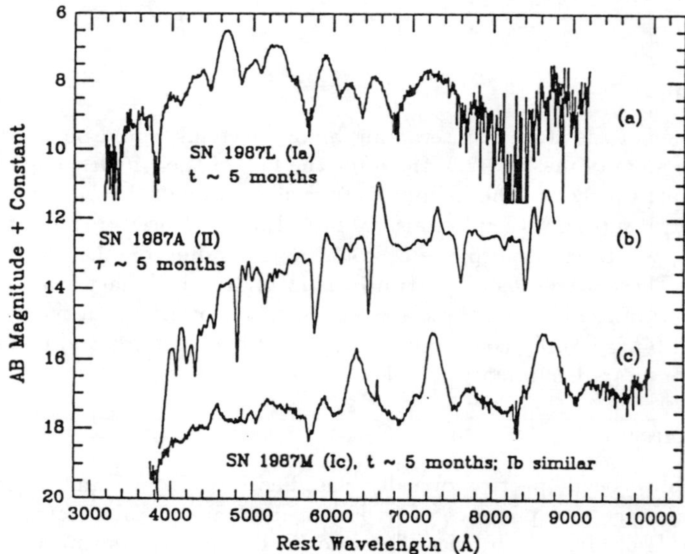

Fig. 2: Spectra of supernovae, showing late–time distinctions between different types and subtypes (Branch *et al.* 1991). Notation is the same as in Figure 1. The SN Ia 1987N (Fig. 1) was spectroscopically similar to SN 1987L, shown above. At even later phases, SN 1987A was dominated by strong emission lines of Hα, [O I], [Ca II], and the Ca II near–infrared triplet, with only a weak continuum.

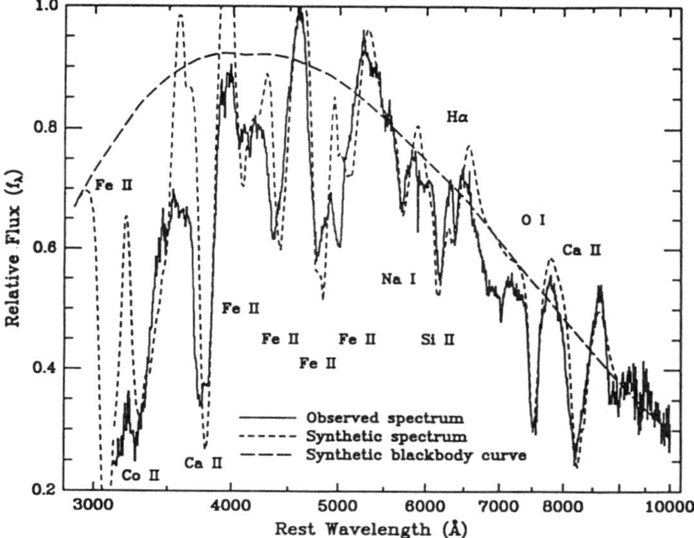

Fig. 3: The observed spectrum of SN Ic 1987M compared with a synthetic spectrum (Jeffery *et al.* 1991).

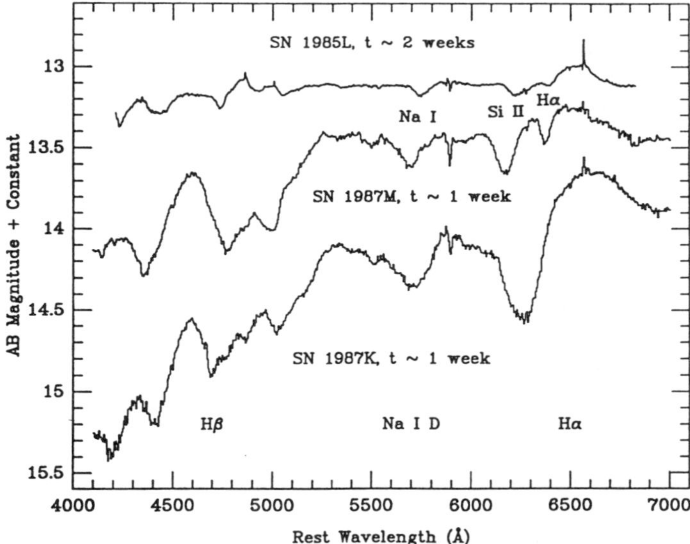

Fig. 4: Comparison of the observed spectra of SN Ic 1987M, SN IIb 1987K, and SN II 1985L (Filippenko *et al.* 1990; Jeffery *et al.* 1991).

$\lambda 5876$, distinguish SNe Ib from SNe Ic at early times, i.e., SNe Ib exhibit absorption lines of [He I], whereas these lines are absent in SNe Ic (Harkness and Wheeler 1990 for a review). Type II supernovae (SNe II) are subclassified into SNe II-P (*plateau*), SNe II-L (*linear*), and SNe II-BL (*bright linear*) according to the light curve shape (Branch *et al.* 1991).

The late–time ($t \sim 5 - 10$ months) optical spectra of SNe provide additional constraints on the classification scheme (Fig. 2; Branch *et al.* 1991). SNe Ia show strong blends of Fe emission lines. SNe Ib and Ic, on the other hand, are dominated by emission lines of intermediate–mass elements such as [O I], [Ca II], and Ca II. SNe II are dominated by the strong Hα emission line.

The supernova classification has become more complicated since the discovery of SN 1987K whose spectral classification changed from Type II to Type Ib/Ic as it aged, thereby being called as SN IIb (Filippenko 1988). Hydrogen feature has also been identified in the early time spectrum of SNe Ic 1987M (Fig. 3: Jeffery *et al.* 1991) and 1991A (Filippenko 1991). Interestingly the early time spectra of SN 1987K are very similar to SNe Ic (1983V and 87M) (Fig. 4: Filippenko *et al.* 1990; Wheeler and Harkness 1990).

The lack of strong hydrogen lines implies that the progenitors of SNe I have lost most of their hydrogen-rich envelope at the time of explosion. Two cases are possible: (1) mass loss over to the companion star during the evolution of a close binary system, and (2) stellar wind type mass loss from a single star. For these cases, we have basically three candidates for the progenitors of SNe I, i.e., (1) white dwarfs, (2) helium stars in binaries, and (3) single Wolf-Rayet stars. Because of the complicated mass loss processes, it has not been easy to identify the exact evolutionary origin of SNe I.

1.2. Low Mass X-Ray Binaries and Binary Pulsars

An unexpectedly large number of low mass binary pulsars (LMBPs) have recently been discovered. The birth rate of LMBPs is now estimated to be about 100 times higher than that of low mass X-ray binaries (LMXBs) in both the Galactic disk (Kulkarni and Narayan 1988; Narayan *et al.* 1990) and the globular clusters (Kulkarni *et al.* 1990; Romani 1990). Since LMXBs have been thought to be the progenitors of LMBPs, this birth rate discrepancy has raised a serious question about the evolutionary origin of LBMPs. Two scenarios have been proposed to resolve this problem: (1) accretion-induced collapse (AIC) of white dwarfs in close binaries (e.g., Bailyn and Grindlay 1990 and references therein), and (2) shortening of the LMXB phase due to the evaporation of the companion star (e.g., Tavani 1990 and references therein). Further, combinations of AIC and the tidal capture of neutron stars have been suggested as an explanation for the very high incidence of LMBPs in globular clusters (Romani 1990; Ray and Kluzniak 1990).

2. Evolution of Accreting White Dwarfs

Isolated white dwarfs are simply cooling stars that eventually end up as invisible frigid stars. The white dwarf in a close binary system evolves differently because the companion star expands and transfers matter over to the white dwarf at a certain stage of its evolution. The mass accretion can *rejuvenate* the cold white dwarf (e.g., Nomoto and Sugimoto 1977), which could lead to a SNe Ia or AIC in some cases.

The mass accretion onto the white dwarf releases gravitational energy at the white dwarf surface. Most of the released energy is radiated away from the shocked region as UV and does not contribute much to heating the white dwarf interior. The continuing accretion compresses the previously accreted matter and releases gravitational energy in the interior. A part of this energy is transported to the surface and radiated away from the surface (radiative cooling) but the rest goes into thermal energy of the interior matter (compressional heating). Thus the interior temperature of the white dwarf is determined by the competition between compressional heating and radiative cooling; that is the white dwarf is hotter if the mass accretion rate $\dot{M}$ is larger, and vice versa (e.g., Nomoto 1982a).

The scenario that possibly brings a close binary system to a SN Ia or AIC is as follows (although the exact evolutionary origin has not been understood): Initially the close binary system consists of two intermediate mass stars ($M \lesssim 8\ M_\odot$). As a result of Roche lobe overflow, the primary star of this system becomes a white dwarf composed of carbon and oxygen (C+O). When the secondary star evolves, it begins to transfer hydrogen-rich matter over to the white dwarf.

When a certain amount of hydrogen $\Delta M_{\rm H}$ is accumulated on the white dwarf surface, hydrogen shell burning is ignited (Fig. 5; Nariai and Nomoto 1979; Nomoto 1982a). Its outcome depends on $\dot{M}$: For slow accretion ($\dot{M} \lesssim 1 \times 10^{-8}\ M_\odot\ {\rm yr}^{-1}$), hydrogen shell burning is unstable and tends to *flash*, which leads to the ejection of most of the accreted matter from the white dwarf; the strongest flash grows into a *nova* explosion (e.g, Nariai *et al.* 1980 and references therein). For these cases, the white dwarf does not become a *supernova* since its mass hardly grows. In other words, it seems rather unlikely that novae are the precursors of supernovae.

For intermediate accretion rates ($3 \times 10^{-6}\ M_\odot\ {\rm yr}^{-1} \gtrsim \dot{M} \gtrsim 1 \times 10^{-8}\ M_\odot\ {\rm yr}^{-1}$), on the other hand, hydrogen flashes and the subsequent helium flashes are of moderate strength, thereby increasing the C+O white dwarf mass toward the Chandrasekhar mass. When the white dwarf mass becomes 1.4 $M_\odot$ and the central density reaches $\sim 3 \times 10^9\ {\rm g\ cm}^{-3}$, explosive carbon burning starts in the white dwarf's center. The precursor systems might be observed as symbiotic stars (Nomoto 1982a,b).

If the accretion rate is higher than $\sim 3 \times 10^{-6}\ M_\odot\ {\rm yr}^{-1}$, the accreted matter is too hot to be swallowed by the white dwarf (Fig. 6; Nomoto *et al.* 1979b). The matter forms a common envelope, which is eventually lost from the system. As a result of mass and angular momentum losses from the system, some binaries form a pair of C+O white dwarfs. Further evolution of such a double white dwarf system

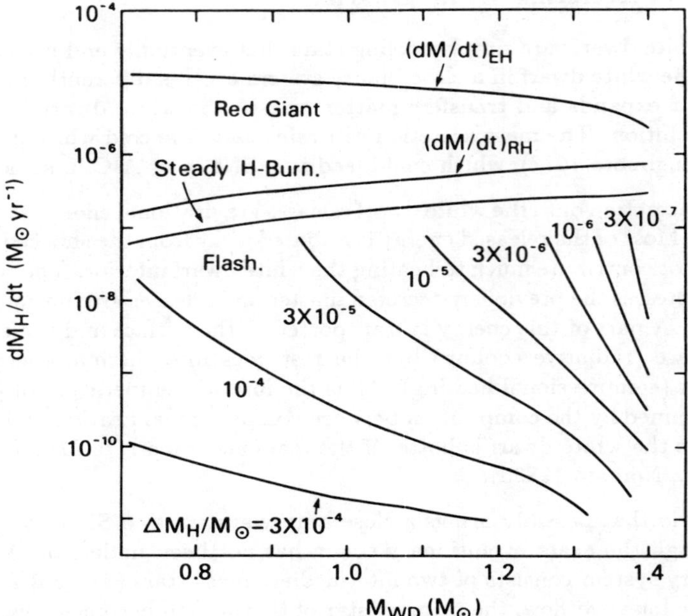

Fig. 5: Types and strength of hydrogen shell burning as a function of accretion rate and the white dwarf mass (Nomoto 1982).

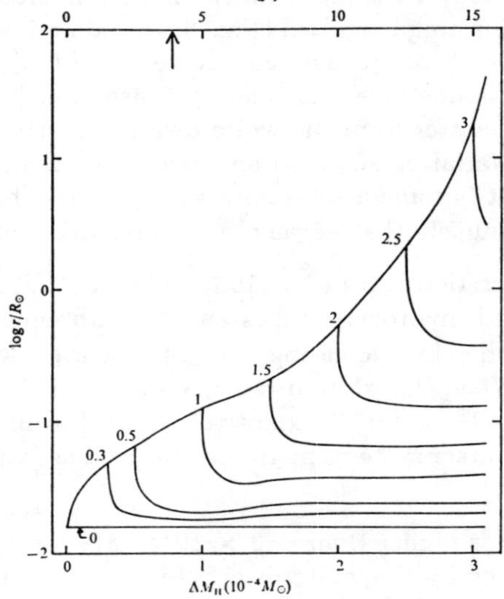

Fig. 6: Increase in the radius of the accreting white dwarf upon the rapid accretion.

is driven by gravitational wave radiation and leads to a Roche lobe overflow of the smaller mass C+O white dwarf (Iben and Tutukov 1984; Webbink 1984). The fate of these merging white dwarfs is not clear yet but would be either a Type Ia supernova explosion or a collapse to form a single neutron star.

It is possible that the accreting C+O white dwarfs could collapse rather than explode, depending on the conditions of the white dwarfs. As described above, compression of the white dwarf by the accreted matter first heats up a surface layer because of the small pressure scale height there. Later, heat diffuses inward (Nomoto *et al.* 1984). The diffusion timescale depends on $\dot{M}$ and is short for larger $\dot{M}$ because of the large heat flux and steep temperature gradient generated by rapid accretion. For example, the time it takes for the heat wave to reach the central region is $\sim 2 \times 10^5$ yr for $\dot{M} \sim 10^{-6}\ M_\odot\ \mathrm{yr}^{-1}$ (Nomoto and Iben 1985) and 5×10^6 yr for $\dot{M} \sim 4 \times 10^{-8}\ M_\odot\ \mathrm{yr}^{-1}$ (Nomoto *et al.* 1984). If the initial mass of the white dwarf, M_{CO}, is smaller than 1.2 $M_\odot$, the entropy in the center increases substantially due to the heat inflow and thus carbon ignites at relatively low central density ($\rho_c \sim 3 \times 10^9$ g cm^{-3}). On the other hand, if the white dwarf is initially more massive than 1.2 $M_\odot$ and cold at the onset of accretion, the central region is compressed only adiabatically and thus is cold when carbon is ignited in the center. In the latter case, the ignition density is as high as 10^{10} g cm^{-3} (e.g., Isern *et al.* 1983) and the white dwarf may well have a solid core. For such a case, it is necessary to determine the critical condition for which a carbon deflagration induces collapse rather than explosion.

3. Type Ia Supernovae

3.1. Carbon Deflagration

When carbon is ignited at the white dwarf's center, carbon burning is so explosive as to incinerate the material into iron-peak elements; the central temperature reaches $\sim 10^{10}$ K. The resulting shock wave is not strong enough to ignite carbon in the adjacent layer; in other words, a *detonation* wave that propagates at supersonic speed does not form. Instead, the interface between the burned and unburned layers becomes convectively unstable. As a result of mixing with the hot material, fresh carbon is ignited. In this way, a carbon burning front propagates outward on the time scale for convective heat transport (Nomoto *et al.* 1976, 1984; Woosley and Weaver 1986a,b). This kind of explosive burning front that propagates at a subsonic speed is called a *convective deflagration* wave. In the standard model, the propagation speed of the convective deflagration wave is on the average about one-fifth of the sound speed. It takes about one second for the front to reach the surface region, which is significantly slower than the supersonic detonation wave. Hence the white dwarf expands during the propagation of the deflagration wave.

Behind the deflagration wave, the material undergoes explosive nuclear burning of silicon, oxygen, neon, and carbon depending on the peak temperatures. In the inner layer, nuclear reactions are rapid enough to incinerate the material into iron-peak elements, mostly ^{56}Ni. When the deflagration wave arrives at the outer layers,

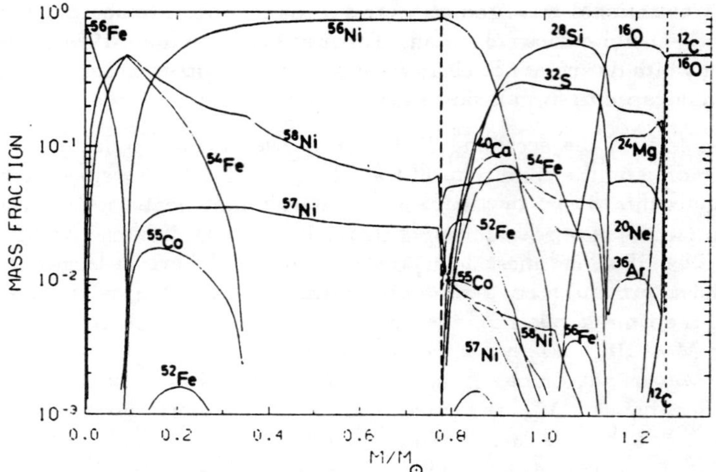

Fig. 7: Composition of a carbon deflagration model W7 for Type Ia supernovae as a function of interior mass (Thielemann *et al.* 1986).

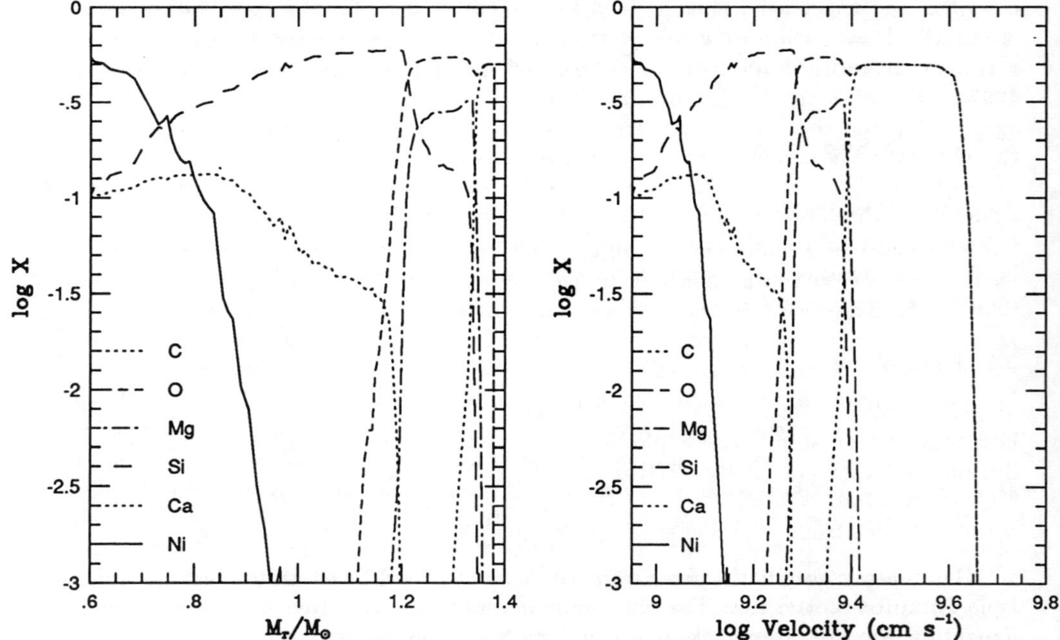

Fig. 8: Composition of a delayed detonation model for Type Ia supernovae as functions of M_r and the expansion velocity (Shigeyama and Nomoto 1991).

the density it encounters has already decreased due to the expansion of the white dwarf. At such low densities, the peak temperature is too low to complete silicon burning and thus only Ca, Ar, S, and Si are produced from oxygen burning. In the intermediate layers, explosive burning of carbon and neon synthesizes S, Si, and Mg. In the outermost layers, the deflagration wave dies and C+O remain unburned. The composition structure after freeze-out is shown in Figure 7 (Thielemann *et al.* 1986).

In the standard carbon deflagration model W7 (Nomoto *et al.* 1984), the amount of ^{56}Ni produced is $M_{\rm Ni} = 0.6\ M_{\odot}$, and the explosion energy is $E =$ (Nuclear energy release – Binding energy of the white dwarf) $= 1.3 \times 10^{51}$ erg. The nuclear energy release is large enough to disrupt the white dwarf completely and no compact star is left behind.

3.2. Delayed Detonation

The outcome of carbon deflagration depends on its propagation speed, which is highly uncertain involving a parameter such as the mixing length of convection. The preceding standard model W7 has been chosen because it accounts well for the observed light curve and spectra at both early and late times of SN Ia.

However, it is a possibility that the propagation speed of the deflagration wave is initially slower than in W7 and then increased at low density layers to induce a transition into a detonation wave (*delayed detonation*; Khokhlov 1991; Woosley 1991). Figure 8 shows a composition structure of the delayed detonation model against M_r and the expansion velocity; here the transition from the deflagration to the detonation is assumed to occur when the density at $M_r = 0.6\ M_{\odot}$ has been decreased to 3×10^7 g cm^{-3} (Shigeyama and Nomoto 1991). The composition structure with respect to M_r is not so different from W7 including some unburned C and O, but Si and Ca layers expand significantly faster than W7. Comparison between the synthetic spectra and the pre-maximum spectra of SN 1990N would be important (Leibundgut *et al.* 1991).

3.3. Light Curve

The explosion energy goes into the kinetic energy of expansion, and without a late time energy source the exploding white dwarf could not be bright. However, during the expansion phase, ^{56}Ni decays into ^{56}Co with a half-life of 6.6 days and ^{56}Co decays into ^{56}Fe with a half-life of 77 days. These radioactive decays produce γ-rays and positrons whose energies power the light curve as follows (Fig. 9).

Gamma-rays originating from radioactive decays are degraded into X-rays by multiple Compton scatterings. The photoelectric absorption of X-rays and the collisional ionization due to energetic electrons eventually heat up the expanding materials and produce the optical light as clearly observed in SN 1987A (e.g., Kumagai *et al.* 1989; Shigeyama *et al.* 1988; Shigeyama and Nomoto 1990).

The light curve powered by the radioactive decays reaches its peak at about 15 days after the explosion and declines because of the increasing transparency of

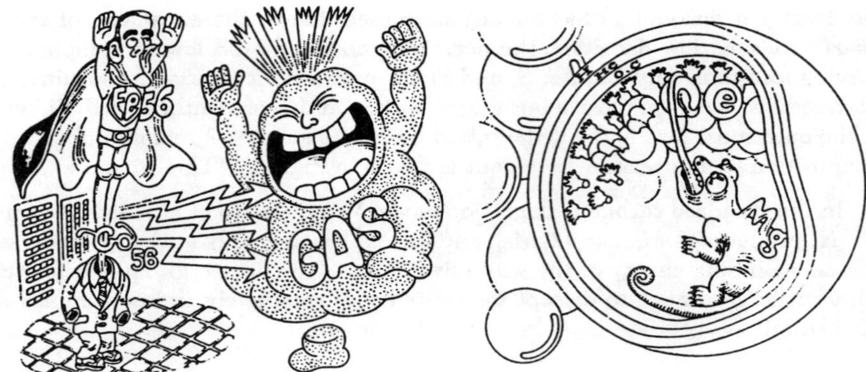

Fig. 9: (a: left) The expanding supernova matter is excited by the decay of ^{56}Co into ^{56}Fe. (b: right) Collapse of an O+Ne+Mg core is induced by electron capture (H. Nomoto 1989).

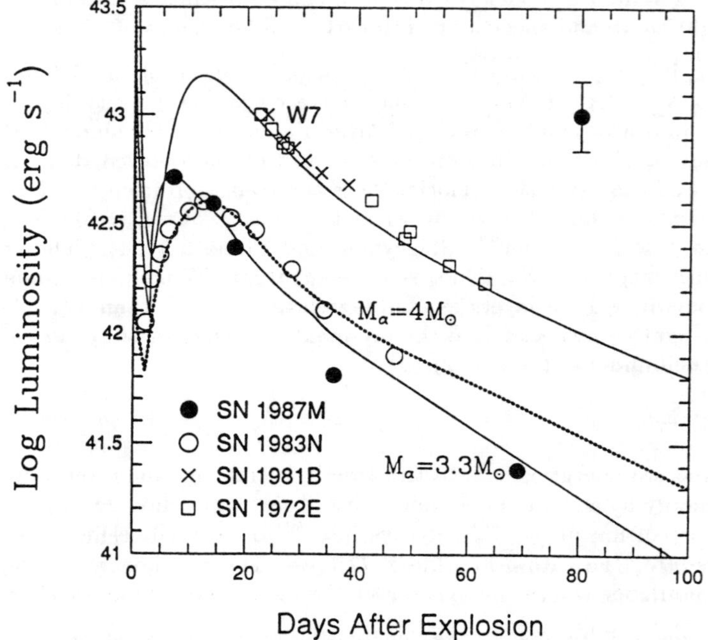

Fig. 10: Approximate bolometric light curve of SN Ic 1987M, and the bolometric light curves of SNe Ia 1972E and 1981B and of SN Ib 1983N. The predicted curves of the 3.3 $M_\odot$ model for SN 1987M, the 4 $M_\odot$ model for SN Ib, and the W7 model for SN Ia are indicated by solid and dotted lines. The error bar illustrates the 2σ photometric uncertainty in the SN 1987M points (Nomoto *et al.* 1990).

the ejecta to γ-rays as well as the decreasing number of radioactive elements. The calculated curve is in good agreement with the observed bolometric light curves of SN 1972E and SN 1981B (Fig. 10; Nomoto *et al.* 1990).

3.4. Spectra

Because SNe Ia do not have a thick hydrogen-rich envelope, elements newly synthesized during the explosion can be observed in the spectra; this enables us to diagnose the internal hydrodynamics and nucleosynthesis in SNe Ia.

Synthetic spectra are calculated based on the abundance distribution and expansion velocities of the standard model and are found to be in excellent agreement with the observed optical spectra of SN 1981B (Branch *et al.* 1985) and 1989B (Harkness 1991). The material velocity at the photosphere near maximum light is $\sim$ 10,000 km s^{-1} and the spectral features are identified as P-Cygni profiles of Fe, Ca, S, Si, Mg, O.

At late times, the spectra are dominated by the emission lines of Fe and Co. The outer layers are transparent and the inner Ni-Co-Fe core is exposed. Synthetic spectra of emission lines of [Fe II] and [Co I] agree quite well with the spectra observed at such phase (Axelrod 1980; Woosley and Weaver 1986b). The agreement implies that both explosion energy and nucleosynthesis in the carbon deflagration model are consistent with the observations of SNe Ia.

4. Accretion–Induced Collapse (AIC) of White Dwarfs

Possible models for AIC previously advanced include solid C+O white dwarfs (Canal *et al.* 1980; Isern *et al.* 1983) and O+Ne+Mg white dwarfs (Nomoto *et al.* 1979a), whose masses could grow to the Chandrasekhar's mass limit for a white dwarf. In the AIC models, collapse of the white dwarf is induced by electron capture that effectively reduces the Chandrasekhar mass limit.

However, since the white dwarf contains nuclear fuel, whether the white dwarf undergoes collapse or explosion depends on which is faster behind the deflagration wave, nuclear energy release or electron capture. The energy generation rate is determined mainly by the propagation velocity of the deflagration wave, $v_{\rm def}$, while the electron capture rate depends on the density. If $v_{\rm def}$ is lower than a certain critical speed, electron capture induces collapse. If on the other hand $v_{\rm def}$ is sufficiently high, complete disruption results. It is important to determine the critical velocity that divides collapse and explosion.

4.1. Solid C+O White Dwarfs

For solid C+O white dwarfs, recent work find that no significant separation between carbon and oxygen occurs during solidification (Barrata *et al.* 1988; Ichimaru *et al.* 1988). This in turn leads to the ignition of explosive carbon burning at much lower densities than in models which postulate chemical separation (Isern *et al.* 1983).

In solid cores, carbon ignition takes place in the pycnonuclear reaction regime (Ogata *et al.* 1990) and develops into explosive burning at $\rho_c \sim 1 \times 10^{10}$ g cm^{-3}. After thermal runaway of carbon burning, it is likely that a conductive deflagration wave propagates in the solid core. A detonation wave would not form because of steep temperature gradient in the central solid region. Convection would not be effective unless the solid core is melted by the heating from nuclear burning or neutrinos (Canal *et al.* 1990a).

The conductive deflagration in the solid C+O white dwarf is calculated assuming a constant ratio of $v_{\rm def}/v_{\rm s}$ for conductive deflagration wave (Nomoto 1986, 1987b; Nomoto and Kondo 1991; Canal *et al.* 1990b).

Figure 11 shows the change in the central density associated with the propagation of the deflagration wave. Three cases with $v_{\rm def}/v_{\rm s}$ = 0.05, 0.03, and 0.01 are calculated and the latter two slow cases undergo collapse. This implies that the critical velocity, $v_{\rm crit}$, that divides collapse and explosion is $v_{\rm crit} \sim 0.04\ v_{\rm s}$ for $\rho_c \sim 10^{10}$ g cm^{-3}. Since the realistic value of conductive deflagration speed is $v_{\rm def} \sim 0.01\ v_{\rm s}$ (Woosley and Weaver 1986a), the collapse is the most likely outcome for the solid white dwarf.

4.2. O+Ne+Mg White Dwarfs

The O+Ne+Mg white dwarfs are formed from stars of main-sequence masses of 8 - 12 $M_\odot$ in close binaries with the initial masses as large as 1.1 - 1.37 $M_\odot$ (Nomoto *et al.* 1979a; Nomoto 1984). After mass accretion from the companion star, the mass of the white dwarf increases toward the Chandrasekhar mass for a certain range of accretion rate (Fig. 14). When ρ_c exceeds 4×10^9 g cm^{-3}, the O+Ne+Mg white dwarf undergoes electron captures ^{24}Mg (e$^-$, ν) ^{24}Na (e$^-$, ν) ^{20}Ne and ^{20}Ne (e$^-$, ν) ^{20}F (e$^-$, ν) ^{20}O (Fig. 9). Electron capture not only reduces the effective Chandrasekhar mass but also releases heat due to γ-ray emission which eventually ignites oxygen deflagration at a certain central density.

In the previous AIC models (Nomoto *et al.* 1979a), oxygen is ignited at $\rho_{\rm ig} \sim 2.5 \times 10^{10}$ g cm^{-3} after the initiation of collapse (Miyaji *et al.* 1980; Nomoto 1987a). At such central densities, electron capture is much faster than oxygen burning, thus promoting further collapse. However, $\rho_{\rm ig}$ has been found to depend on the timescale of semiconvective mixing in the electron capture region (Nomoto 1984b; Mochkovitch 1984; Miyaji and Nomoto 1987). If semiconvective mixing is negligible and the heating due to γ-ray emission is confined to the very central region, oxygen burning is ignited at $\rho_{\rm ig} \sim 9.95 \times 10^9$ g cm^{-3} before collapsing (Miyaji and Nomoto 1987). Hydrodynamical calculation is carried out to see whether this model leads to collapse or explosion (Nomoto and Kondo 1981).

The heat released by electron capture on ^{24}Mg results in the formation of a liquid core even if the white dwarf had initially a solid core (Mochkovitch 1984; Miyaji and Nomoto 1987; Canal *et al.* 1990a). When γ-rays resulting from electron capture on ^{20}Ne ignite explosive oxygen burning, there exists several possible modes of the

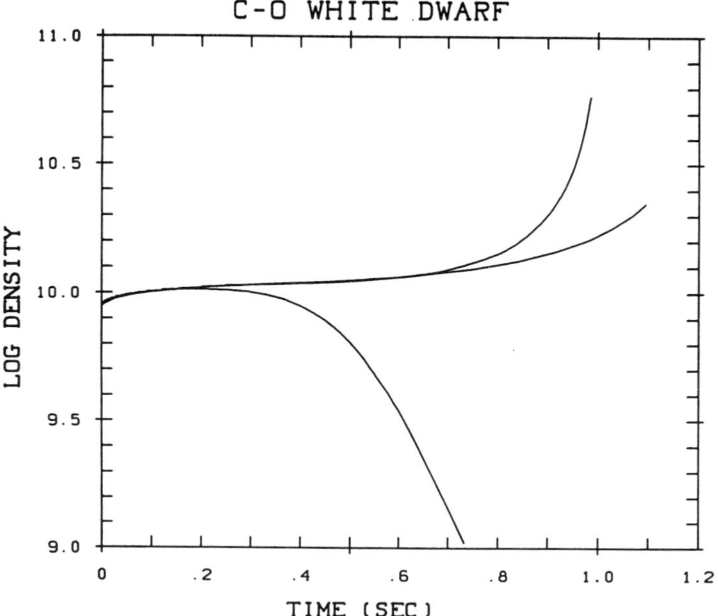

Fig. 11: Change in the central density of the C+O white dwarfs following the propagation of the conductive carbon deflagration wave in the initially solid core. Three cases with $v_{\rm def}/v_{\rm s} = 0.05$, 0.03, and 0.01 are shown and the latter two undergo collapse (Nomoto and Kondo 1991).

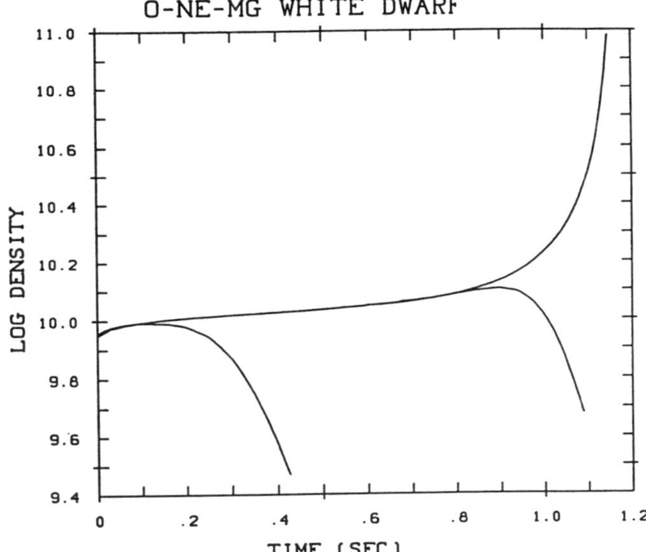

Fig. 12: Same as Fig. 11 but for the O+Ne+Mg white dwarfs following the propagation of the oxygen deflagration wave for three cases with $\ell/\min\,(H_p, r) = 1.4$, 1.0, and 0.7. For the slowest case, the white dwarf undergoes collapse, while faster propagation induces an explosion (Nomoto and Kondo 1991).

subsequent propagation of explosive burning front. Among them the formation of a detonation wave is very unlikely in the present case because negligible semiconvective mixing forms a very steep temperature gradient when explosive oxygen burning starts.

Therefore it is likely that an oxygen deflagration wave forms to propagate at a subsonic velocity. The propagation velocity, $v_{\rm def}$, depends on which mode of heat transport is faster, conductive or convective. For conductive deflagration, we apply $v_{\rm def} = 0.01\ v_{\rm s} \sim 100$ km s^{-1} where $v_{\rm s}$ denotes the local sound velocity, which is a good approximation of $v_{\rm def}$ obtained by numerical calculations (Woosley and Weaver 1986a). For the convective deflagration wave, we apply a time dependent mixing length prescription using the ratio between the mixing length and the pressure scale height (or radial distance) $\alpha = \ell/\min(H_p, r) = 0.7$, 1.4, and 2 (Nomoto *et al.* 1984). For small α, the deflagration speed in the very central region is slower than the conductive deflagration because of small buoyancy force across the burning front; then the minimum $v_{\rm def}$ is set to be 0.01 $v_{\rm s}$.

Whether this will lead to collapse or explosion depends on which is faster behind the deflagration wave, nuclear energy release or electron capture. The energy generation rate is determined mainly by the propagation velocity of the deflagration wave, $v_{\rm def}$, while the electron capture rate depends on the density. If $v_{\rm def}$ is lower than a certain critical speed, electron capture induces collapse. If on the other hand $v_{\rm def}$ is sufficiently high, complete disruption results.

Figure 12 shows the changes in the central density associated with the propagating deflagration front for three cases. It is seen that the slowest case of $\alpha = 0.7$ leads to increasing $\rho_{\rm c}$, i.e., collapse, while the propagation with $\alpha = 2$ results in explosion; the intermediate case with $\alpha = 1.4$ is marginal. The fate of the convective deflagration wave depends mainly on whether $v_{\rm def}$ exceeds $\sim 0.03\ v_{\rm s}$ in the central region at M_r $< \sim 0.1$ - $0.2\ M_\odot$.

Though the determination of $v_{\rm def}$ may require multi-dimensional calculations, the carbon deflagration model for Type Ia supernovae favors $\alpha = 0.7$; the model with α $= 0.7$ can nicely account for many of the observed features of Type Ia supernovae, while the propagation with $\alpha = 0.8$ is a little too fast to be consistent with spectral features of SN Ia (Nomoto *et al.* 1984). For oxygen deflagration, $\alpha < 1$ may also be the case for the same prescription of deflagration; then the collapse of O+Ne+Mg white dwarfs would be the most likely outcome since $\alpha \sim 1.4$ is marginal between collapse and explosion. (If total disruption results from the white dwarf with central density of $\sim 10^{10}$ g cm^{-3}, such an explosion should be extremely rare since the ejection of too much neutron-rich iron-peak elements would not be compatible with solar isotopic ratios. In addition, the explosion with such low energy as $\sim$ a few times 10^{50} ergs due to large neutrino losses does not match any subclass of SN I frequently observed.)

4.3. Conditions for AIC and SN Ia

In applying the above models to the AIC scenario for LMBPs (Michel 1987;

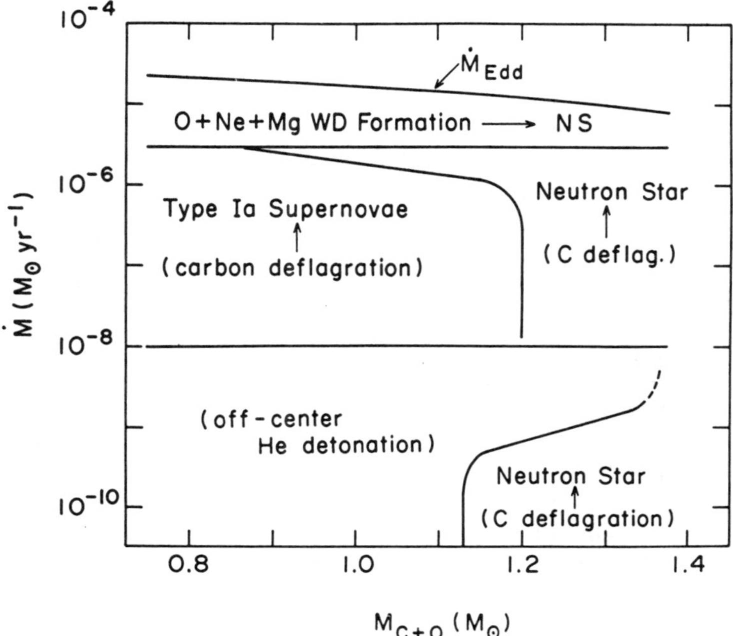

Fig. 13: The final fate of accreting C+O white dwarfs expected for their initial mass and accretion rate $\dot{M}$ (Nomoto and Kondo 1991).

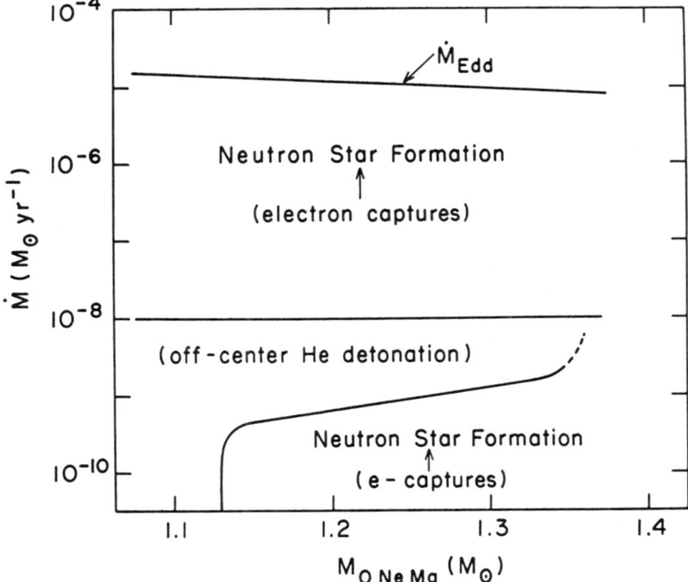

Fig. 14: Same as Figure 13 but for O+Ne+Mg white dwarfs. Collapse is triggered by electron capture on ^{24}Mg and ^{20}Ne (Nomoto and Kondo 1991).

Chanmugam and Brecher 1987; Bailyn and Grindlay 1990; Romani 1990; Ray and Kluzniak 1990),

To make estimates for the occurrence frequency of AIC and SN Ia, we draw boundaries for AIC in a diagram of mass accretion rate ($\dot{M}$) *versus* mass of the white dwarf at the onset of accretion (M_{CO} and M_{ONeMg}) in Figures 13 and 14 (Nomoto 1986; Nomoto and Kondo 1991).

In these figures, the fate of accreting white dwarfs depends mainly on the mass accretion rate as discussed in §2. We note that the boundaries must be regarded as relatively optimistic ones for the growth of white dwarfs since wind-type mass loss associated with shell flashes of hydrogen and helium is not fully taken into account (e.g., Kato and Hachisu 1989).

(1) For $\dot{M} \gtrsim \dot{M}_{det}$, both hydrogen and helium flashes are weak and recur many times to increase the white dwarf mass.

(2) For $\dot{M}_{det} \gtrsim \dot{M} \gtrsim 10^{-9}$ $M_{\odot}$ yr^{-1}, off-center helium detonation prevents the white dwarf mass from growing (Nomoto 1982b; Woosley *et al.* 1986). Here we adopt $\dot{M}_{det} \sim 1 \times 10^{-8}$ $M_{\odot}$ yr^{-1}, since the $^{14}N(e^-, \nu)^{14}C(\alpha, \gamma)^{18}O$ (NCO) reaction ignites weak helium flashes (Hashimoto *et al.* 1986) if the mass fraction of CNO elements in the accreting material exceeds 0.005. For smaller CNO abundances, the NCO reaction is not effective and thus $\dot{M}_{det} \sim 4 \times 10^{-8}$ $M_{\odot}$ yr^{-1} (Nomoto 1982a).

(3) For $\dot{M} \lesssim 10^{-9}$ $M_{\odot}$ yr^{-1}, accretion of hydrogen-rich material gives rise to nova-like explosion which would not allow the white dwarf mass to grow. If the accreting material is helium, on the other hand, the material is too cold to ignite helium burning, thereby increasing the white dwarf mass. Exception is the case with $M_{CO} \lesssim 1.1$ $M_{\odot}$ where pycnonuclear helium burning is ignited (Nomoto 1982a,b).

For $\dot{M} \gtrsim 2 \times 10^{-6}$ $M_{\odot}$ yr^{-1} in Figure 13, we adopt the following scenario. First merging of double C+O white dwarfs forms a thick disk around more massive component (Benz *et al.* 1990). Subsequent heat generation at the boundary layer ignites off-center carbon burning (Mochkovitch and Livio 1990), which burn the entire C+O white dwarf into O+Ne+Mg quietly (Saio and Nomoto 1985). Eventually the O+Ne+Mg white dwarf collapses. This is an optimistic scenario for AIC since such merging white dwarfs are also the possible candidates of SN Ia progenitors (Iben and Tutukov 1984; Webbink 1984).

For C+O white dwarfs, the outcome of accretion depends not only on $\dot{M}$ but also on the initial mass M_{CO} as seen in Figure 13. For $M_{CO} < 1.2$ $M_{\odot}$, substantial heat inflow from the surface layer into the central region ignites carbon at relatively low central density ($\rho_c \sim 3 \times 10^9$ g cm^{-3}), which make Type Ia supernovae. On the other hand, if the white dwarf is sufficiently massive and cold at the onset of accretion, the central region is compressed only adiabatically, thereby being cold when carbon is ignited in the center of density as high as 10^{10} g cm^{-3}.

Figure 14 shows that for a relatively wide parameter range the O+Ne+Mg white dwarf can increase its mass. Since M_{ONM} can be very close to the Chandrasekhar mass, only a small increase in mass is enough to trigger a collapse. Such very massive O+Ne+Mg white dwarfs would give rise to recurrent novae (Nariai and Nomoto 1979).

Figures 13 and 14 clearly show that close binaries with relatively high $\dot{M}$ and high initial white dwarf mass are favored for AIC. This leads to the possibility that LMBPs with relatively long orbital period may originate from AIC (Nomoto 1987b; Romani 1990; Ray and Kluzniak 1990) since mass transfer rate from giant stars may be relatively high. On the other hand, Wheeler (1990) suggested that many of the white dwarfs in cataclysmic variables could possibly increase their masses toward the Chandrasekhar mass ending up with $\dot{M} \sim 10^{-9}$ $M_{\odot}$ yr^{-1} at the lower-right hand corners of Figures 13 and 14; resultant systems could be short orbital period LMBPs. Also if we consider various types of helium star companions (helium main-sequence, helium subgiants, helium white dwarfs, etc.) as well as companions surrounded by a common envelope (Hachisu *et al.* 1989), LMBPs with relatively short orbital period could be formed from AIC.

5. Type Ib/Ic Supernovae

Wolf-Rayet stars with a wide range of masses have been proposed for the progenitors of SNe Ib and Ic, since most of SNe Ib/Ic are associated with star-forming regions (Wheeler and Harkness 1990 for a review). Recently Shigeyama *et al.* (1990), Hachisu *et al.* (1991), Nomoto *et al.* (1990), and Yamaoka and Nomoto (1991) have calculated the progenitor's evolution, nucleosynthesis, Rayleigh-Taylor instabilities, and optical light curves of exploding helium stars. They have suggested that the helium stars of 3–5 $M_{\odot}$ (which form from stars with initial masses $M_{\rm i}$ $\sim$ 12–18 $M_{\odot}$ in binary systems) are the most likely progenitors of typical SNe Ib/Ic and that SNe Ic progenitors may be slightly less massive than those of SNe Ib. Such low mass helium star models can account for the observations that: (1) the light curves of SNe Ic decline faster than SNe Ib, and (2) the early time spectra of SNe Ic show the presence of hydrogen (Jeffery *et al.* 1991), while hydrogen is absent in SNe Ib. It remains an open question how the presence of hydrogen causes the difference between SNe Ib and Ic in their early time spectra.

5.1. Evolution of Interacting Binaries

The difference in the spectral feature between SNe Ic and Ib may be due to the presence of a thin envelope of hydrogen in SNe Ic immediately prior to the explosion. By evolving massive stars in close binary systems, we examine whether hydrogen can be left on the helium stars after mass exchange and wind-type mass loss. Followings are some preliminary results for two cases 13A and 18A, where the initial masses of the primary stars are $M_{\rm i}$ $= 13$ $M_{\odot}$ (13A) and 18 $M_{\odot}$ (18A), and their Roche lobe radii are 50 $R_{\odot}$ (Yamaoka and Nomoto 1991).

After hydrogen exhaustion, the star undergoes Roche lobe overflow forming a helium star of $\sim$ 3.4 $M_{\odot}$ (13A) and $\sim$ 5 $M_{\odot}$ (18A). Figure 15 (upper) shows the composition structure at the onset of mass transfer during core helium burning (up-

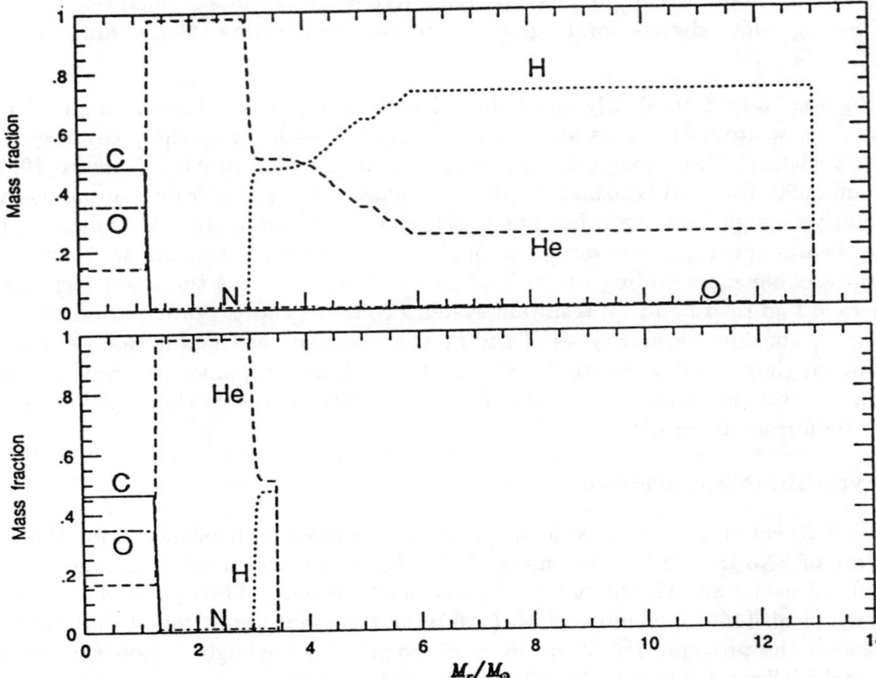

Fig. 15: Change in the composition structure during the binary evolution for the star with $M_{\rm i} = 13\ M_\odot$ (Yamaoka and Nomoto 1991).

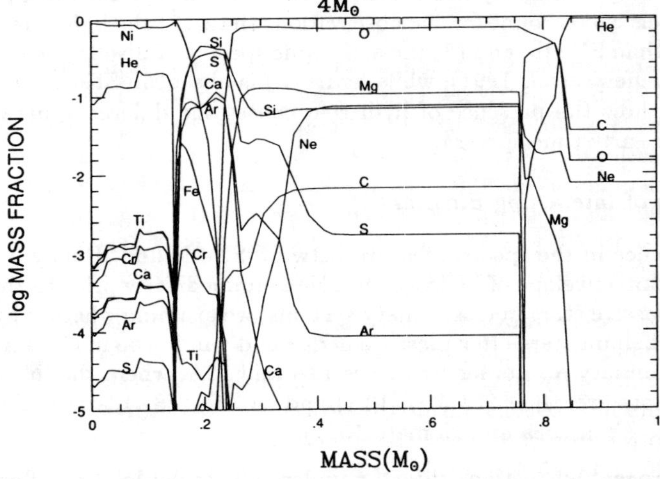

Fig. 16: Explosive nucleosynthesis in the 4 $M_\odot$ helium star (Shigeyama *et al.* 1990). Composition of the innermost 1 $M_\odot$ of the ejecta is shown. (The outermost 1.72 $M_\odot$ helium layer and the 1.28 $M_\odot$ neutron star are not included in the figure.) About 0.15 $M_\odot$ ^{56}Ni and 0.43 $M_\odot$ oxygen are produced.

per) and after the Roche lobe overflow (lower). It is seen that significant amount of hydrogen remains in a relatively thick layer below the surface (0.5 $M_\odot$ for 13A and 1 $M_\odot$ for 18A).

Whether such a hydrogen layer will be further lost from the helium-rich star depends on the Roche lobe radius and wind mass loss rate. If the helium star is detached from the Roche lobe during helium burning, the star loses its masses in a wind. If the mass loss rate depends on the mass such as $\dot{M} \propto M^{2.5}$ (Langer 1989), it may lead to different surface abundances between 13A and 18A. For case 13A hydrogen still may remain in the layer down to $\sim 0.2\ M_\odot$ below the surface, whereas for case 18A all hydrogen may be lost in a wind (Yamaoka and Nomoto 1991). If a common envelope forms or the Roche lobe radius has become small enough for the mass and angular momentum to be lost from the system, then all hydrogen will be lost from the star.

Although more parameter study is needed, the present results suggest that more hydrogen may remain on the stellar surface if the helium star had initially smaller main-sequence mass. Also more compact binary systems may lose more hydrogen through common envelope evolution.

5.2. *Nucleosynthesis*

Assuming that the helium star progenitors of SNe Ib/Ic are formed in close binary systems as described above, Shigeyama *et al.* (1990) performed hydrodynamical calculations of the explosion of helium stars with masses M_α = 3.3, 4, 6, and 8 $M_\odot$. These are presumed to form from the main-sequence stars of masses $M_{\rm i} \sim 13$, 15, 20, and 25 $M_\odot$, respectively. These stars eventually undergo iron core collapse as in SNe II. A shock wave is then formed at the mass cut that divides the neutron star and the ejecta.

Behind the shock wave that propagates outward, materials are processed into nuclear statistical equilibrium (NSE) composition, mostly ^{56}Ni, if the maximum temperature exceeds 5×10^9 K (e.g., Hashimoto *et al.* 1989). As derived from the approximate relation $E = 4\pi r^3/3\ aT^4$ with E being the final kinetic energy of explosion (e.g., Thielemann *et al.* 1990), such a high temperature is realized in a sphere of radius $\sim 3700\ (E/10^{51}\ \mathrm{erg})^{1/3}$ km. With $E = 1 \times 10^{51}$ erg, this region contains a mass $M_{\rm NSE}$ ($\sim 1.44 - 1.46\ M_\odot$ for M_α = 3.3 and 4 $M_\odot$). Then the mass of ^{56}Ni plus neutron-rich iron peak elements is given by $M_{\rm NSE} - M_{\rm NS}$.

The adopted presupernova models (Nomoto and Hashimoto 1988) have the iron core masses as small as 1.18 $M_\odot$ and 1.28 $M_\odot$ for M_α = 3.3 $M_\odot$ and 4 $M_\odot$, respectively, being significantly smaller than 1.4 $M_\odot$ in the 6 $M_\odot$ star due to the larger effect of Coulomb interactions during the progenitor's evolution. If $M_{\rm NS}$ is approximately equal to the iron core mass, the *upper limit* to the possible ^{56}Ni masses are obtained as 0.26 and 0.15 $M_\odot$ for M_α = 3.3 and 4 $M_\odot$, respectively.

Figure 16 shows the abundance distribution after explosive burning for M_α = 4 $M_\odot$ (Shigeyama *et al.* 1990). The masses of oxygen produced in the outer layers

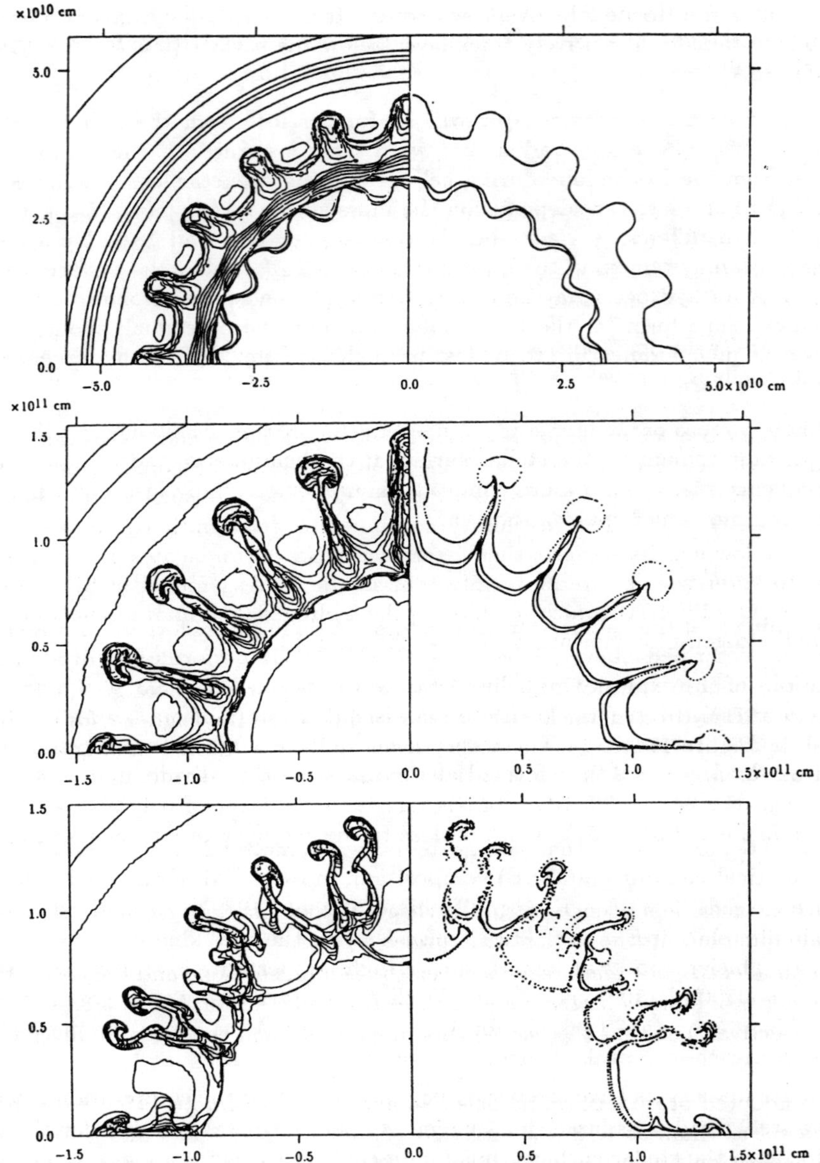

Fig. 17: Rayleigh-Taylor instabilities in the exploding helium stars of $M_\alpha = 6\ M_\odot$ (t = 84 s) (a), 4 $M_\odot$ (t = 200 s) (b), and 3.3 $M_\odot$ (t = 180 s) (c). Shown are the density contour map (*left*) and the marker particles at the composition interfaces, He/C+O, O/Si, and Ni/Si from the outerside (*right*) (Hachisu *et al.* 1991).

are 0.21 and 0.43 $M_\odot$ for $M_\alpha = 3.3$ and 4 $M_\odot$, respectively. These masses could be consistent with those inferred from the late time spectra of SNe Ib/Ic in view of the strong dependence of the oxygen mass on the temperature of the ejecta (e.g., Uomoto 1986).

5.3. *Rayleigh-Taylor Instabilities and Mixing*

As will be shown in §5.4, the helium star models are in good agreement with the observed SNe Ib light curves only if the extensive mixing of ^{56}Ni takes place. Mixing and clumpiness in SNe Ib are also inferred from the late time emission line features (Fransson and Chevalier 1989; Filippenko and Sargent 1989).

Rayleigh-Taylor instabilities in helium stars develop as follows. When the shock wave hits the helium envelope, the expansion of the inner core is largely decelerated, which forms a reverse shock. Then a pressure inversion appears (i.e., the pressure increases outward) in the layer between the forward shock and the reverse shock. The interface between the core and the helium envelope becomes most strongly Rayleigh-Taylor unstable because the density decreases outward steeply and thus $(dP/dr)(d\rho/dr) < 0$. The instability continues to grow until the forward shock reaches the low density surface; then a rarefaction wave propagates inward from the surface to stabilize the interior. Note that in the 20 $M_\odot$ model of SN 1987A by Hachisu *et al.* (1990), the most unstable is the hydrogen/helium interface due to the massive hydrogen-rich envelope as will be shown in §3.

For the helium stars of $M_\alpha = 3.3$, 4, and 6 $M_\odot$, Hachisu *et al.* (1991) have carried out 2D hydrodynamical calculations to follow the Rayleigh-Taylor instability. As seen in Figure 17a–c, the instability leads to only a limited mixing and clump formation for $M_\alpha = 6$ $M_\odot$, while it does induce a large scale mixing for $M_\alpha = 3.3$ and 4 $M_\odot$. For $M_\alpha = 3.3$ $M_\odot$, ^{56}Ni is mixed to the layer of 0.4 $M_\odot$ beneath the surface (Hachisu *et al.* 1991), which is close to the extent of mixing as required from the light curves.

Such a mass dependence of the Rayleigh-Taylor instability can be understood from the difference in the stellar structure as follows.
(1) For smaller M_α the mass ratio between the helium envelope and the core (excluding the neutron star) is larger (i.e., 2.5, 2.7, 1.0, and 0.45 for $M_\alpha = 3.3$, 4, 6, and 8 $M_\odot$, respectively) so that the deceleration of the core and the pressure inversion are larger.
(2) Smaller mass stars have steeper density gradient near the composition interface.
(3) The stellar radius is larger for smaller M_α, so that it takes longer for the shock wave to reach the stellar surface; thus the instability grows for a longer time.

We should emphasize the importance of the density structure rather than stellar mass. For example, a single Wolf-Rayet star which reduces its mass down to 4–5 $M_\odot$ by wind could be a SNe Ib/Ic progenitor (Langer 1989). However, such a star would not undergo extensive mixing despite the small mass, because its helium envelope would be too small to largely decelerate the core.

5.4. *Light Curves*

Figure 10 shows the observed bolometric light curves of SNe Ia 1972E and 1981B (Graham 1987), SN Ib 1983N (Panagia 1987), and the approximate bolometric light curve of SN Ic 1987M constructed from flux-calibrated spectra (Filippenko *et al.* 1990; Nomoto *et al.* 1990). In each case, the observed light curve has been shifted along the abscissa to match the corresponding theoretical curve. The peak bolometric luminosities assume $H_0 = 60$ km s^{-1} Mpc^{-1}.

The previous Wolf-Rayet star models have some difficulties (1) in reproducing the light curves of typical SNe Ib which decline as fast as SNe Ia (Panagia 1987), and (2) in producing enough ^{56}Ni to attain the maximum luminosities of SNe Ib in relatively low mass helium star models (Ensman and Woosley 1988). In particular, Figure 10 demonstrates an important feature of SN Ic 1987M, i.e., its brightness fell somewhat more rapidly than that of SNe Ia and SN Ib 1983N (also SN Ic 1983I reported by Tsvetkov 1985). Maximum brightness of 1987M is not significantly different from those of SNe Ib if we take the extinction estimated by Jeffery *et al.* (1991).

Figure 10 also shows the calculated bolometric light curves of the exploding helium star models with $M_\alpha = 3.3\ M_\odot$ for SN Ic and 4 $M_\odot$ for SN Ib as well as the white dwarf model W7 for SNe Ia (Nomoto *et al.* 1984). The amount of ^{56}Ni is 0.58 $M_\odot$ (W7), 0.15 $M_\odot$ (SN Ic), and 0.15 $M_\odot$ (SN Ib). The helium star models assume uniformly mixed distribution of elements from the center through the layer at 0.2 $M_\odot$ beneath the surface for both cases. Such a mixing may be due to the Rayleigh-Taylor instability during the explosion (Fig. 17; Hachisu *et al.* 1991).

The calculated bolometric light curves of helium stars are powered by the radioactive decays of ^{56}Ni and ^{56}Co (Fig. 9). Maximum brightness is higher if the ^{56}Ni mass is larger and the date of maximum earlier. After the peak, the optical light curve declines at a rate that depends on how fast γ-rays from the radioactive decays escape from the star without being thermalized, thereby declining faster if the ejected mass is smaller and if ^{56}Ni is mixed closer to the surface.

The resulted bolometric light curves of $M_\alpha = 3.3\ M_\odot$ and 4 $M_\odot$ are in good agreement with SN Ic 1987M and SN Ib 1983N, respectively. Compared with the 4 $M_\odot$ model, the light curve of the 3.3 $M_\odot$ model declines faster due to the smaller ejected mass, just as observed in SN Ic 1987M.

6. EVOLUTIONARY ORIGIN OF BINARY PULSARS AND SN IB/IC

The low mass helium stars considered for the progenitors of SNe Ib/Ic would mostly occur in close binaries, because the 12 – 18 $M_\odot$ stars would not lose their entire hydrogen-rich envelope by wind mass loss. Meurs and van den Heuvel (1989) predicted that more than 70 percent of massive star explosions would occur in close binaries. This estimate predicts that the occurrence frequencies of SNe Ib/Ic are higher than SN II, which might be consistent with the increasing number of SNe Ic recently discovered.

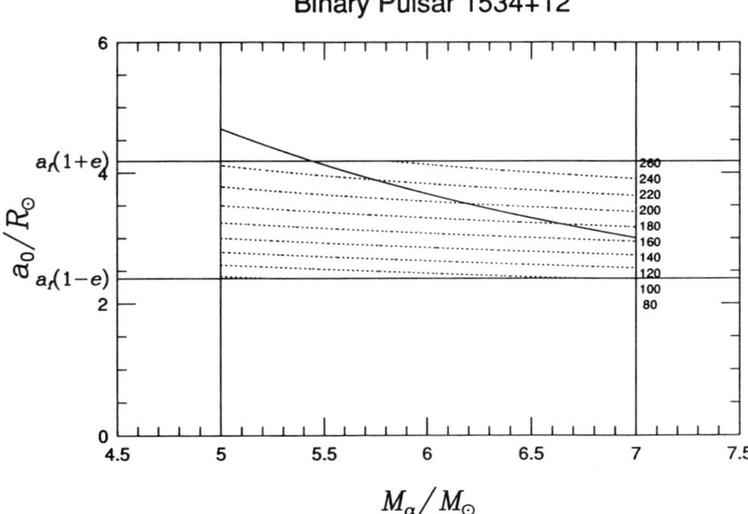

Fig. 18: Kick velocities imparted to the neutron star at the explosion of the helium star 2 of mass M_α as a function of M_α and the initial orbital radius a_0 before the explosion (Yamaoka *et al.* 1991). Here $a_f(1-e) < a_0 < a_f(1+e)$. For the solid line, the radius of the helium star is equal to its Roche lobe radius, so that only the upper-right part of the parameter space is allowed. For PSR 1534+12, $v_{\mathrm{kick}} \approx 170 - 260$ km s^{-1} is necessary to avoid the disruption of the binary system.

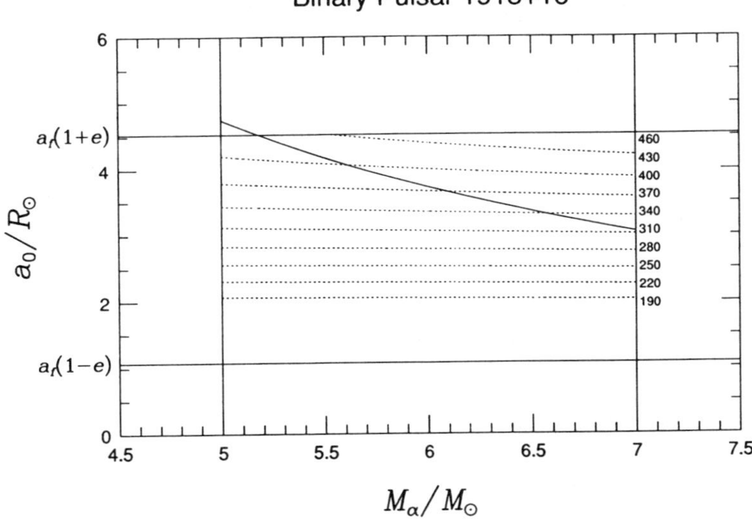

Fig. 19: Same as Fig. 18 but for PSR 1913 + 16, where $v_{\mathrm{kick}} \approx 320 - 460$ km s^{-1} is necessary (Yamaoka *et al.* 1991).

The binary scenario suggests that SNe Ib/Ic might be closely related to the formation of binary pulsars and X-ray binaries. If the binary system is not disrupted by the supernova mass ejection, a neutron star is left to orbit around a various types of a companion star (a main-sequence star or a helium star). Many of them would become Be X-ray binaries.

Recently the masses of the component stars in the binary pulsar system 1534+12 has been determined, which strongly suggests a neutron star companion (Wolszczan 1991). Their masses, eccentricity, semi-major axis, and orbital period are summarized in Table 1 together with those of the first binary pulsar 1913+16 (Taylor 1991). A possible evolutionary scenario for such systems are as follows (e.g., van den Heuvel 1991): (1) Two main-sequence stars 1 and 2, (2) Roche lobe overflow of star 1, which becomes a helium star 1, (3) the first supernova explosion of the helium star 1 to form a neutron star 1, (4) Roche lobe overflow of star 2, which leads to a spiral-in of the neutron star 1 into star 2 and thus to a considerable shrink of the system due to the losses of angular momentum and mass from the system; the system now consists of the recycled neutron star 1 and a helium star 2, (5) the second supernova explosion of the helium star 2; this forms a two neutron star system in an eccentric orbit.

Given the observed orbital parameters in Table 1 and the assumption of a circular orbit for the pre-explosion helium star 2–neutron star 1 system, the mass of the helium star 2 M_α and the possible kick velocity $v_{\rm kick}$ at the explosion can be calculated. If the explosion is spherical (i.e., $v_{\rm kick} = 0$), $M_\alpha \approx 2.1\ M_\odot$ (Wolszczan 1991). This is smaller than the minimum mass of the helium star that can form a neutron star [$\sim 2.5\ M_\odot$ (Nomoto 1984a) – $\sim 2.2\ M_\odot$ (Habets 1986) depending on the treatment of overshooting and semi-convection]. This suggests that either the explosion is not spherical or the exploding star had lost even its helium layer before the explosion.

If we introduce a finite $v_{\rm kick}$ for the explosion of the helium star 2, the kick velocity and its direction can be calculated as functions of assumed M_α and the initial orbital radius a_0 before the explosion where $a_{\rm f}(1-e) < a_0 < a_{\rm f}(1+e)$ (Fig. 18; Yamaoka *et al.* 1991). It should be noted that smaller helium stars have larger radii at the collapse which is $\sim 3\ R_\odot$ for $M_\alpha \approx 3.3\ M_\odot$ (Nomoto and Hashimoto 1988). For the solid line in Figures 18 and 19, the radius of helium star is equal to its Roche lobe radius. As far as the star 2 is a helium star, therefore, M_α should be larger than 5 $M_\odot$ to underfill the Roche lobe. If this is the case, a kick velocity of $v_{\rm kick} \approx 170 - 260$ km s^{-1} is necessary to avoid the disruption of the binary system. The same relation is obtained for 1913 + 16 in Figure 19, where $v_{\rm kick} \approx 320 - 460$ km s^{-1} is necessary (see also Burrows and Woosley 1986).

Two possible extreme cases are: (1) the star 2 is a helium star of more massive than $\sim 5\ M_\odot$, for which the explosion produces a large kick velocity, and (2) the star 2, being initially a helium star of smaller than 5 $M_\odot$, loses its helium envelope to become an almost bare C+O star. The masses of C+O stars are: 6.0, 3.8, 2.1, and 1.8 $M_\odot$ for $M_\alpha = 8, 6, 4, 3.3\ M_\odot$, respectively (Nomoto and Hashimoto 1988). For $M_\alpha \lesssim 4\ M_\odot$, therefore, the explosion of star 2 could be spherical with $v_{\rm kick} = 0$.

If the main difference between SNe Ib and Ic are the absence and presence of hydrogen-rich envelope, the first explosion could be SN Ic because of possible presence of hydrogen, while the second explosion might be SN Ib if star 2 loses even its helium envelope. These binary scenario also suggests that SNe Ic occur more frequently than SNe Ib.

Table 1. Binary Pulsars

PSR	M_p ($M_\odot$)	M_c ($M_\odot$)	e	a_f ($R_\odot$)	P_b (hours)	$P_b/\dot{P}_b$ (years)
1913+16	1.4421±0.0012	1.3875±0.0012	0.617127	2.8	7.752	3×10^8
1534+12	1.32±0.03	1.36±0.03	0.273676	3.28	10.10	2.4×10^8

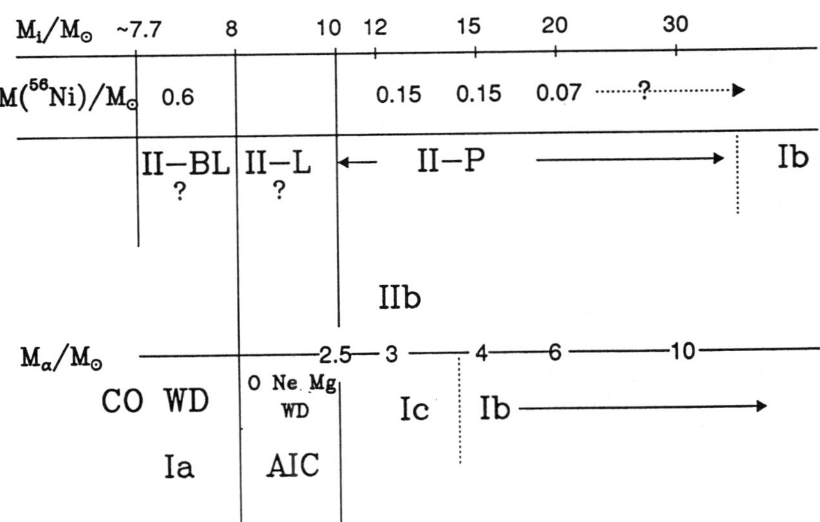

Fig. 20: Hypothetical connection between supernova types and their progenitors for single stars (upper) and close binary stars (lower). M_i and M_α are the initial mass and the helium star mass, respectively. AIC stands for accretion-induced collapse of white dwarfs.

7. Concluding Remarks

Figure 20 summarizes the initial masses M_i of the progenitors for the various types of supernovae currently proposed by several groups (e.g., Branch *et al.* 1991).

The upper and lower rows respectively show the cases of single stars and helium stars of masses M_α (or white dwarfs) in close binary stars. The produced masses of ^{56}Ni are inferred from light curves based on the radioactive decay model.

Single stars more massive than $\sim 8\ M_\odot$ would retain their hydrogen-rich envelope, thereby ending up their lives as SNe II. Among them, SNe II-BL and SNe II-L are tentatively assumed to be the explosions of AGB stars having degenerate C+O cores (carbon deflagration) and O+Ne+Mg cores (electron capture induced collapse), respectively (Swartz *et al.* 1991). Interacting binaries are the likely progenitors of SNe I as has been discussed in the preceding chapters. It must be emphasized that the supernova types vs. progenitor's mass relation presented in Figure 14 is still highly hypothetical and will be tested by future observations of light curve tails and good spectra (Branch *et al.* 1991).

We would like to thank M. Hashimoto, F.-K. Thielemann, A.V. Filippenko, D. Branch, D. Jeffery, and Y. Kondo for collaborative work on the subjects discussed in this paper. This work has been supported in part by the grant-in-Aid for Scientific Research (01540216, 01790169, 02234202, 02302024) of the Ministry of Education, Science, and Culture in Japan, and by the Japan-U.S. Cooperative Science Program (EPAR-071/88-15999) operated by the JSPS and the NSF.

REFERENCES

Axelrod, T. 1980, in *Type I Supernovae*, ed. J.C. Wheeler (Univ. of Texas), p. 80.

Bailyn, C.D., and Grindlay, J.E. 1990, *Ap. J.*, **353**, 159.

Barrat, J.L., Hansen, J.P., and Mochkovitch, R. 1988, *Astr. Ap.*, **199**, L15.

Benz, W., Bowers, R.L., Cameron, A.G.W., and Press, W. 1990, *Ap. J.*, **348**, 647.

Branch, D., Doggett, J.B., Nomoto, K., and Thielemann, F.-K. 1985, *Ap. J.*, **294**, 619.

Branch, D., Nomoto, K., and Filippenko, A.V. 1991, *Comments on Astrophysics.*

Burrows, A., and Woosley, S.E. 1986, *Ap. J.*, **308**, 680.

Canal, R., Garcia, D., Isern, J., and Labay, J. 1990b, *Ap. J. (Letters)*, **356**, L51.

Canal, R., Isern, J., and Labay, J. 1980, *Ap. J. (Letters)*, **241**, L33.

Canal, R., Isern, J., and Labay, J. 1990a, *Ann. Rev. Astr. Ap.*, **28**, 183.

Chanmugam, G., and Brecher, K. 1987, *Nature*, **329**, 696.

Ensman, L., and Woosley, S.E. 1988, *Ap. J.*, **333**, 754.

Filippenko, A.V. 1988, *A. J.*, **96**, 1941.

Filippenko, A.V. 1991, *IAU Cir.* No. 5169.

Filippenko, A.V., Porter, A.C., and Sargent, W.L.W. 1990, *A. J.*, **100**, 1575.

Filippenko, A.V., and Sargent, W.L.W. 1989, *Ap. J. (Letters)*, **345**, L43.

Fransson, C., and Chevalier, R.A. 1989, *Ap. J.*, **343**, 323.

Graham, J.R. 1987, *Ap. J.*, **315**, 588.

Habets, G.M.H.J. 1986, *Astr. Ap.*, **167**, 61.

Hachisu, I., Kato, M., and Saio, H. 1989, *Ap. J. (Letters)*, **342**, L19.

Hachisu, I., Matsuda, T., Nomoto, K., and Shigeyama, T. 1990, *Ap. J. (Letters)*, **358**, L57.
Hachisu, I., Matsuda, T., Nomoto, K., and Shigeyama, T. 1991, *Ap. J. (Letters)*, **368**, L27.
Harkness, R.P. 1991, in *SN 1987A and Other Supernovae*, ed. I.J. Danziger (Garching: ESO), in press.
Harkness, R.P., and Wheeler, J.C. 1990, in *Supernovae*, ed. A. Petschek (Springer-Verlag), p. 1.
Hashimoto, M., Nomoto, K., Arai, K., and Kaminisi, K. 1986, *Ap. J.*, **307**, 687.
Hashimoto, M., Nomoto, K., and Shigeyama, T. 1989, *Astr. Ap.*, **210**, L5.
Hoyle, F., and Fowler, W.A. 1960, *Ap. J.*, **132**, 565.
Iben, I.Jr., and Tutukov, A. 1984, *Ap. J. Suppl.*, **55**, 335.
Ichimaru, S., Iyetomi, H., and Ogata, S. 1988, *Ap. J. (Letters)*, **334**, L17.
Isern, J., Labay, J., Hernanz, M., and Canal, R. 1983, *Ap. J.*, **273**, 320.
Jeffery, D., Branch, D., Filippenko, A.V., and Nomoto, K. 1991, *Ap. J.*, submitted.
Kato, M., and Hachisu, I. 1989, *Ap. J.*, **346**, 424.
Kohkhlov, A.M. 1991, *Astr. Ap.*, in press.
Kumagai, S., Shigeyama, T., Nomoto, K., Itoh, M., Nishimura, J., and Tsuruta, S. 1989, *Ap. J.*, **345**, 412.
Kulkarni, S.R., and Narayan, R. 1988, *Ap. J.*, **335**, 755.
Kulkarni, S.R., Narayan, R., and Romani, R.W. 1990, *Ap. J.*, **356**, 174.
Langer, N. 1989, *Astr. Ap.*, **220**, 135.
Leibundgut, B., Kirshner R.P., Filippenko, A.V., Shields, J.C., Foltz, C.B., Phillips, M.M., and Sonneborn, G. 1991, *Ap. J. (Letters)*, **371**, L23.
Meurs, E.J.A., and van den Heuvel, E.P.J. 1989, *Astr. Ap.*, **226**, 88.
Michel, F.C. 1987, *Nature*, **329**, 310.
Miyaji, S., and Nomoto, K. 1987, *Ap. J.*, **318**, 307.
Miyaji, S., Nomoto, K., Yokoi, K., and Sugimoto, D. 1980, *Pub. Astr. Soc. Japan*, **32**, 303.
Mochkovitch, R. 1984, *Problems of Collapse and Numerical Relativity*, ed. D. Bancel and M. Signore (Dordrecht: Reidel), p. 125.
Mochkovitch, R., and Livio, M. 1990, *Astr. Ap.*, **236**, 378.
Narayan, R., Fruchter, A.S., Kulkarni, S.R., and Romani, R.W. 1990, in *Proc. 11th North American Workshop on CVs and LMXRBs*, ed. C. Mauche, in press.
Nariai, K., and Nomoto, K. 1979, in *IAU Colloquium 53, White Dwarfs and Variable Degenerate Stars*, ed. H.M. Van Horn and V. Weidemann (Rochester: Univ. of Rochester), p.525.
Nariai, K., Nomoto, K., and Sugimoto, D. 1980, *Pub. Astr. Soc. Japan*, **32**, 473.
Nomoto, H. 1989, *Exploring SN 1987A* (in Japanese), (Tokyo: Kodansha).
Nomoto, K. 1982a, *Ap. J.*, **253**, 798.
Nomoto, K. 1982b, *Ap. J.*, **257**, 780.
Nomoto, K. 1984a, *Ap. J.*, **277**, 791.
Nomoto, K. 1984b, *Problems of Collapse and Numerical Relativity*, ed. D. Bancel and M. Signore (Dordrecht: Reidel), p. 89.
Nomoto, K. 1986, *Prog. Part. Nucl. Phys.*, **17**, 249.

Nomoto, K. 1987a, *Ap. J.*, **322**, 206.

Nomoto, K. 1987b, in *IAU Symposium 125, The Origin and Evolution of Neutron Stars*, ed. D.J. Helfand and J.-H. Huang (Dordrecht: Reidel), p. 281.

Nomoto, K., Filippenko, A.V., and Shigeyama, T. 1990, *Astr. Ap.*, **240**, L1.

Nomoto, K., and Hashimoto, M. 1988, *Physics Reports*, **163**, 13.

Nomoto, K., and Iben, I.Jr. 1985, *Ap. J.*, **297**, 531.

Nomoto, K., and Kondo, Y. 1991, *Ap. J. (Letters)*, **367**, L19.

Nomoto, K., Miyaji, S., Sugimoto, D., and Yokoi, K. 1979a, in *IAU Colloquium 53, White Dwarfs and Variable Degenerate Stars*, ed. H.M. Van Horn and V. Weidemann (Rochester: Univ. of Rochester), p.56.

Nomoto, K., Nariai, K., and Sugimoto, D. 1979b, *Pub. Astr. Soc. Japan*, **31**, 287.

Nomoto, K., and Sugimoto, D. 1977, *Pub. Astr. Soc. Japan*, **29**, 765.

Nomoto, K., Sugimoto, D., and Neo, S. 1976, *Ap. Space Sci.*, **39**, L37.

Nomoto, K., Thielemann, F.-K., and Yokoi, K. 1984, *Ap. J.*, **286**, 644.

Ogata, S., Iyetomi, H., and Ichimaru, S. 1991, *Ap. J.*, in press.

Panagia, N. 1987, in *High Energy Phenomena Around Collapsed Stars*, ed. F. Pacini (D. Reidel), p. 33.

Ray, A., and Kluzniak, W. 1990, *Nature*, **344**, 415.

Romani, R.W. 1990, *Ap. J.*, **357**, 493.

Saio, H. and Nomoto, K. 1985, *Astr. Ap.*, **150**, L21.

Shigeyama, T., and Nomoto, K. 1990, *Ap. J.*, **360**, 242.

Shigeyama, T., and Nomoto, K. 1991, in preparation.

Shigeyama, T., Nomoto, K., and Hashimoto, M. 1988, *Astr. Ap.*, **196**, 141.

Shigeyama, T., Nomoto, K., Tsujimoto, T., and Hashimoto, M. 1990, *Ap. J. (Letters)*, **361**, L23.

Swartz, D.A., Wheeler, J.C., and Harkness, R.P., 1991, *Ap. J.*, in press.

Tavani, M. 1991, *Ap. J.*, in press.

Taylor, J. 1991, in *NATO ARW, X-Ray Binaries and the Formation of Binary and Millisecond Radio Pulsars*, ed. E.P.J. van den Heuvel (Kluwer), in press.

Thielemann, F.-K., Hashimoto, M., and Nomoto, K. 1990, *Ap. J.*, **349**, 222.

Thielemann, F.-K., Nomoto, K., and Yokoi, K. 1986, *Astr. Ap.*, **158**, 17.

Tsvetkov, D.Yu. 1985, *Sov. Astr.*, **29**, 211.

Uomoto, A. 1986, *Ap. J. (Letters)*, **310**, L35.

van den Heuvel, E.P.J. 1991, in this volume.

Webbink, R. 1984, *Ap. J.*, **277**, 355.

Wheeler, J.C., and Harkness, R. 1990, *Rep. Prog. Phys.*, **53**, 1467.

Wolszczan, A. 1991, *Nature*, submitted.

Woosley, S.E. 1991, in *Supernovae* (Les Houches, Session LIV), ed. J. Audouze et al. (Elsevier Sci. Publ.), in press.

Woosley, S.E., Taam, R.E., and Weaver, T.A. 1986, *Ap. J.*, **301**, 601.

Woosley, S.E., and Weaver, T.A. 1986a, *Lecture Notes in Physics*, **255**, 91.

Woosley, S.E., and Weaver, T.A. 1986b, *Ann. Rev. Astr. Ap.*, **24**, 205.

Yamaoka, H., and Nomoto, K. 1991, in *SN 1987A and Other Supernovae*, ed. I.J. Danziger (Garching: ESO), in press.

Yamaoka, H., Shigeyama, T., and Nomoto, K. 1991, in preparation.

EVOLUTION OF CLOSE BINARIES AND THE FORMATION OF MILLISECOND RADIO PULSARS

E.P.J. VAN DEN HEUVEL
Institute for Theoretical Physics
University of California
Santa Barbara, CA 93106
and
Astronomical Institute
& Center for High Energy Astrophysics
University of Amsterdam
The Netherlands

ABSTRACT. All millisecond pulsars share the special characteristics of the binary radio pulsars (on average much weaker magnetic fields and much faster spin than the general radio pulsar population), and a large fraction ($\sim$ 40 percent) of millisecond pulsars are found in binaries.

All observational evidence indicates that the millisecond and binary pulsars are relatively old neutron stars that have been spun-up by accretion in binary systems (so-called "recycling").

The two classes of binary radio pulsars (PSR 1913+16 class and PSR 1953+29 class) appear to be the descendants of wide High Mass X-ray Binaries and wide Low-Mass X-ray Binaries, respectively. Low-Mass X-ray Binaries with short orbital periods ($< 0,5d$) are expected to turn into millisecond binary radio pulsars which destroy (evaporate) their low-mass companions, presumably leaving a single millisecond pulsar in most cases.

There is no longer clear evidence that magnetic fields of isolated neutron stars do decay spontaneously. Only neutron stars that have been recycled in binaries show clear evidence for magnetic field decay. These recycled pulsars are expected to highly contaminate (on the 50 percent level or more) the general population of single pulsars in the Galaxy – creating the (probably false) impression that the magnetic fields of isolated neutron stars decay on a relatively short timescale ($\sim 10^7$ yrs).

1. Introduction

Millisecond pulsars and binary radio pulsars have many characteristics in common, which set them apart from the bulk of the "normal" radio pulsars, and suggest that they form a separate class with a common evolutionary origin. Characteristics which they share are: (i) a very high percentage of binaries (for the binary pulsars, of course, 100%); (ii) much shorter pulse periods and (iii) much weaker surface magnetic field strengths B_s, than the bulk of the radio pulsars. Table 1 and Figure 1 illustrate these points, which can be expressed in quantitative terms as follows: – only about 3 per cent of all known

J. Ventura and D. Pines (eds.), Neutron Stars: Theory and Observation, 171–196.

(~520) pulsars are found in binaries, whereas among the pulsars with pulse periods in the millisecond range ($< 12ms$) some 40% are found in binaries (in the galactic disk this percentage is ~75%, in globular clusters ~25%);

- about half of all known binary pulsars have $P < 12ms$, whereas 97 per cent of all pulsars have $P > 30ms$;
- most binary pulsars and all millisecond pulsars have $B_s < 4.10^{10}G$, whereas 96 per cent of all known pulsars have $B_s > 3.10^{11}G$.

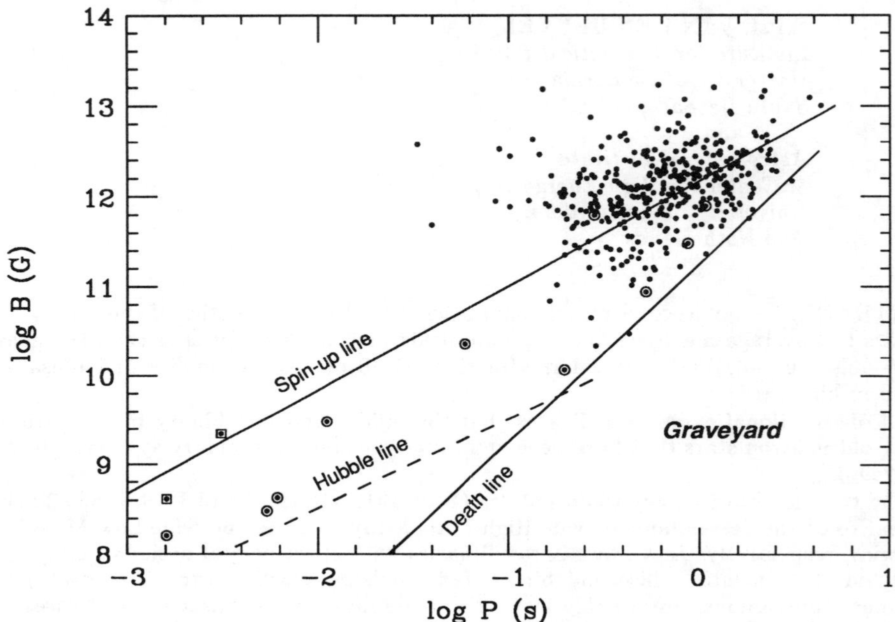

Figure 1. The derived magnetic fields ($\propto \sqrt{P\dot{P}}$) of 403 radio pulsars plotted against their spin periods. Pulsars in binaries are shown encircled and single millisecond pulsars are shown as squares. The "death line" corresponds to a polar cap voltage below which the pulsar activity is likely to switch off. The "Hubble line" represents a spindown age $\tau \equiv P/2\dot{P}$ of 10^{10} yr for a pulsar. The spin-up line represents the minimum spin period to which spin-up may proceed in an Eddington-limited accretion (see text).

Because of these great similarities between the binary and millisecond pulsars we will concentrate in this review on the broader topic of: the origin and evolution of binary and millisecond radio pulsars. In Section 2 we briefly describe the observed characteristics of the two classes of binary systems that contain accreting neutron stars, the High Mass and the Low Mass X-ray Binaries (which we will abbreviate throughout this paper as HMXB and LMXB), and outline how these systems may be related to the binary radio pulsars. After this global introduction we review in Section 3 how the HMXB and LMXB are expected to evolve through the X-ray binary phase and beyond, and the various possible ways in which binary and millisecond pulsars may be produced by this evolution. Also

TABLE 1

Binary Pulsars, Millisecond Pulsars and Pulsars in Globular Clusters (as known on January 31, 1991) (Detailed source references to these pulsars are given in Bhattacharya and van den Heuvel 1991 and Lyne 1991).

PSR	P ms	log $\dot{P}$	log B_s (G)	P_{orb} days	e	$f(M)$ $M_\odot$	M_c $M_\odot$	location
0021-72C	5.76			-	-	-	-	GC 47Tuc
0021-72D	5.36			-	-	-	-	GC 47Tuc
0021-72E	3.536			~ 2				GC 47Tuc
0021-72F	2.624			single				GC 47Tuc
0021-72G	4.040			single				GC 47Tuc
0021-72H	3.210			$\sim$ (?)				GC 47Tuc
0021-72I	3.485			$\sim$ (?)				GC 47Tuc
0021-72J	2.101 or 4.201			0.12				GC 47Tuc
0021-72K	1.786			Binary				GC 47Tuc
0021-72L	4.346			?				GC 47Tuc
0655+64	195.67	-18.17	10.06	1.03	7.5×10^{-6}	0.0712	0.7-0.8	disk
0820+02	864.87	-15.98	11.48	1232.5	0.012	0.003	0.2-0.4	disk
1257 +12	6.22	-	-	-	-	-	-	disk
1310+18	33.16			255				GC M53
1516+02A	5.55			-	-	-	-	GC M5
1516+02B	7.95			6.9				GC M5
1534+12	37.90	-17.61	9.98	0.42	0.2737	0.3146	1.36	disk
1620-26	11.08	-18.09	9.48	191.4	0.025	0.008	0.3-0.6	GC M4
1639+36A	10.38			-	-	-	-	GC M13
1639+36B	30			1.25				GC M13
1744-24A*	11.56	< -19.4	< 8.83	0.075	$< 10^{-3}$	3.2×10^{-4}	> 0.1	GC Ter5
1746-20	288.62			-	-	-	-	GC NGC6440
1802-07	23.10			2.6	0.22	0.0097	> 0.25	GC NGC6539
1820-11	279.83	-14.86	11.79	357.8	0.795	0.068	0.8-1.0	disk
1820-30A	5.44			-	-	-	-	GC NGC6624
1820-30B	378.61			-	-	-	-	GC NGC6624
1821-24	3.05	-17.81	9.34	-	-	-	-	GC M28
1831-00	520.95	-16.84	10.94	1.81	< 0.004	1.2×10^{-4}	0.06-0.13	disk
1855+09	5.36	-19.77	8.48	12.33	2.1×10^{-5}	0.0052	0.2-0.4	disk
1908+00	3.6							GC NGC6760
1913+16	59.03	-17.06	10.35	0.32	0.617	0.1322	1.4	disk
1937+21	1.56	-18.97	8.61	-	-	-	-	disk
1953+29	6.13	-19.53	8.63	117.35	3.3×10^{-4}	0.0027	0.2-0.4	disk
1957+20*	1.61	-19.79	8.21	0.38	$< 2 \times 10^{-5}$	5.2×10^{-6}	~ 0.01	disk
2127+11A	110.66	?[a]		-	-	-	-	GC M15
2127+11B	56.13	-17.05	10.35	-	-	-	-	GC M15
2127+11C	30.53			0.34	0.68	0.15	> 0.9	GC M15
2127+11D	4.65			-	-	-	-	GC M15
2127+11E	4.8			-	-	-	-	GC M15
2303+46	1066.37	-15.24	11.89	12.34	0.658	0.2463	~ 1.5	disk

Notes:

PSR0021-72A and PSR0021-72B are not thought to be located in 47Tuc because of their discordant dispersion measures.

*Companion in the process of evaporation.

[a]Observed $\dot{P}$ is negative, $-2 \times 10^{-17} ss^{-1}$. This apparent $\dot{P}$ is almost certainly due to a gravitational acceleration of the pulsar.

ways in which single millisecond pulsars may be produced are discussed. In Section 4 we discuss the implications of these findings for our understanding of the magnetic-field evolution of neutron stars. For more details we refer to the recent review by Bhattacharya and van den Heuvel (1991).

2. From X-ray Binaries To Binary Radio Pulsars: The "Recycling" Picture

2.1. THE TWO CLASSES OF BINARY X-RAY SOURCES

The X-ray binaries can, broadly speaking, be divided into two groups, the HMXB and LMXB, which differ in a number of important characteristics, as follows (see Table 2): (i) In the HMXB the companion of the X-ray source is a luminous early-type star (spectral type O or B), with a mass typically in the range 10–40 $M_\odot$; in the LMXB it is a faint star of mass $< 1,2 M_\odot$. (ii) In space the HMXB show a strong concentration towards the galactic plane , but not towards the galactic center, as is typical for a very young stellar population. This fits perfectly with the fact that stars with masses $> 10 M_\odot$ do not live for more than $(1-2).10^7$ years. Hence, the HMXB and their neutron stars must have ages less than this value. On the other hand, the LMXB show a strong concentration towards the galctic center, and a much broader distribution around the galactic plane, as is typical for an old stellar population, with age $> (,5-1,5).10^{10}$ years. Indeed, a sizeable number (eleven) of the LMXB are found in globular star clusters, the oldest objects in our Galaxy (ages $(1,5-2).10^{10}$ years). (iii) Each group has its own characteristic type of time variability. In most HMXB the X-ray source is a regular X-ray pulsar, indicating that the neutron star has a strong surface dipole magnetic field (strength $10^{11,5}$ to $10^{13,5} G$), as is supported by the detection of cyclotron features in the spectra of many of these sources (*cf.* Nagase 1989a). On the other hand, very few LMXB are regular X-ray pulsars, but many of them show thermonuclear X-ray flashes (so-called type I X-ray bursts)

TABLE 2
The two groups of strong Galactic X-ray Sources.

Group I	Group II
Optical counterparts massive and luminous early type stars, spectrum O and early B; $L_{opt}/L_x > 1$.	Faint blue optical counterparts $L_{opt}/L_x < 0.1$
Concentrated in space towards the galactic plane: young stellar population, age$< 10^7$ years	Concentrated in space towards the galactic center; fairly wide spread around the galactic plane: old stellar population, age (5–15)$\times 10^9$ years
Type of time variability: Regular X-ray pulsations; no X-ray bursts	Type of time-variability: often X-ray bursts; only in 3 cases regular X-ray pulsations.
Relatively hard X-ray spectra: $kT \gtrsim 15$ keV	Softer X-ray spectra: $kT \lesssim 10$ keV

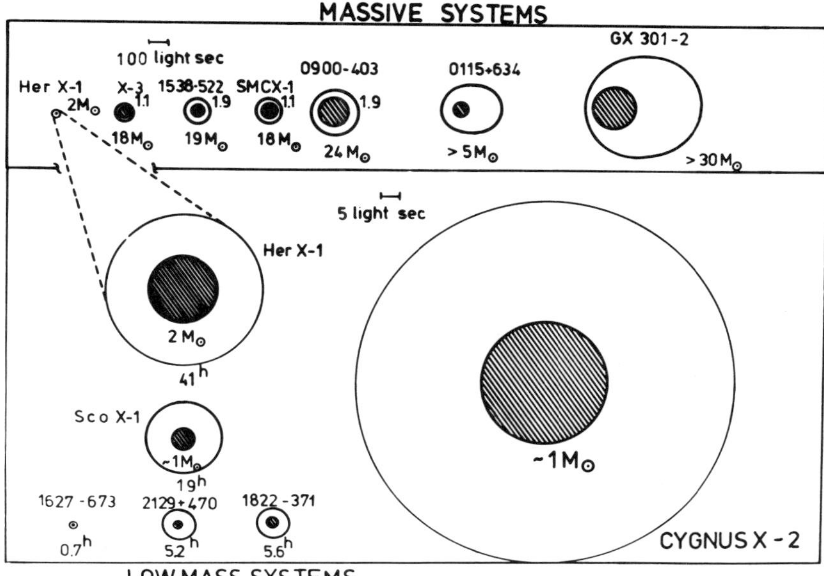

Figure 2. Sketch, to scale, of the orbits and companions (shaded) of a representative number of high-mass (a) and low-mass (b) X-ray binaries. The geometry is based on orbital period and an estimate of the companion mass, derived either from doppler tracking or from its optical characteristics. As indicated in the figure, Hercules X-1 can be considered a member of either group. In the two wide low-mass X-ray binaries the companion is an evolved star (low-mass giant).

of the matter accreted onto the neutron-star surface. As thermonuclear flashes are suppressed if the fieldstrength exceeds $\sim 10^{11}G$ (Lewin and Joss 1983), the surface magnetic fields of the neutron stars in these old systems are expected to be weaker than this value. That these systems contain indeed neutron stars has been demonstrated convincingly by the study of the time evolution of the spectra of X-ray bursts (see, for example, Lewin and Joss 1983). Figure 2 shows the orbits of a representative sample of HMXB and LMXB drawn to scale.

The above summarized characteristics of the two groups of binary X-ray sources seem, at first glance to fit well with the now twenty year old "paradigma" (Gunn and Ostriker 1970) that neutron stars are born with strong surface magnetic fields which decay spontaneously on a timescale of about 10^7 years. In this picture one would indeed expect the neutron stars in HMXB to still have strong magnetic fields $(10^{11} - 10^{13}G)$, whereas the neutron stars in old-population sources (LMXB) should have decayed to low values. However, as we will see later in this paper, this paradigma probably will have to be abandoned, as the evidence for spontaneous magnetic field decay has now become seriously challenged, and the weakness of the fields in LMXB is most probably related to the special history that the neutron stars in these binaries have experienced.

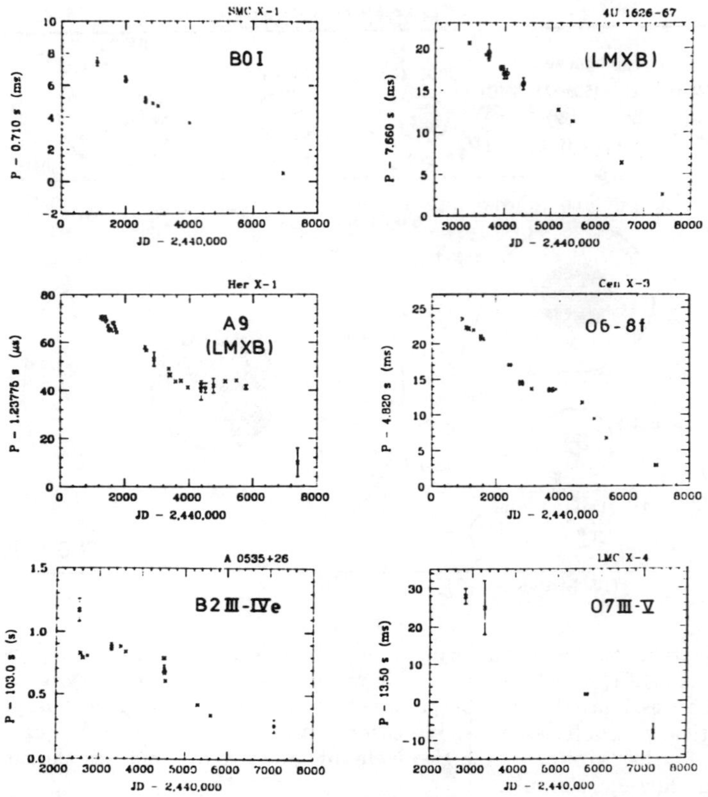

Figure 3. Pulse-period histories of five disk-fed binary X-ray pulsars and one transient X-ray pulsar with a B-emission companion (X0535-26) after Nagase (1989b). The disk-fed pulsars have relatively short pulse periods ($\approx$ 1sec) and show a more or less monotonic decrease in pulse period, so-called "spin up".

A further important observational characteristic of the pulsating sources which are accreting matter from a disk (*e.g.*: Her X-1, Cen X-3, SMC X-1) is that their pulsation periods gradually decrease with time. This so-called "spin-up" behaviour (see Figure 3, after Nagase, 1989b) is due to the fact that the accreted matter carries angular momentum which is fed to the neutron star and causes its rotation to be accellerated. The shortest possible pulse period that can be reached in this process, for a given magnetic field strength B_s and accretion rate $\dot{M}$, is given by the equilibrium spin period $P_{\rm eq}$, defined by (see for example Van den Heuvel 1977, Henrichs 1983, Lamb 1989):

$$P_{\rm eq} = (0.9s) B_{12}^{6/7} M_n^{-5/7} (\dot{M}/\dot{M}_{\rm Edd})^{-3/7} R_6^{16/7} \tag{1}$$

where M_n, R_6 are the mass and radius of the neutron star, in units of a solar mass and of 10^6 cm, respectively, and B_{12} is the surface dipole magnetic field strength of the neutron star in units of $10^{12}G$. Since there is an upper limit to the accretion rate set by the Eddington luminosity: $M = 1,5 \times 10^{-8} M_{\odot}/\mathrm{yr}$, the shortest possible rotation period that can be reached depends, for a standard neutron star of $M_n = 1,0 M_{\odot}$, $R_6 = 1 (R = 10$ km), only on the magnetic field strength:

$$P_{\min} = (0,9s).(B_{12})^{6/7}.(\dot{M}_{\mathrm{acc}}/\dot{M}_{\mathrm{Edd}})^{-3/7} = (2,4ms)B_9^{6/7}.(\dot{M}_{\mathrm{acc}}/\dot{M}_{\mathrm{Edd}})^{-3/7} \quad (2)$$

where B_9 is the surface dipole magnetic field strength in units of 10^9G. This limiting relationship defines the "spin-up line" (Radhakrishnan an Srinivasan 1981, 1984; Alpar *et al.* 1982) in the B-P diagram of radio pulsars, and is indicated in Figure 1.

2.2. THE TWO TYPES OF BINARY RADIO PULSARS

Like the binary X-ray sources, also the binary radio pulsars fall into two groups with distinctly different characteristics, as depicted in Figure 4 and quantitatively indicated in Table 1, as follows(van den Heuvel and Taam 1984):

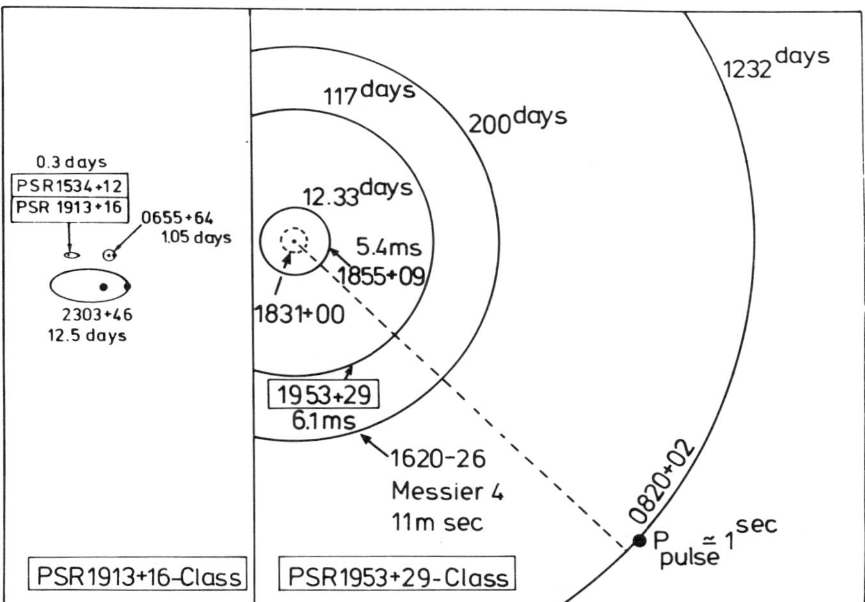

Figure 4. The two classes of binary radio pulsars. Left: The PSR 1913+16-class of systems tend to have narrow and very eccentric orbits; the companion of the pulsar is itself a neutron star or a massive white dwarf. Right: The PSR 1953+29-class systems tend to have wide and circular orbits; here the companion stars always have a low mass, in the range 0.2–0.4 $M_{\odot}$ or smaller – and most probably are helium white dwarfs (see text).

(i) The PSR 1913+16 group. These systems have, in general, narrow orbits and companion masses in the range 0,8 to 1,4 $M_\odot$. Five out of the six known systems of this type have (very) eccentric orbits, and four have orbital periods of the order of one day or less.

(ii) The PSR 1953+29 group. These systems have, on average, much wider orbits, which are circular, and their companion masses tend to be very low, between 0,02 and 0,3 $M_\odot$.

All evidence, from optical identifications or the absence thereof (the distances of the pulsars are known from their dispersion measures), indicates that the companions in both types of systems are themselves compact stars: in group (i) they are most probably neutron stars (in the eccentric-orbit systems) or massive white dwarfs (in the one system with a circular orbit: PSR 0655+64), while in group (ii) they are low-mass white dwarfs.

In fact, they cannot be anything but compact stars, since even a small amount of coronal or stellar wind matter, as is present around all normal stars, is expected to disperse radio pulsar signals from a neutron-star companion beyond recognition.

2.3. ORIGINS OF BINARY RADIO PULSARS: RECYCLING

How is a neutron star that was formed as a companion to a normal star expected to evolve? One expects that first its rotation will rapidly slow down, both as a result of the classical radio pulsar spin-down mechanisms (emission of magnetic dipole radiation and of a pulsar wind), and of torques excerted on the magnetosphere by stellar wind matter of the companion star (*e.g.* see Wang and Robertson 1985, Blondin *et al.* 1990). The neutron star is therefore expected to terminate in the graveyard in the B vs. P diagram within a few million years after it was born, as is indicated by the fully-drawn evolutionary tracks in Figure 5. It will remain in the graveyard region of the B vs. P diagram until its companion starts to overflow its Roche lobe and begins to transfer matter to the neutron star through an accretion disk. The system then has become an X-ray binary and, as shown by Figure 3, the spin period of its neutron star is expected to become shorter on a short timescale (10^3 to 10^4 years). During this spin-up phase, it moves back towards the left in the B-P diagram as indicated by the "spin-up" tracks in Figure 5. During all this time it will not be observed as a radio pulsar, but as an X-ray source. During its stay in the graveyard its magnetic field may or may not have decayed. If it has decayed only little, as is expected to be the case in many HMXB, the pulsar is spun up back to where it came from. If the field decayed substantially (and there is evidence that this has occurred in many binary pulsars, see below) it will be spun up back along a horizontal track that is located much lower in the B-P diagram, as is also indicated in Figure 5.

Only when its companion star has itself terminated its evolution and become a compact star (neutron star or white dwarf), the spun-up neutron star will become observable as a radio pulsar.

The possibility of the production of radio pulsars by the spinning up of an old(er) neutron star by accretion in a binary system was first suggested by Bisnovatyi-Kogan and Komberg (1975) and, independently, by Smarr and Blandford (1976). Radhakrishnan and Srinivasan (1981,1984) and Alpar *et al.* (1982) pointed out that one expects such "recycled" pulsars to be located in the B-P diagram of Figure 5 in the wedge-shaped region between the "spin-up" line and the deathline. Alpar *et al.* (1982), with much foresight, pointed out that if the field of a recycled pulsar does not decay any further, those that have weak surface fields ($< 5.10^9 G$) will take more than a Hubble time to reach the deathline. The "Hubble line" in Figure 5 indicates the position where these spun-up pulsars are located after one Hubble time (about 10^{10} years) of spindown, assuming that they started out on the spin-up line, and have moved to the right in the *B-P* diagram along horizontal tracks (constant *B*). Thus, the recycled pulsars are predicted to be found in the *B-P* diagram in the region bounded by the spin-up line, the deathline and the Hubble line. Figures 1 and 5 show that this prediction of the recycling model fits very well with the observations. With only one exception (PSR 1820-11) all binary radio pulsars are located in the predicted part of the diagram. This, together with the fact that, as mentioned above, all evidence indicates that the companions of the binary pulsars are themselves white dwarfs or neutron stars, leaves little doubt that (with the exception of PSR 1820-11) the observed binary radio pulsars all have been recycled in binaries. This means that indeed they must represent a later stage in the evolution of X-ray binaries. (The only exception, PSR1820-11), may be the only radio pulsar binary in which we are seeing the last-born (new) neutron star. On statistical grounds one would expect one such system to occur for about each dozen recycled ones, see below).

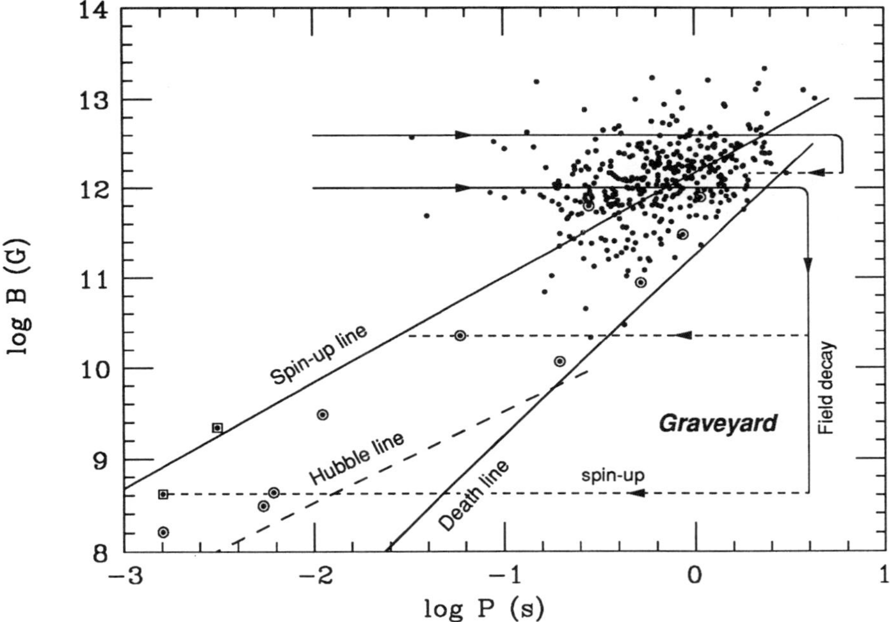

Figure 5. Same diagram as Figure 1 with possible evolutionary tracks of pulsars in binaries indicated. Pulsars are born in the upper left part of the diagram and – if no field decay occurs – move towards the right along horizontal tracks (fully drawn). In the graveyard the field of a single pulsar probably does not decay (see text). In a binary system, the field of pulsars in the graveyard does decay, presumably due to external circumstances (accretion, spin down). Furthermore, the accretion of matter with angular momentum causes the neutron star to be "spun up" towards the left in the diagram along the dashed "spin-up tracks". The shortest period to which a neutron star can be spun up depends on the value of the magnetic field strength: this is indicated by the spin-up line. Only after the companion has itself become a compact star (neutron star or white dwarf) or has disappeared, the spun-up neutron star becomes observable as a radio pulsar: It will then slowly spin-down, *i.e.*, move towards the right again in the diagram.

2.4. RECYCLING AND THE FORMATION OF MILLISECOND PULSARS

Since many of the millisecond radio pulsars ($\sim 40\%$) are in binaries, and since these are exactly found in the predicted region of recycled pulsars in the B-P diagram (Figure 5) it is clear that recycling can indeed produce millisecond pulsars. This is quite an achievement, since a lot of angular momentum has to be added if one whishes to spin a neutron star up to a period of 1,56 millisecond. At the magnetic field strength of $4.10^8 G$ of the $1,56ms$ pulsar this requires the accretion of about $0,15 M_\odot$ (Alpar *et al.* 1982). The accretion rate being limited by the Eddington limiting rate ($1,5.10^{-8} M_\odot/\mathrm{yr}$) this requires accretion for at least 10^7 years (unless a different process took place, such as sudden merging of a neutron star with a white dwarf companion, *cf.* Bonsema and van den Heuvel 1985, van den Heuvel and Bonsema 1984). Since also the single millisecond pulsars are located exactly in the region of the B vs. P diagram predicted for recycled pulsars, it seems most plausible that these as well have been produced by recycling, but have somehow lost their companions. The discovery of the two eclipsing binary millisecond pulsars PSR 1957+20 and PSR 1744-24A (in the globular cluster Ter 5, Lyne *et al.* 1990), in which the companion stars do show clear signs of being evaporated (ablated) under the influence of the irradiation by their pulsar companions shows that indeed in a close binary system a millisecond pulsar may, under certain conditions, be able to evaporate its companion star. In the 1957+20 system the companion has already reached a very low mass ($< 0.02 M_\odot$). Its measured timescale of orbital evolution ($\sim 10^7$ years, Fruchter 1990) shows that the companion will be totally evaporated in about 10^7 years. One thus expects that a single millisecond pulsar will be left, closely resembling the $1,56ms$ pulsar PSR 1937+21. We will now examine in more detail how high-mass and low-mass X-ray binaries may evolve into binary and millisecond radio pulsars.

3. The Final Evolution Of High-mass And Low-mass X-ray Binaries

3.1. STABLE VS. UNSTABLE MASS TRANSFER DRIVEN BY THE NUCLEAR EVOLUTION OF A COMPANION

If mass is transferred from the more massive to the less massive component of a binary, and system mass as well as orbital angular momentum are conserved, the orbit will shrink. In the opposite case, when mass is being transferred from the less massive to the more massive component, the orbit will expand. This can be easily seen from the equation for the orbital change due to mass transfer (see for example van den Heuvel 1983):

$$P/P_o = \left(\frac{(M_1^0.(M - M_1^0))}{(M_1.(M - M_1))}\right)^3 \tag{3}$$

where M is the total mass of the system, M_1 is the mass of one component and $M_2 = M - M_1$ is the mass of the other component. Superscript zero denotes the initial situation, and quantities without superscript indicate the present situation. Due to its interior nuclear evolution the radius of a star increases in the course of time (stars become giants). As a result of this a star in a binary will at a certain time begin to overflow its Roche lobe and start to transfer mass to its companion along the first Lagrangian point L_1. If the mass-losing star is the more massive component, the mass transfer will make the orbit and Roche lobe shrink, while its interior evolution is just causing it to expand. Clearly, this is a very unstable situation, and the star will continue to lose mass. The timescale for this mass transfer is its thermal timescale, and in some cases (stars with convective envelopes) it may become even as short as a dynamical timescale (*cf.* Savonije 1983a for details). The thermal timescale is given by (*cf.* Savonije 1983a, van den Heuvel 1983):

$$\tau_{th} \sim 3.10^7 (M_\odot/M)^2 \text{ years.} \tag{4}$$

For this reason, once in a high-mass X-ray binary the envelope of the companion star overflows the Roche lobe, very high mass transfer rates will ensue, of order 10^{-4} to $10^{-3} M_\odot$/yr. The observable X-ray phase occurs just prior to this phase of full-speed Roche-lobe overflow, and lasts for only a short time, of order 5.10^4 years (Savonije 1983a); it is the phase in which only the tenuous outer atmosphere, but not yet the envelope, overflows the Roche lobe. In this phase of "beginning atmospheric Roche-lobe overflow", the mass transfer rate is still below the Eddington-limit rate but is gradually increasing.

On the other hand, in the low-mass X-ray binaries the mass-losing star is less massive than the neutron star, and the mass transfer will be stable. By transferring some mass the orbit and Roche lobe will gradually widen such that the evolving star will at any time be able to adapt the size of the Roche lobe to its expanding radius.

3.2. THE FATE OF THE HIGH-MASS X-RAY BINARIES

The mass ratio of primary and secondary star in these systems is of the order of ten. Equation (3) shows that in this case after transferring only a few solar masses to the neutron star, the orbital period would have shrunk by orders of magnitude. Clearly, such a large reduction in orbital size is not possible without running into grave difficulties. The general conjecture is that in these systems, once the large-scale mass transfer (on a thermal or dynamical timescale) ensues, the neutron star will be engulfed by the envelope of the massive star, and spiral down into the resulting common envelope due to the large frictional drag which it experiences (Taam *et al.* 1978, see Taam and Bodenheimer, 1989). Two types of outcome of this process are possible:

(i) If the initial orbit is wide ($P > 200d$), the drop in orbital potential energy during the shrinking provides enough energy to expell the envelope, such that the neutron star and the evolved core of the massive star (consisting of helium and possibly heavier elements) remain, in a very tight orbit (P of order hours to days). When the evolved core in this system finally explodes as a supernova, the system may either remain bound, in a very eccentric orbit, or be disrupted. In the first case a system resembling the four binary radio pulsars with eccentric and relatively narrow orbits will remain. In the second case two runaway neutron stars will result, one of them new, and the other recycled. Figure 6 depicts these various outcomes schematically. A third possibility is that the core of the companion remaining after spiral in is just not massive enough to explode as a supernova. In that case it will leave behind a massive white dwarf in a narrow and circular orbit around the neutron star. Such a system resembles the binary radio pulsar PSR 0655+64 (*cf.* van den Heuvel and Taam 1964).

(ii) If the initial orbital period is short ($< 200d$) the neutron star will spiral down into the core of the massive star which then turns into a red supergiant, a so-called Thorne-Zytkow star (Thorne and Zytkow 1977, Biehle 1991, Cannon *et al.* 1991). This star will lose its envelope probably on a short timescale ($\sim 10^5$ years) and a single recycled neutron star will remain, probably hardly rotating any more. Since we are mainly interested here in the origins of the millisecond pulsars, we will not go here into the details of these types of evolution, but note that the class of binary pulsars with relatively tight eccentric orbits and relatively massive companion stars thus appears to be a natural endproduct of the evolution of HMXB with relatively wide orbits. For further details we refer to Bhattacharya and van den Heuvel (1991).

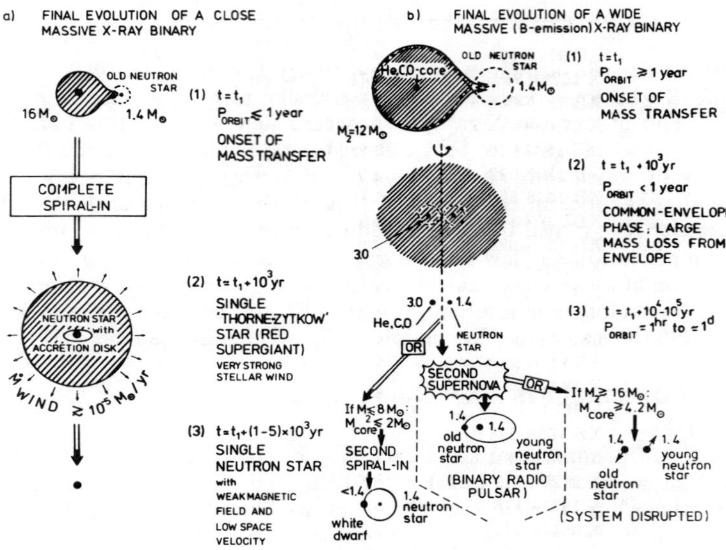

Figure 6. The various possibilities for the final evolution of a High-Mass X-ray Binary. In all cases the onset of Roche-lobe overflow leads to the formation of a common envelope and the occurrence of spiral-in. In systems with orbital periods $\lesssim$ about 1 year there is most probably not enough energy available in the orbit to eject the common-envelope, and the neutron star spirals down into the core of its companion. Subsequently, the envelope is ejected by the liberated accretion energy flux, leaving a single recycled radio pulsar. In systems with orbital periods $\gtrsim$ about a year the common envelope is ejected during spiral-in, and a close binary can be left, consisting of the neutron star and the core (consisting of helium and heavier elements) of the companion. Companions initially more massive than 8–12 $M_\odot$ leave cores that will explode as a supernova, leaving an eccentric-orbit binary pulsar or two runaway pulsars. Systems with companions less than massive than $\approx$ 8–12 $M_\odot$ leave close binaries with a circular orbit and a massive white dwarf companions, similar to PSR 0655+64.

3.3. THE EVOLUTION AND FATE OF LOW-MASS X-RAY BINARIES

3.3.1. Mechanisms Driving The Mass Transfer In Low-mass X-ray Binaries. Table 3 lists the orbital periods of LMXB as given by Parmar and White (1988). The table and Figure 2 show that there are two groups of LMXB: one with short orbital periods ($P < 10 - 12H$), and a group with wider orbits ($P > 12H$). Systems with $P < 10 - 12H$ are so close that a low-mass main-sequence companion star ($< M_\odot$) in it may already overflow its Roche lobe. Table 4 (after Verbunt 1989) gives the radii of hydrogen- and helium- main-sequence stars, as well as of hydrogen and helium degenerate stars, as a function of stellar mass, together with the orbital periods of binaries (with a neutron-star

TABLE 3
The Orbital Periods of LMXB (after Parmar & White 1988)

Source	Period(hr)	Optical	X-ray
XB 1820-303	0.19	-	x
X 1626-673	0.70	o	-
XB 1916-053	0.83	o	x
X 1705-44*	1.30	-	x
MS 1603+26*	1.85	-	x
XB 1323-619	2.9	-	x
XB 1636-536	3.8	o	-
XBT 0748-676	3.8	o	x
XB 1254-690	3.9	o	x
X 1728-169	4.2	o	x
X 1755-338	4.4	o	x
XB 1735-444	4.6	o	-
Cyg X-3	4.8	(IR)	x
XBT 2129+470	5.2	o	x
X 1822-371	5.6	o	x
LMC X-2	6.4	o	x
XB 1658-298	7.2	-	x
XT 0620-003	7.8	o	-
X 2127+119 (M15)	8.5	o	x
X 1957+115	9.3	o	-
X 0547-71*	10.6	-	x
X 1659-487 (G339-4)	14.8	o	-
Cen X-4*	15.1	-	x
Sco X-1	19.2	o	-
X 1813-14*	19.5	-	x
X 1624-490	21	-	x
Cal 83	25	o	-
Her X-1	40.8	o	x
X 0921-630	216	o	x
Cyg X-2	235	o	x

Notes: The letters 'B' and 'T' in the source name indicate a burst source or transient nature, respectively. An 'o' indicates that an orbital periodicity is seen in the optical band, an 'x' that it is seen in the X-ray band.
*After H. Ritter (1990).

TABLE 4
Approximate mass-radius relations for Roche-lobe filling low mass ($\lesssim 1 M_\odot$) stars in thermal equilibrium (during the mass transfer the thermal equilibrium may gradually be disturbed such that deviations from these relations become possible) (after Verbunt 1989):

Main Sequence	$\frac{R_2}{R_\odot} = \frac{M_2}{M_\odot}$	$\mathcal{P} = 9^h \left(\frac{M_2}{M_\odot}\right)$
Helium Main Sequence	$\frac{R_2}{R_\odot} = 0,2\frac{M_2}{M_\odot}$	$\mathcal{P} = 0^h,9 \left(\frac{M_2}{M_\odot}\right)$
Degenerate Star*	$\frac{R_2}{R_\odot} = 0,013(1+X)^{5/3}\left(\frac{M_\odot}{M_2}\right)^{1/3}$	$\mathcal{P} = 48(1+X)^{5/2}\frac{M_\odot}{M_2}$
*here X is the fractional hydrogen abundance.		(sec)

companion mass of 1,4$M_\odot$) in which such stars fill their Roche lobes. It is clear from this table that hydrogen main-sequence stars with masses between 0,2 and 1,0$M_\odot$ fill their Roche lobes in systems with periods between $1,8H$ and $9,0H$, respectively, whereas for helium stars these values are 11 minutes to $1,0H$. In systems with these very short orbital periods the orbits will shrink due to the emission of gravitational radiation (Faulkner 1971). The timescale for orbital decay due to this process is

$$t_{GR} = \frac{(M_1 + M_2)^{1/3}(P/1.6H)^{8.3}}{2^{1/3}M_1M_2}(5.10^7 \text{ yrs}) \qquad (5)$$

For $M_1 \sim M_2 \sim M_\odot$ one observes that for $P < 12H$, t_{GR} is shorter than 10^{10} yr.

Thus, in such systems, even without any evolution of the companion star, continuous mass transfer will ensue due to the shrinking of the orbit (and with it the Roche lobe of the companion). We will further consider this possibility in Section 3.3.5. In systems with periods $> 0,5$ days the orbital angular momentum losses by gravitational radiation are insufficient to drive the mass transfer. For periods between 0,5 and 1,0 days angular momentum losses due to magnetic braking may still help driving the transfer (Verbunt and Zwaan 1981; Verbunt 1989). However, for $P > 1,0d$ the mass transfer can only be driven by the nuclear evolution of the companion star. We will now consider this case in more detail, and in Section 3.3.5 return to the final evolution of systems with $P < 12H$.

3.3.2. The Evolution Of Low-mass X-ray Binaries With Evolved Companion Stars. The evolution of these systems was studied first by Webbink *et al.* (1983) and Taam (1983a). These authors showed that the evolution is in principle very simple. This is due to the relatively simple interior structure of evolved low-mass (solar-type) stars. In such stars the helium core is degenerate and the luminosity is produced by hydrogen burning in a shell around this core. As a result the core mass gradually increases, and the luminosity generated in the hydrogen-burning shell (L_s) gradually increases, which causes the outer radius of the star (R_s) to also gradually increase. For a given initial chemical composition (X, Y, Z), the luminosity and radius of these stars turn out to be a function only of the mass M_c of the degenerate helium core : $L_s(M_c), R_s(M_c)$, independent of the total mass present in the hydrogen-rich envelope (as long as that envelope mass is $> 0,01M_\odot$). Since the core mass M_c determines the stellar radius R_s, and since the Roche-lobe radius of this star is determined by the orbital period (assuming the compact star to have a standard mass, *e.g.* $1,4M_\odot$), the core mass at the moment that the star begins to fill its Roche lobe is entirely determined by the orbital period of the system at that time. Since this core mass determines the evolutionary rate of change of radius at that moment, it determines the ensueing rate of mass transfer. As a result, the initial orbital period P_o (at the onset of the mass transfer) uniquely determines the rate of mass transfer in these systems, as well as the entire further evolution of these systems. Figure 7 shows as an example the evolution of the core mass, radius, luminosity and orbital period of an evolved star in a system that started out with an orbital period of 12,5 days, a neutron-star mass of $1M_\odot$ and a companion mass of $1M_\odot$. The figure shows that the entire evolution of the system takes $8,1.10^7$ years, during which the companion mass decreases to $0,31M_\odot$, the core mass grows from $0,24M_\odot$ to $0,31M_\odot$ and the orbital period increases to 117 days. After an initial episode of slightly super-Eddington mass transfer ($\sim 10^7$ yrs) the mass-transfer rate settles at $\sim 1,0.10^{-8}M_\odot/\text{yr}$, making the system a very bright (near Eddington limit) X-ray source for a period of some 7.10^7 years.

3.3.3. From Wide LMXB To Wide Radio Pulsar Binaries With Circular Orbits. As the giant companion of the system of Figure 7 fills its Roche lobe for $\sim 8.10^7$ years, the strong tidal forces excerted on its convective envelopee by the neutron star are expected to completely circularize the orbit (*cf.* Savonije 1983a). Hence, the system that remains in the end will consist of a neutron star and a low-mass ($\sim 0,3 M_\odot$) helium white dwarf in a wide and circular orbit. Such a system exactly resembles the second class of binary radio pulsars depicted in Figure 4 (the PSR 1953+29 class), as was noticed independently by Joss and Rappaport (1983), Savonije (1983b) and Paczynski(1983), immediately after the discovery of PSR 1953+29 by Boriakoff *et al.* (1983). Important is that, assuming a given initial chemical composition and initial neutron-star mass, the evolution and final binary period of such a system is entirely determined by the initial binary period and companion mass. (For details about the relations between initial and final binary parameters for such systems we refer to Verbunt (1989)). This implies that for the binary pulsars of the PSR 1953+29 class , it is possible to reconstruct, from the present orbital parameters, their orbital periods at the onset of the mass transfer, as well as the rate of mass transfer during the X-ray phase (which was determined by the initial orbital period). The latter can, for $P_o > 2$ days be approximated by (for a $1,0 M_\odot$ companion, $Z = 0,02$)

$$\dot{M} = -8.10^{-10} P_o[d] \tag{6}$$

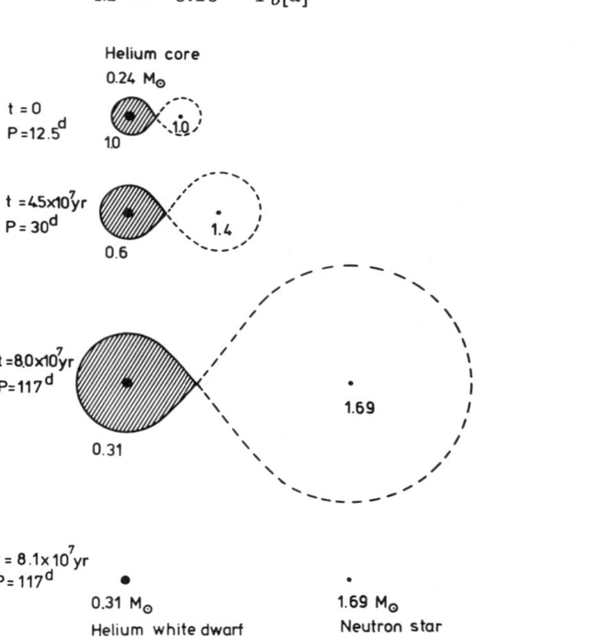

Figure 7. Evolution of a wide low-mass X-ray binary such as Cygnus X-2 into a wide radio-pulsar binary with a circular orbit on a low-mass helium white dwarf companion, such as PSR 1953+29. At the onset of the mass transfer, the low-mass companion is a (sub)giant with a degenerate helium core of 0.24 $M_\odot$. Its light is generated by hydrogen-fusion in a shell around the core. The mass-transfer from the giant to the neutron star is due to the slow expansion of the giant, driven by this hydrogen-shell burning. During the mass transfer the orbit gradually expands (due to angular momentum conservation) and after 8.10^8 yr the system terminates as a wide radio-pulsar binary (after Joss and Rappaport 1983).

Table 5 and Equation (6) show that the mass-transfer rate becomes super-Eddington for $P_o > 12$ days. This implies that for $P_o > 12$ days also mass loss from the systems will have occurred such that the "conservative" evolutionary picture that we adopted will no longer be fully correct. Therefore, for PSR 0820+02, in which the mass-transfer rate was $10^{-7} M_\odot$/yr, the relation between initial and final orbital period of Table 5 cannot be precise, and will be only approximate.

TABLE 5

Initial and final (=present) parameters of three wide binary radio pulsar systems, for an assumed initial companion mass of $0,9 M_\odot$, initial neutron star mass $1,4 M_\odot$ and initial heavy element abundance $Z = 0,02$. Assumed is conservative mass transfer (no mass and orbital angular momentum loss from the system) (*cf.* Verbunt 1989).

System Name	Binary Period (d)		Core Mass ($M_\odot$)		Mass transfer rate ($M_\odot$/yr)	
	P	P_0	M_c^0	M_c^f	$\dot{M}_0$	$\dot{M}_f$
PSR 1913+16	120	6,5	0,22	0,27	$0,8 \times 10^{-8}$	$0,4 \times 10^{-8}$
PSR 1620-62	191	16	0,24	0,29	$1,6 \times 10^{-8}$	$0,6 \times 10^{-8}$
PSR 0820+02	1232	170	0,32	0,38	10^{-7}	$0,5 \times 10^{-7}$

3.3.4. The Final Evolution And Fate Of Low-mass X-ray Binaries With Orbital Periods Between $12H$ *And About 2 Days.* In systems with $P < 12H$ orbital angular momentum losses by "magnetic braking" can drive mass transfer rates one to two orders of magnitude larger than produced by gravitational radiation losses alone (Rappaport *et al.* 1983). Such may be necessary in order to explain why in quite a number of Cataclysmic Variable binaries (abbreviated here as CV) and Low-Mass X-ray binaries with periods $< 12H$, mass-transfer rates considerably larger than the $10^{-10} M_\odot$/yr predicted for GR-losses alone are observed. Assuming orbital angular momentum losses by magnetic braking to be important at $P < 12H$, one can calculate that such losses will also be important in systems with somewhat larger orbital periods, in the range $12H$ to 1-2 days, provided that the low-mass companion does fill its Roche lobe. In such systems, however, the low-mass companion star ($\lesssim M_\odot$) can only come close to filling its Roche lobe if it has evolved away from the main sequence (*i.e.*, if the hydrogen content of its core has been substantially reduced). In such systems therefore one expects that also the internal nuclear evolution of the companion, which is a sub-giant, will contribute to driving the mass transfer. The latter will cause the orbit to expand, while at the same time the angular momentum losses by magnetic braking will cause the orbit to shrink. As a result of the combination of these two opposing effects the orbital period may hardly change during the evolution of these systems. For this reason the low-mass systems that start out with orbital periods in the range about $12H$ to $1 - 2d$ are expected to terminate their evolution with periods in roughly the same range. Precise calculations of the evolution of such systems, taking the combined effects of magnetic braking and internal nuclear evolution into account, were carried ot by Pylyser and Savonije (1988, 1989), and confirm this expectation. Thus, a system like Sco X-1 ($P = 0,8d$) is expected to terminate its evolution with an orbital period in the range $12H$ to $2d$. Similarly, the binary pulsar system PSR 1831+00 ($P = 1,8d$) is expected to have been produced by the evolution of such a system in which the combined effects of magnetic braking and nuclear evolution of the companion (which has left a degenerate helium core of only about $0,2M_\odot$ in this system, *cf.* Dewey *et al.* 1986) have driven the orbital evolution. Had internal nuclear evolution alone driven the orbital evolution of this system (which must have started out with an orbital period of at least 0,5 days in order to leave an helium white dwarf), its final orbital period would have been some ten times longer than its initial orbital period, *i.e.*, longer than 5 days. The only way to explain why PSR 1831+00 has terminated with an orbital period of 1,8 days is, that magnetic braking has played a substantial role in the past evolution of this system. This assertion is a very important one as it provides us with a solid proof that magnetic braking is operating in Nature (gravitational radiation losses are negligible in systems with $P > 0,5d$).

3.3.5. The Final Evolution And Fate Of LMXB With $P < 12H$.

(a) *Effects of gravitational radiation losses and magnetic braking*

As mentioned in Sections 3.3.2 and 3.3.4 the evolution of these systems is expected to be driven by losses of orbital angular momentum by gravitational radiation (which we will abbreviate here as GR) and magnetic braking (abbreviated here as MB). The last-mentioned mechanism is absolutely required to explain the relatively high mass-transfer rates ($> 10^{-9}M_\odot/\mathrm{yr}$) observed in many CV- and LMXB-systems with $P < 12H$. For the sake of argument we first discuss the evolution of a CV- or LMXB-system under the influence of GR-losses alone. Figure 8 depicts the evolution of such a system, in which the companion starts out with a mass of one solar mass. At first, the mass loss causes the orbital period to decrease, since due to the mass loss the radius of the low-mass companion decreases. However, when the timescale of the mass transfer becomes comparable to the thermal timescale of the companion star, this star will no longer be able to remain in thermal equilibrium. Since from its past evolution it contains more heat than a thermal-equilibrium (main-sequence) model of the same mass, its radius will become somewhat inflated. This is the reason why for $M < 0,3M_\odot$ the mass-losing companion no longer follows the main-sequence mass-radius relation given in Table 4. When it reaches a mass of $0,1M_\odot$ the star becomes degenerate and further mass loss will cause its radius to increase (see Table 4). As the star is all the time filling its Roche lobe, an increase in stellar radius implies an increase in orbital radius and orbital period. The minimum orbital period which the system can attain – for the case that only angular momentum losses due to GR-losses occur – is about 80 minutes for normal hydrogen-rich stars (Paczynski and Sienkiewicz 1981). Figure 8 also shows the evolutionary track of a system in which the orbital angular momentum losses by GR were supplemented by angular momentum losses by magnetic braking, driving a ten times larger mass-transfer rate (for examples of calculations of this type we refer to Rappaport, Verbunt and Joss 1983, and Taam 1983b). In this case the deviations from thermal equilibrium are larger, causing the companion to become more inflated, and the period minimum to be located at a somewhat longer orbital period.

(b) *The period gap*

Figure 9 depicts the distributions of the orbital periods of CVs and LMXB as deduced from Ritter's (1990) catalogue. The figure shows that

(i) with only two exceptions CVs have periods longer than 80 minutes (the two ultra-short orbital period systems both appear to have hydrogen-poor companions - which fits with expectations from Table 4);

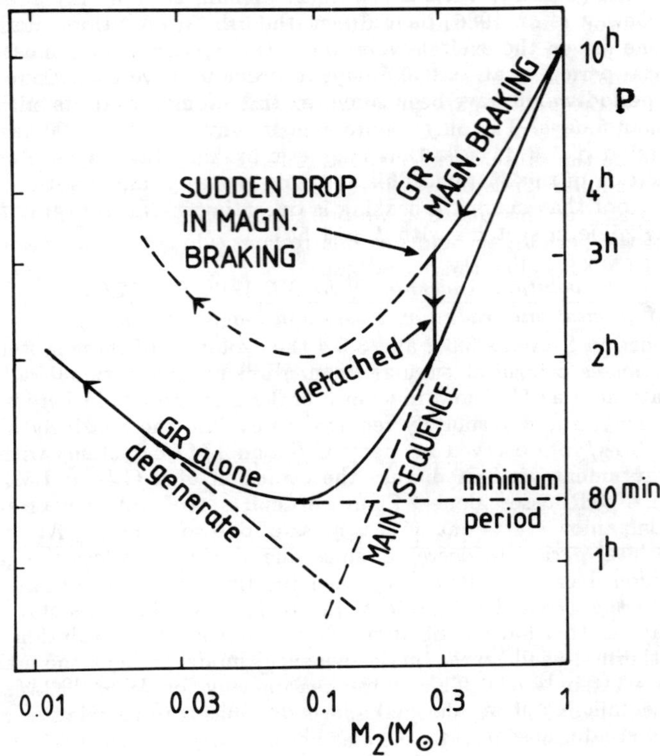

Figure 8. Schematic explanation for the origin of the "Period Gap" of Cataclysmic Variables between 2^H and 3^H (*cf.* Spruit and Ritter 1983). Indicated are evolutionary tracks of CV-binaries and LMXB in which the mass transfer is driven by angular momentum losses by Gravitational Radiation (GR) and/or Magnetic Braking. The system begins in the upper right-hand part of the diagram with $M_2 \lesssim M_\odot$, $P \leq 10^h$. If GR losses alone are operating, the system reaches a minimum orbital period of $\sim$ 80 minutes (companion mass $\sim 0,1 M_\odot$), after which the companion becomes degenerate and the orbit expands upon further mass transfer. If also magnetic braking operates, the mass-transfer rate may be an order of magnitude larger and the companion is driven out of thermal equilibrium, such that its radius becomes inflated. It then moves along the higher track in the diagram. However, in this case, for $M_2 \sim 0,3 M_\odot$ the companion becomes fully convective and the magnetic braking is much reduced. The resulting reduction of the mass transfer rate causes the companion to shrink back to thermal equilibrium and to detach itself from the Roche-lobe. At this point $P_{\rm orb} \simeq 3H$. Subsequent GR losses reduce the orbital period further during this detached phase; when $P_{\rm orb}$ has become $2H$ the companion fills its Roche-lobe again and mass-transfer resumes (after Spruit and Ritter 1983). Thus, between $P_{\rm orb} \approx 2^H$ and 3^H no mass-transferring CVs or LMXB are expected to be found.

(ii) there are hardly any CVs with orbital periods between $2H$ and $3H$: this is the so-called "period gap" for Cataclysmic Variables. On the other hand, for LMXB the figure shows no period gap: in fact, with the exception of again a few ultra-short period systems, which must have hydrogen-poor companions, there are virtually no LMXB below the period gap, *i.e.*, with orbital periods between 80 minutes and $2H$, where some 40 per cent of all CVs are found.

What could be the cause of this striking difference between the period distributions of otherwise very similar kinds of systems?

In order to answer this question we must look at the proposed explanations for the occurrence of the period-gap, and possible reasons why such explanations could lead to an absence of LMXB below the period gap.

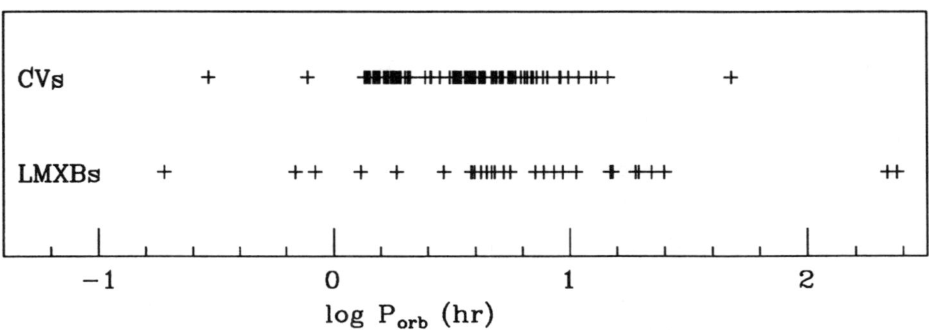

Figure 9. Distributions of the orbital periods of cataclysmic variables (CVs) and low-mass X-ray binaries (LMXBs), after Verbunt (1989). The distributions are roughly similar, but differ in detail. Notably, many CVs have orbital periods between $1^h,3$ and 2^h, where LMXBs are virtually absent.

(c) *Explanations for the origin of the period gap in CV-binaries*

A plausible explanation for the origin of the period gap was put forward by Spruit and Ritter (1982). In this explanation use is made of the fact that below a mass of $0,3M_\odot$ main-sequence stars become fully convective (above this mass they have a radiative core and a convective envelope). Since magnetic fields presumably need anchoring in a radiative core, one expects that at a mass below $0,3M_\odot$ magnetic activity and, therefore, magnetic braking are suddenly decreased. Consequently, in a CV-binary or an LMXB, when the system has evolved along the "magnetic braking" track in Figure 8 down to a companion mass of $\sim 0,3M_\odot$ (orbital period around $3H$), the orbital angular momentum losses by magnetic braking presumably are very much reduced. This causes the mass-transfer rate to suddenly drop by an order of magnitude or more, causing the mass-loss timescale to suddenly become much longer than the thermal timescale of the companion again. This allows the companion star, which was driven far out of thermal equilibrium by its preceding mass loss, to relax back to thermal equilibrium, which causes its inflated radius to shrink to its thermal equilibrium value. This shrinking causes the star to detach itself completely from the Roche lobe, such that the mass loss stops altogether. During the detached phase which now ensues, the orbit will shrink due to the emission of gravitational radiation. When the orbit has shrunken so far that the star (now at its main-sequence thermal-equilibrium radius) again fills its Roche lobe, and the mass transfer resumes, the orbital period is about $2H$. Thus, in a natural way the absence of mass-transferring CV-systems with periods in the range between about $2H$ and about $3H$ is explained. Figure 8 graphically illustrates this explanation.

(d) *Evaporation of companions and the virtual absence of LMXB-systems below the period gap*

To explain why there are virtually no LMXB below the period gap, while there are many CVs in the period range 80 minutes to $2H$, one must pay attention to the following great difference between a CV and an LMXB. Once accretion stops, when the system arrives at the upper end of the period gap, the rapidly spinning magnetized neutron star in an LMXB will turn into a powerful radio pulsar. A CV-binary does not contain a similar "powerhouse" which may strongly affect the structure of a nearby low-mass companion star. On the other hand, the pulsar which appears in the system when an LMXB enters the period gap, may with its powerful energy emission totally evaporate the low-mass companion star ($< 0,3M_\odot$) during the detached phase, when the system is attempting to cross the period gap. We will show below that such a total "evaporation" of the companion may quite well take place. Thus, LMXB with initial orbital periods $< 12H$ may in most cases end their evolution as single millisecond pulsars, similar to PSR 1937+21. An attractive aspect of this proposed evolutionary model (van den Heuvel and van Paradijs 1988) is that kills two birds with one stone: it explains the presence of single millisecond pulsars in the galactic disk, as well as it explains the virtual absence of LMXB with orbital periods below the period gap for CV-binaries. (In globular clusters also other processes, such as exchange collisions and direct collisions with dwarf stars may contribute to the formation of single millisecond pulsars; this may explain the relatively high fraction of single millisecond pulsars in globular clusters , *cf.* Phinney and Sigurdsson 1991, Phinney 1991,1992).

(e) *A simple evaporation model*

For a detailed review of evaporation models for companions of pulsars we refer to Shaham (this volume). Here we give only a simplified treatment (*cf.* van den Heuvel and van Paradijs 1988). In order to see whether a pulsar has, in principle, enough energy available to evaporate its companion, one should compare its rotational energy content with the gravitational binding energy of the companion star. The rotational energy content of the pulsar is

$$E_{\text{rot}} = 0,5I\omega^2 = 2.10^{52} P_{ms}^{-2} \text{ ergs} \tag{7}$$

where P_{ms} is the rotation period of the neutron star in milliseconds. The gravitational binding energy of the companion star is given by:

$$E_b = GM^2/R \sim 10^{48}(M/M_\odot)^2(R/R_\odot)^{-1} \text{ ergs} \tag{8}$$

Furthermore, one should know the efficiency g of the evaporation process, defined as the fraction of the pulsar's rotational energy that can be actually used for the evaporation of the companion. The condition under which the pulsar is able to completely evaporate its companion is:

$$gE_{\text{rot}}/E_b > 1 \tag{9}$$

Using Equations (7) and (8) one observes that a pulsar with $P = 15ms$ (the period which the Crab pulsar presumably had at birth) would in principle be able to evaporate a one solar mass companion, provided that the efficiency factor g is $> 0,0125$. Similarly, a pulsar with $P = 1,5ms$ will be able to evaporate such a companion if $g > 1,25.10^{-4}$. The factor g is the product of the geometrical factor $(R/2a)^2$ and the real efficiency of the evaporation process f, which is the fraction of the pulsar's energy flux falling on the companion's surface, that is used to drive the evaporative wind. Here R is the companion's radius, and a is the orbital radius. The important unknown quantity here is the effiency factor f. (Kluzniak *et al.* 1988, and Ruderman *et al.* 1989a,b) estimate it to be rather low ($< 0,001 to 0,01$), but Krolik and Sincell (1990) expect it to be high (0,01 to 0,1), due to the surface heating produced by the annihilation of the positrons from the pulsar wind. An important other parameter is the timescale of the evaporation process, which can be estimated as follows (*cf.* van den Heuvel and van Paradijs 1988). The pulsar's total energy emission ("luminosity") is:

$$L_p = dE_{\rm rot}/dt = (2R_n^6/3c^3)(B_s^2)(2\pi/P)^4 \tag{10}$$

where R_n is the radius of the neutron star, B_s its surface dipole magnetic field strength, and c the velocity of light. Assuming that the energy is emitted in a spherically symmetric way, the amount of energy available for driving the wind of the companion is:

$$0,5\dot{M}v^2 = f(R/2a)^2(2R_n^6/3c^3)(B_s^2)(2\pi/P)^4 \tag{11}$$

Assuming that, as ususal for stellar winds, the wind velocity is of order of the escape velocity $v_e = (2GM/R)^{0,5}$ from the surface of the companion, the evaporation timescale becomes

$$\tau_{ev} = M/\dot{M} = (3c^3/2R_n^6)(2GM^2/R^3)(a^2/8\pi^4 f)P^4/B_s^2 \tag{12}$$

Thus, the evaporation timescale is proportional to P^4/B_s^2. This implies that for a weak-field pulsar, such as the millisecond pulsars ($B_s \sim 4.10^8 G$) it is some 10^8 times longer than for a strong-field pulsar such as the Crab pulsar. For $M < M_\odot$ the mass-radius relation for main-sequence companions is:

$$R/R_\odot = M/M_\odot \tag{13}$$

and τ_{ev} scales as M^{-1}. For low values of M the star is degenerate and

$$R/R_\odot = 0,013(1+X)^{5/3}(M/M_\odot)^{-1/3} \tag{14}$$

where X is the hydrogen fraction. In the latter case, τ_{ev} scales as M^3, which implies that the smaller the mass becomes, the faster the evaporation proceeds. For $X = 0,75$ the relations (13) and (14) match for $M = 0,075 M_\odot$. For the 1,6 ms pulsar PSR 1957+20 with $M = 0,02 M_\odot$, $a = 2,5 R_\odot$ and $R = 0,166 R_\odot$ (according to Eq.(14)) and $f = 0,1$ one obtains $\tau_{ev} \sim 10^7$ years. This is of the same order as the observed timescale for orbital decay of this system (Fruchter 1990) which presumably is connected with the wind mass loss. This seems to imply that indeed the efficiency factor f of the evaporation process is high, of the order of 0,1, as had been conjectured by Krolik and Sincell (1990). In fact, still higher efficiencies can be reached if one takes into account the strong deformation of the critical equipotential surface (Roche lobe) of the companion due to the pressure of the wind from the nearby pulsar (*cf.* Kluzniak 1991). Concerning the 11ms eclipsing binary pulsar in the globular cluster Ter 5, which has $P_{\rm orb} = 1,8H$ and a companion mass of $\sim 0,1 M_\odot$, one calculates with the above equations that complete evaporation can be achieved provided that $f > 0,01$. Detailed calculations of the evolution of systems with evaporating companions were carried out by Ergma and Fedorova (1990, 1991), to which we further refer.

4. Magnetic field decay revisited

From the above it will be clear that an analysis of the (single) pulsar population in the Galaxy, in which it is assumed that pulsars are simply born as single neutron stars in isolated supernova events, is much too naive. Many pulsars are expected to have been produced by recycling in (mostly) massive binary systems which subsequently were broken up in the second supernova event (see Bailes 1989). The magnetic fields of these recycled neutron stars are expected to be (somewhat) weaker than those of newborn pulsars, as is evidenced by the relatively weak fields observed in some of the eccentric-orbit binary pulsars and also in the PSR0655+64 system, which is the remnant of a massive binary as well (van den Heuvel and Taam 1984). However, other recycled binary pulsars of this category, such as PSR2303+46 appear to have relatively normal field strengths ($10^{12}G$) even though also they are in the "recycled" domain of the B-P diagram. The fact that the PSR0655+64 system is very old ($> 10^9$ yrs, as is evidenced by the cooling age of the white dwarf companion (Kulkarni 1986)), indicates that in neutron stars with surface dipole field strengths $\sim 10^{10}G$ the field does not decay any more. The same is true for the millisecond pulsars, as is evidenced by their large number in the Galaxy ($\sim$ 10 per cent of all pulsars, *cf.* Stokes *et al.*, 1986). This large number can only be explained if their magnetic fields, which range from 2.10^8 to 2.10^9G, do not decay at all (Bhattacharya and Srinivasan 1986, van den Heuvel *et al.* 1986). Further, the presence of a strong ($\sim 10^{12}G$) magnetic field in the old ($\sim 5.10^8$ yrs) neutron star in the Her X-1 system (*cf.* Verbunt *et al.* 1990) indicates that even at $B \sim 10^{12}G$ the surface dipole components of the magnetic fields of neutron stars probably do not decay at all. An additional argument against spontaneous magnetic field decay in isolated neutron stars comes from the observation of cyclotron features in the spectra of gamma-ray burst sources (Murakami *et al.* 1988), which indicate the presence of magnetic fields of about $10^{12}G$. These bursts arise most probably in a population of old neutron stars, which have already reached the graveyard. The fact that these old isolated neutron stars in the graveyard still have retained strong magnetic fields forms an additional argument against the spontaneous B-field decay in isolated neutron stars. The above shows that the "bottom" surface dipole field strengths (*i.e.*, field strengths beyond which no more spontaneous field decay occurs) observed in recycled and non-recycled neutron stars range from 2.10^8 to over $10^{12}G$. This enormous range, of almost a factor 10^4, suggests that there is no unique "bottom" field, but that its value depends on the history of the system. The only thing that differs among the neutron stars in these systems is the total amount of accretion that took place during the X-ray phase (Taam and van den Heuvel 1986; Shibazaki *et al.* 1989; Romani 1990), or the precise rotational history of the neutron star (Srinivasan *et al.* 1990). Taam and van den Heuvel (1986) have shown that the amount of B-field decay observed in accreting systems is a monotonic function of the total amount of accretion that took place onto the neutron star. Thus, it may well be that the field is "killed" by accretion (Shibazaki *et al.* 1989, Romani 1990). Alternatively, the bottom value of the field may be linked to the rotational history of the neutron star, which in a binary is expected to have been totally different than for an isolated neutron star (Srinivasan *et al.* 1990).

In any case, if one assumes that also in the HMXB the B-fields of the recycled neutron stars have, on average, undergone some decay, say, of order of a factor 2 to 3, then after spin-up and release from their binary systems, these neutron stars will remain observable as pulsars some 4 to 9 times longer than the average newborn single neutron star (the recycled ones need 4 to 9 times longer to reach the deathline). Since of order of at least 10 per cent of all neutron stars is expected to have undergone spin-up recycling in a massive binary which subsequently broke up (Meurs and van den Heuvel 1989), one expects the observed population of single pulsars to consist for a large part (on the 50 per cent or more level, see also Bailes 1989) of recycled neutron stars, which are on average characterized by somewhat weaker B-fields than the average pulsar population. Due to this severe contamination with the remnants of accreting binary systems, one may, if one naively analyzes the population of single radio pulsars in terms of a model in which neutron stars are the direct result of isolated (single) supernova events (*cf.* Narayan and Ostriker 1990) get the false impression that the magnetic fields of isolated neutron stars decay spontaneously on a timescale of order 10^7 years. However, taking only field decay induced by binary evolution (*i.e.*, due to accretion or to a peculiar rotational history) into account, and assuming that the fields of isolated-born neutron stars do not decay at all, it is likely to be possible to account for the observed properties of the general pulsar population in the Galaxy (Bailes 1989, Bhattacharya and van den Heuvel 1991). For this reason, the "paradigma" of spontaneous decay of the surface dipole component of the magnetic fields of isolated neutron stars on a relatively short (10^7 yrs) timescale (Gunn and Ostriker 1970) seems no longer necessary. The above implies that the entire statistical analysis of the radio pulsar population will have to be redone on a new basis, in which a large contribution to the observed radio pulsar population from the recycling of neutron stars in binaries is taken into account. A reasonable starting assumption in such a new analysis will be, in view of the above, that neutron star magnetic fields do not decay unless recycling in binaries has taken place. Some model of how the recycling evolution in a binary affects the magnetic field strength will have to be included.

5. Conclusions

(i) A large fraction of the millisecond pulsars ($\sim 40\%$) is found in binaries, and all millisecond pulsars share the special characteristics of the binary radio pulsars (on average much weaker magnetic fields, and much faster spin than the general radio pulsar population). This suggests that binary and millisecond pulsars form one class of objects, with a common evolutionary origin.
(ii) Observational evidence indicates that the binary and millisecond pulsars are relatively old neutron stars that have been spun-up by accretion in binary systems (so-called "recycling").
(iii) The binary radio pulsars fall into two classes: the PSR 1913+16 class (narrow and often eccentric orbits, companion is neutron star or massive white dwarf), and the PSR 1953+29 class (wide and circular orbits, low-mass white dwarf companions). The first class are the decendants of High-Mass X-ray Binaries, the second class of Low-Mass X-ray Binaries with relatively wide orbits (initial orbital period $> 0,5$ days).
(iv) Low-Mass X-ray Binaries with initial orbital periods $< 0,5$ days are expected to turn into millisecond binary radio pulsars when the companion mass has been reduced to $< 0,3M_\odot$. Subsequently, the millisecond pulsar is likely to evaporate its companion, leaving a single millisecond pulsar.
(v) There is no longer clear evidence that the magnetic fields of isolated neutron stars do decay spontaneously. Only neutron stars that have been recycled in binary systems show clear evidence for magnetic field decay. The general galactic radio pulsar population is expected to be highly contaminated (on the 50 per cent level or more) with neutron stars that have been recycled in (mostly massive) binaries, in which decay of the surface dipole component of the magnetic field has taken place, presumably induced either by the accretion or by the special spin evolution that takes place for neutron stars in binaries.

6. Acknowledgements

This research was supported in part by NSF Grant No. PHY89-04035.

7. References

Alpar, M.A., Cheng, A.F., Ruderman, M.A. and Shaham, J. (1982) *Nature*, **300**, 728.

Bailes, M. (1989) *Astrophys. J.*, **342**, 917.

Bhattacharya, D. and Srinivasan, G. (1986) *Current Sci.*, **55**, 327.

Bhattacharya, D. and van den Heuvel, E.P.J. (1991) *Physics Reports* (in press).

Biehle, G.T. (1991) *Astrophys. J.* (in press; Caltech Goldenrod Preprint – 241).

Bisnovatyi-Kogan, G.S. and Komberg, B.V. (1975) *Sov. Astron.* **18**, 217 ((1974) *Astron. Zh.*, **51**, 373).

Blondin, J.M., Kallman, T.R., Fryxell, B.A. and Taam, R.E. (1990) *Astrophys. J.*, **356**, 591.

Bonsema, P.F.J. and van den Heuvel, E.P.J. (1985) *Astronomy & Astrophys.*, 146, L3.

Cannon, R.C., Eggleton, P.P., Zytkow, A.N. and Podsiadlowski, P. (1991), *Astrophys. J.* (in press).

Dewey, R.J., Maguire, C.M., Rawley, L.A., Stokes, G.H. and Taylor, J.H. (1986) *Nature*, **322**, 712.

Ergma, E. and Federova, A.V. (1990) *Astron. Astrophys.* (in press).

Ergma, E. and Federova, A.V. (1991), *Pisma Astron. Zh.* (in press).

Faulkner, J. (1971) *Astrophys. J.*, **170**, L99.

Fruchter, A.S. (1991), private communication.

Gunn, J.E. and Ostriker, J.P. (1970) *Astrophys. J.*, **160**, 979.

Henrichs, H.F. (1983) in W.H.G. Lewin and E.P.J. van den Heuvel (eds.), 'Accretion Driven Stellar X-ray Sources', Cambridge University Press, p. 393.

Joss, P.C. and Rappaport, S.A. (1983) *Nature*, **304**, 419.

Kluzniak, W., Ruderman, M., Shaham, J. and Tavani, M. (1988), *Nature*, **334**, 225.

Krolik, J.H. and Sincell, M.W. (1990), *Astrophys. J.*, **357**, 268.

Kulkarni, S.R. (1986) *Astrophys. J.*, **306**, L85.

Lamb, F.K. (1989) in H. Ögelman and E.P.J. van den Heuvel (eds.), 'Timing Neutron Stars', Kluwer Academic Publishers, Dordrecht, p. 649.

Lewin, W.H.G. and Joss, P.C. (1983) in W.H.G. Lewin and E.P.J. van den Heuvel (eds.), 'Accretion Driven Stellar X-ray Sources', Cambridge University Press, p. 41.

Lyne, A.G., Manchester, R.N., D'Amico, N., Staveley-Smith, L., Johnston, S., Lim, J., Fruchter, A.S. Goss, W.M. and Frail, D. (1990) *Nature*, **347**, 650.

Meurs, E.J.A. and van den Heuvel, E.P.J. (1989) *Astronomy & Astrophys.*, **226**, 88.

Murakami, T., Fujii, M., Hayashida, K., Itoh, M., Nishimura, J., Yamagami, T., Conner, J.P., Evans, W.D., Fenimore, E.E., Klebesadel, R.W., Yoshida, A., Kondo, I. and Kawai, N. (1988) *Nature*, **335**, 235.

Nagase, F. (1989a) in J. Hunt and B. Battrick (eds.), Proc. 23rd ESLAB Symposium on X-ray Binaries (part I), ESA/ESTEC, Noordwijk, The Netherlands, p. 45.

Nagase, F. (1989b) *Publ. Astron. Soc. Japan*, **41**, 1.

Narayan, R. and Ostriker, J.P. (1990) *Astrophys. J.*, **352**, 222.

Paczynski, B. (1983) *Nature*, **304**, 421.
Paczynski, B. and Sienkewicz (1981) *Astrophys. J.*, **248**, L27.
Parmar, A.N. and White, N.E. (1988) in N.E. White and R. Pallavacini (eds.), 'X-ray Astronomy with EXOSAT', *Mem. Astr. Soc. Ital.*, **59**, 147.
Phinney, E.S. (1991) *Proc. Royal Netherlands Acad. Sci.* (in press).
Phinney, E.S. (1992) in E.P.J. van den Heuvel and S.A. Rappaport (eds.), 'X-ray Binaries and the Formation of Binary and Millisecond Radio Pulsars', Kluwer Academic Publishers, Dordrecht (in the press).
Phinney, E.S. and Sigurdsson, S. (1991), *Nature*, **349**, 220.
Pylyser, E.H.P. and Savonije, G.J. (1988) *Astron. Astrophys.*, **191**, 57.
Pylyser, E.H.P. and Savonije, G.J. (1989) *Astron. Astrophys.*, **208**, 52.
Radhakrishnan, V. and Srinivasan, G. (1981) Paper presented at 2nd Asian-Pacific Regional Meeting of the IAU, 24–29 August 1991, Bandung, Indonesia.
Radhakrishnan, V. and Srinivasan, G. (1984) in B. Hidayat and M.W. Feast (eds.), Proc. 2nd Asian-Pacific Regional Meeting of the IAU, 24–29 August 1981, Tira Pustaka, Jakarta, p. 423.
Rappaport, S.A., Verbunt, F. and Joss, P.C. (1983) *Astrophys. J*, **275**, 713.
Ritter, H. (1990) *Astron. & Astrophys. Suppl. Series*, **85**, 1179.
Romani, R.W. (1990) *Nature*, **347**, 741.
Ruderman, M., Shaham, J. and Tavani, M. (1989a) *Astrophys. J.*, **336**, 507.
Ruderman, M., Shaham, J., Tavani, M. and Eichler, D. (1989b) *Astrophys. J.*, **343**, 292.
Savonije, G.J. (1983a), in W.H.G. Lewin and E.P.J. van den Heuvel (eds.), 'Accretion Driven Stellar X-ray Sources', Cambridge University Press, p. 343.
Savonije, G.J. (1983b), *Nature*, **304**, 422.
Shibazaki, N., Murakami, T., Shaham, J. and Nomoto, K. (1989) *Nature*, **342**, 656.
Smarr, L.L. and Blandford, R.D. (1976) *Astrophys. J.*, **207**, 574.
Spruit, H.C. and Ritter, H. (1983) *Astron. Astrophys.*, **124**, 267.
Srinivasan, G., Bhattacharya, D., Muslimov, A.G. and Tsygan, A.I. (1990) *Current Sci.*, **59**, 31.
Stokes, G.H., Segelstein, D.J., Taylor, J.H. and Dewey, R.J. (1986) *Astrophys. J.*, **311**, 694.
Taam, R.E., Bodenheimer, P. and Ostriker, J.P. (1978) *Astrophys. J.*, **222**, 269.
Taam, R.E. and Bodenheimer, P. (1989) *Astrophys. J.*, **337**, 849.
Taam, R.E. and van den Heuvel, E.P.J. (1986) *Astrophys. J.*, **305**, 235.
Taam, R.E. (1983a) *Astrophys. J.*, **270**, 694.
Taam, R.E. (1983b) *Astrophys. J.*, **268**, 361.
Thorne, K.S. and Zytkow, A.N. (1977) *Astrophys. J.*, **212**, 832.
van den Heuvel, E.P.J. (1977) *Annals New York Acad. Sci.*, **302**, 14.
van den Heuvel, E.P.J. (1983) in W.H.G. Lewin and E.P.J. van den Heuvel (eds.), 'Accretion Driven Stell X-ray Sources', Cambridge University Press, p. 303.
van den Heuvel, E.P.J. and Bonsema, P.F.J. (1984) *Astron. & Astrophys.*, **139**, L16.
van den Heuvel, E.P.J. and Taam, R.E. (1984) *Nature*, **309**, 235.
van den Heuvel, E.P.J., van Paradijs, J.A. and Taam, R.E. (1986) *Nature*, **322**, 153.
van den Heuvel, E.P.J. and van Paradijs, J.A. (1988) *Nature*, **334**, 227.
Verbunt, F. (1989) in W. Kundt (ed.), 'Neutron Stars and Their Birth Events', Kluwer Academic Publishers, Dordrecht, p. 179.
Verbunt, F. and Zwaan, C. (1981) *Astron. Astrophys.*, **100**, L7.

Verbunt, F., Wyers, R.A.M.J. and Burm, H. (1990) *Astron. Astrophys.*, **234**, 195.
Wang, Y.M. and Robertson, J.A. (1985) *Astron. Astrophys.*, **151**, 361.
Webbink, R.F., Rappaport, S.A. and Savonije, G.J. (1983) *Astrophys. J.*, **270**, 678.

VAPORIZING NEUTRON STARS IN LOW-MASS X-RAY BINARIES AND THE STATISTICS OF MILLISECOND PULSARS

MARCO TAVANI[1]
University of California
Institute of Geophysics and Planetary Physics, LLNL, CA 94550
and Department of Astronomy, UC Berkeley, CA 94720.

φύσις κρύπτεσθαι φιλεεῖ
Heraclitus

ABSTRACT. Recent data on low-mass X-ray binaries (LMXBs) and msec pulsars (MSPs) pose a challenge to evolutionary theories which neglect the effects of disk and companion irradiation. Here we discuss the main features of a radiation-driven (RD) evolutionary model that may be applicable to several LMXBs. According to this model, radiation from the accreting compact star in LMXBs 'vaporizes' the accretion disk and the companion star by driving a self-sustained mass loss until a sudden accretion-turn off occurs. The main characteristics of the RD-evolution are: (1) the lifetime of RD-LMXB's is of order 10^7 years or less; (2) both the orbital period gap and the X-ray luminosity may be consequences of RD-evolution of LMXB's containing lower main sequence and degenerate companion stars; (3) the companion star may transfer mass to the primary even if it underfills its Roche lobe; (4) the recycled msec pulsar can continue to vaporize the low-mass companion star even after the accretion turn-off produced by a strong pulsar wind; (5) the RD-evolutionary model resolves the apparent statistical discrepancy between the number of MSP's and their LMXB progenitors.

1. Introduction

Heraclitus reminds us that *Nature loves to be cryptic*, and this is particularly true in the case of bright galactic X-ray sources. Although there are approximately 100 known low-mass X-ray binaries (LMXB's) in the Galaxy (and $\sim$ 10 in globular clusters), in many cases the nature of the primary and companion stars, the mass transfer mechanism, and the lifetimes remain elusive. Evolutionary scenarios neglecting the effect of radiation from the primary star have been guiding the study of LMXB's, and have successfully explained the properties of several LMXB's studied *individually* (e.g., [1]). Depending on the nature of the Roche lobe-filling companion star, mass transfer may be driven by angular momentum loss caused by gravitational radiation (GR), magnetic braking (MB), or core expansion for sub-giant companions. In the 'standard model' the mechanism of mass transfer is independent of the nature of the primary star, and works in the same way for binaries containing black holes, neutron stars and white dwarves (cataclysmic variables, CVs). However, several questions arise both from the *statistical* features of LMXB's, and from new data on orbital

[1]Associated to the International Center of Relativistic Astrophysics (ICRA)

J. Ventura and D. Pines (eds.), Neutron Stars: Theory and Observation, 197–202.

timescales: (1) ***Why are the LMXB luminosities close to the Eddington limit of one solar mass primary ? Why have there been no observed steady LMXB's with persistent emission in the range $L_X \sim 10^{36} - 10^{34.5}$ erg s^{-1} ?*** In what follows we will refer to this luminosity range as the 'luminosity gap'. (2) ***Why is the orbital period distribution of LMXB's different from the corresponding distribution of CV's?*** Fig. 1 gives the orbital period distributions for LMXB's and cataclysmic variables (CV's). ***(3) Is the birthrate of millisecond pulsars in the Galaxy and globular clusters the same as the birthrate of their LMXB progenitors ?***

Here we address these questions by showing that radiation from the primary star (operating effectively only for a primary with mass-energy conversion efficiency of order 10 %) may influence binary evolution so as to produce the observed luminosity and orbital period 'gaps' of the LMXB distributions, as well as to reconcile LMXB/MSP birthrates. Some recent observations of $\dot{P}_{orb}/P_{orb}$ for the sources Cyg X-3, X 1822-371, EXO 0748-676 and 4U 1820-30 are interpreted in terms of the RD-evolution model.

2. Radiation-Driven Evolution of LMXB's

The reader is referred to Refs. [2,3,4] for further details of the wind formation mechanism. The RD-*mass loss* rate can be written as $|\dot{m}_{rad}| = 10^{-17} f\hat{L}$ g s^{-1}, where the dimensionless quantity f depends on the details of energy deposition and radiation transport at the base of the corona, and $\hat{L} = \chi\,\Delta\Omega\,L$, where $\Delta\Omega$ is the effective solid angle subtended by the companion star to its primary, L the compact star's luminosity (in erg s^{-1}) and χ an attenuation factor which takes into account possible absorption/screening and scattering/reprocessing effects in the disk or corona surrounding the compact star (typically $\chi \lesssim 0.1$ for X-rays). Under certain conditions, it is possible for the mass loss to be self-sustained. In the simplest model of self-sustained evolution described here, the radiation-driven mass loss (corresponding to a luminosity $L^* \propto \beta|\dot{m}|_{rad}$) can be either of order of $10^{18} - 10^{19}$ g s^{-1} or zero. The transition occurs at a value of the companion's mass m_c (hereafter, *the critical mass*) which corresponds to a sudden decrease of the accretion rate or even a permanent accretion turn-off if a msec pulsar has been produced [2].

2.1 LMXB LIFETIMES

Two features of the self-sustained evolutionary phase make the lifetime of a RD-LMXB substantially shorter than in standard models: the mass loss is driven at a relatively large rate, and one which is weakly dependent on orbital evolution; and the companion may contract inside its Roche lobe and still transfer mass. This last possibility depends crucially upon the quantity β, the amount of mass lost from the binary during the RD-evolutionary phase.

The LMXB lifetime is comprised of two phases, i.e., where the companion fills and underfills its Roche lobe. We have $\tau = \Delta t_{RL} + \Delta t_{NRL}$, where the time the Roche lobe is filled is $\Delta t_{RL} = (m_1/\langle|\dot{m}_{rad}|\rangle_{RL})(1 - m^*/m_1)$; and the time the Roche lobe is underfilled is $\Delta t_{NRL} = (m^*/\langle\dot{m}_{rad}\rangle_{NRL})\ (1 - m_2/m^*)$, with m_1 and m_2 respectively the initial and final companion's masses for $m_2 < m^* \lesssim m_1$, and $\langle|\dot{m}|_{rad}\rangle$ the corresponding average mass loss rate [3]. Although the value of the RD-mass loss rate may depend upon the orbital evolution, here we will adopt the following numerical estimates: $\langle|\dot{m}_{rad}|\rangle_{RL} = \langle|\dot{m}_{rad}|\rangle_{NRL} \simeq 10^{-8}$ M$_\odot$ yr^{-1} $\equiv \dot{m}_8$. For LMXB's with degenerate companions, m_1 can be larger (for an initial $\beta \sim 1/3$) or even equal to m^* (for $\beta \simeq 1$). Where mass transfer is self-sustained and the companion always fills its Roche lobe, $m_2 = m_{c,RL}$, the critical mass being given by $m_{c,RL} \simeq (0.1M/\Sigma)/(1-0.1/\Sigma)$, with $\Sigma = (\xi_{-1} f_{-1} \beta\chi)^{3/2}$, where $\xi_{-1} = \xi/0.1$ (ξ is the conversion of accreted matter into energy), and $f_{-1} = f/(0.1\,\Upsilon)$ with Υ is a dimensionless quantity depending on the irradiating spectrum (typically $\Upsilon \sim 1/2$ for soft gamma-rays and $\Upsilon \sim 10 - 100$ for X-rays). Alternately, after a Roche lobe underfilling evolutionary phase, $m_2 = m_{c,NRL} \simeq 0.6\, m^*[\xi_{-1} f\chi (m^*/0.04\, M_\odot)^{2/3}]^{-3/8}[1 + m^*/M]^{3/8}$, where we used $f \sim \Upsilon$ and $m^* \sim 0.04\, M_\odot$ appropriate to the evolution of binary progenitors of PSR 1957+20 [2]. Applying the relevant formulae given above to LMXB's with degenerate companions, we find the lifetime of a radiation-driven LMXB containing a companion with, say, $m_1 \sim 0.1\, M_\odot$ is $\tau_d \lesssim 10^7$ years for $\Upsilon\chi \sim 1$. In the case of LMXB's with degenerate companions underfilling their Roche lobe (as in the case of the progenitor of PSR 1957+20), $\Delta t_{NRL} \sim 10^6$ years; for example, Cyg X-3 may be in this intermediate phase [5]. For the numerical estimate of the birthrate of very rapid LMXB's containing white dwarfs we will use the lifetime $\tau_{vr,7} = \dot{m}_8^{-1} \tau_{vr}/(10^7 \text{ yrs})$.

Analogously, the inferred lifetime of an LMXB containing a lower main sequence companion with $n = 1/3$ is $\sim 10^7$ yrs [3]. For $\beta = 1$ we obtain $m_c/m^* \simeq 0.33(\xi_{-1} f\chi)^{-3/8} \cdot [(1 + m^*/M)/(m^*/0.4\, M_\odot)]^{1/4}$, and for $\beta = 2/3$ we have $m_c/m^* \simeq 0.11(\xi_{-1} f\chi)^{-3/4} \cdot [(1 + m^*/M)(m^*/0.4\, M_\odot)]^{1/2}$. In what follows, for the lifetime of rapid LMXB's we will use $\tau_r' = \dot{m}_8^{-1} \tau_r/(3 \cdot 10^7 \text{ yrs})$.

2.2 ORBITAL EVOLUTION

The orbital evolution of an LMXB is determined by angular momentum loss (driven by gravitational radiation or a wind), as well as by mass loss from the companion star. For masses and orbital parameters of typical observed LMXB's, mass transfer mechanisms neglecting irradiation effects yield an effective lifetime of order $10^8 - 10^9$. Evolution driven by the combined effects of GR and MB accounts successfully for most of the observed properties of known CV's.

A self-sustained RD-mass loss from the companion (or from the outer edge of the disk) leads to an orbital evolution quite different from the standard model. The irradiated companion can transfer mass even if it does not fill its Roche lobe exactly . We obtain [3]

$$\frac{\dot{P}_{orb}}{P_{orb}} = -3\left[1 - (1-\beta)\tilde{\alpha}(1+q) - \beta q - \frac{1}{3}(1-\beta)\frac{q}{1+q}\right]\frac{\dot{m}_{rad}}{m} + 3\left(\frac{\dot{J}}{J}\right) \tag{1}$$

where m is the mass of the companion star, $\dot{m}_{rad}$ the RD-mass transfer rate, β the fraction of $|\dot{m}|$ which is accreted onto the compact star, $\dot{J}/J = (\dot{J}/J)_{GR} + (\dot{J}/J)_{MB}$ the fractional change of the orbital angular momentum J from gravitational radiation (GR) and magnetic braking (MB), $q = m/M$ the ratio of the companion mass to primary mass M, and $\tilde{\alpha}$ the specific angular momentum parameter defined by $\delta J = \tilde{\alpha}\delta m(1-\beta)a^2 2\pi/P_{orb}$, where P_{orb} is the orbital period. The dimensionless parameter $\tilde{\alpha}$ gives the effective angular momentum loss caused mass loss from the binary. For example, if mass is lost from the binary at the orbital distance a and if the the outflowing gas does not acquire any additional specific angular momentum, then $\tilde{\alpha} = 1$. For a self-sustained value of the mass loss rate the first term is the dominant one. Note that the quantity $\dot{P}_{orb}/P_{orb}$ is expected to be $\sim +10^{-6}$ yr^{-1} for RD-LMXBs with very low mass degenerate companions. For main sequence companions the time derivative of the orbital period may be either positive or negative depending primarily on the value of β and $\tilde{\alpha}$. The associated lifetime turns out to be $\sim 10^7$ years.

2.3 ORBITAL PERIOD 'GAP' OF LMXB's

The LMXB orbital period gap can be 'crossed' relatively rapidly by RD-systems containing degenerate companions, or never crossed and 'repelled' by RD-binaries with main sequence companions. Alternately, even if the mechanism of mass transfer is *not* driven by radiation during the evolution of LMXB's above the period gap (for $P_{orb} \gtrsim 3$ hours), the sudden quenching of mass trasfer occurring at $P_{orb} \sim 3$ hours may yield a millisecond pulsar whose radiation pressure could be large enough to stop any further accretion. Fig. 1 shows the schematic orbital behavior expected in the RD-evolution.

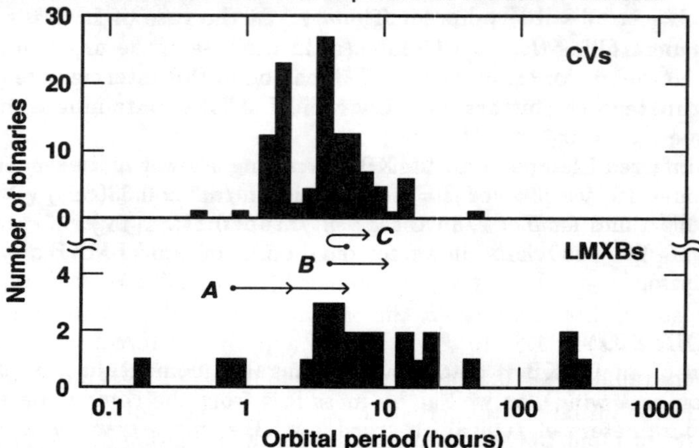

FIGURE 1 Distribution of orbital periods of CVs and LMXBs. Solid curves give the schematic behavior of the orbital RD-evolution. *Line A* represents the rapid expansion of an LMXB with a degenerate companion for β in the range $1/3 \lesssim \beta \lesssim 1$; the timescale may be $\sim 10^6$ yrs. *Line B* and *line C* give the orbital evolution according to Eq. (1) of LMXBs with main sequence companions.

2.4 VAPORIZING MILLISECOND PULSARS

Even in the post-accretion phase, radiation produced by the relativistic wind emanating from millisecond pulsars can continue to vaporize and drive mass loss from their low-mass companions [2,3,4] leading to partial or even total evaporation of the companion star. The ratio of the evaporation time scale $\tau_{evap.} = m_c/|\dot{m}_{rad}|$ and the spin-down time scale defined as $\tau_s = P/\dot{P}$ with P the pulsar spin period and $\dot{P}$ its time derivative is

$$\frac{\tau_{evap}}{\tau_s} \simeq 0.8 \frac{P_{-3}^2}{(f/0.1)I_{45}} \left(\frac{m_c}{0.2\ M_\odot}\right)^{1/3} \frac{1}{\Psi_c^2} \tag{2}$$

where P_{-3} is the spin period in milliseconds, I_{45} the neutron star's moment of inertia in units of 10^{45} g cm^2, Ψ_c the value of $\Psi = R/R_L$ at the accretion turn-off with R_L the Roche lobe radius. In Eq. (2) the total mass has been assumed to be $\sim 1.8\ M_\odot$. Energy deposition in the irradiated companion's atmosphere by secondary radiation produced by the pulsar relativistic wind[2] gives [2] $f \simeq 0.01\Upsilon$. Even though complete evaporation of

---

[2]In the case of PSR 1957+20, whether the ratio $\tau_{evap.}/\tau_s$ is less than unity depends upon currently unknown details of the irradiating spectrum; we obtain $\tau_{evap}/\tau_s \simeq 0.8 f^{-1} I_{45}^{-1}$. For a synchtrotron spectrum at fixed electron energy, $\Upsilon \simeq 10$, and for a time integrated synchrotron spectrum of $e^{\pm}$pairs which lose most of their energy, $\Upsilon \simeq 100$ [2]. It is therefore plausible that the companion of 1957+20 would be completely

low-mass companions is a possible outcome of RD-evolution, only a subclass of MSP's are capable of producing a sufficiently large evaporative wind. The RD-evolution of LMXB's makes plausible the existence of a relatively large number of MSP's in detached binaries containing lower main sequence stars heated and probably slightly evaporated by the pulsar wind. Doppler-induced spread caused by orbital motion in binaries with P_{orb} of order of a few hours makes difficult the detection of this important class of binary millisecond pulsars. However, the detection of MSP's with main sequence companions would provide invaluable information about binary evolution. It could be possible to distinguish between models which neglect companion irradiation and predict binary MSP's with $P_{orb} \simeq 3$ hrs, and the RD-model, which predicts 3 hr $\lesssim P_{orb} \lesssim$ 10 hr.

3. Statistics of Millisecond Pulsars and LMXB's

A recent statistical analysis of galactic MSP's gives a birthrate $BR(MSPs) \simeq 2 \cdot 10^{-5} yr^{-1}$ after taking into account the selection effects associated with radiopulsar surveys [7]. This birthrate is about 1000 times smaller than the galactic supernova formation rate and is too large by a factor $\sim 10 - 100$ to be in agreement with standard evolutionary models of LMXB's. This discrepancy is even more pronounced for the subgroup of 'rapid' low-mass binary pulsar systems (LMBP's), where the orbital period is $P_{orb} \lesssim 10$ days, and the birthrate $BR(rapid\ LMBPs) \simeq 1.5 \cdot 10^{-5}$ yr^{-1}. It is appropriate to consider separately binaries defined as 'very rapid' ($P_i \lesssim 0.5$ days), 'rapid' (0.5 days $\lesssim \mathrm{P_i} \lesssim$ 3 days), and 'slow' ($P_i \gtrsim 3$ days), with P_i the initial orbital period. From the estimates of section 2.1 we can calculate the birthrates of rapid LMXB's evolving due to the radiation-driven mechanism of mass transfer. We obtain [9] $BR(very\ \ rapid\ \ LMXB's) \simeq (7 \cdot 10^{-6}\ \ yr^{-1}) f_w \tau_{vr,7}^{-1}$ and $BR(rapid\ \ LMXB's) \simeq (6 \cdot 10^{-7}\ \ yr^{-1}) f_r \tau_r'^{-1}$, with $f_w \simeq 1/2$ the fraction of very rapid LMXB's with white dwarf and main sequence companions, and f_r the fraction of LMXB's with $P_{orb} \gtrsim 0.5$ days containing sub-giant companions as defined in Ref. [7]. If we compare the birthrate of very rapid LMXB's driven by radiation and of rapid MSP's we find that they are approximately equal. Given the current statistical uncertainties of pulsar surveys and LMXB number estimations the statistical MSP/LMXB discrepancy may therefore be resolved by very rapid RD-LMXB's [9]. We note that since the 'prompt' formation of MSP's after the AIC of white dwarf primaries possibly affects only a fraction of slow binaries, the AIC mechanism is insignificant in this context.

At present it is uncertain whether the MSP/LMXB discrepancy in globular clusters is the same as in the Galaxy [12] or smaller by a factor of ~ 10 [14]. The inferred MSP birthrate is $BR(cluster\ \ MSPs) \simeq (0.2 - 1.7) \cdot 10^{-6}$ yr^{-1}. From the measured value of $\dot{P}_{orb}/P_{orb}$ for 4U 1820-30 (see Table 1) and the large value of the mass loss rate of AC211 in the globular cluster M15, we obtain $BR(HML{-}LMXBs) \simeq 1 \cdot 10^{-6}$ yr^{-1}, if we consider only the two high-mass-loss (HML) binaries 4U 1820-30 and AC211. We note that despite their apparent low X-ray luminosities, LMXB's in globular clusters may have RD-driven orbital timescales of order 10^7 yrs, similarly to the X-ray burster EXO 0748-676 (see Table 1). The RD-LMXB birthrate in globular clusters may well be $\gtrsim$ *a few* 10^{-6} yr^{-1}.

evaporated even if $\Psi_c \simeq 1/3$. However, for the eclipsing pulsar in Ter 5 PSR 1744-24A we have $\tau_{evap.}/\tau_s \gg 1$. We note that the discovery probability of binary pulsars not able to *completely* vaporize their companion is much larger than the probability of discovering systems with $\tau_{evap.}/\tau_s \ll 1$.

Table 1

LMXB	P_{orb} (hours)	$\dot{P}_{orb}/P_{orb}$ (yr^{-1})	timescale (years)	Ref.
Cyg X-3	4.82	$+(2.20 \pm 0.22) \cdot 10^{-6}$	$5 \cdot 10^5$	11
X 1822-371	5.57	$+(3.40 \pm 0.94) \cdot 10^{-7}$	$2.9 \cdot 10^6$	12
EXO 0748-676	3.82	$-(2.02 \pm 0.28) \cdot 10^{-7}$	$5 \cdot 10^6$	13
4U 1820-30	0.18	$-(1.08 \pm 0.19) \cdot 10^{-7}$	10^7	14

4. Observation vs. Theory

The quantity $\dot{P}_{orb}$ has now been measured for Cyg X-3 [11], X 1822-371 [12], EXO 0748-676 [13], and 4U 1820-30 [14], all of which have $P_{orb} \lesssim 5.6$ hours. Table 1 gives the properties of the four LMXB's whose $\dot{P}_{orb}/P_{orb}$ has been measured. *None* of the measured values agrees with the predictions of the standard GR and MB models. The inferred timescale for orbital evolution is about 100 times shorter than expected in all systems. However, the RD-evolutionary mechanism may explain the observations quite naturally [6]. Cyg X-3 fits the evolutionary scenario for a RD-LMXB containing a very low mass white dwarf companion [5,3]. RD-LMXB's with lower main sequence companions may be characterized by a large positive value of $\dot{P}_{orb}/P_{orb}$ if the mass loss from the companion is $\dot{m}_8 \gtrsim 3$; and this may be the case for X 1822-371. Alternately, X 1822-371 may contain a white dwarf companion, and in this case the likely companion mass is $m \simeq (0.1\ M_\odot)\dot{m}_8$. We note that both the sign and magnitude of $\dot{P}_{orb}/P_{orb}$ for Cyg X-3 and X 1822-371 suggest a stable RD-mass transfer with $1/3 \lesssim \beta \lesssim 1$ and $\Psi \lesssim 1$. By contrast, EXO 0748-676 and 4U 1820-30 may belong to a class of RD-LMXB's in which the mass loss from the binary plays a crucial role in determining the orbital evolution [6]. According to the RD-evolution model, EXO 0748-676 and 4U 1820-30 are characterized by a relatively large mass loss from the binary with $\beta \sim 1/10$ and $\Psi \gtrsim 1$. No clear indication of ionized RD-mass loss has been obtained from known LMXB's thus far, with the possible and notable exceptions of Cyg X-3 and the source AC211 in the globular cluster M15. More observational and theoretical work is still necessary to unveil the many cryptic features of LMXB evolution.

References

[1] Rappaport, S., Verbunt, F., and Joss, P.C., 1983, *Ap.J*, **275**, 713.
[2] Tavani, M., 1989, in *Proc. of the 23rd ESLAB Symposium*, ESA SP-296, Vol. 1, p. 241.
[3] Tavani, M., 1990, submitted to *Ap.J.*
[4] Shaham, J., 1990, these Proceedings.
[5] Tavani, M., Ruderman, M. and Shaham, 1989, *Ap. J.(Letters)*, **342**, L31.
[6] Tavani, M., 1990, submitted to *Nature*.
[7] Kulkarni S.R., and Narayan, R., 1988, *Ap.J*, **335**, 755.
[8] Kulkarni, S., Narayan, R., and Romani, R.W., 1990, *Ap.J.*, **356**, 174.
[9] Tavani, M., 1991, *Ap.J. (Letters)*, **366**, L27-L31.
[10] Fruchter, A., Goss, D., 1990, *Ap.J. (Letters)*, **365**, L63-L66.
[11] Kitamoto, S., Miyamoto, S., and Matsui, W., 1987, *P.A.S.J.*, **39**, 259.
[12] Hellier, C., Mason, K.O., Smale, A.P., Kilkenny, D., 1990, *M.N.R.A.S.*, **244**, 39P.
[13] Parmar, A.N., Verbunt, F., Smale, A.P., and Corbet, R.H.D., 1991, *Ap.J.*, **366**, 253.
[14] Tan, J., et al., 1990, submitted to *Ap. J.*

COMPANION WINDS EXCITED BY NEUTRON STAR RADIATION IN LMXBS AND MILLISECOND PULSARS

J. SHAHAM
Physics Department and Columbia Astrophysics Laboratory
Columbia University
538 West 120th Street
New York, NY 10027

I. Abstract. The lecture first describes the basic principles underlying formation of a wind from a stellar atmosphere by external heating. These ideas are then applied to the case of binary companions of neutron stars (NS) which are being heated by the NS radiation in Low Mass X-ray Binaries (LMXBs) and in binary millisecond pulsar systems. Among others, the possibilities of companion evaporation and of self-excited X-ray systems is discussed. Finally, a model for the newly discovered "windy" binary pulsar 1744-24A is presented.

II. Introduction

Several problems associated with our understanding of LMXBs and millisecond pulsars have pointed in the direction of companion winds even before the first such wind was discovered in PSR1957+20:

One problem was that of the overabundance of bright LMXBs shining at, or close to, Eddington luminosities. Theoretical models suggest that the lifetime τ of a LMXB at mass transfer rate $\dot{m}$ is a steep inverse power law in the companion mass m, $\tau \equiv m/|\dot{m}| \propto m^{-\alpha}$ (α is 11/3 for Gravitational Radiation Driven mass transfer from a degenerate companion). Therefore, most observed binary systems in which mass accretion onto a compact object occurs should have low mass-transfer rates. While this seems to be true for Cataclysmic Variables it is manifestly not so for LMXBs, many of which are very bright.

Another problem had to do with the spin-up scenario for millisecond pulsars: if these are indeed spun-up in LMXBs then their periods should slow down as $|\dot{m}|$ drops so that when they emerge as radio pulsars their period p will only be

$$p \sim (6 \text{ msec}) B_9^{6/7} \left(\frac{m}{.1 M_\odot}\right)^{-3/7(\alpha+1)}$$

with B_9 their sufrace dipolar magnetic field in units of $10^9 G$. For $\alpha = 11/3$, $p \propto m^{-2}$, hence a slowdown to p values of order seconds could be expected. How do millisecond pulsars ever emerge from LMXBs?

A related problem was the solitary nature of PSR1937+214: a millisecond pulsar without a companion. If the spin-up scenario was right – how did the companion disappear?

J. Ventura and D. Pines (eds.), Neutron Stars: Theory and Observation, 203–218.

Indications for the very low mass of the companion of Cyg X-3 put it deep inside its Roche Lobe (RL). Can one understand mass transfer without RL overflow? And in 1957+20 itself the companion seemed again to be sunk in its RL. If the only mechanism for mass transfer was RL overflow and 1957+20 was a spun-up pulsar, how was RL overlflow terminated and how did the companion sink in its Lobe?

All of these problems can, in principle, be solved if radiation from the NS can induce a wind from the companion. When accretion from the wind onto the NS suffices to supply the energy of its radiation, bootstrapping can occur and luminosities can be driven up to the Eddington maximum until such time that the companion gets too far away and mass transfer rates drop by several orders of magnitude. The companion need not fill its RL for mass transfer to take place and may, occasionally, fully evaporate to leave behind a solitary millisecond pulsar.

Winds in close binaries with NSs seem, therefore, to play an important role in their behavior and evolution. This is why this topic is part of the present Advance Study Institute.

III. General Concepts – The Sonic Point in Radial Flow

Three equations govern steady wind flow in radial symmetry [1]: the continuity equation

$$\frac{d}{dr}(\rho u r^2) = 0, \tag{3.1}$$

Newton's second law

$$\rho u \frac{du}{dr} = -\frac{dP}{dr} - \rho \frac{d\Phi}{dr} \tag{3.2}$$

and the energy flow equation

$$\frac{1}{r^2}\frac{d}{dr}\left[\rho u r^2 \left(\frac{1}{2}u^2 + \omega + \Phi\right)\right] = Q. \tag{3.3}$$

The symbols in (3.1) – (3.3) are as follows:

- r radial distance from the stellar center
- ρ wind density
- u wind radial velocity
- P pressure
- Φ gravitational potential
- ω enthalpy per unit mass
- Q energy production rate per unit volume.

For an ideal gas the enthalpy function ω is

$$\omega = \frac{\gamma}{\gamma - 1}\frac{P}{\rho} = \frac{\gamma}{\gamma - 1}\frac{kT}{\mu} \tag{3.4}$$

where μ is the mass per pressure particle and γ is the polytropic exponent

$$\gamma = \left(\frac{\partial \log P}{\partial \log \rho}\right)_s .$$

The limit $\gamma \to 1$ would correspond to an isothermal flow (only $\frac{d\omega}{dr}$ enters the equations of motion, hence $\gamma \to 1$ should signal $\frac{dT}{dr} = 0$ by (3.4)).

In general,

$$d\omega = Tds + \frac{1}{\rho}dP$$

where s is the entropy per unit mass. Equation (3.2) can be rewritten as

$$\frac{d}{dr}\left(\frac{1}{2}u^2 + \omega + \Phi\right) = T\frac{ds}{dr} \tag{3.2a}$$

so that

$$T\frac{Ds}{Dt} = uT\frac{ds}{dr} = \frac{Q}{\rho} = \Lambda \tag{3.5}$$

where $\frac{D}{Dt}$ is the comoving derivative and Λ is the energy production rate per unit mass. Equation (3.5) shows the simple fact of life, that as a given chunk of matter moves downstream in the wind, its entropy increases (decreases) by heating (or cooling) if $\Lambda > 0 (\Lambda < 0)$.

The basic point to understand here is why would a wind form in the first place, and at what flux value, F, will it form when it does; we use

$$F \equiv \rho u r^2. \tag{3.6}$$

To get a feeling as to what is involved, consider first the $Q = 0$ case. Then (3.1) (as always) guarantees the constancy of F, (3.3) that of the "energy" per unit mass $\mathcal{E}$,

$$\mathcal{E} = \frac{1}{2}u^2 + \omega + \Phi \tag{3.7}$$

and (3.5) the constancy of the entropy s. From these constraints we can write the following equation for the variation of the Mach number

$$M = |\frac{u}{c}| \tag{3.8}$$

where c is the speed of sound

$$c = \left(\frac{\gamma P}{\rho}\right)^{1/2} : \tag{3.9}$$

$$f(M^2) \equiv \left(\frac{1}{2}M^2 + \frac{1}{\gamma - 1}\right) M^{-\frac{2(\gamma-1)}{\gamma+1}} = \Delta^{-1}(\mathcal{E} - \Phi)r^{\frac{4(\gamma-1)}{\gamma+1}} \tag{3.10}$$

where Δ is a constant which depends on the values of s and F.

If lifting a wind off a stellar surface is what we are after, we are looking for a transonic flow, one that begins with $M = 0$ and ends with $M \to \infty$. Upon analysis of (3.10) we discover, that its l.h.s minimizes as a function of M^2 for $M^2 = 1$; therefore, if one is after a *transonic* solution, the r.h.s. should minimize there too, or else the point at which the r.h.s. minimizes, $r = r_0$, will never be attained, nor will $M = 1$ and the flow will be confined to only one side of r_0 and one side of $M = 1$. r_0 is defined by

$$\frac{d}{dr}\left(\Phi r^{\frac{4(\gamma-1)}{\gamma+1}}\right)\bigg|_{r=r_o} = \frac{4(\gamma-1)}{\gamma+1}\mathcal{E}r_0^{\frac{3\gamma-5}{\gamma+1}} \tag{3.11}$$

From (3.10) and (3.11), at $M = 1$ we should have

$$\Delta = \frac{1}{2} \frac{d\Phi}{dr}\bigg|_{r_0} r_0^{\frac{5\gamma-3}{\gamma+1}} \tag{3.12}$$

Thus, in a transonic flow (3.12) fixes Δ as a function of $\mathcal{E}$ (by (3.11)) and therefore requires a well defined relation between F, s, and $\mathcal{E}$. This is how F can be determined.

For a wind off a star of mass m, $\Phi = -\frac{Gm}{r}$ and (3.11) gives

$$\mathcal{E} = \frac{5-3\gamma}{\gamma-1} \frac{Gm}{4r_0} \tag{3.13}$$

If the stellar radius is R we must clearly have

$$\mathcal{E} < \frac{5-3\gamma}{\gamma-1} \frac{Gm}{4R} \tag{3.14}$$

or else the flow will be either always subsonic or always supersonic. Alternatively, a *shock* could take the flow across the $M = 1$ singularity. If M_1 and M_2 are the Mach numbers on both sides of the shock, and if we denote $M_i^2 - 1 = Y_i$, then the shock conditions require

$$\left(Y_1 + \frac{\gamma+1}{2\gamma}\right)\left(Y_2 + \frac{\gamma+1}{2\gamma}\right) = \left(\frac{\gamma+1}{2\gamma}\right)^2 \tag{3.15}$$

whence $Y_1 Y_2 < 0$.

Two cases require special attention:

(i) $\gamma = 5/3$. This, the most common case of an ideal gas, is in trouble with (3.13). $\mathcal{E}$ has to be negative, but then one cannot get out to infinity!

(ii) $\gamma = 1$. This could be considered as the isothermal flow case, since $p \propto \rho$ means T is constant. (3.11) means $\mathcal{E} \to \infty$, but since (3.11) follows from (3.7) which, in turn, follows from the integration of (3.3), we must start there, because, as $\gamma \to 1$, the limit

$$\lim_{\gamma\to1} \frac{d\omega}{dr} = \lim_{\gamma\to1} \frac{\gamma}{\gamma-1} \frac{k}{\mu} \frac{dT}{dr} = F(r)$$

could be any function $F(r)$ (as long as $\frac{dT}{dr} \to 0$, of course). The correct limit turns out to arise by taking (3.10) to the $\frac{1}{\gamma-1}$ power and then going over to the limit. One gets

$$\frac{e^{M^2}}{M^2} = Ar^4 \exp\left(-\frac{4\Phi}{r_0 \Phi_0'}\right) \tag{3.16}$$

for the transonic solution,

$$A = \frac{1}{r_0^4} \exp\left[\frac{(r^4\Phi)_0'}{r_0^4 \Phi_0'}\right].$$

IV. General Concepts – The Sonic Point with Heating

We now turn to the case $Q > 0$. Equation (3.3) readily integrates to a generalized conservation law,

$$\mathcal{E}(r) = \mathcal{E}_0 + \int_{r_0}^{r} dr \frac{r^2 Q(r)}{F} \equiv \mathcal{E}_0 + G(r)$$
$$= \mathcal{E}_0 + \int_{r_0}^{r} T dS \tag{4.1}$$

It is clear that for heating of a given profile $Q = \tilde{Q} g(r/r_0)$, we shall find

$$F \propto \tilde{Q}$$

so that the wind flux will be proportional to the amount of net heating. The Mach number M varies according to

$$f(M^2) \equiv \left(\frac{1}{2}M^2 + \frac{1}{\gamma - 1}\right) M^{-\frac{2(\gamma-1)}{\gamma+1}} = \Delta^{-1}(\mathcal{E} + G - \Phi) r^{\frac{4(\gamma-1)}{\gamma+1}}, \tag{4.2}$$

an equation similar to (3.10) but with Φ replaced by $\Phi - G$, and with Δ depending on r through the entropy: $\Delta \propto \exp\left(\frac{2(\gamma-1)}{\gamma+1}\frac{s}{k}\mu\right)$. That replacement changes everything, even when $\gamma = 5/3$: A transonic solution readily exists, and (3.11) is replaced by

$$u_0^2 = \frac{1}{2} r_0 \Phi_0' + \frac{1}{2}(\gamma - 1)\frac{r_0 \Lambda_0}{u_0} \tag{4.3}$$

where $u_0 = c_0 \equiv \sqrt{\gamma(p_0/\rho_0)}$ is the flow speed at the sonic point.

With $\Phi = -\frac{Gm}{r}$, (4.3) can be rewritten as [2]

$$\eta_0^2 = 1 + \frac{x}{\eta_0} \tag{4.4}$$

where

$$\eta_0 = \frac{2u_0}{v_e} \qquad x = \frac{4 r_0 \Lambda_0 (\gamma - 1)}{v_e^3} \qquad v_e = \left(\frac{2Gm}{r_0}\right)^{1/2} \tag{4.5}$$

(v_e is the escape velocity from r_0).

(4.4) restricts $\eta_0 \geq 1$; that, in turn, requires the temperature T_0 at the sonic point to satisfy

$$T_0 \geq \frac{\mu v_e^2}{4\gamma k} \tag{4.6}$$

which is a powerful condition on T_0 and r_0.

We can distinguish between the strong heating (SH, $x \gg 1$) and weak heating (WH, $x \ll 1$) cases:

$$\eta_0 \simeq \begin{cases} 1 & x \ll 1, \quad \text{WH} \\ x^{1/3} & x \gg 1, \quad \text{SH} \end{cases} \tag{4.7}$$

From Eqs (3.2a) and (3.5) we see that when the flow speed is still low, $\frac{dT}{dr} \sim \frac{\gamma-1}{\gamma}\frac{\mu}{k}\frac{\Lambda}{u}$ so the first thing that happens is a quick rise in temperature and subsequent decrease in density, while the pressure does not change by much. For SH, the distance δr traveled to reach $T_0 \sim \frac{\mu v_e^2}{4\gamma k}$ satisfies $\frac{\delta r}{R} \sim \frac{1}{x}$.

V. Cooling and the Concept of P_{min}

To model the role played by radiative cooling in the formation of a wind in a stellar atmosphere heated externally at a rate Λ_H, McCray and his collaborators used the expression [3], [4]

$$\Lambda_c = 3.2 \times 10^{12} PT^{-3/2}(T_R - T) \qquad \text{erg/g} \tag{5.1}$$

which approximates free-free (FF) processes in ionized H, excited via photspheric photons at temperature T_R. For fixed P, $-\Lambda_c$ maximizes at $T = 3T_R$, with

$$-\Lambda_{c,max} \cong 1.2 \times 10^{12} PT_R^{-1/2} \qquad \text{erg/g} \tag{5.2}$$

Thermal equilibrium can be maintained as long as there is at least one temperature T for which

$$\Lambda \equiv \Lambda_H + \Lambda_c = 0 \tag{5.3}$$

If Λ_H is constant, (5.3) has generally two solutions, unless

$$P < P_{min} \simeq 10^5 \Lambda_{H,15} T_{R,4}^{1/2} \left(\frac{\alpha_{FF}}{\alpha}\right) \qquad \text{dyne/cm}^2 \tag{5.4}$$

where the α's are the cooling coefficients, ($\alpha_{FF} \equiv 3.2 \times 10^{12}\text{cm}^3(^\circ\text{K})^{1/2}/\text{g}$ for FF alone). Then, there is no solution. As one climbs up in the atmosphere at $u = 0$, (3.2) requires P to drop; but, once P_{min} is reached, $\Lambda = 0$ is impossible to maintain. Around that point is when a flow speed begins to build up and a wind begins to form. When Λ_H arises not from acoustic energy, say (like in the sun) but from external radiative heating, heating may become saturated (due to ionization) once a temperature T_s is reached, with $T_s \gg T_R$ (this condition is not really necessary – but it describes the reality of X-ray heating and simplifies the algebra). Thus [3], [4]

$$\Lambda_H = \begin{cases} \Lambda_{H,0}\left(1 - \frac{T}{T_s}\right) & T \geq T_s \\ 0 & T \geq T_s \end{cases} \tag{5.5}$$

In this case, one may maintain $\Lambda = 0$ below P_{min} with a discontinous jump of the temperature to values very close to T_s. Subsonic flow begins in this region, which becomes an isothermal flow at T_s, subject to (3.16) and (3.17). Across the temperature discontinuity (1 – below, 2 – above),

$$\begin{aligned} P_1 + \rho_1 u_1^2 \equiv P_{min} &= P_2 + \rho_2 u_2^2 = \rho_2 u_2^2(1 + M_2^{-2}) \\ &= \rho_2 u_2 c_2 M_2^{-1}(1 + M_2^2) \end{aligned} \tag{5.6}$$

c_2, the sound speed above the transition to T_s, is the same as it will be at the sonic point because the flow is isothermal. By (4.4), with $x \equiv 0$ for an isothermal flow, we have

$$\sqrt{\frac{kT_s}{\mu}} = c_2 \simeq \frac{1}{2} v_e \qquad (5.7)$$

from which the mass flux density ρu can be estimated [5] as $\sim \frac{2P_{min}}{v_e} \frac{M_2}{1+M_2^2}$.

As a function of $r_0/R \equiv \xi$, the maximum of $\frac{2M_2\xi^{1/2}}{1+M_2^2}$ under (3.16) is 1.1, obtained for $\xi \sim 1$. When $\xi \gg 1$ (a necessity, due to (4.6), if $T_s < \frac{\mu G m}{2\gamma k R}$, for example), it varies as $2\xi^{1/2}e^{-2\xi}$. Thus, the only good-sized wind, that is not just a large atmosphere in equilibrium shedding some of its top, will come when $\xi \sim 1$. Then,

$$\rho u \sim \frac{P_{min}}{v_e}. \qquad (5.8)$$

This is why P_{min} is so important a concept, and its detailed calculation through the calculations of Λ_H and Λ_c is of utmost importance for wind calculations.

For a star of cross sectional area $\mathcal{A}$ which intercepts a total external luminosity $\hat{L}$,

$$\Lambda_H = \frac{\hat{L}}{\rho \mathcal{A} \ell} \qquad (5.9)$$

where $\rho\ell$ is the column density absorbing the radiation,

$$\rho\ell \sim \frac{m_p}{\sigma}$$

with m_p the proton mass and σ the effective heating cross section in the stellar atmosphere. Thus,

$$\Lambda_{H,15} = 4 \times 10^{-16} \frac{\hat{L}}{\mathcal{A}} \left(\frac{\sigma}{\sigma_T}\right) \qquad \text{erg/g} \qquad (5.10)$$

with σ_T the Thomson cross section. Since $\rho u \sim \frac{\dot{m}_\omega}{\mathcal{A}}$ with $\dot{m}_\omega$ the total wind mass outflow per unit time,

$$\dot{m}_\omega \simeq \rho u \mathcal{A} \sim 10^5 \Lambda_{H,15} T_{R,4}^{1/2} \times 2.3 \times 10^{-8} \left(\frac{R}{R_\odot}\right)^{1/2} \left(\frac{M_\odot}{m}\right)^{1/2} \left(\frac{\alpha_{FF}}{\alpha}\right)$$

or

$$\dot{m}_\omega \simeq 1.3 \times 10^{-8} \hat{L}_{36} \left(\frac{\sigma}{\sigma_T}\right) \left(\frac{R}{R_\odot}\right)^{1/2} \left(\frac{M_\odot}{m}\right)^{1/2} \left(\frac{\alpha_{FF}}{\alpha}\right) T_{R,4}^{1/2} \qquad M_\odot/\text{yr} \qquad (5.11)$$

McCray, Sunyaev and their collaborators calculated Λ for Her X-1 in great detail [6], [7], [8]; they found $\frac{\sigma}{\sigma_T} \simeq 15$, $\frac{\alpha_{FF}}{\alpha} \sim .1$, but various authors obtained values in a two-orders-of-magnitude range for $\left(\frac{\sigma}{\sigma_T}\right) \left(\frac{\alpha_{FF}}{\alpha}\right)$! In the end, the issue of having self-excited accretion in Her X-1 was regarded unsettled. I will argue that chances are much better in LMXBs.

It is interesting to write down the expression for the efficiency $\lambda = \frac{E_K}{A\sigma T_R^4}$ of converting the infalling $\hat{L}$ (which determines $\sigma T_R^4 A$) to kinetic energy E_K of a wind. This is

$$\lambda \simeq 2 \times 10^{-3} \left(\frac{\sigma}{\sigma_T}\right) \left(\frac{\alpha_{FF}}{\alpha}\right) \left(\frac{RM_\odot}{mR_\odot}\right)^{-1/2} x^2 \tag{5.12}$$

Naturally, a high λ does not necessarily mean high $\dot{m}$; u_0 may be quite large when $x \gg 1$.

VI. The Slightly Underfilled Roche Lobe

When the companion almost fills its Roche lobe one may use plane parallel geometry: if the z coordinate is measured along the binary line of centers toward the compact star, $z = 0$ being the $L1$ point, then

$$\Phi = \Phi_0 - \frac{1}{2} f z^2 \tag{6.1}$$

is a good approximation as long as $\frac{z}{R} \lesssim 0.2$. Let the stellar surface will be at $z_s = -|z_s|$.

One key feature distinguishes this from the single star situation (i.e., from when the companion is buried inside its lobe): when $z > 0$ the wind is being helped by the gravity of the compact star, hence heating needs only expand the atmosphere to the Roche lobe boundary to form a wind. Indeed, the sonic point condition (4.3) becomes

$$\Lambda_0 = \frac{f u_0 z_0}{\gamma - 1} \tag{6.2}$$

showing that $M = 1$ occurs for positive z_0, hence beyond $L1$ (right *at* $L1$ if $\gamma = 1$).

Two time-scales are important here [4]: that of getting to $L1(\tau_F)$ and that of heating up to $T_s(\tau_H)$, with a ratio

$$\frac{\tau_F}{\tau_H} \simeq 10^4 T_{s,6}^{-3/2} \Lambda_{H,15} |z_s|_{10} \tag{6.3}$$

Also important is the ratio between T_s and $1/2 f R^2/k \equiv T_\Phi$

$$\frac{T_s}{T_\Phi} \cong \left(\frac{m}{M_\odot}\right)^{-1} \left(\frac{R}{R_\odot}\right) T_{s,6} \left(\frac{|z_s|}{0.1R}\right)^{-1} \\ \left(\cong \left(\frac{m}{M_\odot}\right)^{-0.18} T_{s,6} \left(\frac{|z|_s}{0.1R}\right)^{-1} \quad \text{for Main Sequence (MS) companions}\right) \tag{6.4}$$

High values for $\dot{m}$ will naturally be obtained when $T_s \gtrsim T_\Phi$; under such circumstances, the atmosphere could be heated up right off the stellar surface to escape velocities through $L1$. When $\tau_F \gg \tau_H$, the flow will arrive at $L1$ essentially with $M = 1$ and speed $\left(\frac{kT_s}{\mu}\right)^{1/2}$. But the more interesting case is that of $\tau_F \ll \tau_H$. In this case the flow will arrive at $L1$ with a low Mach number and low temperature, possibly close to $3T_R$.

If the pressure at the sonic point is (P_{min}/q) and the temperature $3\theta T_R$ $(q, \theta > 1)$, then

$$\rho u = \frac{\mu P_{min}}{3k\theta T_R q} \cdot \left(\frac{5 \cdot 3k\theta T_R}{3\mu}\right)^{1/2}$$

hence, by (5.4) and (5.10), we find

$$\dot{m}_\omega \simeq 3 \times 10^{-7} \hat{L}_{36} \left(\frac{\sigma}{\sigma_T}\right) \left(\frac{\alpha_{FF}}{\alpha}\right) \left(\frac{1}{q\sqrt{\theta}}\right) \qquad M_\odot/\mathrm{yr} \tag{6.5}$$

A comparison with (5.11) shows a potential for high values for $\dot{m}_\omega$ here. We shall later adopt a representative value of

$$\dot{m}_\omega \simeq 4.5 \times 10^{-7} \hat{L}_{36} \qquad M_\odot/\mathrm{yr} \tag{6.6}$$

based on $q\sqrt{\theta} \sim 10$ and $\left(\frac{\sigma}{\sigma_T}\right) \left(\frac{\alpha_{FF}}{\alpha}\right) \sim 15$.

VII. LMXBs vs. Her X-1

As I mentioned earlier, the possibility of having self excited accretion in Her X-1 is touch-and-go. What is the situation for LMXBs and VLMXBs (Very-Low-Mass-X-Ray-Binaries)?

To scale the binary separation a, let us measure it in units suitable for a $.1M_\odot$ MS companion to fill its Roche lobe around a $1.4M_\odot$ neutron star: $b = \frac{a}{5\times 10^{10}\mathrm{cm}}$. Then, the radiation temperature would be

$$T_R \cong 10^5 b^{-1/2} L_{38}^{1/4} \qquad {}^\circ\mathrm{K} \tag{7.1}$$

$$\Lambda_{15} \simeq 1.3 L_{38} \left(\frac{\sigma}{\sigma_T}\right) b^{-2} \tag{7.2}$$

and

$$P_{min} \cong 4 \times 10^5 L_{38}^{9/8} b^{-9/4} \left(\frac{\sigma}{\sigma_T}\right) \left(\frac{\alpha_{FF}}{\alpha}\right) \qquad \mathrm{dyne/cm^2} \tag{7.3}$$

L_{38} is the NS luminosity in units of 10^{38} erg/sec.

Since the number density n_0 at P_{min} satisfies

$$P_{min} = n_0 k \cdot 3T_R,$$

(7.1) and (7.3) give

$$n_0 \cong 9 \times 10^{15} b^{-7/4} L_{38}^{7/8} \left(\frac{\sigma}{\sigma_T}\right) \left(\frac{\alpha_{FF}}{\alpha}\right) \qquad \mathrm{cm^{-3}} \tag{7.4}$$

The depth at which n_0 will be achieved is given by $R\delta$, with

$$\delta = 1.5 \times 10^{-2} L_{38}^{-7/8} b^{7/4} \left(\frac{\sigma}{\sigma_T}\right)^{-2} \left(\frac{\alpha_{FF}}{\alpha}\right)^{-1} R_{10}^{-1} \tag{7.5}$$

[for a white dwarf companion of same mass, $\delta \propto b^{3/4}$]. The wind flow rate for a companion well inside its lobe and for X-ray heating will be determined by the escape velocity,

$$v_e = 5 \times 10^7 \left(\frac{m}{0.1M_\odot}\right)^{1/2} R_{10}^{-1/2} \qquad \mathrm{cm/sec} \tag{7.6}$$

so that, by (5.8)

$$\dot{m}_\omega \simeq 3.8 \times 10^{-8} L_{38}^{9/8} b^{-9/4} R_{10}^{1/2} \left(\frac{m}{0.1 M_\odot}\right)^{1/2} \left(\frac{\sigma}{\sigma_T}\right) \left(\frac{\alpha_{FF}}{\alpha}\right) \quad M_\odot/\mathrm{yr} \tag{7.7}$$

Now, this value of $\dot{m}_\omega$ is already higher than the value for Her X-1 because T_R is some 10 times higher. However, it could be even higher because at the higher temperatures, while heating by C, N, O, the main absorbers of X-ray heating, remains essentially the same, recombination cooling drops due to complete ionization of hydrogen and $\frac{\alpha_{FF}}{\alpha}$ becomes ~ 1.

VIII. VLMXB Bootstrapped Evolution

This being the case, VLMXBs ($M < 0.08 M_\odot$) have a good chance of emitting self excited X-rays in a bootstrapping process, in which the X-rays are emitted by accreting some of the wind they themselves generate at the companion [2]. The self-excited luminosity will build itself up to the most intense possible, possibly close to the Eddington luminosity. The fraction of the wind leaving the system will determine the evolution of the companion inside its lobe, but when b and R become such that $\dot{m}_\omega$ decreases below the self-sustained value, $\dot{m}_\omega$ will drop abruptly – either to the $\dot{m}_\omega$ expected from the radiation of the neutron star turned Radio pulsar if the companion is inside its Roche lobe, or to the $\dot{m}_\omega$ value of Roche lobe overflow. In either case, a large luminosity gap will form in the LMXB statistics and millisecond pulsars will not spin down due to a dwindling $\dot{m}$. Occasionally, a binary millisecond radio-pulsar with a "windy" companion could emerge [9], [10], some LMXBs could have the companions deep in their RLs [11] and some companions may be fully consumed, leaving behind a solitary millisecond pulsar. We will return to this last point in the next section.

The binary evolution is controlled by the following equations:

$$\dot{\eta}/\eta = \frac{2\dot{m}}{m}\left(1 - h + \frac{h\dot{m}_s}{|\dot{m}|} - \beta\right) \tag{8.1}$$

$$\dot{a}/a = \frac{2\dot{m}}{m}\left(\frac{h\dot{m}_s}{|\dot{m}|} - \beta\right) \tag{8.2}$$

where in (8.1) and (8.2), a is the binary separation and

$$\eta = \frac{R_L}{R_s} \tag{8.3}$$

with R_L the Roche Lobe radius and R_s the stellar one. For R_s one adopts the relation

$$\log\left(\frac{R_s}{R_\odot}\right) = -n \log\left(\frac{m}{M_\odot}\right) \quad , \tag{8.4}$$

β is the fraction of $|\dot{m}|$ that ends up being accreted, and

$$h = \frac{5}{6} - \frac{1}{2}n \tag{8.5}$$

where n is defined in Eq. (8.4). For $n = -.82$, appropriate to, e.g. moderately massive MS stars in thermal equilibirum, $h = 1.24$. For $n = 1/3$, appropriate for degenerate stars, $h = .67$. h is always > 0 but for convective stars, where the adiabatic index n_{ad} is $1/3$, h can have values between 1.24 and .67 depending on whether or not the mass transfer time is longer than thermal times [12]. Under standard evolution or when the star loses mass by external heating, h will be close to 1.24 until nuclear burning shuts off.

Also in (8.1) and (8.2), $\dot{m}_s$ is the mass loss rate expected solely from standard conservative evolution ($\beta \equiv 1$) and Roche-lobe-overflow ($\eta \equiv 1$).

While it seems to be always possible to maintain steady-state Roche Lobe overflow bootstrapping with a degenerate companion, where $h = 2/3$,this is not the case when the companion is a main sequence star and when $h > 1$. If, in Eqs. (8.1) and (8.2), we assume that $|\dot{m}| \gg \dot{m}_s$, we have $\dot{\eta}/\dot{m}\left(\propto \left(\frac{1}{6} + \frac{1}{2}n - \beta\right)\right) < 0$ for any β. Hence, even if at the beginning of the bootstrapped phase $\eta = 1$, as m decreases, η *increases*, meaning $R_L > R_s$: the companion begins to sink inside its Roche lobe. During the bootstrapping episode,

$$a \propto m^{-2\beta} \qquad (\propto m^{-2} \text{ for conservative evolution; } \propto m^{-2/3} \text{ for RL overflow}) \tag{8.7}$$

$$\eta \propto m^{2(1-h-\beta)} \qquad (\equiv 1 \text{ for RL overflow; } m^{-2} \text{ for conservative evolution}) \tag{8.8}$$

$$\Gamma \propto m^{4h+4\beta-10/3} \qquad (\propto m^{2/3} \text{ for RL overflow; } m^{4-2n} \text{ for conservative evolution}) \tag{8.9}$$

Bootstrapping will cease when, at a mass $m = m_c$, Γ becomes too small to trap enough radiation to produce the necessary wind. One finds

$$\nu_c^{4h+4\beta-10/3} = \frac{0.2}{\beta}\left(\frac{f\chi\xi}{0.06}\right)^{-1} M_{1.4}^{2/3}\nu_0^{4(h+\beta-1)} \tag{8.10}$$

where $\nu_c = \frac{m_c}{0.1M_\odot}$, $\nu_0 = \frac{m_0}{0.1M_\odot}$ [m_0 is the mass of the companion at the beginning of bootstrapping, assuming to fill its Roche lobe], $M_{1.4}$ is the neutron stellar mass in $1.4M_\odot$ units, f measures the efficiency of the wind formation process, χ is the fraction of the neutron-stellar radiation in the direction of the companion that is not obstructed and re-directed by matter, and ξ is the fraction of the rest mass energy of accreted matter that is converted to energy at the neutron stellar surface. Clearly, we must have $\frac{0.2}{\beta}\left(\frac{f\chi\xi}{0.06}\right)^{-1}\nu_0^{-2/3} < 1$ or else m_0 itself would already not satisfy the bootstrapping conditions.

IX. The Windy Binary Pulsar 1744-24A: A Test Lab for Accretion onto Millisecond Pulsars?

We heard from Professor Lyne [13] about a new pulsar, 1744–24A, that his group has recently discovered with a period of 11.563 m sec and a binary period of 109 minutes, which exhibits eclipse characteristics reminescent of those of 1957+20, the pulsar with the windy companion. We have discussed 1957+20 in the NATO ASI in Cesme in 1988. However, unlike the relative eclipse stability for 1957+20, intensities and eclipse durations of the new msec pulsar here vary considerably: eclipse durations as long as $\sim 180°$ in orbital phase and even disappearance of pulses for complete orbital periods have been observed. Given the characteristics of the eclipses Tavani and I now suggest that the mass loss in 1744–24A is due again to pulsar radiation, which drives a wind from the illuminated atmosphere of the companion, possibly this time a $\sim 0.1 M_\odot$ cool MS star which slightly underfills its Roche Lobe. Radio signals may be both reflected/refracted and free-free absorbed in the cool streams. Unlike in 1957+20, however, the neutron star might attempt to accrete some of the $\sim 2 \times 10^{-12} M_\odot \mathrm{yr}^{-1}$ mass flow, since the pulsar pressure in this system is only marginally strong enough to expel the flow. In this case the NS could occasionally produce X-ray bursts, as the propeller effect is likely to quench steady accretion; an identification of PSR 1744–24A with the X-ray burster seen in the direction of Ter 5 [14] is possible. The system could have evolved from a neutron star $+0.14 M_\odot$ MS companion binary in a bootstrapped accretion process which spun the pulsar up to a ~ 6 msec period while pushing the companion out. We are now at the end of the evolutionary phase during which the system has been coming back into contact. This scenario would make the system some 10^{10} years old but alternative scenarios with shorter lifetimes are also possible. If our interpretaiton is correct, 1744–24A may turn out to be the best laboratory found to date to study the accretion process on a millisecond period neutron star. Let me now discuss some of the details of this model.

Mass outflowing from an illuminated companion will encounter an opposing pressure from electro magnetic fields and particles in the pulsar wind that will attempt to push it out of the system instead of being accreted [15], [11]. To estimate the upper bound on mass flow that cannot accrete one may compare the magnetic pressure with that of the flowing mass,

$$B^2 = \left(\frac{2GM}{r}\right)^{1/2} \frac{|\dot{m}|}{r^2} \tag{9.1}$$

where B is the magnetic field, r is the distance sfrom the neutron star and we assumed that matter typically moves with Keplerian velocities.

The value of B can be estimated as μ/r^3 out to the light cylinder radius $r_c = \frac{cp}{2\pi}$, with μ the pulsar magnetic moment, p the pulsar period and c the speed of light; and as $4\pi^2\mu/c^2p^2r$ outside of r_c. Since the rhs of Eq. (9.1) varies as $r^{-5/2}$, the value of $|\dot{m}|$ which makes accretion possible, $\dot{m}_2$, is found by using Eq. (9.1) at a, the binary spearation. One then finds

$$\dot{m}_2 = 2.1 \times 10^{-11} \mu_{27}^2 p_0^{-4} M_{1.4}^{-1/2} a^{1/2} \qquad M_\odot \mathrm{yr}^{-1} \tag{9.2}$$

where $\mu_{27} = \frac{\mu}{10^{27} \mathrm{Gcm}^3}$ and $p_0 = \frac{p}{11.563 \text{ msec}}$. We shall consistently assume here that $\mu_{27} \sim 0.5$, the value found for several other millisecond pulsars, notably 1937+21. Note, that for values $|\dot{m}| > \dot{m}_2$, accretion may take place and there is only one solution of Eq. (9.1) inside the binary orbit, at roughly the Alfven radius, r_A.

When $|\dot{m}| < \dot{m}_2$ there are usually two solutions for (9.1), one (again an Alfven radius) inside of r_c and another outside of it, where a bow shock can be formed. These two values of r coincide, at r_c, when $|\dot{m}| = \dot{m}_1$,

$$\dot{m}_1 = 6.5 \times 10^{-13} \mu_{27}^2 p_0^{-7/2} M_{1.4}^{-1/2} \qquad M_\odot \mathrm{yr}^{-1}. \tag{9.3}$$

Unless $r_c > a$, $\dot{m}_1 < \dot{m}_2$. For $|\dot{m}| > \dot{m}_2$ accretion is possible, and for $|\dot{m}| < \dot{m}_1$ expulsion is certain. The mass flow dynamics in the regime $\dot{m}_1 < |\dot{m}| < \dot{m}_2$ depends on additional factors, not the least of which is the past history of $\dot{m}$:

If accretion is *on* and $|\dot{m}|$ begins to drop below $\dot{m}_2$, it will remain *on* as long as $r_A \ll r_c$, because pulsar radiation is likely to be quenched. Only when r_A approaches r_c, i.e., $|\dot{m}|$ approaches $\dot{m}_1$, will expulsion begin. Notice, however, that $\dot{m}_1 \propto r_c^{-7/2}$; if not r_c but some smaller radius, ϕr_c, defines $\dot{m}_1$, then $\dot{m}_1 \propto \phi^{-7/2}$ and even radii close to r_c may raise $\dot{m}_1$ by large factors.

In the other extreme case, when $|\dot{m}|$ begins to exceed $\dot{m}_1$, expulsion persists because pulsar radiation pressure still dominates at the matter injection point [unless these are clumps that now get injected, and they disintegrate further in, at a region where material pressure already dominates]. Only when $|\dot{m}| \gtrsim \dot{m}_2$ may gas accretion really begin. Thus, the accretion-expulsion dynamics exhibits a hysteresis effect.

If the companion of 1744-24A were completely degenerate (there is no reason to actually assume this, since the lower observational bound on its mass [13] is already above the nuclear burning threshold), it would be well inside its Roche lobe and, by (5.11), one might expect a hot ($T \gtrsim 10^7$°K) wind of $3.5 \times 10^{-14} M_\odot \mathrm{yr}^{-1}$ for a H star or $3 \times 10^{-13} M_\odot \mathrm{yr}^{-1}$ for a He one if $\mu_{27} \sim 0.5$. Note, that the luminosity L of a pulsar is expected to be

$$L = 2 \times 10^{33}\ \mu_{27}^2 p_0^{-4} \qquad \mathrm{erg/sec} \tag{9.4}$$

Both these values for the flux and temperature may prove to be incompatible with observations, which seem to indicate a mass-loss rate of order $\gtrsim 10^{-12} M_\odot \mathrm{yr}^{-1}$ and possible free-free absorption in a cool wind.

We shall therefore follow here the suggestion of Professor Lyne and his collaborators that the companion is actually a MS star, almost filling its RL. If the companion were actually to fill the lobe at present, Gravitational Radiation (GR)alone would give $|\dot{m}|/\dot{m}_2 \sim 5\dot{m}_s/\dot{m}_2 \sim 5\nu^2 \mu_{27}^{-2} p^4 M_{1.4}^{5/2} \alpha^{-9/2} \left(\frac{h}{1.24}\right)^{-1} \sim 20(2\mu_{27})^{-2}$ which, at least for the μ values around that of 1937+21, would be inconsistent with the radiation-driven evolution. In this case accretion would have a tendency to be bootstrapped near the Eddington level, and pulsar action would surely be suppressed.

We therefore consider a companion which slightly *underfills* its Roche lobe (and is therefore of mass close to $0.1 M_\odot$).

The induced cool and slow "windy" Roche lobe overflow in this case can be written as (eq. (6.6))

$$\dot{m}_w = 10^{-11} \mu_{27}^2 p_0^{-4} \psi \quad M_\odot \mathrm{yr}^{-1} \tag{9.5}$$

with $\psi = \frac{\Gamma}{0.01}$.

Note, that the energy of the exciting radiation here, sychrotron X-rays produced by relativistic particles in the companion-reflected pulsar magnetic field, is expected to be $0.5\mu_{27}\alpha^{-1}\gamma_5^2$ keV, where γ_5 is the γ factor of the relativistic particles coming from the pulsar in units of 10^5. This radiation should be very very efficient in producing the wind.

Curiously enough, for the value of $\dot{m}_w$ in Eq. (9.5), the ratio

$$\frac{\dot{m}_w}{\dot{m}_2} = 0.5\psi\alpha^{-1/2} M_{1.4}^{1/2}, \tag{9.6}$$

which is independent of μ, is *of order unity*, suggesting the possibility that large changes in wind (and bow shock) profiles will occur; and, since the wind temperature, $T \sim 10^4 T_4^\circ$K, is essentially photospheric, free-free absorption may be very effective with $\tau_{ff} \sim 10 T_4^{-3/2} \dot{m}_{14}^2$, so that large corresponding changes in eclipse durations will occur. One may also expect occasional accretion episodes. In 1957+20, by contrast, $\dot{m}_w/\dot{m}_2 \ll 1$ and no such behavior is expected.

When some material does manage, due to some fluctuation in parameters relevant for wind formation, to overcome pulsar radiation pressure and fall into the neutron stellar lobe, accretion onto the neutron star is still not likely to occur because of the propeller effect [15]: for the $\dot{m}_2$ of Eq. (9.2), the equilibrium spin-up period P_{eq} is longer than the period of 1744–24A,

$$P_{eq} = 43\psi^{-3/7}M_{1.4}^{-5/7} \text{ msec} \qquad (\text{Note}: \mu \text{ independent!})$$

and there is no overcoming of the centrifugal barrier.

We thus argue that, occasionally, $\dot{m}_w/\dot{m}_2 > 1$ and matter finds itself at a distance of 1.8×10^7cm from the neutron star, is being shock-heated there and propelled supersonically out. Whether or not such an episode can produce (e.g. by field-line reconnexion events) the puzzling burst activity observed by Hakucho during Aug. 5 – 21 of 1980 in the direction of Ter 5 is not clear to us at present. We only note here that their rate ($\sim$ .2hr) is consistent with the binary time scale of the system, the total energy output of the 14 observed bursts is of order of the propeller energy for an $\dot{m}_w$ value of order that of Eq. (9.5) and the temporal behaviour of each burst is consistent with the Alfven speed at r_A.

Let us now consider a scenario in which the neutron star has been spun-up by accretion. Unless, at some point, $|\dot{m}| < \dot{m}_1$, accretion would still be going on today; we therefore assume that accretion has gone on at a rate $|\dot{m}|_s$ until some neutron star spin period p_c was reached (which was, of course, no faster than the equilibrium period for the rate $|\dot{m}|_s$), at which mass accretion dropped to a rate $|\dot{m}|_e < \dot{m}_1$. From then on, as long as $\dot{m}_w \ll \dot{m}_2$, no accretion will have occurred again (see (9.6)). Also, p has slowed down to $\sim$ 11msec in a time $\lesssim 10^{10}$yrs. These conditions produce the following constraints on the scenario:

$$0.33 \leq \mu_{27} \leq 10^{-3}\left(\frac{|\dot{m}|_s}{|\dot{m}|_e}\right)\left(\frac{|\dot{m}|_s}{2.3 \times 10^{-11}M_\odot \text{yr}^{-1}}\right)^{1/2} \tag{9.7}$$

and

$$1 \text{ msec} \leq p_c \leq 0.08\left(\frac{|\dot{m}|_s}{|\dot{m}|_e}\right)^{6/7} \text{ msec} \tag{9.8}$$

where the l.h.s. of (9.8) comes from rotational stability requirements. At that point, the companion mass should already be around $0.1M_\odot$ – whether accretion dropped because magnetic breaking ceased [12], or accretion dropped because m_c was reached; in either case the companion will evolve inside its Roche lobe and the binary spearation should begin to decrease towards the present configuration. The time scale for that would be $\frac{0.1M_\odot}{\dot{m}_s}$ and should coincide with the pulsar slowing down time scale of $2.2 \times 10^9\mu_{27}^{-2}$yrs.

Alternatively, the pulsar may not have been spun up by accretion in this system at all, and the companion may have just been captured after the pulsar has been spun up in another binary or in a supernova. Given the high stellar density of the core of Ter 5, this could be the case, but in view of the spin-up scenarios widely accepted for millisecond pulsars, we consider here an evolutionary scenario that does rely on bootstrapping episodes in the same binary. We note, that the position of the pulsar in Ter 5, at some 10 core radii, does not predict ionization of such binary to be very likely.

We begin with a slowly rotating, $1.4M_{\odot}$, pulsar having a $.14M_{\odot}$ main sequence companion orbiting at a spearation $a \sim 10^{11}$cm. In 2.4×10^{9}yrs, GR takes the companion to its Roche lobe, at 6.4×10^{10}cm. Low rate, $3.6 \times 10^{-11} M_{\odot}\mathrm{yr}^{-1}$, GR driven RL overflow begins, which easily dominates the pulsar radiation to commence accretion. Since conditions are right, this immediately triggers bootstrapped accretion near the Eddington level. We assume, for the sake of definiteness, that in the bootstrapping phase all the mass lost from the companion is being accreted onto the neutron star, i.e. $\beta = 1$. We also assume the $\zeta \equiv \frac{f \times \xi}{0.06} = 1$. Similar scenarios are possible for other values of β (but not for all values).

This being a MS companion, the separation a grows to 1.25×10^{11}cm while it loses $0.04M_{\odot}$ and sinks in its Roche lobe, before the critical mass of $0.1M_{\odot}$ is reached, $\lesssim 10^{7}$yrs or so later. Then mass flow rate drops abruptly. By then, the neutron star has been spun up to ~ 6 msec and the pulsar induced wind rate, $\dot{m}_w$, satisfies $\dot{m}_w/\dot{m}_2 \sim 0.02$ and $\dot{m}_w/\dot{m}_1 \sim 1$ so that accretion, at $\sim 10^{-12} M_{\odot}\mathrm{yr}^{-1}$, may or may not still be going on (once it stopped, however, it would not resume at this phase). It is likely, however, that with the aid of the propeller effect accretion is suppressed. This is the beginning of the phase in which we are today, when the companion has been taken back by GR to $a = 6 \times 10^{10}$cm and to almost Roche lobe filling, 8×10^{9}yrs later. During this time the pulsar, if $\mu_{27} = 0.5$, has slowed down to the present spin of $\sim$ 11msec.

This seems to be the simplest scenario, but one can imagine other values of ζ and more massive initial companions, that go in and out several times, undergoing several bootstrapping phases until the present configuration is reached. For example, for $\zeta = 0.5$, the most recent bootstrapping phase, which leads to $m_c = 0.1M_{\odot}$, would begin with a $0.12M_{\odot}$ companion at 6×10^{10}cm, that recedes to 8.5×10^{10}cm while losing $0.028M_{\odot}$ over a period of $\sim 2 \times 10^{9}$yrs. Preceding it, a $0.148M_{\odot}$ Roche-lobe-filling star at $a = 6.54 \times 10^{10}$cm recedes to 9.9×10^{10}cm while losing $0.028M_{\odot}$ and comes to contact again 2.3×10^{9}yrs later. These two stages would have lasted only about half the time the above single process did with $\zeta = 1$. Here, a preceding stage could also exist: a $0.19M_{\odot}$ companion at a separation of 7.4×10^{10}cm, recedes to 1.2×10^{11}cm while losing $.042M_{\odot}$ and comes back into contact 4.7×10^{9}yrs later. A different (but not very different) value of μ_{27} will be needed here to bring the pulsar spin down to 11 msec at the present epoque, or else the intermediate spins are not too fast because μ_{27} would not have decayed yet to its present value.

X. Conclusion

The discovery of the eclipsing msec pulsar in the globular cluster Ter 5 comes only two years after the detection of PSR 1957+20. Both these systems belong to the interesting subset of binary msec pulsars showing evidence of evolution driven by radiation from the neutron star.

In this lecture we described the foundation for such a process and reported some preliminary results concenring the remarkable data from PSR 1744–24A. Even though much work is still to be done concerning both observational and theoretical aspects of the system (such as the phenomenology and detailed mechanisms and time dependence of eclipses as well as binary evolution) we have outlined a possible mechanism which makes possible intermittant pulsar emission and even sporadic episodes of accretion due to gas outflowing from the companion star illuminated by the pulsar wind. Unlike in the case of PSR 1957+20, the peculiar behavior of PSR 1744–24A is caused by the near equality of the ram pressure of the radiation-driven mass loss and the radiation pressure at the companion orbital distance.

Under normal conditions the outflowing gas is not likely to reach the stellar surface even if it is not immediately expelled by the radiation pressure because of the propeller effect. If the companion is a lower main sequence star, it may be slightly underfilling its own Roche lobe and pulsar irradiation of the outer atmosphere is crucial in causing the evaporative wind. Alternatively, an intrinsic wind activity of a low-mass companion star (never observed in other systems) could, in principle, explain the complex eclipse phenomenology.

From the inferred value of the evaporative mass loss from the companions in 1957+20 and 1744-24A, we deduce that the evaporation time scales are larger than the pulsar spin-down time scales. However, even though the companion stars are not likely to be *completely* evaporated in these systems, the characteristics of eclipsing msec pulsars can contribute significantly to clarify the physics involved in the process of formation of single msec pulsar by *complete* companion evaporation. We note that both PSR 1957+20 and PSR 1744–24A belong to a class of *low-flux* evaporating systems, expected to have a relatively long lifetime. From the upper limit on $\dot{p}$ we obtain a spindown time scale of order $p/\dot{p} \gtrsim 7 \times 10^9$ years in 1744–24A, and from the estimated low value of the mass loss we obtain an evaporation time scale $m/|\dot{m}| \gtrsim 10^{12}$years. It is therefore not surprising that 'low-flux' evaporating systems have a relatively high probability of being discovered. Another class of *high-flux* evaporating binaries containing msec pulsars should exist as well: their lifetime is expected to be shorter by several orders of magnitude compared to the low-flux systems, and they may therefore be harder to find.

XI. Acknowledgements

It is a pleasure to thank the directors of this ASI for their kind hospitality and partial support. NASA support for travel, preparing the lecture and manuscript and for carrying out some of the research described here through grant NAGW-1618, is gratefully acknowledged. This is contribution number 450 of the Columbia Astrophysics Laboratory.

XII. References

[1] Holzer, T.E. and Axford, W.I., *Ann. Rev. Astr. Ap.*, **8**, 31-60 (1970).

[2] Ruderman, M., Shaham, J., Tavani, M., and Eichler, D., *Ap.J.*, **343**, 292-312 (1989).

[3] London, R., McCray, R. and Auer, L.H., *Ap.J.*, **243**, 970-982 (1981).

[4] London, R. and Flannery, B.P., *Ap.J.*, **258**, 260-269 (1982).

[5] Basko, M.M., Hatchett, S., McCray, R., and Sunyaev, R.A., *Ap.J.*, **215**, 276-284 (1977).

[6] Basko, M.M., and Sunyaev, R.A., *Astr. Sp. Sci.*, **23**, 117-158 (1973).

[7] Buff, J. and McCray, R., *Ap.J.*, **189**, 147-155 (1974).

[8] McCray, R. and Hatchett, S., *Ap.J.*, **199**, 19s6-205 (1975).

[9] Ruderman, M., Shaham, J., and Tavani, M., *Ap.J.*, **336**, 507-518 (1989).

[10] Kluzniak, W., Ruderman, M., Shaham, J., and Tavani, M., *Nature*, **334**, 225-227 (1988).

[11] Tavani, M., Ruderman, M., and Shaham, J., *Ap.J. (Letters)*, **342**, L31-L34 (1989).

[12] Rappaport, S., Verbunt, F., and Joss, P.C., *Ap.J.*, **275**, 713-731 (1983).

[13] Lyne, A.G., Manchester, R.N., D'Amico, N., Staveley-Smith, L., Johnston, S., Lim, J., Fruchter, A.S., Goss, W.M., and Frail, D., *Nature*, **347**, 650-652 (1990).

[14] Makishima, K. *et al.*, *Ap.J. (Letters)*, **247**, L23-L25 (1981).

[15] Illarionov, A.F. and Sunyaev, R.A., *Astr. Ap.*, **39**, 185-195 (1975).

THE EVOLUTION OF NEUTRON STAR MAGNETIC FIELDS

D. BHATTACHARYA[1,2,*] and G. SRINIVASAN[3]

[1] *Astronomical Institute 'Anton Pannekoek'*
University of Amsterdam
Roetersstraat 15, 1018 WB Amsterdam
The Netherlands

[2] *Centre for High Energy Astrophysics*
NIKHEF-H, Kruislaan 409, 1098 SJ Amsterdam
The Netherlands

[3] *Raman Research Institute*
Bangalore 560080
India

ABSTRACT. In this article we review the arguments for and against the decay of the magnetic fields of neutron stars, in the light of most recent developments. We conclude that the field decay may only occur when the neutron star has been processed in an interacting binary. Some theoretical models for such an evolution are discussed.

1 Introduction

The magnetic fields of neutron stars are strong, and span a wide range of values from $\sim 10^8$ to $\sim 10^{13}$ G. Ever since the discovery of neutron stars questions have been asked about the origin and evolution of these strong fields. Ostriker and Gunn (1969) suggested that the magnetic fields of neutron stars must decay with age due to the ohmic dissipation of currents that maintain these magnetic fields.

This idea has been periodically challanged and supported by different researchers, based on various theoretical and observational considerations. For example, it was pointed out by Baym, Pethick and Pines (1969) that the interior of a neutron star is expected to be very highly conducting, perhaps superconducting, and the magnetic flux passing through the interior will be retained practically for ever. However, observations have shown that *some* field decay does seem to occur. In the present article we shall first discuss the observational

*On leave from Raman Research Institute, Bangalore, India

J. Ventura and D. Pines (eds.), Neutron Stars: Theory and Observation, 219–233.

evidence for field evolution, and then describe some theoretical models that attempt to explain these observations. The classical arguments for field decay are presented in section 2, and the revisions of these ideas necessitated by recent observations are discussed in the following sections.

2 Observational evidence for field decay

Gunn and Ostriker (1970) were the first to argue that there exists observational evidence for field decay in radio pulsars. They plotted the derived surface magnetic fields $B_s \propto \sqrt{P\dot{P}}$ (P=spin period, $\dot{P}$=spindown rate) against the "spindown ages" τ ($\equiv P/2\dot{P}$) of $\sim$ 15 pulsars for which the spindown rates were known at that time. The plot showed a clear decrease in derived B_s with increasing τ. But as pointed out by Lyne, Ritchings and Smith (1975), such arguments for field decay are extremely unreliable, since most of the above trend is artificial, caused by a much larger range in $\dot{P}$ than in P of the observed pulsars.

A more persuasive argument for field decay came with the measurement of the velocities of radio pulsars (Lyne, Ritchings and Smith 1975; Helfand and Tademaru 1977; Lyne, Anderson and Salter 1982; Lyne, Manchester and Taylor 1985). The velocity transverse to the line of sight has now been measured for over $\sim$ 75 pulsars (Lyne, Anderson and Salter 1982; Cordes 1986; Bailes et al. 1989, 1990), and clearly indicates that pulsars are high-velocity objects, with a population average speed of $\sim$ 200 km s^{-1}. As a result of this they are able to migrate to large distances from the plane of our galaxy despite being born very close to it. Thus the distance z of a pulsar from the mean galactic plane can in fact be used as an indicator of its age. Comparing the "kinetic age" t_k, defined as the ratio of the z-distance and the z-velocity, and the spindown age τ of 13 pulsars (fig. 1), Lyne, Anderson and Salter (1982) came to the conclusion that up to a couple of million years the spindown age provides a reasonable estimate of the true age of the pulsar, but beyond this the spindown age quickly becomes a gross overestimate of the true age, exactly as would be expected from a decay of the field strength. If the field strength B_s decreases exponentially, i.e. $B_s = B_s(t=0)\exp(-t/t_d)$, then the spindown age increases as

$$\tau = \left(\tau_0 + \frac{t_d}{2}\right)\exp(2t/t_d) - \frac{t_d}{2}. \tag{1}$$

Eq. (1) shows τ linearly increasing with time for $t \ll t_d$, but a very rapid increase in τ for $t \gtrsim t_d$. Fittting this to the observed data a decay timescale t_d in the range 2–10 Myr was obtained (Lyne, Anderson and Salter 1982; Lyne, Manchester and Taylor 1985), but the large uncertainties in t_k introduced by the unknown radial components of the velocities and the unknown location of a pulsar's birthplace considerably weakens this result.

A similar conclusion can be arrived at from the observed z-distribution of pulsars. Fig. 2 shows the average z-values of 7 groups of 50 pulsars each in different bins of τ. As can be clearly seen, the distribution is confined to within $z \sim$ 500 pc. This, along with the observed population average velocity of $\sim$ 200 km s^{-1} suggests that most pulsars are $\lesssim$ 5 Myr old. The spindown ages, on the other hand, extend more than an order of magnitude beyond this value—up to $\sim 10^8$ yr: suggesting that the spindown ages increase faster than the true age. In fig. 2 the effect of an exponential field decay is also shown. While the z-distribution expected with no field decay (dashed line) is a very poor fit to the observations,

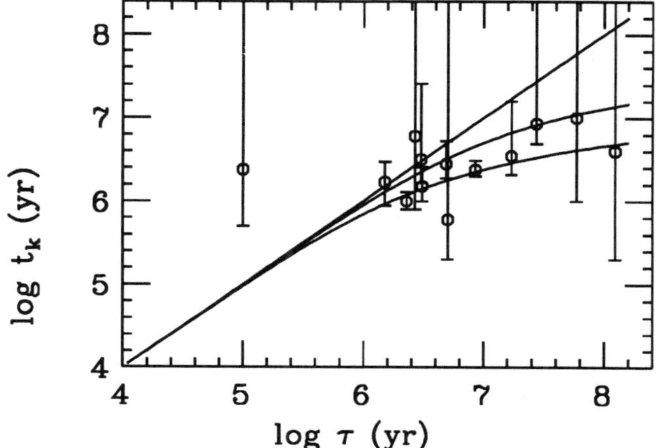

Figure 1: Observed relation between the kinetic ages and the spindown ages of 13 pulsars from Lyne, Anderson and Salter (1982). The three curves correspond to (from top) the expected relation in case of no field decay, and exponential field decay with timescales of 8 Myr and 2 Myr respectively.

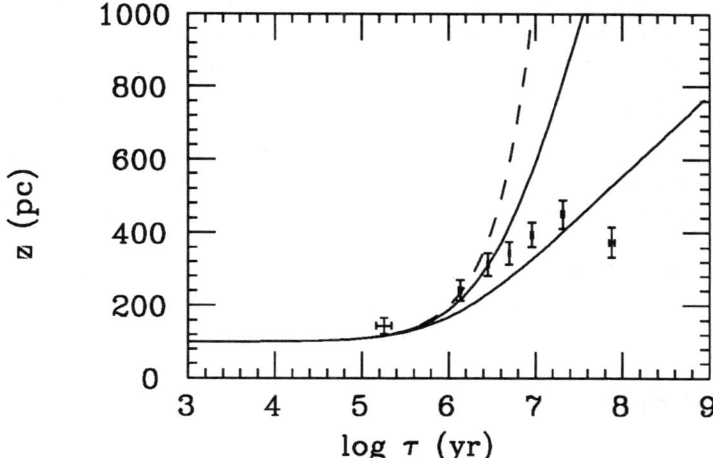

Figure 2: The average distance from the galactic plane of 7 groups of 50 pulsars each, arranged according to their spindown ages. The error bars correspond to $\pm 1\sigma$ errors on the mean values. The dashed curve shows the expected distribution for no field decay, and a population average space velocity of 170 km s^{-1}. The solid lines include exponential field decay with timescales of 8 Myr and 2 Myr respectively.

that including an exponential decay of the field strength (solid line) resembles the data much more closely.

There are several other pieces of supporting evidence in favour of field decay. For example, young pulsars associated with supernova remnants have higher magnetic fields than the old ones, associated with globular clusters or very cool white dwarf companions. Also, the high magnetic field, pulsating X-ray sources have young, massive companion stars while the non-pulsing X-ray sources are associated with old, low-mass donors (see Van Paradijs, this volume).

The distribution of the magnetic fields and periods of the observed pulsars (fig. 3) also provides some evidence for field decay. As can be seen, the number of short period ($P <$ 100 ms) pulsars is particularly lacking among those with magnetic fields between $10^{10.5}$ and 10^{12} G. This is in spite of the fact that pulsars born with low magnetic fields would spend a much longer time at short spin periods than their high-field counterparts. The common interpretation of this has been that these pulsars were in fact born with higher fields, but achieved their present field strengths due to field decay. The curved lines in fig. 3 show such evolutionary tracks for an exponential field decay with a time constant t_d of 5 Myr. The pulsars move horizontally at first till field decay becomes significant at times $t \gtrsim t_d$, following which the trajectories become vertical. Most of these pulsars would cross the "death line" in a few million years and cease to function.

Further evidence for long-term field decay comes from the radio pulsars in binaries (circled points in fig. 3). These objects are expected to have been spun-up due to accretion in the binary systems (see Van den Heuvel, this volume); the maximum possible spin-up for an Eddington-limited accretion being shown as the "spin-up line" in fig. 3. These neutron stars are at least as old as the sum of the "wait period" for the mass transfer to begin after the formation of the neutron star, and the duration of the mass transfer phase. Depending on the kind of binary system the neutron star was born in, its present age can range from a few million years to $> 10^9$ yr. These neutron stars have in fact been revived from the "graveyard" by means of spin-up (see fig. 3). It is significant that most of these old pulsars have magnetic fields clearly much lower than the general population of single radio pulsars (which are much younger, on an average). In addition, the pulsars processed in massive binaries with short evolutionary timescales have, in general, higher magnetic fields than those processed in low-mass binaries with long evolutionary timescales. The derived ages and the magnetic fields of these neutron stars appear to be consistent with an exponential decay with a time constant of ~5 Myr (Taam and Van den Heuvel 1986).

3 The residual field

In the last five years, however, it has become clear that not all is well with such a picture of exponentially decaying magnetic fields. Evidence for long-lived components of the magnetic field (usually called the "residual field") has now emerged from several quarters:

1. The white dwarf companions of at least two binary pulsars have now been detected (Kulkarni 1986). Among these, the companion white dwarf of PSR 0655+64 has a rather low surface temperature, which indicates a "cooling age" in excess of 10^9 yr. The age of the neutron star, the first born member of the binary system (see, e.g.

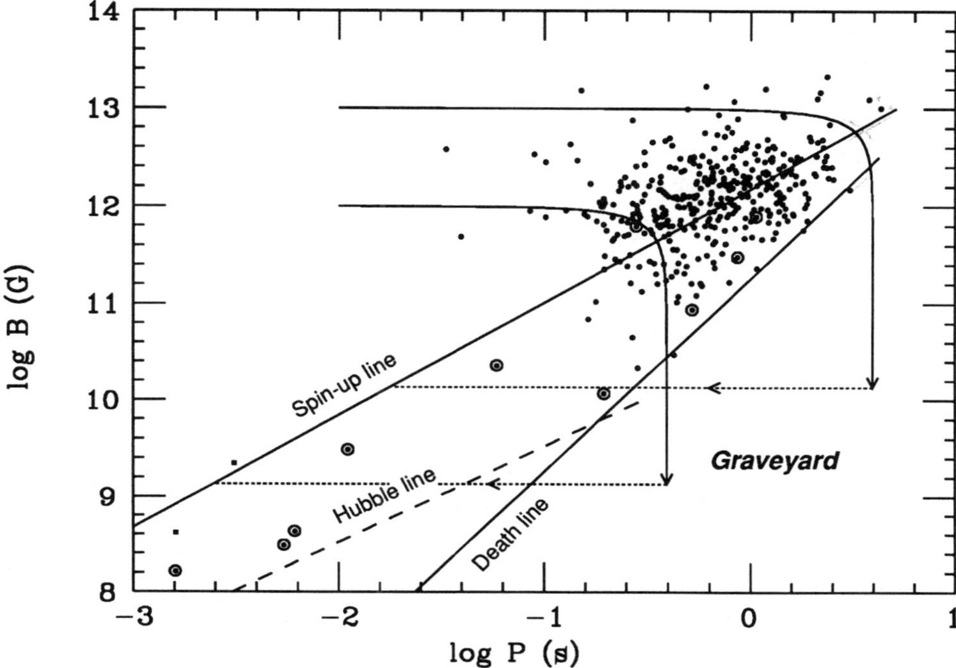

Figure 3: The derived magnetic fields ($\propto \sqrt{P\dot{P}}$) of 403 radio pulsars plotted against their spin periods. Pulsars in binaries are shown encircled and single millisecond pulsars are shown as filled squares. The "death line" corresponds to a polar cap voltage below which the pulsar activity is likely to switch off. The "Hubble line" represents a spindown age $\tau \equiv P/2\dot{P}$ of 10^{10} yr for a pulsar. The two curved lines show typical evolutionary tracks of pulsars, with exponential decay of the magnetic field on a 5 Myr timescale. The horizontal dotted lines depict the spin-up of a dead pulsar in an accretion process, turning it into a functioning radio pulsar again ("recycling"). The "spin-up line" represents the minimum spin period to which such a spin-up may proceed in an Eddington-limited accretion.

Van den Heuvel 1984), must be even larger than this. Nevertheless, this neutron star has retained a magnetic field $\sim 10^{10}$ Gauss. If the magnetic field continued to decay with an e-folding timescale of a few million years, then the expected field at present would be practically zero, and the pulsar activity would have ceased a long time ago. The upper limit on the surface temperature of the white dwarf companion of PSR 1855+09 (Callanan et al. 1989; Kulkarni, Djorgovski and Klemola 1990) also leads one to a similar conclusion.

2. The second piece of evidence comes from the statistics of millisecond pulsars (Bhattacharya and Srinivasan 1986; Van den Heuvel, Van Paradijs and Taam 1986). The prolonged accretion required for spinning a neutron star up to a millisecond period, as well as the nature of their binary companions suggest that the progenitors of millisecond pulsars must be low-mass X-ray binaries. The small volume of the galaxy in which the millisecond pulsars have been detected, and the possible selection effects against their detection suggest that the total number of active millisecond pulsars in the galaxy exceed that of low-mass X-ray binaries by a factor of *at least* 100 (Bhattacharya and Srinivasan 1986; Kulkarni and Narayan 1988; Coté and Pylyser 1989). This means that the active lifetimes of millisecond pulsars must be larger than that of low-mass X-ray binaries by a similar factor. Since the active lifetimes of low-mass X-ray binaries are estimated to be in the range 10^7 to 10^8 yr, that of millisecond pulsars must be larger than 10^9 yr. This would just not be possible if their magnetic fields continue to decay in $\lesssim 10^7$ yr timescale.

3. The gamma-ray bursters are likely to be very old neutron stars, as is indicated by their isotropy around us (see D. Lamb, this volume). The discovery of cyclotron absorption features in some of these gamma-ray bursters (Murakami et al. 1988) suggests that these may have "residual" fields $\sim 10^{12}$ G.

4. It has been estimated that the neutron star in the X-ray binary Her X-1 is at least $\sim 6 \times 10^8$ yr old (Verbunt, Wijers and Burm 1990). Yet this object seems to have retained a magnetic field as high as $\sim 10^{12}$ G.

4 Field decay in binaries?

One would notice from the above examples that they provide not only the evidence for the existence of "residual" fields, but also suggest that the residual fields may have a wide range of values—starting from ~a few times 10^8 G for millisecond pulsars, to $\sim 10^{12}$ G for gamma ray bursters, Her X-1 and 4U 1626−67, with PSR 0655+64 (10^{10} G) falling in between. This, however, encompasses practically the entire range of magnetic fields observed in neutron stars. This confronts us with the question as to what determines the value at which the initial field decay will stop. If the initial field decay is spontaneous, it is hard to see why this should stop at widely different values in different neutron stars. One should therefore turn the question around, and ask what causes the field decay in the first place, and what determines its extent. A promising explanation seems to be that the magnetic field of a neutron star decays significantly only if it has, at some point in its life, been a member of an interacting binary system.

It must be noted that all observed neutron stars which possess significantly low field strengths either are, or bear distinct characteristics of having been, members of accreting binary systems. These neutron stars also exhibit a general correlation between the amount of matter accreted from the companion and the degree of magnetic field decay (Taam and Van den Heuvel 1986). Further, on re-examining the relation between the kinetic ages and the spindown ages of radio pulsars (see above), Bailes (1989) has come to the conclusion that those pulsars which are likely to have been "recycled", exhibit little field decay *after* the end of mass transfer. All this seems to point to the fact that interaction with a binary companion plays a central role in the decay of neutron star magnetic fields.

5 Field decay caused by mass accretion

What physical mechanism may cause a decay of the magnetic field of an accreting neutron star? It was suggested by Bisnovatyi-Kogan and Komberg (1974) that the accreted matter may screen and bury the magnetic field of a neutron star, and thus, even if the intrinsic field of the neutron star does not decay, the exterior dipole field may diminish.

More recently, several authors have attempted to construct models for the decay of the *intrinsic* magnetic field due to accretion. The following physical processes may be relevant in this context (see, e.g. Romani 1990). (*i*) The heating of the crust due to accretion will cause the crustal conductivity to drop, thus hastening ohmic decay of the currents there. (*ii*) If the original flux is entirely confined to the crust, (as it would be if it was generated by thermoelectric effects after neutron star formation—see Blandford, Applegate and Hernquist 1983) then the compression of the current-carrying layers due to the accreted overburden may cause a reduction in the dipole moment by bringing opposing current loops into close contact (Romani 1990). (*iii*) The accreted matter, after settling onto the neutron star and anchoring the magnetic field, may flow towards the equator, dragging field lines and causing reconnection, eventually decreasing the exterior field (Romani 1990). (*iv*) An inverse thermoelectric battery effect may also operate and destroy the crustal flux, aided by thermally-induced field convection transporting flux outwards from deeper layers (Blondin and Freese 1986).

Such mechanisms of field destruction would predict a direct relation between the amount of matter accreted and the amount of flux destroyed. Simple forms of such relation, examined by Shibazaki et al. (1989) and Romani (1990), appear to be able to reproduce the observed field-period combinations of the recycled pulsars. However, one counter-example has recently been pointed out: the neutron star in the X-ray binary 4U 1626-67 may have accreted a large quantity of mass, but has still retained a magnetic field $\sim 10^{12}$ G (Verbunt, Wijers and Burm 1990).

The mechanisms just discussed operate solely in the crust of the neutron star and thus cannot influence the magnetic flux passing through the superconducting interior. A completely different mechanism may, however, operate to relate the evolution of the "core" magnetic field to the accretion process (Srinivasan et al. 1990).

6 Field decay due to spindown

The interior of a neutron star is believed to be composed of superfluid neutrons and superconducting protons. The spin angular momentum of the neutron star is carried by quantized Onsager-Feynman vortex lines in the superfluid neutron component, and any magnetic flux passing through the interior is confined to quantized Abrikosov fluxoids in the superconducting proton component (see Alpar, this volume and Pines, this volume). If the magnetic field of the neutron star has to decay, these fluxoids must be expelled from the superconducting region. Muslimov and Tsygan (1985) pointed out that such an expulsion may indeed occur due to the buoyancy force acting on the fluxoids. The timescale for fluxoid expulsion has been somewhat controversial, but most authors now agree that if the buoyancy force and a drag due to electron scattering are the only forces acting on a fluxoid, the flux can be expelled from the interior in $\lesssim 10^7$ yr (see e.g. Jones 1987,1988).

An additional effect, however, is most likely to frustrate such a flux expulsion. As has been recently pointed out (Sauls 1989; Srinivasan et al. 1990), a strong tendency of pinning may exist between the proton flux tubes and the neutron vortex lines, which would make it impossible for these entities to move independent of each other. Though detailed exploration of the physics of such a pinning is yet to be made, at an intuitive level it seems quite reasonable to expect this effect. In situations encountered in the laboratory, the existence of pinning can be traced to local modifications of the penetration depth, condensation energy, etc. Since the cores of both the fluxoids and the vortices are "normal", they are likely to act as pinning centres. There are several reasons to expect that the pinned and the unpinned configurations will have different energies:

1. Because the proton fluxoid is magnetized, the pressure of matter in the vicinity of the core of a fluxoid is reduced due to the enhanced magnetic pressure, and also due to circulation of supercurrents,

$$\Delta P \sim \frac{1}{2}\rho_{\rm P}\left(\frac{h}{2m}\cdot\frac{1}{r}\right)^2 + \frac{B^2}{8\pi}$$

which implies a density fluctuation $\Delta\rho \sim \Delta P/c_s^2$. Here the subscript p stands for protons, c_s is the sound velocity, and other symbols have their usual meaning.

2. The vortices in the $^3P_2$ core neutron superfluid will be "ferromagnetic" (Sauls 1989), with a vortex magnetization $\sim 10^{11}$ G. This will lead to a density fluctuation in the core of a neutron vortex. There will be additional density fluctuation due to the pressure equilibrium in presence of circulating superfluid as mentioned above.

3. The neutron and proton fluids are coupled to each other via strong interactions. Consequently, the neutron and proton are "dressed" by polarization clouds which involve particles of *both species*. Because the polarization clouds are modified by condensation, the superfluid mass current of the neutron and proton are also modified by strong interactions. An important consequence of such a *superfluid drag* is that the neutron vortex acquires a very strong magnetization, $\sim 10^{15}$ G, four orders of magnitude larger than the magnetic field due to the ferromagnetism of the $^3P_2$ vortex.

Given the above, strong pinning between proton fluxoids and the neutron vortices seem likely. As Sauls (1989) has pointed out, the interaction energy due to proton density perturbations at the centre of a fluxoid may be estimated as follows:

$$\varepsilon_{\rm pin} \sim n_{\rm n} \left(\frac{\Delta_{\rm n}}{E_{\rm Fn}}\right)^2 \left(\frac{\Delta_{\rm p}^2}{E_{\rm Fp}}\right) \left(\xi_{\rm n}^2 \xi_{\rm p}\right) \quad \text{per connection}$$

where Δ is the energy gap, $E_{\rm F}$ the Fermi energy, ξ the coherence length and n is the number density. For typical values of these parameters the pinning energy is $\sim$(0.1–1) MeV per connection. A rough estimate of the pinning *force* per connection, $F_{\rm pin}$, may be obtained by dividing the pinning energy $\varepsilon_{\rm pin}$ by the neutron coherence length $\xi_{\rm n}$, which is a measure of the interaction region. This yields $F_{\rm pin} \sim (0.1 - 1) \times 10^6$ dyne per connection.

Another source of interaction between the fluxoids and the vortices comes from their magnetic fields. The core magnetic fields of both the fluxoids and the vortices are $\sim 10^{15}$ G, and the electromagnetic interaction energy $\varepsilon_{\rm em} \sim B_{\rm v} B_{\rm f} (\pi \lambda_{\rm v}^2) \lambda_{\rm f} / 8\pi$, where λ is the London penetration depth, works out to be $\sim$ 10 MeV per connection (see Ruderman, this volume), but due to a larger scale length involved, the *force* due to such an interaction is of a similar magnitude as the pinning force.

When a neutron star spins down, the neutron vortices in the superfluid decrease in number by moving out and annihilating at the superfluid boundary (see Alpar, this volume). These moving vortex lines will also carry the pinned fluxoids with them, causing an expulsion of the flux from the interior. The amount of flux retained in the interior will thus be determined directly by the spin period of the neutron star. The spin evolution of an isolated neutron star due to dipole braking with $B_s \propto \Omega$ would result in a very slow field evolution $B_s \propto t^{-1/4}$, with a net decay of only about an order of magnitude in a Hubble time (Srinivasan et al. 1990).

Neutron stars in binaries can, however, be spun down to much longer periods, especially during phases of weak accretion from the companion's stellar wind. Spin periods as long as 835 s have been observed among the binary X-ray pulsars. Such a spindown will reduce the number of vortex lines, and consequently the number of fluxoids, by a large factor, causing a substantial field decay. Subsequently strong accretion will spin the neutron star up to a short period (leaving a "recycled pulsar" in the end); but this process will inject new vortices into the core, and all the remaining flux will be trapped there for ever. Thus according to this model the "residual" field of a neutron star will be determined by the *maximum* spin period to which the star is spun down in its binary history (Srinivasan et al. 1990).

7 Comparison with observations

The intimate connection between the evolution of a neutron star in a binary system and the behaviour of its magnetic field thus appears to provide a very promising explanation for the long-term behaviour of the magnetic field. Is it possible to understand the signs of field decay exhibited by the population of *solitary* pulsars based on such a hypothesis? The answer is yes, provided a significant fraction ($\gtrsim$30%) of the the solitary pulsar population is composed of *recycled* pulsars processed in binaries.

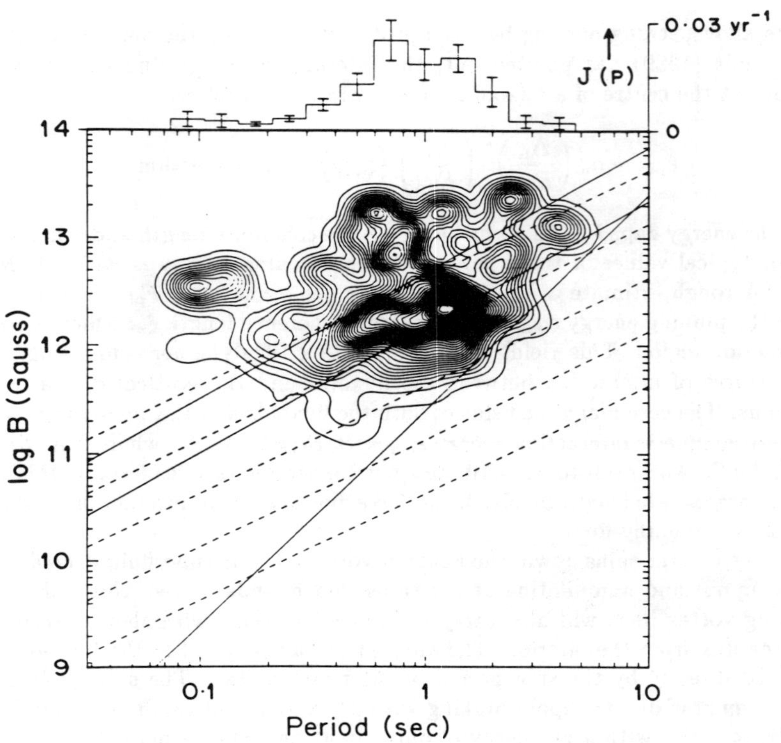

Figure 4: The distribution of pulsar "current" (see text) as a function of magnetic field and spin period. The data are obtained using the pulsars detected in 8 surveys, and correcting for the selection effects in these surveys in a manner described by Narayan (1987), to obtain a representative bias-free sample for the whole galaxy. The three sloping solid lines correspond to (from bottom) the death line, the spin-up line for Eddington accretion rate and the spin-up line for 10 times the Eddington accretion rate. the dashed lines represent constant spindown ages. The horizontal nature of the constant current lines upto nearly the death line suggests that pulsars undergo little field decay during their active lifetime. The histogram at the top is the distribution of current as a function of period, integrated over all magnetic fields.

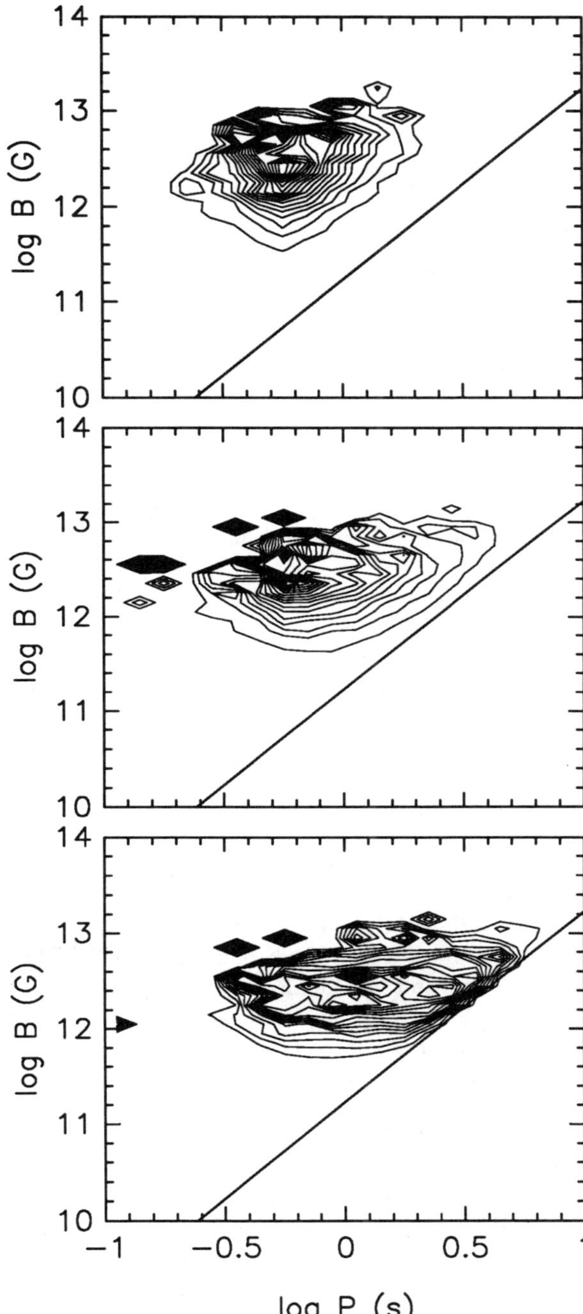

Figure 5: The expected distribution of pulsar current as a function of magnetic field and spin period as obtained from a Monte-Carlo simulation. The three panels include exponential magnetic field decay with time constants of (from top) 2 Myr, 10 Myr and 10^{10} yr respectively. The sloping straight lines in these diagrams represent the "death line" for pulsars (see fig. 3). Comparison with the observed distribution, as depicted in fig. 4, suggests that galactic radio pulsars have field decay timescales much in excess of 10 Myr.

New indication for this has emerged from a recent study of the distribution of the "current" of pulsars as a function of magnetic fields and periods (Deshpande and Srinivasan 1990). The "current" of pulsars due to spindown from a bin of width ΔP around a period P at a magnetic field B is defined as

$$J(B,P) = \frac{1}{\Delta P}\sum_i \dot{P}_i$$

where the sum extends over all pulsars in the galaxy that fall into the appropriate bin. From the observed population of pulsars, the galactic population was constructed by accounting for selection effects as described in Narayan (1987). The result is shown in fig. 4 (details will be published elsewhere). The fact that the constant current lines are more or less horizontal almost all the way to the death line suggests that these pulsars cross the death line essentially by spinning down, with hardly any field decay. For a comparison, we show in fig. 5 the *expected* distribution of $J(B,P)$ for field decay timescales of 2 Myr, 10 Myr and 10^{10} yr. These distributions were obtained from a Monte-Carlo simulation of pulsar evolution. It is clearly seen that to be consistent with observations (namely, fig. 4), a field decay timescale much in excess of 10 Myr will be necessary. In fact, using the current distribution in fig. 4 Deshpande and Srinivasan (1990) estimate that the decay timescale must be greater than 20 Myr.

One thing to note in fig. 4 is the near absence of the contribution to pulsar current of pulsars with magnetic fields below about $10^{11.5}$ G. There is some indication from this statistical study that these pulsars form a separate population, and are restricted below the so-called "spin-up" line (see fig. 6). It is very likely that these pulsars, which have conventionally been thought to have come from higher field pulsars via exponential field decay, have actually undergone "recycling" in binary systems, and their magnetic fields have been reduced beacuse of the processes mentioned above. This immediately raises the question as to whether there are enough progenitor binaries that can provide such a large number of isolated recycled pulsars.

The objects that are likely to produce a significantly large number of recycled pulsars, and also cause the disruption of the binary in a second supernova explosion are neutron star binaries with B/Be star companions. Some of these binaries are visible as transient X-ray sources. Observations by the Ginga satellite indicate that there is a large population of such objects in the galaxy (Koyama 1988). Meurs and Van den Heuvel (1989) have recently estimated this number to be at least $\sim 2\times 10^4$. Since these binaries have a typical lifetime of $\sim 10^7$ yr, they can provide $\gtrsim$10% of the pulsar birthrate in *recycled* pulsars alone. Further, since the active lifetime of a radio pulsar in absence of field decay is inversely proportional to its field strength, these recycled pulsars—despite their lower birthrate—can constitute a huge fraction of the total pulsar population if the processing of these neutron stars in binaries decrease their field strengths by, say, an order of magnitude.

The last piece of observation that remains to be understood is the restriction of pulsar z-distribution to $\lesssim$1 kpc, which would imply lifetimes $\lesssim 10^7$ yr for an average radio pulsar (cf. fig. 2). If field decay is related to binary evolution, then, as mentioned above, one expects a fraction of radio pulsars to have lifetimes much longer than this. With a z-velocity of $\sim$ 100 km s^{-1} such a pulsar would move several kiloparsecs away from the galactic plane, which is not observed. Here we would like to advocate some caution. Most of the distances

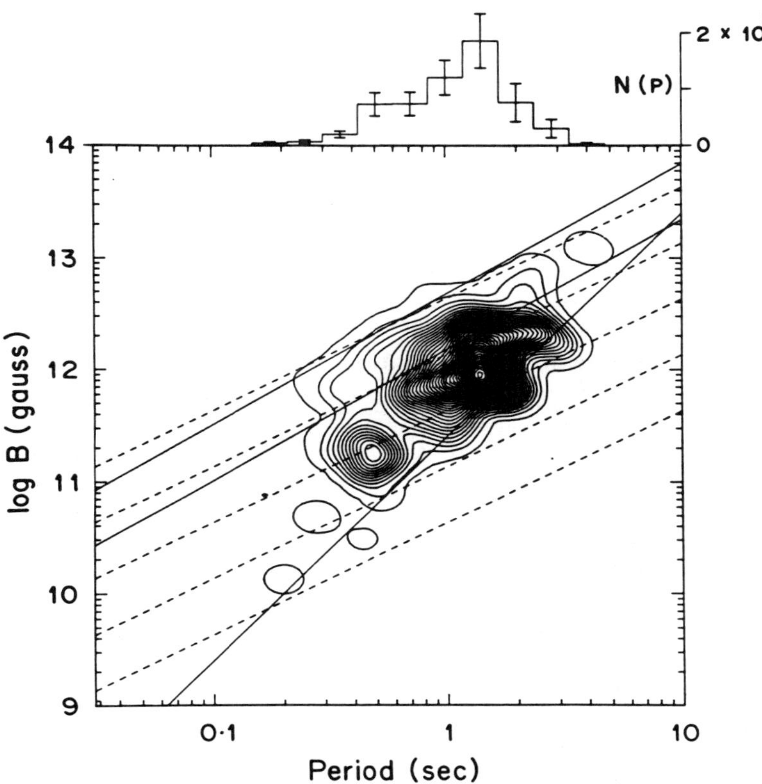

Figure 6: The number distribution of galactic pulsars as a function of magnetic field and spin period, obtained using the same procedure for correcting selection effects in pulsar surveys as has been used in fig. 4. The sloping straight lines have the same meaning as in fig. 4. There is some indication that pulsars with magnetic fields below $10^{11.5}$ G form a distinct population, presumably of recycled pulsars. The histogram at the top is the derived period distribution of pulsars, integrated over all magnetic fields.

to radio pulsars are obtained from their dispersion measures, *i.e.* the column density of free electrons in the line of sight. Recent observations of radio pulsars in globular clusters clearly show that this electron layer has a scale height of $\sim 500 - 1000$ pc (Reynolds 1989; Bhattacharya and Verbunt 1990). This would lead to an underestimate of the distances of radio pulsars far away from the galactic plane. Pulsars currently thought to be located at $\lesssim 1$ kpc may in fact be at higher z. Thus the observed z-distribution of radio pulsars cannot be taken as a contradiction to a longer-than-10^7-yr lifetime for some pulsars (Bhattacharya and Verbunt 1990).

8 Conclusions

The main conslusions we can arrive at from the above discussions are the following:

1. There are strong observational indications that the decay of the magnetic field of a neutron star is closely related to its being processed in a binary system. It seems very likely that this is caused by the expulsion of magnetic flux from the superconducting interior, the fluxoids being dragged by neutron vortices as the star spins down. If so, the field decay will be related to the spindown of the neutron star. The "residual" field strength will in this case be determined by the maximum spindown of the neutron star.

2. The most likely explanation of the signs of field decay exhibited by the population of solitary pulsars is that it contains a large sub-population of recycled pulsars, coming from B/Be star + neutron star binaries.

Acknowledgements

We are grateful to A. A. Deshpande for providing us figures 4 and 6 in advance of publication. In addition, GS would like to thank A. A. Deshpande for numerous illuminating discussions.

References

Bailes, M.: 1989, *Astrophys. J.* **342**, 917

Bailes, M., Manchester, R. N., Kesteven, M. J., Norris, R. P., and Reynolds, J. E.: 1989, *Astrophys. J., Lett.* **343**, L53

Bailes, M., Manchester, R. N., Kesteven, M. J., Norris, R. P., and Reynolds, J. E.: 1990, *Nature* **343**, 240

Baym, G., Pethick, C., and Pines, D.: 1969, *Nature* **224**, 673

Bhattacharya, D. and Srinivasan, G.: 1986, *Curr. Sci.* **55**, 327

Bhattacharya, D. and Verbunt, F.: 1990, *Astron. Astrophys.*, in press

Bisnovatyi-Kogan, G. S. and Komberg, B. V.: 1974, *Sov. Astron.* **18**, 217

Blandford, R. D., Applegate, J. H., and Hernquist, L.: 1983, *Mon. Not. R. astr. Soc.* **204**, 1025

Blondin, J. M. and Freese, K.: 1986, *Nature* **323**, 786

Callanan, P. J., Charles, P. A., Hassal, B. M. J., Machin, G., Mason, K. O., Naylor, T., Smale, A. P., and van Paradijs, J.: 1989, *Mon. Not. R. astr. Soc.* **238**, 25P

Cordes, J. M.: 1986, *Astrophys. J.* **311**, 183

Coté, J. and Pylyser, E. H. P.: 1989, *Astron. Astrophys.* **218**, 131
Deshpande, A. A. and Srinivasan, G.: 1990, in preparation
Gunn, J. E. and Ostriker, J. P.: 1970, *Astrophys. J.* **160**, 979
Helfand, D. J. and Tademaru, E.: 1977, *Astrophys. J.* **216**, 842
Jones, P. B.: 1987, *Mon. Not. R. astr. Soc.* **228**, 513
Jones, P. B.: 1988, *Mon. Not. R. astr. Soc.* **233**, 875
Koyama, K.: 1988, *Comm. Astrophys.* **12**, 287
Kulkarni, S., Djorgovski, S., and Klemola, A.: 1990, submitted to *Astrophys. J.*
Kulkarni, S. R.: 1986, *Astrophys. J.* **306**, L85
Kulkarni, S. R. and Narayan, R.: 1988, *Astrophys. J.* **335**, 755
Lyne, A. G., Anderson, B., and Salter, M. J.: 1982, *Mon. Not. R. astr. Soc.* **201**, 503
Lyne, A. G., Manchester, R. N., and Taylor, J. H.: 1985, *Mon. Not. R. astr. Soc.* **213**, 613
Lyne, A. G., Ritchings, R. T., and Smith, F. G.: 1975, *Mon. Not. R. astr. Soc.* **171**, 579
Meurs, E. J. A. and Van den Heuvel, E. P. J.: 1989, *Astron. Astrophys.* **226**, 88
Murakami, T., Fujii, M., Hayashida, K., Itoh, M., Nishimura, J., Yamagami, T., Conner, J. P., Evans, W. D., Fenimore, E. E., Klebesadel, R. W., Yoshida, A., Kondo, I., and Kawai, N.: 1988, *Nature* **335**, 235
Muslimov, A. G. and Tsygan, A. I.: 1985, *Sov. Astron. Lett.* **11**, 80
Narayan, R.: 1987, *Astrophys. J.* **319**, 162
Ostriker, J. P. and Gunn, J. E.: 1969, *Astrophys. J.* **157**, 1395
Reynolds, R. J.: 1989, *Astrophys. J., Lett.* **339**, L29
Romani, R. W.: 1990, *Nature* **347**, 741
Sauls, J. A.: 1989, in Ögelman, H. and van den Heuvel, E. P. J., eds., *Timing Neutron Stars*, page 457, Kluwer Academic Publishers, Dordrecht
Shibazaki, N., Murakami, T., Shaham, J., and Nomoto, K.: 1989, *Nature* **342**, 656
Srinivasan, G., Bhattacharya, D., Muslimov, A. G., and Tsygan, A. I.: 1990, *Curr. Sci.* **59**, 31
Taam, R. E. and van den Heuvel, E. P. J.: 1986, *Astrophys. J.* **305**, 235
van den Heuvel, E. P. J.: 1984, *J. Astrophys. Astron.* **5**, 209
van den Heuvel, E. P. J., van Paradijs, J. A., and Taam, R. E.: 1986, *Nature* **322**, 153
Verbunt, F., Wijers, R. A. M. J., and Burm, H.: 1990, *Astron. Astrophys.* **234**, 195

ELECTRICAL CONDUCTIVITY OF NEUTRON STAR CORES AND EVOLUTION OF INTERNAL MAGNETIC FIELDS

D.G. Yakovlev
A.F.Ioffe Institute of Physics and Technology
SU 194021, Leningrad, USSR

Abstract

Physical input to the equations which govern the evolution of magnetic fields in non-superconducting parts of neutron star cores is discussed with emphasis to the conductivity of matter. Magnetic fields can be influenced by the effects associated with the balance of forces, by circulation (induced by the heat transport), by anisotropic resistivity of magnetized plasma (Ohmic decay of longitudinal and transverse currents and the Hall drift), etc. The timescales of these processes range from several minutes to the Universe age; the field lifetimes $\sim 10^4 - 10^7$ yr might be expected.

1 Introduction

Evolution of the magnetic fields $\boldsymbol{B}$ in the non-superconducting parts of neutron star (NS) cores, at densities $\rho \sim \rho_0$ ($\rho_0 = 2.8 \times 10^{14}\,\mathrm{g\,cm^{-3}}$ is the standard nuclear matter density), is still unclear. Basically it can be followed by solving a set of well-known equations of magnetic induction, force-balance and heat transfer

$$\frac{\partial \boldsymbol{B}}{\partial t} = \mathrm{curl}\,(\boldsymbol{v} \times \boldsymbol{B}) - c\,\mathrm{curl}\,\boldsymbol{E}\ , \tag{1}$$

$$\boldsymbol{g}\,\rho - \nabla P + \frac{1}{c}\boldsymbol{j} \times \boldsymbol{B} = 0, \tag{2}$$

$$\frac{\partial s}{\partial t} + \boldsymbol{v}\,\nabla s = Q - \mathrm{div}\,\boldsymbol{q}\ , \tag{3}$$

where $\boldsymbol{v}$ is a possible slow (Sec. 3.3) velocity of matter, $\boldsymbol{j}$ the electric current, $\boldsymbol{E}$ the electric field in a plasma-comoving reference frame, P is the pressure, $\boldsymbol{g}$ the gravity, $\boldsymbol{q}$ the thermal current, s is the entropy density, and Q is the heat source. Eqs. (1)-(3) should be supplemented by the transport equations which express $\boldsymbol{E}$ and $\boldsymbol{q}$ through $\boldsymbol{j}$ and temperature gradient ∇T by means of the transport coefficients (electrical resistivity,

J. Ventura and D. Pines (eds.), Neutron Stars: Theory and Observation, 235–242.

thermal conductivity and thermopower). In a magnetized plasma the transport coefficients are generally tensors. For instance, the transport equation for $\boldsymbol{E}$ reads

$$\boldsymbol{E} = \hat{\mathcal{R}}\, \boldsymbol{j} + \hat{\beta} \nabla T, \tag{4}$$

where $\hat{\mathcal{R}}$ and $\hat{\beta}$ are the resistivity and thermopower tensors, respectively. In a coordinate frame with the z-axis parallel to $\boldsymbol{B}$ each tensor is determined by three components; for instance,

$$\hat{\mathcal{R}} = \begin{pmatrix} \mathcal{R}_\perp & \mathcal{R}_H & 0 \\ -\mathcal{R}_H & \mathcal{R}_\perp & 0 \\ 0 & 0 & \mathcal{R}_\parallel \end{pmatrix}. \tag{5}$$

Here $\mathcal{R}_\parallel$ and $\mathcal{R}_\perp$ are the resistivities in the directions parallel and perpendicular to $\boldsymbol{B}$, respectively, and $\mathcal{R}_H$ is the Hall resistivity. Their properties will be discussed in Sec. 2. Instead of $\hat{\mathcal{R}}$ one often introduces the electric conductivity tensor $\hat{\sigma} = \hat{\mathcal{R}}^{-1}$. Notice that in Eq. (4) we have omitted an electric field produced by polarization of charged particles in the presence of gradients of their chemical potentials; this field is often curl-free and does not contribute to magnetic induction (Eq. (1)). Magnetic field $\boldsymbol{B}(\boldsymbol{r},t)$ can be found from the above equations with appropriate initial and boundary conditions.

The present discussion will be restricted in several respects. Firstly, the origin of the magnetic fields in a NS core will not be discussed; the field will be assumed to appear at the early stage of NS evolution. Secondly, the relation of the internal field to the surface field will not be analysed (some aspects of this problem can be clarified from the results of Sang *et al.* 1990, and from the references quoted there). Thirdly, a non-superconducting part of the core will be studied where the magnetic field can exist freely . The evolution of the magnetic fields in superconducting NS interiors has been reviewed by Bhattacharya (1991) (also see Srinivasan *et al.* , 1990).

2 Electric Conductivity of Neutron Star Cores

We shall mainly consider the conductivity of ordinary *npe* dense matter, which consists of neutrons with a few-per-cent admixture of protons and electrons (and all constituents are strongly degenerate). The problem has been studied since the classical work of Baym *et al.* (1969). The main results are summarized below.

(i) *Collision rates.* The conductivity is determined by the momentum transfer rates $J_{\alpha\beta} = J_{\beta\alpha}$ in collisions between particles of species α and β (e.g., Yakovlev and Shalybkov, 1990b), or equivalently by the effective relaxation times $\tau_{\alpha\beta} = n_\alpha m^*_\alpha / J_{\alpha\beta}$ for the particle species α in the $\alpha \to \beta$ collisions, where n_α is the number density of the particles, and m^*_α is their (relativistic) effective mass at the Fermi surface. In *npe* matter, *np* collisions are due to strong interaction, *ep* collisions are Coulombic, and *en* collisions are due to charge - magnetic moment interaction. The collision rates have been calculated by Baym *et al.* (1969), Haensel *et al.* (1990), Yakovlev and Shalybkov (1990 a,b), and Østgaard and Yakovlev (1990). *en* collisions appear to be mainly negligible, while for other collisions (Østgaard and Yakovlev, 1990)

$$J_{ep} = \frac{\pi}{6}\left(\frac{\pi e^2 m_p^*}{\hbar\, p_p}\right)^{3/2} \frac{n_p}{\hbar c^2}(k_B T)^2, \; J_{pn} = \frac{8\pi}{9\hbar^3}(k_B T)^2 (m_n^* + m_p^*)^2 n_p (\sigma_{np})_0 F, \tag{6}$$

where p_p is the Fermi momentum of the protons, and $(\sigma_{np})_0$ is the zero-angle differential cross section of *np*-scattering. F takes into account the effect of neutron superfluidity. When the neutrons are normal ($T > T_c$), one has F =1, and the *pn* collisions are much more frequent than the *ep* ones. When $T < T_c$ and the superfluidity is due to the $^1S_0$ neutron pairing, F may be fitted by an analytic expression

$$F(y) \approx \left(\frac{2+\sqrt{1+y^2}}{3}\right)^{5/2} \exp\left(1-\sqrt{1+y^2}\right), \; y = \frac{T_s}{T}, \tag{7}$$

where T_s is the enegry gap in temperature units. If $T \ll T_c$, the *pn* collisions are strongly suppressed by the superfluidity. The parameters in Eqs. (6) - (7) can be taken from a model of *npe* matter. For a semi-quantitative analysis, one can use the model of free particles (e.g., Shapiro and Teukolsky, 1983). At $\rho \lesssim 2\rho_0$ the model yields the Fermi momenta $p_n = \hbar(3\pi^2\rho/m_n)^{1/3}$, and $p_e = p_p = p_n^2/(2m_n c)$.

(ii) *The longitudinal resistivity* $\mathcal{R}_{\parallel} = \sigma_0^{-1}$ (where σ_0 is the field-free conductiviy) does not depend on the magnetic field; this is a general property of $\mathcal{R}_{\parallel}$. For *npe* matter, $\mathcal{R}_{\parallel}$ is mainly provided by the electrons (Baym *et al.* , 1969)

$$\mathcal{R}_{\parallel} = \frac{m_e^*}{e^2 n_e \tau_{ep}} \approx 1.0 \times 10^{-28}\, T_8^2 \left(\frac{\rho_0}{\rho}\right)^3 \text{ s.} \tag{8}$$

A numerical estimate in (8), as well as the estimates in Eqs. (9), (11), and (12), is done in the free-particle model.

(iii) *The Hall resistivity* is given by a very simple equation (Yakovlev and Shalybkov, 1990 a,b; Østgaard and Yakovlev, 1990)

$$\mathcal{R}_H = \mathcal{R}_{\parallel}\,(\Omega_e \tau_{ep}) = \frac{B}{c e n_e} \approx 7.2 \times 10^{-26}\, B_{12} \left(\frac{\rho_0}{\rho}\right)^2 \text{ s,} \tag{9}$$

where $B_{12} = B/(10^{12}$ G), and $\Omega_e = eB/(m_e^* c)$ is the electron gyrofrequency. $\mathcal{R}_H$ is independent of the neutron superfluidity and also of temperature. It depends linearly on B and exceeds $\mathcal{R}_{\parallel}$ when $\Omega_e \tau_{ep} > 1$, i.e. when $B \gtrsim 10^9 T_8^2$ G. .

(iv) *The transverse resistivity* may be expressed as (Østgaard and Yakovlev, 1990)

$$\mathcal{R}_{\perp} = \mathcal{R}_{\parallel} + \mathcal{R}_B\,, \;\; \mathcal{R}_B = \mathcal{R}_B^{(0)}\, S, \;\; S \equiv \frac{X_n^2}{F}, \tag{10}$$

$$\mathcal{R}_B = \mathcal{R}_{\parallel}\,(\Omega_e \tau_{ep})(\Omega_p \tau_{pn}) = \frac{B^2}{c^2 J_{np}} X_n^2 \approx 1.5 \times 10^{-26} \left(\frac{B_{12}}{T_8}\right)^2 S \left(\frac{\rho_0}{\rho}\right)^{5/3} \text{ s,} \tag{11}$$

where X_n is the mass fraction of normal neutrons among all normal (non-superfluid) particles. $\mathcal{R}_B$ describes an enhancement of the transverse resistivity in a magnetic field. The enhancement occurs because Larmor rotation of charged particles suppresses their motion across $\boldsymbol{B}$. $\mathcal{R}_B$ is clearly provided by the protons and is sensitive to the neutron superfluidity. $\mathcal{R}_B^{(0)}$ is the value of $\mathcal{R}_B$ for the normal neutrons, and the effect of the neutron superfluidity is described by the factor S.

When the neutrons are normal, one has $F = 1$, $X_n \approx 1$, and $S \approx 1$. Then the enhancement term $\mathcal{R}_B^{(0)} \propto T^{-2}$. It dominates ($\mathcal{R}_B^{(0)} \gg \mathcal{R}_{\parallel}$), if $(\Omega_e\tau_{ep})(\Omega_p\tau_{pn}) \gg 1$, i.e. $B \gg 10^{11} T_8^2$ G.

For $T < T_c$ (neglecting the contribution of the electrons to X_n), we have

$$X_n \approx \left(1 + \frac{X_p}{X_n^{(n)}}\right)^{-1}, \quad X_n^{(n)} \approx \left(\frac{1+10\sqrt{1+y^2}}{11}\right)^{1/2} \exp\left(1-\sqrt{1+y^2}\right). \tag{12}$$

where $X_p \approx 0.0057\rho/\rho_0$ is the mass fraction of the protons among nonsuperfluid particles, and $X_n^{(n)}$ is the mass fraction of the normal neutrons among all the neutrons fitted by an analytic equation (again given for the $^1S_0$ pairing)

When $T \ll T_c$, S is very small, $S \propto \exp(-T_s/T)$. Then $\mathcal{R}_B \to 0$ and the enhancement does not operate. This is explained (e.g., Yakovlev and Shalybkov, 1990b) by suppression of the proton contribution to $\mathcal{R}_\perp$ in the case when the proton collisions become ineffective in transferring the proton momentum to neutral particles.

When T decreases below T_c, S changes from 1 to 0 non-monotonously (Østgaard and Yakovlev, 1990). For the $^1S_0$ pairing and $\rho \sim \rho_0$, S initially grows up, reaches maximum $S \approx 20$ at $T \approx T_c/4$, decreases to its initial value $S = 1$ at $T \approx T_c/6$ and then becomes < 1. Hence the enhancement of $\mathcal{R}_\perp$ by the magnetic field is amplified by the moderate superfluidity and suppressed by the strong superfluidity.

In addition to the case of standard *npe* matter Yakovlev and Shalybkov (1990 b) have considered the conduction properties of *npe* matter with Σ^- hyperons which are created when $\rho > 4\rho_0$ (for the free-particle model). Haensel and Jerzak (1989), and Yakovlev and Shalybkov (1990 a,b) have also discussed the case of strange matter. The main results are the following. The longitudinal resistivity $\mathcal{R}_{\parallel} = \sigma_0^{-1}$ of non-superconducting dense matter is independent of B and is of the order of (8) for all compositions of matter. The Hall resistivity $\mathcal{R}_H$ and the transverse resistivity $\mathcal{R}_\perp$ can depend on B and on the composition, and this dependence becomes more complicated when new particles are involved. When $B \gtrsim 10^9 T_8^2$ G, one usually has $| \mathcal{R}_H | \gg \mathcal{R}_{\parallel}$. In strange matter, as in *npe* matter with strongly superfluid neutrons, $\mathcal{R}_\perp \sim \mathcal{R}_{\parallel}$. In $npe\Sigma^-$ matter $\mathcal{R}_\perp \gg \mathcal{R}_{\parallel}$ provided $B \gg 10^{11} T_8^2$ G irrespectively of the neutron superfluidity.

3 Evolution of Magnetic Fields in Neutron Star Cores

Evolution of magnetic fields is governed by the equations formulated in Sec. 1. It is affected by many factors (electric resistivity, thermopower, heat conduction, magnetic forces, etc.) which should have been examined simultaneously but have not been studied in detail even separately. The discussion below is very preliminary.

3.1 EFFECTS OF ELECTRIC RESISTIVITY

It is often thought that the NS magnetic fields are mainly affected by the electric resistivity. This approximation corresponds to $\boldsymbol{v}$ =0, ∇T=0 in Eqs. (1) and (4). Then

$$\frac{\partial \boldsymbol{B}}{\partial t} = -\frac{c^2}{4\pi} \mathrm{curl}\,(\hat{\mathcal{R}}\ \mathrm{curl}\,\boldsymbol{B}\ ). \tag{13}$$

If the resitivity $\hat{\mathcal{R}}$ were isotropic, the magnetic field would suffer Ohmic decay over a timescale $t_i \sim \zeta\sigma_0$ (e.g., Landau *et al.* , 1984), where $\zeta = (4/\pi)(L/c)^2$, L is a typical length-scale ($L \sim$ the stellar radius or the size of a non-superconducting part of the stellar core). Since actually the resistivity is anisotropic, Eq. (13) describes three processes: Ohmic decay of the electric currents parallel to $\boldsymbol{B}$ ($t_{\|} \sim \zeta/\mathcal{R}_{\|}$), Ohmic decay of the perpendicular currents ($t_{\perp} \sim \zeta/\mathcal{R}_{\perp}$), and Hall drift of the magnetic field with velocity $\mathbf{v}_d = -(c/B)\mathcal{R}_H \mathbf{j}_{\perp}$ ($t_H \sim \zeta/\mathcal{R}_H$). Here and in what follows the numerical estimates are performed for *npe* matter with the normal neutrons at $\rho = 1.7\rho_0$ which is the mean density for a NS with $M = M_{\odot}$ and radius of 10 km.

Longitudinal currents decay with the same rate as in a non-magnetized plasma. From Eq. (8) one gets the decay time

$$t_{\|} \sim 2 \times 10^{12}\, L_6^2 T_8^{-2} \text{ yr}, \tag{14}$$

($L_6 = L/10^6$ cm) which is larger than the Universe age. This result (Baym *et al.* , 1969) is widely accepted in the literature.

Transverse currents decay much faster when $\mathcal{R}_{\perp}$ is enhanced by the magnetic field (Sec. 2). For the normal neutrons, the enhancement occurs when $B \gtrsim 10^{11} T_8^2$ G. Then from Eqs. (10) and (11) one gets

$$t_{\perp} \sim 7.5 \times 10^9\, L_6^2 T_8^2 B_{12}^{-2} \text{ yr}. \tag{15}$$

A model calculation of the decay of the transverse currents in a cooling NS with the normal neutrons has been performed by Haensel *et al.* (1990). It has been shown that the field decays to a value of $\sim 10^{12}$ G irrespectively of its initial value B_0 provided $B_0 \gg 10^{12}$ G. Moreover, at $t \gtrsim 10^6$ yr the enhanced decay produces an intense Joule heating, which delays noticeably NS cooling.

Hall drift of the magnetic field is not accompanied by the direct dissipation but changes the field configuration. An appropriate timescale is obtained from Eq. (9)

$$t_H \sim 2 \times 10^9\, L_6^2 B_{12}^{-1} \text{ yr} \tag{16}$$

and is temperature independent. Owing to the drift, spatial field scales L may become smaller. This enhances (indirectly) the Ohmic decay, Eqs. (14) and (15). Recently, Shalybkov and Urpin (privite communication) have obtained model numerical solutions of Eq. (16) with anisotropic resistivity for toroidal axially symmetric fields. The results confirm the above conclusions. In addition it has been shown that the Hall drift makes the field $\boldsymbol{B}$ asymmetric with respect to the magnetic equator of the NS even if it is initially symmetric. This is easily seen from the expression for $\mathbf{v}_d$ presented above.

3.2 HYDROSTATIC EQUILIBRIUM

The main disadvantage of the analysis of Sec. 3.1 is that it ignores the balance of forces, Eq. (2). If initially the balance is not realized, hydrodynamical motions appear and rearrange the field to a balancing state. The rearrangement proceeds with the Alfven velocity v_A. Its timescale is very short,

$$t_A \sim L/v_A \sim 10^2\, L_6 B_{12}^{-1} \text{ s}. \tag{17}$$

For a rapidly rotating NS, t_A is a factor of $(\Omega L/v_A)$ larger, where Ω is the angular frequency of rotation (see, e.g., Flowers and Ruderman, 1977).

The equilibrium field configurations and their stability in NS cores have not yet been studied in detail. One can expect several possibilities: the magnetic field can be expelled out of the NS core; the field can survive but become force-free; the magnetic force is not zero and is either potential or vortex-like.

Let us discuss the last possibility in more detail. One can easily see that the vortex component of the magnetic force cannot be compensated by the main part of the pressure $P = P_0(\rho)$ which depends on the only variable ρ. However it can be balanced by a small thermal part of the pressure, $P_1(\rho, T)$. In the free-particle model with non-relativistic normal neutrons $P_1 \approx 5.5\times 10^{26}\, T_8^2\, (\rho/\rho_0)^{1/3}$ erg cm^{-3}. In this case the force-balanced state of the NS core will be slightly non-isothermal, with relative temperature variation $\varepsilon \sim \delta T/T \sim B^2 \,/\, (8\pi P_1) \sim 6\times 10^{-5} B_{12}^2 T_8^{-2}$.

Let us emphasize that the force-balancing could be significantly affected by the rotation (e.g., Flowers and Ruderman, 1977).

3.3 HEAT-TRANSFER DRIFT

After a quick rearrangement to a force-balanced state, a much more slow process of the heat transport (Eq. (3)) should be stabilized. Since the density and/or the temperature profiles in a force-balanced NS core depend, to some extent, on $\boldsymbol{B}$, and are not spherically symmetric, the heat transport is generally accompanied by slow motions of matter of meridional-circulation type (e.g., Schwarzschild, 1958). The motions do not violate the hydrostatic equilibrium but cause the magnetic field drift (see Eq. (1)). When the magnetic force is vortex-like and is compensated by the temperature dependent part of the pressure (Sec. 3.2), the circulation timescale can be estimated as $t_{cir} \sim t_{th}/\varepsilon$, where t_{th} is the thermal relaxation time. If t_{th} is determined by the thermal conductivity κ, it is estimated as $t_{th} \sim CL^2/\kappa$, where C is the heat capacity per unit volume. For $C \sim 2\times 10^{19} T_8$ erg cm^{-3}K^{-1} and $\kappa \sim 5\times 10^{23}\, T_8^{-1}$ erg s^{-1} cm^{-1} K^{-1} ('quantum-liquid case' of Flowers and Itoh, 1981, which seems to be most appropriate to the model of study), one has $t_{th} \sim 1.3 \times L_6^2 T_8^2$ yr and

$$t_{cir} \sim 2\times 10^4\, T_8^4 L_6^2 B_{12}^{-2} \text{ yr} \tag{18}$$

although this estimate is very uncertain. If it is valid for a cooling NS, the field drift goes faster with time t. The drift becomes significant when $t_{cir} \sim t$. For $B \sim 10^{12}$ G, this happens at $t \sim 10^4 - 10^5$ yr.

The drift itself does not lead to the magnetic field dissipation but can accelerate the dissipation indirectly producing small-scale fields (similar to the Hall drift, Sec. 3.1).

4 Conclusions

Transport properties of matter in the non-superconducting parts of NS cores (Sec. 2) are more or less clear but the magnetic field evolution is not: it can be affected by many interrelated phenomena with different timescales (Sec 3). The most stringent is the balance of forces (Sec. 3.2) which is achieved in a very short time; the most urgent problem is to understand the force-balanced structure of $\boldsymbol{B}$. The second important process can be the field drift (Sec. 3.3) associated with the heat transfer circulation. It is fast when the magnetic force is vortex-like and is compensated by the thermal part of the pressure force. In any case, since the neutrino emissivity depends on $\boldsymbol{B}$ (e.g., Yakovlev and Tschaepe, 1981), and so does the thermal conductivity, it is difficult to avoid the circulation although it can be much slower. The effects of electric conductivity (Sec. 3.1) include Ohmic dissipation of longitudinal and transverse currents, and the Hall drift. Direct action of these effects may be slower than the action of circulation. It should be added that the magnetic field can also be influenced by many other phenomena. Among them are: rotation (e.g., Flowers and Ruderman, 1977), the thermomagnetic effects associated with the thermopower in Eq. (4), convection and various instabilities (e.g., Vandakurov, 1972; Tayler, 1973), anisotropic heat transport and heat release. They require special analysis.

Further detailed study is needed to understand the actual situation. It is clear that the evolution of various magnetic field configurations (toroidal, poloidal, and mixed fields) can be entirely different and may be strongly affected by rotation and the presence of superfluid parts of NS matter.

Acknowledgements. Many of the above problems have been discussed with C. Pethick, D.A. Shalybkov, and V.A. Urpin whose help and advises are greatly appreciated.

References

Bhattacharya, D. (1991) *'The evolution of neutron star magnetic fields'*, this volume.

Baym, G., Pethick, C., and Pines, D. (1969) *'Electrical conductivity of neutron star matter'*, *Nature* **224**, 674 - 675.

Haensel, P. and Jerzak, A.J.: (1989) *'Transport properties of strange matter'*, *Acta Physica Polonica* **B20**, 141.

Haensel, P., Urpin, V.A., and Yakovlev, D.G.: (1990) *'Ohmic decay of internal magnetic fields in neutron stars'*, *Astron. Astrophys.* **229**, 133 - 137.

Flowers, E. and Itoh, N. (1981) *'Transport properties of dense matter. III. Analytic formulae for thermal conductivity'*, *Astrophys. J.* **250**, 750 - 752.

Flowers, E. and Ruderman, M. (1977) *'Evolution of pulsar magnetic fields'*, *Astrophys. J.* **215**, 302 - 310.

Landau, L.D., Lifshitz, E.M., and Pitaevskii, L.P. (1984) *Electrodynamics of Continuous Media*, Pergamon Press, Oxford.

Sang, Y., Chanmugam, G., and Tsuruta, S. (1990) *'Magnetic field decay in cooling neutron stars'*, in W. Kundt (ed.), *Neutron Stars and Their Birth Events*, Kluwer Academic Publishers, Dordrecht, p.p. 127 -131.

Østgaard, E. and Yakovlev, D.G. (1990) *'Electrical conductivity of npe dense matter. Effects of neutron superfluidity and nucleon interaction'*, *Astrophys. Space Sci.* (in print).

Schwarzschild, M. (1958) *Structure and Evolution of the Stars*, Princeton University Press, Princeton, New Jersey.

Shapiro, S.L. and Teukolsky, S.A.: (1983) *Black Holes, White Dwarfs, and Neutron Stars*, Wiley-Interscience, New York.

Srinivasan, G., Bhattacharya, D., Muslimov, A.G., and Tsygan, A.I. (1990) *'A novel mechanism for the decay of neutron star magnetic fields'*, *Current Sci.* **59**, 31 - 38.

Tayler, R.J. (1973) *'The adiabatic stability of stars containing magnetic fields. I. Toroidal*

fields', *Mon. Not. R. Astron. Soc.* **161**, 365 - 380.
Vandakurov, Yu.V. (1972) *'To the theory of stability of stars with toroidal magnetic fields'*, *Sov. Astr.* **16**, 265.
Yakovlev, D.G. and Shalybkov, D.A. (1990a) *'Electrical conductivity and resistance in magnetized cores of neutron stars'*, *Pisma Astron. Zh. (Soviet Astron. Lett.)* **16**, 202 - 207.
Yakovlev, D.G. and Shalybkov, D.A. (1990) *'Electrical conductivity of neutron star cores in the presence of a magnetic field. Papers I and II'*, *Astrophys. Space Sci.*
Yakovlev, D.G. and Tschaepe, R. (1981) *'Synchrotron neutrino-pair radiation in neutron stars'*, *Astron. Nachr.* **302**, 167 - 176.

3. OBSERVATIONS ON BINARIES AND QPO's

NEUTRON STARS IN X-RAY BINARIES

JAN VAN PARADIJS
Astronomical Institute "Anton Pannekoek", University of Amsterdam
& Center for High-Energy Astrophysics, NIKHEF-H, Amsterdam

ABSTRACT. The galaxy contains two major groups of accretion-driven stellar X-ray sources, the high-mass and the low-mass X-ray binaries. In this lecture the properties of these two types of sources are briefly described. The observational differences between the high-mass and the low-mass X-ray binaries are likely related to a large difference in the magnetic fields of the neutron stars in these objects.

1. Introduction

Following early theoretical ideas about accretion onto compact objects as a source of energy for quasars and X-ray sources [342,460,461] the properties of the spectra of the first two optically identified X-ray sources, Sco X-1 [344] and Cyg X-2 [110] led to speculation that these sources are mass-exchanging binaries with a compact component (see, e.g., [355]). In the case of Sco X-1 the spectrum was found to be similar to those of old novae and U Gem type stars, which were known to be binary stars, in particular through the work of Crawford and Kraft in the 1950s and '60s [76,205,206]. The optical spectrum of Cyg X-2 was found to be composite, showing the signatures of both a late-type star and a component of much higher excitation; Cyg X-2 also showed significant radial-velocity variations. However, the single most important characteristic of a binary star, i.e., an orbital periodicity, was not found in either system (cf. [208]) until many years later, in spite of substantial observational effort (see e.g., [146,207], for discussions of early optical observations of X-ray sources, and references).

The idea that the bright galactic X-ray sources are mass-exchanging binary stars was immediately accepted, and became a paradigm of X-ray astronomy, with the discovery of the optical counterpart of Cygnus X-1 [449], and of the regular eclipses of the pulsating X-ray sources Centaurus X-3 [111,350]. The optical identification of the rapidly variable X-ray source Cyg X-1 was established through an accurate position of a radio source [28,150], which underwent a large upward brightness transition, apparently correlated with a major change in the X-ray spectrum of Cyg X-1[383]. The radio position (accurate to about one arcsecond) coincided with that of the bright star HD 226868. This early-type supergiant was soon found to be a single-lined spectroscopic binary, with an orbital period of 5.6 days [449]. The observed mass function, in conjunction with an estimate of the mass of the optical star (from its position in the Hertzsprung-Russell diagram) led to the conclusion that the compact X-ray source in this system has a mass of at least 6 $M_{\odot}$. Thus, the first X-ray source for which the binary nature was reasonably well established, contained an accreting compact object that is likely a black hole (see [273] for a recent

J. Ventura and D. Pines (eds.), Neutron Stars: Theory and Observation, 245–287.

review of the status of black-hole candidates in the stellar mass range).

The observation of variable delays of the pulse arrival times from Cen X-3, in phase with the periodic (2.1 days) eclipses of the X-ray source, showed persuasively that in this system the X rays are generated by accretion onto a strongly magnetized neutron star, rotating at the observed 4.8 s pulse period, in orbit around a companion star. The properties of the X-ray orbit showed that the mass-transferring companion star is very massive (> 10 $M_\odot$), a result which was later confirmed by the optical identification of this source with an O-type giant star [210]. The discovery that Cen X-3 is an X-ray binary star was soon followed by more observations of eclipsing X-ray sources, some of them pulsating, and the identification of these X-ray sources with early-type stars (see, e.g., [239,322,437]). In addition, a general framework for the origin and evolution of a massive X-ray binary, as a rather normal episode in the life of a massive close binary star with successive stages of mass transfer between the two components, was readily accepted [405]. Thus, within a few years the existence of a galactic population of high-mass X-ray binaries (HMXB) was well established.

The clustering of bright X-ray sources within ~30° of the direction of the galactic center without a strong background of unresolved sources showed already quite early that there is a group of sources located in the central regions of the Galaxy [341,353]. It was, therefore, suspected that apart from the above-described HMXB there is a class of low-mass X-ray binaries (see, e.g., [343]), but proof for this basic idea was hard to obtain. Apart from the difficulty of finding orbital periods, the apparent heterogeneity of the properties of low-mass X-ray binaries (LMXB) probably played a role. Compared to the HMXB the first handful of systems now classified as LMXB (Her X-1, Cyg X-3, Sco X-1, Cir X-1)

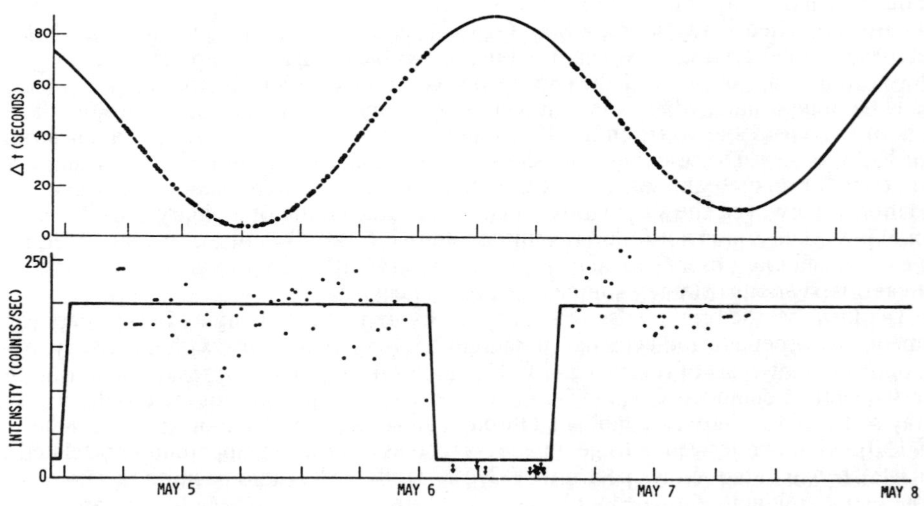

Fig. 1: *The top panel shows the Doppler delay of the arrival times of the 4.8 s X-ray pulsations of Cen X-3 (relative to the times expected for a constant pulse period), as observed with Uhuru in May 1971. The bottom panel shows the X-ray intensity variations during the same time. The maximum delay coincides with the center of the X-ray eclipse (adapted from [350]).*

show rather more diversity than similarity in their properties. As a result, the establishment of two broad groups of galactic X-ray sources has come slowly, and along various roads: only at the end of the 1970s it became clear that with respect to their sky distributions, X-ray spectral characteristics, optical properties, and types of X-ray variability, the LMXB are distinct from the HMXB as a group with "family traits" (see, e.g., [229]). The LMXB comprise the globular cluster X-ray sources, X-ray bursters, soft X-ray transients, and the bright galactic bulge X-ray sources (most of which have recently been found to be QPO sources; see [216,236,409], for recent reviews).

In this paper we briefly describe the main properties of the HMXB and LMXB, and show how the differences between these two groups of X-ray sources may be linked to a difference in the strength of the magnetic fields of the neutron stars they harbour. For the convenience of the reader I have collected in Tables 1 and 2 the X-ray sources for which the classification as an LMXB or an HMXB, respectively, is fairly secure, together with some basic information, and recent references that should facilitate entry into the literature.

Within the limits of this lecture I cannot strive for completeness. For background information on various topics related to X-ray binaries the interested reader is advised to consult individual chapters in the books by Shapiro & Teukolsky [354], Lewin & Van den Heuvel [230], Eggleton & Pringle [97], Frank, King & Raine [103], Pringle & Wade [331], Lewin, Trümper & Brinkmann [234], Mason, Watson & White [265], Pallavicini & White [309], Filipov & White [100], and Oegelman & Van den Heuvel [302]. References on optically identified individual sources can be found in Bradt & McClintock [27] and Van Paradijs [420].

2. Optical Counterparts

The optical counterparts of HMXB have normal early-type spectra, in the sense that they can be MK-classified without particular difficulty, on the basis of ratios of spectral line strengths (see Fig. 2). This is highly informative, since it immediately provides us with at least an approximate idea about the masses, radii, and ages of these stars. Some disturbance of the spectrum, indicative of anisotropic gas flow near the primary may show up as variable emission/absorption components, particularly in Hα, Hβ, He II λ4686, and the CIII-NIII λ4630-50 complex. However, when the latter two lines are strongly in emission (see e.g., [141]) this is likely due to a very early spectral type (Of characteristic) of the primary, and not to the presence of the X-ray source. The reason that the X-ray source does not seem to affect the spectral properties of the primary much, is that the bolometric luminosity of the latter generally exceeds the X-ray luminosity, often by a large margin [420].

With respect to the spectral types of the optical counterparts the HMXB can be divided into two subgroups, as follows. (i) The spectral type is earlier than B2, and the luminosity class is I to III, i.e., the primary star has evolved off the main sequence. These stars are filling, or close to filling their Roche lobes, as is apparent from the amplitudes of their optical light curves. (ii) The primary is a B-emission (Be) star, located in the Hertzsprung-Russell diagram rather close to the main sequence. The orbits of these Be/X-ray binaries are eccentric, and their periods tend to be long. The primaries underfill their Roche lobes.

As first suggested by Maraschi *et al.* [253] the mass transfer in these two groups is driven by different mechanisms. In the first group mass is transferred via a strong stellar wind (in a few short-lived sources Roche-lobe overflow is important). In the Be/X-ray binaries the mass transfer is related to the anisotropic (often highly variable) shedding of mass as observed in all Be stars, which is believed to be the result of their rapid rotation [358]. This inferred difference in mass transfer mechanism is supported by the different

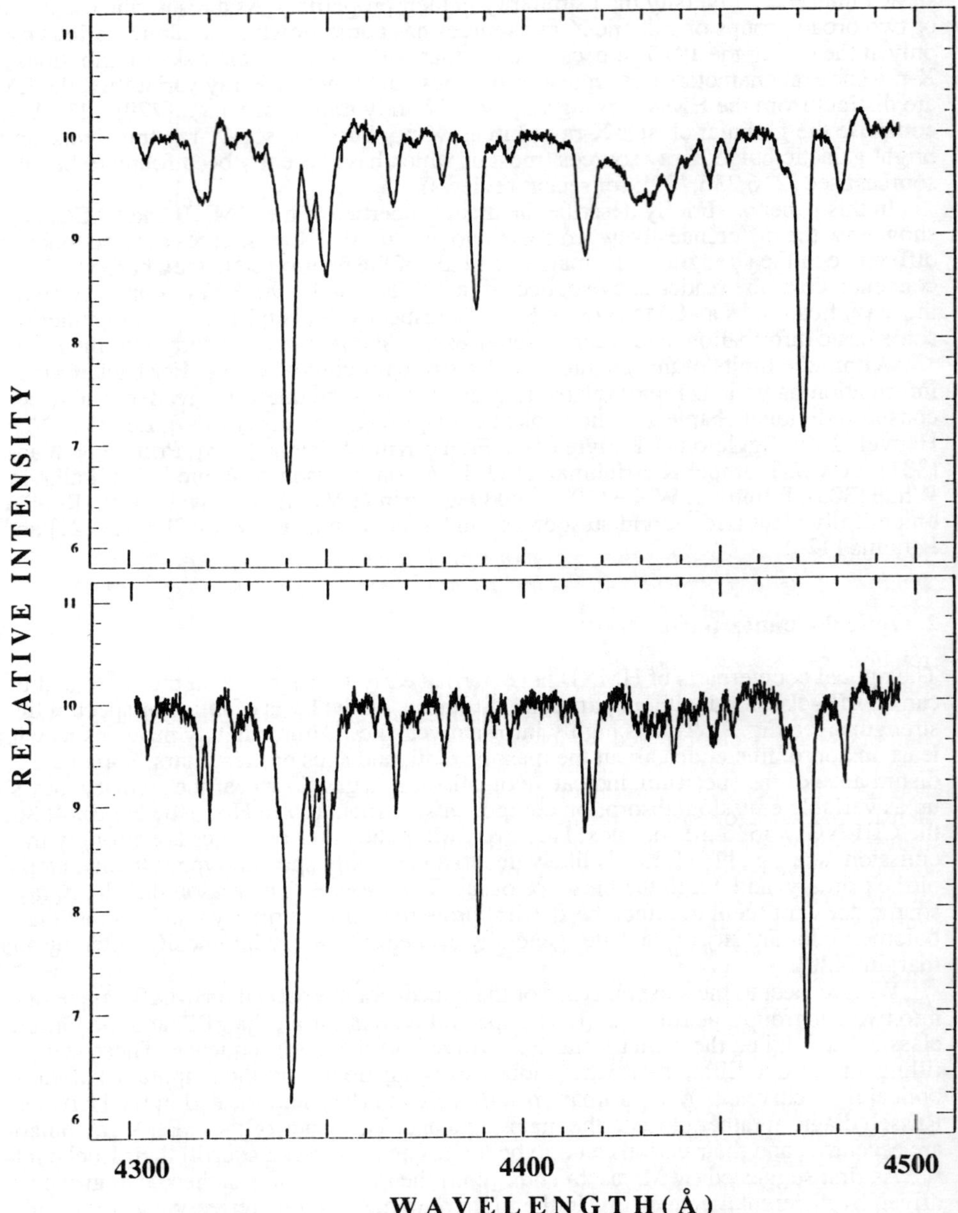

Fig. 2: *Comparison of the optical spectra of HD 77581, the optical counterpart of the HMXB Vela X-1 (top panel), and of the normal star κ Ori (lower panel), which have spectral types B0.5 Ib and B0.5 Ia, respectively. Note the difference in line widths, caused by the larger rotational velocity of HD 77581 (coutrtesy M. van Kerkwijk).*

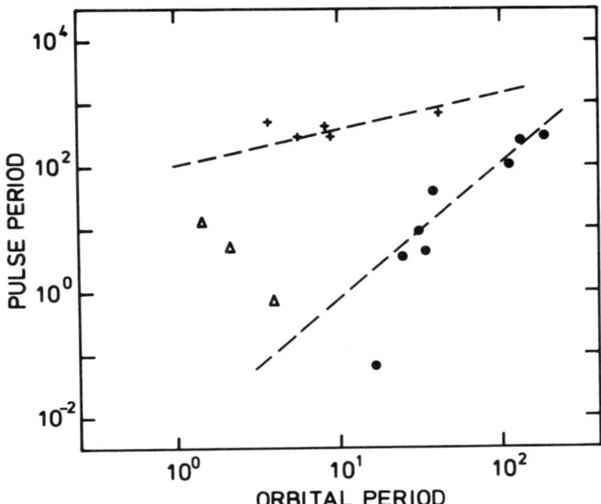

Fig. 3: *Relation between pulse period and orbital period for three types of high-mass X-ray binaries: Be/X-ray binaries (filled circles), non-Be wind driven sources (crosses), and non-Be X-ray sources driven by Roche lobe overflow (triangles) [60,65,445].*

relations between orbital period and X-ray pulse period, first pointed out in by Corbet [60], for these two groups of sources (see [65,407,445]).

Most Be/X-ray binaries are highly variable, or transient. In some of them recurrent outbursts have been observed, which reflect the varying accretion rate onto the neutron star as it moves in its eccentric orbit through regions of varying density around the Be star. In addition, a more sudden turning on and off of the accretion can occur when the wind density becomes too low for the neutron star magnetosphere to be within the corotation radius, so that accretion becomes centrifugally inhibited [365]. However, in many cases outbursts have been observed which are not related to the orbit of the Be/X-ray binary, but are due to a sudden enhancement of the mass loss of the Be star (for recent reviews of various aspects of Be stars see [359]).

The optical counterparts of LMXB are rather faint stars. Their spectra show a few emission lines, particularly Hα, Hβ, He II λ4686, and C III-N III λ4630-50, superposed on a flat (in frequency) continuum. These spectra, which definitely are not those of normal stars, are dominated by the emission from an accretion disk around a neutron star, which radiates mainly through reprocessing of incident X-rays into optical/UV photons. In very few cases the signature of a companion star can be discerned. According to Motch & Pakull [290] the relative strength of the C III-N III emission complex relative to the λ4686 emission provides a good measure of the heavy-element abundances in the accreted matter.

It appears that the optical properties of LMXB are rather uniform, and can in general be reasonably described by average values [420]. The colour indices B-V and U-B (reddening-corrected) have average values of 0.0 ± 0.3 and -0.9 ± 0.2, respectively (errors are one sigma standard deviations), close to those expected for a flat continuum (F_ν = constant). The distribution of the ratio of X-ray to optical luminosity is rather sharply peaked. Expressed in terms of an "optical/X-ray colour index" $B_o + 2.5 \log F_X$ (μJy), the peak occurs at 21.5, corresponding to a ratio of fluxes emitted in X-rays (2-11 keV) and in optical light (3000-7000 Å) of ~350.

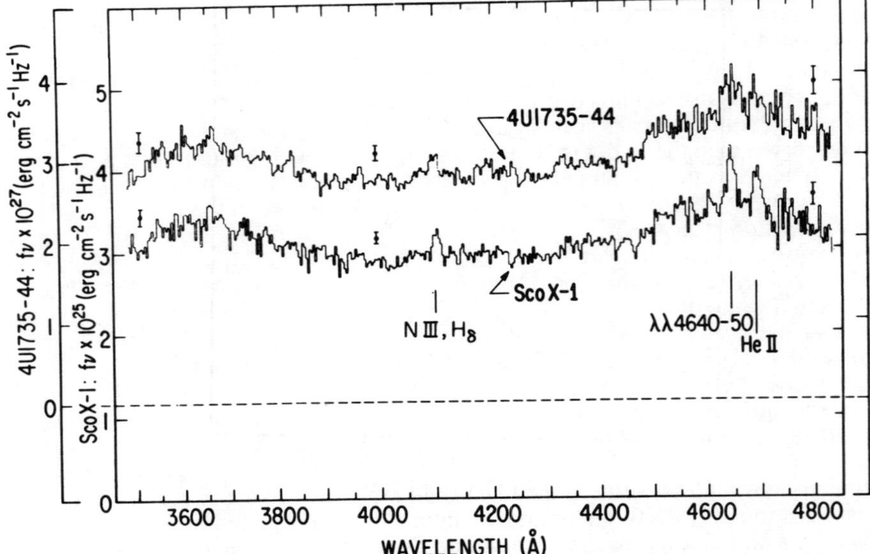

Fig. 4: *Optical spectra of the low-mass X-ray binaries Sco X-1 and 4U/MXB 1735-44 (a burst source). The similarity of these spectra is striking [269].*

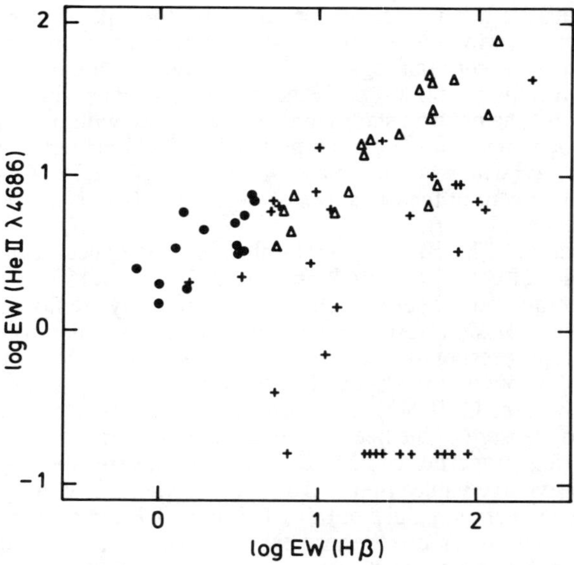

Fig. 5: *Relations between the equivalent widths of the He II λ4686 line and Hβ. Low-mass X-ray binaries are indicated by filled dots, magnetic cataclysmic variables by triangles, and non-magnetic cataclysmic variables by crosses (adapted from [423]).*

Absolute visual magnitudes M_V have been estimated for optical counterparts of X-ray burst sources and some soft X-ray transients, for which reasonable distance determinations can be made. These absolute magnitudes (average value $M_V = 1.0$) scatter over a fairly small interval of ± 1 mag. This small scatter may be related to a rather small range in orbit: periods of the systems used in this average. If the relative shape of the accretion disk (in particular its angular thickness as seen from the neutron star) is assumed to be independent of the orbital period then the X-ray irradiated accretion disk will (for the same X-ray luminosity) be hotter as the orbital period decreases since they are then smaller [T^4 (:) a^{-2} (:) $P^{-4/3}$]. A relatively larger fraction of its emission will then be in the UV (i.e., the bolometric correction increases).

It is of some interest to compare the optical properties of LMXB with the closely related cataclysmic variables (CVs), which are different in that the accreting compact star is not a neutron star but a white dwarf. The spectra of CVs bear a general resemblance to those of LMXB, showing emission lines superposed on a continuum. However, in general the equivalent widths of these lines in LMXB spectra, in particular that of Hβ, are much smaller than those in CV spectra [423]. Since the absolute magnitudes of LMXB and CV differ by values ranging between ~3.5 mag (for nova-like variables, and dwarf novae in outburst) and at least 6 mag (for dwarf novae in quiescence) the luminosities in Hβ are substantially higher in the LMXB than in CV. In general, the equivalent width of HeII λ4686 varies somewhat less between the LMXB and CV (except the AM Her systems); this suggests that in the LMXB the luminosity in this line is enhanced by X-ray reprocessing by a similar factor as the continuum flux.

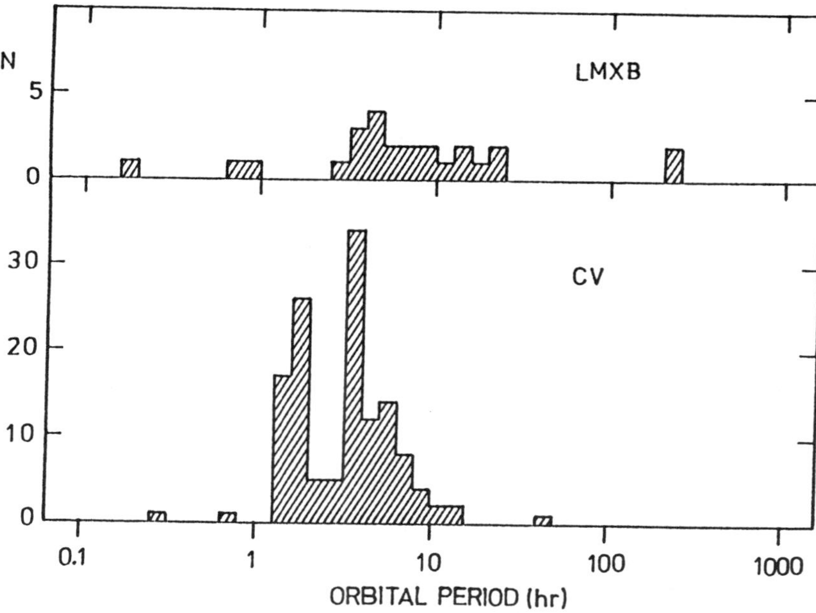

Fig. 6: *Distributions of well-defined orbital periods for low-mass X-ray binaries and cataclysmic variables. Data have been taken from [336].*

The orbital-period distributions of CV and LMXB (Fig. 6) are somewhat different. Compared to the CV there is a relatively larger fraction of LMXB with periods above about half a day; this may partly reflect the fact that CV systems with long orbital periods may be classified as symbiotic stars. Also, there are no LMXB in the period range between 1 and 2 hours (i.e., below the period gap), which is well populated by the CV. This is perhaps the result of evaporation of the companion stars by the large luminosity from the rapidly rotating neutron star (spun up by accretion torques), which becomes active as a millisecond radio pulsar after the mass transfer stops when the system has reached the upper edge of the period gap [340,408].

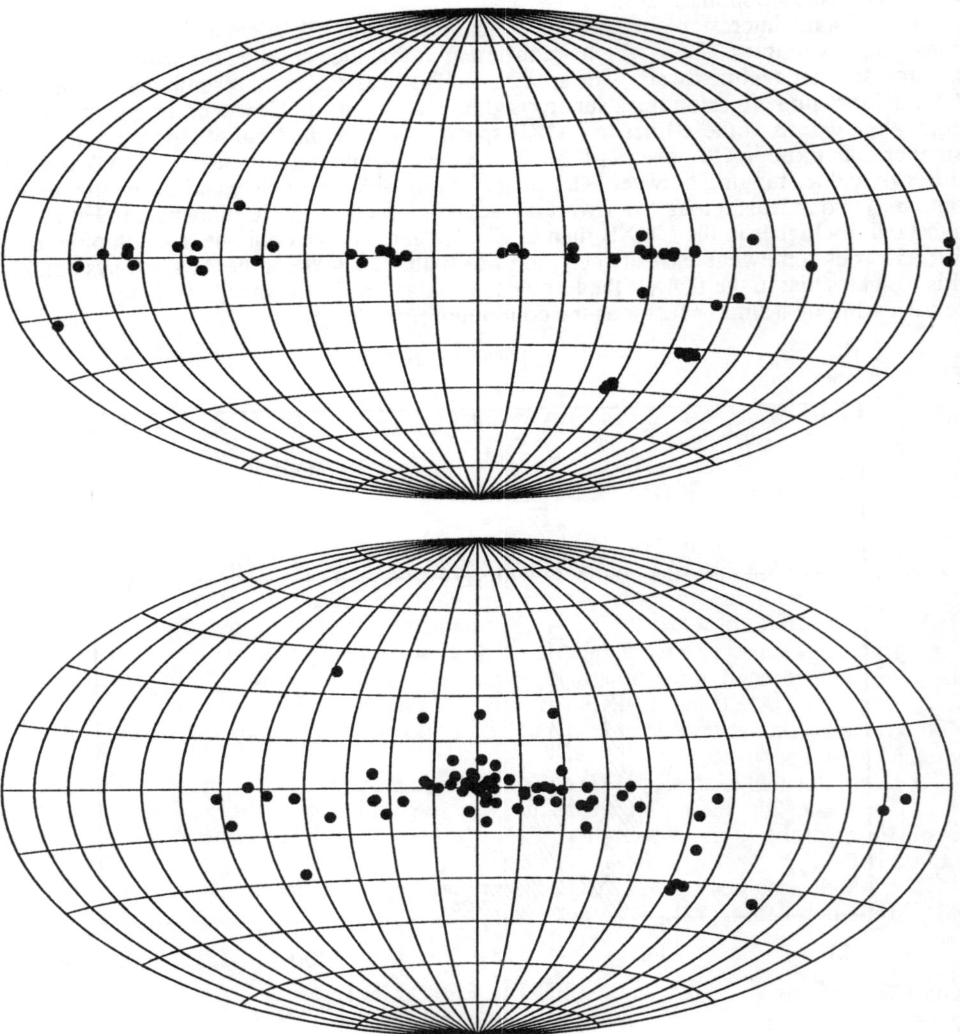

Fig. 7: *Sky maps (in galactic coordinates) of the high-mass X-ray binaries (top panel) and low-mass X-ray binaries (bottom panel); the latter also include the globular-cluster sources.*

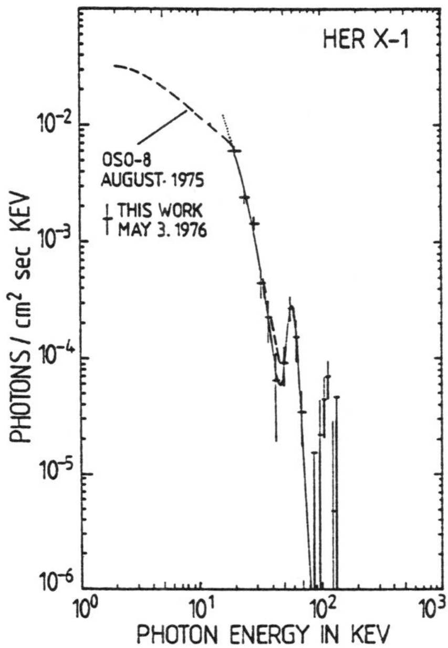

Fig. 8: *Cyclotron line in the hard X-ray spectrum of Her X-1 [397].*

Galactic Distributions of HMXB and LMXB

The sky distributions of the HMXB and LMXB are shown in Fig. 7. The galactic HMXB are distributed along the galactic plane, with an average latitude $<b^{II}> = -1.7 \pm 8.3°$ (if we leave out X Per and the other very nearby high-latitude Be/X-ray systems found by Tuohy et al. [401] we find $<b^{II}> = 0.1 \pm 1.9°$). The galactic LMXB (not including the globular-cluster sources) have a wider latitude distribution ($<b^{II}> = 0.1 \pm 9.1°$), and are also more concentrated in the general direction of the galactic center. These distributions fit the idea that the HMXB and LMXB are parts of a very young galactic population of massive stars (population I), and of a much older population (population II, and old disk population), respectively.

A recent detailed analysis [418] of the kinematic properties of the optically identified HMXB indicates that these objects are runaway stars; this is perhaps the result of asymmetries of the supernova explosions in which the (now) accreting neutron stars were formed [418].

The radial velocities of the LMXB optical counterparts support their membership of an old galactic population [72].

X-ray Pulsations

Almost all HMXB show X-ray pulsations, which indicates that the accreting compact stars in these systems are strongly magnetized neutron stars (for reviews of various aspects of X-ray pulsars see e.g., [188,296,334]). Strong magnetic fields (a few 10^{12} G) have also been inferred from the presence of cyclotron lines in the hard X-ray spectra of some X-ray pulsars (see [198] for a review of this subject).

Observed pulse periods range over a factor ~10^4, between 61 msec and 835 s. From EXOSAT observations of the transient source EXO 2030+475, which covered a large range in luminosity, Parmar *et al.* [315] found a correlation between the pulse profile and the X-ray luminosity of the source; according to their analysis the emission changed from a fan beam to a pencil beam pattern as the luminosity decreased (see also [452]).

Pulse-arrival time measurements for pulsating HMXB, in combination with radial-velocity observations of their massive companions, have provided invaluable information on the masses of neutron stars. In Fig. 4 our present knowledge on neutron-star masses is summarized. These results are consistent with a standard neutron star mass between 1.2 and 1.6 $M_\odot$; however, mass differences of more than 0.5 $M_\odot$ cannot at present be excluded. The limiting factor in the accuracy of neutron star masses are the optical radial-velocity data (which so far have mainly been based on photographic and image-tube observations). Significant progress in this area in the form of much improved error ranges for these masses, and an extension of the sample of observed neutron stars, appears definitely possible with presently available CCD spectrographs.

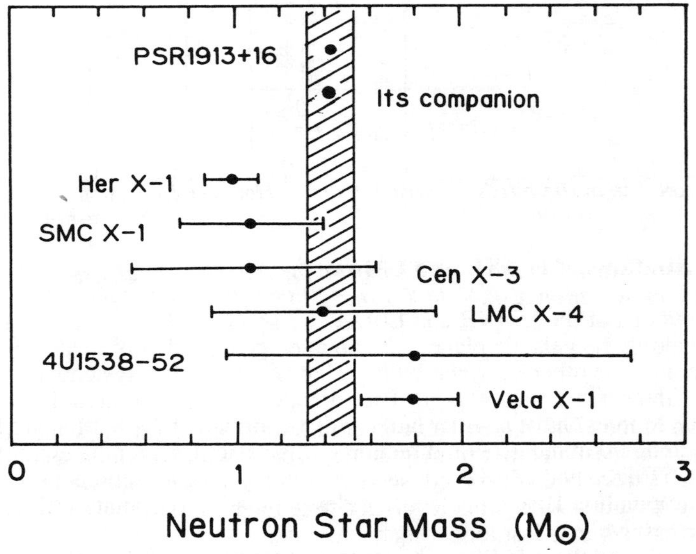

Figure 9: *Summary of our present knowledge of neutron star masses from observations of binary X-ray pulsars and the binary radio pulsar PSR 1913+16 (adapted from [296]).*

X-ray pulse periods show long-term changes which are caused by a combination of external accretion torques, and internal torques due to the coupling of the crust and the core of the neutron star. In general, the sources in which the accretion takes place via an accretion disk show a fairly regular long-term spin up of the pulsar, with typical time scales of roughly 10^4 years. Sources with stellar-wind accretion show much more irregular spin period fluctuations (see [214,296] for detailed reviews, and references).

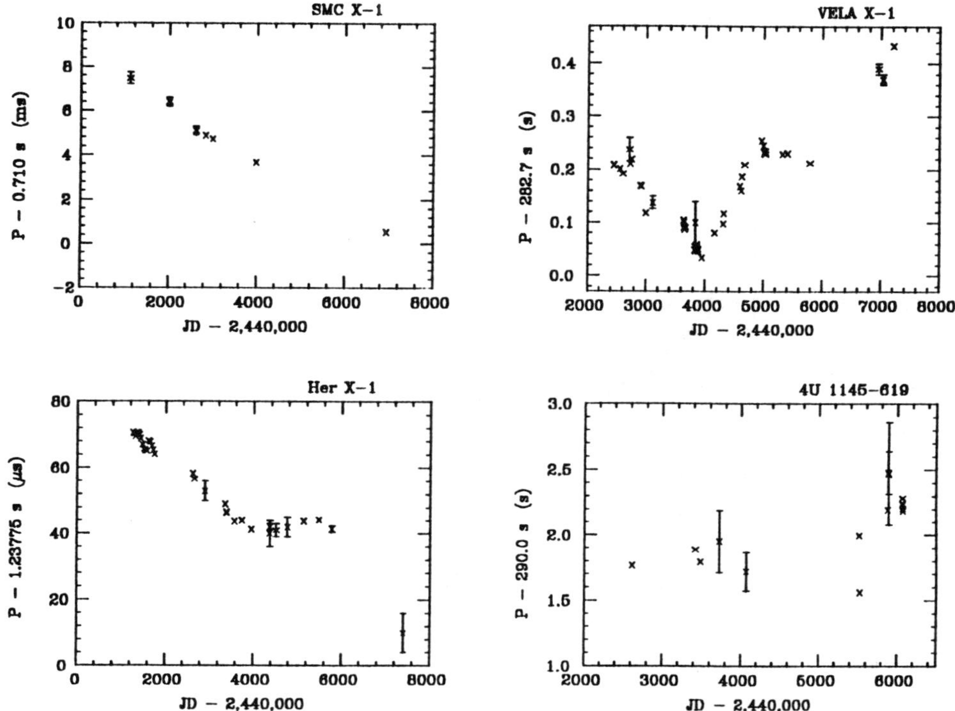

Fig. 10: *Long-term variation of the pulse period for four X-ray pulsars. SMC X-1 and Her X-1 accrete via an accretion disk, Vela X-1 and 1145-619 are wind accretors (after [296]).*

X-ray bursts

A large fraction of the LMXB (see Table 2) emit X-ray bursts, during which the X-ray flux rises by typically at least an order of magnitude, usually within about a second (but rise times up to ~ 10 s have been observed, see e.g., [293]). This is followed by a decay, generally to the pre-burst X-ray flux level, in a time interval between ~ 10 s and about a minute. In rare cases bursts last longer than a minute.

Two types of X-ray bursts can be distinguished [160], called type I and II. The type I bursts show a distinct softening of the X-ray spectrum during the decay of the burst. Their recurrence times are generally of the order of hours and longer, but on occasion time intervals between bursts as short as five minutes have been observed. The spectral development in type II bursts is much less pronounced than that in the type I bursts.

The type II bursts have been observed only from the Rapid Burster [226] (in a few other sources similar events may have been observed, see [231] for references). The time intervals between these bursts can be as short as ~10 s, and as long as ~1h. They come in a characteristic pattern such that the total energy in a burst is approximately proportional to the time interval to the following burst [226]: the Rapid Burster behaves like a relaxation oscillator. This strongly indicates that the type II bursts are the result of an accretion instability [226].

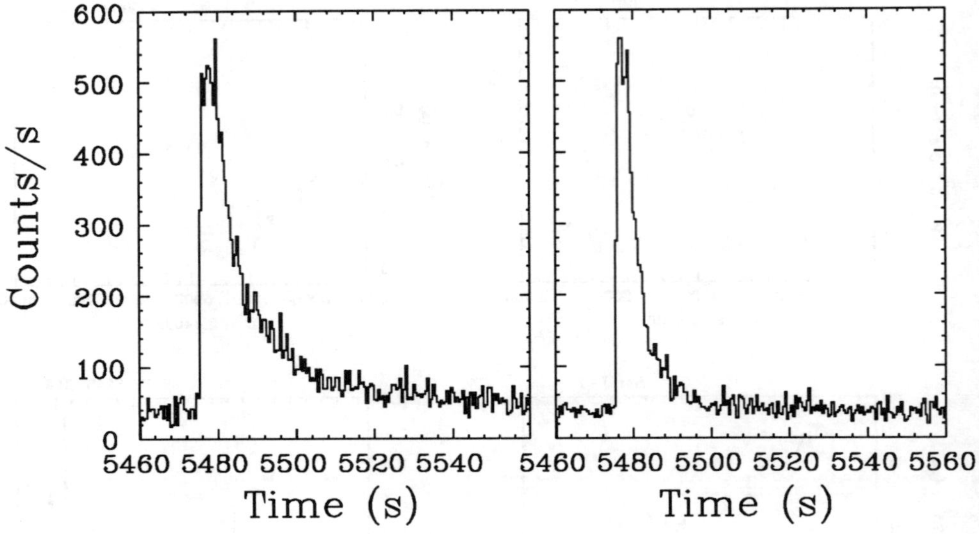

Fig. 11: *Type I X-ray burst from 1702-42 as observed with Exosat in the 1.2 - 5.3 keV band (top) and the 5.3 - 19.0 keV band (bottom); the softening of the X-ray burst spectrum is apparent as a longer tail in the low-energy burst profile (courtesy T. Oosterbroek).*

The sky distribution of X-ray bursters [228] is strongly concentrated to the center of the Galaxy (all known luminous LMXB in globular clusters are X-ray bursters). From this one can infer that X-ray bursters are located at average distances of ~ 8 kpc, and that the total energy and maximum luminosity in type I bursts are of the order of 10^{39} erg, and 10^{38} erg s^{-1}, respectively.

The ratio, α, of the total energy emitted in the persistent flux, to that emitted in bursts, is typically of the order 10^2. This is nicely accounted for by the thermonuclear-flash model of type I X-ray bursts, put forward by Woosley & Taam [458] and Maraschi & Cavaliere [254], and worked out in substantial detail in [11, 105,106,186,379,459]. In this model, after each type I X-ray burst, a fresh layer of matter accretes onto the surface of a neutron star. When a sufficient amount of matter has accumulated on the neutron star surface, critical conditions may develop at the base of this envelope, causing unstable helium burning. The sudden release of nuclear energy gives rise to an X-ray burst. In this model α is the ratio between the gravitational binding energy to the available thermonuclear energy per gram of accreted material (i.e., of order $10^{2\pm0.5}$).Typical values for the rise time, decay time, and recurrence time, and for the maximum luminosity and integrated energy for type I X-ray bursts, are well reproduced by this model.

The radiation observed during an X-ray burst originates directly from the surface of the neutron star (with possibly some modification due to e.g., Compton scattering in intervening hot plasma). Swank *et al.* [374] found that for a particular burst they observed the X-ray spectrum was best fit by a blackbody spectrum, with a temperature that decreased during the decay of the burst. The blackbody radius they found during burst decay (assuming a distance of 10 kpc) was ~15 km. Later work [158,159,419] showed that in all cases the observed spectra were well described by a Planck function, and that the blackbody radii fitted the idea that the type I bursts originate from the surface of a neutron star. These results form the basis for attempts to study the mass-radius relation of neutron stars from observations of the spectral development of X-ray bursts (see, e.g. [429]).

There is a global correlation between the properties of the X-ray bursts and the X-ray luminosity (i.e., mass accretion rate). Following results obtaine by Gottwald *et al.* [117] for the transient source 0748-676, Van Paradijs *et al.* [428] used as a measure of the latter the ratio γ of the persistent X-ray flux to the peak flux of bursts that show photospheric

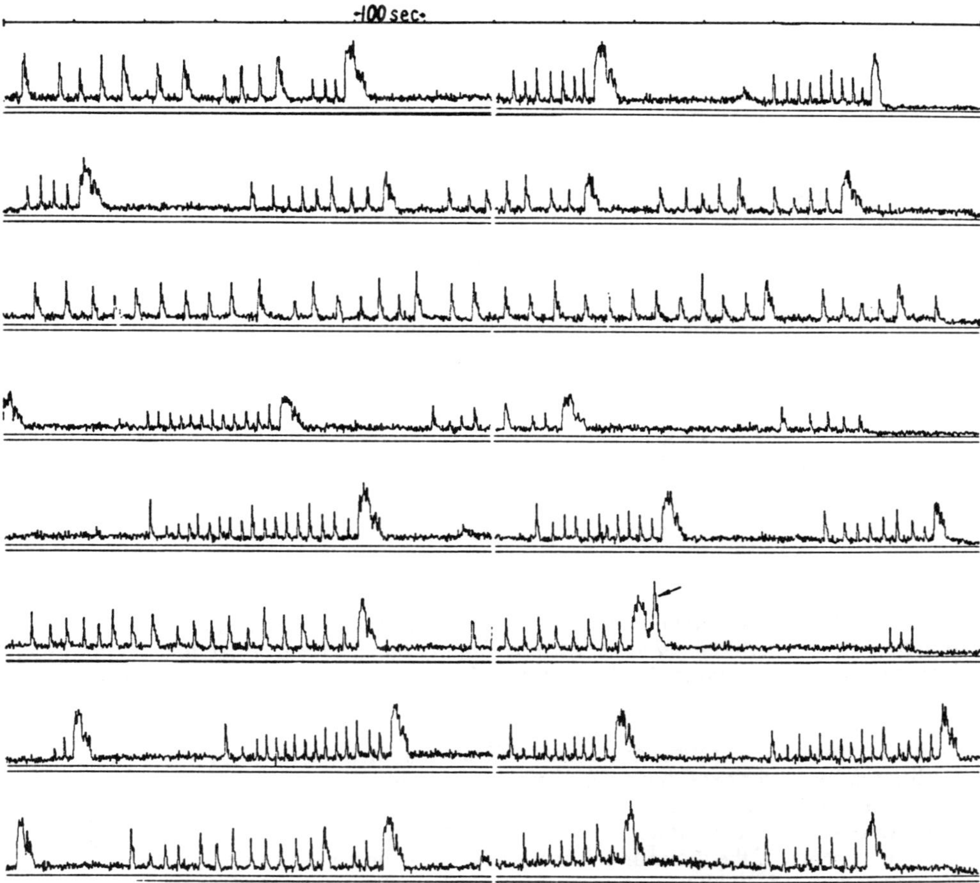

Fig. 12: *Series of type II X-ray bursts from the Rapid Burster observed with SAS-3. Each line represents about 24 minutes of data (from [225]).*

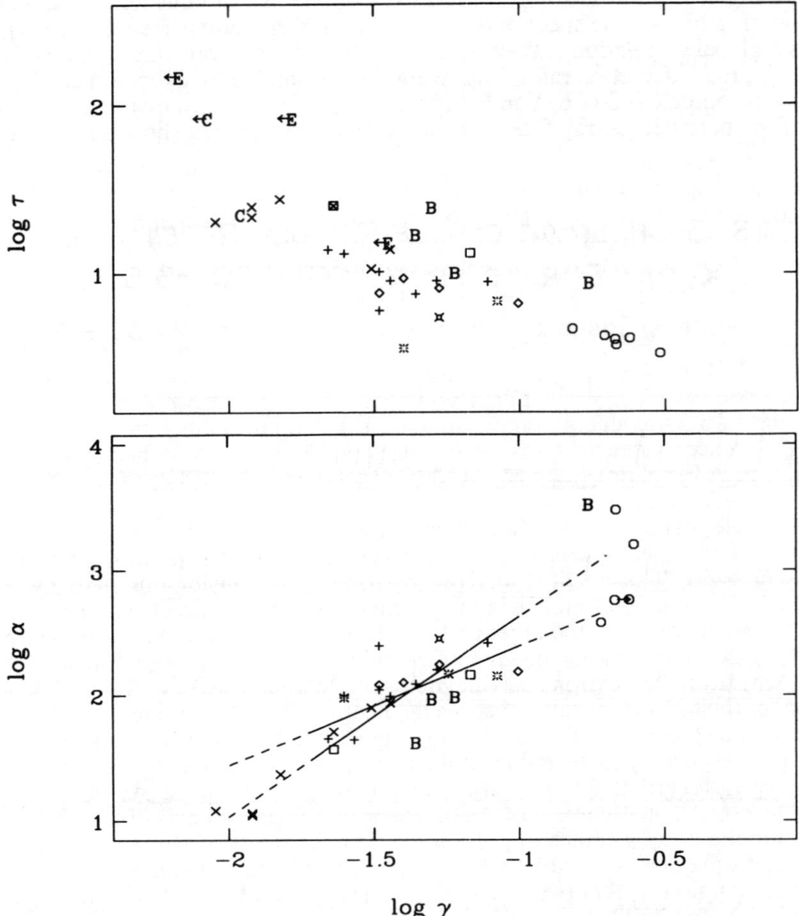

Fig. 13: *Relations between the luminosity parameter γ (see text) and the burst duration τ (top panel), and between γ and the α ratio (bottom panel), for a collection of sources which have produced X-ray bursts with photospheric radius expansion (after [428]).*

radius expansion (theoretical arguments indicate that then the burst luminosity equals the Eddington limit, not expected to vary much between different burst sources), and found that the burst duration τ and the above ratio α are strongly correlated with γ (see Fig. 13). The decrease of the burst duration with γ indicates the decreasing importance of hydrogen in the energetics of the thermonuclear flashes, as the persistent luminosity increases. The strong increase of a with γ implies that independent of the accretion rate, after a given waiting time a source produces an X-ray burst with approximately the same energy. This result is not accounted for by present thermonuclear-flash models.

X-Ray Spectra

X-ray pulsations occur only rarely in LMXB. This suggests that the magnetic fields of the neutron stars in these systems are generally much weaker than in HMXB. (The alternative that the magnetic and rotational axes of the neutron stars are aligned, is discussed below).

Not a single source is known that shows both pulsations and bursts. Apparently, the presence of a strong magnetic field suppresses the instability of the nuclear reactions that gives rise to bursts (as expected from models for thermonuclear flashes, see e.g., [187]). This mutual exclusion of bursts and pulsations indicates that it is a weaker magnetic field, and not only alignment of the field axis, which distinguishes the neutron stars in LMXB from those in HMXB. An additional argument that alignment does not play a crucial role in this context can be inferred from a comparison of the X-ray spectra of the HMXB and LMXB.

The X-ray spectra of HMXB (most of which are pulsars) are generally much harder than those of LMXB [184,384,453]. This distinction i4 present for both the steady and transient X-ray sources [55]. As is illustrated in Fig. 13 the difference in spectral hardness persists into the hard X-ray energy range, up to ~10^2 keV. From this figure it appears that the average difference in spectral hardness, as measured by the ratio of the count rates observed with the A4 experiment on HEAO-1 [223] and with the SSI on Ariel-5 [275,443], is about a factor 10.

It is remarkable that the few LMXB which show pulsations (GX 1+4, Her X-1, and 1627-673) have X-ray spectra which are as hard as those of HMXB (almost all of which are, likewise, pulsars). This result strongly suggests that the division into hard and soft X-ray spectra is related to a difference in the geometry of the accretion flow. For neutron-star magnetic fields of the order of 10^{12} G, and sub-Eddington accretion rates, the accretion flow is dominated by the magnetic field within a relatively large distance (of the order of 10^3 km) from the neutron star (magnetospheric radius, see e.g., [140,214]); a large fraction of the inflowing matter reaches the neutron star via an accretion column on a relatively small area (near the polar caps). For magnetic fields $< 10^9$ G the magnetospheric radius becomes comparable to the radius of the neutron star; one then expects that the accreting material is distributed over a larger fraction of the neutron star surface.

It should be noted that alignment of the magnetic and rotational axis of the neutron star may also lead to the disappearance of pulsations, even for fields of 10^{12} G. However, since the magnetospheric radius will not be much affected by this alignment one does not expect the accretion flow within ~10^3 km of the neutron star to be much affected by the alignment; it is therefore unlikely that alignment alone can explain the systematic difference in the hardness of the X-ray spectra of pulsating and non-pulsating sources. Thus, the differences in these X-ray spectra support the inference, from the distribution of pulsars and bursters among the HMXB and LMXB, that the magnetic fields of the neutron stars in LMXB are systematically weaker than those of the neutron stars in HMXB.

In a number of the brightest LMXB a correlation has been observed between the X-ray intensity and the hardness of the X-ray spectrum (see e.g., [351]). In addition, these sources have, at times, been observed in a different spectral state, in which the X-ray spectrum is relatively hard (but still much softer than the X-ray spectra of pulsars), and does not change much with source intensity. These different spectral states (which, after the location of the sources in an X-ray spectral hardness-intensity diagram, are called the normal branch, and the horizontal branch) are correlated with different fast-variability behavior, with high-frequency intensity-dependent QPO occurring in the HB state, and low-frequency QPO (~5 Hz, independent of source intensity) occurring in the NB state [134,413]. In some sources, a third spectral state (the "flaring branch" in the spectral hardness vs. intensity diagram) has been distinguished, which also is correlated with QPO behavior [135]. After the topology of these spectral branches in the hardness-intensity

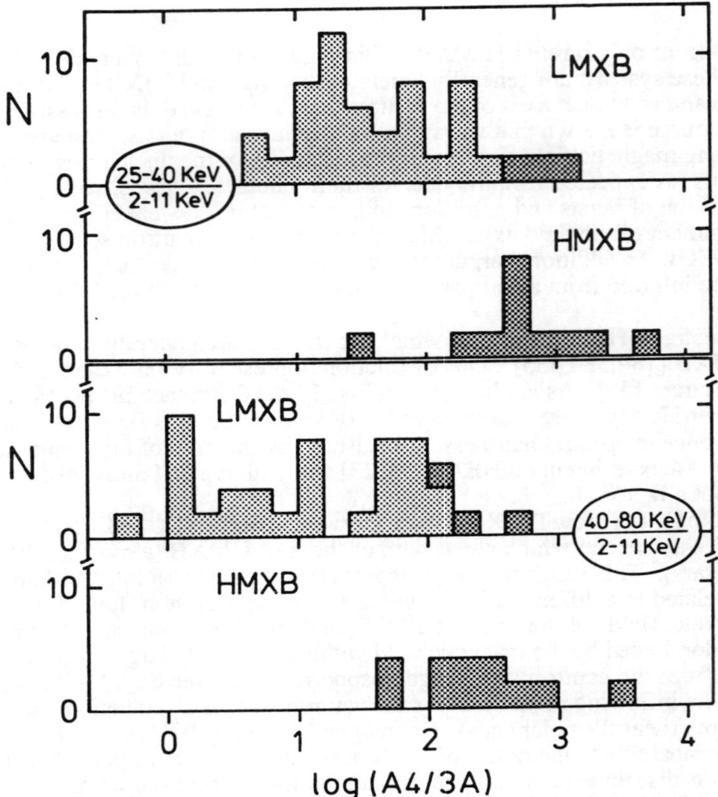

Figure 14: *Distributions of the ratios of count rates, as observed with the HEAO-1 A4 experiment [223] in the 25-40 keV and 40-80 keV bands, to that observed with Ariel 5 in the 2-11 keV band [275,443], shown separately for the high-mass and low-mass X-ray binaries. X-ray pulsars are indicated by a darker shade.*

diagram, these sources are called "Z sources". From the correlated optical/UV behaviour of Cyg X-2 along these branches Hasinger *et al.* [137] concluded that the mass transfer rate increases as one moves from the horizointal branch along the normal branch to the flaring branch.

Recently, Hasinger and Van der Klis [136] have shown that with respect to their spectral behaviour another group of LMXB can be distinguished, the so-called "atoll sources", which are characterized by so-called "island" and "banana" states in an X-ray colour-colour diagram. These sources have different QPO properties from the Z sources, and as a group are substantially less luminous than the latter. The island and banana states are likely governed by the mass accretion rate; this is supported by the recently discovered dependence of the (type I) burst properties on these source states [415].

For reviews of QPO in LMXB with emphasis on observations and data analysis we refer to [236,409]. The origin of the different spectral states, and the connection to the QPO properties are presently a subject of an intense study ; for recent reviews with emphasis on models for QPO we refer to [216].

Magnetic Field Decay?

The strong correlation between X-ray pulsars and HMXB on the one hand, and between X-ray bursters and LMXB on the other hand, and, in addition, the striking correlated difference in the persistent X-ray spectra of HMXB and LMXB (except for the LMXB pulsars), persuasively argues for a systematic difference in the magnetic field strengths of the neutron stars in these two groups of X-ray sources.

There are two possible ways to understand this difference. In the first place the magnetic fields of the (generally old) neutron stars in LMXB may be much weaker than those of the (young) neutron stars in HMXB, because they have always been very weak. This difference might be related to a difference in the formation mechanism of neutron stars in HMXB and LMXB, *viz*. via the normal evolution of a massive star, and via the accretion-induced collapse of a white dwarf, respectively (see e.g., [40]). However, this idea is hard to reconcile with: (i) observed periods and period derivatives of new-born radio pulsars; (ii) the observed weak magnetic fields of the binary radio pulsars PSR 1913+16 and PSR 0655+64, which are late stages in the evolution of massive binary stars (see, e.g., [404]); (iii) the statistics of CVs and LMXB in globular clusters [434].

It appears that the simplest description of the properties of LMXB vis a vis the HMXB is provided by the assumption that the magnetic fields of all neutron stars decay. This would be in agreement with a simple interpretation of kinematic data of new-born radio pulsars [242]. However, as argued recently by Bhattacharya and Verbunt [18], this kinematic evidence for field decay is not as unequivocal as generally believed so far.

It is possible that decay of a magnetic field takes place only in accreting neutron stars. As shown by Taam & Van den Heuvel [380] one cannot distinguish between spontaneous field decay, and decay as a consequence of accretion; in particular the inferred magnetic fields of neutron stars correlate well with the total amount of accreted material.

Observations of millisecond radio pulsars in binaries provide good evidence that any decay of the field of neutron stars does not continue indefinitely. From the colours of the optical counterpart of PSR 0655+64 (a white dwarf) Kulkarni [212] estimated that the (cooling-) age of this system is ~2 10^9 years, yet the magnetic field of neutron star (as inferred from its period derivative) is a few 10^{10} Gauss. A similar conclusion follows from the observed number of millisecond pulsars in binaries which are thought to have descended from LMXB [17,213,406]. These results indicate that when the magnetic field has decayed to a bottom value of the order of 10^{9-10} G (this value may be different for different sources) the decay time increases substantially (from ~10^7 to more than 10^9 years).

It is of interest to note that for two of the three (old) LMXB which show pulsations (Her X-1, GX 1+4) the accretion life time can be calculated from the system parameters; these lifetimes turn out to be very short (of the order of 10^7 years) compared to the expected accretion lifetime of an LMXB (of the order of 10^9 years). This coincidence might suggest that in these systems the neutron star magnetic field is high because the neutron star was formed recently through the accretion induced collapse of a white dwarf during the same stage of mass trtansfer in which we now observe the system as a source of X rays. However, Verbunt *et al*. [435] have shown that this view is hard to reconcile with evolutionary constraints on these systems.They argue that the bottom fields for old neutron stars may in some cases be as high as 10^{12} G.

We are presently lacking a clear view of the evolution of neutron star magnetic fields. An interesting recent development is the suggestion [362] that the magnetic-field history of a neutron star is strongly coupled to the history of its spin rate (see the contribution by Bhattacharya to this Volume).

Table 1 Low-mass X-ray Binaries

Source	Type	RA DEC	ℓ^{II} b^{II}	P_{orb}(hr) P_{puls}(s)	Optical Counterpart	V	B–V	U–B	E_{B-V}	F_X (μJy)	Radio	References
0512–401	GB	05 12 27.9 –40 06 02 [143]	244.5 –35.0		NGC1851 [143]					6 [314,433]		[2,13]
0521–720 LMC X-2		05 21 18.0 –72 00 26 [27]	283.1 –32.7	 [37,75]	*22 [181]	18.5	0.0	–0.8	0.3	26 [420]		[22]
0543–682 CAL 83		05 43 48 –68 23 34 [361]	278.6 –31.3	24.96 [361]	*V [71,361]	16.3- 17.3	0.0		0.3	1 [241]		[74]
0547–711 CAL 87		05 47 26.8 –71 09 50 [71]	281.8 –30.7	10.62 [35]	*X [308]	18.8	–0.15		0.3	0.2 [241]		
0614+091	B?	06 14 22.8 +09 09 22 [27]	200.9 – 3.4		V1055 Ori [91]	18.5	0.3	–0.5		50 [420]		
0620–003 V616 Mon	TU	06 20 11.1 –00 19 11 [27]	210.0 – 6.5	7.75 [272]	V616 Mon [95]	11.2	0.2	–0.8	0.4	50000 [420]	[109,153, 305]	[183]
0748–676	TB	07 48 25.0 –67 37 32 [439]	280.0 –19.8	3.82 [313]	UY Vol	17.1	0.1			30 [117]		[292,349]

0918–549		09 18 54.7 –54 59 37 [47]	275.9 –3.8		*X [47]	21.0	0.3	–0.9	0.3	10 [420]		
0921-630		09 21 25.1 –63 04 48 [27]	281.8 – 9.3	216.18 [69]	V395 Car	15.3	0.4	–0.7	<0.3	2.3 [420]		[267]
1254–690	B	12 54 21.0 –69 01 08 [27]	303.5 – 6.4	3.93 [66,288]	GR Mus [123]	19.1	0.3		0.35	29 [420]		
1323–62	B	13 23 16.8 –61 52 36 [316]	307.1 + 0.2	2.93 [316]					~5:	7 [420]		
1354–64 Cen X-2?	TU	13 54 27.5 –64 29 29 [200]	310.0 – 2.8		BW Cir [200]	16.9	1.1	–0.1	~4:	120 [200]		
1455–314 Cen X-4	TB	14 55 19.6 –31 28 09 [27]	332.2 +23.9	15.10 [48,73]	V822 Cen [38]	12.8	0.05	–0.9	0.1	4200 [420]	[148,153]	[425]
1516–569 Cir X-1	B	15 16 48.5 –56 59 13 [27]	322.1 + 0.0	398.4 [189]	star [284]					5-3000 [420]	[326,371]	[387,388]
1524–617 TrA X-1	T	15 24 05.8 –61 42 35 [27]	320.3 – 4.4		KY TrA *N [295]	17.5B [295]			0.7	800 [420]		
1543–624		15 43 34.1 –62 24 51 [27]	321.8 – 6.3		*6 [6]	~20(B)			0.5	40 [420]		

Source	Type	RA DEC	ℓ^{II} b^{II}	P_{orb}(hr) P_{puls}(s)	Optical Counterpart	V	B–V	U–B	E_{B-V}	F_X (μJy)	Radio	References
1543–475	TU	15 43 33.9 –47 30 54 [318]	330.9 + 5.4		star [44]	14.9	0.6		0.7	15000 [199]		[199,416]
1556–605		15 56 45.8 –60 35 52 [27]	324.1 – 5.9		*X [291]	18.6- 19.5	0.4	–0.7	0.6	26 [420]		
1608–52	TBA	16 08 52.2 –52 17 43 [27]	330.9 – 0.9		star [124]	I≤18.2			1.5:	1100 [420]		[136,320]
1617–155 Sco X-1	Z	16 17 04.5 –15 31 15 [27]	359.1 +23.8	18.90 [116,219]	V818 Sco [344]	12.2	0.2	–0.8	0.3	9000 [420]	[1,151, 155]	[136,236, 409]
1624–49		16 24 17.8 –49 04 46 [27]	334.9 – 0.3	22.10 [185]					~7: [447]	50 [443]		
1627–673	P	16 27 14.7 –67 21 18 [27]	321.8 –13.1	0.69 P:7.7 [277,296]	KZ TrA *4 [178]	18.5	0.1	–1.2	0.1 [424]	20 ˈ420]		[224]
1630–472	T	16 30 19.4 –47 17 24 [312]	336.9 + 0.3							400 [27]		[312]
1636–536	BA	16 36 56.4 –53 39 18 [27	332.9 – 4.8	3.80 [430]	V801 Ara *3 [178]	17.5	0.7	–0.7	0.6	260 [420]		[136,235, 317,415]

1642–455 GX 340+0	Z	16 42 09.3 –45 31 15 [144]	339.6 – 0.1						12: [144]	450 [443]	[126]	[136,236, 321,409]
1656+354 Her X-1	P	16 56 01.6 +35 25 05 [27]	58.2 +37.5	40.80 P:1.24 [82,296]	HZ Her [92]	13.2	–0.1	–0.8	0.0	160 [420]		[296,398]
1658–298	TB	16 58 55.4 –29 52 28 [27]	353.8 + 7.3	7.11 [56]	V2134 Oph *T [94]	18.3	0.45	–0.4	0.3	40 [420]		
1659–487 GX339-4		16 59 02.0 –48 43 07 [27]	338.9 – 4.3	14.83 [162]	V821 Ara *V [94]	15.5	0.8	–0.1	1.2	400 [420]		[64,179, 286]
1702–429	B	17 02 41.0 –42 58 09 [144]	343.9 – 1.3		 [144]				2.5: [144]	100 [303]		
1702–363 GX349+2	Z	17 02 22.9 –36 21 23 [59]	349.1 + 2.7		*6 [8,59]	18.7	1.5		1.3: [456]	700 [443]	[59,109, 126]	[136,236, 409]
1705–440	B	17 05 17.9 –44 02 13 [27]	343.8 - 2.3		 [335]				1.5: [218]	280 [443]		
1705–250	T	17 05 10.4 –25 01 38 [27]	358.6 + 9.1		V2107 Oph [122]	15.9	0.6		0.5	1200 [420]		
1715–321	TB	17 15 32.3 –32 07 34 [27]	354.1 + 3.1		 [335]					28 [443]		

Source	Type	RA DEC	ℓ^{II} b^{II}	P_{orb}(hr) P_{puls}(s)	Optical Counterpart	V	B–V	U–B	E_{B-V}	F_X (μJy)	Radio	References
1724–307	GB	17 24 20.1 –30 45 39 [143]	356.3 + 2.3		Ter 2 [125]				0.7-1.2 [252]	15 [433]		
1728–247 GX 1+4	P	17 28 57.9 –24 43 35 [27]	1.9 + 4.8	304d? P:114 [78,296]	V2116 Oph *GF [93]	19.0			1.7 [125]	125 [420]		[274]
1728–337 GX 354-0	B	17 28 39.2 –33 47 55 [143]	354.3 – 0.2						3.8: [144]	110 [433]		[12]
1728–169 GX 9+9	A	17 28 50 –16 55 32 [27]	8.5 + 9.0	4.20 [145,347]	V2216 Oph [93]	16.8	0.3	–0.7	0.3 [298]	250 [420]		
1730–335 Rapid Burster	GB	17 30 06.6 –33 21 13 [143]	354.8 – 0.2		Lil 1 [240]				3.3: [252]	<0.1 - 250 [27,368]		[89,226, 369]
1731–260	TB	17 31 06.8 –26 03 10 [373]	1.1 3.6						3.2:	110 [373]		
1732–30	GB	17 32 34.8 –30 27 03 [314]	357.6 0.9		Ter 1 [248]				1.5 [448]	16 [356,433]		[356]
1735–444	B	17 35 19.3 –44 25 20 [27]	346.1 – 7.0	4.65 [403]	V926 Sco *5 [178]	17.5	0.2	–0.8	0.15 [132,424]	200 [420]		[403,427]

1741–293 MXB1743–29?	TB	17 41 38 –29 19 53 [174]	359.6 – 0.1			26 [174]	[174,227]
1742–294 MXB1742-29?	B?	17 42 56 –29 29 24 [356]	359.6 – 0.4			180 [443]	[227,446]
1742–289 MXB1743-29?	T	17 42 26.3 –28 59 57 [27]	359.9 0.0			2000 [27]	[31,81,98]
1743–28	B	17 43.6 –28 30 [227]	0.5 – 0.1				
1744–300	B	17 44 13.6 –30 01 29 [357]	359.3 – 1.0				
1744–265 GX3+1	AB	17 44 48.9 –26 32 50 [144]	2.3 + 0.8		3.7 [144]	400 [443]	[136,233, 236,409]
1745–247	GB	17 45 00 –24 45.8 [448]	3.8 + 1.7			16 [433]	[173,248]
1745–203	GT	17 45 55.0 –20 21 07 [143]	7.7 + 3.8	NGC 6440 [102,259]	1.1 [298]	180 [102]	
1746–331		17 46 33.2 –33 11 03 [357]	356.9 – 3.1			6 - 15 [357]	

Source	Type	RA DEC	ℓ^{II} b^{II}	P_{orb}(hr) P_{puls}(s)	Optical Counterpart	V	B–V	U–B	E_{B-V}	F_X (μJy)	Radio	References
1746–370	GB	17 46 48.5 –37 02 18 [143]	353.5 – 5.0		NGC 6441 [143]				0.50 [378]	32 [433]		
1747–214	BT	17 47 25 –21 24 46 [311]	19.8 22.7							70 [311]		
1755–338		17 55 21.5 –33 48 14 [27]	357.2 – 4.9	4.46 [454]	V4134 Sgr [264]	18.5	0.8		0.3 [298]	60 [420]		[264]
1758–250 GX 5–1	Z	17 58 03.1 –25 04 43 [144]	5.1 – 1.0		 [335]				7.8: [144]	1050 [443]	[29,126]	[136,236,278, 409,412,413]
1758–205 GX9+1	A	17 58 33.5 –20 31 44 [144]	9.1 + 1.2		 [335]				3.7: [144]	600 [443]		[136,409]
1811–171 GX13+1	A	18 11 37.2 –17 10 16 [27]	13.5 + 0.1		 [93]					300 [443]	[107,126]	[101,107, 136,409]
1812–12	B	18 12.4 –12 06.0 [294]	18.1 + 2.3							30 [443]		[294]
1813–140 GX 17+2	ZB	18 13 10.9 –14 03 15 [27]	16.4 + 1.3		G star? [93,232]	17.5 [255]	1.3	1.0	2.5: [351,377]	600 [443]	[126,150, 319]	[377]

1820–303	GBA	18 20 27.8 –30 23 17 [143]	2.8 – 7.9	0.19 [285,366]	NGC 6624				0.3 [402]	220 [433]	[150]	[129,136,236, 329,376,409]
1822–371		18 22 22.7 –37 08 04 [27]	356.9 – 11.3	5.57 [261,450]	V691 CrA [123]	15.3	0.1	–0.9	0.1-0.2 [262,263]	9 [420]		[263]
1837+049 Ser X-1	B	18 37 29.5 +04 59 20 [27]	36.1 + 4.8		MM Ser *DS [394]	19.2B [394]		–0.5	≥0.4 [394]	225 [420]		
1850–087	GB	18 50 21.2 –08 46 04 [143]	25.4 – 4.3		NGC 6712 [77,220, 246]	~21 [77,300]			0.42 [77]	7 [314,433]		[245,314]
1905+000	B	19 05 53.4 +00 05 18 [46]	35.0 – 3.7		star [46]	20.5	0.5	–0.5	0.5 [46]	23 [443]		[50]
1908+005 Aql X-1	TB	19 08 42.9 +00 30 05 [27]	35.7 – 4.1		V1333 Aql [392]	14.8 [41]	1.5	–0.4	0.4 [392]	1300 [420]		[41,202,425]
1916–053	B	19 16 08.0 –05 19 42 [27]	31.4 – 8.5	0.83 [127,348, 360,451]	V1405 Aql [348,442]	21.0 [127]	0.4	–0.5		20 [420]		[328,375]
1957+115		19 57 02.2 +11 34 16 [257]	51.3 – 9.3	9.33 [391]	V1408 Aql [72]	18.7 [72]	0.3	–0.6	0.3-0.5 [257]	38 [420]		[36]
2000+25	TU	20 00 42.9 +25 05 44 [304]	63.4 – 3.1		QZ Vul *N [400]	17.5B [304]			1.1-1.4 [24,400]	11000 [382]	[153]	[24,400]

Source	Type	RA DEC	ℓ^{II} b^{II}	P_{orb}(hr) P_{puls}(s)	Optical Counterpart	V	B–V	U–B	E_{B-V}	F_X (μJy)	Radio	References
2023+338	T	20 22 06.3 +33 42 18 [440]	73.2 − 2.2		V404 Cyg [440]	12.7 [34]	1.5	0.3	1.0 [42]	20000 [382]	[156,157]	[42,400]
2030+407 Cyg X-3		20 30 37.6 +40 47 13 [27]	79.9 + 0.7	4.79 [414]	V1521 Cyg [441]				6.3 [282]	260 [420]	[149,281, 283]	[23,266, 283,372]
2127+119	GB	21 27 32.9 +11 56 51 [9]	65.0 − 27.3	8.54 [142,172, 297]	AC211 [9]	15.8 -16.4 [10]	−0.1	−1.2	0.1 [20]	5 [433]		[90,432]
2129+470		21 29 36.2 +47 04 08 [27]	91.6 − 3.0	5.24 [270,271]	V1727 Cyg [393]	16.6 [271]	0.4	−0.5	0.5-0.8 [393]	4 [420]		[49,108,395]
2142+380 Cyg X-2	ZB(?)	21 42 36.9 +38 05 28 [27]	87.3 −11.3	236.2 [68]	V1341 Cyg [110]	14.7 [243]	0.5	−0.2	0.45 [51]	480 [420]	[152,154]	[135,137,138, 154,431,438]

Table 2 High-mass X-ray Binaries

Name(s)	RA DEC	ℓ^{II} b^{II}	Type	Optical Counterpart	Spectral Type	V	B–V	U–B	E_{B-V}	F_X (μJy)	P_{pulse} (sec)	P_{orbit} (days)	radio	references
0050–727 SMC X-3	00 50 19.5 –72 42 24 [27]	302.9 –44.7	T	*4 [52]	O9 III-Ve	~14	–0.3	–1.0	0.03	<1 - 5 [27]				[52]
0053–739 SMC X-2	00 52 53.1 –73 57 19 [27]	302.6 –43.4	T	*5 [52]	B1.5 Ve	16.0	–0.3	–0.5	0.03	<1 - 7 [27]				[52]
0053+604	00 53 40.3 +60 26 47 [27]	123.6 – 2.1		γ Cas	B0.5 III-Ve	1.6-3.0	–0.15	–1.08	0.05	5 - 11 [27]				[104,390]
0103–762	01 07 45.1 –75 00 38 [401]	301.9 –41.1		star [401]	Be	17				2.3 [457]				
0114+650	01 14 41.8 +65 01 32 [27]	125.7 + 2.6		LS I+65°010 [91]	B0.5 IIIe	11.0	1.2	0.1	1.6	4 [27]				
0115+634	01 15 13.8 +63 28 38 [27]	125.9 + 1.0	TP	V635 Cas [180]	OBe	14.5 -16.3	1.4	0.3	1.7	<2-350 [27]	3.61 [296]	24.3		[330,399]
0115–737 SMC X-1	01 15 45.6 –73 42 22 [27]	300.4 –43.6	P	Sk 160 [345]	B0 Ib	13.3	–0.14	–0.98	0.03	0.5 - 57 [27]	0.71 [296]	3.89		[4,396]

Name(s)	RA DEC	ℓ^{II} b^{II}	Type	Optical Counterpart	Spectral Type	V	B–V	U–B	E_{B-V}	F_X (μJy)	P_{pulse} (sec)	P_{orbit} (days)	radio	references
0236+61	02 36 40.6 +61 00 54 [120]	135.7 + 1.1	T	LS I+61°303 [32,133]	B4.5 IIIe	10.7	0.8	0.75		0.2 [332]		26.45	[386]	[16,79,121, 164,165,276, 338,386]
0331+530	03 31 14.9 +53 00 24 [201]	146.1 – 2.2	TP	BQ Cam [201]	Be	15.1 -15.4	1.6 -2.3		1.9	<0.5-1250 [364]	4.4 [296]	34.25		[14,63,163, 364]
0352+309	03 52 15.1 +30 54 01 [27]	163.1 –17.1	P	X Per	O9 III-Ve	6.0-6.6	0.29	–0.82	0.5	<9 - 37 [27]	835 [296]			[280,323,337]
0532–664 LMC X-4	05 32 47.3 –66 24 13 [27]	276.3 –32.5	P	Sk-Ph [45]	O7 III-V	14.0	–0.1	–1.1	0.1	<3 - 60 [27]	13.5 [296]	1.40		[139,191, 217,324]
0535–668	05 35 42.4 –66 53 39 [27]	276.9 –32.2	TP	*Q [182]	B2 IIIe	12.3 -14.9	0.1	–0.9	0.1	<0.01-180 [27]	0.069 [296]	16.7		[61,86]
0535+262	05 35 48.0 +26 17 18 [27]	181.4 – 2.6	TP	V725 Tau [238]	O9.7 IIe	8.9 -9.6	0.45 -0.62	–0.54	0.8	<3-2800 [27]	104 [296]	111 55.7?		[176,244,258]
0538–641 LMC X-3	05 38 39.7 –64 06 34 [27]	273.6 –32.1	B	*1 [411]	B3 Ve	16.7- 17.5	–0.3- 0.0		0.1	<1.7-44 [27]				[21,70,170 211,426]
0540–697 LMC X-1	05 40 05.5 –69 46 04 [27]	280.2 –31.5		*32 [67]	O7-9 III	14.5	0.29	–0.70	0.32	3 - 25 [27]		~4		[15,96,167, 307]

0544–665	05 44 15.6 –66 34 59 [181]	276.5 –31.4		*1 [181]	B1 Ve	15.4	–0.23	–0.96	0.1	1.8 [181]			[410]
0556+286	05 52 44.3 +28 46 41 [401]	181.3 + 1.9		HD249179	Be	9.2				1.1 [457]			
0726–260	07 26 50.0 –26 00 13 [363]	240.3 – 4.1		star [363]	B0-1	11.6	0.3	–0.6	0.6	1.2 - 4.7 [443,457]			
0739–529	07 46 09.8 –53 12 28 [401]	266.4 –13.7		HD63666 SAO235515	Be	7.62	0.02	–0.24		0.7 [457]			
0749–600	07 55 27.7 –60 57 54 [401]	274.0 –16.2		HD65663 SAO250018		7.0				0.7 [457]			
0900–403 Vela X-1	09 00 13.2 –40 21 25 [27]	263.1 + 3.9	P	HD77581 GP Vel	B0.5 Ib	6.9	0.47	–0.51	0.7	2 - 1100 [27]	283 [296]	8.96	[25,83,84, 87,197,421, 396]
1024.0 –5732	10 24 05.4 –57 33 24 [39]	284.5 + 0.2	P	Wack 2134 [389]	O5: [39]	12.7	1.5		1.8	1 - 10 [39]	0.061 [39]		
1034–56	10 28 28.3 –56 49 14 [401]	284.6 + 0.7		HD91188 SAO238130	Be	6.64	–0.10	–0.56		3.3 [443]			
1118–615	11 18 45.2 –61 38 31 [27]	292.5 – 0.9	PT	He3-640 [175]	O9.5 III-Ve	12.1	0.96	–0.30	1.2	0.1 - 70 [27,289]	405 [296]		[289]

Name(s)	RA DEC	ℓ^{II} b^{II}	Type	Optical Counterpart	Spectral Type	V	B–V	U–B	E_{B-V}	F_X (μJy)	P_{pulse} (sec)	P_{orbit} (days)	radio	references
1119–603 Cen X-3	11 19 01.9 –60 20 57 [27]	292.1 + 0.3	P	V779 Cen [33]	O6.5 II-III	13.3	1.07	–0.04	1.4	10 - 312 [27]	4.84 [296]	2.089 [193]		[53,128,192] 196,396]
1145–619	11 45 33.6 –61 55 44 [27]	295.6 – 0.2	PT	Hen 715 HD102567 [26]	B0-1 Ve	9.3	0.18	–0.81	0.25 -0.49	4 - 1000 [27]	292 [296]	187.5		[57,85,91, 330]
1145.1 –6141	11 45 02.3 –61 40 33 [27]	295.5 – 0.0	P	star [166]	B2Iae	13.1	1.5	0.15	1.7	4 - 40 [166]	298 [296]	10.76? [169]		[169,171]
1223–624 GX 301–2	12 23 49.7 –62 29 37 [27]	300.1 – 0.0	P	BP Cru Wra 977 [26]	B1-1.5 Ia	10.8	1.76	0.42	1.8	9-1000 [27]	696 [296]	41.5		[339,346,455]
1249–637	12 39 53.2 –62 47 06 [401]	302.0 – 0.2		HD110432 SAO252002	Be	5.1 [299]	0.27	–0.99		2.2 [457]				
1253–761	12 35 59.8 –75 05 43 [401]	302.1 –12.5		HD109857 SAO256967	Be	6.5 [299]	0.08	–0.24		0.6 [457]				
1255–567	12 51 39.6 –56 53 50 [401]	303.4 + 5.7		μ^2 Cru HD112091	Be	5.17 [299]	–0.12	–0.51		0.8 [457]				
1258–613 GX 304–1	12 58 11.8 –61 19 58 [27]	304.1 + 1.2	P	MMV *2 [260]	B2 Vne	13.5- 14.2	1.7	0.8	2.0	6-200 [27]	272 [296]	133?		[62,325,327, 330]

1417–624	14 17 25.5 –62 28 11 [27]	313.0 – 1.6	P	*7 [7]	Hα em.	17	1.5	1.0	1.0	2 - 43 [27]	17.6 [296]		
1538–522	15 38 38.6 –52 13 37 [27]	327.4 + 2.1	P	QV Nor *12 [6]	B0 Iab	14.4	1.9	0.6	2.1	<3 - 30 [27]	529 [296]	3.73	[251,306,310]
1553–542	15 53 55.6 –54 16 15 [27]	327.9 – 0.9	P							27 [27]	9.3 [296]	30.6	[193]
1555–552	15 50 26.4 –55 10 54 [401]	327.0 – 1.2		HD141926 SAO243098	Be	8.6				1.7 [457]			
1657–415	16 57 16.8 –41 35 59 [27]	344.4 + 0.3	P							4 - 42 [27]	38 [296]		
1700–377	17 00 32.7 –37 46 29 [27]	347.8 + 2.2		HD153919 V884 Sco	O6.5f	6.6	0.25	–0.72	0.52	<11 - 110 [27]		3.41	[43,87,118, 130,131, 190,422]
1722–363	17 22 33 –36 22 05 [444]	351.5 +0.6	P							0.2 - 3 [381]	413 [296]		[381,385]
1833–076 Sct X-1	18 33.8 –07 38.9 [27]	24.5 – 0.2	P							1.6 - 2.7 [203,443]	111 [203]		[203,204]
1836–04	18 36.0 –04.5 [203]	27.9 – 1.9	P							2.5 [203]	81.1 [203]		

Name(s)	RA DEC	ℓ^{II} b^{II}	Type	Optical Counterpart	Spectral Type	V	B–V	U–B	E_{B-V}	F_X (μJy)	P_{pulse} (sec)	P_{orbit} (days)	radio	references
1843+00	18 43.0 +0.9 [203]	33.1 + 1.7	P							33 [203]	29.5 [203]			
1843–024	18 43.0 –02.4 [203]	30.2 + 1.1	P							6 [203]	94.8 [203]			
1907+097	19 07 15.1 +09 44 54 [27]	43.7 + 0.5	P	 [129]	B I	16.4	3.2		3.3	4 - 275 [27]	438 [296]	8.38		[58,250, 417]
1909+048	19 09 21.2 +04 53 54 [27]	39.7 – 2.2		SS433 [209]	pec V1343 Aql	14.2	2.1	0.6	2.6	2 - 10 [27]		13.1	[436]	[5,99,194, 221,222,256, 268,370]
1936+541	19 31 42.6 +53 46 12 [401]	85.9 +15.9		DM+53°2262	Be	9.8				0.7 [457]				
1956+350 Cyg X-1	19 56 28.9 +35 03 55 [27]	71.3 + 3.1	B	HD226868 V1357 Cyg [332]	O9.7 Iab	8.9	0.78	–0.30	1.06	235-1320 [27]		5.6	[30,147, 150]	[30,88,112, 113,147,150, 195,279,301]
2030+375	20 30 22.1 +37 28 00 [287]	77.2 – 1.3	PT	star [287]	Be	19.7	3.3		3.6	<0.5-1400 [315]	41.8 [296]	45.6-47.5		[3,54,177, 287,315]
2137+57 Cep X-4?	21 36.0 +57.1 [204]	99.1 + 3.4	PT							<6 - 100 [204]	66.2 [296]			[204]

2202+501	21 59 44.1 +49 55 35 [401]	97.3 – 4.0	DM+49°3718	Be	8.8				0.7 [457]	
2206+543	22 06 07.4 +54 16 23 [363]	100.6 – 1.1	star [363]	B1	9.9	0.2	–0.6	0.5	0.6 - 5.5 [443,457]	[363]
2214+589	22 24 47.8 +60 58 59 [401]	106.4 + 3.1	GG3 71 [114]	Be	11				0.5 [457]	

References

[1] Ables, J.G. 1969, *Proc. Astron. Soc. Australia* **1**, 237.
[2] Alcaino, G. *et al.* 1990, *Astron. J.* **99**, 817.
[3] Angelini, L. *et al.* 1989, *Astroph. J.* **346**, 906.
[4] Angelini, L. *et al.* 1990, *Astroph. J.* (accepted).
[5] Antokhina, E.A., Cherepashchuk, A.M. 1988, *Sov. Astron. J.* **31**, 295.
[6] Apparao, K.M.V. *et al.* 1978, *Nature* **271**, 225.
[7] Apparao, K.M.V. *et al.* 1980, *Astron. Astroph.* **89**, 249.
[8] Augusteijn, Th. *et al.* 1990, *Astron. Astroph.* (in preparation).
[9] Auriere, M. *et al.* 1984, *Astron. Astroph.* **138**, 415.
[10] Auriere, M. *et al.* 1986, *Astron. Astroph.* **158**, 158.
[11] Ayasli, S., Joss, P.C. 1982, *Astroph. J.* **256**, 637,
[12] Basinska, E.M. *et al.* 1984, *Astroph. J.* **281**, 337.
[13] Baylin, C.D. *et al.* 1988, *Astroph. J.* **331**, 303.
[14] Bernacca, P.L. 1984, *Astron. Astroph.* **132**, L8.
[15] Bianchi, L., Pakull, M.W. 1985, *Astron. Astroph.* **146**, 242.
[16] Bignami, G.F. *et al.* 1981, *Astroph. J.* **246**, L85.
[17] Bhattacharya, D., Srinivasan, G. 1986, *Current Sci.,* **55**, 327.
[18] Bhattacharya, D., Verbunt, F. 1990, *Astron. Astroph.* (in press)
[19] Bignami, G.F. *et al.* 1981, *Astroph. J.* **247**, L85.
[20] Bingham, W. *et al.* 1984, *Mon. Not. Roy. Astron. Soc.* **209**, 765.
[21] Bochkarev, N.G. *et al.* 1988, *Sov. Astron. J.* **32**, 405.
[22] Bonnet-Bidaud, J.M. *et al.* 1989, *Astron. Astroph.* **213**, 97.
[23] Bonnet-Bidaud, J.M., Chardin, G. 1989, *Phys. Rept.* **170**, 325.
[24] Borisov, N.V. *et al.* 1989, *Proc. 23rd ESLAB Symposium*, ESA SP-296, p. 305.
[25] Boynton, P.E. *et al.* 1986, *Astroph. J.* **307**, 545.
[26] Bradt, H.V. *et al.* 1977, *Nature* **269**, 21.
[27] Bradt, H.V., McClintock, J.E. 1983, *Ann. Rev. Astron. Astroph.* **21**, 63.
[28] Braes, L.L.E., Miley, G.K. 1971, *Nature* **232**, 246.
[29] Braes, L.L.E. *et al.* 1972, *Nature* **236**, 392.
[30] Braes. L.L.E., Miley, G.K. 1976, *Nature* **264**, 731.
[31] Branduardi, G. *et al.* 1976, *Mon. Not. Roy. Astron. Soc.* **175**, 47P.
[32] Brodskaya, E.J., Shajn, P.F. 1958, *Izw. Krim. Astroph. Obs.* **20**, 299.
[33] Brucato, R.J. *et al.* 1972, *Astroph. J.* **175**, L137.
[34] Buie, M.W., Bond, H.E. 1989, *IAU Circular* 4786.
[35] Callanan, P.J. *et al.* 1989, *Mon. Not. Roy. Astron. Soc.* **241**, 37P.
[36] Callanan, P.J. *et al.* 1989, *Proc. 23rd ESLAB Symposium*, ESA SP-296, p. 331.
[37] Callanan, P. *et al.* 1990, *Mon. Not. Roy. Astron. Soc.* (in press).
[38] Canizares, C. *et al.* 1980, *Astroph. J.* **236**, L55.
[39] Caraveo, P.A. *et al.* 1989, *Astroph. J.* **338**, 338.
[40] Chanmugan, G., Brecher, K. 1987, *Nature*, **329**, 696.
[41] Charles, P.A.. *et al.* 1980, *Astroph. J.* **237**, 154.
[42] Charles, P.A. *et al.* 1989, *Proc. 23rd ESLAB Symposium*, ESA SP-296, p. 103.
[43] Cherepashchuk, A.M., Khruzina, T.S. 1981, *Sov. Astron. J.* **25**, 697.
[44] Chevalier, C. 1989, *Proc. 23rd ESLAB Symposium*, ESA SP-296, p. 341.
[45] Chevalier, C., Ilovaisky, S.A. 1978, *ESO Messenger* **5**, 4.
[46] Chevalier, C. *et al.* 1985, *Astron. Astroph.* **147**, L3.
[47] Chevalier, C., Ilovaisky, S.A. 1987, *Astron. Astroph.* **172**, 167.
[48] Chevalier, C. *et al.* 1989, *Astron. Astroph.* **210**, 114.
[49] Chevalier, C. *et al.* 1989, *Astron. Astroph.* **217**, 108.
[50] Chevalier, C., Ilovaisky, S.A. 1990, *Astron. Astroph.* **228**, 115.

[51] Chiappetti, L. *et al.* 1983, *Astroph. J.* **265**, 354.
[52] Clark, G.W. *et al.* 1978, *Astroph. J.* **221**, L37.
[53] Clark, G.W. *et al.* 1988, *Astroph. J.* **324**, 974.
[54] Coe, M.J. *et al.* 1988, *Mon. Not. Roy. Astron. Soc.* **232**, 865.
[55] Cominsky, L. *et al.* 1978, *Astroph. J.* **224**, 46.
[56] Cominsky, L., Wood, K.S. 1989, *Astroph. J.* **337**. 485.
[57] Cook, M.C., Warwick, R.S. 1987, *Mon. Not. Roy. Astron. Soc.* **225**, 369; **227**, 661.
[58] Cook, M.C., Page, C.G. 1987, *Mon. Not. Roy. Astron. Soc.* **225**, 381.
[59] Cooke, B.A., Ponman, T.J. 1990, *Astron. Astroph.* (in press)
[60] Corbet, R. H. D. 1984, *Astron. Astroph.***141**, 91.
[61] Corbet, R.H.D. *et al.* 1983, *Mon. Not. Roy. Astron. Soc.* **212**, 565.
[62] Corbet, R.H.D. *et al.* 1986, *Mon. Not. Roy. Astron. Soc.* **221**, 961.
[63] Corbet, R.H.D. *et al.* 1986, *Astron. Astroph.* **162**, 117.
[64] Corbet, R.H.D. *et al.* 1987, *Mon. Not. Roy. Astron. Soc.* **227**, 1055.
[65] Corbet, R. H. D. 1987, in ref. [359], p. 63.
[66] Courvoisier, T. J.-L. *et al.* 1986, *Astroph. J.* **309**, 265.
[67] Cowley, A.P. *et al.* 1978, *Astron. J.* **83**, 1619.
[68] Cowley, A.P. *et al.* 1979, *Astroph. J.* **231**, 539.
[69] Cowley, A.P. *et al.* 1982, *Astroph. J.* **256**, 605.
[70] Cowley, A.P. *et al.* 1983, *Astroph. J.* **272**, 118.
[71] Cowley, A.P. *et al.* 1984, *Astroph. J.* **286**, 196.
[72] Cowley, A.P. *et al.* 1988, *Astroph. J.* **333**, 906.
[73] Cowley, A.P. *et al.* 1988, *Astron. J.* **95**, 1231.
[74] Crampton, D. *et al.* 1987, *Astroph. J.* **321**, 745.
[75] Crampton, D. *et al.* 1990, *Astroph. J.* **355**, 496.
[76] Crawford, J. A., Kraft, R. P. 1956, *Astroph. J.* **123**, 44.
[77] Cudworth, K.M. 1988, *Astron. J.* **96**, 105.
[78] Cutler, E.P. *et al.* 1986, *Astroph. J.* **300**, 551.
[79] D'Amico, N. *et al.* 1987, *Astron. Astroph.* **180**, 114.
[80] Davidsen, A. *et al.* 1977, *Astroph. J.* **211**, 866.
[81] Davies, R.D. *et al.* 1976, *Nature* **261**, 476.
[82] Deeter, J.E. *et al.* 1981, *Astroph. J.* **247**, 1003.
[83] Deeter, J.E. *et al.* 1987, *Astroph. J.* **314**, 634.
[84] Deeter, J.E. *et al.* 1989, *Astroph. J.* **336**, 376.
[85] De Loore, C. *et al.* 1981, *Astron. Astroph.* **104**, 150.
[86] Densham, R.A. *et al.* 1982, *Mon. Not. Roy. Astron. Soc.* **205**, 1117.
[87] Dolan, J.F., Tapia, S. 1988, *Astron. Astroph.* **202**, 124.
[88] Dolan, J.F., Tapia, S. 1989, *Astroph. J.* **344**, 830.
[89] Dotani, T. *et al.* 1990, *Astroph. J.* **350**, 395.
[90] Dotani, T. *et al.* 1990, *Nature* **347**, 534.
[91] Dower, R.G. *et al.* 1978, *Nature* **273**, 364
[92] Doxsey, R,E. *et al.* 1973, *Astroph. J.* **182**, L25.
[93] Doxsey, R.E. *et al.* 1977, *Nature* **270**, 586.
[94] Doxsey, R. *et al.* 1979, *Astroph. J.* **228**, L67.
[95] Duerbeck, H.W. 1987, *Space Sci. Rev.* **45**, 1.
[96] Ebisawa, K. *et al.* 1989, *Publ. Astron. Soc. Japan* **41**, 519.
[97] Eggleton, P. P. & Pringle, J. E. 1985, *Interacting Binaries*, NATO ASI Series C. Vol. 150 (Reidel).
[98] Eyles, C.J. *et al.* 1975, *Nature* **257**, 291.
[99] Filipenko, A.V. *et al.* 1988, *Astron. J.* **96**, 242.
[100] Filipov, L., White, N. E. (eds) 1988, Proc. COSPAR/IAU Symposium *The Physics of Compact Objects: Theory versus Observations* (Pergamon Press).

[101] Fleischman, J.R. 1985, *Astron. Astroph.* **153**, 106.
[102] Forman, W. *et al.* 1976, *Astroph. J.* **207**, L25.
[103] Frank, J., King, A.R. & Raine, D. J. 1985, *Accretion Power in Astrophysics*, (Cambridge Univ. Press).
[104] Frontera, F. *et al.* 1987, *Astroph. J.* **320**, L127.
[105] Fujimoto, M. Y., Hanawa, T. & Miyaji, S. 1981, *Astroph. J.* **247**, 267.
[106] Fushiki, I., Lamb, D. Q. 1987, *Astroph. J.* **323**, L55.
[107] Garcia, M.R. *et al.* 1988, *Astroph. J.* **328**, 552.
[108] Garcia, M. *et al.* 1989, *Astroph. J.* **341**, L75.
[109] Geldzahler, B.J. 1983, *Astroph. J.* **264**, L49.
[110] Giacconi, R. *et al.* 1967, *Astroph. J.* **148**, L129.
[111] Giacconi, R. *et al.* 1971, *Astroph. J.* **167**, L67.
[112] Gies, D.R., Bolton, C.T. 1982, *Astroph. J.* **260**, 240.
[113] Gies, D.R., Bolton, C.T. 1986, *Astroph. J.* **304**, 371; **304**, 389.
[114] Gonzales, G., Gonzales, G. 1956, *Bol. Obs. Tonantzintla y Tacubaya* **2**, No. 15, 16.
[115] Gorayo, P.S., Tur, N.S. 1988, *Astroph. Space Sci.* **145**, 263.
[116] Gottlieb, E.W. *et al.* 1975, *Astroph. J.* **195**, L33.
[117] Gottwald, M. *et al.* 1986, *Astroph. J.* **308**, 213.
[118] Gottwald, M. *et al.* 1986, *Mon. Not. Roy. Astron. Soc.* **222**, 21P.
[119] Gottwald, M. *et al.* 1989, *Astroph. J.* **339**, 1044.
[120] Gregory, P.C. *et al.* 1979, *Astron. J.* **84**, 1030.
[121] Gregory, P.C. *et al.* 1989, *Astroph. J.* **339**, 1054.
[122] Griffiths, R.E. *et al.* 1978, *Astroph. J.* **221**, L63.
[123] Griffiths, R.E. *et al.* 1978, *Nature* **276**, 247.
[124] Grindlay, J.E., Liller, W. 1978, *Astroph. J.* **220**, L127.
[125] Grindlay, J.E. *et al.* 1980, *Astroph. J.* **240**, L121.
[126] Grindlay, J.E., Seaquist, E.R. 1986, *Astroph. J.* **310**, 172.
[127] Grindlay, J.E. *et al.* 1988, *Astroph. J.* **334**, L25.
[128] Gruber, D.E. 1988, *Astroph. J.* 328, 265.
[129] Haberl, F. *et al.* 1987, *Astroph. J.* **314**, 266.
[130] Haberl, F. *et al.* 1989, *Astroph. J.* **343**, 409.
[131] Hammerschlag-Hensberge, G. *et al.* 1978, *Astron. Astroph.* **64**, 399.
[132] Hammerschlag-Hensberge, G. *et al.* 1982, *Astroph. J.* **254**, L1.
[133] Hardorp, J. *et al.* 1959, *Luminous Stars in the Northern Milky Way I* (Hamburger Sternwarte - Warner and Swasey Observatory).
[134] Hasinger, G. 1987, *Astron. Astroph.* **186**, 153.
[135] Hasinger, G. 1987, *Proc. IAU Symposium* **125**, 333.
[136] Hasinger, G., Van der Klis, M. 1989, *Astron. Astroph.* **225**, 79.
[137] Hasinger, G. *et al.* 1990, *Astron. Astroph.* **235**, 131.
[138] Hasinger, G. 1988, *Adv. Space Res.* **8(2)**, 377.
[139] Heemskerk, M., Van Paradijs, J. 1989, *Astron. Astroph.* **223**, 154.
[140] Henrichs, H. F. 1983, in ref. [230], p. 393.
[141] Hensberge, G. *et al.* 1973, *Astron. Astroph.* **29**, 69.
[142] Hertz, P. 1987, *Astroph. J.* **315**, L119.
[143] Hertz, P., Grindlay, J.E. 1983, *Astroph. J.* **275**, 105.
[144] Hertz, P., Grindlay, J.E. 1984, *Astroph. J.* **282**, 118.
[145] Hertz, P., Grindlay, J.E. 1988, *Astroph. J.* **331**, 764.
[146] Hiltner W. A., Mook D. E. 1970, *Ann. Rev. Astron. Astroph.* **8**, 139.
[147] Hjellming, R.M. 1973, *Astroph. J.* **182**, L29.
[148] Hjellming, R.M. 1979, *IAU Circular* 3369.
[149] Hjellming, R.M. 1988, in: *'Galactic and Extragalactic Radio Astronomy'*, eds. G.L. Verschuur & K.I. Kellerman, p. 381 (Springer Verlag).

[150] Hjellming, R.M., Wade, C.M. 1971a, *Astroph. J.* **168**, L21.
[151] Hjellming, R.M., Wade, C.M. 1971b, *Astroph. J.* **164**, L1.
[152] Hjellming, R.M., Blankenship, L.C., 1973, *Nature Phys. Sci.* **243**, 81.
[153] Hjellming, R.M. *et al.* 1988, *Astroph. J.* **335**, L75.
[154] Hjellming, R.M. *et al.* 1990, *Astron. Astroph.* **235**, 147.
[155] Hjellming, R.M. *et al.* 1990b, *Astroph. J.* (submitted)
[156] Hjellming, R.M. *et al.* 1989, *IAU Circular* 4790.
[157] Hjellming, R.M., Han, X.H. 1989, *IAU Circular* 4879.
[158] Hoffman, J. A. *et al.* 1977a, *Mon. Not. Roy. Astron. Soc.* **179**, 57P.
[159] Hoffman, J. A. *et al.* 1977b, *Astroph. J.* **217**, L23.
[160] Hoffman, J. A. *et al.* 1978, *Nature*, **271**, 630.
[161] Hoffman, J.A. *et al.* 1980, *Astroph. J.* **240**, L27.
[162] Honey, W.B. *et al.* 1988, *IAU Circular* 4532.
[163] Honeycutt, R.K., Schlegel, E.M. 1985, *Publ. Astron. Soc. Pacific* **97**, 300.
[164] Howarth, I.D. 1983, *Mon. Not. Roy. Astron. Soc.* **203**, 801.
[165] Hutchings, J.B., Crampton, D. 1981, *Publ. Astron. Soc. Pacific* **93**, 486.
[166] Hutchings, J.B. *et al.* 1981, *Astron. J.* **86**, 871.
[167] Hutchings, J.B. *et al.* 1983, *Astroph. J.* **275**, L43.
[168] Hutchings, J.B. *et al.* 1985, *Astroph. J.* **292**, 670.
[169] Hutchings, J.B. *et al.* 1987, *Publ. Astron. Soc. Pacific* **99**, 420.
[170] Hutchings, J.B., Cowley, A.P. 1990, *Proc. IAU Symposium* **148** (in press).
[171] Ilovaisky, S.A. *et al.* 1982, *Astron. Astroph.* **114**, L7.
[172] Ilovaisky, S.A. *et al.* 1987, *Astron. Astroph.* **179**, L1.
[173] Inoue, H. *et al.* 1984, *Publ. Astron. Soc. Japan* **36**, 855.
[174] In 't Zand, J. et al. 1990, *Proc. 28th Cospar Meeting* (in press).
[175] Janot-Pacheco, E. *et al.* 1981, *Astron. Astroph.* **99**, 274.
[176] Janot-Pacheco, E. *et al.* 1987, *Astron. Astroph.* **177**, 91.
[177] Janot-Pacheco, E. *et al.* 1988, *Astron. Astroph.* **202**, 81.
[178] Jernigan, J.G. *et al.* 1977, *Nature* **270**, 321.
[179] Jernigan, J.G. *et al.* 1978, *Nature* **272**, 701.
[180] Johnston, M. et al. 1978, *Astroph. J.* **223**, L71.
[181] Johnston, M.D. *et al.* 1979, *Astroph. J.* **233**, 514.
[182] Johnston, M.D. *et al.* 1980, *Nature* **285**, 26.
[183] Johnston, H.M. *et al.* 1989, *Astroph. J.* **345**, 492.
[184] Jones, C. 1977, *Astroph. J.* **214**, 856.
[185] Jones, M.H., Watson, M.G. 1989, *Proc. 23rd ESLAB Symposium*, ESA SP-296, p. 439.
[186] Joss, P.C. 1978, *Astroph. J.* **225**, L123.
[187] Joss, P. C., Li, F. K. 1980, *Astroph. J.* **238**, 287.
[188] Joss, P. C., Rappaport, S. A. 1984, *Ann. Rev. Astron. Astroph.*, **24**, 537.
[189] Kaluzienski, L.J. *et al.* 1976, *Astroph. J.* **208**, L71.
[190] Kaper, L. *et al.* 1990, *Nature* **347**, 652.
[191] Kelley, R. et al. 1983, *Astroph. J.* **264**, 568.
[192] Kelley, R.E. et al. 1983, *Astroph. J.* **268**, 790.
[193] Kelley, R.L. *et al.* 1983, *Astroph. J.* **274**, 765.
[194] Kemp, J.C. *et al.* 1986, *Astroph. J.* **305**, 805.
[195] Kemp, J.C. *et al.* 1987, *Sov. Astron. J.* **31**, 170.
[196] Khruzina, T.S., Cherepashchuk, A.M. 1986, *Sov. Astron. J.* **30**, 295
[197] Khruzina, T.S., Cherepashchuk, A.M. 1986, *Sov. Astron. J.* **30**, 422
[198] Kirk, J. & Trümper, J. 1983, in ref. [230], p. 261.
[199] Kitamoto, S. *et al.* 1984, *Publ. Astron. Soc. Japan* **36**, 799.
[200] Kitamoto, S. *et al.* 1990, *Astroph. J.* **361**, 590.
[201] Kodaira, K. *et al.* 1985, *Publ. Astron. Soc. Japan* **37**, 97.

[202] Koyama, K. *et al.* 1981, *Astroph. J.* **247**, L27.
[203] Koyama, K. *et al.* 1990, *Nature* **343**, 148.
[204] Koyama, K. *et al.* 1990, *Astroph. J.* (submitted).
[205] Kraft, R. P. 1962, *Astroph. J.* **135**, 408.
[206] Kraft, R. P. 1964, *Astroph. J.* **139**, 457.
[207] Kraft, R. P. 1973, *Proc. I.A.U. Symposium* **55**, 36.
[208] Kraft, R. P., Miller, J.S. 1969, *Astroph. J.* **155**, L159.
[209] Krumenaker, L.E. 1975, *Publ. Astron. Soc. Pacific* **87**, 185
[210] Krzeminski, W. 1974, *Astroph. J.* **192**, L135.
[211] Kuiper, L. *et al.* 1988, *Astron. Astroph.* **203**, 79.
[212] Kulkarni, S. 1986, *Astroph. J.* **306**, L85.
[213] Kulkarni, S., Narayan, R. 1988, *Astroph. J.* **335**, 755.
[214] Lamb, F. K. 1988, in ref. [302], p. 649.
[215] Lamb, F.K. 1989, *Proc. 23rd ESLAB Symposium*, ESA SP-296, p. 215.
[216] Lamb, F.K. 1990, *Astroph. J.* in press.
[217] Lang, F.L. *et al.* 1981, *Astroph. J.* **246**, L21.
[218] Langmeier, A. *et al.* 1987, *Astroph. J.* **323**, 288.
[219] LaSala, J., Thorstensen, J.R. 1985, *Astron. J.* **90**, 2077.
[220] Lehto, H. *et al.* 1990, *Nature* **347**, 49.
[221] Leibowitz, E.M. 1984, *Mon. Not. Roy. Astron. Soc.* **210**, 279.
[222] Leibowitz, E.M. *et al.* 1984, *Mon. Not. Roy. Astron. Soc.* **206**, 751.
[223] Levine A. M. *et al.* 1984, *Astroph. J. Suppl.*, **54**, 581.
[224] Levine, A. *et al.* 1988, *Astroph. J.* **327**, 732.
[225] Lewin, W.H.G. 1977, *American Scientist* **65**, 605.
[226] Lewin, W.H.G. *et al.* 1976, *Astroph. J.* **207**, L95.
[227] Lewin, W.H.G. *et al.* 1976, *Mon. Not. Roy. Astron. Soc.* **177**, 83P.
[228] Lewin, W.H.G. *et al.* 1977, *Nature* , **267**, 28.
[229] Lewin, W. H.G. & Clark, G. W. 1980, *Ann. New York Ac. Sci.*, **336**. 451.
[230] Lewin, W.H.G. & Van den Heuvel, E.P.J. (eds), 1983, *Accretion Driven Stellar X-ray Sources* (Cambridge University press).
[231] Lewin, W. H. G. & Joss, P. C. 1983, in ref. [230], p. 41.
[232] Lewin, W.H.G., Van Paradijs, J. 1985, *Astron. Astroph.* **142**, 361.
[233] Lewin, W.H.G. *et al.* 1987, *Mon. Not. Roy. Astron. Soc.* **226**, 383.
[234] Lewin, W. H. G., Trümper, J. & Brinkmann, W. (eds) 1986, *"The Evolution of Galact. X-ray Binaries"*, NATO ASI Series C, Vol. 167 (Reidel).
[235] Lewin, W.H.G. *et al.* 1987, *Astroph. J.* **319**, 893.
[236] Lewin, W.H.G. *et al.* 1988, *Space Sci. Rev.* **46**, 273.
[237] Li, F.K. *et al.* 1978, *Nature* **276**, 799.
[238] Li, F.K. *et al.* 1979, *Astroph. J.* **228**, 893.
[239] Liller, W. 1973, *Astroph. J.* **184**, L23.
[240] Liller, W. 1977, *Astroph. J.* **213**, L21.
[241] Long, K.S. *et al.* 1981, *Astroph. J.* **248**, 925.
[242] Lyne, A. G. *et al.* 1982, *Mon. Not. Roy. Astron. Soc.* **201**, 503.
[243] Lyutyi, V.M., Shugarov, S. Yu. 1979, *Sov. Astron. J. Lett.* **5**, 206.
[244] Lyutyi, V.M. *et al.* 1989, *Sov. Astron. J. Lett.* **15**, 182.
[245] Machin, G. *et al.* 1989, *IAU Circular* 4818.
[246] Machin, G. *et al.* 1990, *Mon. Not. Roy. Astron. Soc.* **246**, 237.
[247] Magnier, E. *et al.* 1989, *Mon. Not. Roy. Astron. Soc.* **237**, 729.
[248] Makishima, K. *et al.* 1981, *Astroph. J.* **247**, L23.
[249] Makishima, K. *et al.* 1983, *Astroph. J.* **267**, 310.
[250] Makishima, K. *et al.* 1984, *Publ. Astron.Soc. Japan* **36**, 679.
[251] Makishima, K. *et al.* 1987, *Astroph. J.* **314**, 619.

[252] Malkan, M. *et al.* 1980, *Astroph. J.* **237**, 432.
[253] Maraschi, L. *et al.* 1975, *Nature*, **259**, 292.
[254] Maraschi, L., Cavaliere, A. 1977, *Highlights of Astronomy*, 4, part 1, p. 127.
[255] Margon, B. 1976, *Astroph. J.* **179**, 585.
[256] Margon, B. 1984, *Ann. Rev. Astron. Astroph.* **22**, 507.
[257] Margon, B. *et al.* 1978, *Astroph. J.* **221**, 907.
[258] Margoni, R. *et al.* 1988, *Astron. Astroph.* **195**, 148.
[259] Markert, T.H. *et al.* 1975, *Nature* **275**, 32.
[260] Mason, K.O. *et al.* 1978, *Mon. Not. Roy. Astron. Soc.* **184**, 45P.
[261] Mason, K.O. *et al.* 1980, *Astroph. J.* **242**, L109.
[262] Mason, K.O., Cordova, F.A. 1982, *Astroph. J.* **255**, 603.
[263] Mason, K.O., Cordova, F.A. 1982, *Astroph. J.* **262**, 253.
[264] Mason, K.O. *et al.* 1985, *Mon. Not. Roy. Astron. Soc.* **216**, 1033.
[265] Mason, K. O. Watson, M. G. & White, N. E. (eds) 1986, *The Physics of Accretion Onto Compact Objects,* Lecture Notes in Physics, Vol. 266, (Springer).
[266] Mason, K.O. *et al.* 1986, *Astroph. J.* **309**, 700.
[267] Mason, K.O. *et al.* 1987, *Mon. Not. Roy. Astron. Soc.* **226**, 423.
[268] Mazeh, T. *et al.* 1987, *Astroph. J.* **317**, 824.
[269] McClintock, J.E. *et al.* 1978, *Astroph. J.* **223**, L75.
[270] McClintock, J.E. *et al.* 1981, *Astroph. J.* **243**, 900.
[271] McClintock, J.E. *et al.* 1979, *Astroph. J.* **258**, 245.
[272] McClintock, J.E., Remillard, R. 1986, *Astroph. J.* **308**, 110.
[273] McClintock, J.E. 1988, in ref. [100], p. 191.
[274] McClintock, J.E., Leventhal, M. 1989, *Astroph. J.* **346**, 143.
[275] McHardy, I.M., *et al.* 1981, *Mon. Not. Roy. Astron. Soc.* **197**, 893.
[276] Mendelson, H., Mazeh, T. 1989, *Mon. Not. Roy. Astron. Soc.* **239**, 723.
[277] Middleditch, J. *et al.* 1981, *Astroph. J.* **244**, 1001.
[278] Mitsuda, K. *et al.* 1989, *Publ. Astron. Soc. Japan* **41**, 557.
[279] Miyamoto, S., Kitamoto, S. 1989, *Nature* **342**, 773.
[280] Moffat, A.F.J. *et al.* 1973, *Astron. Astroph.* **23**, 433.
[281] Molnar, L.A. et al. 1985, in: *'Radio Stars'*, eds. R.M. Hjellming & D. Gibson, p. 329 (Reidel)
[282] Molnar, L.A. *et al.* 1988, *Bull. A.A.S.* **20**, 736.
[283] Molnar, L.A. *et al.* 1988, *Astroph. J.* **331**, 494.
[284] Moneti, A. 1990, *ESO Messenger* **58**, 7.
[285] Morgan, E. *et al.* 1990, *Astroph. J.* (in press).
[286] Motch, C. *et al.* 1983, *Astron. Astroph.* **119**, 171.
[287] Motch, C., Janot-Pacheco, E. 1987, *Astron. Astroph.* **182**, L55.
[288] Motch, C. *et al.* 1987, *Astroph. J.* **313**, 792.
[289] Motch, C. *et al.* 1988, *Astron. Astroph.* **201**, 63.
[290] Motch, C., Pakull, M.W. 1989, *Astron. Astroph.* **214**, L1.
[291] Motch, C. *et al.* 1989, *Astron. Astroph.* **219**, 158.
[292] Motch, C. *et al.* 1989, *Proc. 23rd ESLAB Symposium*, ESA SP-296, p. 545.
[293] Murakami, T., Inoue, H., Koyama, K. *et al.* 1980, *Astroph. J.* **240,** L143.
[294] Murakami, T. *et al.* 1983, *Publ. Astron. Soc. Japan* **35**, 531.
[295] Murdin, P. *et al.* 1977, *Mon. Not. Roy. Astron. Soc.* **178**, 27P.
[296] Nagase, F. 1989, *Publ Astron. Soc. Japan* **41**, 1.
[297] Naylor, T. *et al.* 1988, *Mon. Not. Roy. Astron. Soc.* **233**, 285.
[298] Neckel, T., Klare, G. 1980, *Astron. Astroph. Suppl.* **42**, 251.
[299] Nicolet, B. 1978, *Astron. Astroph. Suppl.* **34**, 1.
[300] Nieto, J.L. *et al.* 1990, *IAU Circular* 4944.
[301] Ninkov, Z. *et al.* 1987, *Astroph. J.* **321**, 425; 438.

[302] Oegelman, H., Van den Heuvel, E.P.J. (eds), 1989, *Timing Neutron Stars*, (Kluwer Acad. Publ.).
[303] Oosterbroek, T. *et al.* 1990, *Astron. Astroph.* (submitted).
[304] Okumura, S., Noguchi, T. 1989, *IAU Circular* 4589.
[305] Owen, F.N. *et al.* 1976, *Astroph. J.* **203**, L15.
[306] Pakull, M.W. *et al.* 1983, *Astron. Astroph.* **122**, 79.
[307] Pakull, M.W., Angebault, L.P. 1986, *Nature* **322**, 511.
[308] Pakull, M.W. *et al.* 1988, *Astron. Astroph.* **203**, L27.
[309] Pallavicini, R., White, N.E. (eds) 1988, *X-ray Astronomy with EXOSAT*, (Memorie della Societa' Astronomica Italiana).
[310] Parkes, G.E. *et al.* 1978, *Mon. Not. Roy. Astron. Soc.* **184**, 73P.
[311] Parmar, A.N. *et al.* 1985, *IAU Circular* 4058.
[312] Parmar, A.N. *et al.* 1986, *Astroph. J.* **304**, 664.
[313] Parmar, A.N. *et al.* 1986, *Astroph. J.* **308**, 199.
[314] Parmar, A.N. *et al.* 1989, *Astron. Astroph.* **222**, 96.
[315] Parmar, A.N. *et al.* 1989, *Astroph. J.* **338**, 359; 373.
[316] Parmar, A.N. *et al.* 1989, *Astroph. J.* **338**, 1024.
[317] Pedersen, H. *et al.* 1982, *Astroph. J.* **263**, 340.
[318] Pedersen, H. *et al.* 1983, *IAU Circular* 3858.
[319] Penninx, W. *et al.* 1988, *Nature* **336**, 147.
[320] Penninx, W. *et al.* 1989, *Astron. Astroph.* **208**, 146.
[321] Penninx, W. *et al.* 1990, *Mon. Not. Roy. Astron. Soc.* (in press).
[322] Penny, A. J. *et al.* 1973, *Mon. Not. Roy. Astron. Soc.***163**, 7p.
[323] Penrod, D.G., Vogt, S.S. 1985, *Astroph. J.* **299**, 653.
[324] Pietsch, W. *et al.* 1985, *Space Sci. Rev.* **40**, 371.
[325] Pietsch, W. *et al.* 1986, *Astron. Astroph.* **163**, 93.
[326] Preston, R.A. *et al.* 1983, *Astroph. J.* **268**, L23.
[327] Priedhorsky, W.C., Terrell, J. 1983, *Astroph. J.* **273**, 709.
[328] Priedhorsky, W., Terrell, J. 1984, *Astroph. J.* **280**, 661.
[329] Priedhorsky, W., Terrell, J. 1984, *Astroph. J.* **284**. L17.
[330] Priedhorsky, W.C., Holt, S.S. 1987, *Sp. Sci. Rev.* **45**, 291.
[331] Pringle, J. E. & Wade, R. A. (eds) 1985: *Interacting Binary Stars*, (Cambridge Univ. Press).
[332] Rappaport, S.A. *et al.* 1971, *Astroph. J.* **168**, L17.
[333] Rappaport, S.A. *et al.* 1978, *Astroph. J.* **224**, L1.
[334] Rappaport, S. & Joss, P. C. 1983, in ref. [230], p. 1.
[335] Reid, C.A. *et al.* 1980, *Astron. J.* **85**, 1062.
[336] Ritter, H. 1990, *Astron. Astroph. Suppl.*, **85**, 1179.
[337] Robba, N.R., Warwick, R.S. 1989, *Astroph. J.* **346**, 469.
[338] Rössiger, S. 1978, *IAU Circular* 3210.
[339] Rothschild, R.E., Soong, Y. 1987, *Astroph. J.* **315**, 154.
[340] Ruderman, M. A. *et al.* 1989, *Astroph. J.* **336**, 507.
[341] Ryter, C. 1970, *Astron. Astroph.* **9**, 288.
[342] Salpeter, E.E. 1964, *Astroph. J.* **140**, 796.
[343] Salpeter, E. E. 1973, *IAU Symposium*, **55**, 135.
[344] Sandage, A.R. *et al.* 1966, *Astroph. J.* **146**, 316.
[345] Sanduleak, N. 1968, *Astron. J.* **73**, 246.
[346] Sato, N. *et al.* 1986, *Astroph. J.* **304**, 241.
[347] Schaefer, B. 1990, *Astroph. J.* **354**, 720.
[348] Schmidtke, P. 1988, *Astron. J.* **95**, 1528.
[349] Schmidtke, P., Cowley, A.P. 1987, *Astron. J.* **93**, 374.
[350] Schreier E. *et al.* 1972, *Astroph. J.* **172**, L112.

[351] Schulz, N. *et al.* 1989, *Astron. Astroph.* **225**, 48.
[352] Schwartz, D.A. *et al.* 1980, *Astron. J.* **85**, 549.
[353] Setti, G., Woltjer, L. 1970, *Ap. Space Sci.*, **9**, 185.
[354] Shapiro, S. L. & Teukolsky, S. A. 1983: *Black Holes, White Dwarfs and Neutron Stars,* (John Wiley and Sons).
[355] Shklovskii, J. 1967, *Astroph. J.* **148**, L1.
[356] Skinner, G.K. *et al.* 1987, *Nature* **330**, 544.
[357] Skinner, G.K. *et al.* 1990, *Mon. Not. Roy. Astron. Soc.* **243**, 72.
[358] Slettebak, A. 1987, in ref. [359], p. 24.
[359] Slettebak, A. & Snow, T. P. (eds) 1987: *Physics of Be Stars*, (Cambridge Univ. Press).
[360] Smale, A.P. *et al.* 1988, *Mon. Not. Roy. Astron. Soc.* **232**, 647.
[361] Smale, A.P. *et al.* 1988, *Mon. Not. Roy. Astron. Soc.* **233**, 51.
[362] Srinivasan, G. *et al.* 1990, *Curr. Sci.* **59**, 31.
[363] Steiner, J.E. *et al.* 1984, *Astroph. J.* **280**, 688.
[364] Stella, L. *et al.* 1985, *Astroph. J.* **288**, L45.
[365] Stella, L. *et al.* 1986, *Astroph. J.* **308**, 669.
[366] Stella, L. *et al.* 1987a, *Astroph. J.* **312**, L17.
[367] Stella, L. *et al.* 1987b, *Astroph. J.* **315**, L49.
[368] Stella, L. *et al.* 1988a, *Astroph. J.* **324**, 361.
[369] Stella, L. *et al.* 1988b, *Astroph. J.* **324**, 379.
[370] Stewart, G.C. *et al.* 1987, *Mon. Not. Roy. Astron. Soc.* **228**, 293.
[371] Stewart, R.T. *et al.* 1990, *Mon. Not. Roy. Astron. Soc.* (submitted).
[372] Strom, R.G. *et al.* 1989, *Nature* **337**, 234.
[373] Sunyaev, R. *et al.* 1989, *Proc. 23rd ESLAB Symposium*, ESA SP-296, p. 641.
[374] Swank, J. H. *et al.* 1977, *Astroph. J.* **212**, L73.
[375] Swank, J.H. *et al.* 1984, *Astroph. J.* **277**, 724.
[376] Sztajno, M. *et al.* 1983, *Astroph. J.* **267**, 713.
[377] Sztajno, M. *et al.* 1986, *Mon. Not. Roy. Astron. Soc.* **222**, 499.
[378] Sztajno, M. *et al.* 1987, *Mon. Not. Roy. Astron. Soc.* **226**, 39.
[379] Taam, R.E. 1982, *Astroph. J.* **258**, 761.
[380] Taam, R. E., Van den Heuvel, E. P. J. 1986, *Astroph. J.* **305**, 235.
[381] Takeuchi, Y. *et al.* 1990, *Publ. Astron. Soc. Japan* **42**, 287.
[382] Tanaka, Y. 1989, *Proc. 23rd ESLAB Symposium*, ESA SP-296, p. 3.
[383] Tananbaum, H. *et al.* 1972, *Astroph. J.* **177**, L5.
[384] Tananbaum, H. 1973, *IAU Symposium*, **55**, 9.
[385] Tawara, Y. *et al.* 1989, *Publ. Astron. Soc. Japan* **41**, 473.
[386] Taylor, A.R., Gregory, P.C. 1982, *Astroph. J.* **255**, 210.
[387] Tennant, A.F. *et al.* 1986, *Mon. Not. Roy. Astron. Soc.* **221**, 27P.
[388] Tennant, A.F. 1987, *Mon. Not. Roy. Astron. Soc.* **230**, 403.
[389] The, P.S. 1966, *Contr. Bosscha Obs.* No. 35.
[390] Thom, C. *et al.* 1986, *Astron. Astroph.* **165**, L13.
[391] Thorstensen, J.R. 1987, *Astroph. J.* **312**, 739.
[392] Thorstensen, J.R. *et al.* 1978, *Astroph. J.* **220**, L131.
[393] Thorstensen, J.R. *et al.* 1979, *Astroph. J.* **233**, L57.
[394] Thorstensen, J.R. *et al.* 1980, *Astroph. J.* **238**, 964.
[395] Thorstensen, J.R. *et al.* 1988, *Astroph. J.* **334**, 430.
[396] Tjemkes, S.A. *et al.* 1986, *Astron. Astroph.* **154**, 77.
[397] Trümper, J. et al. 1978, *Astroph. J.* **219**, L105.
[398] Trümper, J. *et al.* 1986, *Astroph. J.* **300**, L63.
[399] Tsunemi, H., Kitamoto, S,. 1988, *Astroph. J.* **334**, L21.
[400] Tsunemi, H. *et al.* 1989, *Astroph. J.* **337**, L81.
[401] Tuohy, I.R. *et al.* 1988, in: "*Physics of neutron Stars and Black Holes*", ed. Y. Tanaka,

(Universal Academic Press, Tokyo), p. 93.
[402] Vacca, W.D. *et al.* 1986, *Mon. Not. Roy. Astron. Soc.* **220**, 339.
[403] Van Amerongen, S. *et al.* 1987, *Astron. Astroph.* **185**, 147.
[404] Van den Heuvel, E. P. J. 1984, *J. Astroph. Astron.* **5**, 209.
[405] Van den Heuvel, E. P. J., Heise, J. 1972, *Nature Phys. Sci.*, **239**, 67.
[406] Van den Heuvel, E. P. J. *et al.* 1986, *Nature*, **322**, 153.
[407] Van den Heuvel, E. P. J., Rappaport, S. 1987, in ref. [359], p. 291.
[408] Van den Heuvel, E. P. J., Van Paradijs, J. 1988, *Nature*, **334**, 227.
[409] Van der Klis, M. 1989, *Ann. Rev. Astron. Astroph.* **27**, 517.
[410] Van der Klis, M. *et al.* 1983, *Mon. Not. Roy. Astron. Soc.* **203**, 279.
[411] Van der Klis, M. *et al.* 1985, *Astron. Astroph.* **151**, 322.
[412] Van der Klis, M. *et al.* 1985, *Nature* **316**, 225.
[413] Van der Klis, M. *et al.* 1987, *Astroph. J.* **313**, L19.
[414] Van der Klis, M., Bonnet-Bidaud, J.M. 1989, *Astron. Astroph.* **214**, 203.
[415] Van der Klis, M. *et al.* 1990, *Astroph. J.* **360**, L19.
[416] Van der Woerd, H. *et al.* 1989, *Astroph. J.* **344**, 320.
[417] Van Kerkwijk, M. *et al.* 1987, *Astron. Astroph.* **209**, 173.
[418] Van Oyen, J. 1989, *Astron. Astroph.* **217**, 115.
[419] Van Paradijs, J. 1978, *Nature*, **274**, 650.
[420] Van Paradijs, J. 1983: in *Accretion Driven Stellar X-ray Sources*, ed. W.H.G. Lewin & E.P.J. van den Heuvel (Cambridge University Press), p. 189.
[421] Van Paradijs, J. *et al.* 1977, *Astron. Astroph. Suppl.* **30**, 195.
[422] Van Paradijs, J. *et al.* 1984, *Astron. Astroph. Suppl.* **55**, 7.
[423] Van Paradijs, J. & Verbunt, F. 1984, in *High Energy Transients in Astrophysics*, ed. S.E. Woosley, AIP Proc. Vol. **115**, p. 49.
[424] Van Paradijs, J. *et al.* 1986, *Astron. Astroph. Suppl.* **63**, 71.
[425] Van Paradijs, J. *et al.* 1987, *Astron. Astroph.* **182**, 47.
[426] Van Paradijs, J. *et al.* 1987, *Astron. Astroph.* **184**, 201.
[427] Van Paradijs, J. *et al.* 1988, *Astron. Astroph.* **192**, 147.
[428] Van Paradijs, J. *et al.* 1988, *Mon. Not. Roy. Astron. Soc.* **233**, 437.
[429] Van Paradijs, J., Lewin, W.H.G. 1989, in ref. [309], p. 213.
[430] Van Paradijs, J. *et al.* 1990, *Astron. Astroph.* **234**, 181.
[431] Van Paradijs, J. *et al.* 1990, *Astron. Astroph.* **235**, 156.
[432] Van Paradijs, J. *et al.* 1990, *Publ. Astron. Soc. Japan* **42**, No. 5 (in press).
[433] Verbunt, F. *et al.* 1984, *Mon. Not. Roy. Astron. Soc.* **210**, 899.
[434] Verbunt, F. & Meylan, G. 1988, *Astron. Astroph.* **203**, 297.
[435] Verbunt, F. *et al.* 1990, *Astron. Astroph.* **234**, 195.
[436] Vermeulen, R,. 1989, Ph. D. Thesis, Univ. Leiden.
[437] Vidal, N. V. 1973, *Astroph. J.* **182**, L77.
[438] Vrtilek, S.D. *et al.* 1990, *Astron. Astroph.* **235**, 162.
[439] Wade, R.A. *et al.* 1985, *Publ. Astron. Soc. Pacific* **97**, 1092.
[440] Wagner, R.M. *et al.* 1989, *IAU Circular* 4783.
[441] Wagner, R.M. *et al.* 1989, *Astroph. J.* **346**, 971.
[442] Walter, F. *et al.* 1982, *Astroph. J.* **253**, L67.
[443] Warwick, R.S. *et al.* 1981, *Mon. Not. Roy. Astron. Soc.* **197**, 865.
[444] Warwick, R.S. *et al.* 1988, *Mon. Not. Roy. Astron. Soc.* **232**, 551.
[445] Waters, L., Van Kerkwijk, M. 1989, *Astron. Astroph.* **223**, 196.
[446] Watson, M.G. *et al.* 1981, *Astroph. J.* **250**, 142.
[447] Watson, M.G. *et al.* 1985, *Space Sci. Rev.* **40**, 195.
[448] Webbink, R.F. 1983, *Proc. IAU Symposium* **113**, 541.
[449] Webster, B.L., Murdin, P. 1971, *Nature*, **237**, 35.
[450] White, N.E. *et al.* 1981, *Astroph. J.* **247**, 994.

[451] White, N.E., Swank, J.H. 1982, *Astroph. J.* **253**, L61.
[452] White, N. E. *et al.* 1984, *Astroph. J.* **270**, 711.
[453] White, N. E., Marshall, F. E. 1984, *Astroph. J.* **281**, 354.
[454] White, N.E. *et al.* 1984, *Astroph. J.* **283**, L9.
[455] White, N.E., Swank, J.H. 1984, *Astroph. J.* **287**, 856.
[456] White, N.E. *et al.* 1988, *Astroph. J.* **324**, 363.
[457] Wood, K.S. *et al.* 1984, *Astroph. J. Suppl.* **56**, 507.
[458] Woosley, S., Taam, R. E. 1976, *Nature*, **263**, 101.
[459] Woosley, S.E., Weaver, T.A.1984, in *High Energy Transients in Astrophysics*, ed. S.E. Woosley, AIP Proc. **115**, 273.
[460] Zel'dovich, Ya. B. 1964, *Sov. Phys. Dokl.* **9**, 246.
[461] Zel'dovich, Ya. B., Guzeinov, O.H. 1965, *Astroph. J.* **144**, 840.

OPTICAL LIGHT CURVES OF X-RAY BINARIES

JAN VAN PARADIJS
Astronomical Institute "Anton Pannekoek", University of Amsterdam
& Center for High-Energy Astrophysics, NIKHEF-H, Amsterdam

ABSTRACT. As discussed in the accompanying lecture notes on "Neutron stars in X-ray binaries", two broad groups of X-ray binaries can be distinguished on the basis of their X-ray spectra, their X-ray variability characteristics and their optical properties; these are the high-mass X-ray binaries (HMXB) and the low-mass X-ray binaries (LMXB). In this lecture the orbital variations of the optical brightness of these systems are discussed in the framework of a relatively simple model, in which contributions to the optical emission from the companion and the accretion disk, their mutual eclipses, and reprocessing of X rays in matter surrounding the X-ray source, are included.

1. Optical Light Curves of High-mass X-ray Binaries

1.1. Introduction

The optical emission of high-mass X-ray binaries (HMXB) does not, in general, appear to be much affected by the presence of an accreting compact star; the reason for this is that the bolometric luminosity of the massive companion star generally exceeds the X-ray luminosity, often by a large margin [154, 160]. The optical counterparts of many HMXB show brightness variations in phase with the orbital period, with amplitudes typically ≤ 0.1 magnitude, superposed on which is apparently irregular variability (amplitude a few percent). These orbital light curves often show a double-waved shape, with two maxima and two minima per orbital cycle. From a comparison with optical radial-velocity curves or with Doppler delay curves of the X-ray pulse arrival times it appears that these minima occur when the binary components are in conjuction.

Such double-waved (so-called ellipsoidal) light curves are caused by the rotational and tidal distortion of the companion star, which fills (or nearly fills) a critical surface, and a non-uniform distribution of its surface brightness (gravity darkening). As discussed in Sect. 1.2. the distortion of the companin which, together with the inclination of the orbit, determines the shape and amplitude of the light curve, depends on its relative size (i.e., the extent to which it fills its critical surface), and on the mass ratio of the binary components.

For some systems purely ellipsoidal light curves are inadequate to describe the observed brightness variations: X-ray heating and the presence of an accretion disk around the compact component can have an appreciably effect.

In Sect. 1.2. I describe a relatively simple geometric model for ellipsoidal light curves, incorporating effects related to the presence of the X-ray source (X-ray heating, accretion disk). Observations of orbital light curves of individual HMXB are discussed in Sect. 1.3.

J. Ventura and D. Pines (eds.), Neutron Stars: Theory and Observation, 289–317.

1.2. Description of the model

1.2.1. *Ellipsoidal variations*

The geometric model is based on the following assumptions. (i) The gravitational potential of the two components of the binary can be described by that of two point masses. (ii) The companion (i.e., non-compact) star is co-rotating with the orbital angular velocity; its rotation axis is perpendicular to the orbital plane. (iii) The orbit is circular. We introduce a Cartesian coordinate system (x, y, z) with its origin at the center of mass of the companion, the x-axis pointing toward the compact component, the y-axis in the orbital plane and the z-axis perpendicular to the orbital plane.

Under these assumptions the forces acting on a fluid element of the companion star can be written as the gradient of a potential function ψ, given by:

$$\psi(x, y, z) = GM_1/r_1 + GM_2/r_2 + 1/2\ \omega^2(x^2+y^2) - \omega^2 x a M_2/(M_1 + M_2) \qquad (1.1).$$

Here M_1 and M_2 are the masses of the companion and the compact star, respectively, and a is the distance between the stars. The other quantities are given by:

$$r_1{}^2 = x^2 + y^2 + z^2 \qquad (1.2a),$$
$$r_2{}^2 = (x - a)^2 + y^2 + z^2 \qquad (1.2b),$$
$$\omega^2 = G(M_1 + M_2)/a^3 \quad \text{(orbital angular velocity)} \qquad (1.2c).$$

The above expression for ψ can be written conveniently in dimensionless form by adopting a polar coordinate system (r, θ, ϕ), and using the distance a as the unit of length. Defining:

$$x = r \cos\theta \quad = r\lambda \qquad (1.3a),$$
$$y = r \sin\theta \sin\phi \quad = r\mu \qquad (1.3b),$$
$$z = r \sin\theta \cos\phi \quad = r\nu \qquad (1.3c),$$

we can write the potential in dimensionless form as:

$$\Omega = r^{-1} + q\,(1 - 2\lambda r + r^2)^{-1/2} - q\lambda r + 1/2\,(1+q)\,r^2\,(1 - \nu^2) \qquad (1.4),$$

where

$$\Omega = (a\psi/GM_1) - M_2{}^2/[2M_1(M_1 + M_2)] \qquad (1.5).$$

Here q is the mass ratio: $q = M_2/M_1$. For a given value of q the shape of an equipotential surface is fully determined by Ω (see Fig. 1). There exists a critical value Ω_c of Ω for which the equipotential surface consists of two closed surfaces, one around each binary component, which are linked at a single point on the X-axis, the first Lagrangean point L_1. For values of Ω larger than Ω_c the equipotential surface consists of two separate closed surfaces, one around each component. Since at L_1 the gradient of the potential along the X-axis vanishes (Ω is saddle-shaped near L_1) matter is free to leave the companion there and be transferred to the compact star (or lost from the system). It is therefore assumed that the companion star cannot exceed this critical surface (Roche lobe).

The distribution of the emergent flux across the stellar surface is determined by Von Zeipel's gravity darkening "law", which states that the local radiation flux is proportional to the local gravity acceleration. In short, this can be demonstrated as follows (cf. [29, 105]). For a star in hydrostatic equilibrium the pressure and density are related through

$$\nabla P = \rho\ \nabla\ \psi \qquad (1.6),$$

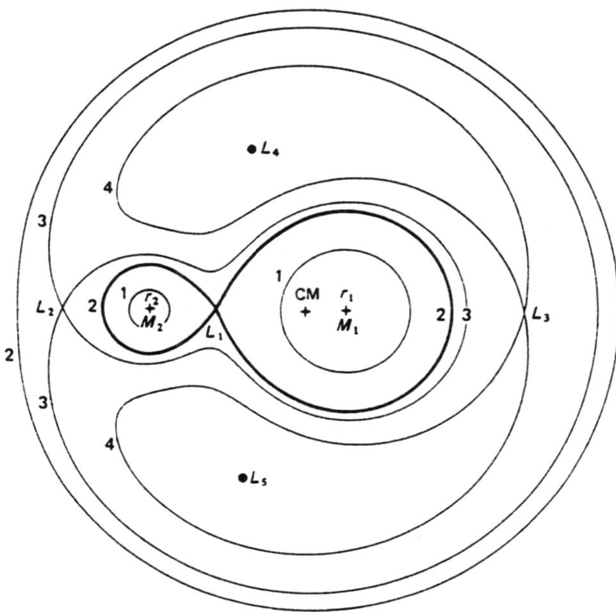

Fig. 1: Cuts in the orbital plane of Roche equipotential surfaces for a binary system with mass ratio $q = 0.2$. The equipotential surfaces are indicated with 1-4 in order of increasing value of ψ; CM indicates the center of mass of the system. At the inner Lagrangean point L_1 the potential has a saddle point. (from [51]).

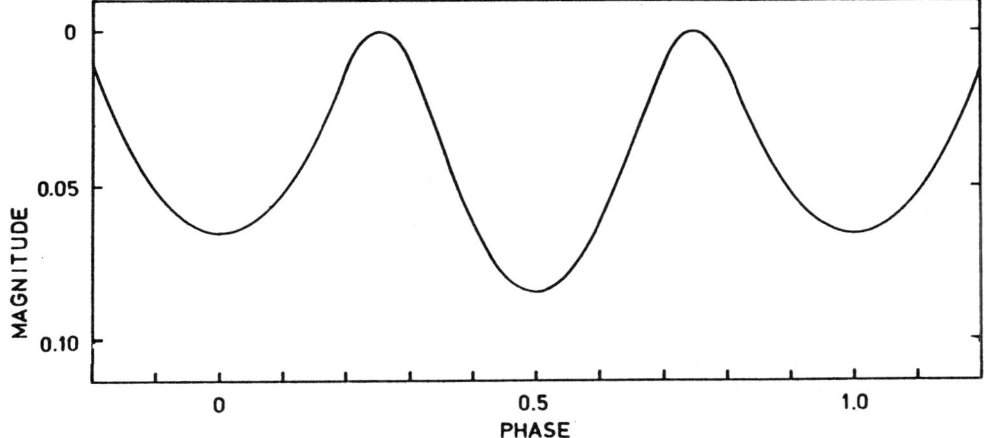

Fig. 2: *Theoretical ellipsoidal light curve for a Roche lobe filling primary star (mass ratio $q = 0.076$, orbital inclination 90°). Adapted from [183].*

where ∇P is the local gradient of the total pressure, ρ is the density, and the potential gradient $\nabla\psi$ includes all external forces acting on the fluid element. Since the density is a scalar quantity, the local gradients of P and ψ are everywhere parallel, which implies that P is constant on an equipotential surface. Therefore, P is a function of ψ only, i.e. $P = P(\psi)$, and in view of eq. (1.6) we also have $\rho = \rho(\psi)$. For a given equation of state this leads to the conclusion that also the temperature is a function of ψ only, since $T = T(P,\psi) = T(\psi)$. In radiative equilibrium the radiative flux is given by:

$$H = \sigma T_e^4 = -[(16\,\sigma T^3)/(3\,\kappa\rho)]\ \nabla T = -[(16\,\sigma T^3)/(3\,\kappa\rho)]\ (\partial T/\partial Y)\ \nabla\psi \qquad (1.7),$$

where $\kappa = \kappa(\rho, T)$ is the Rosseland absorption coefficient, and σ the Stefan-Boltzmann constant. Since T, κ, ρ and $\partial T/\partial\psi$ are all functions of ψ only it follows that the radiative flux is proportional to the local gradient of the potential, i.e. the local gravity acceleration g. It should be noted that based on eq. (1.7) it can be demonstrated that the condition of radiative equilibrium is violated, and that slow meridional circulation currents will develop. The velocities of these currents are small enough that hydrostatic equilibrium is not affected [105, 130].

We note that in the case the outer envelope is not in radiative equilibrium (as assumed above), but instead is in convective equilibrium, the relation between T_{eff} and g is approximately given by $T_{eff} \propto g^{0.08}$ [86].

To relate observed light curves to this geometric model one makes the assumption that the shape of the companion star is that of an equipotential surface ψ = constant. The general features of an ellipsoidal light curve (see Fig. 2) can be understood from the roughly pear-like shape of the equipotential surfaces: near conjunctions the (projected) stellar disk is smallest, at quadratures it is largest. This explains the double-waved shape of the light curve. At superior conjunction of the primary the L_1 point is directed towards the observer; since near L_1 the surface gravity acceleration and, therefore, the surface brightness is minimal, the corresponding minimum in the light curve is the deepest of the two.

In order to numerically calculate a light curve a grid of surface elements is defined on the surface of the companion, for each of which the gravity acceleration, and the effective temperature have a locally well defined value. The light curve is then defined by scaling parameters (M_1, a, and the bolometric luminosity L_1 of the companion), parameters that define the shape of the companion (q, Ω), and the inclination i of the orbital plane.

For a discussion of the assumptions underlying this model, and its limitations, and of approximative ways to treat the effect of non-circularity of the orbit and non-synchronous rotation of the companion, I refer to [147, 181, 183]. Results of detailed calculations have been described in [5, 11, 122, 139, 147, 179, 180, 182].

1.2.2. *Effects related to the presence of the X-ray source*

X rays incident on the atmosphere of the primary star can have a substantial effect on the temperature structure of its outer layers, and on the properties of the stellar wind (through ionization of the atomic species responsible for the radiative acceleration in the wind; see, e.g. [87, 138]). In a simplifying first-order approximation one assumes that a fraction η (the X-ray albedo) of the infalling X-ray flux is reflected, and a fraction $(1-\eta)$ absorbed by the atmosphere of the primary. The absorbed X rays are reradiated as lower energy (UV and optical) photons. In equilibrium the additionally emitted flux equals the absorbed part of the incident X-ray flux. In the "deep heating" approximation this reprocessing of X rays is described as an increase of the local effective temperature according to

$$\sigma T_e^4 = \sigma T_{e,0}^4 + (1-\eta)\ L_X \cos\chi\,/(4\pi d^2) \qquad (1.8).$$

Here $T_{e,0}$ is the undisturbed local effective temperature (calculated from gravity darkening),

L_X is the luminosity of the X-ray source (assumed to be radiating isotropically), χ is the angle between the surface normal and the direction from the surface element to the X-ray source, and d is the distance between the surface element and the X-ray source. Of course, the temperature is affected only for those surface elements for which $\cos \chi > 0$.

The presence of an accretion disk affects the light curve is several ways. Firstly, the disk will act as an independent source of optical light, which in the absence of mutual eclipses will decrease the amplitude of the light curve. Secondly, mutual eclipses of the companion star and the disk may occur, which will increase the amplitude of the light curve. Finally, the disk will cast an X-ray shadow on the companion, so that the effect of X-ray heating may be substantially decreased. In view of our meagre knowledge of the structure of X-ray irradiated accretion disks it is probably in most cases not justified to model the disk in any detail beyond a simple circular flat cylinder (this is not the case for LMXB in which the disk dominates the optical emission, see Sect. 2). For details about the numerical treatment of the mutual eclipses of the disk and the primary, and of the X-ray shadowing by the disk see [52, 92, 147].

1.3. Observations of optical light curves of HMXB

In this Section I discuss some examples of orbital brightness variations of HMXB, in the light of the model described in Sect. 1.2. This section is not intended to be comprehensive; for additional references on individual sources the reader should consult Tables 1 and 2 in my lecture notes on "Neutron stars in X-ray binaries" in the present Volume.

1.3.1. *Vela X-1 = GP Vel*

The shape of the average light curve of this source is qualitatively similiar to a purely ellipsoidal light curve, with the deepest of the two minima occurring when the companion is in superior conjunction (see Fig. 3). However, a more detailed investigation shows that the average light curve of GP Vel cannot be understood in terms of purely ellipsoidal variations [147]. This is probably due to the fact that in the eccentric orbit of this system ($e \sim 0.09$) one or more of the assumptions underlying the model, are not valid. In particular, the assumption that the companion star instantaneously fills a (varying) equipotential surface may be wrong. It is possible that in an eccentric orbit the shape of the companion star is not even a unique function of orbital phase at all. The variable gravitational disturbance of the compact object may, e.g., excite high-order pulsation modes in the envelope of the companion star, with characteristic time scales less than the orbital period and longer damping times. Superposition of such excited oscillations could result in apparently irregular brightness variations. In addition, the companion may show pulsation-like variability which is not related to the presence of the compact star, as has been seen in many other early-type supergiants [88].

1.3.2. *Cen X-3/ V779 Cen and SMC X-1*

Although the average light curves of Cen X-3 and SMC X-1 are double-waved (see Fig. 4 and 5), they cannot be well described in terms of ellipsoidal variations alone [147]. The observed relative depths of the two minima are reversed compared to ellipsoidal light curves. X-ray heating and the presence of an accretion disk (indicating that Roche-lobe overflow contributes significantly to the mass transfer in this system) play a significant role in these systems. When we use known values of the system parameters, the average light curves of Cen X-3 and SMC X-1 can be satisfactorily accounted for by the model described in Section 2.2.

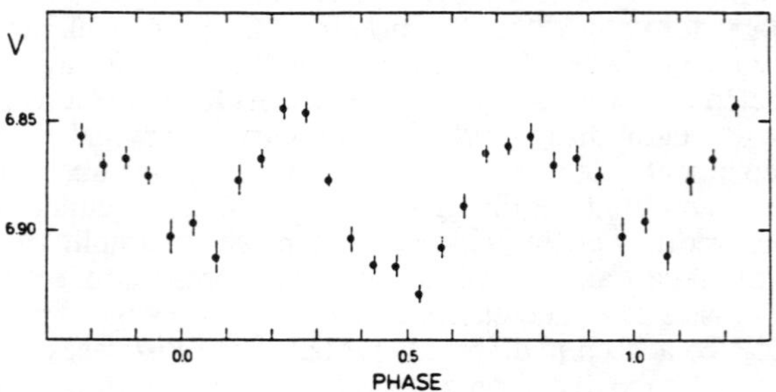

Fig. 3: *Average V band light curve of Vela X-1 [147].*

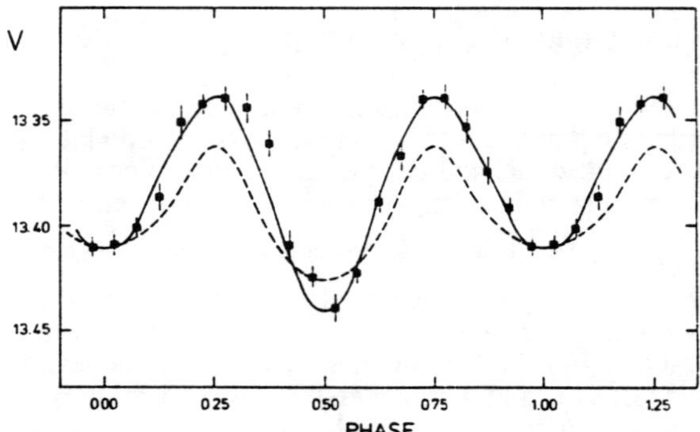

Fig. 4: *Average V band light curve of Cen X-3. The full curve represents a theoretical light curve with the effects of X-ray heating and an accretion disk included. The dashed curve includes only ellipsoidal variations [147].*

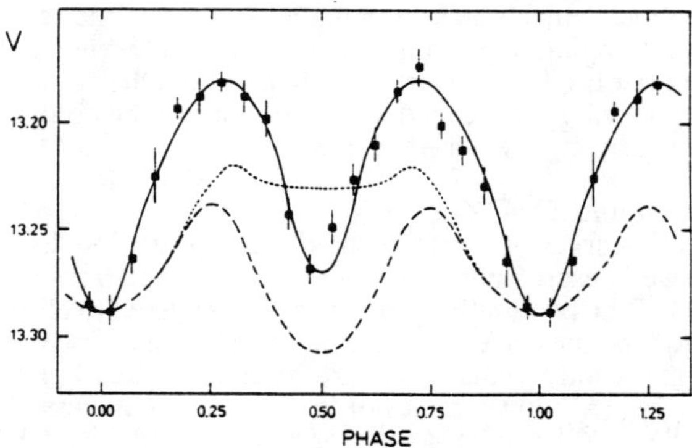

Fig. 5: *Average V band light curve of SMC X-1. The full curve represents a theoretical light curve with the effects of X-ray heating and an accretion disk included. The dashed curve includes only ellipsoidal variations, the dotted one in addition X-ray heating [147].*

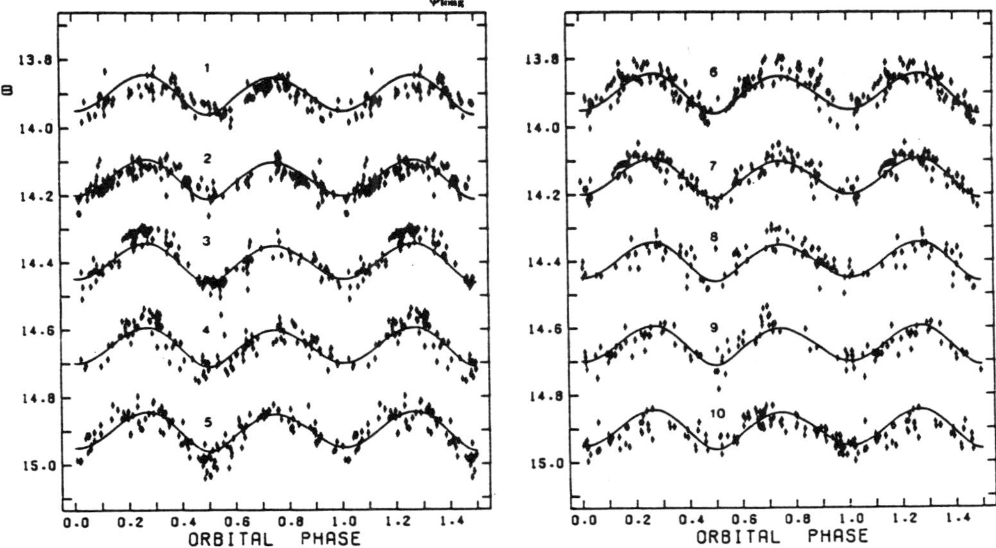

Fig. 6: *B band light curves of LMC X-4 in ten consecutive phase bins of the 30.4 day precessional cycle [64].*

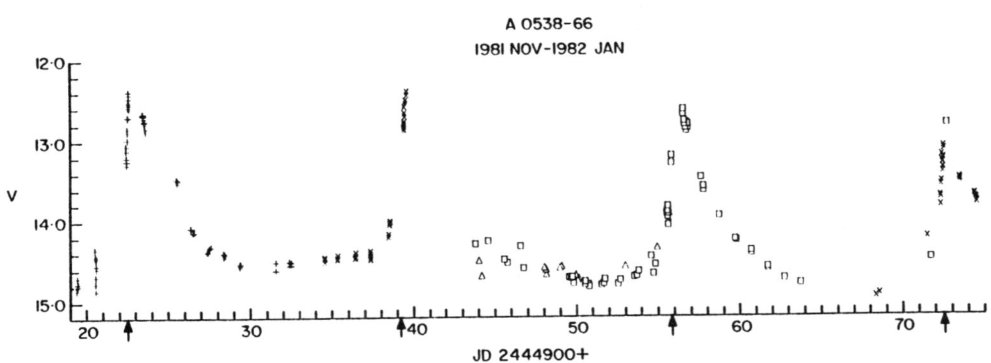

Fig. 7: *Light curve of A0538-66, covering four consecutive 16.6 day cycles [46].*

1.3.3. *LMC X-4*

In addition to regular X-ray eclipses (orbital period 1.4 d) this X-ray source shows a periodic (30.4-day) on-off cycle [83]. This long-term periodicity is also visible in the optical brightness variations [74], in the form of a backward-moving (in orbital phase) feature in the double-waved orbital light curve (see Fig. 6). The 30.4-day X-ray on-off cycle and the corresponding long-term variations of the optical light curve have been qualitatively explained by precession of a tilted accretion disk, analogous to Her X-1 [52] and SS 433 [85, 89]. This qualitative explanation has been supported by a detailed analysis based on an extension of the model described in Section 1.2 that incorporates a tilt and precession of the accretion disk [64].

1.3.4. *SS 433/ V1343 Aqr*

This source is famous because of the jets which move out at about a quarter of the speed of light, and precess with a period of ~163 days [89]. The orbital period of SS 433 is 13.2 days, as evidenced by radial velocity variations [43]. The optical brightness of SS 433 varies regularly with this orbital period [54, 80], however, the detailed shape of the light curve varies during the 163 day precession period [81]. The orbital light curve has been interpreted in terms of a distorted primary star and a tilted precessing accretion disk, with mutual eclipses between these two [1, 85]. Unlike in most other HMXB (with the exception of LMC X-3, see [73, 156]) the brightness of the accretion disk is not small compared to that of the primary star; this reflects the high mass transfer rate ($\sim 10^{-4}\ M_{\odot}\ yr^{-1}$) in SS 433.

1.3.5. *A0538–66*

This system consists of a rapidly spinning neutron star (rotation period 69 ms [132]) in a highly eccentric 16.6 day orbit around a $\geq 10\ M_{\odot}$ B-type star. The system shows long periods of activity, in which near periastron the X-ray luminosity exceeds the Eddington limit by a substantial factor [23, 172], and the optical brightness increases by about an order of magnitude ([46]; see Fig. 7). These active periods are alternated by long quiescent intervals in which no X rays are detected; the optical star is then relatively faint, and shows no orbital brightness modulation [149]. The long-term variation in the outburst activity is probably caused by the irregular shedding by the B star of a dense equatorial envelope; X-ray outbursts occur as the neutron star penetrates this envelope. Howarth *et al.* [70] have argued that the optical outbursts are caused by the corresponding increased X-ray heating of the primary.

1.3.6. *LMC X-3*

LMC X-3 is one of the only three known good candidates for a black hole in the stellar mass range [38, 116]. Orbital variations have been detected in the radial velocity and the optical brightness, but not in X rays. On a time scale of weeks the X-ray flux changes substantially, and the average optical brightness is correlated with these changes [156]. Recently, Hutchings and Cowley [73] found that the average X-ray flux varies regularly with a period of ~200 days, possibly reflecting the precession of a tilted accretion disk. The orbital light curve can be reasonably described with the geometric model discussed in section 1.2.; the required system parameters are consistent with the idea that the compact star in LMC X-3 is a black hole [82].

1.3.7. *GX 301–2/ Wray 977*

The orbit of the binary X-ray pulsar GX 301-2 is eccentric ($e \sim 0.5$); the orbital period is 41.5 days [175]. Near periastron the system shows periodic outbursts [106, 124, 166]. By combining published optical photometry [13, 61, 129, 152, 153] with the results of his own observations Pakull [117] found evidence for a periodic brightness variation at the orbital period. The average light curve shows a single maximum which occurs near periastron, and has an amplitude of 0.03 mag. The increase of X-ray luminosity during the recurrent outbursts is insufficient to account for this optical maximum through increased reprocessing of X rays. Pakull *et al.*[119] found that "ellipsoidal" brightness variations caused by variable distortion of the supergiant companion during its eccentric orbit can explain this single-peaked light curve if it is assumed that the supergiant fills its critical lobe near periastron.

1.3.8. *LS I +61°301*

The early-type star LS I +61°303 [62] is the massive component of a Be/X-ray binary [10], associated with the variable radio source GT 0236+610, which shows periodic (26.5 days) outbursts of non-thermal radio emission [141, 142]. These outbursts have been explained as a result of supercritical accretion (however, the X-ray luminosity is below the Eddington limit by many orders of magnitude). The orbital motion, suspected to cause the periodic radio outbursts was confirmed by Hutchings and Crampton [72] who found that the radial-velocity variations also show the 26.5 day periodicity. Mendelson and Mazeh [104] found that the optical brightness of LS I +61°303 also varies with the orbital period; the light curve has an amplitude of ~ 0.04 mag. They investigated the possibilities that the optical variability is related to variable synchrotron radiation (the favoured mechanism of the radio emission), and heating of the accretion disk; both models have problems.

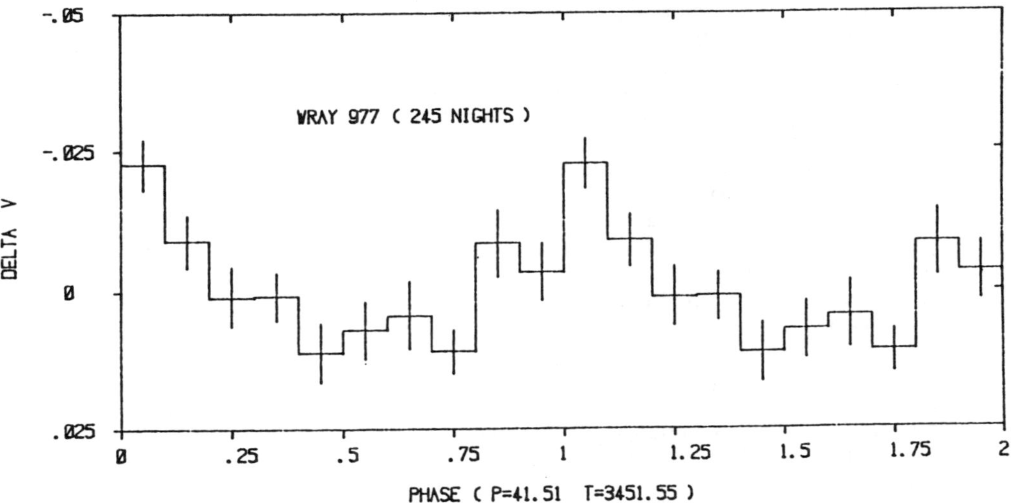

Fig. 8: *Average V band light curve of Wra 977 [119].*

2. Optical Light Curves of Low-mass X-ray Binaries

2.1. Introduction

Several lines of evidence suggest that the optical emission of low-mass X-ray binaries is dominated by reprocessing of X rays in material surrounding the X-ray source [154]. These include: (i) the delay and smearing of optical bursts which are coincident with X-ray bursts; (ii) the correlation of the optical and X-ray fluxes, over a very large range of variation, observed during soft X-ray transient outbursts; (iii) the operation of the Bowen mechanism in LMXB.

For many optically identified LMXB the optical brightness shows a fairly regular orbital modulation. This indicates that the spatial distribution of the reprocessing material is not axisymmetric around the X-ray source. The phase relation of this optical brightness variation with X-ray eclipses and X-ray "dips" shows that (i) X-ray heating of the companion star star contributes significantly, and (ii) X-ray reprocessing in the accretion disk dominates the optical emission.

It is perhaps worth noting here that in luminous LMXB the energy generated in the disk, by conversion of gravitational potential energy of the inwardly drifting material, provides an insignificant contribution to the total optical luminosity of the disk. This follows from a comparison of the absolute magnitudes of cataclysmic variables and LMXB at similar values of the mass transfer rate ($\sim 10^{-8}\ M_{\odot}\ yr^{-1}$), which are $M_V \sim +4.5$ and $\sim +1.5$, respectively [154, 165]. In case internal energy generation dominates it follows from energy conservation that the radial variation of the (local) effective temperature in the disk is approximately given by $T_e(r) \propto r^{-3/4}$[131]. The radial dependence of X-ray heating is proportional to $L_X r^{-2}$, and therefore heating becomes relatively more important compared to internal energy generation as r increases. In the case that X-ray heating dominates, T_e and the disk thickness h are linked in view of the dependence of X-ray heating of the disk on the shape of the latter through the cos χ factor in Eq. (2.8). By making the assumption that at a given radial distance the vertical structure of the disk is isothermal Vrtilek *et al.* [161] were able to derive the radial distribution of both T_e and of the disk thickness, h(r). They found $T_e(r) \propto r^{-3/7}$ (i.e., the temperature distribution is flatter, as expected).

The geometric model for the orbital variation of the optical brightness of HMXB, described in Section 1.2., can be appied to LMXB as well, although with rather different system parameters. The emission of the companion star itself (in the absence of X-ray heating) contributes only negligibly to the optical brightness; as a consequence ellipsoidal variations are absent. The light curve is determined predominantly by X-ray heating of the companion star (in as far as this star is not in the X-ray shadow of the disk), emission from the disk, and mutual eclipses of the disk and the companion star.

In section 2.2. and 2.3. I describe the observed orbital light curves of LMXB, separately for systems whose orbital inclinations are sufficiently high that X-ray eclipses and "dips" are observed and for low-inclination systems without such features in their X-ray light curve. The amplitude of the optical light curve depends quite strongly on the orbital inclination (see Fig. 9). For low inclination angles, as apparent from the lack of X-ray eclipses and X-ray "dips", the light curves are approximately sinusoidal, with an amplitude of a few tenths of a magnitude. For systems with somewhat higher inclination angles, as evident from the presence of periodic X-ray "dips" but absence of X-ray eclipses, the amplitude of this sinusoidal light curve increases to ~ 0.5 mag. At the highest inclinations the amplitude reaches ~ 1.5 mag. For these systems the optical light curve can be decomposed into the sine wave that is also observed for systems at lower inclinations, and a rather sharp cusp superposed on the minimum of the sine wave. From a comparison of the phases of the optical and X-ray intensity curves it appears that the cusp (and therefore the minimum of the sine wave component) occurs at superior conjunction of the

X-ray source. This indicates that the cusp in the optical light curve is due to the eclipse of the luminous accretion disk. The relative phasing for the non-eclipsing system 1755-338 of the sinusoidal optical light curve and the X-ray dips [93] confirms that this picture is also valid for the lower-inclination systems. The correlation of the amplitude of the sinusoidal component with inclination angle is likely due to the fact that as the inclination angle decreases the average brightness of the accretion disk increases (larger projection factor, less self shielding), and the relative importance of the variable component (due to eclipses of the disk and the heating of the companion star) decreases. For the non-eclipsing systems the amplitude of the light curve predominantly reflects the ratio of solid angles (as seen from the X-ray source) of the companion star and the acretion disk.

In section 2.4. I give a brief discussion of the optical light curves of soft X-ray transients during quiescence; it turns out that when the X rays are turned off the optical light curves of LMXB are dominated by ellipsoidal variations due to the distortion of the low-mass companion star.

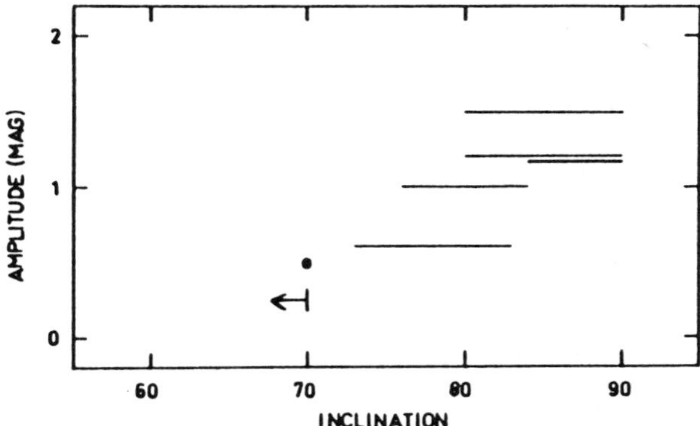

Fig. 9: *Dependence of the amplitude of optical light curves of LMXB on the orbital inclination [157]. Eclipsing systems are represented by horizontal lines which indicate the allowed range of inclination. Three sources which show dips but not eclipses, are assumed to have i = 70°. Sources which do not show eclipses or dips are likely to have smaller inclinations. Systems in which the companion star contributes significantly to the optical brightness have not been included in this figure.*

2.2. Orbital light curves of low-mass X-ray binaries with high orbital inclination

2.2.1. *0547-711/ CAL 87*

Pakull *et al.* [118] identified the optical counterpart of CAL 87, and found that its light curve showed an ~ 1 magnitude deep eclipse of the accretion disk by the companion star. This was confirmed by Callanan *et al.* [19] who derived an orbital period of 10.6 hours. They also found a small secondary eclipse near phase 0.5, and a significant orbital variation of B-V. The optical light curve of CAL 87 is similar to that of 1822-37 (see Figure 10), in spite of the fact that the X-ray spectra of the two systems are very different. This suggests that the disk structure inferred for 1822-37 by modelling the optical light curve (see above) is representative for that in all LMXB. From radial-velocity variations of CAL 87 Cowley *et al.* [41] inferred that the compact object in this system is likely a black hole.

2.2.2. ***0748–676***
The X-ray intensity curve of this transient source shows bursts, eclipses and dips [57, 120]; the orbital period of 3.8 hours decreases on a time scale of ~6 10^6 yrs [136]. The optical counterpart of 0748–676, which in quiescence is fainter than 23rd magnitude [162], reached 17th magnitude for at least 3 years after an outburst which occurred in 1985. There is a tight correlation between the long-term X-ray and optical flux variations of 0748–676, which can be explained as a result of the blackbody temperature variation of the accretion disk caused by variable X-ray heating [112]. Optical photometry has been reported in [44, 112, 127, 157]. The shape of the orbital light curve depends on the average brightness. In the "high" state (V ~ 17.0) the light curve shows a rather narrow eclipse which sits on top of a much broader absorption between phases 0.6 and 1.2 [112]. This light curve is similar to that of 2129+47 (and Her X-1) except for a smaller amplitude (Fig. 11). Motch et al. [112] interpreted this light curve as a combination of eclipses of the bright inner part of the accretion disk by the companion star and a dark bulge at the outer rim of the disk. In the "low" state (V ~ 17.5) the light curve is fairly flat outside the eclipse; the disappearance of the broad absorption may reflect a decrease of the height of the bulge on the disk edge.

2.2.3. ***0921–631***
Cowley *et al.* [37] found an orbital period of of 9.0 days from radial-velocity variations; the kinematic properties of 0921–630 indicate that it is a halo object at a distance of the order of 10 kpc. The optical spectrum shows both emission and absorption lines; the strength of both varies as a function of orbital phase. The variations of B and B–V are strongly correlated; the system is reddest near minimum brightness. This is due to the relatively larger contribution of the evolved (red) companion star at minimum light, as also indicated by the observation that then the late-type absorption features in the spectrum are strongest. Initially the optical light curve of 0921-630 appeared to show deep recurring eclipses which recur at long intervals [17, 27]. According to Branduardi-Raymont *et al.* [18] the optical light curve does not contain a stable eclipse; however, dips (up to 1 mag deep) occur preferentially around superior conjunction of the X-ray source. Superposed on the average optical light curve are large (~ 0.5 mag) irregular variations. The X-ray intensity undergoes a long partial eclipse at the expected time of superior conjunction of the X-ray source, indicating that 0921–630 is an ADC source (like 1822–371 and 2129+470) [95].

2.2.4. ***1254–691***
The X-ray intensity curve of the burst source 1254–691 shows X-ray dips which recur at intervals of 3.9 hours [35]. This (orbital) periodicity has also been found in optical brightness variations. The minimum of the sinusoidal optical light curve (Fig. 12) occurs about 0.2 orbital cycles after the dips. This fits the idea that the optical brightness variations of LMXB with moderate or small orbital inclinations are dominated by aspect changes of the X-ray heated secondary star [111].

2.2.5. ***1656+354/ Her X-1.***
The optical light curve of Her X-1 (amplitude ~ 1.5 mag in B) shows a single minimum per orbital cycle[6]; also the colours B-V and U-B vary, with amplitudes of ~0.3 and ~1.0 magnitudes, respectively. The optical variability is dominated by X-ray heating of the companion. In addition to the 1.7 day orbital cycle the X-ray intensity of Her X-1 shows a long-period (35 day) on-off cycle [53]. This long-term periodicity also appears in the optical light curve, in the form of a feature which drifts continuously toward earlier orbital phase throughtout the 35 day cycle [15, 24]. This long-term X-ray and optical variation can be explained by assuming that Her X-1 contains a precessing tilted accretion disk [52], which acts as a (variable) source of optical radiation, gives a variable X-ray heating of the companion star, and a variable occultation of that star (see Sect. 1.2.).

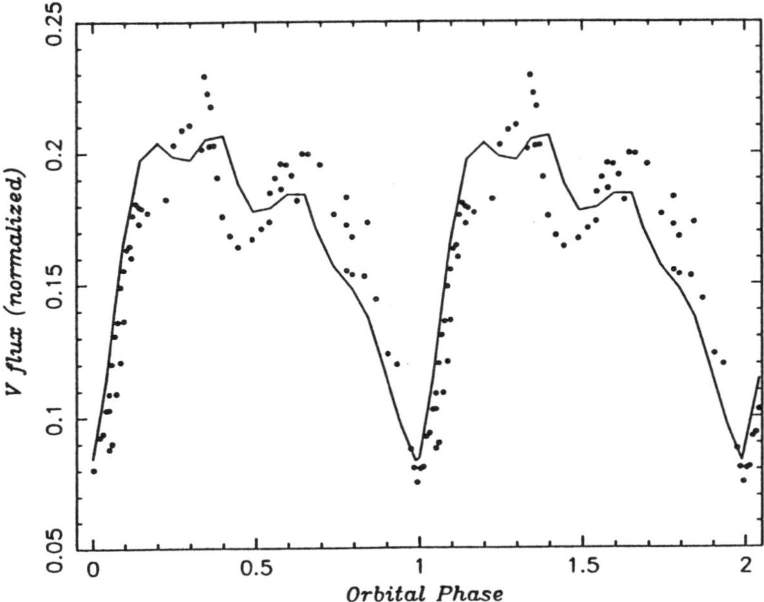

Fig. 10: *V band light curve of CAL 87, superposed on which is a plot of the light curve of 1822-371 19].*

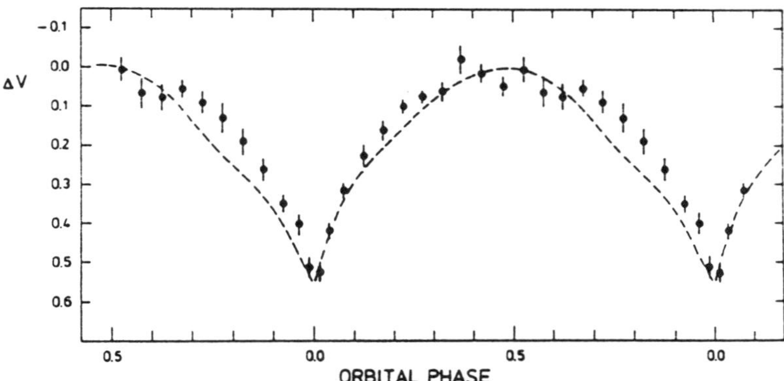

Fig. 11: *Average V band light curve of 0748-676, compared with that of 2129+47, after scaling the amplitude of the latter (dashed curve) [157].*

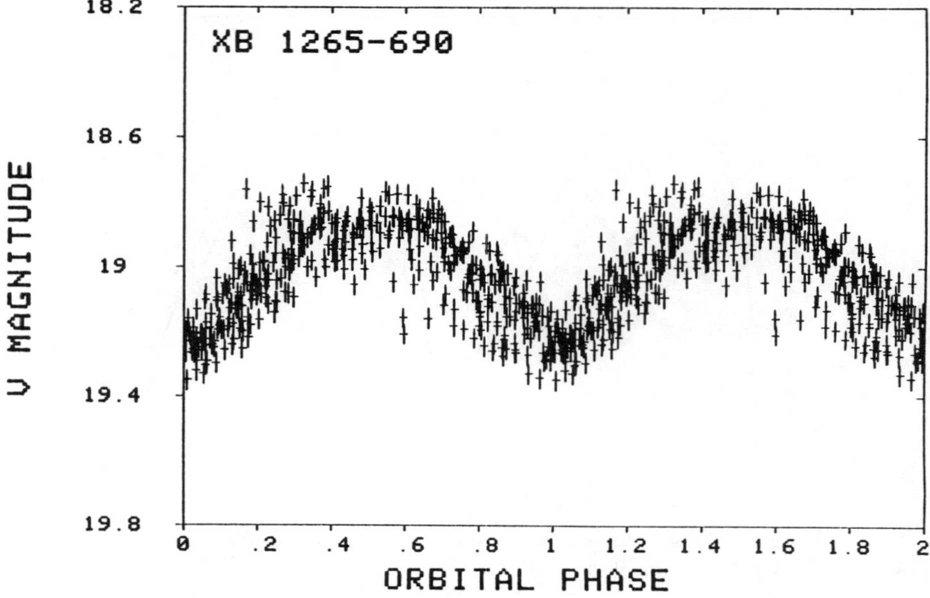

Fig. 12: *V band light curve of 1254-691 [111].*

2.2.6. *1755–338*

The 4.4-hour orbital X-ray intensity curve of 1755–33 shows dips, but not eclipses [177]. The optical brightness varies in an approximately sinusoidal fashion (amplitude ~ 0.4 mag) with the minimum occurring about 0.15 orbital cycles after the minimum in the X-ray dips ([93]; see Fig. 14). In view of the phase relation between X-ray dips and eclipses for systems in which both are present in the X-ray intensity curve, this result indicates that the minimum of the optical light curve coincides with the superior conjunction of the compact star. It is very likely that the same is true for the optical light curves of low-inclination systems which show neither dips not eclipses in their X-ray intensity curves.

2.2.7. *1822–371*

Optical brightness variations of this source were discovered by Mason *et al.* [91] who found that every 5.57 hours the light curve shows a very broad minimum preceded by a slow ingress, which together affect the light curve over about half of the orbital cycle. They suggested that this slow ingress represents additional absorption related to the accretion stream from the secondary. The X-ray intensity curve contains a dip superposed on a relatively smooth modulation, which White *et al.* [173] interpreted in terms of an accretion disk corona (ADC) model. In this model the central X-ray source cannot be observed as it is shielded by the accretion disk. The observed X-rays have been scattered over the edge of the disk by an extended cloud of highly ionized gas blown off the surface of the accretion disk. This extended scattering region is partially occulted by the secondary once per orbital cycle. Mason and Cordova [92] have modelled the optical, UV, and IR brightness variations, based on an extension of the ADC model for the X-ray intensity curve, with an azimuthally structured disk, in which contributions to the intensity come from the following

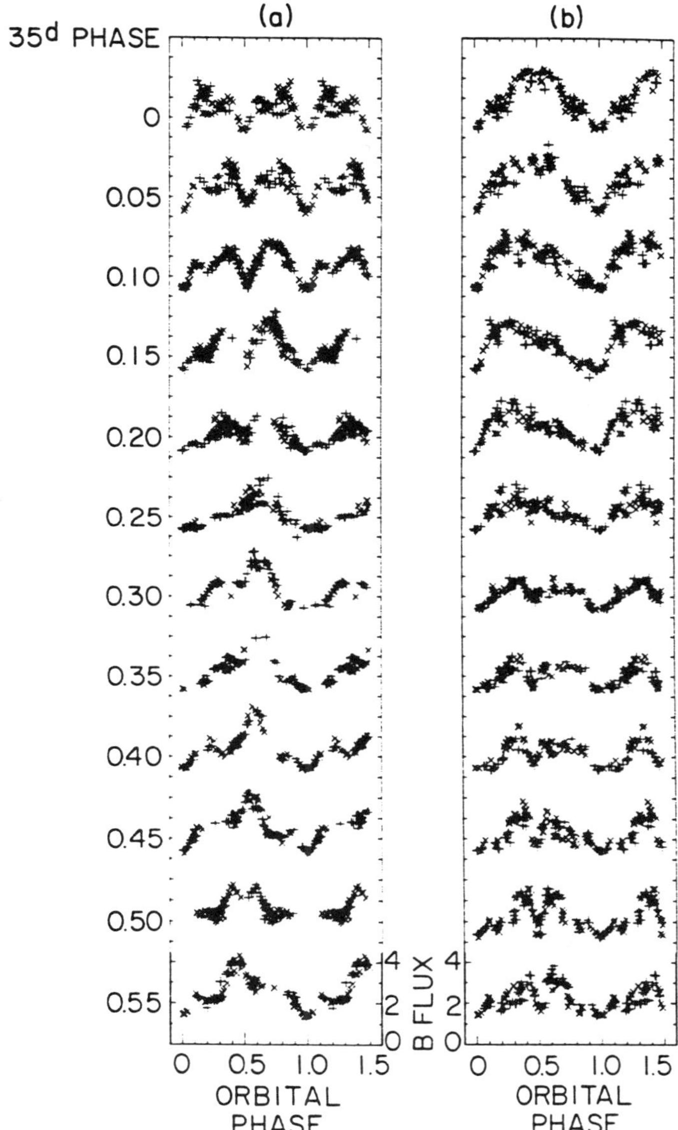

Fig. 13: *B band light curve of Her X-1 in consecutive phase intervals of the 35 day precessional cycle [52]. The curves in the left column have been folded modulo the 1.7 day orbital period, and data have been combined according to the symmetry rule $B(\phi, \psi) = B(-\phi, -\psi)$, where ϕ and ψ are the orbital and precessional phases, respectively. The curves in the right column have been sampled along a diagonal in the (ϕ, ψ) plane and represent the light curves that would be measured by an observer moving around the binary system in 35 days together with the precession of the disk.*

components: (i) the accretion disk; (ii) the inward facing side of the rim of the disk; (iii) the outer side of the rim; (iv) the X-ray heated side of the secondary (see Fig. 15). The relative contributions from these components depend on the orbital inclination, and light curves as different as those of 1822–37 and 2129+47 (see below) can be successfully described with this model.

2.2.8. *1916–053*

The X-ray intensity curve of 1916–053 displays erratic dips, which recur at intervals of ~ 50 minutes; estimates of the average recurrence period of these dips range between 2982 and 3005 seconds [135, 164, 174]. The optical counterpart shows regular brightness variations [126]; however, the period of these variations (3027.5 s) differs from the X-ray period [60]. Grindlay *et al.* [60] have interpreted this difference as evidence that 1916-053 is part of a hierarchical triple system (see also [59]). This interpretation requires the orbital period to be that of the optical variations. White [171] and Smale *et al.* [136] consider it more likely that the X-ray period is the orbital period, since in 0748–676 [120] and 1658-298 [31] the X-ray dips follow the eclipse period. They suggest that the optical modulations may be related to the superhumps observed during superoutbursts of the SU UMa class of dwarf novae.

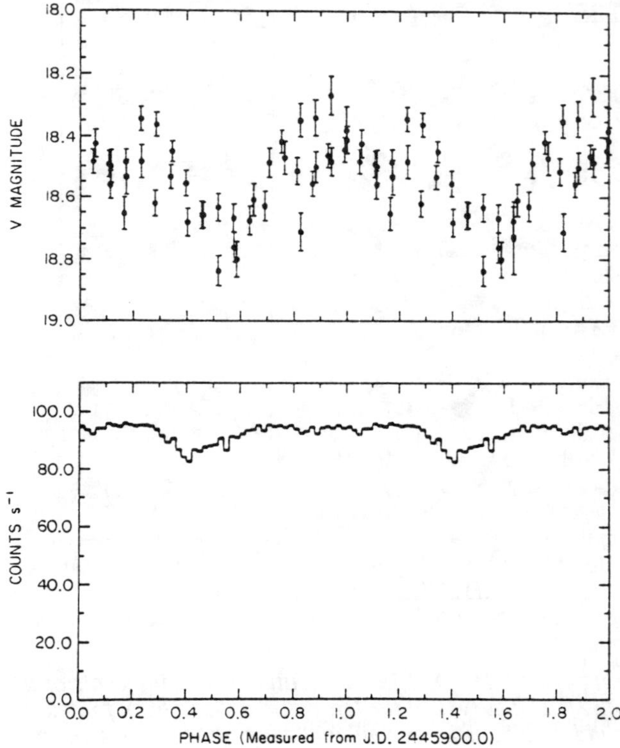

Fig. 14: *Orbital variation of the V band intensity and X-ray intensity of 1755-338, showing the phase relation between the approximately sinusoidal optical light curve and the X-ray dips [93].*

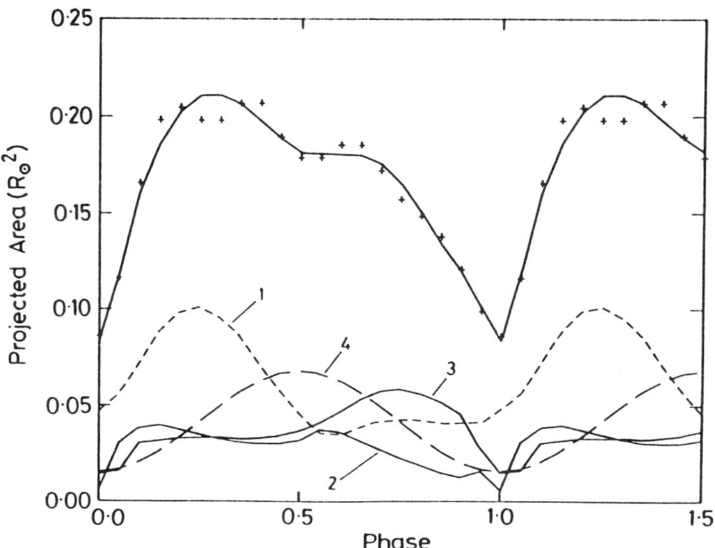

Fig. 15: *Optical light curve of 1822-371 and best-fit theoretical light curve consisting of four components (see text) which are also shown individually [92].*

2.2.9. *2129+470*

X-ray observations before 1983 showed that 2129+470 has a 5.2 hour orbital period [100]; the shape of the X-ray intensity curve was independent of X-ray energy, and can be understood as a result of a partial eclipse of an X-ray scattering accretion disk corona. The optical brightness and colours showed a 5.2 hour modulation with amplitudes of ~1.5 mag in B, 0.25 mag in B–V, and 0.5 mag in U–B ([99, 144]; see Fig. 16). The light and colour curves are similar to those of Her X-1, but quite different from those of 1822–37; the optical emission is dominated by X-ray heating of the companion star. The reason that the X-ray light curve of Her X-1 is so different is probably that the accretion disk in that system is tilted. Pietsch *et al.* [123] found that 2129+470 was not detectable when they observed it in 1983 with Exosat. Coordinated optical observations showed that the large-amplitude optical variations had also disappeared. Further observations showed that in this X-ray off state the spectrum of the optical counterpart is that of an F5-8 main-sequence star; the absorption lines in this F-type spectrum spectrum do not show a detectable radial-velocity variation. This F-type star is not the companion star, since it is too large too fit inside the Roche lobe at the 5.5 hour orbital period. Further optical photometry [25, 146] confirms that there are no detectable brightness variations in the off state (upper limit to the amplitude 0.015 mag in V). This lack of detectable ellipsoidal variations (expected amplitude ≥ 0.1 mag, see [146]) indicates that the off-state optical counterpart is substantially fainter than the F star, from which it is located at most 0.3" away [25]; it has been suggested that the F star and the X-ray binary for a hierarchical triple system [25, 50].

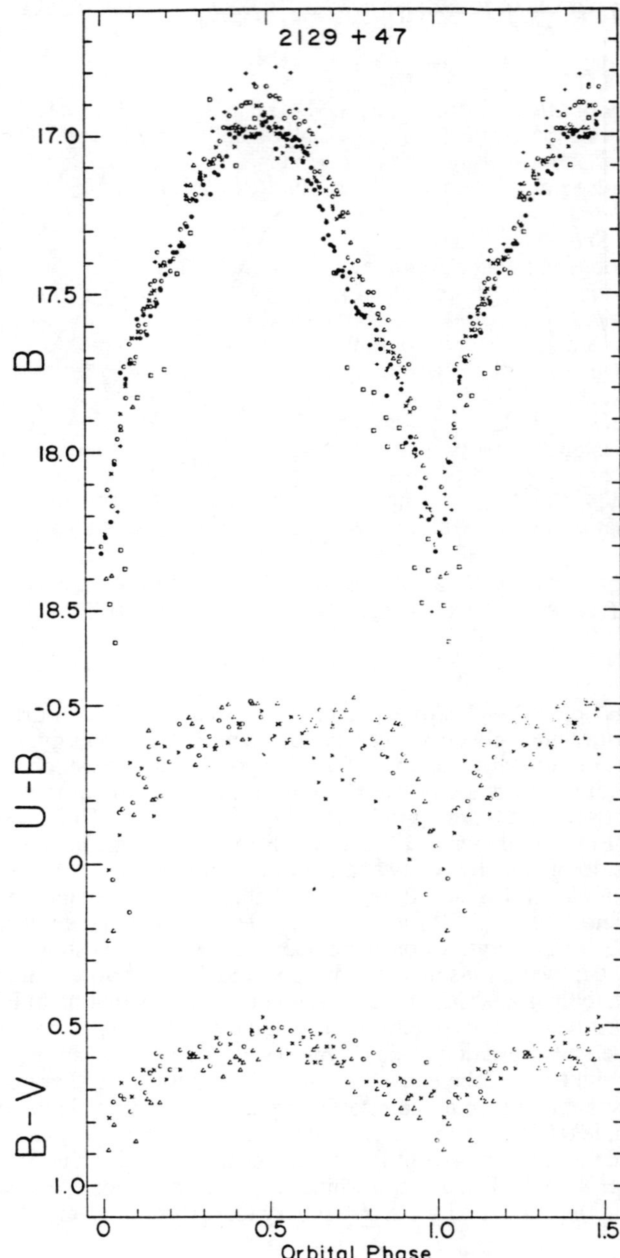

Fig. 16: *Optical light and colour curves of 2129+470 [99].*

2.3. Orbital light curves of low-mass X-ray binaries with low orbital inclination

2.3.1. *0543–681/ CAL 83*

The radial velocity of CAL 83 shows this LMXB is a member of the LMC [45]; it has an extremely soft X-ray spectrum, and an unusually small X-ray to optical luminosity ratio. Optical photometry shows that the orbital period is 1.05 days; the light curve is approximately sinusoidal and has a (full) amplitude of 0.23 mag [134].

2.3.2. *1617-155/ Sco X-1*

In spite of much observational effort it took about a decade after the optical identification of Sco X-1 to prove the binary nature of this source. Gottlieb *et al.* [56] found a regular ~ 0.2 magnitude brightness variation with a period of 18.9 hours, from an analysis of more than one thousand Harvard archival photographic plates (Fig. 17). Superposed on this periodic variation, which was subsequently also found in variations of the emission-line radial velocities [42, 84], are much larger irregular variations which make it difficult to detect the orbital light curve in short data stretches. From the phase relation between the optical light curve and the emission-line radial-velocity curve it appears that minimum light occurs when the emission-line object is in superior conjunction. In view of the phase relations between the orbital light curves and X-ray eclipses, observed for high-inclination systems, this result indicates that the optical emission lines originate in the accretion disk.

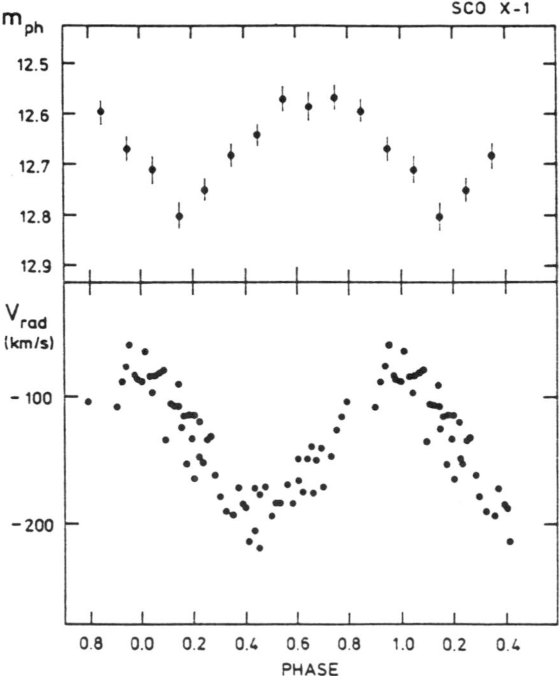

Fig. 17: *Average optical light curve and He II λ4686 radial-velocity of Sco X-1 [56].*

2.3.3. *1636–536*

Based on high-speed photometry performed to detect optical bursts coincident with X-ray bursts Pedersen *et al.* [121] found that the optical brightness of 1636–53 varies with a period of 3.8 hours. Subsequent observations confirmed this periodicity [128], and showed that the average orbital light curve is approximately sinusoidal [133]. From an analysis of all published photometry Van Paradijs *et al.* [158] found that there are substantial night-to-night changes in the shape of the orbital light curve, likely caused by substantial long-term changes in the spatial distribution of the reprocessing material in the binary system.

2.3.4. *1659–487/ GX 339–4*

This source has been considered a black-hole candidate in view of its very soft X-ray spectrum during the X-ray high state (cf. [176]), and its rapid X-ray variabiltity (the latter phenomenon is not, however, accepted as a good argument any more, see [137]).
GX 339-4 has been observed in at least three X-ray states, called the high, low and off states, with corresponding average visual magnitudes V ~ 17, 15.4 and 18.7 to 20, respectively ([34], and references therein). Both optical and X-ray quasi-periodic oscillations have been observed, with periods in the range between 50 ms and 20 s [76, 108, 109]. Radial-velocity variations have been reported by Cowley *et al.* [39]. During the off state and the high state periodic optical brightness variations, with a period of 14.8 hours, have been observed [20, 69]. The off-state light curve (amplitude 0.3 magnitudes) is fairly smooth (Fig. 18), suggesting that there may be a significant contribution from heating of the secondary.

2.3.5. *1728–169/ GX 9+9*

GX 9+9 is a luminous galactic-bulge source, which Hasinger and Van der Klis [63] showed is a member of the "atoll" group of LMXB. Hertz and Wood [66] detected a 4.2 hour periodicity in its X-ray flux, which is also present in optical brightness variations; the sinusoidal light curve has a (full) amplitude of ~0.2 mag [125].

2.3.6. *1735–444*

McClintock and Petro [98] suggested a periodicity (at 4.3 hours) in the brightness variation of the optical counterpart of the burst source 1735-44, based on one night of continuous observations. The orbital period (4.65 hours) was found by Corbet *et al.* [33] and confirmed by Van Amerongen *et al.* [150]. The average light curve is approximately sinusoidal and has a full amplitude of 0.2 mag. Superposed on this average light curve are irregular (~0.1 mag. amplitude) variations on time scales between ~ 1 hour and a few days.

2.3.7. *1957+115*

This is one of the sources proposed as a possible black-hole candidate on the basis of its ultra-soft X-ray spectrum [176]. Its optical brightnes is variable [110], with a periodicity of 9.33 hours [145]; the light curve is sinusoidal with a peak-to-peak amplitude of 0.23 mag.

2.3.8. *2030+407/Cyg X-3*

Cyg X-3 is a very interesting source of X rays and possibly photons of ultra-high energy (see, however, [12]). It shows a regular periodicity of 4.79 hours in its X-ray variations, generally accepted to be the orbital period [157]. Cyg X-3 is also a remarkable radio source from which very strong outbursts have been observed [58, 67]. According to Molnar *et al.* [107] quiescent variability of the radio source contains a periodic component at 4.95 hours, i.e., different from the X-ray period. The visual extinction of Cyg X-3, as estimated from the low-energy cut off in the X-ray spectrum and the neutral-hydrogen (21 cm) column density is ~ 19 magnitudes [28, 168]; as a consequence there is no optical counterpart

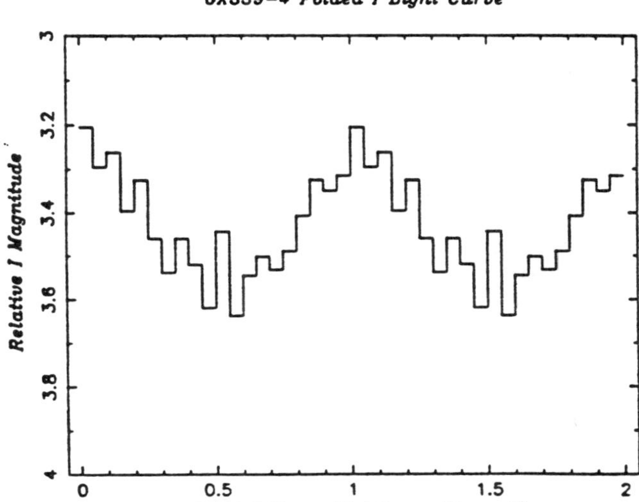

Fig. 18: *Optical light curve of GX 339-4 [20].*

known. However, in view of the much reduced infrared extinction Cyg X-3 has been detected as a bright source at 2.2 and 1.6 μ [8], and at near-infrared wavelengths (I = 17; [167]). Simultaneous X-ray and infrared (2 μ) observations [9, 90] showed that at times the 2 μ intensity variations are very similar to those in the X-ray band. However, co-variability is sometimes absent; in particular the infrared source has at times been found to show no significant orbital variation. Mason *et al.* [94] found that during a simultaneous X-ray/infrared observation the infrared variations tracked those in X rays (although at a much reduced amplitude); in addition, the infrared intensity curve showed rapid strong flares which lasted for a few minutes. I band CCD photometry [163] showed that the near-infrared brightness of Cyg X-3 undergoes large (3 magnitude) long-term variations, and confirmed large night-to-night variations of the amplitude of the orbital modulation of the infrared brightness. It is unlikely that the infrared variability of Cyg X-3 can be explained by the geometric model described in section 1.2.

2.3.9. *2129+119/ AC 211*

The X-ray burst source in M15 [47, 159] is the only LMXB in a globular cluster which has been unambiguously optically identified (with the star AC 211; see [2]), through positional coincidence, UV brightness and variability [3, 4], and the detection of the 8.55 hour orbital period in both X rays [65] and in optical photometric [75] and spectroscopic [114] data. The spectroscopic characteristics of AC 211 are complex, comprising substantial variations of the mean radial velocity of the He I absorption lines, erratic variations of the absorption and emission line profiles, and P Cyg profiles of the Balmer lines. So far there is not a single generally accepted explanation for the spectroscopic behaviour of AC 211. The optical brightness of AC 211 shows a variable, sometimes strong (up to 2 magnitudes in the U band) orbital modulation, which however disappears at times. The ratio of optical to X-ray flux is relatively high for a LMXB; this is unlikely to be caused by shielding of the central X-ray source by the accretion disk (leaving only X rays scattered by an accretion disk corona observable, see section 2.2.7.). Fabian *et al.* [49] have suggested that the very low heavy-element abundances in M15 give rise to a very large ADC which redirects (through scattering) an unusually large fraction of the X-ray luminosity towards the disk, causing excess X-ray heating.

2.3.10. *2142+380/ Cyg X-2*

Variability of the radial velocity and optical brightness of Cyg X-2 was detected soon after its optical identification in 1967, but the orbital period (9.84 days) was not found until 1979 [36]. The long orbital period of Cyg X-2 implies that the companion is a giant star; this star is sufficiently bright in the optical passband that its absorption lines are visible in the optical spectrum. The optical brightness of Cyg X-2 varies regularly with the orbital period [36]. The light curve shows a double-waved lower envelope whose minima occur at the conjunctions; superposed on this ellipsoidal variation are irregular brightness increases (amplitudes up to ~ 1 magnitude), likely caused by variable X-ray heating of the accretion disk. As the brightness excess increases the star gets bluer [55].

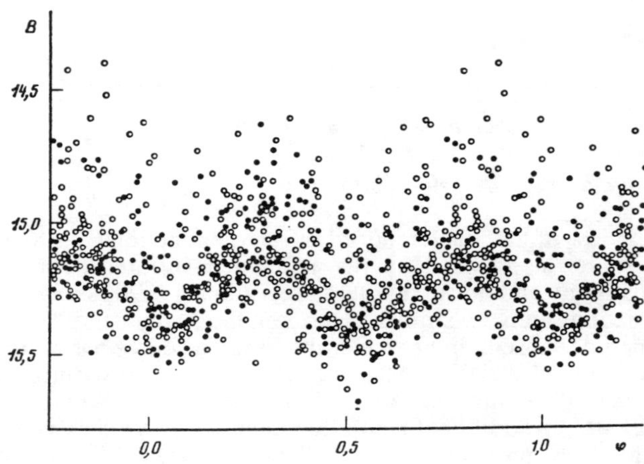

Fig. 19: *Optical light curve of Cyg X-2 [55].*

2.4. Soft X-ray transients in quiescence

A minor fraction of the known X-ray binaries are transient sources, i.e., most of the time they are below the level of detectability, they then turn on and rise to a maximum X-ray brightness in typically a few days, and afterwards decay on a time scale of weeks to months below the level of detectability [30, 177]. Some transients show recurrent outbursts. In addition, some X-ray sources turned off after they had been known as persistent bright sources for many years (see, e.g., sections 2.2.5. and 2.2.9.). Like is the case for X-ray binaries in general, the transient sources fall into two groups with hard and soft X-ray spectra [30], which are called the 'hard X-ray transients' and 'soft X-ray transients', respectively. They are high-mass and low-mass X-ray binaries which show very large variations in the rate of mass accretion onto the neutron star.

Near maximum brightness the X-ray properties of soft X-ray transients (e.g., X-ray spectra, occurrence of X-ray bursts) are similar to those of 'normal' low-mass X-ray binaries. In several cases an optical counterpart has been detected, which together with the X-ray source showed a large increase in brightness (e.g., [22, 170]). Also the optical spectrum near maximum and during the initial decay of the outburst is typical for that of a persistent LMXB.

In the optical spectra of some soft X-ray transients late-type spectral features have been observed after the X-ray flux had reached a very low level [115, 143, 155]. The optical counterpart had by then returned to a faint (~ pre-outburst) flux level. These observations, which indicate that in soft X-ray transients a compact star is accompanied by a low-mass star, have been important in the development of our ideas on the structure of LMXB.

In several cases the optical counterparts of soft X-ray transients in quiescence have shown ellipsoidal brightness variations, indicating that then optical emission related to mass transfer (X-ray heating, accretion disk) is not very important any more.

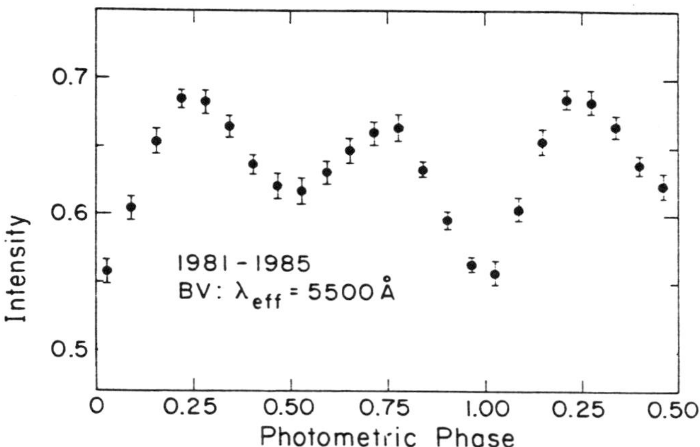

Fig. 20: *Average quiescent optical light curve of the soft X-ray transient A0620-000 [102].*

2.4.1. *A0620-00*

During its outburst in 1975 A0620-00 reached a peak X-ray brightness about three times that of Sco X-1, the brightest known X-ray source. The corresponding optical nova reached a peak magnitude V ~ 11.2 (see [170] for a summary of the X-ray and optical outbursts). In quiescence the optical spectrum is that of a K3-7 dwarf star, with superposed Balmer emission lines [102, 113, 115, 170]. Optical brightness variations in quiescence show that the orbital period is 7.8 hours [101]. The orbital radial-velocity variations of the K dwarf correspond to a mass function in excess of 3 $M_{\odot}$, to which corresponds a firm lower limit of the mass of the compact object of 3.2 $M_{\odot}$ [77, 102].The quiescent orbital light curve is dominated by ellipsoidal modulations, with a contribution from a faint accretion disk and a 'bright spot' similar to those found in the light curves of some cataclysmic variables [102].

2.4.2. *Cen X-4*

Cen X-4 showed X-ray outbursts in 1969 [32] and 1979 [79]. During the second outburst X-ray bursts were observed [97]. The optical counterpart [22] then reached a peak magnitude V ~ 13. Spectroscopic observations during the decay of the outburst revealed spectral features of a K type star [155]. Chevalier *et al.* [26] discovered the orbital period of 15.1 hours from optical photometry in quiescence. The orbital light curve (see also [40,

103]) shows two (unequal) maxima and minima per orbital cycle, which reflects ellipsoidal variations and a contribution similar to that of the 'bright spot' in cataclysmic variables (see the discussion in [103]). In addition, Chevalier *et al.* [26] find evidence for intermittent optical activity in quiescence, probably related to mass transfer. The radial-velocity curve of Cen X-4 [40, 103] is consistent with this interpretation. The mass of the companion star in Cen X-4 is peculiarly low (~ 0.1 $M_\odot$; [40, 103]). A detailed comparison of the properties of Cen X-4 and A0620-00 has been made by McClintock and Remillard [103].

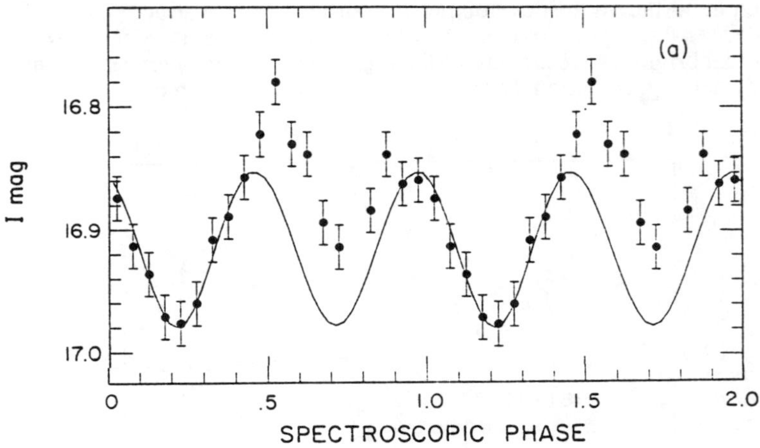

Fig. 21: *Average quiescent optical light curve of the soft X-ray transient Cen X-4 [103].*

2.4.3. *Her X-1; 2129+470*

Magnitude estimates from archival photographic plates, taken over almost a century, showed that during long periods HZ Her, the optical counterpart of Her X-1, is in an optically-low state, indicating that the X-ray emission responsible for the heating of the secondary star had then become much weaker. During part of this low state HZ Her showed a double-waved light curve [78, 169], reflecting ellipsoidal variability. At other times the quiescent light curve showed a single deep minimum, which can be ascribed to the occultation of luminous matter near the compact star [48, 71].

During a similar low state of 2129+470 the expected ellipsoidal variations were not detected (see section 2.2.9). Apparently, the off-state optical brightness of this source is much fainter than that of the quiescent optical counterpart; this has been interpreted as evidence that the system is a hierarchical triple system.

2.4.4. *GS2000+25*

GS2000+25 was discovered as a bright X-ray transient with Ginga in April 1988 [148]. At maximum the corresponding optical nova reached V ~ 16.5. With respect to its X-ray outburst light curve, its X-ray and optical spectra, and the variable radio emission, GS2000+25 was very similar to A0620-000 [14, 68, 140]. During the outburst evidence was found for a 8.3 hour modulation of the optical brightness. In quiescence the optical brightness varies with a ~4.1 hour period; it is possible that this corresponds to an ellipsoidal modulation at the 8.3 hour period [21].

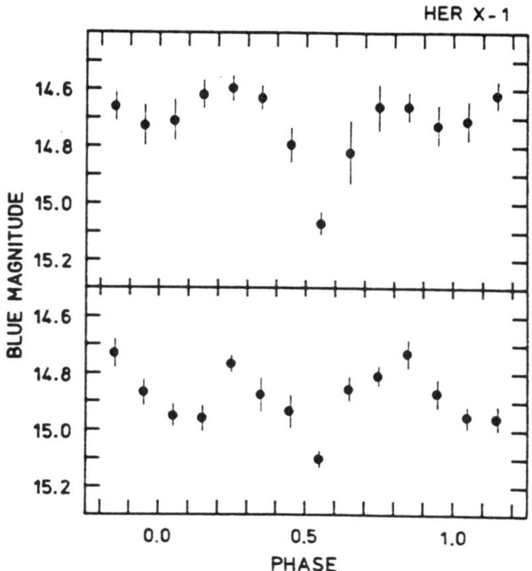

Fig. 22: *Average B band light curve of Her X-1 during extended low states according to [71] (upper curve) and [77] (lower curve).*

References

[1] Antokhina, E.A., Cherepashchuk, A.M. 1987, *Soviet Astron. J.* **31**, 295.
[2] Auriere, M., Cordoni, J.P. 1981, *Astron. Astroph. Suppl.* **46**, 347.
[3] Auriere, M. *et al.* 1984, *Astron. Astroph.* **138**, 415.
[4] Auriere, M. *et al.* 1986, *Astron. Astroph.* **158**, 158.
[5] Avni, Y., Bahcall, J.N. 1975, *Astroph. J.* **197**, 675.
[6] Bahcall, J.N., Bahcall, N.A. 1973, *Astroph. J.* **178**, L1.
[7] Bahcall, J.N. *et al.* 1976, *Publ. Astron. Soc. Pacific* **87**, 141.
[8] Becklin, E.E. *et al.* 1972, *Nature Phys. Sci.* **239**, 130.
[9] Becklin, E.E. *et al.* 1974, *Astroph. J.* **192**, L119.
[10] Bignami, G.F. *et al.* 1981, *Astroph. J.* **247**, L85.
[11] Bochkarev, N.G. *et al.* 1988, *Sov. Astron. J.* **32**, 405.
[12] Bonnet-Bidaud, J.M., Chardin, G. 1988, *Phys. Rept.* **170**, 325.
[13] Bord, D.J. *et al.* 1976, *Astroph. J.* **203**, 689.
[14] Borisov, N.V. *et al.* 1989, *Proc. 23rd ESLAB Symposium*, ESA SP-296, p. 305.
[15] Boynton. P.E. *et al.* 1973, *Astroph. J.* **186**, 617.
[16] Bradt, H.V., McClintock, J.E. 1983, *Ann. Rev. Astron. Astroph.* **21**, 13.
[17] Branduardi-Raymont, G. *et al.* 1981, *Space Sci. Rev.* **30**, 279.
[18] Branduardi-Raymont, G. *et al.* 1983, *Mon. Not. Roy. Astron. Soc.* **205**, 423.
[19] Callanan, P. *et al.* 1989, *Mon. Not. Roy. Astron. Soc.* **241**, 37P.
[20] Callanan, P.J. *et al.* 1990, *Proc. 11th North-American Workshop on CVs and LMXBs* (in press).
[21] Callanan, P., Charles, P. 1990, *Gemini* **30**, 14.
[22] Canizares, C.R. *et al.* 1980, *Astroph. J.* **236**, L55.

[23] Charles, P.A. *et al.* 1983, *Mon. Not. Roy. Astron. Soc.* **202**, 657.
[24] Chevalier, C., Ilovaisky, S.A. 1973, *Nature Phys. Sci.* **245**, 87.
[25] Chevalier, C. *et al.* 1989, *Astron. Astroph.* **210**, 114.
[26] Chevalier, C. *et al.* 1989, *Astron. Astroph.* **217**, 108.
[27] Chevalier, C., Ilovaisky, S.A. 1981, *Astron. Astroph.* **94**, L3.
[28] Chu, K.W., Bieging, J.H. 1973, *Astroph. J.* **179**, L121.
[29] Clayton, D.D. 1968, *Principles of Stellar Evolution and Nucleosynthesis* (McGraw Hill).
[30] Cominsky, L. *et al.* 1978, *Astroph. J.* **224**, 46.
[31] Cominsky, L., Wood, K.S. 1984, *Astroph. J.* **283**, 765.
[32] Conner, J.P. *et al.* 1969, *Astroph. J.* **157**, L157.
[33] Corbet, R.H.D. *et al.* 1987a, *Mon. Not. Roy. Astron. Soc.* **222**, 15P.
[34] Corbet, R.H.D. *et al.* 1987b, *Mon. Not. Roy. Astron. Soc.* **227**, 1055.
[35] Courvoisier, T.J.-L. *et al.* 1986, *Astroph. J.* **309**, 265.
[36] Cowley, A.P. *et al.* 1979, *Astroph. J.* **231**, 539.
[37] Cowley, A.P. *et al.* 1982, *Astroph. J.* **256**, 605.
[38] Cowley, A.P. *et al.* 1983, *Astroph. J.* **272**, 118.
[39] Cowley, A.P. *et al.* 1987, *Astron. J.* **93**, 195.
[40] Cowley, A.P. *et al.* 1988, *Astron. J.* **95**, 1231.
[41] Cowley, A.P. *et al.* 1990a, *Astroph. J.* **350**, 288.
[42] Crampton, D. *et al.* 1976, *Astroph. J.* **207**, 907.
[43] Crampton, D. *et al.* 1980, *Astroph. J.* **235**, L131.
[44] Crampton, D. *et al.* 1986, *Astroph. J.* **306**, 599.
[45] Crampton, D. *et al.* 1987, *Astroph. J.* **321**, 745.
[46] Densham, R.H. *et al.* 1983, *Mon. Not. Roy. Astron. Soc.* **205**, 1117.
[47] Dotani, T. *et al.* 1990, *Nature* **347**, 534.
[48] Dvorak, T.Z. 1976, *IAU Inf. Bull. Variable Stars* No. 1082.
[49] Fabian, A.C. *et al.* 1987, *Mon. Not. Roy. Astron. Soc.* **225**, 29P.
[50] Garcia, M. *et al.* 1989, *Astroph. J.* **341**, L75.
[51] Frank, J. *et al.* 1985, *Accretion Power in Astrophysics* (Cambridge Univ. Press).
[52] Gerend, D., Boynton, P. 1976, *Astroph. J.* **209**, 562.
[53] Giacconi, R. *et al.* 1973, *Astroph. J.* **184**, 227.
[54] Gladyshev, S.A. *et al.* 1979, *Astron. Tsirk.* No. 1086.
[55] Goranskii, V., Lyutyi, V.M. 1988, *Sov. Astron. J.* **31**, 193.
[56] Gottlieb, E.W. *et al.* 1975, *Astroph. J.* **195**, L33.[56]
[57] Gottwald, M. *et al.* 1986, *Astroph. J.* **308**, 213.
[58] Gregory, P.C. *et al.* 1972, *Nature Phys. Sci.* **239**, 114.
[59] Grindlay, J.E. 1990, *Proc. 23rd ESLAB Symposium*, ESA SP-296, p. 121.
[60] Grindlay, J.E. *et al.* 1988, *Astroph. J.* **334**, L25.
[61] Hammerschlag-Hensberge, G. *et al.* 1976, *Astron. Astroph.* **49**, 321.
[62] Hardorp, J. *et al.* 1959, *Luminous Stars in the Northern Milky Way I* (Hamburger Sternwarte - Warner and Swasey Observatory).
[63] Hasinger, H., Van der Klis, M. 1989, *Astron. Astroph.* **225**, 79.
[64] Heemskerk, M., Van Paradijs, J. 1989, *Astron. Astroph.* **223**, 154.
[65] Hertz, P. 1987, *Astroph. J.* **315**, 429.
[66] Hertz, P., Wood, K.S. 1988, *Astroph. J.* **331**, 764.
[67] Hjellming, R.M. 1973, *Science* **182**, 1089.
[68] Hjellming, R.M. *et al.* 1988, *Astroph. J.* **333**, L75.
[69] Honey, W.B. *et al.* 1991, *Mon. Not. Roy. Astron. Soc.* (in preparation).
[70] Howarth, I. *et al.* 1984, *Mon. Not. Roy. Astron. Soc.* **207**, 287.
[71] Hudec, R., Wenzel, W. 1976, *Bull. Astron. Inst. Chech.* **27**, 325.
[72] Hutchings, J.B., Crampton, D. 1981, *Publ. Astron. Soc. Pacific* **93**, 486.

[73] Hutchings, J.B., Cowley, A.P. 1990, *IAU Symposium* **148**, in press.
[74] Ilovaisky, S.A. *et al.* 1984, *Astron. Astroph.* **210**, 251.
[75] Ilovaisky, S.A. *et al.* 1987, *Astron. Astroph.* **179**, L1.
[76] Imamura, J. *et al.* 1990, *Astroph. J.* **365**, 312.
[77] Johnston, H. *et al.* 1989, *Astroph. J.* **345**, 492.
[78] Jones, C. *et al.* 1973, *Astroph. J.* **182**, L109.
[79] Kaluzienski, L.J. *et al.* 1980, *Astroph. J.* **241**, 779.
[80] Kemp, J.C. *et al.* 1980, *Astroph. J.* **238**, L133.
[81] Kemp, J.C. *et al.* 1986, *Astroph. J.* **305**, 805.
[82] Kuiper, L. *et al.* 1988, *Astron. Astroph.* **203**, 79.
[83] Lang, F. *et al.* 1981, *Astroph. J.* **246**, L21.
[84] LaSala, J., Thorstensen, J.R. 1985, *Astron. J.* **90**, 2077.
[85] Leibowitz, E.A. 1984, *Mon. Not. Roy. Astron. Soc.* **210**, 279.
[86] Lucy, L.B. 1967, *Zeitschr. f. Astroph.* **65**, 89.
[87] MacGregor, K.B., Vitello, P.A.J. 1982, *Astroph. J.* **259**, 267.
[88] Maeder, A. 1980, *Astron. Astroph.* **90**, 311.
[89] Margon, B. 1984, *Ann. Rev. Astron. Astroph.* **22**, 507.
[90] Mason, K.O. *et al.* 1976, *Astroph. J.* **207**, 78.
[91] Mason, K.O. *et al.* 1980, *Astroph. J.* **242**, L109.
[92] Mason, K.O., Cordova, F.A. 1982, *Astroph. J.* **262**, 253.
[93] Mason, K.O. *et al.* 1985, *Mon. Not. Roy. Astron. Soc.* **216**, 1033.
[94] Mason, K.O. *et al.* 1986, *Astroph. J.* **309**, 700.
[95] Mason, K.O. *et al.* 1987, *Mon. Not. Roy. Astron. Soc.* **226**, 423.
[96] Mason, K.O. *et al.* 1985, *Mon. Not. Roy. Astron. Soc.* **216**, 1033.
[97] Matsuoka, M. *et al.* 1980, *Astroph. J.* **240**, L137.
[98] McClintock, J.E., Petro, L. 1981, *IAU Circular* No. 3615.
[99] McClintock, J.E. *et al.* 1981, *Astroph. J.* **243**, 900.
[100] McClintock, J.E. *et al.* 1982, *Astroph. J.* **258**, 245.
[101] McClintock, J.E. *et al.* 1983, *Astroph. J.* **266**, L27.
[102] McClintock, J.E., Remillard, R.A. 1986, *Astroph. J.* **308**, 110.
[103] McClintock, J.E., Remillard, R.A. 1990, *Astroph. J.* **350**, 386.
[104] Mendelson, H., Mazeh, T. 1989, *Mon. Not. Roy. Astron. Soc.* **239**, 723.
[105] Mestel, L. 1965, in: *Stellar Structure*, ed. L.H. Aller & D.B. McLaughlin (Univ. Chicago Press), p.465.
[106] Mitani, K. *et al.* 1984, *Astroph. Space Sci.* **103**, 345.
[107] Molnar, L.A. *et al.* 1984, *Nature* **310**, 662.
[108] Motch, C. *et al.* 1982, *Astron. Astroph.* **109**, L1.
[109] Motch, C. *et al.* 1985a, *Space Sci. Rev.* **40**, 219.
[110] Motch, C. *et al.* 1985b, *Space Sci. Rev.* **40**, 239.
[111] Motch, C. *et al.* 1987, *Astroph. J.* **313**, 792.
[112] Motch, C. *et al.* 1990, *Proc. 23rd ESLAB Symposium*, ESA SP-296, p. 545.
[113] Murdin, P. et al. 1980, *Mon. Not. Roy. Astron. Soc.* **192**, 709.
[114] Naylor, T. *et al.* 1987, *Mon. Not. Roy. Astron. Soc.* **233**, 285.
[115] Oke, J.B. 1977, *Astroph. J.* **217**, 181.
[116] Paczynski, B. 1983, *Astroph. J.* **273**, L81.
[117] Pakull, M.W. 1982, *Proc. Workshop on Accreting neutron Stars*, Garching, MPE Report 177, p. 53.
[118] Pakull, M.W. *et al.* 1988, *Astron. Astroph.* **203**, L27.
[119] Pakull, M.W. *et al.* 1991, in preparation.
[120] Parmar, A.N. *et al.* 1986, *Astroph. J.* **308**, 199.
[121] Pedersen, H. *et al.* 1981, *Nature* **294**, 725.
[122] Petro, L., Hiltner, W.A. 1974, *Astroph. J.* **190**, 661.

[123] Pietsch, W. *et al.* 1986, *Astron. Astroph.* **157**, 23.
[124] Priedhorsky, W.C., Terrell, J. 1983, *Astroph. J.* **273**, 709.
[125] Schaefer, B. 1990, *Astroph. J.* **354**, 720.
[126] Schmidtke, P. 1988, *Astron. J.* **95**, 1528.
[127] Schmidtke, P., Cowley, A.P. 1987, *Astron. J.* **92**, 374.
[128] Schoembs, R. *et al.* 1987, *ESO Messenger* **48**, 6.
[129] Schulz-Lüppertz, E. 1979, Ph.D. Thesis, University of Tübingen.
[130] Schwarzschild, M. 1958, *Structure and Evolution of the Stars* (Dover Publ.).
[131] Shakura, N.I., Sunyaev, R.A. 1973, *Astron. Astroph.* **24**, 337.
[132] Skinner, G.K. *et al.* 1982, *Nature* **297**, 568.
[133] Smale, A.P., Mukai, K. 1988, *Mon. Not. Roy. Astron. Soc.* **231**, 663.
[134] Smale, A.P. *et al.* 1988a, *Mon. Not. Roy. Astron. Soc.* **233**, 51.
[135] Smale, A.P. *et al.* 1988b, *Publ. Astron. Soc. Japan* **41**, 607.
[136] Smale, A.P. *et al.* 1990, *Proc. 23rd ESLAB Symposium*, Vol. 1, p. 607.
[137] Stella, L. *et al.* 1985, *Astroph. J.* **288**, L45.
[138] Stevens, I.R., Kallman, T.R. 1990, *Astroph. J.* **365**, 321.
[139] Strittmatter, P.A. *et al.* 1973, *Astron. Astroph.* **25**, 275.
[140] Tanaka, Y. 1989, *Proc. 23rd ESLAB Symposium*, ESA SP-296, p. 3.
[141] Taylor, A.R., Gregory, P.C. 1982, *Astroph. J.* **255**, 210.
[142] Taylor, A.R., Gregory, P.C. 1984, *Astroph. J.* **283**, 273.
[143] Thorstensen, J.R. *et al.* 1978, *Astroph. J.* **220**, L131.
[144] Thorstensen, J.R. *et al.* 1979, *Astroph. J.* **233**, L57.
[145] Thorstensen, J.R. *et al.* 1987, *Astroph. J.* **312**, 739.
[146] Thorstensen, J.R. *et al.* 1988, *Astroph. J.* **334**, 430.
[147] Tjemkes, S.A. *et al.* 1986, *Astron. Astroph.* **154**, 77.
[148] Tsunemi, H. *et al.* 1989, *Astroph. J.* **337**, L81.
[149] Van Amerongen, S. *et al.* 1986, *IAU Inf. Bull. Variable Stars* No. 2901.
[150] Van Amerongen, S.F. *et al.* 1987, *Astron. Astroph.* **185**, 147.
[151] Van der Klis, M., Bonnet-Bidaud, J.M. 1989, *Astron. Astroph.* **214**, 203.
[152] Van Genderen, A.M. 1973, *IAU Inf. Bull. Var. Stars* No. 856.
[153] Van Genderen, A.M. 1977, *Astron. Astroph.* **54**, 733.
[154] Van Paradijs, J. 1983, in: *Accretion-driven Stellar X-ray Sources*, eds. W.H.G. Lewin & E.P.J. van den Heuvel, (Cambridge University Press), p. 192.
[155] Van Paradijs, J. *et al.* 1980, *Astroph. J.* **241**, L161.
[156] Van Paradijs, J. *et al.* 1987, *Astron. Astroph.* **184**, 201.
[157] Van Paradijs, J. *et al.* 1988, *Astron. Astroph. Suppl.* **76**, 185.
[158] Van Paradijs, J. *et al.* 1990a, *Astron. Astroph.* **234**, 181.
[159] Van Paradijs, J. *et al.* 1990b, *Publ. Astron. Soc. Japan* **42**, in press.
[160] Van Paradijs, J. 1991, this Volume.
[161] Vrtilek, S.D. *et al.* 1990, *Astron. Astroph.* **235**, 162.
[162] Wade, R.A. *et al.* 1985, *Publ. Astron. Soc. Pacific* **97**, 1092.
[163] Wagner, R.M. *et al.* 1989, *Astroph. J.* **346**, 971.
[164] Walter, F. *et al.* 1982, *Astroph. J.* **253**, L67.
[165] Warner, B. 1987, *Mon. Not. Roy. Astron. Soc.* **227**, 23.
[166] Watson, M.G. *et al.* 1982, *Mon. Not. Roy. Astron. Soc.* **199**, 915.
[167] Weekes, T.C., Geary, J.C. 1982, *Publ. Astron. Soc. Pacific* **94**, 708.
[168] Westphal, J.A. *et al.* 1972, *Nature Phys. Sci.* **239**, 114.
[169] Whelan, J. 1973, *Astroph. J.* **185**, L127.
[170] Whelan, J.A.J. *et al.* 1977, *Mon. Not. Roy. Astron. Soc.* **180**, 657.
[171] White, N.E. 1989, *Astron. Astroph. Rev.* **1**, 85.
[172] White, N.E., Carpenter, G.F., 1978, *Mon. Not. Roy. Astron. Soc.* **183**, 11P.
[173] White, N.E. *et al.* 1981, *Astroph. J.* **247**, 994.

[174] White, N.E., Swank, J.H. 1982, *Astroph. J.* **253**, L61.
[175] White, N.E., Swank, J.H. 1984, *Astroph. J.* **287**, 856.
[176] White, N.E., Marshall, F.E. 1984, *Astroph. J.* **281**, 354.
[177] White, N.E. *et al.* 1984, *Astroph. J.* **283**, L9.
[178] Whelan, J., Wickramasinghe, D.T. 1976, *Mon. Not. Roy. Astron. Soc.* **174**, 29.
[179] Wickramasinghe, D.T. 1975, *Mon. Not. Roy. Astron. Soc.* **173**, 21.
[180] Wickramasinghe, D.T., Whelan, J. 1975, *Mon. Not. Roy. Astron. Soc.* **172**, 175.
[181] Wilson, R.E. 1979, *Astroph. J.* **234**, 1054.
[182] Wilson, R.E., Devinney, E.J. 1971, *Astroph. J.* **166**, 605.
[183] Zuiderwijk, E.J. 1979, Ph.D. Thesis, University of Amsterdam.

OBSERVATIONS OF Z AND ATOLL SOURCES

M. van der Klis
Astronomical Institute "Anton Pannekoek"
University of Amsterdam
Roetersstraat 15
1018 WB Amsterdam
The Netherlands

ABSTRACT. The correlations observed in bright low-mass X-ray binaries between X-ray colours and rapid X-ray variability (including quasi-periodic oscillations) indicate the existence of two types of sources, the Z and the atoll sources, each showing several different modes of behaviour or "states". The X-ray intensity, the X-ray spectrum, the optical, UV and radio emission and the X-ray burst properties all correlate with this phenomenological classification, but X-ray intensity does so only approximately, showing that it is not strictly proportional to accretion rate. The observational evidence underlying this classification is summarized, a tentative interpretation in terms of differences in accretion rate and magnetic field strength of neutron stars is given and discussed within the framework of binary evolution, and the relation of the Z and atoll sources to other LMXB is considered.

1. Introduction

The EXOSAT observations of bright low-mass X-ray binaries (LMXB) have provided us with a new way of looking at the entire body of information that we have about these systems. It has been known since the early days of X-ray astronomy that in bright LMXB X-ray spectral hardness sometimes varies in correlation with X-ray intensity. The large number (~50) of long (> 1/3 day up to 4 days), often uninterrupted observations performed with EXOSAT of bright LMXB (see Sternberg *et al.* 1986) allowed the realization that this property is only one aspect of an intricate phenomenology that involves at least two distinct classes of sources each exhibiting several different modes of behaviour (Hasinger and van der Klis 1989, hereafter HK89). In addition to the long observing times, a decisive factor in disentangling this phenomenology was the large area (1600 cm^2) of the EXOSAT Medium Energy detectors (Turner *et al.* 1981), which allowed to perform very accurate X-ray spectral hardness measurements and to study the time variability of these bright sources down to time scales of milliseconds. This latter capability led to the discovery of the quasi-periodic oscillations or QPO (van der Klis *et al.* 1985a) which turned out to be an important diagnostic.

J. Ventura and D. Pines (eds.), Neutron Stars: Theory and Observation, 319–346.

As we see it now, bright LMXB include two subtypes, named the Z and the atoll sources, which can be distinguished on the basis of the correlated variations of their X-ray colours and their rapid X-ray variability (QPO and other aperiodic phenomena). The Z sources show three distinct states, whereas the atoll sources exhibit a more gradual but also very systematic variation in their properties between two extremes. In both cases, source state can be classified on the basis of the same two characteristics as are used to classify source type, namely, X-ray colours and rapid X-ray variability.

Many other observational characteristics are found also to be correlated with the source types and -states defined in this way, but it is interesting to note, that X-ray intensity does not correlate strictly in either Z or atoll sources. If we accept the interpretation that it is (in a given source) the accretion rate that determines the source state, as seems very likely, then this means that X-ray intensity is *not* a good measure for accretion rate in these sources, as has long been, often implicitly, assumed.

HK89 have proposed that the differences between Z and atoll sources are caused by a difference in *two* basic source characteristics, Z sources having both higher accretion rates *and* higher neutron-star magnetic-field strengths than atoll sources and that this explains why Z sources show two different types of QPO whereas no similar QPO are seen in the atoll sources. There also exist bright LMXB which are not of one of the two main types; their relation to Z and atoll sources is an as yet nearly unexplored area.

In this paper, which has some tutorial sections that experienced readers may wish to skip, I review our present knowledge about the correlations between the observable properties of the bright LMXB, with some emphasis on the empirical approach and the data analysis techniques used to uncover these correlations. The limitation to bright sources is of a purely observational nature; many of the techniques used require the X-ray count rates to be at least ~100 c/s, somewhat depending on the source-state variability time scales. In practice, this leads to a limiting flux of $\sim 10^{-9}$ erg/cm^{2}s, or luminosity of $\sim 10^{37}$ erg/s at 10 kpc.

2. X-ray colour-colour and hardness-intensity diagrams

The amount of information in an astronomical X-ray spectrum obtained with a proportional counter is relatively limited. The bad spectral resolution ($\Delta E/E = 0.2 \times (E/6 \text{ keV})^{-1/2}$ for the EXOSAT ME argon detectors; Turner *et al.* 1981) combines with a limited effective spectral range (~1–20 keV) and calibration uncertainties (of the order of 1%) to constrain the amount of information about the continuum spectral shape that can be extracted to a relatively small number of parameters. In practice, spectral modeling does not provide a unique description of the data (Section 4).

A model-independent way to characterize X-ray spectra is through "hardness ratios", also called "X-ray colours". Traditionally these are calculated as the ratio of the X-ray count rates in two photon-energy bands, without any corrections for detector response. Background subtraction is routinely performed, where if necessary corrections for detector deadtime are taken into account: if during the source measurement a fraction f_s of the photons goes undetected due to detector deadtime then the observed count rate $s = (1 - f_s)(B + S)$, where B is the incident background count rate and S the incident source count rate. A background measurement (performed either simultaneously with another detector or before or after the source observation) will yield a count rate of $b = (1 - f_b)B$, where f_b is the dead-time fraction during the background observation. The incident source count rate that is used in the hardness ratio is then calculated as $S = s/(1 - f_s) - b/(1 - f_b)$.

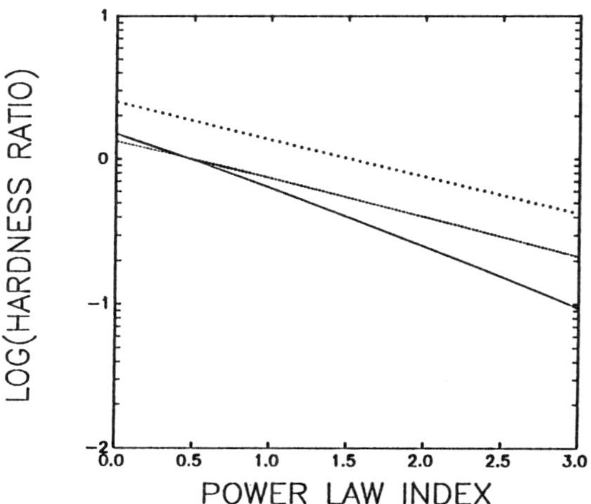

Figure 1. Dependence of the hardness ratio of a pure power law spectrum on the photon power law index for some commonly used energy ranges: 3–6/1–3 keV (drawn), 6–10/3–6 keV (dashed), 10–20/6–10 keV (dotted). Graph is for illustrative purposes only as the effects of detector quantum efficiency were ignored ($f_{34} = f_{12}$ in Eq. 2).

A hardness ratio is a rough measure of the steepness of the spectrum in the photon-energy range considered. If the photon number spectrum is given by $n(E)$, *i.e.*, the number of photons with energies between E and $E+dE$ incident on the detector per cm^2 per second is $n(E)dE$, then the hardness ratio H calculated between energy ranges $[E_1, E_2]$ and $[E_3, E_4]$ with $E_1 < E_2 \leq E_3 < E_4$ is

$$H = \int_{E_3}^{E_4} n(E)f(E)dE \bigg/ \int_{E_1}^{E_2} n(E)f(E)dE, \tag{1}$$

where $f(E)$ is the quantum efficiency of the detector (the fraction of incident photons of energy E that is detected), and where the effect of the finite spectral resolution, which will cause cross-talk between adjacent spectral ranges, is neglected. For a power-law spectral shape $n(E) = CE^{-\alpha}$ we have

$$H = \frac{\int_{E_3}^{E_4} E^{-\alpha} f(E)dE}{\int_{E_1}^{E_2} E^{-\alpha} f(E)dE} = \frac{f_{34}}{f_{12}} \; \frac{1 - (E_3/E_4)^{-\alpha+1}}{1 - (E_1/E_2)^{-\alpha+1}} \left(\frac{E_4}{E_2}\right)^{-\alpha+1}, \tag{2}$$

where f_{34} and f_{12} are weighted averages of the detector quantum efficiency in the two bands, that in general depend on α. We see that if $E_3/E_4 = E_1/E_2$ and $f(E)$ is constant *within* each of the two energy ranges, then $\log H$ is linearly related to the power law index α. In practice, these conditions are not exactly fulfilled, but the relation between $\log H$ and α does not deviate much from a linear one (Figure 1).

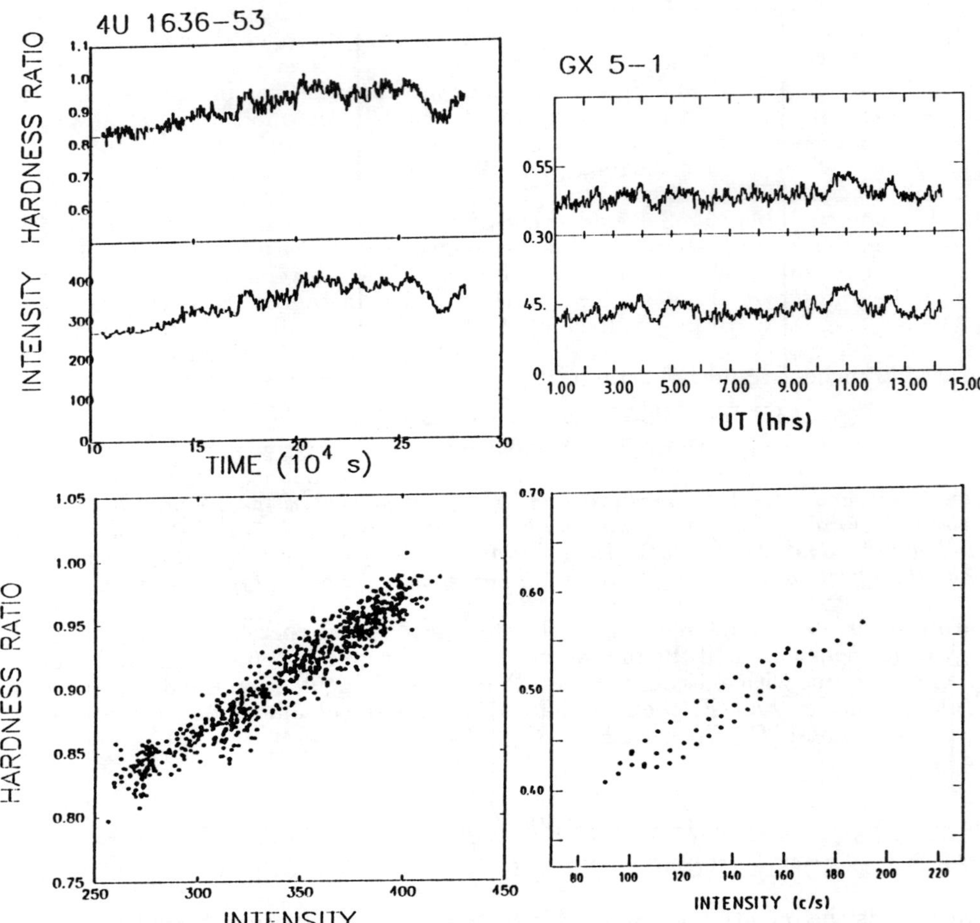

Figure 2. Time variations in X-ray intensity and X-ray spectral hardness and corresponding hardness-intensity diagrams in an atoll (left; EXOSAT ME data of 4U 1636-53) and a Z (right; EXOSAT GSPC data of GX 5-1, after van der Klis et al. 1987a) source. X-ray burst were removed from the light curve of 4U1636-53. GX 5-1 was in the NB during this observation. Note the similarity of the light curves and the correlation between hardness and intensity.

Bright LMXB usually show variations in X-ray intensity and X-ray spectral hardness on time scales of several hours to days (Figure 2). Two types of diagram are commonly used (in addition to plots *vs.* time) to characterize these variations: the hardness-intensity diagram and the colour-colour diagram. In these diagrams (explained below) the variations of the source result in a pattern whose shape is an important diagnostic of source type and -state.

A **hardness-intensity diagram** is a plot of a hardness ratio H *vs.* a total count rate (or "intensity") I, usually the summed count rate of the two bands used to calculate H. Hardness-intensity diagrams have been used since the very earliest days of X-ray astronomy to characterize the spectral variations of X-ray sources as a function of their brightness, where the hardness was in the early days often expressed in terms of the temperature of a best-fit thermal brehmsstrahlung spectrum (*e.g.*, Mason *et al.* 1976, White *et al.* 1976, 1978, 1980, Parsignault and Grindlay 1978, Holt 1980, Charles *et al.* 1980, Branduardi *et al.* 1980, Ponman 1982, Sztajno *et al.* 1983, Basinska *et al.* 1984, Shibazaki and Mitsuda 1984). Before EXOSATs launch in May 1983, it had already been well established that in a LMXB a positive correlation is often observed between spectral hardness and intensity when these two quantities vary in time. In some cases, inverse correlations and "messy" diagrams had also been seen; and in a few instances (Sco X-1: White *et al.* 1976, Cyg X-2: Branduardi *et al.* 1980, GX 5-1: Shibazaki and Mitsuda 1984) multi-valued dependencies of H on I had been reported. In these cases, two separate "branches" could be distinguished in the hardness-intensity diagram.

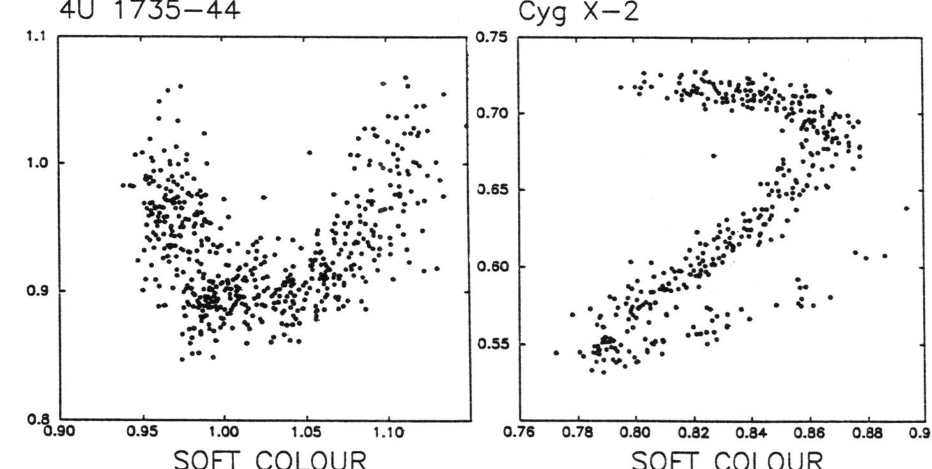

Figure 3. The colour-colour diagram of an atoll (left; EXOSAT ME data on 4U 1735-44) and a Z (right; EXOSAT ME data on Cyg X-2) source. Simple hardness-intensity correlations as in Fig. 2 are only part of the full pattern.

An X-ray **colour-colour diagram** is a plot of a hardness ratio in a higher-energy spectral range (or "hard colour") *vs.* one in a lower range ("soft colour"). The advantages of the X-ray colour-colour diagram over the hardness-intensity diagram are similar to those of the optical $U-B$, $B-V$ diagram from which they borrow their name. (Note, however, that the word "colour" as it is used here does not imply a logarithmic quantity such as an optical magnitude difference.) Such diagrams only consider spectral *shape*, and therefore the distance to the source and variations in the effective detector area do not affect them. Originally, colour-colour diagrams were introduced in X-ray astronomy (Ostriker 1977, Markert *et al.* 1977, White and Marshall 1984) to distinguish between different classes of X-ray sources on the basis of the position of each source in the diagram. It was found that different populations of X-ray binaries show up in quite different places in the diagram; the bright LMXB cover a relatively small part of it, in particular if plausible corrections are applied for the effect of the energy-dependent insterstellar absorption (Schulz *et al.* 1989). All the discussion below concerning the movement of these sources through the diagram is constrained to this small part.

If an X-ray colour-colour diagram is used to characterize the spectral variations of a bright LMXB, then the shape of the pattern in a diagram where a hardness ratio in a low-energy (*e.g.*, 1–6 keV) band is plotted in the positive X direction and in a higher-energy (*e.g.*, 6–20 keV) band in the positive Y direction, is often qualitatively similar to that in a hardness-intensity diagram of the same data (*e.g.*, Schulz *et al.* 1989, Tan *et al.* 1990; note that some authors have chosen to plot colour-colour diagrams differently, *e.g.*, White and Marshall 1984). However, there are clear exceptions. In particular, in some bright LMXB it has been found that the variations of the source are more reproducible from one epoch to the next in a colour-colour diagram than in a hardness-intensity diagram, and sometimes source states that are distinct in the colour-colour diagram, are tangled up in the hardness-intensity diagram (Cyg X-2: Hasinger *et al.* 1990; 4U 1636-53: van der Klis *et al.* 1990; GX 349+2: compare Ponman *et al.* 1988 and HK89).

Because of this better reproducibility and state definition, and also because of the above-mentioned independence of source distance and collimator efficiency effects, most recent work has preferred the use of the colour-colour diagrams for source type and -state determination. As the colour-colour diagram is a strictly empirical way to classify the spectral behaviour of the sources, there has been no point in being dogmatic about the choice of energy bands, and different choices have been made in different instances. Obviously, if it is required to compare several different sources, it may be desirable to stick to the same bands; however, the available data do not always allow this; in some cases it may be helpful to select those bands that in a given source lead to the clearest separation between adjacent branches.

After the discovery of the QPO in GX 5-1, Sco X-1 and Cyg X-2 in 1985 (van der Klis *et al.* 1985a, Middleditch and Priedhorsky 1986, Hasinger *et al.* 1986) exactly the sources of which it was known at the time that the hardness-intensity diagrams were multi-valued, it was quickly realized that the properties of their QPO related to the position of the source in the hardness-intensity diagram (*e.g.*, van der Klis *et al.* 1985b, Priedhorsky *et al.* 1986). This motivated further studies of the hardness-intensity diagrams and colour-colour diagrams of QPO sources, culminating in the identification of the three-branched Z pattern in Cyg X-2 (Hasinger 1987a) and the realization that part of this three-branched pattern had already been observed in GX 5-1 and Sco X-1. Further work on the extensive body of EXOSAT observations of bright LMXB turned up more diagrams that were conceivably part of a Z-shaped pattern (Lewin *et al.* 1987, Schulz *et al.* 1989, Penninx *et al.* 1989), but did not lead to a conclusive classification of all observed colour-colour diagrams. By considering the properties of the **rapid X-ray variability** (Section 3) in relation to the position of the source in the colour-colour diagram, HK89 classified the phenomenology of most bright LMXB contained in the EXOSAT data. In the course of this work they found that there existed a second type of bright LMXB (the atoll sources) in addition to the sources showing a Z-shaped diagram.

It is important to note that the classification of the behaviour of the sources in the colour-colour diagram was not possible on the basis of these diagrams alone, but required the knowledge of the rapid X-ray variability. One reason for this is that not every LMXB displays its entire pattern of behaviour in every given observation. Sometimes this is a matter of bad luck on the part of the observer, in other cases a source may, apparently permanently, be constrained to subsections of the entire pattern (see below). Also, as we shall see in Section 2.2, some sources were just moving too slowly in the colour-colour diagram to obtain the full pattern in a typical observing time. In all such instances, a look at the power spectrum of the rapid variability (see Section 3) can be extremely illuminating.

We now turn to a description of the properties of the colour-colour diagram and hardness-intensity diagram of the two classes of sources. A comparison of the Z-atoll classification with earlier attempts at classifying LMXB and a discussion of the possible underlying physics is deferred to Section 8.

2.1 X-RAY COLOUR-COLOUR DIAGRAMS OF Z SOURCES

Six sources were classified by HK89 as "Z sources": GX 5-1, GX 17+2, GX 340+0, GX 349+2, Cyg X-2 and Sco X-1. These sources show three-branched, roughly Z-shaped hardness-intensity and colour-colour diagrams (Figure 4). The three branches are called the horizontal branch (HB), the normal branch (NB) and the flaring branch (FB) for various historical reasons (see van der Klis 1989a). The HB–NB junction is sometimes called the "apex" (or "hard apex") and the NB–FB junction the "antapex" (or "soft apex").

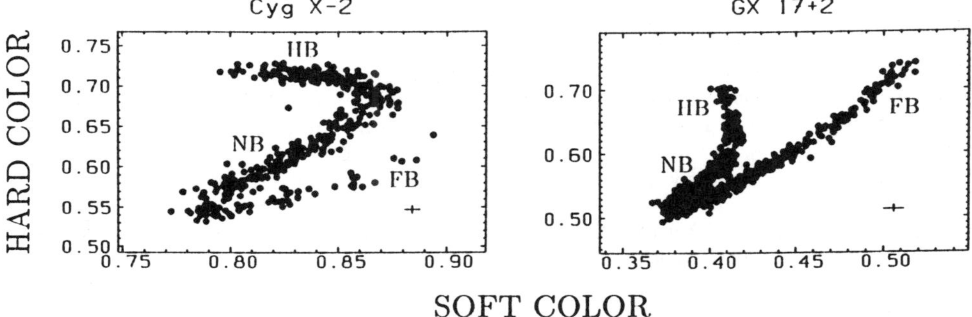

Figure 4. Examples of colour-colour diagrams of Z sources.

As its spectrum varies, a Z source moves through this pattern on time scales of hours to about a day. It never directly jumps from one branch to the other, but always moves gradually from one branch to the next via apex or antapex (Branduardi *et al.* 1980, Shibazaki and Mitsuda 1984, Priedhorsky *et al.* 1986), which suggests that a single parameter maps one-to-one onto the Z pattern. Usually, it is assumed that this parameter is the accretion rate. The branches are narrow, their observed widths being only slightly in excess of those expected from the statistical errors. There is no evidence for any periodicity or other regularity in this motion along the

Z (van der Klis *et al.* 1991).

The shape of the Z pattern depends on the choice of the energy bands. However, even if care is taken to select identical energy bands to calculate the hardness ratios, it is found that the shape of the distorted Z can be rather different between various individual sources. Cyg X-2, GX 5-1 and GX 340+0 are characterized by extended HBs, that are approximately horizontal when the X-ray colour plotted along the Y-axis is in the 5–20 keV range, whereas in GX 17+2 and Sco X-1 in a similar diagram the HB is short and, in contradiction to its historically motivated name, nearly vertical. Observations with the Hakucho and Ginga satellites (Shibazaki and Mitsuda 1984, Tan *et al.* 1990) of GX 5-1 have shown a HB that extends over more than a factor of 2 in X-ray intensity, and at its low-intensity end shows an upward bend. GX 5-1 apparently never enters the FB (van der Klis *et al.* 1991), and judging from its short NB as compared to other Z sources (Schulz *et al.* 1989) may never even attain the antapex, whereas by contrast GX 349+2 has always been found in the FB or the adjacent antapex region (HK89).

There usually is a good correspondence between the hardness-intensity diagram and the colour-colour diagram. However, this is not true for Cyg X-2, in which source the FB extends in the hardness-intensity diagram in the direction of *decreasing* intensity and in the colour-colour diagram often in the direction of *increasing* hardness in both bands (Hasinger *et al.* 1990). Something similar is also the case in GX 340+0, but in this source the FB "inversion" in the hardness-intensity diagram, which happens halfway up the FB, is matched in the colour-colour diagram (Penninx *et al.* 1991). Note that several different locations in the Z (likely, several values for accretion rate) can correspond to the same X-ray intensity.

The stability of the pattern in the diagrams on longer time scales appears to be usually quite good. In GX 5-1, no deviations in excess of 10% in intensity and 20% in hardness were detected during three years of monitoring with the Ginga All Sky Monitor (van der Klis *et al.* 1991). An exception again is Cyg X-2, in which source the entire pattern appears to move through the diagram (Hasinger 1987b). Two Ginga observations of this source showed several more detailed differences in the pattern than just a translation (Hasinger *et al.* 1990). Possibly, this could be described as due to the movement of the pattern as a whole, which happens on time scales that may not be very much longer than those on which the Z itself is traced out.

Since the report by HK89, further confirmation of the Z pattern has come from several Ginga observations of the six sources identified by HK89 (Mitsuda *et al.* 1988, Mitsuda and Dotani 1989, Wood *et al.* 1989, Hasinger *et al.* 1990, Penninx *et al.* 1990, 1991, Tan *et al.* 1990, van der Klis *et al.* 1991) but no new Z sources were found.

2.2 X-RAY COLOUR-COLOUR DIAGRAMS OF ATOLL SOURCES

Of the sixteen bright LMXB observed with EXOSAT that were analyzed by HK89, ten sources did not fit in with the pattern of correlated X-ray colour-colour and rapid X-ray variability behaviour defined by the Z sources. Just like the Z sources, however, these sources (GX 3+1, GX 9+1, GX 9+9, GX13+1, 4U 1636-53, 4U 1705-44, 4U 1820-30, 4U 1728-33, 4U 1735-44 and 4U 1608-52) showed well-defined correlations between their X-ray colour-colour diagrams and the power spectra of their rapid intensity variations. These were used to define a second type of source, the atoll source.

One of the most important observational differences in the X-ray colour behaviour between Z and atoll sources is not apparent when glancing at a completed hardness-intensity diagram or colour-colour diagram: atoll sources usually move much slower in such diagrams than Z sources. Because of this, it is not completely established for each individual atoll source exactly what pattern it traces out in the diagrams in the long run. As a general rule, however, atoll sources appear to exhibit an elongated, upwardly curved branch in both hardness-intensity diagram and colour-colour diagram. In their motion through the diagram sources follow this branch; they do not jump from one end to the other. This once again suggests accretion rate as the parameter determining the position of the source in the branch. Source motion through the diagram is much slower at the left end of the branch, where it seems that a source can hover for at least several, and perhaps many, days, than at the right end, where the time scale on which the source moves through the pattern is of the order of a day or less.

As the EXOSAT observations usually lasted a couple of days at most, this led, after a source had been observed several times at different epochs, to combined diagrams of the type shown in Figure 5, an elongated upwardly curved branch which came to be called the "banana" branch together with one or more "islands", corresponding to observations during which the source was in the part of the diagram where source motion was slow, so that it did not move much during an entire observation. These islands, which were easily identifiable from the presence of a very strong power spectral component called HFN (see Section 3.2.2), always occurred at lower X-ray intensity than the banana branch, and were nearly always located to the left of the banana in the colour-colour diagram.

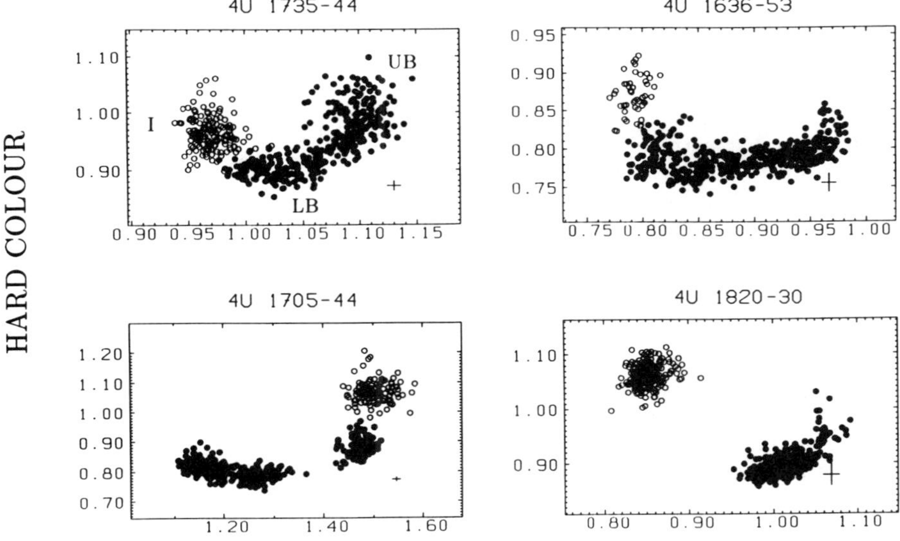

Figure 5. Examples of colour-colour diagrams of atoll sources. Open circles correspond to island states as determined from the HFN strength (see 3.2.2).

As the separate islands and banana branch are caused by observational sampling of a continuous sequence of behaviour, the distinction between island and banana state is somewhat arbitrary; usually the presence of strong ($\gtrsim 10\%$ rms) HFN (see Section 3.2) in addition to little motion in the colour-colour diagram is taken as a criterion of an island state. The right-hand end of the upwardly curved banana branch in some cases bends back to the left upon itself. In those cases, a distinction is made between the "lower" (LB) and the "upper" (UB) banana, located below and above the leftward bend of the branch. The entire pattern is sometimes reminiscent of a geographical map of an atoll, which explains the name of these sources. Contrary to the Z sources, in the atoll sources it is usually found that the width of the branch much exceeds that expected from statistics.

So, in a given source, the banana branch state corresponds to higher X-ray intensity (and likely, higher accretion rate) than the island state. This difference also roughly holds between sources: in some of the brighter atoll sources, only a banana branch and no islands are observed, whereas in some of the fainter ones only islands are seen. The qualitative correspondence between colour-colour diagram and hardness-intensity diagram is usually good, with the islands located to the left of the banana branch in both types of diagram. Exceptions do occur: in 4U 1636-53 one of two island states observed with EXOSAT showed a large drift in intensity during which the colours remained approximately stable (van der Klis *et al.* 1990). Different locations in the pattern (LB and UB; island and left end of the LB) can correspond to the same X-ray intensity.

An interesting case is presented by the Tenma observations of the transient 4U 1608-52 (Mitsuda *et al.* 1989). This source was classified as an atoll source by HK89 on the basis of two EXOSAT "island" observations in 1984 and 1986. As noted by HK89, the 20–22 April 1983 Tenma observations appeared to show a banana, whereas the May–June 1984 observations might show what an island state looks like in the hardness-intensity diagram when the observations are long enough. (The first part of these observations might also correspond to the left end of the banana.) Indeed, Mitsuda *et al.* (1989) reported that the source variability time scale was shorter in the 1983 observations (10 minutes) than in the 1984 observations (a day). In accordance with this, Makishima (1986) finds stronger VLFN (see Section 3.2) in the 1983 than in the 1984 data. His data do not allow to measure the HFN. In the 1984 EXOSAT observation of 4U 1608-52, which took place in July 1984 about ten days after the last Tenma one, strong HFN, as expected in an island state was indeed observed (HK89).

It is important to reiterate at this point that it was not always possible to make the distinction between Z and atoll sources nor indeed between an island state and a banana state on the basis of the colour-colour diagrams alone. In fact, some of the bent-back upper banana branches had previously been identified with HB–NB patterns. The additional information about the rapid variability was crucial. In the following Sections, I describe the way in which the various other source characteristics (Section 3: QPO and noise; Section 4: X-ray spectra; Section 5: Optical and UV emission; Section 6: Radio emission; Section 7: X-ray bursts) relate to the position of the source in its pattern in the colour-colour diagram. The basic method used to select data is always the same: define boxes in the colour-colour diagram and consider the properties of all data obtained when the source was within each box.

3. Quasi-periodic oscillations and noise

The main diagnostic of the rapid X-ray variability of the bright LMXB is the power spectrum of the X-ray intensity variations. I reviewed the analysis techniques associated with the power spectral approach extensively in van der Klis (1989b) and the explanations given there will not be repeated here. For detailed reviews of QPO properties, see Lewin *et al.* 1988, Stella 1988, van der Klis 1989a; for an account of QPO models, see Lamb 1989, and Lamb, this volume. The basic approach is as follows: Data are selected according to their position in the colour-colour diagram. All time segments of data within a given predetermined box in the colour-colour diagram that are of sufficient length are Fourier transformed and the power spectra calculated from the resulting Fourier transforms are averaged. The power spectral component that is due to the counting statistics (Poisson noise modified by detector characteristics, in particular dead time; in practice this is in most cases very nearly a white noise component) is subtracted from the power spectrum and the resulting "background subtracted" power spectral shape is studied by fitting several-component models to it. (Sometimes, the "Poisson level" that should be subtracted from the power spectrum is not sufficiently accurately predictable due to lack of knowledge of instrumental characteristics. Then it is included in the fits as a free parameter.) Fit procedures usually assume Gaussian statistics of the power estimates. Note that this is only correct if a sufficient number of power estimates was averaged (see van der Klis 1989b).

To detect a broad power spectral component in the presence of counting statistics noise at a n sigma level of confidence, we require $r > (\Delta\nu/T)^{1/4}(2n/I)^{1/2}$, where r is the "fractional rms variation" of the variability component (the root-mean-square amplitude of the mean-subtracted signal expressed as a percentage of the mean count rate, hereafter % rms), $\Delta\nu$ its width in the power spectrum, T the observing time and I the count rate. Typical values for EXOSAT observations ($T\sim10^4$ s, $I\sim10^3$ c/s) lead to a 3σ detection level of a 10 Hz wide feature of $\sim$1.4% rms. As the components that we are interested in usually have a fractional rms strength of a few percent, it is obvious that count rates above 100 c/s, and preferably higher, are definitely required to do anything useful.

The highest and lowest frequencies that are represented in the power spectrum depend on the time resolution of the data and the length of the individual data segments. In order to be able to measure the rapid variability components in the power spectrum that are of importance in classifying source type and state, it is desirable to cover a range of at least 10^{-3} to 10^{+3} Hz (so, "rapid" means: on time scales of less than 1000 s). If the lower bound is higher, VLFN and LFN (see below) become difficult to characterize correctly, if the upper bound is lower this is true for the HFN.

We now turn to a description of the power spectral properties of Z and atoll sources in their respective states. In Section 3.3, I discuss the photon energy dependence of the QPO and noise components in these sources.

3.1 POWER SPECTRA OF Z SOURCES

The following power spectral components are distinguished in the Z sources (in order of increasing frequency):

3.1.1 *VLFN (very-low frequency noise)*, a power-law shaped component $p(\nu) = C\nu^{-\alpha}$, with $p(\nu)$ the power density, C a normalization constant and α the power-law index. VLFN is clearly distinct from the LFN (van der Klis *et al.* 1987a, see Section 3.1.2). α is usually in the range 1.5–2.0 with some tendency towards a steepening slope in the sense HB–NB–FB. The strength of this component varies between 0.4–1.6% rms on the HB, increasing via 0.6–2.2% in the NB up to 2–6.4% rms on the FB. It is observed in all Z sources in all three branches and usually dominates the power spectra at the lowest frequencies (below 0.01–1 Hz depending on the presence or absence of LFN, see next section).

3.1.2 *LFN (low-frequency noise)*, a broad noise component fitted by HK89 to a function $p(\nu) = C\nu^{-\alpha}\exp(-\nu/\nu_{cut})$, where ν_{cut} is a cut-off frequency. This component is associated with the HB QPO: the two components are either both present or both absent (van der Klis *et al.* 1987a). Often, but not always, there is a more detailed correlation in their strengths (*e.g.*, GX 5-1: van der Klis *et al.* 1985a, but see also, *e.g.*, Cyg X-2: Hasinger *et al.* 1986). It is observed in the HB and in some cases in the upper half of the NB. The strength of this component is 3–9% rms in the HB and $\lesssim$1.5% rms in the upper NB. The component is sometimes "red" (*i.e.*, monotonically decreasing with frequency), and in those cases produces a broad hump in the VLFN power law most evident between 0.1 and 10 Hz, and sometimes forms a very broad peak with a maximum around ~1 Hz ("peaked LFN"). In the former case, $\alpha \approx 0.3$ and ν_{cut} ~10 Hz, in the latter these values are ~−0.5 to −1.0 and ~2 Hz, respectively. LFN shape appears to correlate with HB slope: peaked LFN appears in short "vertical" HBs (GX 17+2 in Figure 4). The component has been fit with a variety of different functions by various authors. The exponential and Lorentzian shapes favoured early on (see van der Klis 1989a for references) do not fit the component if a sufficiently wide frequency range is considered. Penninx *et al.* (1990, 1991) have recently preferred to fit a function of the shape $p(\nu) = C[\nu^{-\gamma} + \nu^{-\psi}]^{-1}$ to the LFN of GX 17+2 and GX 340+0. LFN has been seen in all Z sources but GX 349+2, which has never been observed in the HB.

3.1.3 *HB QPO (horizontal branch quasi-periodic oscillations).* These oscillations were first found in the HB (van der Klis *et al.* 1985a) but are now known to persist at a lower strength down to about halfway the NB (Hasinger *et al.* 1990). They form a peak in the power spectrum that fits a Lorentzian shape: $P(\nu) = C[(\nu - \nu_{QPO})^2 + (\Delta\nu)^2]^{-1}$, with a centroid frequency ν_{QPO} and a FWHM (the full width of the peak measured at a level halfway down from the maximum) $\Delta\nu$.

Sometimes a second harmonic at $2\nu_{QPO}$ is also detected (Mitsuda *et al.* 1988, Penninx *et al.* 1990, Langmeier *et al.* 1990, Tan *et al.* 1990, see also Stella *et al.* 1986, 1987a). ν_{QPO} of the fundamental peak varies between 15 and 55 Hz on the HB and is strictly correlated with position in the branch (increasing to the right). This leads to a correlation between ν_{QPO} and X-ray intensity which suggests a magnetospheric origin for these QPO (Alpar and Shaham 1985). When the source moves (from the HB) into the apex region and then down the NB, the QPO frequency apparently sticks to the value it had at the apex (Hasinger *et al.* 1990, Tan *et al.* 1990). The HB QPO can occur in the NB *simultaneously* with the NB QPO (below), establishing beyond doubt that these are two different phenomena. The peak width $\Delta\nu$ is between 0.1 and 0.3 ν_{QPO} for the fundamental; the second harmonic peak has been reported to have widths of 2 times the width of the fundamental, which favours frequency drift over life time broadening as the broadening mechanism in the context of an oscillating shot model (Shibazaki and Lamb 1987), up to 8 times, which is hard to explain. The strength of the fundamental QPO peak is between 2 and 9% rms in the HB, increasing with QPO frequency, and drops to 1.5–3% in the NB. The power contained in the harmonic can be as large as that in the fundamental. As noted above (3.3.2), HB QPO and LFN are related. On time scales comparable to the QPO period, it has not been possible to measure this correlation (Norris *et al.* 1990), on longer time scales, it has (van der Klis *et al.* 1985a, Mitsuda 1989). HB QPO have been detected in 4 of the 6 Z sources. The exceptions are GX 349+2, which is always in the FB or the antapex region (HK89), and Sco X-1, which only rarely enters its short, vertical HB (Hasinger *et al.* 1989).

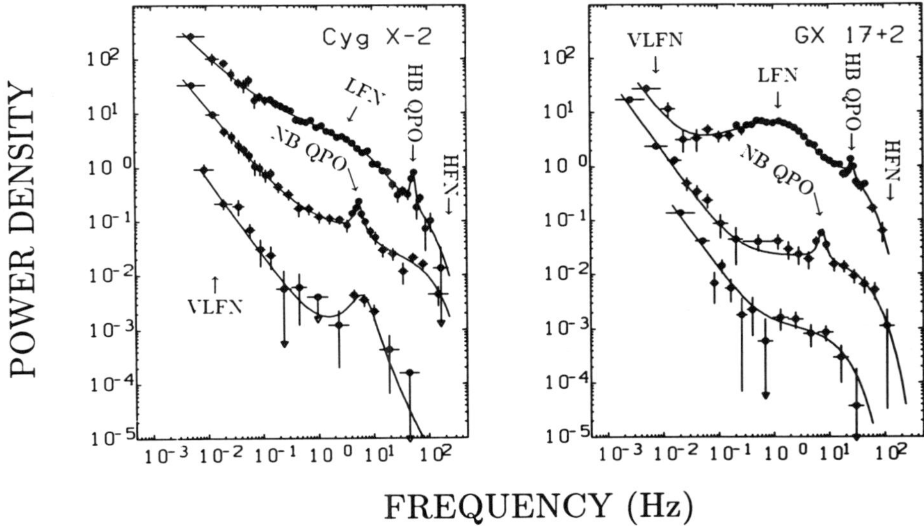

Figure 6. Examples of power spectra of Z sources.

3.1.4 *NB QPO (normal branch quasi-periodic oscillations)*, a Lorentzian shaped peak with a centroid frequency of about 6 Hz and a width (FWHM) of 2–3 Hz first seen in Sco X-1 (Middleditch and Priedhorsky 1986). This QPO peak is observed in all 5 Z sources that show a NB when the source is in the middle or lower parts of it. Its strength is usually 3–4% rms. The QPO are reported to be strongest and narrowest in the middle of the NB and there is some indication that the frequency may slightly increase from the middle of the NB outward to apex and antapex ((Middleditch and Priedhorsky 1986, van der Klis *et al.* 1987c, Hasinger *et al.* 1990, Mitsuda and Dotani 1989, Penninx *et al.* 1990, 1991, Tan *et al.* 1990). No harmonics have been reported.

3.1.5 *FB QPO (flaring branch quasi-periodic oscillations)*, a QPO phenomenon that is apparently closely related to NB QPO. FB QPO form a Lorentzian peak with a centroid frequency between 10 and 30 Hz when the source is in the FB. Its frequency is correlated with the position of the source in the branch, increasing when it moves away from the antapex. Peak width also increases in this sense, from 6 up to tens of Hz; halfway up the FB the peak becomes so wide that it can no longer be measured separately from the HFN (below). When the source moves down from the FB into the antapex region, FB QPO frequency gradually decreases from the value of 10 Hz it has in the lowest part of the FB to match the 6 Hz of the NB QPO, with which it merges smoothly. This QPO peak has been observed with certainty in Sco X-1 and GX 17+2 (Middleditch and Priedhorsky 1986, Penninx *et al.* 1990) and broad excesses probably due to this same peak (moving rapidly in frequency as the source moves up and down the flaring branch) were also seen in Cyg X-2, GX349+2 and GX 340+0 (*e.g.*, HK89, Hasinger *et al.* 1990, Penninx *et al.* 1991).

3.1.6 *HFN (high-frequency noise)*, a component often detected up to frequencies of ~100 Hz (HK89) that gradually decreases in strength from the HB (10–4% rms) via the NB (5–<2% rms) to the FB (4–1.5% rms). This component fits a similar shape as LFN: $p(\nu) = C\nu^{-\alpha}\exp(-\nu/\nu_{cut})$, with ν_{cut} usually between 30 and 100 Hz and $\alpha{\sim}0$. (HK89 uniformly adopted $\alpha \equiv 0$ for the HFN in Z sources.) Note that HFN strength systematically decreases from HB via NB to FB, while at the same time the VLFN increases. HFN is clearly distinct from LFN; in the HB both components are detected at the same time. This component has been seen in all Z sources; there is some evidence that its shape may in some cases be peaked (Penninx *et al.* 1990).

Physical models have been proposed for two sets of power spectral components of Z sources: HB QPO/LFN seem likely to be due to an interaction of the accretion stream with a small (~10^{7-8} cm) neutron star magnetosphere (neutron-star magnetic-field strength ~10^{9-10} G; Alpar and Shaham 1985, Lamb *et al.* 1985, Lamb 1989), whereas for the explanation of NB/FB QPO radiation-pressure-limited accretion flows near to the Eddington limit have been gaining ground (Hasinger 1987a, 1988a, b, Hasinger *et al.* 1989, Lamb 1986, 1989, van der Klis *et al.* 1987c, see in particular Fortner *et al.* 1989). VLFN is often implicitly assumed to be due to external variations in mass-transfer rate. No models have been proposed for HFN.

3.2 POWER SPECTRA OF ATOLL SOURCES

The power spectra of atoll sources can be described in terms of just two components which are observed in nearly all of them (HK89):

3.2.1 *VLFN (very-low frequency noise)*, a component that is very similar to Z-source VLFN with a power law shape: $p(\nu) = C\nu^{-\alpha}$ increasing in strength from $\lesssim 1\%$ rms and usually undetectable in the island state to usually ~2.5% in the left end of the LB and ~4% further up the banana. Power law slope α is a bit less than in the Z sources and varies between 1 and 1.5%. VLFN has been seen in all atoll sources in the banana state.

3.2.2 *HFN (high-frequency noise)*, a very strong component whose shape was described by HK89 by the function $p(\nu) = C\nu^{-\alpha}\exp(-\nu/\nu_{cut})$, that just like in the Z sources decreases in strength when the VLFN increases, from a whopping 10–20% rms in the island state via 6–<3% rms in the LB to usually <2% rms in the UB. Atoll source HFN often has a tendency to be "peaked" (show a local maximum) such as also sometimes seen in Z-source LFN and sometimes shows additional "wiggles"; in most cases $\alpha < 0$ and ν_{cut} between 15 and 30 Hz. Dotani *et al.* (1989) fit the HFN in 4U 1820-30 with two Lorentzians with different centroid frequencies and widths and report that this provides a better fit than the function used by HK89. HFN has been detected in all atoll sources but GX 13+1.

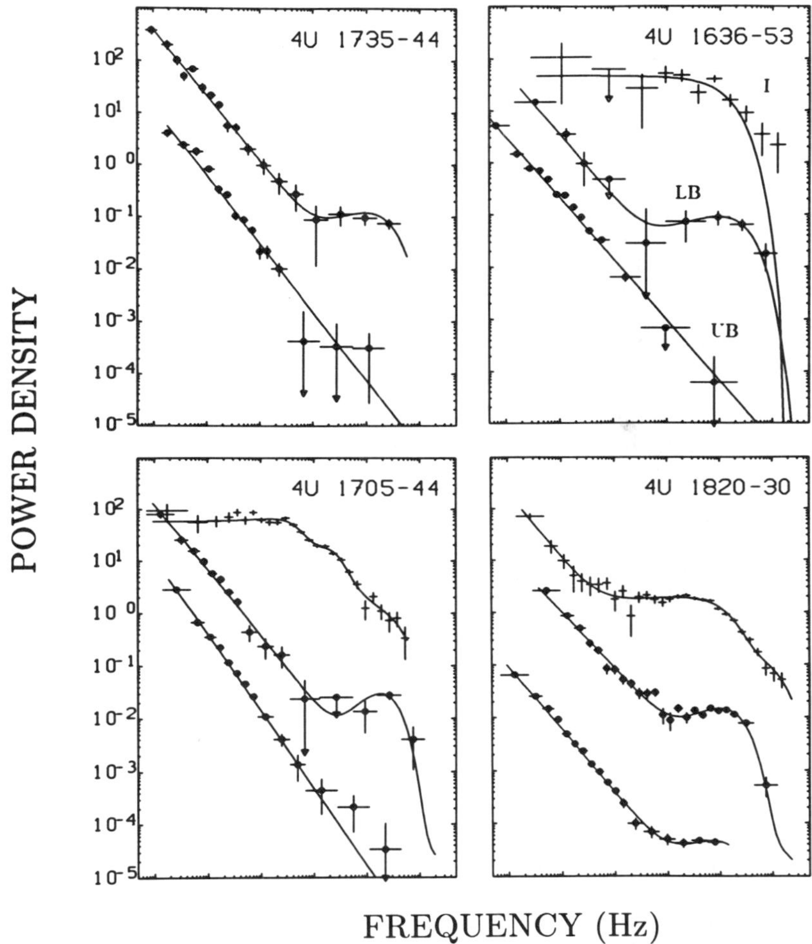

Figure 7. Examples of power spectra of atoll sources. Dots: banana state; crosses: island state. Frequency scale as Fig. 6.

The interplay between these two components leads to a variety of power spectral shapes (Figure 7). The peaked HFN components have sometimes been described as QPO but are clearly quite different from both HB QPO and NB/FB QPO in peak width and shape, and in correlation with other source characteristics such as LFN and X-ray intensity and colours.

There are no direct indications that the VLFN and HFN components in atoll sources can be identified with those in the Z sources. The main indication that they may be the same thing (apart from a general, but not exact, similarity in shape) is the similar dependencies of their strengths on source state (and likely, accretion rate) in both cases (VLFN increases, HFN decreases with accretion rate).

No explanations have been proposed for atoll source power spectral shapes; see, however, the remark in Section 3.1 about Z source VLFN which also applies here.

3.3 PHOTON ENERGY DEPENDENCE OF THE RAPID VARIABILITY

Two different quantities are necessary to completely characterize the photon energy dependence of the rapid variability components described above: the strength of the component as a function of photon energy, and its phase as a function of photon energy. The first quantity, sometimes described as the X-ray spectrum of the variability component, can be measured by constructing power spectra of data obtained in different photon energy bands and determining the strength of the component in each power spectrum. The second quantity is usally discussed in terms of the "time lag" of the variability in one photon energy band with respect to that in another band. It can be measured by constructing cross-correlation functions or cross-spectra between the data in the two bands (see van der Klis *et al.* 1987b, van der Klis 1988, Lewin *et al.* 1988 for a summary of these techniques). Note that, as the power spectral components that we are interested in are broad, different time lags can apply at different variability frequencies in the same component. The interpretation of such frequency-dependent lags may be nothing else than a difference in the shape of the modulation at different photon energies (van der Klis *et al.* 1987b). The available information about this aspect of the rapid variability is far from complete, in particular in the case of the atoll sources. In the following I summarize what is known; for a discussion see Section 8.

3.3.1 *VLFN.* At frequencies below 1 mHz, the VLFN can be identified with the variations in source flux that trace out the Z and atoll patterns in the hardness-intensity diagram. So, if a positive correlation is seen between X-ray intensity and spectral hardness, as is the case in the NB and the FB, and often in the banana branch, then this low-frequency end of the VLFN will necessarily show an X-ray spectrum that is harder than the average X-ray spectrum. In an HB where spectral hardness does not vary, and in the part of the banana branch where this is the case, one would expect a VLFN X-ray photon energy spectrum similar to the average X-ray spectrum. Various reports about the energy dependence of the VLFN are scattered throughout the literature; often, the information that is supplied is not sufficient to determine the source state, or the frequency ranges over which the power was integrated were chosen such that other power spectral components in addition to the VLFN (Z source LFN, atoll source HFN) were included. In general, the data seem to be in accordance with the idea that the VLFN, also above 1 mHz, has an energy dependence that is similar to that expected if the intensity variations that it represents occur along the branch on which it is observed in both Z and atoll sources (GX 5-1: van der Klis 1987, Tan *et al.* 1990, Sco X-1: van der Klis *et al.* 1987c; GX 3+1: Lewin *et al.* 1987, Makishima *et al.* 1989; GX 340+0: Van Paradijs *et al.* 1988).

3.3.2 *HB QPO and LFN* both show a similar dependence on photon energy (van der Klis 1987, Tan *et al.* 1990, Penninx *et al.* 1991): the fractional amplitude clearly increases from ~3% near 2 keV to 10% near 20 keV. A small time lag of the signal in a hard (*e.g.*, 5–20 keV) with respect to that in a soft (*e.g.*, 1–5 keV) band exists. It decreases as a function of QPO frequency from 4 to 0.0 ms in Cyg X-2 (Hasinger 1987b, van der Klis *et al.* 1987b) and varies between 3.0 and 0.0 ms in GX 5-1 (van der Klis *et al.* 1987b, Mitsuda *et al.* 1988). In the LFN, time lags of the opposite sign are observed: the 1–10 Hz variability shows a "soft lag" of up to 12 ms (van der Klis *et al.* 1987b) and even larger values (up to 90 ms) are reported by Mitsuda *et al.* (1988). Explanations of the QPO time lags in terms of inverse Compton scattering (Hasinger 1987b, van der Klis *et al.* 1987b, Stollman *et al.* 1987, Brainerd and Lamb 1987, Wijers *et al.* 1987, Bussard *et al.* 1988) have been proposed; it is not clear that the required scattering cloud parameters can be obtained in reality.

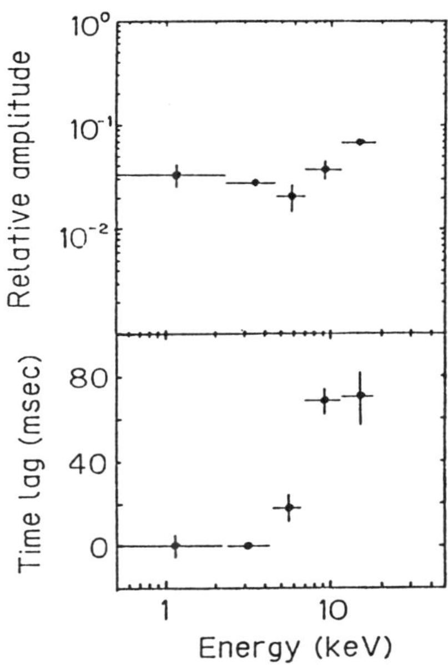

Figure 8. Time lags and relative spectrum of NB QPO in Cyg X-2.

3.3.3 *NB/FB QPO.* In Cyg X-2, Mitsuda and Dotani (1989) observed the energy dependence shown in Figure 8 in the NB QPO. The QPO show a minimum in relative strength and a phase jump of 150^0 near 5 keV. This was interpreted by Lamb (1989) as evidence for a quasi-periodic "rocking" of the X-ray spectrum around a pivot point at this energy caused by quasi-periodic variations in Comptonization optical depth in accordance with numerical modeling by Fortner *et al.* (1989). In GX 5-1, the NB QPO (Mitsuda 1989) show a similar phase jump near 2.5 keV. No reports have been made about the energy dependence of FB QPO.

3.3.4 *HFN.* The energy dependence of the peaked HFN in 4U 1820-30 and GX 3+1 has been studied by Makishima *et al.* (1989), Dotani *et al.* (1989) and Mitsuda (1989). No time lags were detected. The spectrum of the HFN was harder than the average spectrum, and the ***absolute*** rms variation in the HFN roughly fitted a blackbody spectrum with a temperature of 2.2 keV.

4. X-ray spectra

As noted in Section 2, spectral modeling does not lead to unique representations. Often, several different models fit the data (see, *e.g.*, Hasinger *et al.* 1990, Vacca *et al.* 1987) and even within one model more than one set of parameter values may fit (Vacca *et al.* 1987). Testing of specific physically motivated models has led to the rejection of some models (see, *e.g.*, White *et al.* 1986, 1988), but it is quite possible that more complicated models than can be rigorously tested using proportional counter data may in fact be needed for a physically realistic description (see, *e.g.*, Vrtilek *et al.* 1988, Taam and Mészáros 1987).

In recent years, there has been a number of claims of physically motivated models that might fit the spectra of bright LMXB. Some of these models involve two different components (there exist two schools of thought here, see Mitsuda *et al.* 1984, 1989 on one hand and White *et al.* 1986, 1988 on the other), in which one component is identified with the emission of the neutron star surface or the neutron star/accretion disk boundary layer, and the other with emission from the inner accretion disk. Others seek to describe the spectrum through more or less realistic modeling of Comptonization of an assumed input spectrum in some idealized geometry (Lamb 1989, Fortner *et al.* 1989, Schulz and Wijers 1989, Damen *et al.* 1990, Ponman *et al.* 1990). There exists no consensus as to what is the correct spectral model.

One might hope that the clarification afforded by the Z/atoll classification of source types and -states might help in finding the correct model(s), as it can remove the confusion caused by comparing or even averaging spectra obtained in different source states or from sources of different type. However, so far such deeper insight has not emerged. Schulz *et al.* (1989), Lamb (1989) and Schulz and Wijers (1989) all suggest from approximate parametrizations that the Z pattern can be reproduced by parameter changes in a variety of existing spectral models, and Hasinger *et al.* (1990) have shown by direct fits to spectra of Cyg X-2 in various different sections of the Z that the two competing two-component models can both provide a description of the data. For the final verdict the results of systematic surveys including proper X-ray spectral fits to LMXB spectra sorted according to source type and -state must be awaited.

I note that the case of the Z sources appears particularly tantalizing: The fact that the Z patterns are observed shows that two subtle, but clearly observable, transitions are in fact detected in the presently available X-ray spectra at two specific values (corresponding to apex and antapex) of the accretion rate; it should be possible to devise a model for the accretion flow and its X-ray emission that convincingly predicts these transitions.

5. Optical and UV emission

In correlated optical and UV/X-ray (and radio, see Section 6) observing campaigns of Cyg X-2 (Hasinger *et al.* 1990) and Sco X-1 (Wood *et al.* 1989), clear correlations of the optical and UV properties of these sources with source state were seen. In general, the optical and UV flux increases in the order HB–NB–FB. Hasinger *et al.* 1990 argue that this provides evidence for an increase of accretion rate in this sense. According to these authors, the accretion disk, whose optical and UV emission due to reprocessing of X-rays dominates that of the system, has a "better view" of the X-ray source than we have on Earth (our line of sight being obstructed in a source-state dependent way), so that optical and UV flux are better indicators for the X-ray luminosity, and therefore for the accretion rate, than is the X-ray intensity.

Among a large body of optical monitoring data of bright LMXB (in particular, Sco X-1), sometimes with simultaneous X-ray coverage, that existed before the Z/atoll classification was proposed (*e.g.*, Evans *et al.* 1970, Mook *et al.* 1975, Canizares *et al.* 1975, Bradt *et al.* 1975, Willis *et al.* 1980, Ilovaisky *et al.* 1980) only a small part can be interpreted in terms of this classification. This is mostly due to the fact that X-ray hardness data are usually lacking.

The simultaneous X-ray-optical observations by Corbet *et al.* (1989) of the atoll source 4U 1735-44 show a clear upper and lower banana branch. The optical and X-ray flux are correlated along the branch, strengthening the proposed (HK89) positive correlation of accretion rate with X-ray intensity on this branch.

6. Radio emission

As pointed out by Penninx (1989), Z sources tend to be brighter radio sources (typically a few to a few tens of mJy) than atoll sources, most of which are undetected (typically <0.4 to at most a few mJy). A series of correlated X-ray/radio observations of Z sources has shown that in general sources are most radio bright and/or show most radio flares in the HB. In the NB, all Z sources may have roughly the same radio luminosity (Penninx 1989).

It has been suggested (Hasinger *et al.* 1990) that the strong magnetospheric interactions responsible for the HB QPO that take place in the HB are also responsible for the particle acceleration causing the HB radio emission. Recent observations (Hjellming *et al.*, in prep.) suggest that in GX 5-1 the radio flux may be consistently lower in the HB than in the NB.

7. X-ray bursts

It was shown by Damen *et al.* (1989) that the duration τ of an X-ray burst correlates with the (variable) position of its cooling track in the X-ray luminosity *vs.* best-fit blackbody temperature diagram, parametrized by the temperature $T_{0.1}$ at the point in the burst decay where $L_x = 0.1L_{Edd}$. This was a surprising result, as $T_{0.1}$ is believed to be determined by the radiation transfer of the burst flux in the neutron star atmosphere, whereas τ is determined by the properties of the deep layer in the neutron star envelope where the thermonuclear flash takes place (in particular, by its chemical composition). The confusion was compounded by the fact that neither $T_{0.1}$ nor τ correlated well with the persistent intensity (although over a wide range in X-ray intensity and taking together many burst sources, such a correlation can be found [Van Paradijs *et al.* 1988]).

Van der Klis *et al.* (1990) have shown that in 4U 1636-53 τ and $T_{0.1}$ correlate much better with accretion rate as parametrized by atoll source state, than with X-ray intensity, up to then usually taken as a measure for accretion rate. This suggests that whatever mechanism causes τ and $T_{0.1}$ to correlate, it is driven by accretion rate (see van der Klis *et al.* 1990 for possible ways in which this could arise). Inversely, the correlation provides further support for the interpretation of atoll source state in terms of accretion rate. Similar results have been obtained for the atoll source 4U 1705-44. The properties of 4U 1735-44 and 4U 1820-30 are also in accordance with this correlation (van der Klis *et al.* 1991, in prep.).

8. Discussion

8.1 CLASSIFICATION OF LOW-MASS X-RAY BINARIES

Various attempts at classifying LMXB based on X-ray properties that did not consider rapid variability (Parsignault and Grindlay 1978, Ponman 1982), including those based on detailed X-ray spectral fits (White and Mason 1985, White *et al.* 1988) or hardness-intensity diagrams and colour-colour diagrams of EXOSAT data (Schulz *et al.* 1989), in general led to a division in a high X-ray luminosity (hereafter "hi") and a low X-ray luminosity ("lo") group. With few exceptions, the "hi" group includes the bright galactic bulge sources ("GX sources"). The "lo" group includes X-ray burst sources, sources showing periodic X-ray dips and sources which are located in globular clusters (these groups partly overlap). Table 1 lists the brightest persistent LMXB in order of average X-ray intensity, together with the classification of each source according to above-mentioned authors. It can be seen, that there is a good correspondence between them, with in general the GX sources ending up in the "hi" group.

By contrast, in the Z/atoll classification of HK89, only four of the GX sources are grouped together with Sco X-1 and Cyg X-2 in what might be identified as the "hi" group (the Z sources), and four other GX sources with classical X-ray burst sources such as 4U 1636-53 in the "lo" group (atoll sources). As will be clear from the previous sections, I feel that the earlier classifications, which could not take advantage of our present knowledge of the rapid variability, are unsatisfactory, as they group together sources with very different QPO and noise characteristics, colour-colour diagrams and radio properties.

Apart from the array of diagnostics described above, there is another source characteristic that might be related to the classification: orbital period. As seen in Table 1, of the six Z sources two (maybe three) orbital periods are known and all are in excess of 19 hrs, of the ten atoll sources, four orbital periods are known, and all are shorter than 5 hrs. Obviously, the numbers are small, but the result is suggestive of an evolutionary connection (see Section 8.3).

Table 1. Bright low-mass X-ray binaries[a]

Source	I_x [b] (μJy)	P_{orb} [c] (hr)	Type[d] PG	P	W	S	HK	Phenomena[e]
Sco X-1	12400	19.2	hi	—	hi	hi	Z	QPO
GX 5-1	1200	—	hi	hi	—	hi	Z	QPO
GX 349+2	780	—	hi	hi	hi	hi	Z	QPO
GX 17+2	680	19.8?	hi	hi	hi	hi	Z	QPO,(bu)
GX 9+1	650	—	hi	hi?	hi	hi	A	—
GX 340+0	490	—	lo?	—	—	hi	Z	QPO
GX 3+1	430	—	hi	hi	—	hi	A	QPO,(Bu)
Cyg X-2	430	235.	hi?	—	hi	hi	Z	QPO,(bu),Mo
GX 13+1	340	—	hi	hi	—	hi	A	—
GX 9+9	290	4.2	hi	lo?	—	hi	A	Mo
4U 1820-30	260	0.2	lo	lo	hi/lo	lo	A	QPO,(Bu),Mo
4U 1705-44	260	—	—	lo	hi	lo	A	Bu
4U 1636-53	220	3.8	lo	lo	lo	lo	A	Bu
Ser X-1	200	—	lo	lo	lo	lo	—	Bu
4U 1728-33	170	—	lo	lo	lo	lo	A	Bu
4U 1735-44	160	4.6	lo	lo	—	lo	A	Bu

[a] All variable objects in 3A catalogue (McHardy *et al.* 1981, Warwick *et al.* 1981) with an average flux $\geq 100\mu$Jy not identified with an early type star (excluding the peculiar – possibly ADC – source Cyg X-3, the galactic center source GCX-1 and the black hole candidate GX 339-4).

[b] Mean intensity converted from Ariel V ASM counts into μJy (2–11 keV) according to 1 ASM c/s = 2.6 μJy (Bradt and McClintock 1983).

[c] See Parmar and White (1988).

[d] PG: Parsignault and Grindlay (1978); P: Ponman (1982); W: White and Mason (1985), White *et al.* (1988); S: Schulz *et al.* (1989); HK: HK89, A: atoll

[e] Bu: regular X-ray bursts; (Bu): has shown an episode of regular X-ray bursts; (bu): occasional X-ray bursts reported; Mo: shows periodic X-ray modulation.

8.2 A TENTATIVE INTERPRETATION OF THE Z/ATOLL CLASSIFICATION

A tentative interpretation of the Z/atoll classification was put forward by HK89. They suggest that the sources they classified as Z sources have *both* higher neutron-star magnetic-field strengths than atoll sources, *and* higher accretion rates. The higher neutron-star magnetic-field strength explains why HB QPO/LFN, which are likely a magnetospheric phenomenon, occur in Z sources only. The higher accretion rate can explain why NB/FB QPO, which were proposed to be caused by an instability in the accretion flow that occurs near to the Eddington limit, have also only been observed in Z sources, and why Z sources are brighter.

HK89 argue that it is hard to see how other differences such as orbital inclination (a gradual range in properties would be expected rather than a division in two groups), or neutron star rotation rate (HB QPO are expected in magnetospheric accretion at any rotation rate), or a difference in *just* accretion rate (why no HB

QPO in the lower accretion-rate atoll sources, and where are the sources that switch between Z and atoll behaviour as their accretion rate changes), or *just* neutron-star magnetic-field strength (why are Z sources brighter), could consistently explain all the observed differences.

They also suggest that the fact that Z sources show three states with two distinct transitions between them, whereas atoll sources show no such transitions but only a gradual range of properties, may be explained by two important physical changes that can occur in the accretion stream of a Z source, but not of an atoll source: (i) the "engulfing" of the magnetosphere by the inner disk when with increasing accretion rate the disk would swell up due to radiation pressure and (ii) the onset of mass ejection as the source reaches the Eddington limit. These events could be identified with apex and antapex, respectively. Such a qualitative picture of the HB as the state of magnetospheric disk accretion, the NB of more spherical inflow and the FB of super-Eddington outflow is in accordance with the "unified model" of QPO sources proposed by Lamb (1989).

In these terms, atoll sources are characterized by considerably sub-Eddington non-magnetospheric disk accretion. I note that the HFN, observed in Z sources as well as in atoll sources, and inversely correlated to accretion rate in both source types, which includes the highest observed variability frequencies in any of these sources, may be an important universal diagnostic. Similar noise components are also observed in disk-fed X-ray pulsars and in Cyg X-1 (Belloni and Hasinger 1990): the component may be a general feature of disk accretion.

8.3 A LINK WITH LOW-MASS X-RAY BINARY EVOLUTION

If we accept the interpretation of the differences between the Z and the atoll sources in terms of a difference in *two* basic source characteristics, accretion rate and neutron-star magnetic-field strength, then it remains to be explained why among the brightest LMXB these two characteristics correlate. The answer to this may be related with LMXB evolution.

As is well known, Webbink *et al.* (1983) proposed that for Eddington accretion rates to be sustained in LMXB, the mass transfer should likely be driven by the evolutionary expansion of a low-mass star away from the main sequence towards the red giant stage (hereafter "evolved companion"), rather than by gravitational radiation and magnetic braking in a system containing a main-sequence or degenerate dwarf companion. They specifically mentioned that the bright galactic bulge sources (the "GX" sources or the ones classified "hi" in Table 1) should be the ones to be identified with the systems containing evolved companions. Although systems with an evolved companion can both expand and contract in their evolution (Pylyser and Savonije 1988a, b), high accretion rates are expected to correlate to longer orbital periods than are consistent with a main sequence companion ($\sim$10 hr) in this picture.

This presents a problem for any classification scheme that includes GX 9+9 (P = 4.2 hr) among the "hi" sources. After the discovery of the QPO in GX 5-1, and their interpretation as a magnetospheric phenomenon, it was proposed that the presence of a neutron-star magnetic field strong enough to form a magnetosphere (10^{9-10} G) might be related to the high accretion rates possible from an evolved companion (van der Klis *et al.* 1985a, see also Lewin and Van Paradijs 1985, Van Paradijs and Lewin 1986, Lewin and Van Paradijs 1986) and that consequently QPO and evolved companions should correlate. This idea subsequently lost its appeal when GX 9+1, GX 9+9 and GX 13+1, usually classified in the "hi" class and therefore assumed to have evolved companions, did not show QPO, and when a short orbital period as well as QPO were discovered in 4U 1820-30 (Stella *et al.* 1987b, d).

In the Z/atoll scheme, the short period of GX 9+9 and the absence of HB QPO from GX 9+1, GX 9+9 and GX 13+1 (and GX 3+1, whose previously reported QPO [Lewin *et al.* 1987] do not resemble HB QPO and were reinterpreted by HK89 as peaked HFN) are no problem, but on the contrary expected consequences of the membership of these sources of the (small-companion, low neutron-star magnetic-field strength) atoll class. For the Z sources, the previously supposed link between evolved companions, high accretion rates and higher neutron-star magnetic-field strengths now once more appears to hold.

However, the mechanism originally proposed for the evolutionary link between these properties, accretion-induced collapse of a white dwarf in high-accretion-rate systems that produces young neutron stars with less decayed fields in the evolved-companion systems, has meanwhile become much less attractive, as the evidence for spontaneous gradual field decay seems to have evaporated (see Bhattacharya, this volume). As noted by HK89, the observational evidence suggests that we may have to look for another way to link these properties. The mechanism for spin-rate-change-induced field decay proposed by Srinivasan *et al.* (1990) might provide such a way. In this mechanism, the residual (core) field strength is proportional to the lowest frequency to which the neutron star has been spun down in its life (it is not important whether it has been spun back up later). If a neutron star would be spun down by propellor braking to lower spin rates during the onset of mass transfer in a system driven by gravitational radiation and magnetic braking (for example, because the low mass-transfer stage lasts longer) than in a system driven by evolutionary expansion of the companion, then this migh provide the required link between neutron-star magnetic-field strength and companion type.

8.4 OTHER SOURCES

As can be seen in Table 1, the Z/atoll classification encompasses all the brightest persistent LMXB with the exception of Ser X-1 (which is probably an atoll source, Hasinger and van der Klis, priv. comm.), Cyg X-3 (see below), GCX-1 (source near galactic center) and GX 339-4 (a black-hole candidate). Among these $\sim$20 brightest LMXB, no X-ray dip sources occur (Cyg X-3 may be an accretion-disk corona source, which is not the same [White and Mason 1985]; in general ADC sources need to be treated separately in any classification as the fact that we do not see the X-ray source directly strongly modifies the observed characteristics). The absence of dippers among the brightest sources suggests that the structure in the accretion stream that is causing the dips (outer disk rim [White and Mason 1985], deflected stream [King, this volume]) is shadowed in these sources, possibly by the ADC (or absent). There is only one known globular cluster source among the brightest LMXB.

So, as to classify X-ray sources count rates $\gtrsim$100 c/s are required, HK89 could not classify any of the X-ray dip sources as either Z or atoll, and classified only one globular cluster source (4U 1820-30, as atoll source). On the basis of the interpretation given by HK89 for the differences between Z and atoll sources (Section 8.2), one might expect most burst sources to belong to the class of the atoll sources (if burst models predicting an absence of bursts in high accretion rate neutron stars [Joss 1979] are correct – note, however, that GX 17+2 has shown X-ray bursts [Tawara *et al.* 1984]).

Indeed, the burst source in the globular cluster M15 (Van Paradijs *et al.* 1990) and the burst source 4U 1702-42 (Oosterbroek *et al.* 1990) were recently classified as atoll sources in the island and the banana state, respectively, on the basis of their X-ray colour-colour diagram and rapid X-ray variability. Note, that the M15 source has an orbital period of 8.5 hr.

Even if binary evolution causes a correlation between accretion rate and neutron-star magnetic-field strength, with Z sources representing the high, and atoll sources the low niche in this correlation, one might look for peculiar sources that, through some tortuous evolutionary path, came to occupy the two remaining niches (high $\dot{M}$ low B, and *vice versa*). A high $\dot{M}$, low B source would be predicted to show NB/FB QPO, but no HB QPO; for a low $\dot{M}$, high B source the opposite would be expected. As the persistently bright sources are mostly covered, a candidate high $\dot{M}$, low B source should be transient or strongly variable. Cir X-1, which has shown NB/FB-like QPO and hardness-intensity diagram behaviour (Tennant 1988), may fill this ticket. No examples spring to mind of high B, low $\dot{M}$ sources – the way to find one would be to follow a soft transient that displays Z behaviour when it is bright, down its decay. I note that such a source would be a prime candidate for the discovery of millisecond X-ray pulsations ($B{\sim}10^{9-10}$, $\dot{M} \ll L_{Edd}$).

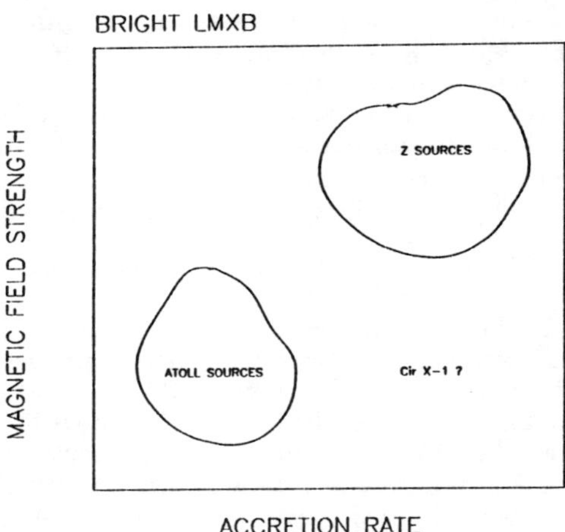

Figure 9. Z, atoll and "other"sources in the $\dot{M}$, B plane.

It is far from clear yet, from empirical considerations as well as from the point of view of binary evolution, how well binary period might be correlated to the Z/atoll distinction. To test this, the unclassified long-period LMXB deserve further attention. The 9.8-day period source X 0921-63 (an ADC source) could be a Z source edge-on. Although it is too weak in X-rays to classify it, it might turn out to be a radio source similar to other Z sources. The 21-hr X-ray dip source X 1624-490 (the "big dipper") is another interesting case: if an orbital period $\gtrsim$5 hrs implies an evolved companion, then apparently such a companion does not always lead to high accretion rates (assuming on the basis of its location that the distance to the source is roughly that to the galactic bulge) or the absence of X-ray dips.

9. References

Alpar, M.A., Shaham, J., 1985, *Nature* **316**, 239.

Basinska, E.M., Lewin, W.H.G., Sztajno, M., Cominsky, L.R., Marshall, F.J., 1984, *Astrophys. J.* **281**, 337.

Belloni, T., Hasinger, G., 1990, *Astron. Astrophys.* **230**, 103.

Bradt, H.V.D., McClintock, J.E., 1983, *Annu. Rev. Astron. Astrophys.* **21**, 13.

Bradt, H.V., *et al.*, 1975, *Astrophys. J.* **197**, 443.

Brainerd, J., Lamb, F.K., 1987, *Astrophys. J. (Letters)* **317**, L33.

Branduardi, G., Kylafis, N.D., Lamb, D.Q., Mason, K.O., 1980, *Astrophys. J.* **235**, L153.

Bussard, R.W., Weisskopf, M.C., Elsner, R.F., Shibazaki, N., 1988, *Astrophys. J.* **327**, 284.

Canizares, C.R., *et al.*, 1975, *Astrophys. J.* **197**, 457.

Charles, P.A., Thorstensen, J.R., Bowyer, S., *et al.*, 1980, *Astrophys. J.* **237**, 154.

Corbet, R.H.D., Smale, A.P., Charles, P.A., Lewin, W.H.G., Menzies, J.W., Naylor, T., Penninx, W., Sztajno, M., Thorstensen, J.R., Trümper, J., Van Paradijs, J., 1989, *Monthly Notices Roy. Astron. Soc.* **239**, 533.

Damen, E., Jansen, F., Penninx, W., Oosterbroek, T., Van Paradijs, J., Lewin, W.H.G., 1989, *Monthly Notices Roy. Astron. Soc.* **237**, 523.

Damen, E., Wijers, R.A.M.J., Van Paradijs, J., Penninx, W., Oosterbroek, T., Lewin, W.H.G., Jansen, F., 1990, *Astron. Astrophys.* **233**, 121.

Dotani, T., Mitsuda, K., Makishima, K., Jones, M.H., 1989, *Proc. Astron. Soc. Japan* **41**, 577.

Evans, W.D., Belian, R.D., Conner, J.P., Strong, I.B., Hiltner, W.A., Kunkel, W.E., 1970, *Astrophys. J.* **162**, L115.

Fortner, B., Lamb, F.K., Miller, G.S., 1989, *Nature* **342**, 775.

Hasinger, G., 1987a, *Astron. Astrophys.* **186**, 153.

Hasinger, G., 1987b, *IAU Symp.* **125**, 333.

Hasinger, G., 1988a, in *Physics of Neutron Stars and Black Holes*, Y. Tanaka (ed.), Tokyo: Universal Academy Press, p. 97.

Hasinger, G., 1988b, *Advances in Space Research* **8(2)**, 377.

Hasinger, G., van der Klis, M., 1989, *Astron. Astrophys.* **225**, 79. [HK89]

Hasinger, G., Langmeier, A., Sztajno, M., Trümper, J., Lewin, W.H.G., White, N.E., 1986, *Nature* **319**, 469.

Hasinger, G., Priedhorsky, W.C., Middleditch, J., 1989, *Astrophys. J.* **337**, 843.

Hasinger, G., van der Klis, M., Ebisawa, K., Dotani, T., Mitsuda, K., 1990, *Astron. Astrophys.* **235**, 131.

Holt, S.S., 1980, in *X-ray Astronomy*, R. Giacconi and G. Setti (eds.), Reidel - Dordrecht, p. 237.

Ilovaisky, S.A., *et al.*, 1980, *Monthly Notices Roy. Astron. Soc.* **191**, 81.

Joss, P.C., 1979, *Comments on Astrophys.* **8**, 109.

Lamb, F.K., 1986, in *The Evolution of Galactic X-ray Binaries, NATO ASI Series C* **167**, 151.

Lamb, F.K., 1989, Proc. 23rd ESLAB Symp. *Two Topics in X-Ray Astronomy, ESA SP-* **296**, 215.

Lamb, F.K., Shibazaki, N., Alpar, M.A., Shaham, J., 1985, *Nature* **317**, 681.

Langmeier, A., Hasinger, G., Trümper, J., 1990, *Astron. Astrophys.* **228**, 89.

Lewin, W.H.G., *et al.*, 1987, *Monthly Notices Roy. Astron. Soc.* **226**, 383.

Lewin, W.H.G., Van Paradijs, J., van der Klis, M., 1988, *Space Sci. Rev.* **46**, 273.

Lewin, W.H.G., Van Paradijs, J., 1986, *Comments on Astrophys.* **11**, 217.

Lewin, W.H.G., Van Paradijs, J., 1985, *Astron. Astrophys.* **149**, L27.

Makishima, K., 1986, *ISAS Research Note* **313**.

Makishima, K., Ishida, M., Ohashi, T., Dotani, T., Inoue, H., Mitsuda, K., Turner, M.J.L., Hoshi, R., 1989, *Proc. Astron. Soc. Japan* **41**, 531.

Markert, T.H., Canizares, C.R., Clark, G.W., Hearn, D.R., Li, F.K., Sprott, G.F., Winkler, P.F., 1977, *Astrophys. J.* **218**, 801.

Mason, K.O., Charles, P.A., White, N.E., Culhane, J.L., Sanford, P.W., Strong, K.T., 1976, *Monthly Notices Roy. Astron. Soc.* **177**, 513.

McHardy, I.M., Lawrence, A., Pye, J.P., Ponds, K.A., 1981, *Monthly Notices Roy. Astron. Soc.* **197**, 893.

Middleditch, J., Priedhorsky, W.C., 1986, *Astrophys. J.* **306**, 230.

Mitsuda, K., 1989, Proc. 23rd ESLAB Symp. *Two Topics in X-ray Astronomy, ESA SP-* **296**, 197.

Mitsuda, K., Dotani, T, 1989, *Proc. Astron. Soc. Japan* **41**, 557.

Mitsuda, K., Dotani, T., Yoshida, A., 1988, in: *Physics of Neutron Stars and Black Holes*, Y. Tanaka (ed.), Universal Academy Press, p. 133.

Mitsuda, K., Inoue, H., Koyama, K., Mkishima, K., Matsuoka, M., Ogawara, Y., Shibazaki, N., Suzuki, K., Tanaka, Y., 1984, *Proc. Astron. Soc. Japan* **36**, 741.

Mitsuda, K., Inoue, H., Nakamura, N., Tanaka, Y., 1989, *Proc. Astron. Soc. Japan* **41**, 97.

Mook, D.E., *et al.*, 1975, *Astrophys. J.* **197**, 425.

Norris, J.P., Hertz, P., Wood, K.S., Vaughan, B.A., Michelson, P.F., Mitsuda, K., Dotani, T., 1990, *Astrophys. J.* **361**, 514.

Oosterbroek, T., Penninx, W., van der Klis, M., Van Paradijs, J., lw, 1990, *Astron. Astrophys.* , in prep.

Ostriker, J.P., 1977, *Annals N.Y. Acad. Sciences* **302**, 229.

Parmar, A.N., White, N.E., 1988, *MEMORIE della Società Astr. Italiana* **59**, 147.

Parsignault, D.R., Grindlay, J.E., 1978, *Astrophys. J.* **225**, 970.

Penninx, W., 1989, Proc. 23rd ESLAB Symp. *Two Topics in X-ray Astronomy, ESA SP-* **296**, 185.

Penninx, W., Hasinger, G., Lewin, W.H.G., van der Klis, M., 1989, *Monthly Notices Roy. Astron. Soc.* **238**, 851.

Penninx, W., Lewin, W.H.G., Mitsuda, K., van der Klis, M., Van Paradijs, J., Zijlstra, A.A., 1990, *Monthly Notices Roy. Astron. Soc.* **243**, 114.

Penninx, W., Lewin, W.H.G., Tan, J., Mitsuda, K., van der Klis, M., Van Paradijs, J., 1991, *Monthly Notices Roy. Astron. Soc.* , in press.

Ponman, T., 1982, *Monthly Notices Roy. Astron. Soc.* **201**, 769.

Ponman, T.J., Cooke, B.A., Stella, L., 1988, *Monthly Notices Roy. Astron. Soc.* **231**, 999.

Ponman, T.J., Foster, A.J., Ross, R.R., 1990, *Monthly Notices Roy. Astron. Soc.* **246**, 287.

Priedhorsky, W., Hasinger, G., Lewin, W.H.G., Middleditch, J., Parmar, A., Stella, L., White, N., 1986, *Astrophys. J.* **306**, L91.

Pylyser, E., Savonije, G.J., 1988a, *Astron. Astrophys.* **191**, 57.

Pylyser, E.H.P., Savonije, G.J., 1988b, *Astron. Astrophys.* **208**, 52.
Schulz, N.S., Hasinger, G., Trümper, J., 1989, *Astron. Astrophys.* **225**, 48.
Schulz, N.S., Wijers, R.A.M.J., 1989, Proc. 23rd ESLAB Symp. *Two Topics in X-Ray Astronomy, ESA SP-* **296**, 601.
Shibazaki, N., Lamb, F.K., 1987, *Astrophys. J.* **318**, 767.
Shibazaki, N., Mitsuda, K., 1984, in: *High Energy Transients in Astrophysics*, S.E. Woosley (ed.) *AIP Conf. Proc.* **115**, 63.
Srinivasan, G., Bhattacharya, D., Muslimov, A.G., Tsygan, A.I., 1990, *Current Science* **59**, 31.
Stella, L., 1988, *MEMORIE della Società Astr. Italiana* **59**, 185.
Stella, L., Chiapetti, L., Ciapi, A.L., Maraschi, L., Tanzi, E.G., Treves, A., 1986, Proc. of 4th *Marcel Grossmann Meeting on General Relativity*, R. Ruffini (ed.), Elsevier Science Publ., p. 861.
Stella, L., Parmar, A.N., White, N.E., 1987a, *Astrophys. J.* **321**, 418.
Stella, L., Priedhorsky, W., White, N.E., 1987b, *Astrophys. J. (Letters)* **312**, L17.
Stella, L., Haberl, F., Lewin, W.H.G., Parmar, A.N., van der Klis, M., Van Paradijs, J., 1987c, *Astrophys. J. (Letters)* **327**, L13.
Stella, L., White, N.E., Priedhorsky, W., 1987d, *Astrophys. J. (Letters)* **315**, L49.
Sternberg, J.R., White, N.E., Barr, P., and Osborne, L., 1986, *The EXOSAT observation log*, ESA/ESTEC, Noordwijk.
Stollman, G.M., Hasinger, G., Lewin, W.H.G., van der Klis, M., Van Paradijs, J., 1987, *Monthly Notices Roy. Astron. Soc.* **227**, 7p.
Sztajno, M., Basinska, E.M., Cominsky, L.R., Marshall, F.J., Lewin, W.H.G., 1983, *Astrophys. J.* **267**, 723.
Taam, R.E., Mészáros, 1987, *Astrophys. J.* **322**, 329.
Tan, J., Lewin, W.H.G., Penninx, W., Van Paradijs, J., van der Klis, M., Mitsuda, K., 1990, in prep.
Tawara, Y., Hirano, T., Kii, T., Matsuoka, M., Murakami, T., 1984, *Proc. Astron. Soc. Japan* **36**, 861.
Tennant, A.F., 1988, *Monthly Notices Roy. Astron. Soc.* **230**, 403.
Turner, M.J.L., Smith, A., Zimmermann, H.U., 1981, *Space Sci. Rev.* **30**, 513.
Vacca, W.D., Sztajno, M., Lewin, W.H.G., Truemper, J., Van Paradijs, J., Smith, A., 1987, *Astron. Astrophys.* **172**, 143.
Van der Klis, M., 1987, Proc. *The Physics of Accretion onto Compact Objects, Lecture Notes in Physics* **266**, 157.
Van der Klis, M., 1988, *Advances in Space Research* **8(2)**, 383.
Van der Klis, M., 1989a, *Annu. Rev. Astron. Astrophys.* **27**, 517.
Van der Klis, M., 1989b, *Timing Neutron Stars*, Ögelman and van den Heuvel (eds.), Kluwer, p. 27.
Van der Klis, M., 1990, Proc. 1990 Vulcano Meeting *Frontier Objects in Astrophysics and Particle Physics*, in press. Preprint Astronomical Inst. Univ. Amsterdam AIAP-1990-014
Van der Klis, M., Hasinger, G., Damen, E., Penninx, W., Van Paradijs, J., Lewin, W.H.G., 1990, *Astrophys. J. (Letters)* **360**, L19.
Van der Klis, M., Hasinger, G., Stella, L., Langmeier, A., 1987b, *Astrophys. J.* **319**, L13.
Van der Klis, M., Jansen, F., Van Paradijs, J., Lewin, W.H.G., Sztajno, M., Trümper, J., 1987a, *Astrophys. J.* **313**, L19.
Van der Klis, M., Jansen, F., Van Paradijs, J., Lewin, W.H.G., van den Heuvel, E.P.J., Trümper, J.E., Sztajno, M., 1985a, *Nature* **316**, 225.

Van der Klis, M., Jansen, F., Van Paradijs, J., Lewin, W.H.G., Trümper, J.E., Sztajno, M., 1985b, *IAU Circ.* **4140**.
Van der Klis, M., Kitamoto, S., Tsunemi, H., Miyamoto, S., 1991, *Monthly Notices Roy. Astron. Soc.* , in press; AIAP preprint 1990-15.
Van der Klis, M., Stella, L., White, N., Jansen, F., Parmar, A.N., 1987c, *Astrophys. J.* **316**, 411.
Van Paradijs, J., Penninx, W., Lewin, W.H.G., 1988, *Monthly Notices Roy. Astron. Soc.* **233**, 437.
Van Paradijs, J., Lewin, W.H.G., 1986, in *The Evolution of Galactic X-ray Binaries, NATO ASI Series C* **167**, 187.
Van Paradijs, J., Dotani, T., Tanaka, Y., Tsuru, T., 1990, *Proc. Astron. Soc. Japan* , in press.
Verbunt, F., Zwaan, C., 1981, *Astron. Astrophys.* **100**, L7.
Vrtilek, S.D., Swank, J.H., Kallman. T.R., 1988, *Astrophys. J.* **326**, 186.
Warwick, R.S., Marshall, N., Fraser, G.W., Watson, M.G., Lawrence, A., *et al.*, 1981, *Monthly Notices Roy. Astron. Soc.* **197**, 865.
Webbink, R.F., Rappaport, S., Savonije, G.J., 1983, *Astrophys. J.* **270**, 678.
White, N.E., Charles, P.A., Thorstensen, J.R., 1980, *Monthly Notices Roy. Astron. Soc.* **193**, 731.
White, N.E., Marshall, F.E., 1984, *Astrophys. J.* **281**, 354.
White, N.E., Mason, K.O., 1985, *Space Sci. Rev.* **40**, 167.
White, N.E., Mason, K.O., Sanford, P.W., Ilovaisky, S.A., Chevalier, C., 1976, *Monthly Notices Roy. Astron. Soc.* **176**, 91.
White, N.E., Mason, K.O., Sanford, P.W., Johnson, H.M., Catura, R.C., 1978, *Astrophys. J.* **220**, 600.
White, N.E., Peacock, A., Hasinger, G., Mason, K.O., Manzo, G., Taylor, B.G., Branduardi-Raymont, G., 1986, *Monthly Notices Roy. Astron. Soc.* **218**, 129.
White, N.E., Stella, L., Parmar, A.N., 1988, *Astrophys. J.* **324**, 363.
Wijers, R.A.M.J., Van Paradijs, J., Lewin, W.H.G., 1987, *Monthly Notices Roy. Astron. Soc.* **228**, 17p.
Willis, A.J., *et al.*, 1980, *Astrophys. J.* **237**, 596.
Wood, K.S., Hertz, P., Norris, J.P., Vaughan, B.A., Michelson, P.F., Mitsuda, K., Dotani, T., 1989, Proc. 23rd ESLAB Symp. *Two Topics in X-Ray Astronomy, ESA SP-* **296**, 689.

4. PULSAR AND X-RAY EMISSION

AN EMPIRICAL THEORY OF PULSAR EMISSION

Joanna M. Rankin†
Physics Department
University of Vermont
Burlington, Vermont 05405 USA

ABSTRACT. A system of profile classification is reviewed which serves as a starting point for studying the emission characteristics of pulsars. Two types or mechanisms of pulsar radiation are identified which combine geometrically to produce five major species of profile. The core emission, which forms a pencil beam of radiation, is apparently produced close to the stellar surface within the entire polar-cap region by low γ particles. The conal emission, which consists of a hollow-conical beam, then seems to be emitted at heights of 10 to 20 stellar radii by currents of high γ particles travelling along some of the most peripheral of the "open" field lines.

1. THEORETICAL AND OBSERVATIONAL APPROACHES

The problem of pulsar emission has presented great challenges for both theorists and observers. In part because of preoccupation with technical difficulties, efforts have tended to fragment into distinct observational and theoretical areas which have not always been complementary.

Some theorists have complained that too much emphasis was being given to emission region models because only they could be directly compared with the observations. Certainly the primary work, they would say, is in building global pulsar models on the solid foundation of basic physics. Curt Michel (1990), in particular, has expressed strong cautions about attempts to "verify" emission region theories which are not grounded on global models. He compares these to "snake-oil cures"—that is, attempts at treatment of a disease which are not grounded immediately on an epidemiology of the disease and ultimately on basic biochemical research.

Observers, however, have had a very different problem: Not even the most credible emission-region theories have been able to make meaningful predictions about specific individual pulsars. While certainly quantitative, their level of detail can only begin to account for the emission characteristics of pulsars as a class. Of course, we observers have no option but to observe specific individual pulsars. This severely limits what can be learned through comparing the observations and theories, because these theories cannot be falsified empirically owing to their lack of specificity.

This latter concern has motivated the analytical work which is described below. In a

†Current Address: Raman Research Institute, C. V. Raman Avenue,
Sadashivanagar, Bangalore 560080 India

J. Ventura and D. Pines (eds.), Neutron Stars: Theory and Observation, 349–362.

number of instances historically, a phenomenological approach has proved a useful means of gaining physical insight in situations which were too complex to approach from theoretical first principles. The work on stellar classification and its physical interpretation by Hertzsprung and Russell represent one such example.

A system of classification of pulsar emission phenomena has been developed which serves both as an overall description or "Empirical Theory" and as a source of insight into how individual pulsars differ from others. In the sections that follow, we shall see that, for radio pulsars, this insight has primarily been of benefit in helping to understand the basic emission geometry of individual stars.

Table 1
Pulsar Classification System

Profile Species	TYPE	%	Emission type	PRIMARY IDENTIFICATION CRITERIA	
				Spectral Evolution	Polarization
Core Single	S_t	34	Core (and conal at high freqs.)	· profile has a SINGLE COMPONENT near 400 MHz and usually develops a CONAL OUTRIDER PAIR above 1400 MHz. S_t profiles thus evolve to triple or even double forms at very high frequencies, given the steeper spectrum of core emission.	· Circular: moderate in degree, often with an antisymmetric change of hand; moderate symmetric (non-reversing) also observed · Linear: nearly complete to unpolarized · PA traverse: unsystematic; does not appear to follow single-vector (R&C) model.
Triple	T	25	Central core component plus conal component pair(s)	· profile consists of both a CENTRAL CORE COMPONENT and one CONAL OUTRIDER PAIR. T profiles tend to evolve into double forms at very high frequencies, given the steeper spectrum of core emission.	· Circular: moderate in degree, often with an antisymmetric change of hand; moderate symmetric (non-reversing) also observed; 50-70% circular is observed in some stars · Linear: moderate to virtually unpolarized; polarization-mode depolarization at wings
Five Component	M	9		· profile consists of both a CENTRAL CORE COMPONENT and TWO CONAL OUTRIDER PAIRS. M profiles tend to evolve into featureless "boxy" forms at very high frequencies, as the components merge and the core emission subsides.	· PA traverse: S-shaped, steep, usually near 180° for M, less for T; closely following the single vector (R&C) model · Corellation: if circular sense change is lh to rh, linear traverse is usually cw (neg.), and vice versa
Conal Single	S_d	15	Conal	· profile has a SINGLE COMPONENT near 400 MHz, which BROADENS AND BIFURCATES progressively at lower frequencies, due to the high-altitude spreading of the conal emission beam. ↕ *closely connected* ↕	· PA traverse: shallow (≤90°), closely following the single vector (R&C) model · Linear: moderate to virtually unpolarized; polarization-mode depolarization at wings · Circular: small and unsystematic ↕ *closely connected* ↕
Conal Double	D	10		· profile has a TWO COMPONENTS virtually throughout spectrum, which slowly broaden at lower frequencies, due to the high-altitude spreading of the conal emission beam.	· PA traverse: S-shaped, ~180°, closely following the single vector (R&C) model · Linear: moderate to virtually unpolarized; polarization-mode depolarization at wings · Circular: small and unsystematic
OneSided Triple	$T\frac{1}{2}$	3	as above	· one-sided "triple" pulsars are similar to triple (T) pulsars pulsars in all respects, except that only one conal outrider in apparent. It is not clear whether the outrider is actually	
Conal Triple	cT	3	conal	· pulsars with cT profiles are closely related to those with M profiles, except that the sight line to crosses the polar-cap region at the periphery of the conal emission beams;	
OneSided Double	$D\frac{1}{2}$	>1	conal	*No pulsars belonging to this species have been identified although there is no* a priori *why they should not if the excitation of the conal emission beam is "patchy"—and*	
Others?	?	1?	?	?	?

2. Pulsar Profiles: Patterns of Frequency Evolution

The system of classification which is summarized in Table 1 considers the morphological characteristics of polarized average profiles with particular attention to their formal evolution with radio frequency as well as certain pulse-sequence properties, mode changing, drifting subpulses and pulse nulling. It has been described in a series of published papers (Rankin 1983*a*, *b*; 1986; Rankin *et al.* 1989) and is the fundamental empirical foundation for this study.

Classification of pulsar characteristics has long been recognized as a potential source of physical insight into the emission process. Inspired by Radhakrishnan and Cooke's (1969; hereafter RC) hollow-cone emission model, Huguenin, *et al.* (1971) first proposed

Pulsar Classification System (cont'd)

Geometry	Beam size (°)	Acceler. Potential B_{12}/P^2	Profile Mode Changes	Sub-pulse Modulation	Null-ing	Age $<\log_{10} t>$ (10^6years) $<\lvert z\rvert>$
Uncertain: the linear angle traverse apparently provides little reliable information about where the sight line crosses the emission beam	$\frac{2.45°}{P^{1/2}\sin\alpha}$ (FWHM)	large, ≥2.5; typically 10	not observed	little; steady, flat fluctuation spectrum	none	young 6.1 160 pc
The sight line crosses the polar-cap region somewhat peripherally —*i.e.*, away from the center of the conal emission beam	• core beam diam. 2.45° • conal beam radii 4.3° and 5.9°, discrete values for M	full range; large to small	yes (always?)	stationary subpulse modulation, associated with the conal components	yes	middle age to old 6.6 210 pc
The sight line crosses the polar-cap region near the magnetic axis—*i.e.*, near the center of the conal emission beam	• both scale as $P^{-1/2}$ and $\sin^{-1}\alpha$	small, generally less than 2.5				old 7.3 ≤400 pc
The sight line crosses the polar-cap region near the periphery of the conal emission beam ↕ *closely connected* ↕	not well known	small, ≤ 2.5; lowest known value 0.16	maybe; difficult to discern	yes, orderly ***drifting*** subpulses		old 7.6 335 pc
The sight line crosses the polar-cap region near the magnetic axis—*i.e.*, near the center of the conal emission beam				little or no drifting; stationary subpulse modulation		old 7.7 260 pc
missing or whether it is simply merged or overlain with the core component.		as with triple				
thus, the the inner conal beam appears as a central conal component.		≤2.5	yes	***drifting*** subpulses	yes	old 7.7 ≤400 pc
there are many indications that the illumination is at least non-uniform.		*presumably very much like double (D) pulsars*				
?	?	?	?	?	?	?

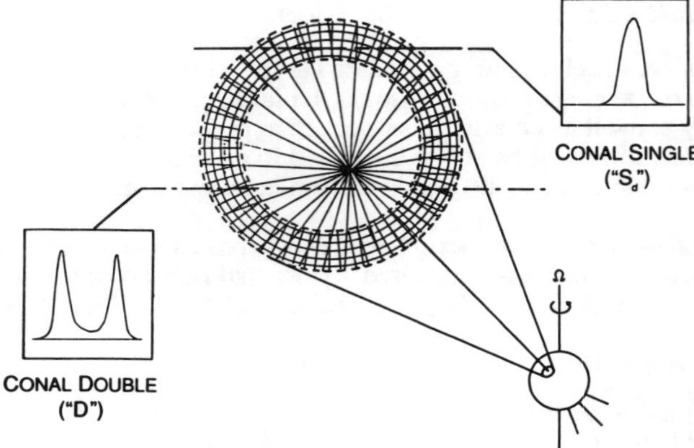

Figure 1—*Schematic diagram showing the geometry of conal beaming. Single or double profiles result depending upon the centrality of the sight line traverse.*

a classification scheme for average profiles, and Backer (1976) then greatly elaborated it. This latest system builds directly upon these early ideas.

On the basis of their profile properties, five major categories of pulsar are delineated, the core-single (**S**$_t$), conal single (**S**$_d$), double (**D**), triple (**T**), and five-component (**M**). A sixth class, the conal triple (**cT**) is also known which is closely related to the **M** stars. These species in turn are found to entail two distinct types or mechanisms of radiation, *core* emission and *conal* emission. Depending upon a) the relative strengths of the core and conal contributions to the overall emission, and b) the orientation of our sight line to the emission beam, one of the several characteristic profiles is observed. The possibilities for an entirely conal geometry are depicted in Figure 1.

One of the species, the five-component (**M**) profile, deserves special mention. More than 15 pulsars have been identified which appear to be members of the **M** profile species. These profiles all have a single, central core component and two pairs of conal outriders. The outer component pairs (I and V, and II and IV, respectively) apparently result from a double-conal emission beam, and the central core component (III) from a smaller core beam within it. This geometry is diagrammed in Figure 2. They emit both core and conal radiation in roughly comparable amounts. Pulsars of the **M** class exhibit the highest degree of profile complexity known. Indeed, no convincing examples of pulsars with four or with more than five components have been identified.

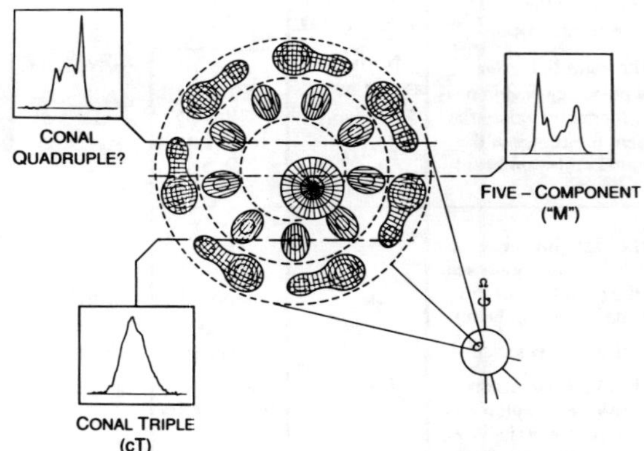

Figure 2—*Schematic diagram showing the geometry of both the core beam and the double-conal beam. Central traverses of the sight line give a five-component (**M**) profile; more oblique traverses provide other, entirely conal possibilities. Several examples of conal triple (c**T**) profiles have been identified, but no conal quadruple (**Q**) stars are known. Furthermore, core emission together with a single conal zone produces a triple (**T**) profile, and core emission by itself a core-single (**S**$_t$) profile.*

Summarizing the material in Table 1 and anticipating some of the conclusions below, let us compare the characteristics of the core and conal radiation. Core emission is primary in some 70% of all pulsars. It is prominent in core-single (S_t), triple (**T**), and five-component (**M**) pulsars, associated with younger pulsars, and emitted at low altitude. Its modulation tends to be steady, and it is frequently marked by sense-reversing (antisymmetric) circular polarization. Finally, it typically has a softer spectrum and thus tends to diminish at high frequency.

Conal emission, by contrast, is primary in about 30% of all pulsars. It is most prominent in conal single (S_d), double (**D**), and conal triple (c**T**) pulsars, associated with the older stars, and emitted at altitudes of 10 to 20 stellar radii. Conal emission exhibits complex, quasi-periodic modulation patterns, characteristic linear polarization, and a harder spectrum.

3. Physical basis of Speciation

The foregoing classification scheme has clear physical consequences, and these circumstances provide clues to the evolutionary significance of the various species. Figure 3 gives the magnetic field as a function of spindown age for some 150 pulsars with well determined classifications; note the tendency of the various species to "clump" in certain regions of the diagram. This tendency is even more marked in Figure 4, where the "acceleration potential", proportional to B_{12}/P^2, is plotted against spindown age. The "linear" form

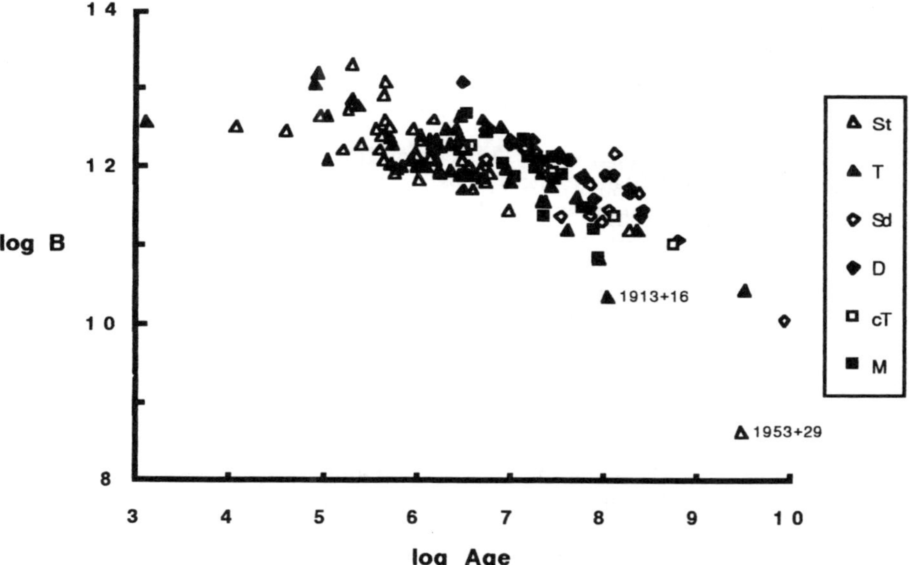

Figure 3.—*Inferred magnetic field versus spindown age for some 150 pulsars. Note the segregation of the S_t pulsars at the left of the diagram and the S_d, **D**, c**T**, and **M** stars at the right. The triple (**T**) pulsars are found throughout the entire range of values.*

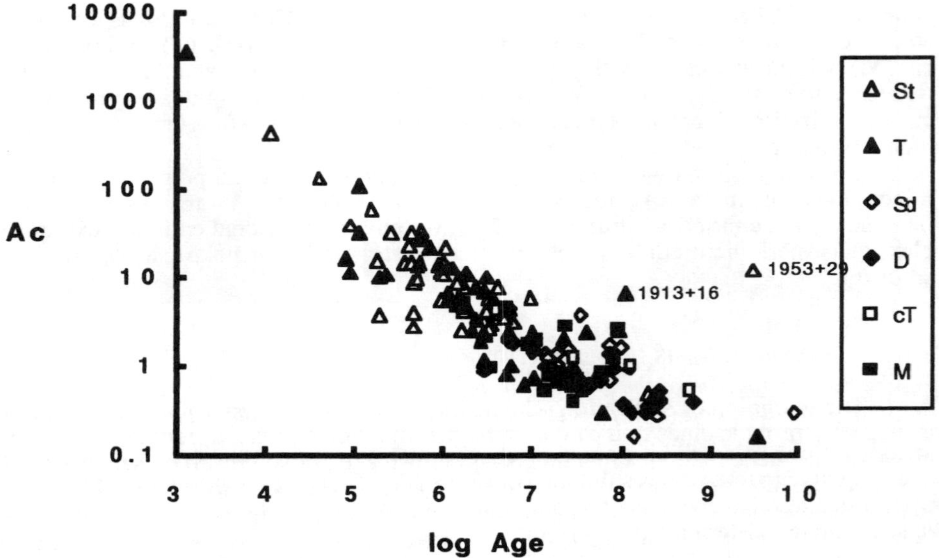

Figure 4.— *"Acceleration parameter" B_{12}/P^2 versus spindown age for some 150 pulsars*

of the plot is expected in that the two quantities are correlated. Note, however, how the $\mathbf{S_t}$ pulsars have B_{12}/P^2 values greater than about 2.5, whereas the $\mathbf{S_d}$, **D**, **cT**, and **M** pulsars almost always fall below 2.5. The triple (**T**) pulsars, by contrast, seem to populate the entire range of B_{12}/P^2 values.

Furthermore, the various species differ significantly in age. The $\mathbf{S_t}$ group is by far the youngest with a mean log age of 6.1, whereas the $\mathbf{S_d}$, **D**, **cT**, and **M** pulsars are all old with mean log age values of about 7.5. The **T** pulsars again fall in between with a mean value of about 6.6. These differences in age are reiterated in Figure 5, which gives the galactic z-height distribution for $\mathbf{S_t}$ pulsars in contrast with the combined group of $\mathbf{S_d}$ and **D** stars. Note that the $\mathbf{S_t}$ pulsars have a scale height $<|z|>$ of only about 160 parsecs, whereas the other group has a value almost twice as great.

4. Interpulsars

Twelve pulsars are now known which exhibit core emission in their main-pulse and /or interpulse profiles. These pulsars are very important because only for these few stars do we have any direct information about the angle, α, between their rotation and magnetic axes. Study of these interpulsars having core emission has resulted in the conclusion that most, but not all, have magnetic axes which are nearly orthogonal to their rotation axes (Rankin 1990*a*).

Six of the stars have core components whose width (i.e., full width at half maximum) can be measured with reasonable accuracy and interpolated to 1 GHz. When fitted against

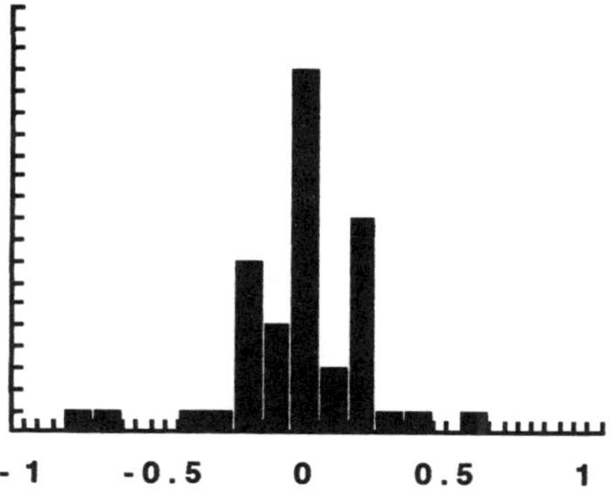

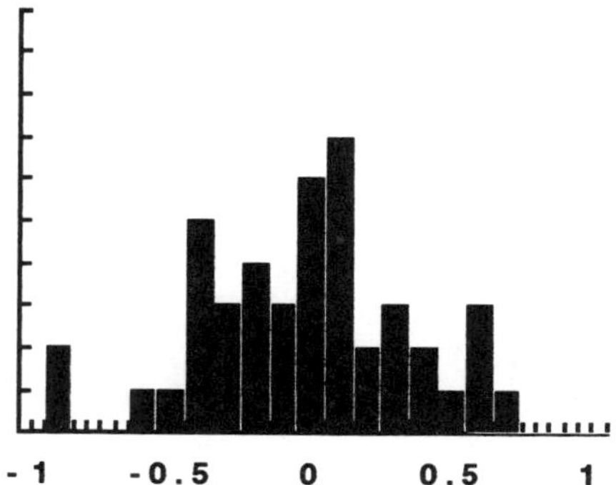

Figure 5.—*Histogram of the z-distribution, in kiloparsecs, of (right) pulsars with core-single (S_t) profiles, and (left) pulsars with conal-single (S_d) and double (D) profiles.*

period, these values exhibit a surprisingly accurate power-law relation. A least-squares fit to these values yields the result that $W = 2.45°/P^{0.50}$.The core components of most other interpulsars also have widths comparable with those given by the above relation, but cannot be so accurately determined.

This core width-period relationship has a very simple interpretation in erms of the magnetic field structure of the pulsar. Assuming a dipolar magnetic field, the angle ρ between the field-line tangent and the magnetic axis can easily be evaluated. Taking the 1-GHz width of the core component as twice this value

$$W_{core} = 2\rho \approx 2.49° \, (r/R)^{1/2}/P^{1/2},$$

where r is the emission height, measured from the center of the star, and R the stellar radius, assumed to be 10 km.

Comparing the empirical relationship for W with the geometrically-derived expression for W_{core} above, both equations have a $P^{-1/2}$ term, and thus it appears that the period dependence of the core width is geometrical in origin. However, the two expressions can only be reconciled numerically if the ratio of the emission height, r, to the stellar radius, R, is about unity. This then suggests that the core emission comes from near the stellar surface.

Most pulsars, of course, do not have interpulses, and in general, we must consider how an emission beam of angular radius ρ about the magnetic axis projects onto the sight line direction. Simple geometrical arguments (see Rankin 1990*a*) suggest that the above relation can be generalized as follows to any pulsar with a core component

$$W_{core} = 2.45° \, P^{-1/2}/\sin\alpha. \qquad (1)$$

Apparently, this simple relationship describes the angular width of core emission beams at 1 GHz. The relationship depends only on the pulsar period, which determines the height of the velocity-of-light cylinder and thus the polar-cap radius, and the angle α which enters in considering how much of the angular rotation cycle the core beam occupies.

5. Geometry of Core Emission

The 1-GHz, half-power width of pulsars with core-single (S_t) profiles are plotted as a function of period in Figure 6. The points corresponding to the interpulsars (filled sym

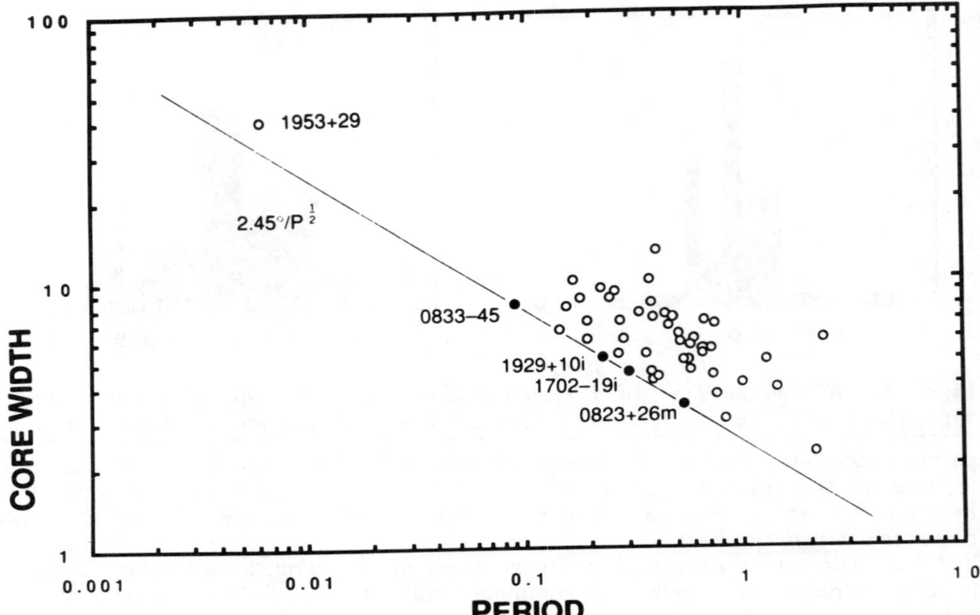

Figure 6.—*The half-power profile width of core-single (S_t) pulsars as a function of period. The symbols of a few prominent pulsars are labelled, and those with interpulses indicated by a filled symbol. (The suffixes "i", "m", and "p" indicate interpulse, main pulse and postcursor, respectively). The indicated curve is 2.45° $P^{-1/2}$ (see text).*

bols) as well as several other stars are labelled, and a line showing the fitted widths of the interpulsars indicated. Note the minimum width defined by the four interpulsars; several other pulsars—most notably the 6-msec pulsar 1953+29—have core widths just in excess of the interpulsar minimum, but none have smaller widths.

The widths of the central core components of pulsars with triple (**T**) and five-component (**M**) profiles are plotted in Figure 7. Pulsars with triple profiles are indicated by triangular symbols and the **M** stars with squares. Filled symbols are used to indicate the interpulsars which, along with several other prominent stars, are explicitly identified. A curve indicating the fitted widths of the interpulsars is again superposed. The core width values of the seven triple interpulsars (filled triangles) all fall accurately on this curve within their measurement or estimation errors. (By contrast, the core width value for 0826-34m, which has an aligned geometry, falls furthermost from the curve.)

We can interpret the core-width/period relationship in Figures 6 and 7 as deriving primarily from the polar-cap geometry, and as such it should be well described by eq. (1) above. Only two factors apparently determine the 1-GHz width of a core component: a) the angular radius of the polar-cap emission region at the stellar surface, which goes as 2.45° $P^{-1/2}$, and b) the angle between the magnetic axis and the rotation axis, α.

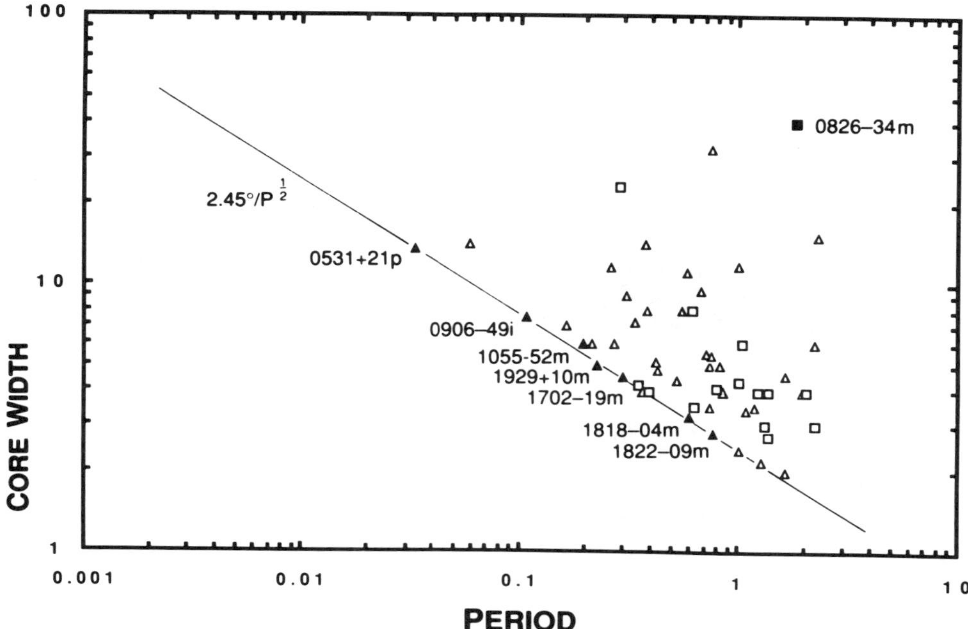

Figure 7.—*The half-power core-component widths of triple (**T**) pulsars (open triangle symbol) and five-component (**M**) pulsars (open square symbol) as a function of period as in Figure 6.*

6. Estimation of α Values

Assuming that the angular width of all core emission is determined by the angular extent of the open field lines at the stellar surface, then eq. (1) can be inverted to estimate the magnetic orientation angle α. Histograms of the α values are then given in Figure 8. a) The angles for the $\mathbf{S_t}$ stars range between some 15° and 90° and exhibit a median value of about 35°. b) The α values for the **T** (filled bars) and **M** (striped bars) stars range more widely than for the $\mathbf{S_t}$ stars. The median value is still about 35°, but the distribution is flatter and a larger fraction are very nearly aligned.

The population of $\mathbf{S_t}$ pulsars has been divided into two subgroups according to their spindown age. The younger group (solid bars) has a mean log age of 5.6, whereas the older one (striped bars) has a mean log spindown age of 6.7. The older group shows a weak tendency toward alignment, which is probably not significant at this level of analysis. The α distributions for the **T** (filled bars) and **M** pulsars are very similar despite differences in mean log age (6.6 and 7.3, respectively). These results are summarized in Figure 9; clearly, we find no support here for the proposition that the magnetic axes of pulsars tend to align with age.

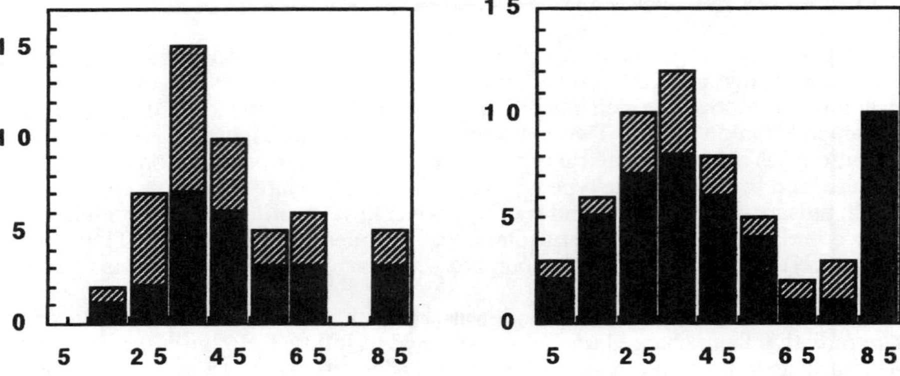

Figure 8.—*Histograms of inferred* α *values in degrees for (left) core-single* (S_t) *pulsars, and (right) triple* (***T***) *and five-component* (***M***) *pulsars. The* S_t *population is divided into two equal groups, a younger group (solid bars) whose mean log spindown age is 5.6, and an older group (striped bars) whose mean log age is 6.7. Similarly, the* ***T*** *(solid bars) and* ***M*** *(striped bars) pulsars have mean log ages of 6.6 and 7.3, respectively.*

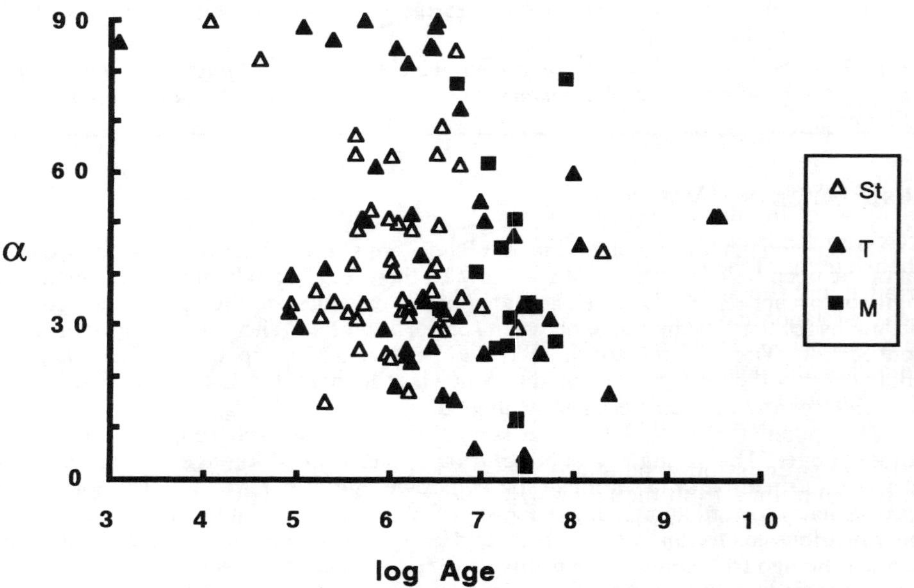

Figure 9.—*Magnetic inclination angle* α *values plotted against spindown age for populations of* S_t*,* ***T****, and* ***M*** *pulsars*

7. CIRCULAR POLARIZATION AND THE FIGURE OF THE POLAR-CAP REGION

A phenomenological study of the circular polarization associated with pulsar emission has shown that most is observed in core components—that is, in core-single (S_t) profiles and in the central components of triple (T) and five-component (M) profiles (Radhakrishnan and Rankin 1990). Two extreme types of circular signature are identified in the observations: a) an antisymmetric type wherein the circular polarization changes sense in mid-pulse, and b) a symmetric type wherein it is predominantly of one sense.

In pulsars with triple (T) and five-component (M) profiles, the antisymmetric type is usually correlated with the sense of rotation of the linear position angle. Transitions from positive (LH) to negative (RH) are found to accompany negative (clockwise) rotations of the position angle, and vice versa.

In the general framework of models in which the radio emission is produced by curvature radiation from charge bunches constrained to follow field lines, the linear polarization is intrinsic to the emission mechanism, and is, furthermore, a purely geometric property independent of the polarity of the magnetic field or of the sign of the charges. The correlation we find then requires that the antisymmetric circular polarization is also a purely geometric property of the emission process. Curvature radiation will have significant net circular polarization if there are gradients in the emissivity over angular scales comparable with the emission cone of a single charge (i.e. $\gamma \sim 1$, where γ is the Lorentz factor of a charge bunch). The observation of significant circular polarization therefore implies that $\gamma \lesssim 20$ for the core emission. Furthermore, no net circular polarization is produced if the emissivity is circularly symmetric about the magnetic dipole axis. The sign of the correlation is consistent with an emission region more extended in longitude than in latitude (referred to the rotation axis).

8. GEOMETRY OF CONAL EMISSION

We saw in eq. (1) above that width of the the 1-GHz core components can be described in terms of the rotation period P and the angle between the rotation and magnetic axes α. α can then be determined in any pulsar with a core component.

On the basis of this knowledge of α, the radius of the conal emission zone can be determined from the width of the conal component pair and the sweep rate of the linear polarization angle. The geometry is depicted in Figure 10.

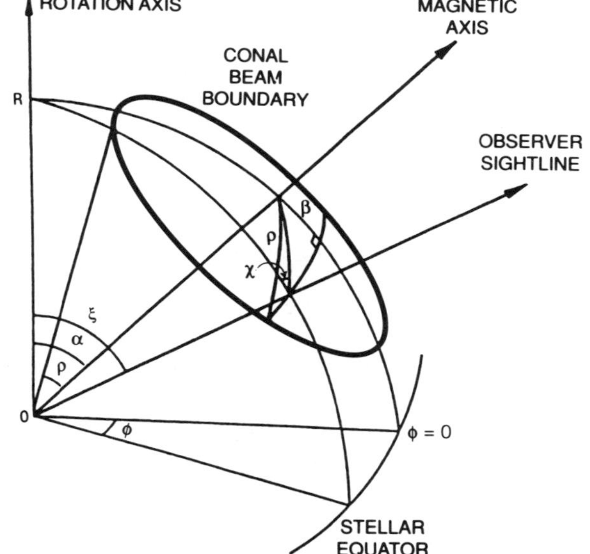

Figure 10.—*Geometry of the conal emission region*

The full 1-GHz width of the conal profile pair (to the outside half-power points), $\Delta\psi$, is

$$\Delta\psi = 4 \sin^{-1}\{\sin(\rho/2+\beta/2)\sin(\rho/2-\beta/2)/\sin\alpha \sin\zeta\}^{1/2}$$

where ρ is the radius of the conal emission zone and β the "impact angle" of the sight line with the magnetic axis. Solving for ρ, we have

$$\rho = \cos^{-1}\{\cos\beta - 2\sin\alpha\sin\zeta\sin^2(\Delta\Psi/4)\} \qquad (2)$$

The angle β can then be calculated via the relation that

$$|d\chi/d\varphi|_{\varphi_o} = \sin\alpha/\sin\beta,$$

where χ is the polarization angle and φ is the longitude.

On the premise that the core emission is produced close to the ($R \simeq 10$ km.) surface of the neutron star, the 1-GHz emission height of the conal radiation can be estimated via the following relation, which assumes a purely dipolar geometry in the emitting region

$$h_{1GHz} = 10 \text{ km.}[2\rho/W_{core}]^2 = 10 \text{ km.}[2\rho/(2.45°P^{-1/2}/\sin\alpha)]^2$$

Using the above relationships, the conal emission geometry of five-component (**M**) pulsars can be calculated with some confidence, owing to the presence of both a core component and two pairs of conal components. The results of the calculations are shown in Figure 11. The 1-GHz angular radii of the inner and outer conal zones exhibit a very regular behavior and indeed also scale as $P^{-1/2}$ as follows:

$$\rho_{inner} = 4.3° P^{-1/2}, \text{ and}$$
$$\rho_{outer} = 5.9° P^{-1/2}.$$

Because both the core and conal width scale as $P^{-1/2}$, the emission height is found to be independent of the period. The curves fitted to the 1-GHz inner and outer conal emission heights then correspond to values of:

h_{inner} = ~120 km. or ~12 stellar radii, and

h_{outer} = ~210 km. or ~21 stellar radii ,

where we have assumed, for the sake of calculation, that the inner and outer conal zones are emitted along the same peripheral field lines at different heights. Clearly, other relationships between the two emission zones are possible.

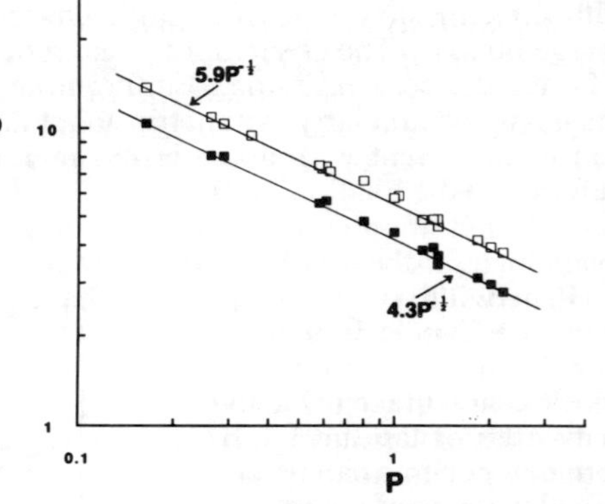

Figure 11.—*Conal radius ρ as a function of period P for the inner and outer conal zones of five-component (M) pulsars. The fitted curves $4.3° P^{-1/2}$ and $5.9° P^{-1/2}$ are indicated.*

In any case, let us here compare these results for **M** stars with other pulsars having conal emission, looking first at those pulsars with core-single ($\mathbf{S_t}$) and triple (**T**) profiles. Exactly the same procedure can be followed: first compute the orientation of the magnetic axis α by evaluating the width of the core component and then calculate ρ, the radius of the conal emission beam using eq. (2) above.

The results for $\mathbf{S_t}$ (open triangles) and **T** (solid triangles) pulsars are given in Figure 12. Notice the overall $P^{-1/2}$ trend in the curve, with significant departures. Note, however, that the $\mathbf{S_t}$ stars generally lie close to the lower curve; it is the **T** stars which have the least orderly behavior.

Finally, a few points representing $\mathbf{S_d}$, **D**, and **cT** pulsars are also plotted on the diagram. None of these pulsars have core components and so the technique described above is not applicable. Instead, α values (as well as the other observational data) have been taken

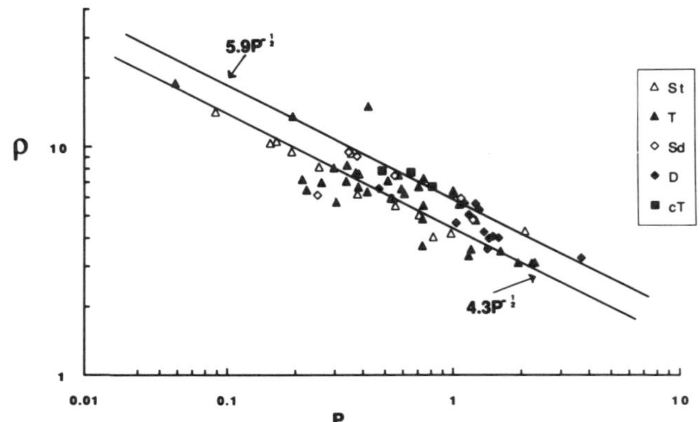

from the analysis of Lyne and Manchester (1988), as it is just for such cone-dominated profiles that their analysis is best motivated. Note that the conal single and double pulsars have ρ values which generally lie between the two solid curves, and that the conal triple values fall essentially on the outer curve as expected.

Figure 12.—*Conal emission radius ρ as a function of period for populations of* S_t*, and* ***T*** *pulsars. These values are computed from the conal analysis as described in the text. ρ values are also shown for a small group of* S_d*,* ***D****, and pulsars c****T*** *stars; here the α values and other observational data are taken from the analysis of Lyne and Manchester (1988). Curves corresponding to the inner and outer conal zones of the* ***M*** *stars, 4.3* $P^{-1/2}$ *and 5.9* $P^{-1/2}$*, are indicated.*

9. Discussion

We reemphasize that pulsars exhibit two modes or mechanisms of pulsar emission. Core emission is found in a majority (60-70%) of mostly younger pulsars, whereas conal emission predominates in the profiles of the remaining older pulsars.

The width of core components have a regular dependence on rotation period and on the magnetic inclination angle α. The magnitude of these widths suggests that the core radiation is emitted close to the neutron-star surface throughout the entire polar-cap region. Furthermore, its circular polarization properties appear compatible only with a radiating population of low γ particles in an asymmetric emission region—*i.e.*, one which is longitudinally extended. The conal emission, by contrast, seems to be emitted at heights of some 100-200 km. This value results from consideration of 1-GHz profiles; the lower frequency emission will then come from larger heights and vice versa.

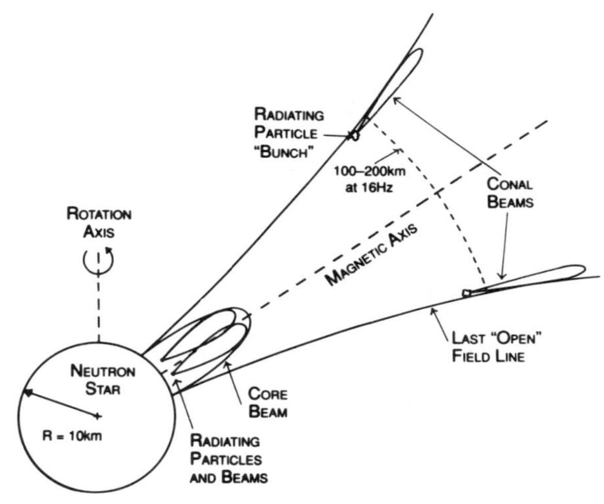

Figure 13.—*Geometry of the emission region*

There is thus still good reason to associate the conal emission with relatively high γ particles, and this is further supported by its linear polarization, which (apart from mode changes) closely follows the single-vector (RC) model.

Figure 13 gives a cartoon of the core and conal emission geometry.

The classification of pulsar profiles appears to give useful information about the geometry of the emission region. The orderly sequence of profile types as well as their generally orderly evolution with frequency suggests, among other things, that the pulsar magnetic field is usually highly dipolar. Similarly, there appears to be little observational evidence for "patchy" or incomplete cones of emission as suggested by Lyne and Manchester (1988).

Finally, the five-component **M** pulsars exhibit the highest degree of profile complexity and appear to express the full potentialities of the pulsar emission process. The inner conal zone has geometrical properties very similar to the type of conal emission exhibited by core-single ($\mathbf{S}_t$) pulsars at high frequency, and the the outer conal zone seems to have much in common with the conal radiation pattern seen in conal triple (c**T**), conal single ($\mathbf{S}_d$) and double (**D**) pulsars. Triple (**T**) pulsars are apparently less regular in their behavior and warrant further detailed study.

ACKNOWLEDGEMENTS

This work was partially supported by grants from the National Science Foundation (AST 89-17722) and the Vermont EPSCoR Program. Arecibo Observatory is operated by Cornell University under contract to the National Science Foundation.

10. REFERENCES

D. C. Backer 1976, "Pulsar Average Wave Forms and Hollow-Cone Beam Models" *Ap. J.* **209**, 895.

Huguenin, G. R., Manchester, R. N., and Taylor, J. H. 1971, "Properties of Pulsars" *Ap. J.* **169**, 97.

Lyne, A. G., and Manchester, R. N. 1988, "The Shape of Pulsar Radio Beams" *M. N. R. A. S.* **234**, 477.

Michel, F. C. 1990, *Proceedings of IAU Colloquium #128 "The Magnetospheric Structure and Emission Mechanisms of Radio Pulsars, Hankins, T. H. et al. eds. (Univ. Zielona Gora Press, Poland; in press).*

Radhakrishnan, V., and Cooke, D. J. 1969, "Magnetic Poles and the Polarization Structure of Pulsar Radiation" *Astrophys. Lett.* **3**, 225.

Radhakrishnan, V., and Rankin, J. M. 1990, "Toward and Empirical Theory of Pulsar Emission. V. On the Circular Polarization in Pulsar Emission" *Ap. J* **352**, 258.

Rankin, J. M. 1983a, "Toward and Empirical Theory of Pulsar Emission. I. Morphological Taxonomy" *Ap. J* **274**, 333.

Rankin, J. M. 1983b, "Toward and Empirical Theory of Pulsar Emission. II. On the Spectral Behavior of Component Width" *Ap. J* **274**, 333.

Rankin, J. M. 1986, "Toward and Empirical Theory of Pulsar Emission. III. Mode Changing, Drifting Subpulses, and Pulse Nulling" *Ap. J* **301**, 901.

Rankin, J. M. 1990*a*, "Toward and Empirical Theory of Pulsar Emission. IV. Geometry of the Core Emission Region" *Ap. J* **352**, 247.

Rankin, J. M., Stinebring, D. R., and Weisberg, J. M. 1989, "Arecibo 21-cm Polarimetry of 55 Pulsars: A Guide to Classification" *Ap. J* **346**, 869.

PLASMA PHYSICS OF ACCRETING NEUTRON STARS

PRANAB GHOSH AND FREDERICK K. LAMB
Departments of Physics and Astronomy
University of Illinois at Urbana-Champaign
1110 W. Green St., Urbana, IL 61801
U.S.A.

ABSTRACT. Accretion of plasma by a neutron star is thought to be the fundamental process that powers many X- and γ-ray sources in the cosmos. Examples of such sources include the X-ray pulsars found in high- and low-mass binary systems and the luminous X-ray stars found in low-mass binary systems in the galactic bulge. These lectures provide an introduction to some of the plasma concepts and phenomena that are needed to understand such sources. The capture of material from the wind or from the atmosphere or envelope of a binary companion star is described and the resulting types of accretion flows discussed. The reasons for the formation of a magnetosphere around the neutron star are explained. The qualitative features of the magnetospheres of accreting neutron stars are then described and compared with the qualitative features of the geomagnetosphere. The conditions for stable flow and for angular and linear momentum conservation are explained in the context of accretion by magnetic neutron stars and applied to obtain rough estimates of the scale of the magnetosphere. Accretion from Keplerian disks is then considered in some detail. The radial structure of geometrically-thin disk flows, the interaction of disk flows with the neutron star magnetosphere, and models of steady accretion from Keplerian disks are described. Accretion torques and the resulting changes in the spin frequencies of rotating neutron stars are considered. The predicted behavior is then compared with observations of accretion-powered pulsars. Magnetospheric processes that may accelerate particles to very high energies, producing GeV and, perhaps, TeV γ-rays are discussed. Finally, the mechanisms that decelerate and eventually stop accreting plasma at the surfaces of strongly magnetic neutron stars are described.

1. Introduction

This article provides an introduction to some of the plasma concepts and phenomena needed to understand accretion by magnetic neutron stars. It does not provide a complete and balanced review of the subject but is instead an informal account of progress on several problems that have interested us in recent years. These range from the flow patterns of gravitationally-captured plasma to the formation and structure of neutron star magnetospheres, from the

J. Ventura and D. Pines (eds.), Neutron Stars: Theory and Observation, 363–444.

magnetohydrodynamic interaction of accreting plasma with the stellar magnetic field to the kinetic theory of the interaction of infalling protons with plasma at the stellar surface. Although there has been a substantial amount of work on these problems, many important questions remain unanswered. One goal of this article is to interest more workers in these problems, with the hope that they may help to solve them.

Accretion of plasma by magnetic neutron stars is still a relatively new topic in astrophysics. Study of this subject began in the late 1960s and was spurred by the discovery in the early 1970s of periodic oscillations in the intensities of the bright X-ray stars Cen X-3 (Giacconi *et al.* 1971; Schreier *et al.* 1972) and Her X-1 (Tananbaum *et al.* 1972), and the interpretation of these objects as strongly magnetic ($B \sim 10^{12}$ G), rotating neutron stars accreting plasma from their binary companions (Pringle and Rees 1972; Davidson and Ostriker 1973; Lamb, Pethick, and Pines 1973). About 30 of these accretion-powered pulsars are now known. Most are in high-mass ($M_{\rm tot} \gtrsim 15\,M_\odot$) binary systems located in the disk of the galaxy, but a few have been found in low-mass ($M_{\rm tot} \lesssim 3\,M_\odot$) systems located in the galactic bulge. For a recent review of accretion-powered pulsar observations, see Nagase (1989).

Research on plasma accretion by magnetic neutron stars received new impetus with the discovery in 1985 of several types of quasi-periodic oscillations (QPOs) in the X-ray intensities of many of the luminous X-ray sources in the galactic bulge (van der Klis *et al.* 1985; Hasinger *et al.* 1986; Middleditch and Priedhorsky 1986) and the suggestion that the QPOs of one type, the so-called horizontal branch oscillations (HBOs), are produced by interaction of a small, rapidly-rotating magnetosphere with density and magnetic field inhomogeneities in the Keplerian accretion disk that feeds plasma to the neutron star (Alpar and Shaham 1985; Lamb *et al.* 1985; Shibazaki and Lamb 1987). For recent reviews of QPO observations, see Lewin, van Paradijs, and van der Klis (1988), Stella (1988), and van der Klis (1989, 1991). For discussions of QPO models, see Lamb (1988b, 1989b, 1991).

The different types of galactic binary systems that are thought to contain accreting neutron stars are listed in Table 1.1, together with the kinds of X-ray variability expected in each case. At least some γ-ray burst sources are also thought to be accreting magnetic neutron stars (see Lamb 1988a; Murakami *et al.* 1988; Fenimore *et al.* 1988).

As Ögelman (1989) and Lamb (1989a) have discussed, the evidence that accretion-powered pulsars are strongly magnetic accreting neutron stars is compelling. The stability of the periodic X-ray intensity oscillations rules out the possibility that these sources are black holes, while the high inferred spin rates ($\sim$1–10 Hz) and luminosities ($\sim 10^{37}$–10^{38} ergs s^{-1}) of many are inconsistent with degenerate dwarf models, which have maximum spin rates $\sim$0.1 Hz and maximum X-ray luminosities $\sim 4 \times 10^{36}$ ergs s^{-1}(Kylafis and Lamb 1979). Furthermore, the changes in stellar spin rate inferred from observed changes in the frequencies of the X-ray oscillations agree quantitatively with models of the torques produced by accretion onto neutron stars with dipole magnetic fields $\sim 10^{11}$–10^{13} G, but are several orders of magnitude larger than the changes in spin rate that would be expected for degenerate dwarfs with the same luminosities (Lamb, Pethick, and Pines 1973; see also Ghosh and Lamb 1979b and references therein). Finally, the observed X-ray pulse shapes and spectra agree qualitatively with magnetic neutron star models and indicate surface field strengths as high as 10^{12}–10^{13} G (Kirk and Trümper 1983; Mészáros 1982, 1986). At least half a dozen accretion-powered

TABLE 1.1

Examples of Accreting Neutron Stars

Galactic Location	Mass of System	Magnetic Field Strength (G)	Type of X-ray Intensity Variation Expected[a]
Disk	High	10^{11}–10^{13}	Periodic and quasi-periodic oscillations
Bulge	Low	10^{11}–10^{12}	Periodic and quasi-periodic oscillations
Bulge	Low	10^{8}–10^{9}	Quasi-periodic oscillations and bursts
Bulge	Low	$\lesssim 10^{7}$	Small-amplitude variations, flares, and bursts

[a]Based on current ideas of how the various types of X-ray intensity variations are produced. The quasi-periodic oscillations referred to are the so-called horizontal branch oscillations (HBOs) discussed by Lamb (1988b, 1989b, 1991). For the purpose of this table, the HBOs are assumed to be a magnetospheric phenomenon. The X-ray bursts referred to are the so-called Type I bursts, which are thought to be caused by thermonuclear flashes in matter that has accreted onto the surface of the star (see Lewin and Joss 1983; Taam 1985).

pulsars have now been observed to show X-ray spectral features that appear to be cyclotron scattering lines (Trümper *et al.* 1978; Wheaton *et al.* 1979; Rose *et al.* 1979; Voges *et al.* 1982; Clark *et al.* 1990; Mihara *et al.* 1990; Tanaka 1991). If they are, the indicated surface magnetic field strengths are $\sim 10^{12}$ G.

The evidence that the luminous QPO sources are also accreting neutron stars is strong, although not as compelling as the evidence that accretion-powered pulsars are neutron stars. The arguments that these presumed neutron stars have significant magnetospheres is indirect, but highly suggestive. At present, the most promising model for the HBOs is the beat-frequency modulated accretion (BFMA) model (Alpar and Shaham 1985; Lamb *et al.* 1985; Shibazaki and Lamb 1987). In this model, interaction of the neutron star magnetosphere with the inner edge of the accretion disk causes the accretion rate and hence the stellar luminosity to oscillate at the beat frequency given by the difference between the orbital frequency of inhomogeneities in the inner disk and the rotation frequency of the neutron star. The BFMA model gives oscillation frequencies that agree with observed HBO frequencies for neutron star spin frequencies $\sim$100–1000 Hz and surface magnetic field strengths $\sim 10^9$ G, values expected on the basis of prior evolutionary arguments which suggested that the luminous low-mass X-ray binaries are progenitors of the millisecond rotation-powered pulsars (Alpar and Shaham 1985; see van den Heuvel 1991). Magnetic field strengths of this order are also consistent with recent work which suggests that the magnetic dipole moment of a neutron star does not decay below $\sim 10^8$–10^9 G on the relevant evolutionary time scales (Kulkarni 1986;

van den Heuvel, van Paradijs, and Taam 1986; Bailes 1989). The BFMA model appears even more appealing as an explanation for the HBOs following the recent discovery of quasi-periodic oscillations in the X-ray intensity of the 7 s accretion-powered pulsar EXO 2030+375 (Angelini, Stella, and Parmar 1989) at a frequency and with a strength which indicate that it is probably caused by modulation of the accretion rate at the beat frequency.

The remainder of this article is organized as follows. In §2 we describe the flow of the accreting plasma when it is well outside the magnetosphere of the neutron star. The interaction between the accreting plasma and the neutron star magnetic field and the structures of the quite different magnetospheres that form in different types of accretion flows are discussed in §3. In §4, several aspects of accretion from a Keplerian disk are considered in more detail. Disk accretion is emphasized in this article because observations suggest that most accretion-powered pulsars and all the luminous QPO sources are fed by Keplerian or near-Keplerian plasma flows. §5 describes how studies of the spin rates of accreting neutron star X-ray sources can provide information on the structure of the magnetosphere and the accretion flows in these sources. (The theory of quasi-periodic intensity oscillations in accreting neutron star X-ray sources, including the role of magnetospheric plasma phenomena, is discussed in a companion article; see Lamb 1991). In §6 we describe magnetospheric processes that may accelerate particles to high energies, producing GeV and, perhaps, TeV γ-rays. Finally, in §7 we discuss the deceleration of accreting plasma near the surface of a strongly magnetic neutron star. We do not discuss the microscopic processes by which accreting neutron stars emit X-rays, nor do we review the very large body of relevant observations. Reviews of these topics may be found in the articles by Giacconi (1975), Mészáros (1982, 1986), White (1982), Bradt and McClintock (1983), Kirk and Trümper (1983), Joss and Rappaport (1984), and Lewin, van Paradijs, and van der Klis (1988).

2. Exterior Plasma Flows

The flow pattern of accreting plasma near the neutron star magnetosphere is determined in part by the accretion flow far from the star, where the effect of the stellar magnetic field is negligible. We refer to the accretion flow in this region as the *exterior flow*. In the present section we describe mechanisms by which a neutron star can capture plasma from its binary companion, and discuss the basic types of exterior flows that can result (see also Treves, Maraschi, and Abramowicz 1988). It will be helpful, in this discussion, to have in mind the most important macroscopic length scales in binary systems containing magnetic neutron stars. These are listed in Table 2.1, along with their typical values.

In high-mass binary systems, the radius R of the neutron star is typically much smaller than the radius r_m of the magnetosphere, which in turn is much smaller than the accretion capture radius r_a. Moreover, the accretion capture radius is typically much smaller than binary separation a. Hence, the interaction of the accreting plasma with the neutron star magnetosphere can be treated, to a first approximation, separately from processes by which plasma is lost from the companion star and captured by the neutron star. Similarly, the interaction of the accreting plasma with the neutron star surface can be treated, to a first approximation, separately from the interaction with the magnetosphere. The

TABLE 2.1

Length Scales in Accreting Neutron Star Binaries[a]

Length	Symbol	Typical value (cm)
Binary separation	a	$\sim 10^{10}$–10^{12}
Accretion capture radius	r_a	$\sim 10^{10}$
Magnetospheric radius	r_m	$\sim 10^{7}$–10^{9}
Neutron star radius	R	$\approx 10^{6}$

[a]The value of the accretion capture radius quoted is for capture from a wind with a velocity $v_w = 10^8\ \mathrm{cm\,s^{-1}}$. The range of magnetospheric radii quoted are for mass accretion rates $\dot{M}$ in the range 10^{17}–$10^{18}\ \mathrm{g\,s^{-1}}$, a neutron star mass $M = 1\,M_\odot$, and a surface dipole magnetic field component in the range 10^9–10^{13} G.

thermal state of plasma in the exterior flow can, however, be strongly affected by the radiation produced near the neutron star surface.

In low-mass binary systems, on the other hand, the neutron star magnetic field may ometimes be so weak that the radius r_m of the magnetosphere is not much larger than the radius R of the neutron star. In this case, the interaction of accreting plasma with the magnetosphere cannot be accurately treated separately from the interaction with the neutron star surface. The radius of the magnetosphere is, however, typically much smaller than the outer radius of the exterior Keplerian disk flow. Thus, the interaction of accreting plasma with the magnetosphere typically can be treated separately from plasma transfer and capture processes.

The most luminous accreting neutron stars have total X-ray luminosities $\sim 10^{35}$–$10^{38}\ \mathrm{ergs\,s^{-1}}$. This X-ray emission is powered largely by release of gravitational binding energy as plasma accretes to the surface of the star. The energy released by adding a unit mass to the star is just the chemical potential ϕ_s of the star. Thus, if the mass flux to the stellar surface is $\dot{M}$, the rate of energy release (the so-called accretion luminosity) is

$$L = \dot{M}\phi_s\,. \tag{2.1}$$

The energy released as a star adjusts to the addition of plasma is negligible compared to the gravitational binding energy, for stars more than a few hours old. The energy released in nuclear reactions in the surface layers of a neutron star is also small compared to the gravitational binding energy, when averaged over time. If these processes are neglected, the chemical potential ϕ_s of the star is equal to the gravitational potential $\phi_g(r)$ evaluated at the radius of the stellar surface. For current equations of state, general relativistic corrections to the Newtonian expression for ϕ_g can be as large as 10–20%. If these are also neglected, equation (2.1) becomes

$$L \approx \dot{M}\,GM/R = 1.33\times 10^{37}\,\dot{M}_{17}\,(M/M_\odot)\,R_6^{-1}\ \ \mathrm{ergs\,s^{-1}}\,, \tag{2.2}$$

where M is the mass of the neutron star and R is its radius. Here and below, numerical subscripts on quantities indicate that they are to be evaluated in terms of the corresponding power-of-ten of the cgs unit. Thus, for example, in equation (2.2), $\dot{M}_{17} \equiv \dot{M}/10^{17}\,\mathrm{g\,s^{-1}}$ and $R_6 \equiv R/10^6$ cm.

The luminosities of the most luminous accreting neutron stars imply mass accretion rates $\sim 10^{-11}$–$10^{-8}\,M_\odot\,\mathrm{yr}^{-1}$. For a neutron star with an extensive magnetosphere, the energy released near the stellar surface completely dominates that released near the magnetospheric boundary, since $\phi_g(r) \propto 1/r$. Near the star, the kinetic energy acquired by the accreting plasma in falling into the deep gravitational potential is converted first into heat and then into radiation (very little is conducted into the star, even if it is cold). The luminosity of the star in the 0.1–100 keV X-ray band depends on the efficiency with which the kinetic energy of the infall is converted into X-rays, but cannot exceed the accretion luminosity. In practice, most of the accretion luminosity emerges as X-radiation, except perhaps in stars that are completely swamped by accreting plasma.

2.1 CAPTURE OF PLASMA

Only neutron stars in close binary systems with a companion star of mass $\gtrsim 15\,M_\odot$ or $\lesssim 2\,M_\odot$ are expected to be observed as X-ray stars, since only companion stars with masses in these ranges provide plasma to the neutron star at sufficient rates to power luminous X-ray emission for an appreciable length of time (see Taam 1984; van den Heuvel 1989, 1991). Companion stars in the upper mass range transfer plasma to the neutron star via a radiation-pressure-driven stellar wind or by having their atmosphere overflow the gravitational potential well confining the star. Under some conditions, massive companions may transfer plasma by both mechanisms simultaneously. Companion stars in the lower mass range transfer plasma by envelope overflow or, possibly, via a self-excited wind, if their outermost layers are sufficiently heated by absorbing soft X-rays coming from the neutron star. The atmosphere or envelope of the companion star may overflow its gravitational potential well if the star expands as a result of nuclear evolution or mass loss. In short-period systems, mass transfer is driven by loss of orbital angular momentum via gravitational radiation or mass loss. These possible mechanisms of mass transfer are summarized in Table 2.2.

2.1.1 *Capture from a wind.* When the companion star loses plasma via a wind, some of it will be deflected and focused by the gravity the neutron star, and then compressionally heated, cooled, and captured. The velocity of the radiatively-driven wind from a massive star is expected to be $\sim 10^8\ \mathrm{cm\,s^{-1}}$ at the orbit of the neutron star (see Waters and van Kerkwijk 1989). Moreover, the orbital velocity of the neutron star is typically $\gtrsim 10^7\ \mathrm{cm\,s^{-1}}$. Thus, the velocity of the neutron star relative to the wind plasma is expected to be hypersonic.

The traditional picture of accretion from a homogeneous medium onto a star moving at a hypersonic velocity v_0 (Hoyle and Lyttleton 1939) assumes that fluid elements of the accreting plasma follow Keplerian orbits. All fluid elements with the same impact parameter then collide at the same point behind the star, where they share their momentum. If it is further assumed that fluid elements with an impact parameter less than a certain value are accreted, whereas those with larger impact parameters are not, the resulting mass accretion rate is

$$\dot{M}_0 \equiv \rho_0 v_0 \pi r_a^2 \qquad (2.3)$$

TABLE 2.2

Mass Transfer Mechanisms in Close Binary Systems

Mass of System	Source of Accreting Plasma	Mean Angular Momentum	Peak Angular Momentum
High	Radiation-driven wind	Low	Low to high
	Overflowing atmosphere	High	High
Low	Self-excited wind	Low	Low to high
	Overflowing envelope	High	High

where ρ_0 is the mass density far from the neutron star and

$$r_a = \frac{\xi 2GM}{v_0^2} \approx 10^{10} \left(\frac{M}{M_\odot}\right) \left(\frac{10^8 \,\mathrm{cm\,s^{-1}}}{v_0}\right)^2 \tag{2.4}$$

is the so-called accretion capture radius. Here ξ is a dimensionless number that depends on one's assumptions about the gas dynamics of the capture process but is presumed to be of order unity.

For example, if one assumes that the separatrix between the fluid elements that are captured and those that are not occurs where the inward velocity falls below the escape velocity (Hoyle and Lyttleton 1939), $\xi = 1$. Other assumptions lead to other values for ξ. Bondi and Hoyle (1944) have suggested that this picture can be extended to subsonic velocities by using the interpolation formula

$$r_a = \frac{\xi 2GM}{(v_0^2 + c_s^2)}, \tag{2.5}$$

where c_s is the sound speed far from the star and ξ is between 1 and 2.

The change in the spin rate of the neutron star caused by accretion of plasma can be determined from the change in the mass and angular momentum of the star. Quite generally, the time rate of change of the angular momentum $\mathbf{J}_s$ of the star is given by

$$\dot{\mathbf{J}}_s = \mathbf{N}_s\,, \tag{2.6}$$

where

$$\mathbf{N}_s \equiv -\int_S (\mathbf{r} \times \mathbf{\Pi}) \cdot \hat{\mathbf{n}}\, dS \tag{2.7}$$

is the torque on the star, S is a surface enclosing the star, $\mathbf{\Pi}$ is the momentum flux density tensor, and $\hat{\mathbf{n}}$ is a unit vector normal to the surface and oriented outward.

From the assumed symmetry and steadiness of the hypersonic accretion flow just described, it is clear that the plasma accreted by the star can add no angular momentum. Hence the angular momentum $\mathbf{J}_s$ of the star does not change with time, that is

$$\dot{\mathbf{J}}_s = 0\,. \tag{2.8}$$

Even so, the ***spin rate*** of the star usually changes, since the moment of inertia of the star generally changes as mass is added. The change in spin rate can be either positive or negative, depending on the structure of the star (see §5.2).

Despite the important simplifying assumptions involved, this picture has often been applied to capture of plasma by a neutron star immersed in a stellar wind, with

$$v_0 = (v_w^2 + v_{\rm orb}^2)^{1/2} \tag{2.9}$$

in terms of the wind velocity v_w at the orbit of the neutron star and the orbital velocity $v_{\rm orb}$ of the neutron star. The results can be very misleading, for several reasons:

- Capture depends in a fundamental way on the physical processes that occur in the region where the streamlines intersect, since unless energy is lost, the plasma remains unbound. Yet these physical processes (formation of a shock wave, cooling by radiation, etc.) are not included in the analysis.
- A stellar wind is not a homogeneous medium. At a minimum there are density and velocity gradients on length scales comparable to the distance from the mass-losing star. In fact, there is strong evidence for large velocity and density gradients on much smaller length scales, as discussed below.
- The structure of the wind is likely to be affected by tidal distortion of the mass-losing star, by focusing of the wind by the gravitational field of the neutron star, and by ionization, heating, and cooling of the wind plasma by X-radiation from the neutron star.
- The dimensions of the accreting object (star or magnetosphere) and the boundary conditions imposed on the flow at the central object obviously can affect the capture process, but are not considered in the analysis.

In the last decade a so, a variety of efforts have been made to address some of these questions, first with analytical methods and, more recently, with numerical simulations. Although progress has been made, important issues remain unresolved. Here we briefly summarize the current situation (for more detailed discussions, see Ho 1988, Livio 1991, and Matsuda *et al.* 1991).

Density and velocity gradients in the wind will cause the capture cross section to be deformed, creating circulation in the flow and leading to accretion of angular momentum by the star. Early efforts to treat such effects analytically (Illarionov and Sunyaev 1975; Shapiro and Lightman 1976; see also Gething 1951 and Dodd and McCrea 1952) considered a constant velocity wind with a small density variation of the form

$$\rho(x) = (1 + \epsilon_\rho x/r_a)\rho_0\,, \tag{2.10}$$

where the coordinate x increases in a direction normal to the velocity. The estimates (2.3)–(2.5) for the mass flux and accretion capture radius were found to be correct, within the accuracy of the calculation. In estimating the angular momentum flux, the contribution to $\mathbf{\Pi}$ of stresses other than the material stress and the deformation of the accretion capture cross section caused by the density gradient were both neglected. The result obtained for the angular momentum captured per unit mass captured,

$$\ell \equiv N_s/\dot{M}\,, \tag{2.11}$$

was

$$\ell_w = \epsilon_\rho r_a v_0\,. \tag{2.12}$$

Shapiro and Lightman estimated that for a constant velocity, homogeneous stellar wind, the size of the density gradient at the orbit of the neutron star is

$$\epsilon_\rho \approx (v_{\rm orb}/v_w)(r_a/a)\,, \tag{2.13}$$

where a is the binary separation. It is interesting to parameterize the angular momentum transferred to the neutron star per unit mass captured from the wind using the expression

$$\boldsymbol{\ell}_w = \eta_w r_a^2\,\boldsymbol{\Omega}_{\rm orb}\,, \tag{2.14}$$

where $\boldsymbol{\Omega}_{\rm orb}$ is the orbital angular velocity of the system and η_w is a dimensionless number. The results found by Shapiro and Lightman then correspond to η_w positive and of order unity.

Subsequently, Davies and Pringle (1980) analyzed capture of material from a wind by a point mass in a simplified 2D geometry using the Hoyle-Lyttleton approach. They concluded that, to first order in r_a/a, no angular momentum is captured. Wang (1981) argued that the zero value of ℓ_w found by Davies and Pringle is a result of their assumption of flow along an accretion line to a point mass. He therefore concluded that if accretion occurs in a column rather than along a line, finite angular momentum capture is possible. Soker and Livio (1984) then used a Hoyle-Lyttleton-type approach to study accretion from a hypersonic flow in 3D. They found that the accreting plasma is confined to a thin "accretion layer" and that an upstream density gradient produces a finite value of ℓ_w, due to the finite pressure in the "accretion layer", but that the specific angular momentum is small compared to ℓ_{w0}. Ho (1988) has argued that the Hoyle-Lyttleton approach gives a reasonable estimate of the mass capture rate but that it is unrealistic to expect that it can give accurate estimates of quantities, such as the angular momentum capture rate, that are sensitive to the detailed structure of the accretion flow.

Wang (1981), Anzer, Börner, and Monaghan (1987), and Ho (1988) have recalculated the angular momentum capture rate when there are density and velocity gradients in the upstream flow and only material stresses are included in the integral (2.7). For capture from a spherically symmetric wind leaving the companion star, the expected sizes of the density and velocity gradients are

$$\epsilon_\rho = (v_{\rm orb}/v_w)\,(r_a/a)\,(1+q) \tag{2.15}$$

and

$$\epsilon_v = (v_{\rm orb}/v_w)\,(r_a/a)\,(1-q)\,, \tag{2.16}$$

where $q \equiv (d\ln v_w/d\ln r)$ is to be evaluated at the orbit of the neutron star. For a wind that is still accelerating near the neutron star, $q > 0$, indicating that the acceleration increases ϵ_ρ but decreases ϵ_v. Of course, $q = 0$ for a wind that is coasting. These authors find that when the effect on the angular momentum capture rate of the deformation of the capture cross section is included, ℓ_w is similar in magnitude ***but of opposite sign*** to the estimate made by Shapiro and Lightman (1976), that is, η_w in equation (2.14) is of order unity, but negative. Ho (1988) has argued that pressure stresses may make a contribution to ℓ_w that is of the same order as the material stress, for the usual choice of the surface S in the integral (2.7), but he found it difficult to estimate the sign of the pressure stress contribution. Under some conditions, viscous shear and magnetic stresses may also be important in equation (2.7). Thus, Ho has concluded that the sign as well as the magnitude of η_w may depend on the detailed properties of the flow at the capture radius.

In recent years, several groups have carried out numerical studies of mass capture from a wind. On the basis of their numerical experiments, Hunt (1971) and Shima *et al.* (1985) concluded that the estimates (2.3)–(2.5) for the mass flux and accretion capture radius are probably correct to within a factor of order unity, at least if the upstream flow is uniform and the flow pattern is steady. Livio *et al.* (1986a,b) and Soker *et al.* (1986) studied uniform and nonuniform upstream flows using a 3D particle-in-cell code. From a series of simulations, they concluded that the the mass flux and accretion capture radius are close to the values given by equations (2.3) and (2.5), not only when the upstream flow is uniform, but also when there is a density or pressure gradient.

Livio *et al.* (1986a,b) and Soker *et al.* (1986) also computed the angular momentum captured by the central object when the upstream flow has a density or pressure gradient. They found that the central object accretes some angular momentum, but concluded that most of it is due to fluid from the upstream side accreting directly onto the central object. As a result, the angular momentum accreted is only a few percent of the angular momentum flowing through the circular Bondi-Hoyle cross-section, implying a value for $\ell_w \ll \epsilon_\rho r_a v_0$.

Using a higher-resolution 2D smoothed-particle code, Anzer, Börner, and Monaghan (1987) studied the effect of an upstream velocity gradient on the mass capture process and concluded that such a gradient causes accretion of some angular momentum. However, their flows were unsteady. Matsuda, Inoue, and Sawada (1987) studied mass capture from a thermally-driven wind using a higher resolution 2D code and found that the flow pattern did not approach a steady state in some cases. Instead, the shock cone oscillated from side to side, producing strong circulation that reversed quasi-periodically. Taam and Fryxell (1988) and Fryxell and Taam (1988) used a 2D, piecewise-parabolic method with an upstream density gradient and found quasi-periodic reversals in the circulation of the flow, accompanied by spikes in the mass accretion rate at the times when the circulation was small.

Ho (1988) suggested that the oscillating behavior found in the 2D calculations might be a consequence of the artificially large interaction between the passing wind and the material circulating around the star that is inherent in the 2D approximation. The absence of oscillations in the 3D calculations could thus be either because the large effective viscosity produced by the necessarily relatively coarse zoning of the 3D calculations suppresses this phenomenon, or because the phenomenon is an artifact of the 2D approximation. In an effort to determine whether the reversals of circulation found in the 2D simulations are an artifact or genuine, Sawada *et al.* (1989) carried out similar 2D and 3D simulations. They were able to follow the time development of the 2D simulations but not the 3D simulations, due to the very large computational resources that would have been required. Sawada *et al.* therefore adopted a local time-stepping technique, which assumes the existence of a steady flow, in order to obtain 3D flow solutions. Sawada *et al.* found that the mean specific angular momentum of the accreted plasma was considerably less in the 3D simulations than in the corresponding 2D simulations. In some cases, their 2D simulations did develop flow oscillations.

The most detailed simulations to date are those reported by Blondin *et al.* (1990) and Matsuda *et al.* (1991). Blondin *et al.* simulated the gas flow in the orbital plane of a high mass X-ray binary system in 2D using an explicit, Eulerian version of the piecewise parabolic method. They systematically investigated the effects of the gravitational field of the central object and X-ray ionization and heating on the large-scale structure of the wind, using a spherical grid centered on the mass-losing star. By limiting the spatial resolution near the compact object, it was possible to follow the large-scale properties of the flow for many flow times across the binary separation. Twelve zones were used across the accretion capture radius and the region immediately surrounding the accreting object was modeled by setting the normal velocity inside a central region was to zero at each time step. In the absence of X-radiation, Blondin *et al.* found that the shock cone swings back and forth quasi-periodically. They interpreted this "flip-flop" behavior as a consequence of the asymmetry of the upstream flow.

When X-ray ionization and heating and radiative cooling were included, Blondin *et al.* found that the bow shock oscillated back and forth with a period corresponding to $\sim$10 flow times across the binary separation. Inside the bow shock, a disk flow developed, in which an appreciable quantity of plasma circulated around the compact object at Keplerian or near-Keplerian speeds. The disk and wake changed quasi-periodically from a quasi-steady state in which the circulation of the disk flow was in the same sense as the binary orbital motion to a state in which the circulation was in the opposite sense. The time-averaged mass capture rate was only $\sim$30% of the rate predicted by equations (2.3)–(2.5), although the instantaneous mass capture rate reached the predicted rate during the accretion rate peaks that occurred whenever the circulation of the disk reversed.

The results of Fryxell and Taam mentioned above suggested that the accretion flow may sometimes be unsteady even if upstream density and velocity field are unifrom. Matsuda *et al.* therefore explored accretion from a *uniform* upstream flow, using 2D and 3D cell and finite-volume methods (see also Ho *et al.* 1989). Most of the computational burden of the simulations carried out by Fryxell and Taam and Sawada *et al.* came from their use of curvilinear coordinates that placed small cells near the accreting object. The Courant condition then forces the use of very short time steps. To avoid this difficulty, Matsuda *et al.* (1991) used Cartesian coordinates, sacrificing spatial resolution near theaccreting object in order to be able to follow the time evolution of 3D flows. They did not consider the effects of X-radiation produced by the accreting object.

Matsuda *et al.* (1991) found that a steady flow develops in 2D and 3D simulations when the radius of the accreting body is relatively large and/or the zoning used in the simulation is relatively coarse. In contrast, whenever the radius of the accreting body is relatively small ($\lesssim 0.06$–$0.12\, r_a$) and the zoning is fine enough, the circulation of the flow oscillates quasi-periodically, accompanied by peaks in the mass accretion rate, even though the upstream flow is uniform. The development of quasi-periodic oscillations even when the upstream flow is uniform appears at odds with the argument of Blondin *et al.* that the oscillations are initiated by asymmetries in the upstream flow.

Matsuda *et al.* (1991; see also Livio 1991) argue that the flow oscillations begin once a quasi-rigid plasma "atmosphere" develops around the accreting object, and that the flow then resembles uniform flow past a rigid, gravitating sphere, which seems to be unsteady whenever the viscosity is sufficiently low (see Shima *et al.* 1985; Fryxell, Taam, and McMillan 1987; Matsuda *et al.* 1989). They therefore conclude that the flow oscillation is not caused by density or velocity gradients and is unrelated to the accretion-line instabilities discussed by Cowie (1977) and Soker (1991). However, a simple physical explanation for the origin of the oscillations remains to be given.

The time-average specific angular momentum of the plasma captured in most of the numerical experiments is rather uncertain, but the results to date suggest that the long-term average value of $|\ell_w|$ is $\ll r_a v_0$. In the 2D simulations of Taam and Fryxell (1988) and Blondin *et al.* (1990), for example, the time-average value of the angular momentum accreted by the compact object corresponds to a value of $|\ell_w|$ that is $\ll r_a v_0$, although the instantaneous value of $|\ell_w|$ is $\sim$0.15 when a circulating "disk" flow is well-developed. Matsuda *et al.* (1991) computed both the mass capture and angular momentum capture rates, but reported them in unspecified units. For their flows that developed oscillations, the time-average angular momentum capture rate was much less than the peak rates. The long-term-average value of $|\ell_w|$ for capture from a wind is uncertain in part because it has only been possible to run simulations for relatively short times.

Observations of the pulse frequencies of accreting neutron stars in high mass X-ray binary systems have provided evidence that the accretion flows in these systems do indeed have strong circulations that reverse on short time scales. For example, the spin-rate changes observed in Vela X-1 by Boynton *et al.* (1984) on time scales of days imply that the specific angular momentum of the captured plasma is large enough to produce Keplerian or near-Keplerian azimuthal velocities near the magnetospheric boundary, and that the sense of the circulation revereses on time scales at least as short as a few days. Similar behavior has been observed in GX 301−2 by Sato *et al.* (1986), in 4U1538–52 by Makishima *et al.* (1987), and in X Per by Murakami *et al.* (1987).

Ho and Arons (1987) have suggested that large-amplitude fluctuations in the density and velocity of a radiation-driven wind can occur as the result of overstable oscillations produced by the coupling between the mass capture rate and the wind velocity via the effect of X-radiation from the neutron star on the radiation force in the spectral lines. They argue that η_w could be as large as 0.5 for such a flow.

2.1.2 *Overflow of atmosphere or envelope.* When the companion star loses plasma by overflowing its gravitational potential well, relatively cold plasma from the atmosphere or envelope flows slowly over the gravitational saddle point between the companion star and the neutron star, and is immediately captured by the neutron star. The mass transfer rate is determined by the evolution of the binary system, which can be affected by nuclear burning in the companion star, gravitational radiation of orbital angular momentum, magnetic braking of the orbital motion, and tidal coupling between the neutron star and accreting plasma and the companion star.

For low-mass systems, the moment arm of plasma flowing over the saddle point between the stars is $\sim a$ and the specific angular momentum of the plasma captured by the neutron is therefore

$$\ell_o = \eta_o a^2 \, \Omega_{\rm orb} \, , \tag{2.17}$$

where η_o is a dimensionless number of order unity. The precise value of η_o depends on the mass ratio and other properties of the system. Note that ℓ_o is at least $(a/r_a)^2$ times larger than ℓ_w, for $|\eta_w| \lesssim 1$.

2.2 DISK AND RADIAL FLOWS

The character of the accretion flow near the neutron star magnetosphere depends strongly on the amount of circulation. We define a *Keplerian disk flow* as one in which the angular velocity Ω at each radius is closely equal to the angular velocity $\Omega_K(r)$ of a circular Keplerian orbit at the same radius. A *radial flow* is one in which the angular velocity is much less. Of course intermediate cases, in which the angular velocity is comparable but not closely equal to $\Omega_K(r)$, are also possible.

Accreting plasma with specific angular momentum ℓ relative to the neutron star will tend to orbit it at the radius

$$r_\ell \equiv \ell^2/GM \tag{2.18}$$

of the circular Keplerian orbit with the same specific angular momentum. Thus, if

$$\ell \leq \ell_{\rm crit} \equiv r_m^2\, \Omega_K(r_m) = (GMr_m)^{1/2}\,, \tag{2.19}$$

then $r_\ell < r_m$ and accreting plasma will fall directly onto the magnetosphere. If on the other hand $\ell > \ell_{\rm crit}$ and the motion of accreting plasmaing plasma will orbit outside the magnetosphere, provided that the length scales on which the density and velocity of the plasma vary are larger than the critical impact parameter

$$b_{\rm crit} \equiv (1 + r_a/r_m)^{1/2} r_m \tag{2.20}$$

separating orbits that intersect the magnetosphere from those that do not (Ho 1988).

For plasma captured from a wind, equation (2.14) for the mean specific angular momentum of the plasma gives

$$r_{\ell w} = \eta_w^2 r_a^4 \Omega_{\rm orb}^2/GM \sim \eta_w^2\, 10^5 \ \ {\rm cm}\,, \tag{2.21}$$

where in the last expression on the right we have assumed a wind velocity $v_w = 10^8\ {\rm cm\,s^{-1}}$ and a binary orbital period $P_{\rm orb} = 1\,{\rm day}$. Equation (2.21) shows that η_w must be $\gtrsim 10^2$ in order for an extensive disk to form around the magnetosphere of a strongly magnetic neutron star, which extends to a radius greater than 10^8 cm (cf. Table 2.1). Thus, an extensive disk cannot form if the wind is homogeneous, but may form if the flow has a large circulation, as suggested by the observations of spin-rate changes of wind-fed neutron stars discussed in §2.1.1.

If no disk forms, the radial component v_r of the inflow velocity will approach the free-fall velocity

$$v_{\rm ff}(r) \equiv (2GM/r)^{1/2} = 0.94 \times 10^9\, (M/M_\odot)^{1/2}\, r_9^{-1/2} \ \ {\rm cm\,s^{-1}} \tag{2.22}$$

near the magnetosphere, and the flow time scale will be comparable to the free-fall time

$$t_{\rm ff}(r) \equiv r/v_{\rm ff}(r) = 1.06\, (M/M_\odot)^{-1/2}\, r_9^{3/2} \ \ {\rm s}\,. \tag{2.23}$$

Assuming that viscous shear stresses in the flow are unimportant, the specific angular momentum of the accreting plasma will remain approximately constant as the plasma falls toward the neutron star, and the angular velocity Ω of the accreting plasma will therefore be $\approx \ell/r^2$. If a disk forms, outward transport of angular momentum will cause it to expand radially. If the disk becomes extensive, it may significantly affect mass capture from the wind, as discussed in §2.1.1.

For plasma overflowing the gravitational potential well of the companion star,

$$r_{\ell o} = \eta_o^2 a^4 \Omega_{\rm orb}^2 / GM \sim 10^{10}\,\eta_o^2 \quad {\rm cm}\,, \tag{2.24}$$

where we have used equation (2.16) and in the last expression on the right we have assumed a binary separation $a = 10^{11}$ cm and a binary orbital period $P_{\rm orb} = 1$ day. Thus, plasma supplied by atmospheric or envelope overflow will always form a Keplerian disk around the neutron star magnetosphere (the magnetospheric radius is expected to be $\lesssim 10^9$ cm, even for a stellar magnetic field $\sim 10^{13}$ G).

Figure 2.1 shows stages in the formation of a Keplerian disk by plasma overflowing the gravitational potential lobe of the companion star. The flow around the neutron star is not initially circular but becomes circular within a few orbital periods. The resulting ring then spreads radially inward and outward as the shear stress transports angular momentum outward. The outer radius r_d of the Keplerian disk depends on whether the neutron star is a source or sink of angular momentum and on whether the tidal interaction of the outer part of the disk with the companion star (which can transfer angular momentum from the disk to the orbital motion of the system) is strong or weak. Normally, r_d will exceed the radius r_ℓ. If the tidal coupling of the plasma in the outer part of the disk to the orbital motion is just strong enough, the flow may settle down to a steady state in which plasma enters the disk from the stream and then spirals slowly inward as its angular momentum is transported outward by shear stresses and tidal coupling to the orbital motion.

3. Magnetospheres

We use the term *magnetosphere* to refer to the volume around the neutron star in which its magnetic field strongly affects the flow of mass, energy, and angular momentum. This volume can profoundly affect the observed properties of accreting neutron stars. For example, the intensity of the X-radiation seen by a distant observer will oscillate periodically if the stellar magnetic field is not symmetric about the rotation axis, if the field is strong enough to channel the accretion flow, and if the resulting beamed radiation produced near the star can escape without being isotropized. Such periodic intensity oscillations are especially important because they reflect directly the stellar rotation rate.

The very existence of a magnetosphere changes the rotation-rate behavior of an accreting neutron star, by increasing the radius where the accreting plasma couples to the star and changing the nature of the coupling. A magnetosphere also makes possible new, potentially important radiation processes, such as cyclotron emission and scattering. The interaction of the stellar magnetic field with the accretion flow can create large electrical potential drops in the system, which may accelerate charged particles to very high energies, leading to production of γ-rays.

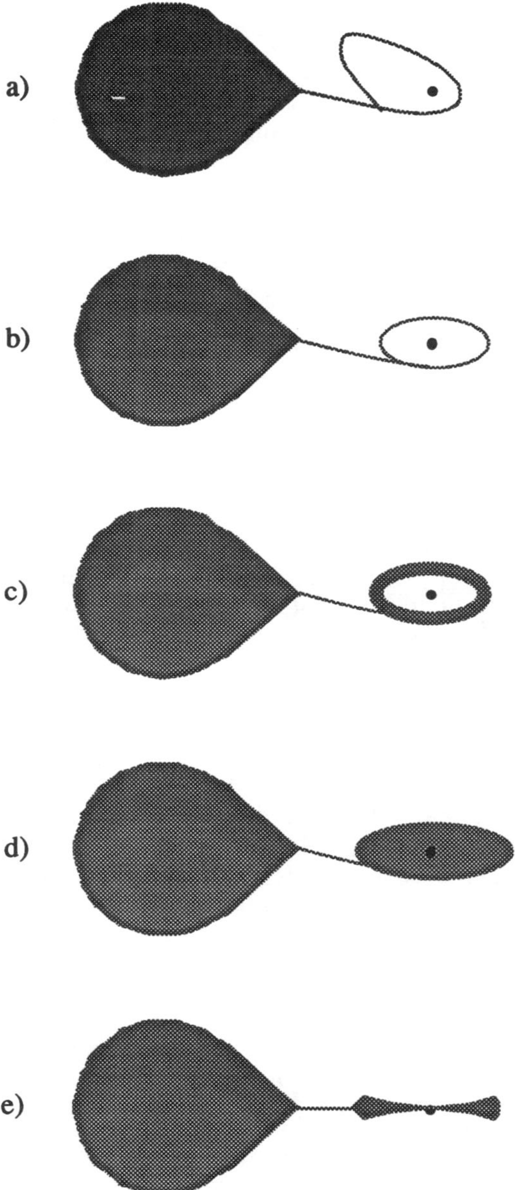

Fig. 2.1.—Schematic picture of the formation of a Keplerian accretion disk around a neutron star, showing (a) collision of the initial gas stream with itself, (b) formation of a circular ring, (c) spreading of the ring inward and outward, and (d) development of a steady disk flow. Panel (e) shows a side view of the steady flow.

The shapes of pulses, the spectra of the radiation, and the time scales of quasi-periodic and random variations in intensity are all affected by the structure of the magnetosphere.

Aside from its astronomical importance, the study of neutron star magnetospheres is also a fascinating plasma physics problem, which provides a check on our understanding of plasma and magnetospheric physics under conditions that differ by many orders of magnitude from those encountered in terrestrial plasmas and planetary magnetospheres.

In the present section we describe the qualitative features of the interaction between the accretion flow and the stellar magnetic field, the characteristic features of accretion flows that lead to formation of a magnetosphere, and the physical principles that determine the scale of the magnetosphere.

3.1 FORMATION OF A MAGNETOPSHERE

The structure of the outer magnetosphere depends on the angular velocity of the plasma near the magnetosphere as well as the angular velocity of the neutron star. A useful measure of the dynamical importance of the circulation of the accreting plasma is the dimensionless number (Elsner and Lamb 1977)

$$\omega_p \equiv \Omega(r_m)/\Omega_K(r_m)\,, \tag{3.1}$$

which is less than or equal to 1. Here $\Omega(r_m)$ is the angular velocity of the plasma at the magnetospheric boundary and $\Omega_K(r_m)$ is the Keplerian angular velocity there. Circulation is dynamically unimportant for approximately radial accretion flows (those with $\omega_p \ll 1$) but important for orbital flows (those with $\omega_p \approx 1$). For Keplerian accretion disks, $\omega_p = 1$. In most systems, the angular velocity of the plasma at the magnetospheric boundary appears to be comparable to the Keplerian velocity there, and one has to consider the dynamical effects of this circulation. The radial accretion problem, on the other hand, is defined by the assumption that circulation is dynamically unimportant.

Despite their profound differences, disk and radial accretion flows have several important features in common:

- Both are inward flows controlled primarily by gravity, which causes the plasma streamlines to converge on the neutron star. Because of this convergence, both flows tend to trap the stellar magnetic field and sweep it inward toward the neutron star, at least to a first approximation.
- In both types of flows, the material stress ρv^2 increases more slowly with decreasing radius than does the stress of the stellar magnetic field. In radial flows, for example, the material stress scales approximately as $r^{-5/2}$. In disk flows, the material stress scales approximately as $r^{-5/2}$ in the middle, gas-pressure-dominated region (see §4.1). In the inner region, the scaling of the material stress with r depends on the model. For most models, the material stress optically-thick, In contrast, increases more slowly with decreasing r than $r^{-5/2}$. In fact, in the optically-thick, radiation-pressure-dominated inner disk model of Shakura and Sunyaev (1973), the material stress actually decreases with decreasing r. In contrast, the magnetic stress of a dipolar stellar magnetic field scales approximately as r^{-6}.
- The electrical conductivity of the accreting plasma in both types of flows is very high. As a result, the dynamo number is very large, that is, the diffusion

time scale $t_d \equiv r^2/\eta_m$ is much larger than the flow time scale $t_f \equiv r/v$. Here η_m is the magnetic diffusivity. Motion across field lines therefore occurs slowly on the time scale of the inflow, and the tendency of the flow to confine the stellar magnetic field in the sectors where it is converging is realized to an important degree.

If the intrinsic (undistorted) magnetic field of the neutron star has non-zero higher multipole moments, the stress of the intrinisic magnetic field of the star will increase even more steeply with decreasing radius than r^{-6}, near the stellar surface. Even if the intrinsic stellar magnetic field is purely dipolar, the confinement of the stellar magnetic field by the flow (screening) and the resulting higher magnetic multipole moments created by the electrical currents in the flow and the electrical currents induced in the star will cause the stress of the magnetospheric magnetic field to increase more steeply with decreasing radius than r^{-6}. As a result, even for relatively weak neutron star magnetic fields ($B \sim 10^8$ G), there is a volume around the neutron star in which the magnetic field strongly affects the flow of mass, energy, and angular momentum.

Because the magnetic field of the neutron star is comparatively weak at the radii where plasma is captured by the gravitational field of the star, and because the subsequent flow of the highly conducting plasma converges toward the neutron star, a negligible fraction of the magnetic flux threading the star is expected to thread the accreting plasma at the time it is captured. This fact, and the high electrical conductivity of the accreting plasma, means that the plasma does not have ready access to the stellar surface. The flux of mass into the magnetosphere due to microscopic processes, such as collisional diffusion or loss-cone entry, is generally negligible compared to the total mass fluxes inferred from the luminosities of neutron stars in X-ray binary systems (Elsner and Lamb 1984). Thus, the large inward mass fluxes must be the result of macroscopic processes. An important problem in both disk and radial flow geometries is to understand how plasma eventually reaches the surface of the neutron star.

Fortunately, the radius of the magnetosphere is typically much smaller than the radius at which plasma is captured (cf. Table 2.1), so that the structure of the magnetosphere can be discussed, to a first approximation, independently of the capture process. In accretion-powered pulsars (but not in QPO sources) the magnetospheric radius is estimated to be much larger than the stellar radius, so that the interaction of the flow with the magnetospheric boundary can also be treated, to a first approximation, separately from the interaction of the flow with the stellar surface.

As discussed further below, the material stresses associated with plasma circulation and flow into the magnetosphere typically dominate the plasma thermal pressure at the magnetospheric boundary. The scale and structure of the magnetosphere therefore depends on the accretion flow pattern, which is generally quite complex, and can vary greatly for different types of flows. The dynamical character of the interaction between the accretion flow and the stellar magnetic field makes it difficult to model.

A quantitative description of the magnetospheres of accreting neutron stars is particularly difficult to construct for the conditions expected in accretion-powered pulsars and QPO sources. In these sources, the magnetic field of the neutron star is thought to be misaligned with its spin axis. As a result, the rotation of the star causes the magnetic field at a given point in space to vary explicitly with time, which imposes an explicit time dependence on the interaction between the star and

TABLE 3.1
Magnetospheres of the Earth and Neutron Stars

Aspect	Earth	Neutron Stars
Exterior flow	Always similar (solar wind)	Highly varied ($\neq$ solar wind)
Effect of gravity	Very small	Very large
Mass influx	Very small	Very large
Stresses caused by plasma entry	Small	Large
Plasma density	Low	High
Collison rates	Low	Low to high
Effects of cooling	Small	Very large
Effects of radiation	Very small	Very large
Effects of rotation	Small	Typically large

the flow even if the flow far away from the star is time-independent. In addition, the stellar spin axis is likely to be approximately parallel to the rotation axis of the binary system. Hence the accreting plasma most likely circulates in a plane perpendicular to the stelalr spin axis, making the interaction of the stellar magnetic field with the accretion flow asymmetrical and hence three-dimensional. Finally, the accretion flow may be unsteady, with large-amplitude fluctuations. Although some attempts to analyze the explicitly time-dependent, three-dimensional case of misaligned spin and magnetic axes have been made (see, *e. g.*, Anzer and Börner 1980), quantitative models have been developed only for steady radial or axisymmetric flows.

3.2 COMPARISON WITH THE GEOMAGNETOSPHERE

Some workers have tried to apply models of the geomagnetosphere directly to accreting magnetic neutron stars. This approach fails because of the profound differences between the magnetospheres of accreting neutron stars and the geomagnetosphere, some of which are listed in Table 3.1 and illustrated in Figure 3.1.

The most important differences between the geomagnetosphere and the magnetospheres of accreting neutron stars can be traced to one or more of the following factors (cf. Table 3.1):

- The strong gravitational field of the neutron star, which creates a converging flow while the accreting plasma is still far away from the magnetosphere and continues to control the flow of plasma into the magnetosphere and toward the stellar surface.

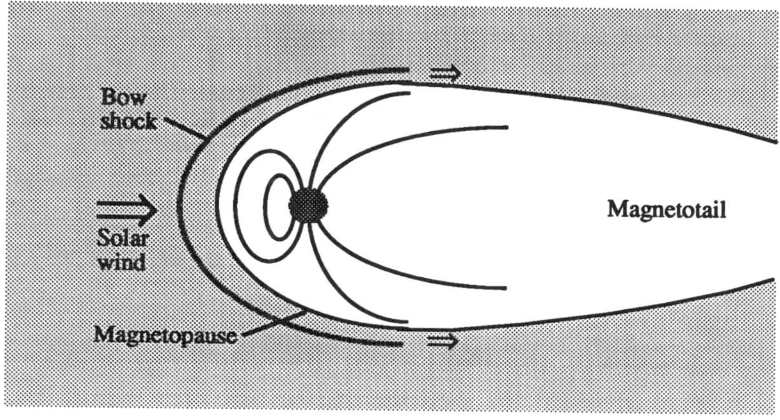

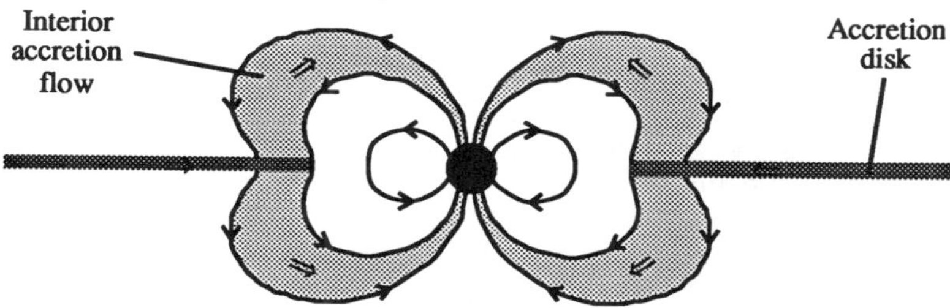

Fig. 3.1.—Side views of the geomagnetosphere (top) and the magnetosphere of a neutron star accreting from a Keplerian disk (bottom), illustrating the profound differences in the exterior plasma flows and the resulting structures of the two magnetospheres. Almost all of the plasma that interacts with the geomagnetosphere flows on by, whereas essentially all of the plasma that interacts with the magnetosphere of an accreting neutron star enters it. Also, the geomagnetosphere is always immersed in the solar wind, whereas the magnetospheres of different neutron stars are immersed in a variety of flows, ranging from nearly radial to Keplerian. Moreover, the flow around a given neutron star magnetopshere may vary dramatically with time.

- The highly variable nature of the plasma flow toward the neutron star magnetosphere, which can cause the structure of the magnetosphere to change dramatically with time.
- The large flux of plasma into the neutron star magnetosphere, which stresses the outer magnetosphere and produces a relatively high plasma density throughout.
- The high plasma density near the neutron star, which makes the plasma collisional or quasi-collisional on the scale of the magnetosphere.

- The strong magnetic field of the neutron star, which typically causes charged particles to lose their energy by cyclotron emission in much less than one transit time across the magnetosphere.
- The intense radiation field from near the surface of the neutron star, which cools the plasma in the magnetosphere in much less than an infall time and can produce radiation pressure forces even larger than gravity.
- The typically strong circulation of the accreting plasma and the rapid rotation of the neutron star, which stress the outer magnetosphere and can strongly affect the plasma flow pattern.

The fact that all or almost all of the plasma that interacts with a neutron star magnetosphere enters it and the highly variable nature of accretion flows deserve further comment.

Because almost all of the plasma that interacts with the geomagnetosphere flows on by (see Fig. 3.1), the characteristic radius of the nose of the geomagnetosphere can be estimated simply by balancing a static plasma pressure, set equal to the ram pressure of the solar wind, against the static pressure of the confined magnetic field of the earth. In contrast, essentially all of the plasma that interacts with the magnetosphere of an accreting neutron star penetrates it and reaches the stellar surface within one free-fall or transit time (again see Fig. 3.1). The material stresses associated with the flow of plasma into a neutron star magnetosphere are therefore generally as large as the stresses outside, and play an important role in determining the scale and structure of the magnetosphere. As a result, determining the structure of the magnetosphere of an accreting neutron star is a dynamic rather than a static problem, even to a first approximation.

A further important difference between neutron satr magnetospheres and the geomagnetosphere is that the geomagnetosphere is always immersed in a single, albeit somewhat variable, environment, namely the solar wind. In contrast, accreting neutron stars are found in a wide variety of environments, ranging from nearly radial flows to geometrically thin Keplerian disks, none of which are similar to the solar wind. Indeed, as noted in §2.1, even the flow near a given neutron star may change dramatically with time. Thus, one expects large variations in the structure of the magnetosphere from one accreting neutron star to another and in the structure of the magnetosphere of a single neutron star with time.

Although the structure of the magnetosphere of an accreting neutron star is generally quite complex, for some purposes it is both useful and sufficient to have an order-of-magnitude estimate of the radius inside which the stellar magnetic field strongly affects the motion of accreting plasma. In the next three subsections we review such estimates. Of course, neutron star magnetospheres are not spherical, so that any single length scale can only be a very rough guide to their dimensions.

3.3 STABLE MAGNETOSPHERIC FLOW

To be credible, an estimate of the radius of the magnetosphere of an accreting neutron star must take into account the fact that essentially all of the plasma approaching the star flows through the magnetosphere to the stellar surface. For this reason, estimates made by simply balancing the thermal pressure of static plasma outside the magnetosphere against the pressure of a static magnetic field inside are not credible. Indeed, if the plasma were to remain outside the magnetosphere, no plasma could reach the stellar surface and hence, whatever else it might be, such a star could not be a luminous X-ray source. If, as expected, the plasma does *not* remain in a static configuration outside, then the assumption of *static* pressure equilibrium is not consistent.

There is ample evidence that the flow of accreting plasma in accretion-powered pulsars is guided by the stellar magnetic field toward the magnetic poles (see Mészáros 1982, 1986; Kirk and Trümper 1983). This is natural if the accreting plasma has become threaded by the stellar magnetic field. Assuming that threading occurs readily, one can construct a first estimate of the scale of the magnetosphere based on *the* fundamental characteristic of accretion, namely plasma *flow*, and the fact that the stellar magnetic field can control the flow only if it is strong enough (see Lamb, Pethick, and Pines 1973; Elsner and Lamb 1977; Ghosh, Lamb, and Pethick 1977). Two forms of the argument are possible, one global and the other local. Both consider plasma flow inside the magnetosphere (which I call *magnetospheric flow*) and hence can be applied for any exterior flow.

3.3.1 Global argument. The global argument (see Fig. 3.2a) starts with the fact that the inward falling plasma cannot be deflected by the stellar magnetic field in a time shorter than the time taken by an Alfvén wave propagating in the stellar field to travel from the accreting plasma to the surface of the star and back. Thus, the stellar magnetic field can significantly affect the motion of accreting plasma at radius r only if

$$t_A(r) < t_f(r)\,, \tag{3.2}$$

where

$$t_A(r) \equiv -\int_r^R \frac{dr}{v_A} \tag{3.3}$$

is the time for an Alfvén wave to travel from r to the stellar surface at R and

$$t_f(r) \equiv \int_r^R \frac{dr}{v_r} \tag{3.4}$$

is the time for the flow to travel from r to the stellar surface. The Alfvén velocity in equation (3.3) is given by

$$v_A = \frac{B_p}{\sqrt{4\pi\rho}}\,, \tag{3.5}$$

where B_p is the poloidal component of the stellar magnetic field and ρ is the density of the plasma *on the stellar field lines*.

Both integrals (3.3) and (3.4) are typically dominated by the domain where r is large, since both v_r and v_A are usually smallest there. Thus, $t_f(r) \approx r/v_r(r)$ and $t_A(r) \approx r/v_A(r)$. On substituting these expressions into inequality (3.2), one finds that an *upper bound* on the radius of the magnetosphere is given implicitly by the inequality

$$\rho(r)\,v_r^2(r) \leq \frac{B_p(r)^2}{4\pi}\,, \tag{3.6}$$

where ρ is the mass density on the stellar field lines. Note that condition (3.6) does not guarantee that the magnetic stress is large enough to control the flow, if other components of the flow velocity are larger than v_r. Thus, the actual radius of the magnetosphere may be less than the radius given by condition (3.6).

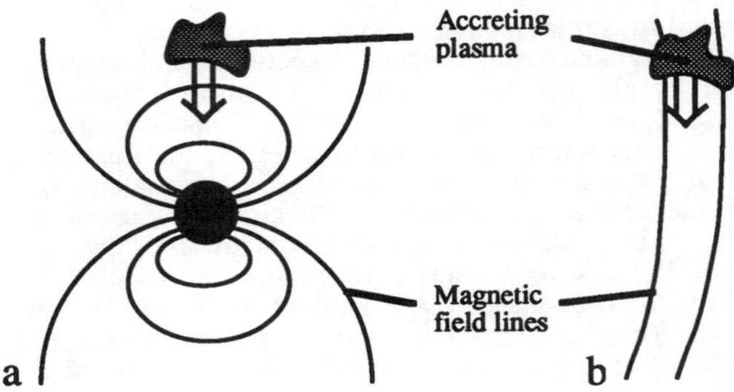

Fig. 3.2.—Schematic illustration of global (a) and local (b) MHD stability arguments that set an upper bound on the extent of the magnetosphere of an accreting neutron star. Both arguments are based on the fact that the disturbance of the stellar magnetic field by the accreting plasma propagates at the Alfvén speed.

3.3.2 Local argument. The local form of this argument (see Fig. 3.2b) may be stated as follows. The magnetic field of the star can control the flow only if the flow is stable. In order to be stable, field-aligned flow must be sub-Alfvénic in the poloidal field of the star (Lamb and Pethick 1974; Williams 1975; Ghosh, Lamb, and Pethick 1977; Adam 1978). The condition that it be sub-Alfvénic is

$$\rho(r)\, v_p(r)^2 \leq \frac{B_p(r)^2}{4\pi}, \tag{3.7}$$

where v_p and B_p are the poloidal components of the bulk flow velocity and the magnetospheric magnetic field, respectively.

Inequality (3.7) is typically also hardest to satisfy at large r. For example, if the accreting plasma is falling freely inward along the lines of a dipolar field, $\rho v_p^2 \propto r^{-7/2}$, whereas $B_p^2 \propto r^{-6}$. Thus, inequality (3.7) implicitly sets an upper bound on the radius of the magnetosphere on each flux tube that carries accreting plasma.

The global and local arguments agree if $v_r \sim v_p$. The radius defined by these arguments is called the ***Alfvén radius*** and is usually denoted r_A.

3.4 INTERACTION WITH DISK FLOWS

The converging streamlines of a Keplerian accretion flow tend to sweep the magnetic field of the neutron star inward. Cross-field motion due to microscopic diffusion is negligible, so that even the relatively slow inward drift of the disk plasma is sufficient to confine much of the magnetic flux of the star to its vicinity. However, the inward flow is typically confined by gravity to a solid angle $\sim 10^{-2}$ ster. As a result, the stellar magnetic field is likely to be much more compressed in the disk plane than in directions away from it, and hence the magnetosphere is not spherical. Instead, the magnetosphere is pinched inward in the disk plane but balloons outward in directions away from the plane. Some lines of the stellar magnetic field may be open.

3.4.1 ***"Diamagnetic disk" models.*** Several authors (Aly 1980; Riffert 1980; Kundt and Robnik 1980; Anzer and Borner 1983; White and Stella 1987) have assumed that a geometrically thin Keplerian disk flow will interact with the magnetic field of the neutron star only by excluding it from the volume occupied by the disk. In order for this to happen, the disk flow would have to be stable and the plasma in the disk perfectly diamagnetic. Despite the absence of any couple between the disk and the star in this "diamagnetic disk" picture, the disk is usually assumed to have an inner edge, from which plasma falls freely to the stellar surface. The radius of the inner edge of the disk is typically estimated by balancing the thermal pressure of the plasma at the midplane of a steady disk flow model against the magnetic pressure of an undistorted dipolar magnetic field (see, for example, White and Stella 1987).

Although the diamagnetic disk picture provides some insight into the topology of the magnetic field near a Keplerian disk flow, it has several major defects as a model for neutron star X-ray sources. First, the very physical processes that cause the plasma in the disk to drift inward by creating a substantial shear stress will also lead to a large magnetic diffusivity (Ghosh and Lamb 1978, 1979a). These processes include turbulent motions, and magnetic field amplification and reconnection. Instabilities of the disk-magnetic field interface, such as the Kelvin-Helmholtz instability, will also assist the magnetospheric magnetic field to mix into the disk. Reasonable estimates of the rate at which the magnetospheric magnetic field diffuses into the disk indicate that the field will thread through the disk in a time much shorter than the radial drift time, as discussed below. If these estimates are correct, the assumption of a diamagnetic disk is self-contradictory and such a disk can never form. Even if the threading of the disk plasma by the magnetospheric magnetic field is not complete, the interaction between the disk plasma and the magnetospheric magnetic field will create a couple between the disk and the star that radically changes the character of the flow.

A second defect of the diamagnetic disk picture, if applied consistently, is that plasma in the disk never reaches the stellar surface on time scales of interest (assuming these are much shorter than the time scales on which electrical currents in the neutron star decay). This can be seen by considering the initial value problem of the formation and evolution of a Keplerian disk from a stream of accreting plasma approaching the neutron star (cf. Fig. 2.1). In accordance with the diamagnetic disk picture, we assume that the incoming flow is stable and perfectly diamagnetic, and that the boundary between the Keplerian flow and the magnetic field can be determined by balancing the static thermal pressure of the plasma in the flow against the pressure of the surrounding magnetic field. Because of the extreme weakness of the stellar magnetic field at the radius r_ℓ where the material from the stream first orbits the neutron star, the pressure balance surface is initially far from the center of the plasma ring, as shown in the top panel of Figure 3.3. As the plasma streaming into the ring diffuses inward and outward, the ring spreads, the pressure at the inner edge of the ring increases, and the inner edge of the pressure balance surface moves inward, as shown in the middle panel of Figure 3.3.

Since, in accordance with the diamagnetic disk picture, there is no couple between the magnetic field and the flow, angular momentum can be removed from the Keplerian flow only by internal shear stresses. However, as shown below in §4.1, removal of angular momentum by shear stresses internal to a geometrically thin Keplerian flow can never proceed fast enough to cause the angular velocity of the flow to deviate significantly from the Keplerian value or to cause the radial velocity to approach free-fall. Thus, the plasma in the disk will continue to diffuse

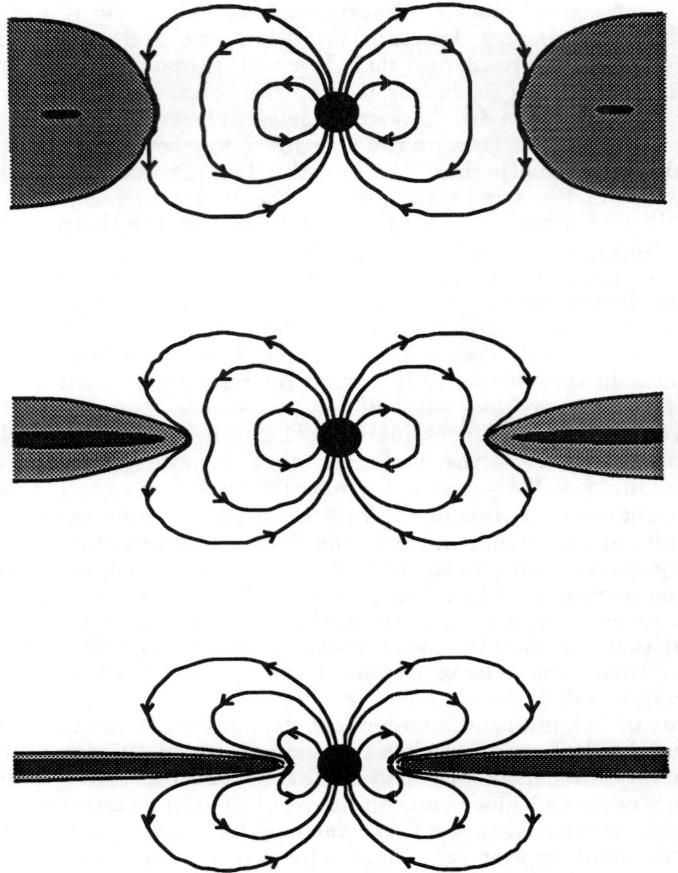

Fig. 3.3.—Schematic side views of the time evolution of a stable, perfectly diamagnetic Keplerian disk flow around an aligned rotator, showing the initial interaction of the plasma ring with the stellar magnetic field (top), the inward and outward diffusion of the plasma in the ring (middle), which causes the inner edge of the pressure balance surface to move inward, and the gradual accumulation of plasma near the star (bottom), which increasingly pinches the stellar magnetic field between the inner edge of the flow and the star. A steady flow never develops. We expect a stable, perfectly diamagnetic Keplerian disk flow around a non-aligned rotator to undergo a similar time evolution.

slowly inward indefinitely, maintaining its Keplerian azimuthal velocity. As the density and pressure at the inner edge of the disk increase, the inner edge of the pressure balance surface will continue to move inward, as shown in the bottom panel of Figure 3.3.

As the pressure of the magnetic field confined between the inner edge of the disk and the star increases, ever increasing plasma pressure is required to move the inner edge of the pressure balance surface further inward. Even if the intrinsic magnetic field of the star is purely dipolar (and it may not be), once the pressure

TABLE 3.2
Important Time Scales in Disk Accretion[a]

Time Scale	Symbol	Typical Value (s)
Kelvin-Helmholtz growth time	τ_{KH}	$\sim 10^{-2}$
Magnetic flux reconnection	τ_R	$\sim 10^{-1}$
Keplerian orbital period	τ_K	$\sim 10^{-1}$
Turbulent diffusion through disk	τ_D	~ 1
Radial drift	τ_r	$\sim 10^3$
Doubling of spin rate	τ_s	10^9–10^{13}

[a]Near the inner edge of the disk. Accretion flow and neutron star parameters are the same as in Table 2.1.

balance surface is within a distance $\sim R$ of the stellar surface, the higher multipole moments of the pinched stellar magnetic field will become very important, causing the magnetic pressure to rise even more steeply as the inner radius of the pressure balance surface decreases. Thus, the structure of the Keplerian flow continues to evolve and is very unlikely to be similar to that in models of steady-state disk flows. Since no plasma ever reaches the neutron star, the mass accretion rate to the stellar surface is zero, there is no energy release at the stellar surface, and the luminosity comes only from the disk.

This discussion shows that attempts to determine the radius of the inner edge of the Keplerian disk, from which plasma supposedly "falls" from the disk to the star, by balancing the thermal pressure at the midplane of a steady-state flow model against the magnetic pressure of an undistorted dipolar magnetic field are deeply misleading, for several reasons: (1) The tacit assumption that the pressure at the inner edge of the time-dependent diamagnetic inflow is the same as the pressure in the midplane of a steady-state Keplerian disk flow is unjustified, and is probably wrong. (2) The tacit assumption that the magnetic pressure of the pinched stellar magnetic field is the same as the pressure of an undistorted dipolar field at the same radius is incorrect. (3) The assumption that plasma deviates significantly from Keplerian motion at the pressure balance surface violates angular momentum conservation, since by assumption there is no couple between the disk and the star and since the shear stress in a geometrically thin flow is too small to cause any significant deviation from Keplerian motion. On the other hand, if there *is* a significant couple between the disk and the star, then the use of static pressure balance to determine the inner edge of the Keplerian disk flow is unfounded.

3.4.2 *The couple between disk and star.* Consider now the physical processes that create a couple between the disk flow and the neutron star. Typical values of the coupling times due to some of these processes, as well as other important time scales in disk accretion, are listed in Table 3.2.

Near the inner edge of the Keplerian flow, the interface between the disk flow and the magnetospheric magnetic field is Kelvin-Helmholtz unstable (Ghosh and Lamb 1978, 1979a). Penetration of the Keplerian flow by the magnetospheric magnetic field is assured if unstable modes grow to an amplitude comparable to the semi-thickness h of the disk. Modes with wavelengths λ greater than h can achieve such an amplitude while still in the linear regime. Since the growth times of the Kelvin-Helmholtz modes in the linear regime can be estimated analytically, using the MHD dispersion relation appropriate to the disk-magnetosphere interface, Ghosh and Lamb concentrated on long-wavelength modes. They found that near the inner edge of the disk, the growth time for long-wavelength Kelvin-Helmholtz modes is $\sim 10^{-5}$ times the radial drift time. Thus, there appears to be plenty of time for these modes to grow to sufficiently large amplitude to greatly disturb the disk surface and allow the magnetospheric magnetic field to mix with the disk plasma before the plasma drifts significantly inward. This mixing produces a couple between the disk and the star.

The disk and the star will also be coupled by turbulent diffusion of the magnetospheric field into the disk, as shown in Figure 3.4. This process is important in the region where the kinetic energy density of convective or turbulent motions in the disk exceeds the energy density of the magnetospheric magnetic field just outside the disk. In this region one expects the magnetospheric magnetic field to be entrained by the convective motions and carried into the disk. In order to make an estimate of the efficiency of this process, Ghosh and Lamb (1978, 1979a) adopted a mixing length approach, assuming a diffusion coefficient $\sim 0.1\, u_t \ell_t$, where u_t is the turbulent velocity and ℓ_t is the length scale of the largest eddies. They found that the magnetospheric magnetic field would diffuse through the disk in a time $\sim 10^{-3}$ times the radial drift time. Thus, there appears to be ample time for the magnetic field to diffuse into the disk. Once inside the disk, magnetic field lines that were formerly above the disk can reconnect to magnetic field lines that were formerly below the disk, creating an additional couple between the disk and the neutron star.

A third process that couples the disk and the star is reconnection of the magnetospheric magnetic field to magnetic fields in the disk (Ghosh and Lamb 1978, 1979a). The disk flow amplifies and reconnects any seed field in the stream of plasma that feeds it, creating closed loops of magnetic flux within the disk. These flux loops are continually distorted by convection and turbulence within the disk, and by reconnection (see Stella and Rosner 1984, and references therein). Even if there were no magnetic field initially threading the disk flow, turbulent diffusion of the magnetospheric magnetic field into the disk flow will create such a field. Once there are magnetic flux loops within the disk, either because magnetic fields were present in the incoming stream or because the magnetospheric magnetic field earlier diffused into the disk, it will be energetically favorable for the magnetospheric field to reconnect to the disk fields. The time scale for this to occur near the inner edge of the disk is $\sim 10^{-3}$ times the radial drift time. In the usual reconnection picture, almost all of the magnetic flux is conserved, but the topology is radically changed. As a result, field lines that initially formed closed loops in the disk now connect to the star. This produces a couple between the star and the disk flow.

In summary, even if the Keplerian disk flow were not initially coupled to the stellar magnetic field, a significant couple would develop rapidly compared to the radial drift time. Thus, any realistic model of disk accretion by a magnetic neutron star must take into account the coupling between the disk and the star produced by the stellar magnetic field.

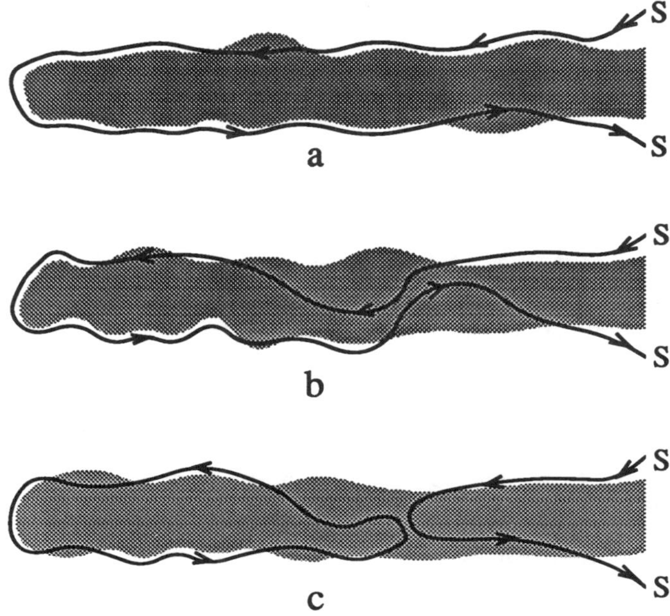

Fig. 3.4.—Side view of the inner edge of a Keplerian disk flow (shaded) being threaded by the magnetospheric magnetic field via turbulent diffusion and reconnection. Field lines that connect to the neutron star are indicated by the letter S. The stellar magnetic field first diffuses into the disk (a-b) and then reconnects (c), linking magnetic field lines that were formerly above and below the disk through it (right) and creating a magnetic island (left) within the disk. Such an island will subsequently be distorted and subdivided by turbulence, convection, and reconnection within the disk. Afterwards, its orientation could favor reconnection to the magnetospheric magnetic field just outside the disk.

3.4.3 *Radius of the inner edge of the disk.* As noted above, if the neutron star did not have a magnetic field, or if the magnetic field did not couple the disk and the star, the plasma in the disk would remain in near-Keplerian motion, diffusing slowly inward as the weak shear stresses within the disk slowly transported angular momentum outward. In order for the Keplerian flow to have an inner edge other than at the stellar surface, angular momentum must be extracted from the Keplerian flow by some mechanism other than viscous shear stresses. The interaction with the magnetic field of the neutron star can be such a mechanism.

Consider, for simplicity, a slowly rotating neutron star. The coupling between the star and the disk via the stellar magnetic field increases rapidly as the flow nears the star. This is true even if the stellar magnetic field is dipolar and the flow does not appreciably screen the stellar field, simply because of the steep increase of B^2 with decreasing radius. At a certain distance from the star, the flow of angular momentum out of the disk via magnetic stresses becomes larger than the radially outward flow of angular momentum via the the shear stresses in the disk. The Keplerian disk does not end at this radius, however, because the angular momentum flux out of the disk is still far too small to cause a significant departure from

TABLE 3.3

Static and Dynamic Stresses in Accretion from a Keplerian Disk[a]

Stress Ratio	Flow in the Disk	Flow in the Magnetosphere
$\rho v_p^2 : P$	$\sim 10^{-4}$	$\sim 10^4$
$\rho v_p v_\phi : P$	~ 1	$\sim 10^3$
$\rho v_\phi^2 : P$	$\sim 10^4$	$\sim 10^4$

[a] Accretion flow and neutron star parameters are the same as in Table 2.1. Here ρ is the mass density of the flow, v_p is the poloidal velocity, v_ϕ is the azimuthal velocity, and P is the thermal pressure.

Keplerian motion. The Keplerian disk ends when the magnetic coupling to the star is large enough to remove the angular momentum of the flow in a radial distance Δr small compared to r. This is what marks the end of the Keplerian disk flow and the beginning of the magnetospheric flow.

This analysis shows that the inner radius of the Keplerian disk is determined by conservation of angular momentum (Ghosh and Lamb 1978, 1979a). The radius at which the magnetic stress removes the angular momentum of the Keplerian flow in a radial distance Δr ($\ll r$) is given implicitly by the azimuthal stress balance condition

$$\frac{B_p B_\phi}{4\pi} 4\pi r^2 \Delta r \approx \dot{M} r v_K \,, \tag{3.8}$$

where B_p and B_ϕ are the poloidal and toroidal components of the magnetospheric field, Δr is the radial distance over which the magnetospheric field and the disk interact, $\dot{M}$ is the mass flux through the inner disk, and $v_K = r\Omega_K$ is the azimuthal velocity of the flow in terms of the Keplerian angular velocity Ω_K. Equation (3.8) implicitly determines the radius r_0 of the inner edge of the Keplerian disk.

The shape and structure of the magnetosphere is affected by the large material stresses associated with the orbital motion of the disk plasma, the flow of plasma into the magnetosphere, and the tendency of plasma within the magnetosphere to corotate with the neutron star. The relative sizes of these stresses just outside and just inside the magnetosphere are compared with one another and with the thermal pressure in Table 3.3. In the disk, the radial ram pressure is typically much less than the thermal pressure. However, the off-diagonal part of the material stress is typically comparable to the thermal pressure, while the diagonal part of the material stress in the azimuthal direction is large compared to the thermal pressure. Just inside the magnetosphere, all the dynamical stresses are much larger than the thermal pressure. The structure of the magnetosphere is clearly a dynamical problem, not a static one.

3.5 INTERACTION WITH RADIAL FLOWS

We define an approximately radial accretion flow as one for which $\omega_p \ll 1$. Approximately radial accretion by neutron stars has been the focus of a substantial

TABLE 3.4
Important Time Scales in Radial Accretion[a]

Time Scale	Symbol	Typical Value (s)
Compton cooling	t_C	10^{-6}–10^{-2}
Electron-ion energy exchange	$t_{\rm ei}$	$\sim 10^{-1}$
Free-fall (dynamical)	$t_{\rm ff}$	$\sim 10^{-1}$
Bremsstrahlung cooling	$t_{\rm br}$	10^{-1}–1

[a]Near the magnetospheric boundary. Accretion flow and neutron star parameters are the same as in Table 2.1.

theoretical effort (Lamb 1975a,b; Elsner 1976; Elsner and Lamb 1976, 1977, 1984; Arons and Lea 1976a,b, 1980; Michel 1977a,b,c; Wang and Welter 1982; Burnard, Lea, and Arons 1983; Wang and Nepveu 1983; Wang, Nepveu, and Robertson 1984; Wang and Robertson 1984), motivated in part by its comparative tractability. Under some conditions, accretion flows near neutron stars that are capturing plasma from a stellar wind may be approximately radial, as discussed in §2.2. Investigations of approximately radial flows have usually assumed that the neutron star is rotating sufficiently slowly that centrifugal forces can be neglected. Even though the dynamical effects of the circulation of the accreting plasma are, by definition, negligible in approximately radial flows, the difference between the angular velocity of the neutron star and that of the plasma just outside the magnetosphere may strongly affect the way in which the accreting plasma enters the magnetosphere.

3.5.1 *Cooling regimes.* The character of approximately radial accretion flows depends sensitively on whether there is a substantial flow of plasma to the stellar surface at the time in question. If there is little or no flow to the stellar surface, plasma near but outside the magnetosphere cools predominantly via electron-ion energy exchange and bremsstrahlung radiation. If, on the other hand, there *is* a substantial flow of plasma to the stellar surface, the X-rays produced there rapidly cool the electrons in the accreting plasma by the inverse Compton process, and the electrons then rapidly cool the ions by collisional energy exchange.

Whereas cooling by bremsstrahlung emission typically takes much longer than the free-fall time to the stellar surface, inverse Compton cooling of plasma exposed to X-rays from the stellar surface typically takes much less than a free-fall time, as indicated in Table 3.4. Thus, when X-rays are being emitted, plasma accreting to the stellar surface cools as it falls. For electrons moving in the magnetospheric field, the cyclotron cooling time is shorter still, while ion cyclotron cooling becomes important near the stellar surface. As long as the accreting plasma remains optically thin, cooling occurs so rapidly that neither the electrons nor the ions will mirror in the magnetospheric field (Lamb, Pethick, and Pines 1973).

3.5.2 *Evolution of radial flows.* As discussed in §2.2, approximately radial inflow is possible only if plasma is captured from a stellar wind. Consider a wave front of plasma falling approximately radially from the accretion capture radius r_a toward

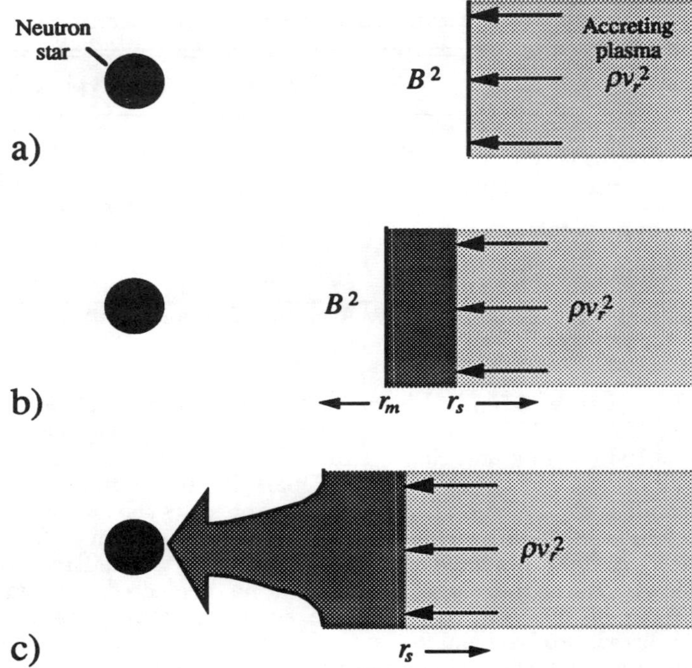

Fig. 3.5.—Schematic view of stages in the interaction between an initially radial inflow and the magnetic field of the neutron star, showing the initial compression of the stellar magnetic field by the infalling, diamagnetic plasma (a), the accumulation of shock-heated plasma outside the magnetopause (b), and the large-scale change in the structure of the magnetosphere as the plasma cools and the magnetosphere becomes unstable (c). The radius of the magnetopause is denoted by r_m and the radius of the shock front, by r_s.

the neutron star. At first, the magnetic field ahead of the wave front is very weak, and hence the infalling plasma readily sweeps it inward, toward the star. However, as the wave front compresses the magnetic field, the field becomes increasingly rigid. If, as expected, the plasma approaching the star is highly conducting and therefore cannot readily penetrate the stellar magnetic field, the rising back-pressure of the magnetic field will decelerate the plasma. When this happens, a shock wave forms near the plasma-magnetic field interface (the so-called *magnetopause*) and begins to propagate upstream through the accreting plasma, heating it and slowing its inward motion (see Fortner, Lamb, and Zylstra 1984). Post-shock electron and ion temperatures may reach $\sim 10^{10}$ K for a shock radius $r_s \sim 10^8$ cm. After the plasma passes through the shock front, its inward motion is slowed further by a gradient in the plasma pressure between the shock front and the magnetic field. Eventually the magnetic field becomes rigid enough to completely halt the inflow of the heated plasma near the initial magnetospheric radius r_{m0}.

This sequence of events is shown schematically in Figures 3.5a and 3.5b. In Figure 3.5a, the plasma is driving the magnetic field inward. As a result, the location of the interface between the plasma and the stellar magnetic field, the so-called magnetopause, is not determined by static pressure balance. As the increasingly compressed stellar magnetic field becomes stiffer, it eventually slows the inflow, and a shock wave forms. A brief interval of quasi-static evolution follows, as plasma accumulates between the outward-moving shock front at r_s and the inward-moving magnetopause at r_m, as shown in Figure 3.5b. During this phase little plasma reaches the stellar surface, and the neutron star is not a luminous X-ray source, although some emission of ~100 keV hard X-rays is expected. Even so, the magnetopause continues to move inward as plasma accumulates outside.

As discussed below, very little of the shock-heated plasma is expected to penetrate the stellar magnetic field on a free-fall time, and hence at first very little plasma reaches the stellar surface to produce cooling X-rays. Moreover, thermal conduction can probably be neglected, since even a very small magnetic field in the accreting plasma will suppress it. Thus, the shock-heated plasma is expected to cool primarily via electron-ion energy exchange and bremsstrahlung radiation. Since the bremsstrahlung cooling time is typically longer than the time required for the magnetosphere to assume its equilibrium shape (a time of order the free-fall time, which is greater than or comparable to the time for an Alfvén wave to cross the magnetosphere), there is a brief interval when the shape of the magnetosphere is determined by a quasi-static balance between the pressure of plasma and swept-up magnetic fields outside the magnetosphere and the pressure of the stellar magnetic field inside.

Relative motion between the plasma just outside and just inside the magnetopause will tend to drive the magnetopause Kelvin-Helmholtz unstable, mixing the accreting plasma into the magnetosphere. If a sufficient quantity of plasma enters the magnetosphere and falls to the stellar surface, releasing X-rays, inverse Compton cooling will accelerate the cooling of the plasma near the magnetopause. After approximately one cooling time, the plasma near the magnetopause has become sufficiently dense that the magnetopause tends to become Rayleigh-Taylor unstable. At this point a large-scale flow of plasma into the magnetosphere develops, as shown schematically in Figure 3.5c. Once this large inward flow toward the stellar surface has begun, the large-scale structure of the magnetosphere is distorted by the stresses associated with the copious flow of plasma into the magnetosphere and static pressure balance is no longer a good guide to the location and shape of the magnetopause. When the increased inflow of plasma approaches the stellar surface, the star begins to emit an increased flux of X-rays, speeding cooling. If the infalling plasma is threaded rapidly by the stellar magnetic field, ordered flow cannot extend outside the Alfvén radius r_A defined in §3.3.

Clearly, the time development of approximately radial accretion flows and the resulting structure of the magnetosphere depend sensitively on whether or not the plasma-field interface is hydromagnetically stable. It is therefore useful to consider models in which the magnetopause is presumed stable separately from those in which it is unstable.

3.5.3 *Stable magnetopause.* When the magnetosphere is hydromagnetically stable and in quasi-static equilibrium, the shape of the magnetosphere is determined by conservation of linear momentum. During this phase, the accreted plasma is separated from the interior of the magnetosphere by a current layer. Electron and proton temperatures $T \sim 10^{10}$ K and particle densities $n \sim 10^{15}\ \mathrm{cm}^{-3}$ are typical of the conditions expected outside the magnetosphere for accretion rates typical of the luminous neutron star X-ray sources. Lower particle densities may be of interest for neutron star models of flaring X-ray and γ-ray sources (Lamb *et al.* 1977). Under these conditions, both the electrons and protons are collisionless, with mean free paths $\gtrsim r_{m0}$.

The current layer separating the accreted plasma from the interior of the magnetosphere is expected to have a thickness $\Delta r_m \sim a_i$, where a_i is the ion Larmor radius in the magnetospheric field (Lamb, Pethick, and Pines 1973). Since a_i is typically $\sim$1 cm, Δr_m is completely negligible compared to r_{m0}, and the gravitational force acting on the plasma in the boundary layer can be neglected in calculating the balance of linear momentum across the magnetopause, which may therefore be expressed as

$$\left(\frac{B_t^2}{8\pi}\right)_{\mathrm{in}} = \left(\frac{B_t^2}{8\pi}\right)_{\mathrm{out}} + P_{\mathrm{out}}, \tag{3.9}$$

where B_t is the component of the magnetic field tangential to the boundary and P is the plasma thermal pressure. The subscripts indicate whether the terms are to be evaluated just inside or just outside the magnetopause.

Applying conservation of linear momentum across the initially-thin shocked plasma layer (including the effect of the gravitational force on the plasma within the layer), the initial plasma thermal pressure just outside the magnetopause is

$$P_{\mathrm{out}} \approx \rho v_r^2, \tag{3.10}$$

where v_r is the (supersonic) inward radial velocity of the accreting plasma approaching the shock front. Thus, the *initial* radius r_{m0} of the magnetopause is the so-called Chapman-Ferraro radius r_{CF}, defined implicitly by (Chapman and Ferraro 1931)

$$\left(\frac{B_t^2}{8\pi}\right) = \rho v_r^2 . \tag{3.11}$$

For a purely radial inflow, the shape and location of the initial plasma-magnetic field interface $r_{m0}(\theta)$ is given by equation (3.11). As discussed below in §3.7, typical values of r_{CF} are $\sim 10^7$–10^9 cm for accretion-powered pulsars, much greater than the radius $R \approx 10^6$ cm of the neutron star. At these radii, the stellar magnetic field is dominated by its dipole component, and there are point cusps above the magnetic poles.

As accretion continues, plasma accumulates between the shock front and the magnetopause, and hence the radius of the shock front tends to increase. Simultaneously, the shocked plasma cools, the pressure at the magnetopause increases, and hence r_m decreases. If gravitational forces were absent, the magnetopause would remain stable as the plasma cools, since it is convex toward the plasma. However, gravitational forces *are* present, and hence the boundary becomes Rayleigh-Taylor unstable as soon as the plasma cools sufficiently (see below).

Appreciable entry of plasma by *any* mechanism causes copious emission of X-rays from the stellar surface and extremely rapid cooling of the plasma at the magnetospheric boundary by the inverse Compton process, as discussed above. Thus, in order to compete effectively with the Kelvin-Helmholtz and Rayleigh-Taylor instabilities, other entry mechanisms must create an average inflow velocity comparable to the free-fall velocity at the magnetospheric boundary, since the average inflow velocity produced by hydromagnetic instabilities will be of this order in any case.

What processes might compete with hydromagnetic instabilities? Particles can enter the magnetosphere via loss-cone entry through the polar cusps and microscopic diffusion across the magnetopause. However, these processes are not important for conditions typical of the luminous binary X-ray sources (Elsner and Lamb 1984). According to current theoretical models of magnetic flux reconnection, plasma entry into the magnetosphere via reconnection cannot compete with the Kelvin-Helmholtz and Rayleigh-Taylor instabilities unless the magnetic field within the accreting plasma is comparable in scale and strength to the magnetospheric magnetic field; otherwise magnetic flux linkage will not proceed fast enough nor go deep enough into the magnetosphere for reconnection to be competitive. Among other things, this requires that the strength of the magnetic field in the plasma accreting onto a given neutron star have a particular value at the capture radius, otherwise the magnetic field at the magnetopause will either be too weak or too small-scale for entry via flux reconnection to compete with entry via the Kelvin-Helmholtz and Rayleigh-Taylor instabilities. Given the available evidence concerning magnetic fields and plasma flow patterns in neutron star binaries, it appears very unlikely that plasma entry into the magnetosphere via magnetic flux reconnection can be competitive (Elsner and Lamb 1984).

In summary, the plasma entry mechanisms that operate when the magnetopause is hydromagnetically stable cannot compete with plasma entry via hydromagnetic instabilities, except perhaps in very exceptional circumstances.

3.5.4 *Unstable magnetopause.* Consider now the mixing of plasma into the magnetosphere produced by the Rayleigh-Taylor and Kelvin-Helmholtz instabilities.

Although there is general agreement on the conditions under which the magnetospheric boundary is Rayleigh-Taylor unstable, the wavelengths of the modes that are most important for transporting plasma into the magnetosphere when this happens remain uncertain. Lamb (1975a,b), Elsner (1976), and Elsner and Lamb (1976, 1977) emphasized the likely importance of relatively long-wavelength modes and rapid diffusion of plasma onto magnetospheric magnetic field lines, whereas Arons and Lea (1976a,b, 1980) assumed that only short-wavelength modes can exist and that the plasma would diffuse only slowly onto field lines. In evaluating the results of a numerical study, Wang and Nepveu (1983) conjectured that mass transport into the magnetosphere via long-wavelength modes is favored, both because the growth of long-wavelength modes tends to saturate more slowly than that of shorter wavelength modes and because there are numerous effects, including rotation, viscosity, and magnetic shear, that preferentially stabilize shorter-wavelength modes. However, in a later numerical study, Wang, Nepveu, and Robertson (1984) found that the shortest wavelengths consistent with the effective viscosity of the plasma flow were likely to be the most important (these could, however, be comparable to r_m, if the pressure scale-height in the shocked plasma is $\sim r_m$).

If the most important nonlinear modes turn out to have wavelengths that are a substantial fraction of r_m, the problem of flow into the magnetosphere becomes much more complex than assumed by Arons and Lea, because tangential pressure gradients and material stresses will distort the shape of the outer magnetosphere, while the magnetopause itself (here defined as the region that contains screening currents and in which the magnetic field transfers momentum to the inflowing plasma) may broaden to an appreciable fraction of r_m. If relatively long-wavelength modes are the most important for plasma flow into the magnetosphere, there are a number of important and interesting observational consequences. For example, in persistent X-ray sources in which plasma enters the magnetosphere predominantly as a result of the Rayleigh-Taylor instability, one may expect relatively large amplitude X-ray intensity fluctuations on time scales comparable to the dynamical time at the magnetospheric boundary. Mass transport into the magnetosphere by relatively long-wavelength modes also implies relatively large fluctuations in the corresponding accretion torque, with important observational consequences. Finally, the growth of such modes will produce plasma sheets or filaments that have a much greater chance of surviving intact deep into the magnetosphere, where their breakup would lead to plasma inflow over a substantial fraction of the stellar surface and hence to the formation of complex pulse waveforms at keV energies (Elsner and Lamb 1976).

The possibility that the Rayleigh-Taylor instability occurs preferentially near the polar cusps is of interest because X-ray emission from accretion-powered pulsars is localized on the stellar surface, as indicated by the pulsed emission. Even though the cusp region is inherently more stable than the equatorial magnetopause, most of the accreting plasma might still enter the magnetosphere there if the luminosity is low and the cusps are preferentially illuminated by X-rays from the stellar surface. However, entry near the cusps cannot be the dominant entry process if the star has a persistent luminosity greater than $\sim 10^{36}\,\mathrm{ergs\,s^{-1}}$ and, as in the presently known pulsing sources, a substantial fraction of the magnetopause is illuminated (Elsner and Lamb 1984). Under these conditions, plasma near the magnetopause cools in a time much shorter than the free-fall time there, and hence there is no time for an extensive plasma atmosphere to accumulate outside the magnetosphere. Instead, almost everywhere outside the magnetosphere plasma will fall into it via Rayleigh-Taylor instability of the magnetopause before it has time to flow a significant distance around the magnetospheric boundary. Plasma above the cusp axis, which is stable, can enter the magnetosphere by flowing a short distance sideways to a point where the magnetopause is unstable.

For approximately spherically-symmetric radial accretion by neutron stars with moderate rotation rates, the relative motion of the magnetosphere and the surrounding plasma suppresses the Rayleigh-Taylor instability. Burnard, Lea, and Arons (1983) have studied such stars and concluded that the accreting plasma is mixed into the magnetosphere by the Kelvin-Helmholtz instability. However, Wang and Welter (1982) concluded that even a very weak magnetic field within the plasma may suppress this instability, and suggested that mixing of moving plasma into the magnetosphere is not the result solely of the Kelvin-Helmholtz instability. In a later numerical study of the nonlinear development of the Kelvin-Helmholtz instability, Wang and Robertson (1984) found that the tendency for neighboring vortices to coalesce causes the longest wavelength modes to dominate the mixing process. They also found that any component of the magnetic field parallel to the flow does have a stabilizing effect. Wang and Robertson concluded that the thickness of the Kelvin-Helmholtz mixing layer is likely to grow until it becomes comparable to the scale-height of the plasma outside the magnetopause, or until it becomes comparable to the scale-length over which the field-line orientation changes near the magnetospheric boundary.

3.6 EFFECTS OF STELLAR ROTATION

Accretion-powered pulsars are frequently characterized as short period ($P < 10\,\mathrm{s}$), intermediate ($10\,\mathrm{s} \leq P \leq 100\,\mathrm{s}$), or long-period ($P > 100\,\mathrm{s}$) rotators. However, the the ***dynamical*** importance of the neutron star's spin depends on the size of the magnetosphere compared to the so-called ***centrifugal radius***

$$r_c = (GM/\Omega_s)^{1/3} = 1.5 \times 10^8\, P^{2/3} (M/M_\odot)^{1/3} \quad \mathrm{cm}\,, \tag{3.12}$$

where the centrifugal force acting on plasma corotating with the star just balances gravity. Here P is the stellar rotation period in seconds. The dynamical importance of the stellar spin therefore depends on the mass accretion rate and the magnetic moment of the star, as well as the star's spin rate.

For plasma within the magnetosphere that has been brought into corotation with the star, centrifugal force becomes comparable to gravity when r_c approaches r_m or, equivalently, when the so-called ***fastness parameter*** (Elsner and Lamb 1977; Ghosh and Lamb 1979a)

$$\omega_s \equiv \Omega_s / \Omega_K(r_m) \tag{3.13}$$

approaches 1. Here the magnetospheric radius r_m is to be interpreted as the angular momentum balance radius r_0, for Keplerian accretion flows, or the Chapman-Ferraro radius $r_{\rm CF}$, for approximately radial flows. The star's rotation is dynamically unimportant for $\omega_s \ll 1$. Note that $\omega_s = (r_m/r_c)^{3/2}$. Values of ω_s inferred from observation range from $\sim 10^{-3}$ to ~ 1.

3.7 SCALE OF THE MAGNETOSPHERE

The various order-of-magnitude estimates of the scale of the magnetosphere discussed in the preceding sections, and the types of flows to which they are applicable and inapplicable, can be summarized as follows.

3.7.1 *Stable magnetospheric flow.* An estimate of the radius inside which ordered, field-aligned flow is possible can be made based on the MHD stability arguments summarized in §3.3. As explained there, field-aligned flow is stable inside the Alfvén radius r_A, given implicitly by the condition

$$\frac{B_p^2}{4\pi} = \rho v_p^2\,. \tag{3.14}$$

This condition recognizes the fundamental characteristic of accretion, namely inward flow of plasma, and can be applied when the flow outside the magnetosphere is radial, Keplerian, or anything in between.

As an example of the application of condition (3.14), suppose we assume as a rough approximation that (1) the poloidal component of the magnetic field in the outer magnetosphere is given approximately by

$$B_p(r) = \mu / r^3\,, \tag{3.15}$$

where μ is the stellar dipole moment, (2) the inward poloidal velocity v_p of the accreting plasma at radius r is approximately equal to the free-fall velocity $v_{\rm ff}(r)$

there, and (3) the mass density ρ at radius r in the outer magnetosphere is equal to the mass density

$$\rho_{\rm ff}(r) \equiv \dot{M}/4\pi r^2 v_{\rm ff}(r) \tag{3.16}$$

that would be there if the flow were spherically-symmetric inflow at the free-fall velocity. Then $\rho v_p^2 \propto r^{-5/2}$ whereas $B_p^2 \propto r^{-6}$, and the Alfvén radius is

$$r_{A_1} \approx \left(\frac{\mu^4}{GM\dot{M}^2}\right)^{1/7} \approx \left(\frac{\mu^4 GM}{L^2R^2}\right)^{1/7} = 4\times 10^8 \left[\frac{\mu_{30}^4}{L_{37}^2 R_6^2}\left(\frac{M}{M_\odot}\right)\right]^{1/7} \text{cm}. \tag{3.17}$$

In the last two expressions on the right we have used equation (2.2) and in the last expression μ_{30} is the magnetic moment in units of 10^{30} gauss cm^3, L_{37} is the accretion luminosity in units of 10^{37} ergs s^{-1}, and R_6 is the neutron star radius in units of 10^6 cm.

As a second example of the application of condition (3.14), suppose that (1) the poloidal component of the magnetic field in the outer magnetosphere is approximately dipolar, (2) the flow in the outer magnetosphere is field-aligned, so that

$$\rho v_p/B_p = \text{const}. \tag{3.18}$$

on each flux tube, (3) the inward poloidal velocity v_p of the accreting plasma at r is again approximately equal to the free-fall velocity $v_{\rm ff}(r)$, and (4) the accreting plasma falls on an area A at the stellar surface, which is a fraction $f \equiv A/4\pi R^2$ of the whole surface. Then $\rho v_p^2 \propto r^{-7/2}$ while B_p^2 is again $\propto r^{-6}$. The resulting estimate of the Alfvén radius is

$$r_{A_2} \approx \left(\frac{\mu^4 f^2}{GM\dot{M}^2R^2}\right)^{1/5} \approx \left(\frac{GM\mu^4 f^2}{L^2R^4}\right)^{1/5} \approx 7\times 10^8 \left[\frac{\mu_{30}^4 f_{-2}^2}{L_{37}^2 R_6^4}\left(\frac{M}{M_\odot}\right)\right]^{1/5} \text{cm}, \tag{3.19}$$

where $f_{-2} \equiv f/10^-2$. Equation (3.19) is particularly useful for setting an upper bound on the size of the magnetosphere if observational data provide an estimate of f (see, for example, Tuohy *et al.* 1981).

Note that if $f \approx R/r_{A_2}$, as expected on dimensional grounds if threading occurs readily (Lamb, Pethick, and Pines 1973), the estimate r_{A_2} is the same as the estimate r_{A_1}.

3.7.2 *Angular momentum conservation.* For Keplerian disk flows, the radius at which the Keplerian disk ends is determined by angular momentum conservation, as explained in §3.4. The radius r_0 of the inner edge of the Keplerian disk is therefore given implicitly by

$$\gamma B_p^2 r^2 = \dot{M} v_K\,, \tag{3.20}$$

where

$$\gamma \equiv (B_\phi/B_p)(\Delta r/r) \tag{3.21}$$

in terms of the mean azimuthal magnetic pitch B_ϕ/B_p in the region of radial width Δr at the inner edge of the Keplerian disk where the stellar magnetic field strongly interacts with the flow, and $v_K = (GM/r)^{1/2}$ is the Keplerian azimuthal velocity.

As an example of how condition (3.20) can be applied, suppose that the neutron star is slowly rotating ($r_c \gg r_m$). Suppose also that all important length scales in the inner part of the Keplerian disk flow are approximately proportional to r. This is in fact the case for the standard "α-model" of geometrically thin Keplerian disk flows around neutron stars, in the so-called gas-pressure-dominated (GPD) region, which extends outward from a transition radius r_t, which is $\sim 10^8$ cm for neutron stars accreting at rates near their Eddingon critical rates (see Treves, Maraschi, and Abramowicz 1988 and §4.1). Then, since the star is not rotating and the stellar magnetic field has no length scale either, the interaction between the Keplerian flow and the stellar magnetic field *in the strong interaction region* has no length scale. As a result, the mean azimuthal magnetic pitch B_ϕ/B_p in the strong interaction region must be independent of r, while the width Δr of the interaction region must scale approximately as r. Thus, γ is independent of r, and

$$r_0 \approx \gamma^{2/7}\left(\frac{\mu^4}{GM\dot{M}^2}\right)^{1/7} \approx \gamma^{2/7}\left(\frac{\mu^4 GM}{L^2R^2}\right)^{1/7}$$

$$= 4\times 10^8\,\gamma^{2/7}\left[\frac{\mu_{30}^4}{L_{37}^2R_6^2}\left(\frac{M}{M_\odot}\right)\right]^{1/7}\ \text{cm}. \tag{3.22}$$

What is a reasonable estimate for γ? The width Δr of the region where the stellar magnetic field strongly interacts with the Keplerian flow is limited by the fact that $B_p^2 r^2 \propto r^{-4}$ whereas $\dot{M}v_K \propto r^{-1/2}$. Thus, from the steep dependence of B_p on r alone, one would expect $\Delta r \approx 0.3\,r$. The actual width may be smaller, due to toroidal screening currents generated in the flow both in the disk plane and above and below it and toroidal screening currents flowing within the magnetosphere above and below the disk. It seems unlikly that Δr can be less than $\sim 0.01\,r$. The mean azimuthal magnetic pitch B_ϕ/B_p is limited to values $\sim$1–3 by the expansion of the magnetosphere that occurs when the magnetospheric magnetic field is sheared, as discussed in more detail in §4.2. Thus, $\gamma^{2/7}$ is probably in the range 0.3–0.7. With these assumptions, one finds $r_0 \propto \dot{M}^{-2/7}$.

To get some idea of the sensitivity of the scaling of r_0 with $\dot{M}$ and other parameters, consider a second example. In the inner, radiation-pressure-dominated (RPD) region of the standard, optically-thick α-disk, the disk thickness is approximately independent of radius (again see Treves, Maraschi, and Abramowicz 1988 and §4.1). Suppose that as a result, the width Δr of the interaction region is approximately independent of r and equal to a fraction of the disk thickness h. Assuming that the mean azimuthal magnetic pitch B_ϕ/B_p in the interaction region is also independent of r, this gives $\gamma \approx f\,(h/r)$, where $f \equiv (\Delta r/h)(B_\phi/B_p)$, and

$$r_0 \approx \left(\frac{f^2h^2\mu^4}{GM\dot{M}^2}\right)^{1/9} \approx \left(\frac{f^2h^2\mu^4 GM}{L^2R^2}\right)^{1/9} \tag{3.23}$$

With these assumptions, $r_0 \propto \dot{M}^{-2/9}$.

Note that in these two examples, the exponent of $\dot{M}$ varies from about 0.2 to about 0.3. Obviously, more careful estimates of the radius r_0 of the inner edge of the Keplerian flow can be made, based on more detailed modeling of the structure of the inner disk and of the disk-magnetosphere interaction, as described in §4. Several models of the inner disk lead to scalings not very different from the estimate (3.22). However, *an important complication arises for neutron stars rotating sufficiently*

TABLE 3.5
Physical Principles and Characteristic Radii

Physical Principle	Relevant Condition	Characteristic Radius
MHD stability of field-aligned flow	$B_p^2/8\pi = \rho v_p^2$	Maximum radius r_A of stable magnetospheric flow
Angular momentum conservation	$\gamma B_p^2 r^2 = \dot{M} v_K$	Inner disk radius r_0 for Keplerian disk flows
Linear momentum conservation	$B_t^2/8\pi = \rho v_r^2$	Initial magnetopause radius $r_{\rm CF}$ for nearly radial flows

rapidly that the corotation radius r_c is not large compared to r_0. For such stars, other lengths (such as $r_0 - r_c$) enter the problem, and may cause the scaling of r_0 to deviate significantly from the slow-rotator scaling (3.22).

3.7.3 *Linear momentum conservation.* For radial flows, the initial shape and location of the magnetopause is determined by linear momentum conservation, as explained in §3.5. The initial radius $r_{\rm CF}$ of the magnetopause is given implicitly by

$$\left(\frac{B_t^2}{8\pi}\right) = \rho v_r^2 \,. \tag{3.24}$$

However, once the magnetopause becomes unstable and there is a large flux of plasma into the magnetosphere, material stresses will distort the magnetopause and hence condition (3.24) will give only a rough estimate of the location of the magnetospheric boundary.

As an example of the application of condition (3.24), suppose that (1) the magnetic field at the radius r_m of the magnetospheric boundary varies approximately as r_m^{-3}, (2) the accreting plasma is falling freely toward the neutron star, and (3) only a thin layer of shocked plasma has accumulated outside the magnetopause. Then the Chapman-Ferraro radius is

$$r_{\rm CF} \approx \left(\frac{\mu^4}{GM\dot{M}^2}\right)^{1/7} \approx \left(\frac{\mu^4 GM}{L^2 R^2}\right)^{1/7} = 4 \times 10^8 \left[\frac{\mu_{30}^4}{L_{37}^2 R_6^2}\left(\frac{M}{M_\odot}\right)\right]^{1/7} \text{ cm} \,. \tag{3.25}$$

For easy reference, the physical principles discussed here and the characteristic radii that they give are listed in Table 3.5.

4. Disk Accretion by Magnetic Stars

In considering disk accretion by magnetic neutron stars, one could address the observed X-ray pulsing behavior directly by trying to model an oblique rotator, and there have been some attempts to do this (see, *e. g.*, Anzer and Borner 1980). However, two features make the oblique rotator much more difficult to handle than the aligned rotator. First, the rotation of the star imposes an explicit time dependence on the flow. Second, the oblique rotator has no symmetry, so one is forced to attempt three-dimensional modeling. Even when the rotation axes of the disk and star are parallel, the magnetic field in the disk plane is not axisymmetric, the magnetic stresses acting on the disk plasma are explicitly time-dependent, and magnetic pressure gradients affect the azimuthal, as well as the radial, motion. These difficulties have so far prevented detailed quantitative calculations of accretion by oblique rotators.

In contrast to the oblique rotator, the aligned rotator allows the possibility of a stationary flow and, if the flow is axisymmetric, two-dimensional modeling. Moreover, solutions for the poloidal magnetic field around a perfectly conducting rigid plate with a circular hole and a magnetic dipole centered in the hole have a topology near the inner edge of the plate that is the same as that of the aligned rotator, for magnetic field inclination angles $\lesssim 60°$. (Aly 1980; Arons 1987). For these reasons, we consider only axisymmetric disk accretion by an aligned rotator in the discussion that follows.

4.1 DISK STRUCTURE

In an extensive Keplerian accretion disk, the properties of the flow well inside the outer radius r_d of the disk are relatively insensitive to the properties of the stream of plasma flowing into the disk, since the radial diffusion of the plasma causes it to "forget" many of the details of the flow at the outer boundary. Most discussions of disk accretion by magnetic neutron stars therefore assume that the flow near the magnetosphere is steady and Keplerian.

The geometrical thickness of accretion disks depends on the mass flux $\dot{M}$. If $\dot{M}$ is $\ll \dot{M}_E$, where $\dot{M}_E$ is the mass flux that would produce an accretion luminosity (see eq. [2.2]) equal to the Eddington critical luminosity

$$L_E \equiv 4\pi G M m_p c/\sigma_T = 1.3 \times 10^{38}\,(M/M_\odot) \quad \mathrm{ergs\,s^{-1}} \,. \tag{4.1}$$

geometrically-thin models (with heights $h \ll r$) are possible. However, if $\dot{M} \approx \dot{M}_E$, the height becomes comparable to the radius. Finally, if $\dot{M} \gg \dot{M}_E$, only geometrically-thick solutions are possible.

At present there is no truly satisfactory theory of Keplerian disk flows. The diverse time scales and complexity of such flows is too great for current supercomputers and analytical methods. As a result, almost all modeling is based on relatively simple, semi-analytical solutions for geometrically-thin Keplerian flows. The most widely used model of such flows is the so-called *α-model* introduced by Shakura (1972) and Shakura and Sunyaev (1973).

4.1.1 *The α-model of accretion disks.* The key assumptions of the α-model are as follows (see Pringle 1981; Treves, Maraschi, and Abramowicz 1988):

- The accretion disk flow is steady.
- The flow is axially symmetric about the neutron star, and hence $\partial/\partial\phi \equiv 0$.
- The flow is symmetric with respect to the plane $z = 0$, and hence the vertical structure of the flow depends only on $|z|$.

- The flow is geometrically thin, *i. e.*, its semi-thickness h at radius r is small compared to r.
- The shear stress is related to the thermal pressure ρc_s^2 by

$$-\eta r \frac{\partial \Omega}{\partial r} \equiv \alpha \rho c_s^2 \,, \tag{4.2}$$

where c_s is the sound speed and η is the effective dynamic viscosity, which is thought to be due to turbulence and/or magnetic field amplification and dissipation in the disk (the kinetic viscosity is negligible). The dimensionless quantity α, which is assumed constant in space and time, is a free parameter of the model (however, the model is self-consistent only if $\alpha \lesssim 1$).
- The azimuthal motion is closely Keplerian, *i. e.*, v_ϕ at radius r is very close to v_K.
- The velocity field satisfies $|v_z| \ll |v_r| \ll |v_\phi|$, and $|v_r| \ll c_s$.
- The radial pressure gradient and heat flux are negligible.
- The vertical optical depth to absorption or scattering is greater than 1.

The discussion that follows includes references to the radial drift time scale $t_r \equiv |r/v_r|$ and the hydrodynamic time scale $t_H \equiv h/c_s$.

4.1.2 *Vertical structure.* Plasma in the α-disk is confined vertically by gravity. Since $|v_r|$ and $|v_z|$ are small compared to c_s, the disk plasma is in hydrostatic equilibrium in the z-direction. Thus,

$$-\rho\, g_z \approx -\partial p/\partial z \,, \tag{4.3}$$

where g_x is the z-component of the gravitational acceleration. The left side of equation (4.3) is $\sim \rho(h/r)(v_K^2/r)$ while the right side is $\sim \rho c_s^2/h$. Hence, equation (4.3) implies

$$h \approx (c_s/v_K)\, r \,. \tag{4.4}$$

At all but the smallest radii, estimates of the vertical heat flux carried by radiation lead to temperatures sufficiently small that $c_s \ll v_K$, so that the initial assumption that the flow is geometrically thin turns out to be self-consistent.

4.1.3 *Radial drift velocity.* The inward radial velocity of the flow in the α-disk can be estimated from angular momentum conservation, since plasma in the disk can fall inward only as it loses its angular momentum, which is carried outward by the viscous shear stress. The net transport of angular momentum through the disk is given by the difference between transport inward by the material stress, $\sigma_m \approx \rho v_r v_\phi$, and transport outward by the shear stress, $\sigma_v \approx \eta r \partial\Omega/\partial r$. Thus, for a steady flow, angular momentum conservation implies

$$4\pi\eta h r^3 \frac{\partial \Omega}{\partial r} + \dot{M} r^2 \Omega = \dot{J}_s \,, \tag{4.5}$$

where the first term on the left side describes the outward transport of angular momentum by the shear stress and the second, the inward transport by the material stress. $\dot{M}$ is the inward mass flux while $\dot{J}_s$ is the net inward flux of angular

TABLE 4.1
Radial Structure of the α-Disk

Name	Dominant Pressure	Dominant Opacity
Outer disk	Gas pressure	Absorption
Middle disk	Gas Pressure	Electron scattering
Inner disk	Radiation pressure	Electron scattering

momentum. Neither $\dot{M}$ nor $\dot{J}_s$ are determined by the model. Assuming that the flow is steady, they are independent of radius and time, and are usually presumed to be fixed by the boundary conditions at the outer and inner edges of the disk flow.

At radii r much greater than the radius r_0 of the inner edge of the disk, both terms on the left side of equation (4.5) scale approximately as $r^{1/2}$. Hence, at sufficiently large radii, the two terms on the left side are each much larger than the term on the right side, and equation (4.5) becomes

$$-\alpha\rho c_s^2 - \rho v_r v_\phi \approx 0\,. \tag{4.6}$$

Therefore, the radial drift velocity v_r at large radii is

$$v_r \approx -\alpha(c_s^2/v_K^2)\,v_K \approx -\alpha(h/r)^2\,v_K\,, \tag{4.7}$$

where in the last expression on the right we have used equation (4.4). This gives the estimate

$$t_r \approx (v_K^2/\alpha c_s^2)\,t_H \gg t_H \tag{4.8}$$

for the radial drift time.

4.1.4 *Radial structure.* The α-disk has three distinct regions, which are listed in Table 4.1. The transition between the middle, gas-pressure-dominated (GPD) region and the inner, radiation-pressure-dominated (RPD) region occurs very gradually near the transition radius (see Treves, Maraschi, and Abramowicz 1988)

$$r_t \approx 10^8\,\alpha^{2/21}(M/M_\odot)^{1/3}R_6^{16/21}(L/L_E)^{16/21}\quad\text{cm}\,, \tag{4.9}$$

where R_6 is the stellar radius in units of 10^6 cm. Thus, the magnetospheres of neutron stars with strong magnetic fields, such as the accretion-powered pulsars, are immersed in the GPD region, whereas the magnetospheres of stars with weak fields, like those thought to be present in the QPO sources, are immersed in the RPD region (White and Stella 1987).

The structure and behavior of the α-disk in the outer and middle GPD regions is comparatively well understood. Cooling is efficient in this region, and hence the disk is quite thin ($h/r \sim 10^{-2}$–10^{-3}). The structure and behavior of the inner,

TABLE 4.2

Stability Properties of Models of the Inner Region of Accretion Disks

Disk Model	Thermal Modes	Viscous Modes
1T Opt thick GPD, shear stress $\propto P_{\mathrm{tot}}$	Stable	Stable
1T Opt thick RPD, shear stress $\propto P_{\mathrm{tot}}$	Unstable	Unstable
1T Opt thick RPD, shear stress $\propto P_{\mathrm{gas}}$	Stable	Stable
2T Opt thin GPD, Compt soft photons	Stable?[a]	Stable
2T Opt thin GPD, Compt brems	Stable?[a]	Stable

[a]The stability properties of these models are not yet certain (see Pringle 1976; White and Lightman 1989, 1990; Lightman 1990).

RPD region is much less clear. In particular, α-models of the RPD region which are optically-thick in the vertical direction are thermally and viscously unstable, if the shear stress is, as originally proposed, proportional to the *total* pressure (Pringle, Rees, and Pacholczyk 1973; Lightman and Eardley 1974; Shakura and Sunyaev 1976; Pringle 1976; Piran 1978). Because of this, a variety of alternative models of the RPD region have been explored. One of the earliest is the so-called β-model proposed by Cunningham (1973), which assumes that the shear stress in the RPD is proportional to the gas pressure alone. This model is both thermally and viscously stable. However, whether the shear stress can be proportional to the gas pressure in the RPD region remains uncertain, although Stella and Rosner (1984) have argued that the shear stress produced by small-scale magnetic fields in the disk is proportional to the gas pressure. Models in which the shear stress is proportional to the product of various fractional powers of the gas pressure and the total pressure have also been investigated (see Taam and Lin 1984, Szuszkiewicz 1990, and references therein).

An alternative approach has been to construct models of the inner disk that are *optically thin* in the vertical direction, in contrast to the models just described. Optically-thin models are much hotter than their optically-thick counterparts. As a result, gas pressure dominates the radiation pressure. Also, the ions are typically much hotter than the electrons. Examples of such "two-temperature" disk models include the model proposed by Shapiro, Lightman, and Eardley (1976), in which the electrons are cooled by Comptonization of an external source of soft photons, and so-called "Comptonized bremsstrahlung models" (see White and Lightman 1989 and references therein), in which the dominant cooling mechanism is emission of bremsstrahlung which at high accretion rates is Comptonized by the electrons in the disk. In these models the α prescription for the shear stress has generally been retained. In recent years, the effects of electron-positron pairs on such hot disk models have also been considered (Liang 1979; Kusunose and Takahara 1988, 1989, 1990; Tritz and Tsuruta 1989; White and Lightman 1989, 1990; Bjornsson 1990; Bjornsson and Svensson 1991). Some essential features of the disk models described in this section are summarized in Table 4.2.

4.1.5 *Slim and thick disk models.* As noted earlier, if the mass flux through the disk approaches $\dot{M}_E$, the radiation pressure in the RPD region becomes so great that the inner disk expands vertically and can no longer be treated as geometrically thin (see Treves, Maraschi, and Abramowicz 1988). Efforts to extend the theory of accretion disks beyond the geometrically-thin approximation have been made by Paczyński and Bisnovatyi-Kogan (1981), Muchotrzeb and Paczyński (1982), and Abramowicz *et al.* (1988). In the resulting "slim" accretion disk models, the vertical structure is averaged in the same way as in geometrically thin disks, but the effects of radial gradients in the fluid variables are included in the radial structure equations. In particular, advection and radial pressure forces are included in the momentum equation, and advective transport of heat is included in the energy equation. With the resulting equations, a consistent treatment of moderately high accretion rates ($\dot{M} \approx \dot{M}_E$) is possible. True to their name, slim disks have vertical thicknesses h less than, but comparable to, the local value of the radius r.

At very high accretion rates ($\dot{M} \gg \dot{M}_E$), radiation forces cause the height of an accretion disk to become comparable to its radius. In such geometrically thick ($h \sim r$) accretion disks, radial and vertical gradients of the pressure and other fluid variables are generally comparable everywhere. Two possibilities for geometrically thick disks have been studied. One possibility is that radiation forces drive a strong wind, which determines the structure of the disk (see Frank 1979 and references therein). An alternative possibility is that radiation pressure and centrifugal force closely balance gravity, so that there is little mass loss. Solutions are possible in which the specific binding energy of the inner region of the disk is small compared with GM/r and the energy radiated per unit mass of accreting matter is small compared to that in geometrically-thin disks (Kozlowski, Jaroszyński and Abramowicz 1978; Abramowicz, Jaroszyński, and Sikora 1978; Jaroszyński, Abramowicz, and Paczyński 1980).

Constructing dynamical models of geometrically thick disks is much more difficult than for geometrically-thin or slim disks. However, in geometrically-thick disks, the shear stresses determine the angular momentum and enthalpy distributions, and hence the shape of surfaces of constant pressure. As a result, it is possible to construct global analytical models of the outer surface and the distribution of the luminosity and angular momentum on the surface without reference to the internal structure or the viscosity (Paczyński and Wiita 1980; Jaroszyński, Abramowicz, and Paczyński 1980; Abramowicz, Calvani, and Nobili 1980; Wiita 1982). Numerical models of geometrically-thick disks with $\alpha \ll 1$ have been constructed by Pachzyński (1980) and Różyczka and Muchotrzeb (1982).

4.1.6 *Effect of disk-magnetosphere interaction.* Using their model of disk accretion by magnetic stars, Ghosh and Lamb (1979a) calculated the changes in the structure of the inner disk produced by its interaction with a neutron star magnetosphere, when the inner edge of the disk lies in the GPD region, and found that the changes are relatively small, except when the neutron star is rotating rapidly. Most other authors have simply assumed that the interaction with the neutron star magnetic field leaves the disk structure unaltered from that around a nonmagnetic star, but this is certainly not a safe assumption in general.

4.2 DISK-MAGNETOSPHERE INTERACTION

In this section we discuss some of the consequences of the coupling between the disk flow and the stellar magnetic field described in §3.4. For the purposes of this discussion, we assume that (1) the neutron star has a dipole magnetic field, (2) the star is rotating with angular velocity Ω_s about its dipole axis, (3) the rotation axis is perpendicular to the plane of the disk, and (4) the flow is steady and has axial symmetry everywhere. We use a cylindrical polar co-ordinate system (ϖ,ϕ,z) centered on the neutron star and oriented such that the polar (z-) axis lies along the rotation axes of the star and disk.

The matter in the neutron star and the disk has a high electrical conductivity. Therefore, as this matter moves in directions perpendicular to the magnetic field, it creates a $\mathbf{v} \times \mathbf{B}$ polarization electric field. The potential difference V across the system created by the cross-field motion of disk plasma could be as large as

$$V = 300\,\frac{1}{c}\int \mathbf{v} \times \mathbf{B} \cdot d\boldsymbol{\ell} \approx E_r\,\Delta r \sim 10^{13}\ \text{volts}\,. \tag{4.10}$$

The presence of such large potential differences may play a role in creating the TeV γ-rays reported from some accretion-powered pulsars, as described in §6.

If there is sufficient plasma between the star and the disk, the electrical potential difference can drive appreciable electrical currents within the system. The resulting $\mathbf{j} \times \mathbf{B}$ forces act on the matter in the star, the magnetosphere, and the disk to reduce their relative motion. The character of this electrodynamic interaction depends on the distribution of electrical resistance along the current paths. Ghosh and Lamb (1979a) examined this question and concluded that the conductivity along field lines in the magnetosphere is likely to be high and that cross-field currents are more likely to flow in the disk and the star. They therefore arrived at a system of field-aligned currents that is not unlike the polar-cap ionospheric current system that is thought to couple the rotation of Jupiter's ionosphere to its magnetospheric sheath (see Kennel and Coroniti 1975, 1977).

An example of this type of current system is shown in Figure 7. The azimuthal motion of the disk plasma relative to the star generates electric fields that drive field-aligned currents within the magnetosphere and cross-field currents within the star and the disk. The poloidal components of the magnetospheric currents create a toroidal magnetic field, which acts to synchronize the orbital motion of the disk plasma and the rotation of the star. In an analogous manner, the radial inward motion of the plasma across the poloidal component of the magnetic field within the disk generates a toroidal electric field that drives toroidal currents, partially screening the disk from the stellar magnetic field and pinching the magnetic field lines inward in the disk plane. The toroidal components of the field-aligned currents within the magnetosphere also contribute to the total poloidal magnetic field.

Despite its high electrical conductivity, the plasma between the disk and the star cannot support arbitrarily large electrical currents. Since the plasma in this region is essentially electrically neutral, the currents driven by electric fields are conduction currents. Such currents drive instabilities that dissipate them, if the magnitude of the relative velocity between the oppositely charged species exceeds the local sound speed (see Spicer 1982). How significant is this limitation on the strength of the electrodynamic coupling beween the disk and the star, for accretion-powered pulsars and QPO sources?

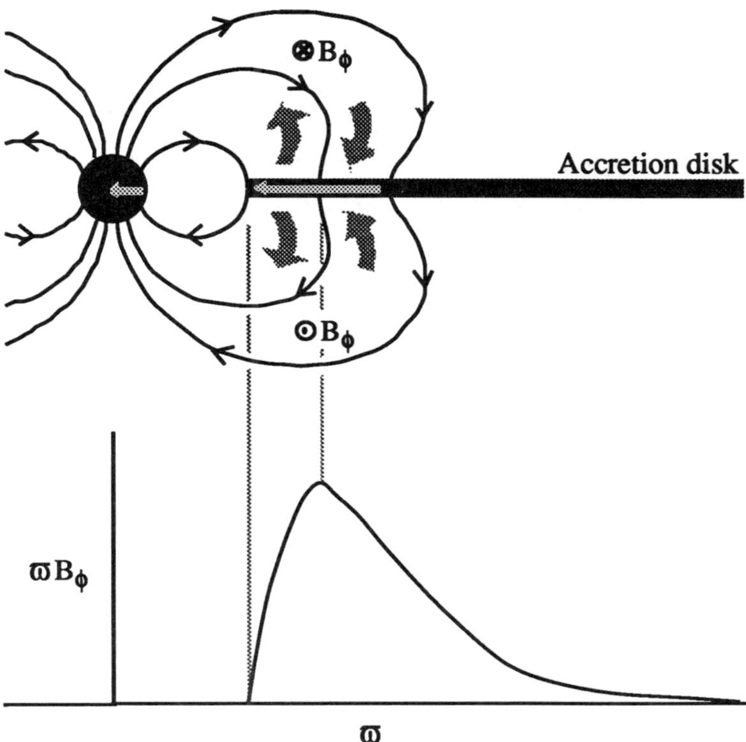

Fig. 4.1.—Side view of an aligned rotator (top) interacting with disk plasma that has an angular velocity greater than the star. The azimuthal motion of the plasma in the disk plane into the page generates a radial electric field within the disk that drives field-aligned currents (shaded arrows) within the magnetosphere and cross-field currents (light arrows) within the disk and the neutron star. The poloidal components of the magnetospheric currents create a toroidal magnetic field that points into the page above the disk and out of the page below. The resulting torque density ϖB_ϕ on the star (bottom) acts to try to spin up the star. The density of the poloidal field-aligned current within the magnetosphere vanishes on the poloidal flux surface where the torque density is a maximum. The toroidal components of the currents within the disk and the field-aligned currents within the magnetosphere create a poloidal magnetic field that points radially inward above the disk and radially outward below it, pinching the total poloidal field inward in the disk plane.

This question was examined by Ghosh and Lamb (1979a) and has recently been reconsidered by Zylstra (1988), based on new observational evidence concerning the plasma density surrounding neutron star X-ray sources. Zylstra estimated the magnetic field gradient $(\Delta B/\Delta r)_{max}$ that could be supported by the maximum stable current density. The undistorted stellar magnetic field near the inner edge of the disk is expected to be approximately dipolar in both pulsars and QPO sources, corresponding to stellar dipole magnetic moments $\sim 10^{30}$ G cm^3 and $\sim 10^{27}$ G cm^3, respectively. Spectral and temporal evidence indicates that the magnetospheres of the neutron stars in the QPO sources are surrounded by a dense central corona with an electron number density of at least $10^{16} - 10^{18}\,\mathrm{cm}^{-3}$ within $\sim 10^7 - 10^8$ cm from the neutron star. The evidence concerning the plasma density near the inner edge of Keplerian disks in accretion-powered pulsars is not as clear. However, observations of Her X-1 indicate the presence of a central corona with an electron number density $\gtrsim 10^{12}\,\mathrm{cm}^{-3}$ within $\sim 10^9$–10^{10} cm of the neutron star. Plasma with a comparable density is probably present between the disk and the star in other accretion-powered pulsars. Coronal temperatures are expected to be ~ 10 keV.

Based on this evidence, Zylstra (1988) estimated that electrical currents flowing within magnetosphere could support magnetic field gradients at radius r as large as

$$(\Delta B/\Delta r)_{max} \sim 10^4\text{–}10^8\, n_{12} r_8^4 (B_0/r)\,, \tag{4.11}$$

where B_0 is the magnitude of the undistorted dipole field at r. The smaller estimate is for QPO sources whereas the larger one is for accretion-powered pulsars. These estimates indicate that in both types of source, electrical currents flowing between the disk and the star could support very large magnetic field gradients. Of course, magnetic field gradients are also limited by other effects, such as magnetohydrodynamic instabilities, so that gradients as large as those that could be supported by the plasma between the disk and the star are unlikely to arise.

The change in the structure of the magnetosphere and the azimuthal magnetic pitch that occurs as the accretion disk pinches and twists the stellar magnetic field has been studied by Zylstra (1988) and Zylstra, Lamb, and Aly (1991), who used an iterative numerical scheme to determine the two-dimensional structure of the magnetospheric magnetic field, which was assumed to be in force-free equilibrium. Several distributions of the poloidal magnetic flux on the disk were explored. Given the symmetry of the problem, the poloidal flux distribution can be written in terms of a poloidal flux function ψ, which is constant on a given magnetic fieldline. Distributions of ψ at the disk plane were chosen to model strong and weak screening by toroidal electrical currents in the disk. One of the poloidal flux distributions Zylstra *et al.*explored is shown in Figure 4.2a. The effect of the azimuthal electrical currents can be judged by comparing the screened field with the unscreened field.

Various models of the interaction of the azimuthal motion of the disk plasma with the magnetic field were explored by specifying different distributions of the *torque density* $n \equiv \varpi B_\phi$ on the disk plane. In axisymmetric systems, the torque density is a function only of the poloidal magnetic flux function ψ. In their calculations, Zylstra *et al.*specified only the *shape* of the torque density function (TDF). The magnitude of the TDF was determined as part of the calculation. Given the distribution of ψ on the disk, knowledge of the TDF specifies the distribution of B_ϕ on the disk.

The sign of $n(\psi)$ is expected to reflect the sign of the difference between the orbital angular velocity of the disk flow and the rotational angular velocity of the star. The magnitude of $n(\psi)$ depends on the effective electrical conductivity of the disk, magnetosphere, and star, and the largest twists that the magnetic field can support in equilibrium. Figure 4.2b shows two of the TDFs Zylstra, Lamb, and Aly considered, corresponding to the interaction between the Keplerian disk flow and the star that is expected for slow rotators (curve SR) and fast rotators (curve FR). For the FR curve, the corotation radius (at which n vanishes) is only slightly larger than the radius of the inner edge of the disk.

Zylstra *et al.* find that as the magnetospheric field is twisted, the surfaces of constant poloidal flux expand outward away from the disk and star. As a result of

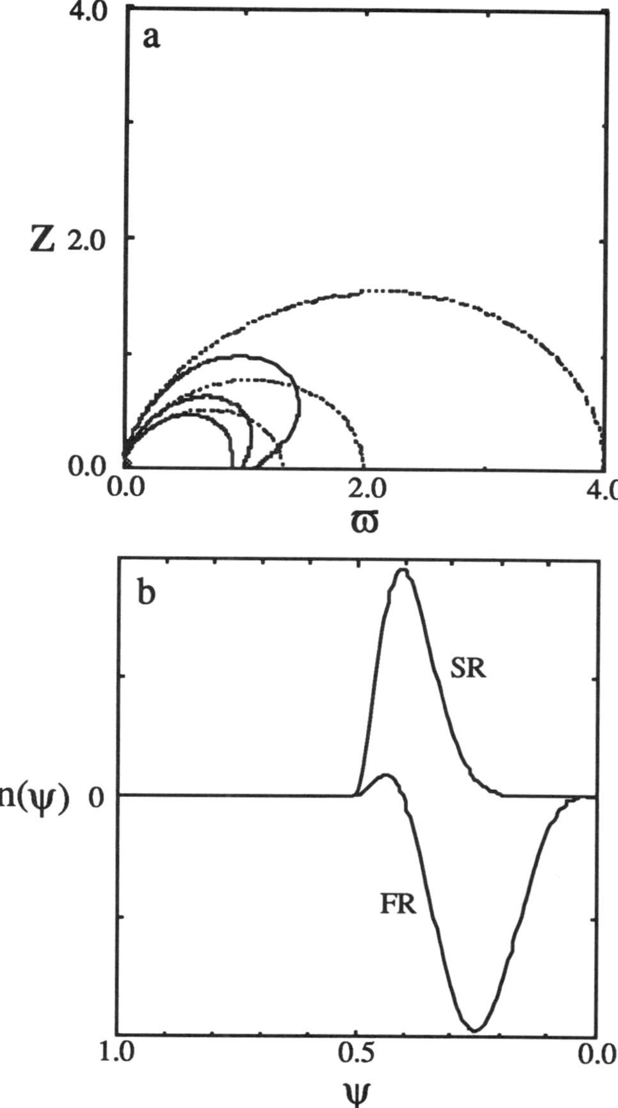

Fig. 4.2.—(a) Side view of the star, accretion disk, and magnetosphere, showing constant-flux surfaces of the undistorted dipolar magnetic field of the star (dashed curves) and the magnetospheric field produced by the currents in the star and the accretion disk (solid curves). The small, shaded quadrant in the lower left shows the size of the star. (b) Torque density functions $n(\psi)$ corresponding to a slow rotator (SR) and a fast rotator (FR). Note that the flux function ψ increases to the left. The equator of the star is at $\psi = 1.0$ while the inner edge of the disk is at $\psi = 0.5$. The fast rotator curve models a star with a spin rate such that the corotation point in the disk is at $\psi \approx 0.6$.

this expansion of the field above and below the disk, the poloidal component of the magnetic field develops an inward pinch in the disk plane as the twist increases, even if the poloidal flux distribution on the disk is unscreened (*i. e.*, the same as for an undistorted dipolar magnetic field). The expansion of the magnetic field is not uniform throughout the inner magnetosphere. Rather, the expansion is large for some field lines but relatively modest for others. This is illustrated in Figure 4.3, which shows the structure of the untwisted field and the maximally expanded field for a solution sequence based on the FR torque density function shown in Figure 4.2b. In this sequence, the stellar radius R is 0.875 times the radius ϖ_0 of the inner edge of the disk. As a result of the expansion of the magnetosphere, the magnitude of B_ϕ on and near the surfaces of the disk first increases and then decreases, while the magnitude of B_ϕ in regions far from the disk increases monotonically.

Zylstra, Lamb, and Aly also find that the structure of the magnetospheric field is relatively insensitive to the ratio of the stellar radius to the inner radius of the disk. Thus, for example, the maximum magnetic pitch $|B_\phi/B_\theta|_{max}$ at the surface of the disk depends only weakly on the ratio R/ϖ_0. It is 0.12 for $R/\varpi_0 = 0.875$ and 0.13 for $R/\varpi_0 = 0.100$. The appropriately scaled torque on the star also depends only weakly on the ratio R/ϖ_0. The reason is that the magnetic field is twisted primarily near the inner edge of the disk rather than near the star.

4.3 STEADY FLOW MODELS

Based on concepts similar to those just described, Ghosh and Lamb (1978, 1979a,b; hereafter GL) developed a quantitative model of disk accretion by an aligned rotator by solving the two-dimensional, non-ideal hydromagnetic equations, assuming that the effective conductivity of the disk plasma is relatively low as a result of the magnetic field diffusion and reconnection processes discussed in section §3.4.

4.3.1 *Structure of steady flow solutions.* The region where the stellar magnetic field threads the disk constitutes a *transition region* between undisturbed disk flow far from the star and the flow inside the magnetosphere. The interaction between the *azimuthal* motion of the flow in the transition region and the stellar magnetic field generates radial currents that tend to twist the field, as described in §4.2, creating an azimuthal field component. The interaction between the *radial* motion of the plasma in the transition region and the magnetic field generates azimuthal currents which tend to confine the poloidal field, creating a radial field component.

GL showed that their 2D MHD model exhibits boundary layer behavior at the radius given implicitly by the angular momentum conservation condition (3.8). That is, the velocity and magnetic fields in the model change on a length scale $\Delta\varpi \ll \varpi$ near the radius ϖ_0 given implicitly by equation (3.8). Thus, the transition region divides naturally into two zones: a broad outer zone, where the motion is Keplerian, and a narrrow inner zone, where the angular velocity falls sharply from the local Keplerian value to the stellar angular velocity at the radius ϖ_{co}. GL defined the radius ϖ_0 of the boundary between these two zones as the radius where $\partial\Omega/\partial\varpi = 0$. Thus, at $\varpi = \varpi_0$, the shear stress vanishes.

In the *inner transition zone*, between ϖ_{co} and ϖ_0, the interaction of the azimuthal flow with the magnetic field significantly reduces the angular velocity of the inflowing plasma is reduced from its Keplerian value just outside ϖ_0 to the corotational value at ϖ_{co}. In addition, azimuthal currents flowing in the disk plane

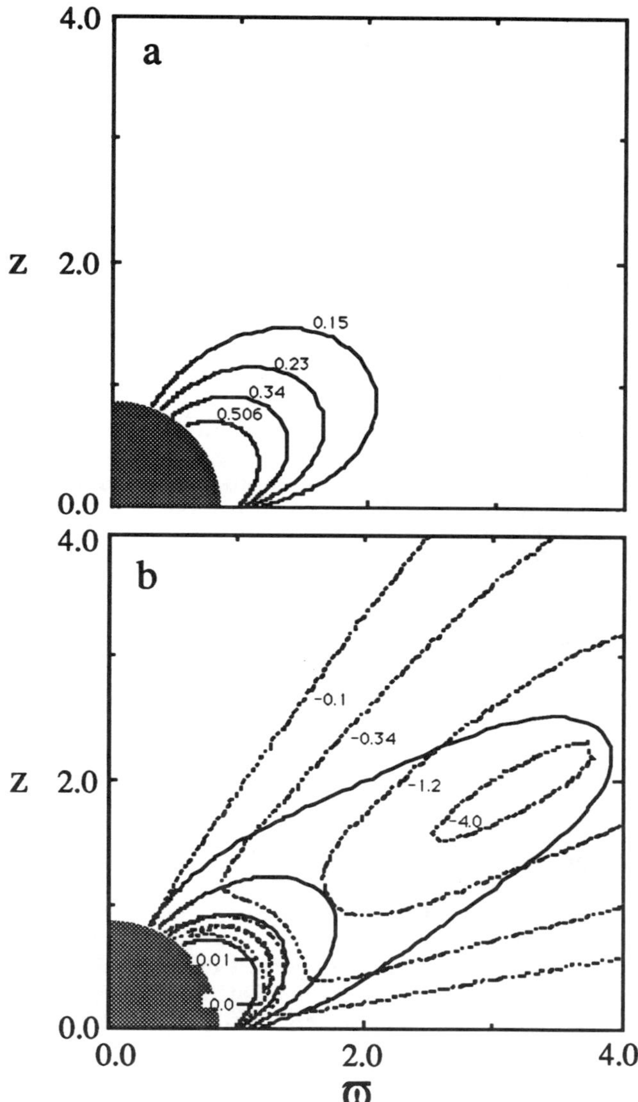

Fig. 4.3.—Side view of the star, accretion disk, and magnetosphere, showing the structure of the magnetosphere when the currents in the disk strongly screen a relatively weak stellar magnetic field. The radius of the star is 0.85 times the radius of the inner edge of the disk (the star is indicated by the shaded quadrant in the lower left). (a) Selected surfaces of constant poloidal magnetic flux when the field is not sheared. The poloidal flux surfaces are labelled with the value of the poloidal magnetic flux function ψ. (b) The same surfaces (solid curves) and surfaces of constant magnetic pitch (dashed curves) when the field is maximally expanded.

and in the accreting plasma above and below the disk screen the poloidal component of the stellar magnetic field on the radial length scale

$$\Delta\varpi \approx \frac{c^2}{4\pi\sigma_{\text{eff}} v_{\varpi 0}}, \tag{4.13}$$

where $v_{\varpi 0}$ is the radial velocity of plasma in the boundary layer. These screening currents reduce the poloidal field by a factor $\sim$5 between ϖ_{co} and ϖ_0. As a result of the drop in the angular velocity of the plasmabetween ϖ_{co} and ϖ_0, centrifugal support is lost and the plasma begins to fall rapidly toward the neutron star. Neglecting any effects of the stellar spin on this motion, GL estimated the radial velocity within the boundary layer to be

$$v_{\varpi 0} \approx (\Delta\varpi/\varpi)^{1/2} v_{\text{ff}}(\varpi_0), \tag{4.14}$$

using the radial component of the plasma momentum equation.

In the *outer transition zone*, outside ϖ_0, the disk structure is similar to that of an undisturbed disk, except that angular momentum is removed from or fed into the disk by the stellar magnetic field, which is twisted by its interaction with the disk plasma. In addition, the total energy dissipation rate is augmented by resistive dissipation of the electrical currents flowing in the disk. GL found that the couple between the outer transition zone and the star is a substantial fraction of the total couple. Since the couple falls off gradually, the outer boundary of the outer transition zone is purely conventional.

GL showed that if the stellar magnetic field threads the outer transition zone, steady flow is possible only if the effective electrical conductivity σ_{eff} for radial currents in the disk has the value

$$\sigma_{\text{crit}} = \frac{c^2}{4\pi} \left| \frac{\gamma_\phi}{rh(\Omega - \Omega_s)} \right| . \tag{4.12}$$

GL argued that magnetic flux reconnection in the disk flow is inherently a self-adjusting process that might give an effective conductivity equal to σ_{crit}. They therefore adopted $\sigma_{\text{eff}} = \sigma_{\text{crit}}$ as a working hypothesis, assuming for simplicity that the conductivity tensor is isotropic.

The plasma in the disk is confined in the vertical direction by the pressure of the magnetospheric field rather than gravity inside the radius r_p, which is somewhat larger than r_0. Aly and Kuiper (1990; see also Kuiper 1990) have suggested that inside r_p, plasma orbiting in the disk plane will break up into isolated clumps or blobs, which will then spiral inward due to their interaction with the stellar magnetic field. We emphasize that the time-averaged structure of this flow is necessarily similar to the structure in the same region of the GL model, and that the innermost radius at which the orbital motion of the blobs is Keplerian is, when calculated correctly, the same as the innermost radius of the Keplerian flow in the GL picture.

The field lines that thread the boundary layer form an *accretion bundle*, defined as the field lines along which plasma accretes to the surface of the star. The motion of the plasma in the accretion bundle at the disk plane and just above and below it has a significant cross-field component. Therefore the field lines in this region are only approximate streamlines of the flow. As plasma falls closer to the star, the cross-field motion is diminished by the increasing stress exerted by the magnetic field, while the field-aligned motion is accelerated by the component of the gravitational force that is along the field line. As a result, the motion of the accreting plasma can approach free fall. GL showed that in their model, the magnetospheric flow is sub-Alfvénic in the magnetospheric field even if it is free fall, and hence is MHD stable.

4.3.2 *Location of the inner edge of the disk.* The radius ϖ_0 of the inner edge of the Keplerian disk flow is an excellent diagnostic of conditions in the inner disk, since it depends on the structure of the inner disk and affects several directly observable properties as well as the evolutionary behavior of accreting neutron stars (Ghosh and Lamb 1991). GL showed that equations (4.12)–(4.14), together with the angular momentum conservation condition (3.8), are sufficient to estimate ϖ_0. For the accretion-powered pulsars investigated by GL, these equations give a result very similar to the simple expression (3.22). GL therefore proposed the use of expression (3.22), on the grounds that the uncertainties in the model are larger than the very small deviation from equation (3.22) predicted by the model. The boundary layer in the GL model is relatively narrow, with $\Delta\varpi \approx 0.03\,\varpi_0$, and hence $\gamma^{2/7} \approx 0.5$.

Following the investigation of QPOs from bright low-mass X-ray binaries, it became clear (see §1) that the accreting neutron stars in these systems have weak magnetic fields. As a result, the inner edges of the accretion disks in these systems are in the so-called "inner" region of the disk (see §4.1), unlike the inner edges of the disks in the accretion-powered pulsars investigated by GL, which are in the "middle" region of the disk (White and Stella 1987). However, the approach used by GL is easily generalized to other models of the inner disk. When the inner edge of the accretion disk lies well within a given region of the disk, the resulting estimate for ϖ_0 scales as various fractional powers of the key accretion variables $\dot{M}$, μ, and M, that is

$$\varpi_0 \propto \dot{M}^a \mu^b M^c \,, \tag{4.15}$$

where the values of the exponents a, b, c depend on the particular model of the inner disk and boundary layer. The values of these exponents are listed in Table 4.3 for the geometrically-thin inner disk models that were described in §4.1 and listed in Table 4.2

Table 4.3 lists the exact values of the exponents for the optically-thick GPD "middle" disk structure, rather than the slightly different approximate values given in GL, to emphasize the fact that the physical arguments leading to the scaling given in GL are fundamentally different from the physical arguments leading to the expression (3.17) for the Alfén radius in spherical accretion, which happens to have almost the same scaling. As should be clear from the discussion in §3, the similarity of the scalings that arise in these two very different physical situations is *not* an indication that the basic physical processes are similar. This is a point on which there is considerable confusion in the literature.

Table 4.3 shows that the scaling of ϖ_0 with the key accretion variables when the Keplerian flow ends in the "inner" disk is generally quite different from the scaling when the Keplerian flow ends in the "middle" disk. Moreover, different models of the "inner" disk produce quite different scalings of ϖ_0 with the key accretion variables. These scalings can be explored observationally in several ways (see Ghosh and Lamb 1991). In particular, it may be possible to determine the scaling of the Keplerian frequency at ϖ_0 in different systems. For example, studies

TABLE 4.3

Scaling of the Innermost Radius of the Disk

	$\varpi_0 \propto \dot{M}^a \mu^b M^c$		
Disk Model	Value of a	Value of b	Value of c
1T Opt thick GPD	−0.25	0.58	−0.21
1T Opt thick RPD*	−0.15	0.51	−0.13
2T Opt thin GPD Compt soft photon	−1.70	0.80	0.73
2T Opt thin GPD Compt brems	−0.48	0.57	0.05

*Scalings are the same for both prescriptions of the viscous shear stress listed in Table 4.2 (Ghosh and Lamb 1991), since the radius of the inner edge of the disk does not depend on the viscous shear stress (GL).

of the variation of the angular acceleration of accretion-powered pulsars with X-ray intensity can provide information on the scaling of the Keplerian frequency, and hence ϖ_0, with $\dot{M}$ (see §5). As another example, accurate measurements of the frequencies of horizontal-branch of QPOs—which are thought to reflect the difference between the stellar spin frequency and the Keplerian orbital frequency at ϖ_0 (see Lamb 1991 and references therein)—may provide data on the variation of ϖ_0 with $\dot{M}$. Finally, studies of the so-called "spin-up line" for recycled pulsars in the pulsar P-$\dot{P}$ diagram—which is thought to reflect the Keplerian orbital frequency at ϖ_0 when the star is accreting at the Eddington critical rate $\dot{M}_E$ (see van den Heuvel 1991 and references therein)—may provide information on the scaling of ϖ_0 with μ.

5. Accretion Torques and Stellar Response

If the magnetic field of the neutron star is strong enough to force the accreting plasma to flow toward the magnetic poles, the radiation produced near the star will be anisotropic. If in addition the star's magnetic axis is not aligned with its spin axis, a distant observer will see periodic oscillations in the X-ray intensity with a frequency equal to the spin frequency of the star. Thus, the X-ray oscillation frequency reveals the neutron star spin rate. The accretion torque can then be studied by measuring the changes in the spin rate it produces.

Changes in the stellar spin rate are important because they can be measured very accurately and offer the possibility of direct, quantitative comparisons with theoretical predictions. These comparisons provide insights about the physics of the accretion flow near the magnetosphere and the internal properties of neutron stars. Thus, for example, observations of long-term changes in the spin rate provide information about the time-averaged circulation of the accreting plasma, the strength of the dipole component of the stellar magnetic field, and the size of the star's moment of inertia. Measurements of short-term fluctuations in the spin rate probe the stability of the accretion flow and the internal dynamical properties of the neutron star (Lamb, Pines, and Shaham 1978).

In the present section we describe how the torque produced by accretion of matter from a Keplerian disk can be calculated and summarize the results of the calculation for the two-dimensional, steady flow solutions discussed in §4.3. We then discuss the changes in the angular momentum, spin rate, and rotational energy of the star produced by the accretion torque. Finally, we briefly compare the predicted changes in the spin rate with current observations.

5.1 ACCRETION TORQUES

5.1.1 *Early work.* The accretion torque produced by capture of plasma from a wind was first considered by Davidson and Ostriker (1973). They considered only material stress, tacitly assuming that all other stresses can be neglected and that the capture cross section is unaffected by density and velocity gradients in the wind. Davidson and Ostriker argued that the variation in wind density across the capture cross section causes the accreted matter to have a mean specific angular momentum of order ℓ_w (eq. [2.12]). However, as discussed in §2.1, the presence of a density gradient deforms the capture cross section in such a way that the mean specific angular momentum of the captured matter actually has the opposite sign, if only the material stress is included. Moreover, other stresses may make important contributions to the accretion torque. The sign, magnitude, and time-dependence of the accretion torque produced by capture of plasma from a stellar wind is a complex problem that has only been partially solved.

The accretion torque produced by capture of plasma from a Keplerian disk is in some ways simpler. This problem was considered by Pringle and Rees (1972). They estimated the torque by considering only the contribution of the material stress at the inner edge of the disk, tacitly assuming that all other stresses are negligible. As a result, they found a torque of order

$$N_0 \equiv \dot{M}(GMr_0)^{1/2}\,, \tag{5.1}$$

where r_0 is the radius of the inner edge of the disk. Pringle and Rees suggested that r_0 might be located at the point where the static pressure of an undistorted dipole magnetic field balances the thermal pressure of the gas at the midplane of an unperturbed steady disk flow, a procedure that we have argued in §3.4 is not meaningful for disk flows interacting with a stellar magnetic field. They also mentioned that the the off-diagonal magnetic stress B_rB_ϕ might play a role in determining r_0.

Lamb, Pethick, and Pines (1973) wrote down a general equation describing the flow of angular momentum into the stellar magnetosphere and noted that not only material, but also magnetic and viscous stresses could make important contributions to the torque on the star. They pointed out that for slow rotators accreting from a disk, the angular velocity of the flow necessarily passes through a maximum at a radius near the inner edge of the disk, and noted that at this radius the shear stress vanishes. By placing the surface S of the integral (2.7) at the radius r_0 of the angular velocity inflection point and noting that the magnetic couple to the region of the disk beyond this point can only increase the torque by an amount of order unity, they showed that the torque on a slowly rotating star accreting from a disk is approximately equal to N_0. In evaluating N_0, Lamb, Pethick, and Pines based their estimate of r_0 on MHD stability arguments like those outlined in §3.3, but used a flow velocity larger than that later found by GL. Compared with the angular momentum conservation condition, which we have argued in §3.4 is a more accurate way of estimating r_0, their argument somewhat underestimates r_0 and hence also the torque.

Ghosh, Lamb, and Pethick (1977) sought to make the torque argument of Lamb, Pethick, and Pines more precise for steady, axisymmetric flows. These authors showed that if the width Δr_m of the region where the magnetic field couples to the disk is small compared to r, then the accretion torque is closely equal to N_0, for all ω_s for which there is a steady accretion flow. Thus, for disks rotating in the same sense as the star, a narrow transition region would imply a spin-up torque approximately equal to N_0, even on rapidly rotating stars. This result strongly suggested that the transition region for disk flows is in fact broad, since otherwise the typical torque on Her X-1, which is 30–40 times smaller than N_0, would be extremely difficult to understand.

GL showed that in their model of disk accretion, the torque on the star produced by the magnetic couple between the disk plasma in the outer transition zone and the star is comparable to the torque produced by the material falling onto the star from the inner transition zone. The couple between the plasma in the outer transition zone and the star increases the spin-up torque on slowly rotating stars, but reduces the spin-up torque as the spin rate of the star increases. For very fast spin rates ($\omega_s \approx 1$), the torque produced by the couple between the outer transition zone and the star can exceed that of the matter falling onto the star, causing the net torque to become negative, braking the star's rotation. This result does not contradict the lower bound on the torque found earlier by Ghosh, Lamb, and Pethick, because the transition region in the GL model is broad rather than narrow.

5.1.2 *Calculating the accretion torque.* The accretion torque produced by flow of matter onto a neutron star is given by the integral (2.7) of the angular momentum flux density over any surface S that encloses the star, if we neglect the very short time required for the stress to be communicated from the surface S to the star. For a time-independent, axisymmetric flow, this integral may be written in the form (Lamb 1977)

$$N = \int_S \left(-\rho\varpi^2\Omega\mathbf{v}_p + \varpi\frac{B_\phi\mathbf{B}_p}{4\pi} + \eta\varpi^2\nabla\Omega \right) \cdot \hat{\mathbf{n}}\, dS\,, \tag{5.2}$$

where again ϖ is the cylindrical radius. Equation (5.2) displays explicitly the various stresses that contribute to the torque. Here Ω and v_p are the angular and poloidal velocities of the flow, B_ϕ and B_p are the poloidal and azimuthal components of the magnetic field, η is the effective dynamic viscosity, and $\hat{\mathbf{n}}$ is a unit vector normal to S and oriented outward. The three terms on the right side of equation (5.2) represent, in turn, the contributions of the material, magnetic, and viscous stresses to the accretion torque. The relative sizes of these three contributions depend on the location and shape of the surface used to evaluate the integral. However, for any surface S close to the surface of the neutron star, the angular momentum flux is carried almost entirely by the magnetic stress.

In evaluating expression (5.2), it is convenient to choose for S a surface composed of three parts: (1) an approximately cylindrical surface S_1, which coincides with the poloidal magnetic flux surface that leaves the disk at the radius ϖ_0 where the angular velocity goes through its maximum, and which extends slightly above and below the disk flow, (2) a surface S_2 consisting of two circular, approximately horizontal surfaces having a circular hole of radius ϖ_0 at the center, which connect to the top and bottom of S_1 at ϖ_0 and extend outward just above and below the flow to large ϖ, and (3) a surface S_3 composed of two separate, hemispherical surfaces that meet the sheets of S_2 at large ϖ and close the surface S above and below the star.

The total torque on the star is the sum

$$N_s = N_1 + N_2 + N_3 \tag{5.3}$$

of the torque contributions from each of the three surfaces S_1, S_2, and S_3. The contribution N_1 is the torque that is eventually communicated to the star by the magnetic field lines that interact with the flow in the inner transition zone, whereas the contribution N_2 is the torque that is communicated to the star by the magnetic field lines that interact with the flow in the outer transition zone. The contribution N_3 is negligible, since the magnetic stress falls off much more rapidly than r^{-3}.

The torque contribution N_1 is given by the material stress on S_1, since both the viscous and the magnetic stresses on S_1 are negligible, owing to the way in which this surface was chosen. The angular velocity Ω_0 of the plasma at ϖ_0 must lie between the angular velocity $\Omega_K(\varpi_0)$ and the stellar angular velocity Ω_s. Thus, for a fast rotator, $\Omega_0 \approx \Omega_K(\varpi_0)$. Even for a slow rotator, Ω_0 will be a substantial fraction (~0.5) of $\Omega_K(\varpi_0)$. Thus,

$$N_1 \approx -\int_S \rho v_r \cdot \varpi_0^2\, \Omega_K(\varpi_0) \cdot 2\pi\varpi_0 \cdot 2h \approx N_0\,. \tag{5.4}$$

The torque contribution N_2 is given by the magnetic stress on S_2, since the material stress on S_2 is zero (no matter crosses it) while the viscous stress is expected to be quite small. Thus

$$N_2 \approx \int_{S_2} (\varpi B_z B_\phi / 4\pi)\, dS\,. \tag{5.5}$$

5.1.3 *GL accretion torque.* Using their flow model to evaluate equation (5.5), GL found that N_s could be written

$$N_s \approx n(\omega_s)\, N_0\,, \tag{5.6}$$

where n is a dimensionless function that depends primarily on the fastness parameter ω_s, as indicated by the explicit dependence on ω_s displayed in equation (5.6). For stars rotating slowly in the same sense as the disk flow ($\omega_s \ll 1$), GL found $n(\omega_s) \approx 1.4$. They also found that the dimensionless torque function n decreases with increasing ω_s, vanishing at a certain critical fastness ω_c and becoming negative for $\omega_s > \omega_c$. Finally, for ω_s greater than a certain maximum fastness $\omega_{\max}$

(typically ≈ 0.95) GL could find no stationary solutions to the two-dimensional flow equations, and hence could not calculate the torque on the star using the approach described here. A simple expression for $n(\omega_s)$ that agrees approximately with their numerical results over the whole range of ω_s from 0 to $\omega_{\max}$ is

$$n(\omega_s) \approx 1.4 \left(\frac{1 - \omega_s/\omega_c}{1 - \omega_s} \right). \tag{5.7}$$

GL found $\omega_c \sim 0.35$ from their model, but stressed that the actual value was relatively uncertain. Subsequent work (see below) indicates that ω_c is unlikely to be less than 0.2, but could perhaps be as large as $\sim$0.9.

The behavior of n as a function of ω_s can be understood as follows. The accretion torque is the sum of the torque N_1 produced by accretion of the angular momentum of the matter that falls onto the star and the torque N_2 contributed by the twisted field lines from the star that interact with the disk in the outer transition zone. N_1 always acts to spin up a star rotating in the same sense as the disk flow, whereas N_2 can have either sign, since the azimuthal pitch of the stellar field lines that interact with the Keplerian flow in the outer transition zone must change sign at the corotation radius ϖ_c, where the angular velocity of the disk plasma is the same as that of the star. The contribution to the torque from the field lines threading the disk between ϖ_0 and ϖ_c is positive whereas the contribution from the field lines threading the disk outside ϖ_c is negative. For slow rotators, ϖ_0 is $\ll \varpi_c$, the positive contribution to N_2 dominates the negative contribution, and N_2 adds a further spin-up torque, approximately equal to $0.4\,N_0$, to the spin-up torque N_0 contributed by the term N_1. For moderately fast rotators, ϖ_0 is $\sim\varpi_c$, the negative contribution to N_2 dominates the positive contribution, and N_2 contributes a spin-down torque that partly cancels N_1. For very fast rotators, ϖ_0 is $\approx \varpi_c$, the spin-down torque contributed by N_2 dominates the spin-up torque contributed by N_1 and there is a net spin-down torque on the star.

The maximum strength of the net spin-down torque depends sensitively on the azimuthal magnetic pitch B_ϕ/B_p in the outer transition zone. In order to calculate the azimuthal pitch caused by interaction of the magnetosphere with the Keplerian disk flow in the outer transition zone, GL used a simple time scale argument to estimate the rate at which B_ϕ is generated by shearing of the polodial component of magnetic field, with the result $\gamma_a|\Omega_K - \Omega_s|B_\phi$, where γ_a is a numerical factor of order unity. However, as pointed out by Lamb (1978, 1984), the induction equation implies that B_ϕ is actually generated at the rate $\gamma_a|\Omega_K - \Omega_s|B_z$. Thus, whereas the time scale argument of GL leads to the expression

$$|B_\phi| = \left(\frac{\gamma_a}{\xi}\right) 2h(4\pi\rho)^{1/2}|\Omega_K - \Omega_s| \tag{5.8}$$

for the azimuthal component of the magnetic field, the induction equation leads to the result (Wang 1987; Zylstra 1988; Miller and Lamb 1991; Ghosh and Lamb 1991)

$$|B_\phi| = \left(\frac{\gamma_a}{\xi}\right)^{1/2} \left[2h(4\pi\rho)^{1/2}\right]^{1/2} |\Omega_{\rm K} - \Omega_s|^{1/2}|B_z|^{1/2}. \tag{5.9}$$

In both equations (5.8) and (5.9), ξ is a numerical factor of order 0.1–1 that characterizes the magnetic field reconnection rate in the outer magnetosphere (which was assumed to be $\xi v_A/2h$). Use of the corrected expression (5.9) leads to smaller magnetic pitches in the outer transition zone (Lamb 1978, 1984; Wang 1987; Miller and Lamb 1991; Ghosh and Lamb 1991).

GL calculated the azimuthal component of the field-aligned currents in the outer magnetosphere and the resulting screening of the poloidal magnetic field at large distances using expression (4.11) for the effective electrical conductivity with the azimuthal magnetic pitch given by expression (5.8) for B_ϕ. The magnitude of the resulting poloidal screening field exceeded the unscreened poloidal field at large distances from the inner edge of the disk. Since such "overscreening" is unphysical, GL set the poloidal field to zero everywhere outside the radius r_s where the poloidal magnetic field produced by the screening currents exactly cancelled the unscreened field. When instead the corrected expression (5.9) for B_ϕ is used, B_ϕ decreases with increasing distance from the star in a plausible manner (Wang 1987; Campbell 1987) and the poloidal magnetic field decreases with increasing radius faster than the unscreened stellar field but is never overscreened (Ghosh and Lamb 1991).

Expression (5.9) for B_ϕ gives a spin-down torque on fast rotators that is smaller than the spin-down torque estimated by GL, and a value for ω_c that is correspondingly larger. Wang (1987) has obtained an expression for the torque that gives a spin-down torque on fast rotators ($\omega_s \approx 1$) that is several times larger in magnitude than the spin-up torque on moderately slow rotators ($\omega_s \approx 0.1$). However, Wang's expression for the accretion torque diverges as the stellar spin rate goes to zero ($\omega_s \to 0$) and therefore cannot be correct, at least in this limit. Wang's expression also does not take into account screening of the poloidal field by currents flowing within the magnetosphere, without which a description of the interaction of the stellar magnetic field with the accretion disk is not self-consistent. Calculations that treat the generation of the azimuthal component of the magnetic field and screening of the poloidal component of the magnetic field self-consistently are underway and will be reported elsewhere (Ghosh and Lamb 1991).

Arons and several collaborators (see Arons *et al.* 1984; Arons 1987) have suggested that interaction of the stellar magnetic field with the disk flow will cause plasma to be mixed with the field via the Kelvin-Helmholtz instability, loading the field lines that pass near the surface of the disk with plasma. Then, if the magnetic stress on this plasma is great enough to accelerate it to a slightly super-Keplerian velocity, the plasma will flow outward along the top and bottom surfaces of the disk from some radius $\varpi_{\rm wind}$ ($> \varpi_0$), opening up the loaded field lines and carrying away angular momentum from the system. This has been proposed as a mechanism for creating a spin-down torque on the neutron star.

Several points are worth noting about this suggestion. First, it is not clear that an ordered poloidal field will persist in the disk once the stellar magnetic field has been opened up, since the field lines carrying the outflowing plasma are then no longer connected to the neutron star. Of course, a disordered magnetic field that extends above the disk might still generate outflow. Second, the angular momentum carried outward by the flow is, in the first instance, removed from the disk, not the star. As mentioned several times previously, loss of angular momentum from the disk, even a relatively large loss, will not cause the plasma in the disk to depart significantly from Keplerian motion, unless the loss is so great that it removes all the angular momentum of the plasma in a radial distance small compared to r. In the picture in question, the magnetic stress that creates the outflow is being generated by the orbital motion at a slightly smaller radius, so it seems unlikely that all the angular momentum of the flow can be removed in a short distance by this mechanism. As long as the plasma at ϖ_0 has a near-Keplerian azimuthal velocity, the term N_1 in the sum (5.3) will contribute a strong spin-up torque, comparable to N_0.

Could this strong spin-up torque nevertheless be overwhelmed by a spin-down torque N_2, due to coupling between the region of outflowing plasma and adjacent closed field lines that thread the surface S_2 and connect to the star? The following argument suggests that this is unlikely. First, note that the spin-up torque N_1 is the maximum torque that can be transmitted from the accreting plasma to the star by the substantial fraction (~0.3–0.8) of the star's poloidal magnetic flux that interacts with the accreting plasma inside ϖ_0. Now suppose that a much smaller fraction of the star's poloidal magnetic flux interacts with the disk between ϖ_0 and ϖ_{wind}. This magnetic flux will contribute to N_2. Suppose further that N_2 is indeed a spin-down torque. In order for it to be larger in magnitude than the spin-up torque N_1, the small fraction of the star's poloidal magnetic flux that contributes to N_2 would have to transmit a torque greater than the maximum torque that can be transmitted by the much larger fraction of the star's poloidal magnetic flux that contributes to N_1, which seems unlikely. Suppose instead that the fraction of the star's poloidal magnetic flux that interacts with the disk between ϖ_0 and ϖ_{wind} is comparable to or even somewhat greater than the fraction of the star's poloidal magnetic flux that interacts with the disk inside ϖ_0. In this case, most of the star's poloidal magnetic flux outside ϖ_0 interacts with the Keplerian disk flow between ϖ_0 and ϖ_{wind}. Only a small fraction of this poloidal magnetic flux is likely to interact strongly with the outflow from the disk outside ϖ_{wind}. Since this small fraction can transmit at most only a comparatively small torque on the star, the spin-down torque produced by the magnetic couple to the plasma leaving the disk will be much less than the spin-down torque produced by the magnetic couple to the disk plasma within ϖ_{wind}, and the wind spin-down model reduces to the GL model.

5.2 RESPONSE OF THE STAR

Changes in the spin rate and rotational energy of the neutron star can be related to ℓ, the angular momentum added to the star per unit mass of accreted matter (Ghosh, Lamb, and Pethick 1977). For simplicity, in this section we assume that the star responds as a rigid body (but see Lamb, Pines, and Shaham 1978).

5.2.1 *Angular momentum, spin frequency, and rotational energy.* The rate of change of the angular momentum of the star is

$$\dot{J}_s \equiv \frac{d}{dt}(I\Omega_s) = \dot{M}\ell\,, \tag{5.10}$$

where I is the moment of inertia of the star and Ω_s is its angular velocity. It follows that the rate of change of Ω_s is

$$\dot{\Omega}_s = (\zeta - 1)\left(\frac{d\ln I}{d\ln M}\right)\left(\frac{\dot{M}}{M}\right)\Omega_s\,, \tag{5.11}$$

where $J_s = I\Omega_s$, R_g is the radius of gyration of the star, and

$$\zeta \equiv \frac{\ell}{R_g^2\Omega_s}\left(\frac{d\ln M}{d\ln I}\right) = \frac{\ell}{\Omega_s}\frac{dM}{dI} \tag{5.12}$$

is a dimensionless parameter that compares the angular momentum added to the star by accretion of the mass dM with the specific angular momentum required to maintain the star at the same angular velocity, after the moment of inertia I has been changed by adding the mass dM to the star. In contrast, the rate of change of the rotational energy $E_{\rm rot}$ of the star is

$$\dot{E}_{\rm rot} \equiv \frac{1}{2}\frac{d}{dt}\left(I\Omega_s^2\right) = 2(\zeta - \tfrac{1}{2})\left(\frac{d\ln I}{d\ln M}\right)\left(\frac{\dot{M}}{M}\right)E_{\rm rot}\,. \tag{5.13}$$

For $\zeta < 1$, the star loses rotational energy and spins down, whereas for $\zeta > 1$, it gains rotational energy and spins up. For $\frac{1}{2} < \zeta < 1$, *the star gains rotational energy but spins down.*

For neutron stars with masses in the range 0.2–1.6 $M_\odot$, the structure factor $(d\ln I/d\ln M)$ is positive and $\sim$1. For such intermediate-mass stars, $\zeta \approx \ell/R_g^2\Omega_s \gg 1$, unless ℓ is less than or comparable to the small quantity $R_g^2\Omega_s$. In the following, we assume $\zeta \gg 1$. The first term in the expression $(\zeta - 1/2)$ on the right side of equations (5.11) and (5.13) is then the dominant one.

5.2.2 *Change in the stellar spin period.* Note first that a slowly rotating ($\omega_s \ll 1$) star rotating in the same sense as the circulation of the accretion flow and with a magnetosphere of appreciable size will be rapidly spun-up by accretion, *even if the angular velocity Ω of the flow outside the magnetosphere is less than the angular velocity Ω_s of the star.* To see this, note that the contribution of material stresses to ℓ gives

$$\zeta \approx r_m^2\Omega/R_g^2\Omega_s\,. \tag{5.14}$$

Thus for $r_m \gg R$, one can have $\zeta \gg 1$ even if $\Omega \ll \Omega_s$. The reason is that the much larger moment arm of the matter at r_m gives it a much greater specific angular momentum than that of the matter in the star, even though its angular velocity is much less. Thus, when it is added to the star, the star spins up.

For approximately radial flows, the rate of change of the pulse period can be written in the form (Ghosh and Lamb 1979b)

$$-\dot{P} = f_1(MR/I)\,\ell(L,\ldots)\,P^2L\,, \tag{5.15}$$

where the function f_1 depends on MR/I and the mean specific angular momentum ℓ of the flow generally depends on L and other properties of the source, as indicated by the notation $f_1(MR/I)$ and $\ell(L,\ldots)$. Given the equation of state, R and I are functions only of M, so that f_1 actually depends only on the mass of the neutron star. Note that equation (5.15) does not depend on the stellar magnetic moment μ or spin period P, because the magnetosphere plays no role in determining ℓ. Given the present uncertainty about ℓ_w, the mean angular momentum of matter captured from a stellar wind, and its time-dependence, little more can be said about secular changes in the stellar spin rate to be expected for capture of plasma from a wind.

For disk flows, Lamb, Pethick, and Pines (1973) used their expression for the accretion torque to write an equation of the form

$$-\dot{P}/P = f_2(\mu, M)\, PL^{6/7} \tag{5.16}$$

for the rate of change of the pulse period of a slow rotator rotating in the same sense as the disk flow. Note that given the two intrinsic properties μ and M of the star, equation (5.16) gives a single curve in the $(\dot{P}/P,\ PL^{6/7})$-plane. If M and μ are allowed to vary, equation (5.16) generates a two-parameter family of curves.

In the more complete theory of GL, which allows one to calculate the accretion torque for finite values of the fastness ω_s, the equation that corresponds to equation (5.16) can be written in the form

$$-\dot{P}/P = n(\omega_s)\, f_2(\mu, M)\, PL^{6/7}\,, \tag{5.17}$$

which depends on the fastness parameter ω_s as well as μ and M. The reason is that ω_s measures the dynamical effect of stellar rotation on the flow near the inner edge of the disk. For example, the ratio of the centrifugal force to the gravitational force for matter corotating with the star near the inner edge of the disk is just ω_s. Similarly, the ratio of the radius ϖ_0 of the inner edge of the disk to the radius ϖ_c of the corotation point in the disk, which is a measure of the relative size of the magnetic braking torque, is just $\omega_s^{2/3}$. Thus, one expects the spin-up rate to vary with ω_s on quite general grounds, although the particular functional form of the variation necessarily depends on details of the model. Equation (5.16) is independent of ω_s because it is the slow rotation limit ($\omega_s \to 0$) of equation (5.17).

An important general point made by Ghosh and Lamb (1979b) is that in comparing theories of the rate of change of the pulse period produced by disk flows with observations, one should plot $\log(-\dot{P})$ vs. $\log PL^{3/7}$, rather than $\log(-\dot{P}/P)$ vs. $\log PL^{6/7}$. The reason is that otherwise the dependence of the right side of equation (5.17) on ω_s introduces uneccessary scatter in the plot. One can see this by noting that since

$$\omega_s = f_3(\mu, M)/PL^{3/7}\,, \tag{5.18}$$

the right side of equation (5.17) is not simply a function of μ, M, and $PL^{6/7}$, but also depends on $PL^{3/7}$. Therefore, plotting $\log(-\dot{P}/P)$ vs. $\log PL^{6/7}$ will cause neutron stars that are identical except for their spin periods and/or luminosities to scatter over the plane, rather than falling on a single curve. Stated differently, fast rotators occur everywhere in the $(\dot{P}/P,\ PL^{6/7})$-plane.

One can recover a two-parameter family of curves by multiplying both sides of equation (5.17) by P, with the result

$$-\dot{P} = n(\omega_s)\, f_2(\mu, M)\, (PL^{3/7})^2 = f_2(\mu, M)\, f_4(PL^{3/7})\,, \tag{5.19}$$

which depends on P and L only through the combination $PL^{3/7}$. Thus, given the two intrinsic properties μ and M of the star, $\dot{P}$ depends only on the single parameter $PL^{3/7}$. Therefore, measurements of P, $\dot{P}$, and L for neutron stars that have different spin periods and/or different luminosities but are otherwise identical, or measurements of the P, $\dot{P}$, and L of a given neutron star made at different times, when its period and luminosity have different values, will fall on a single curve in a plot of $\log(-\dot{P})$ vs. $\log PL^{3/7}$. As shown by equation (5.18), surfaces of constant fastness are vertical lines in such a plot, with the slow rotation limit ($\omega_s \rightarrow 0$) corresponding to $PL^{3/7} \rightarrow \infty$. As a result, fast and slow rotators are separated in a systematic way, with fast rotators to the left and slow rotators to the right. In the slow-rotation limit, $\log(-\dot{P})$ is a straight line of slope +2.

The dependence of the $\dot{P}$ equation (5.19) on both P and L can produce complex spin-rate behavior (see Elsner, Ghosh, and Lamb 1978). However, the behavior of the spin rate is quite simple in two special cases of astrophysical interest, namely: (1) a star of given spin period that accretes matter at a time-varying rate and (2) a star whose spin changes as it accretes matter at a constant rate.

Consider first the pulse period behavior of a star rotating in the same sense as the disk flow and accreting at a constant rate $\dot{M}$. Such a star can experience either spin up or spin down, depending on its spin period. If P is large, ω_s is small compared to unity, and the star experiences a strong spin-up torque, causing P to decrease. As P falls, ω_s increases, and the spin-up torque decreases, vanishing at the critical spin period $P_c(\dot{M})$ at which $\omega_s = \omega_c$. If on the other hand P is small but still greater than $P_{\min}(\dot{M})$, at which $\omega_s = \omega_{\max}$, the star experiences a spin-down torque. As P increases, ω_s decreases, and the magnitude of the spin-down torque diminishes, vanishing at the critical spin period $P_c(\dot{M})$. Thus, the spin period of such a star will approach the critical period P_c that corresponds to its accretion rate, and will then maintain that period. For P less than $P_{\min}$, accretion, if it occurs at all, is not steady.

The behavior of a star of given spin period accreting at a time-varying rate is somewhat different. If $\dot{M}$ is sufficiently large, ω_s is small compared to unity, and the star is a slow rotator. The star therefore experiences a strong spin-up torque. As $\dot{M}$ decreases, ω_s increases, and the spin-up torque falls, vanishing at the critical accretion rate $\dot{M}_c(P)$, at which $\omega_s = \omega_c$. For accretion rates less than $\dot{M}_c(P)$, the star experiences a spin-down torque, which increases steadily in magnitude with decreasing $\dot{M}$until $\dot{M}$ reaches the minimum accretion rate $\dot{M}_{\min}$ consistent with steady accretion. At this accretion rate, $\omega_s = \omega_{\max}$. If accretion continues at all for accretion rates less than $\dot{M}_{\min}$, it is not steady.

5.3 COMPARISONS WITH OBSERVATION

Starting with the earliest theoretical work, the qualitative agreement between the predicted and measured changes of the spin rates of accretion-powered pulsars has been cited as evidence that these sources are neutron stars. Thus, Pringle and Rees (1972) noted the agreement of their torque estimate with the observed spin-up rate of Cen X-3, assuming it was a neutron star. Lamb, Pethick, and Pines (1973) considered both white dwarfs and neutron stars, and argued on the basis of the observed spin-up rates of Cen X-3 and Her X-1 that both are neutron stars and not white dwarfs.

Elsner and Lamb (1976) compared the slow-rotator spin-up equation of Lamb, Pethick, and Pines (1973) with the more detailed $\dot{P}$ observations then available and concluded that agreement was satisfactory for Cen X-3 but not for Her X-1, which has a mean rate of change of its spin period about 30–40 times smaller than expected for a slow rotator. They argued strongly that the then newly-discovered long-period sources are neutron stars. Elsner and Lamb also emphasized the importance of the predicted spin-up time scale for slow rotators as a discriminant between neutron stars and white dwarfs, pointed out that the torque on fast rotators should be smaller, and suggested that this might explain the slow rate of change of the spin period of Her X-1. Rappaport and Joss (1977) and Mason (1977) compared the larger quantity of $\dot{P}$ data available by the time they did their studies with the accretion torque theory of Lamb, Pethick, and Pines, plotting the left side of Lamb, Pethick, and Pines' spin-up equation (here eq. [5.16]) against the right side. The results of both studies indicated that the theory of Lamb, Pethick, and Pines agrees qualitatively with the observed properties of the sources considered, if these sources are neutron stars accreting from Keplerian or near-Keplerian flows. Ghosh and Lamb (1979b) compared their more quantitative predictions of the torque produced by accretion from a disk with observations of nine pulsing X-ray stars, and concluded that their model was consistent with the $\dot{P}$ data then available. Elsner, Ghosh, and Lamb (1980) pointed out that bright transient pulsars provide the best opportunity to test accretion torque theories.

Recently, Parmar *et al.* (1989) have analyzed EXOSAT pulse period and luminosity data from an outburst of the transient X-ray source EXO 2030+375. This source has a pulse period of 42 s and a binary period ~47 days. During the outburst, the luminosity declined steadily by an order of magnitude, providing the first opportunity for to test accretion torque models quantitatively as suggested earlier by Elsner, Ghosh, and Lamb. Parmar *et al.* find that the dependence of $\dot{P}$ on L is consistent with the GL torque theory. At high luminosities, a fit to a power-law gives $-\dot{P} \propto L^{1.16\pm0.12}$, within 2.5σ of the expected relationship $-\dot{P} \propto L^{6/7} \propto L^{0.86}$. At low luminosities, the $\dot{P}$ vs. L curve is somewhat uncertain, due to an ambiguity in the neutron star orbital solution, but suggests that the torque is indeed negative at low luminosities, as predicted by the GL model. Pulse timing observations of any future outburst of EXO 2030+375 would immediately resolve the ambiguity in the neutron star orbit, allowing the uncertainty in the $\dot{P}$ vs. L data of Parmar *et al.* to be greatly reduced and providing a crucila test of accretion torque theory.

Still more recently, Angelini, Stella, and Parmar (1989) have reported the discovery of relatively narrow peak in the power spectrum of the intensity time series in EXO 2030+375, which they interpret as evidence for the existence of quasi-periodic oscillations (QPOs) in this source. The frequency of this oscillation and its high luminosity strongly suggest that it represents the beating beween the stellar spin frequency and the orbital frequency of plasma at the inner edge of the disk, which Lamb *et al.* (1985) predicted would lead to just such a feature in the power spectra of at least some accretion-powered pulsars. In addition to the QPO peak, Angellini *et al.* have reported a broad feature, which suggest is produced by the orbital frequencies of the plasma near the inner edge of the accretion disk and which may, therefore, be used to determine $\Omega_K(\varpi_0)$. Note that if Ω_s and $\Omega_K(\varpi_0)$ are both known, then ω_s is determined. Furthermore, if ω_s is known, the observed spin-rate changes can then be compared directly with the dimensionless torque curve $n(\omega_s)$, providing a quantitative test of accretion torque theory. Angelini *et al.* conclude that their data are in satisfactory agreement with the torque curve predicted by GL. Future close studies of bright, transient accretion-powered pulsars will provide more stringent tests of accretion torque theory and guidance concerning the physics of the inner disk and the outer magnetosphere.

A more detailed discussion of the observed spin-rate behavior of accreting neutron stars and a comparison of this behavior with the predictions of accretion torque theory may be found in the review by Nagase (1989).

6. Particle Acceleration in the Magnetosphere

The outer magnetosphere of an accreting neutron star, where the magnetic field of the star interacts strongly with a relatively dense plasma flow, is a natural site for acceleration of charged particles by a variety of mechanisms. Moreover, the presence of a strong magnetic field and accreting plasma may allow the energy of accelerated particles to be converted efficiently into γ-rays. Processes that may produce a substantial luminosity of MeV γ-rays are of special current interest, since the *Gamma-Ray Observatory* is scheduled for launch within the next few months. In addition, there have been numerous reports of occasional detection of much higher-energy radiation from accretion-powered neutron stars, including radiation at TeV and even PeV energies.

In this section we briefly mention some of the acceleration mechanisms that may operate in the magnetospheres of accreting neutron stars, summarize the basic features of the TeV radiation reported from neutron stars, and describe some of the constraints on models of TeV emission imposed by these reports. Finally, we suggest that magnetospheric flares may be an important source of high-energy photons.

6.1 ACCELERATION MECHANISMS

Most of models of particle acceleration that have been proposed invoke either shock acceleration or acceleration by macroscopic electric fields. These models may be divided roughly into four categories.

6.1.1 *Acceleration in shock waves.* Shock acceleration of protons has been suggested by several authors. In these scenarios, the well-studied Fermi mechanism for accelerating galactic cosmic rays is applied to accreting neutron stars. Examples include acceleration in a standing shock wave in the accretion flow near the stellar surface or the magnetospheric boundary (Eichler and Verstrand 1985; Kazanas and Ellison 1986) and acceleration in shock waves within outflowing "jets" of energetic protons (Kiraly and Mészáros 1988). Another possibility that has been discussed is acceleration of protons by fluctuating magnetic fields within the magnetosphere (Katz and Smith 1988).

6.1.2 *Acceleration in the outer magnetosphere.* As shown in §4.2, the "unipolar inductor" produced by the magnetic coupling between the spinning neutron star and the plasma in an accretion disk, almost all of which orbits the star with a different angular velocity, naturally generates very large potential differences along the magnetic field lines in the outer magnetosphere. Acceleration by such magnetic-field-aligned electric fields may be an effective mechanism for producing highly energetic charged particles, which may then produce MeV or even TeV radiation (Lamb 1989a). Particle acceleration may take place quasi-steadily in regions where the electric field is high or episodically in large magnetospheric flares. Flaring within the magnetospheres of accreting neutron stars has recently been discussed by Kuipers (1990).

6.1.3 *Acceleration in vacuum gaps.* Cheng and Ruderman (1989) have suggested that the "vacuum gap" model of particle acceleration in rotation-powered pulsars (Cheng, Ho, and Ruderman 1986a,b) can be applied to accretion-powered neutron stars. In this scenario, large regions above and below the inner edge of the accretion disk are assumed to be devoid of accreting plasma. A further assumption of the model is that the density of charged particles in these regions is equal to the Goldreich-Julian density, so that there are only electrons in one part of each region and only protons in an adjacent part. A vacuum then develops at the interface between the electron region and the proton region, because there are not enough charged particles to support the electrical current driven by the differential rotation of the disk and the star. As a result, the full potential drop along the field lines that connect the disk and the star appears across the vacuum region, creating a strong electric field there. This electric field then accelerates any protons that make there way into the region.

6.1.4 *Acceleration in the accretion disk.* A different unipolar-inductor acceleration mechanism has been proposed by Chanmugam and Brecher (1985), who base their proposal on the model of Lovelace (1976). In Lovelace's model of magnetized accretion disks, the differential rotation of the disk plasma twists the large-scale magnetic field that is assumed to thread the disk, generating a potential difference between the inner and outer edges of the disk. Chanmugam and Brecher suggest that this may lead to acceleration of protons to TeV energies.

6.2 TEV GAMMA-RAYS

Detection of occasional TeV and even PeV radiation has been reported from about a dozen accretion-powered neutron stars, including Her X-1, and from about half a dozen rotation-powered neutron stars, including the Crab pulsar (see Weekes 1988; Chadwick *et al.* 1990; de Jager 1991). In many cases, the reported detections have not yet been confirmed. The duty cycles of the emission appear to be quite small ($\lesssim 10^{-2}$). Moreover, there are indications that the radiation reported at PeV energies does not have the signature of γ-rays (Cronin 1990). Nevertheless, there are enough confirmed reports of TeV radiation from neutron stars that it seems worthwhile to consider what processes might produce such energetic radiation. Models of TeV γ-radiation from rotation-powered pulsars have been reviewed extensively by Ruderman (see Ruderman 1989, 1990). Here we restrict our discussion to models of TeV γ-ray emission from accretion-powered neutron stars.

The main arguments for interpreting the reported TeV radiation from neutron stars as γ-rays has been summarized by Ruderman (1990). In order for the radiation to preserve an intensity oscillation with a period close to the neutron star spin period (~seconds) even after flight times much greater than 10^4 years, as claimed, and to arrive from the direction of the neutron star, the radiation must be neutral, almost massless relative to its energy, and able to penetrate at least $10^{-2}\ \mathrm{g\,cm^{-2}}$ of interstellar matter without degradation. Among known particles, only photons and neutrinos meet these requirments, but TeV neutrinos would not cause the reported air showers.

Any mechanism for producing the presumed TeV γ-radiation from accreting neutron stars must meet the following basic observational requirements:

- It must be capable of producing γ-rays with energies of 10^{12} eV or higher.
- It must be capable of producing a γ-ray luminosity at least $\sim 10^{-2}$ times the X-ray luminosity of accretion-powered pulsars.

- It should account for the episodic character of the emission, which sometimes lasts $\sim 10^3$ s, and be compatible with the low observed duty cycles of the radiation ($\sim 10^{-2}$ to $\sim 10^{-7}$).
- It should explain the periodicity of the γ-ray emission in accretion-powered pulsars, which is reportedly close to but not exactly equal to the X-ray period, with typical differences of about 1 part in 10^3.
- If the radiation is beamed, it should be beamed in the disk plane, at least in Her X-1 and Vela X-1, since the TeV radiation reported from Her X-1 varies in phase with the 35-day cycle while the TeV radiation reported from Vela X-1 appears to be eclipsed by the neutron star's companion.

These observational requirements impose strong constraints on models of the emission.

In order to produce TeV γ-rays, it is likely that charged particles must be accelerated to even higher energies. Several theoretical considerations are relevant to such models:

- Near accreting neutron stars, acceleration of electrons to energies as high as 10^{12}–10^{13} eV is sometimes, though not always, prevented by the "drag" produced by inverse Compton scattering of X-rays coming from the star. For this reason, most workers have focused on acceleration of protons.
- Proton acceleration cannot occur too far from the star, since the total luminosity available at the radius $R_{\rm acc}$ at which particle acceleration occurs is unlikely to exceed $(R/R_{\rm acc})\,L_x$, where L_x is the X-ray luminosity.
- The acceleration process must be consistent with the known environments of neutron stars. For example, observations indicate that the plasma particle density at 10^8–10^9 cm from the neutron star in Her X-1 is $\gtrsim 10^{12}$ cm^{-3} (Becker *et al.* 1977; Bai 1980).
- If protons are accelerated by an electric field, the acceleration must be parallel to the magnetic field, since the proton gyroradius is typically small compared to the acceleration length. Electric fields that exceed the magnetic field ($E > B$), allowing protons to move across magnetic field lines, are likely only in regions that are small compared to the acceleration length.
- Protons that have been accelerated to $\sim 10^{13}$ eV can convert their energy fairly efficiently to TeV γ-rays, if they traverse a plasma target of appropriate thickness. According to Stenger (1984), the optimal column density for the target is ~ 50 g cm^{-2}. For such a target, each proton will produce about five γ-rays, each having an energy $\sim 3\%$ of the initial proton energy (see Slane and Fry 1989). Since the absorption length for such γ-rays is ~ 75 g cm^{-2}, few are absorbed in the target.
- Production of TeV γ-radiation must occur sufficiently far from the neutron star that the magnetic field strength is less than 10^6 G; otherwise such energetic γ-rays will be converted to electron-positron pairs after traveling only a short distance.

One process that may meet these requirements is electrodynamic acceleration of protons in the outer magnetosphere, above and below the inner edge of an accretion disk.

6.3 MAGNETOSPHERIC ACCELERATION

The interaction of the disk with the stellar magnetic field described in §4 suggests several scenarios for accelerating charged particles in the outer magnetosphere. Here we briefly discuss three possibilities. Particle acceleration in the outer magnetosphere and associated production of γ-rays is discussed in more detail by Lamb *et al.* (1991).

A side view of the disk and magnetosphere is shown in the upper panel of Figure 6.1. The radial velocity v_r of the disk plasma pinches the poloidal magnetic field lines B_p inward, while the azimuthal velocity v_ϕ of the plasma in the disk twists the magnetic field, generating the azimuthal component B_ϕ. An equivalent circuit is shown in the bottom panel of Figure 6.1. The rotation of the neutron star and the orbital motion of the disk plasma through the magnetic field generate the electromotive forces labelled $\mathcal{E}$. These forces drive field-aligned currents within the highly-conducting magnetosphere, here represented by the poloidal currents $\mathcal{J}_1$, $\mathcal{J}_2$, and $\mathcal{J}_3$, and the azimuthal currents $\mathcal{J}_\phi$. The circuit is closed by cross-field currents flowing in the disk and the star. The flow of these currents through the resistances $\mathcal{R}_{\mathrm{acc}}$ in the accretion funnel and $\mathcal{R}_{\mathrm{acc}}$ and $\mathcal{R}_{\mathrm{acc}}$ in the disk produce potential drops that oppose the electromotive forces produced by the motion of the star and the disk plasma. Also important but not explicitly shown are the inductances of the current loops.

As described in §4.3, twisting of the magnetospheric field causes the magnetosphere to balloon outward. At the same time, a localized region of high azimuthal magnetic pitch develops above and below the disk plane (see Fig. 4.3). This region of high pitch produces a region of high azimuthal current density. If the current density becomes sufficiently large, current-driven instabilities will produce an anomalously high resistivity in this region (see Spicer 1982). Depending on whether or not the instability is self-adjusting, the free energy of the twisted magnetic field may be released fairly steadily or in a large flare. While the resistance is high, a large potential drop will develop across the resistive region. The resulting strong electromotive force

$$\mathcal{E}_{\mathrm{tot}} \approx \mathcal{E}_{\mathrm{disk}} - \mathcal{E}_{\mathrm{star}} \approx \frac{1}{c}(\Omega_{K0} - \Omega_s) B_0 r_0^2 \sim 10^{15}\mathrm{Volts} \tag{6.1}$$

may accelerate charged particles along the magnetic field. The time development of such a circuit can often be described by equivalent circuits with lumped circuit elements, like those shown in the bottom panel of Figure 6.1.

A second possibility is that double layers may develop within the magnetosphere (again see Spicer 1982). Such laminar space-charge layers trap a large fraction of the current-carrying electron population and accelerate the remainder. They can accelerate particles to high energies and can be modeled by an equivalent circuit having a capacitor and a resistor parallel to the magnetic field lines; however, the capacitance and resistance must be treated as highly nonlinear functions of the electric field in the layer.

A third possibility is that the energy stored in the twisted stellar magnetic field is released by reconnection, as suggested by GL. In fast rotators, the azimuthal field reverses sign at $r_c \approx r_0$. The resulting current sheet may tear, releasing the stored magnetic energy. Reconnection near the inner edge of the disk has recently been discussed in more detail by Aly and Kuipers (1990) and Kuipers (1990).

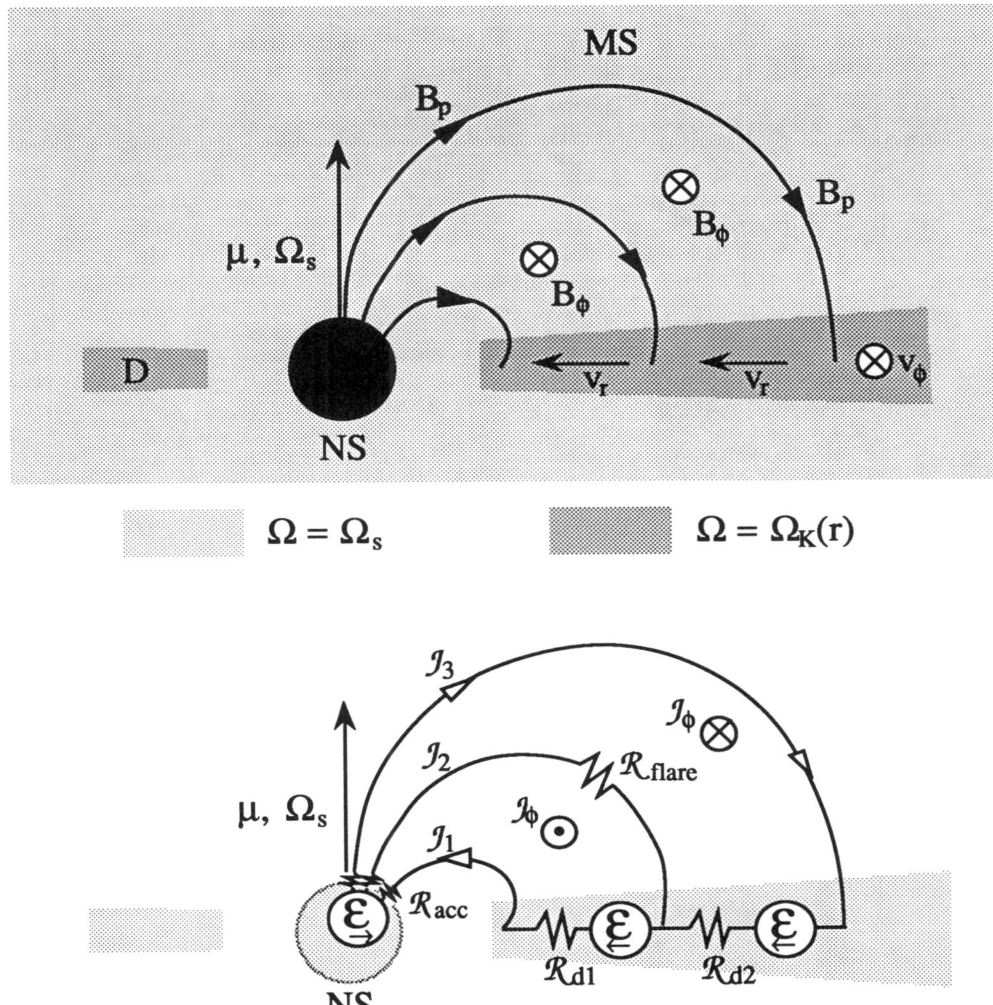

Fig. 6.1.—*Top*: Side view of a magnetic neutron star NS interacting with an acccretion disk D. Plasma in the lightly-shaded region corotates with the star, whereas plasma in the the heavily-shaded region circles the star in Keplerian orbits. *Bottom*: An equivalent circuit, showing the electromotive forces $\mathcal{E}$ produced by the rotation of the neutron star and the orbital motion of the disk plasma through the magnetic field, the resulting poloidal currents $\mathcal{J}_1$, $\mathcal{J}_2$, and $\mathcal{J}_3$, the corresponding azimuthal currents $\mathcal{J}_\phi$, and the circuit resistances $\mathcal{R}_{\rm acc}$ in the accretion funnel and $\mathcal{R}_{\rm acc}$ and $\mathcal{R}_{\rm acc}$ in the disk. A steep rise in the resistance within the magnetosphere ($\mathcal{R}_{\rm flare}$) will create a large potential drop. The resulting strong electric field can accelerate charged particles along the magnetic field.

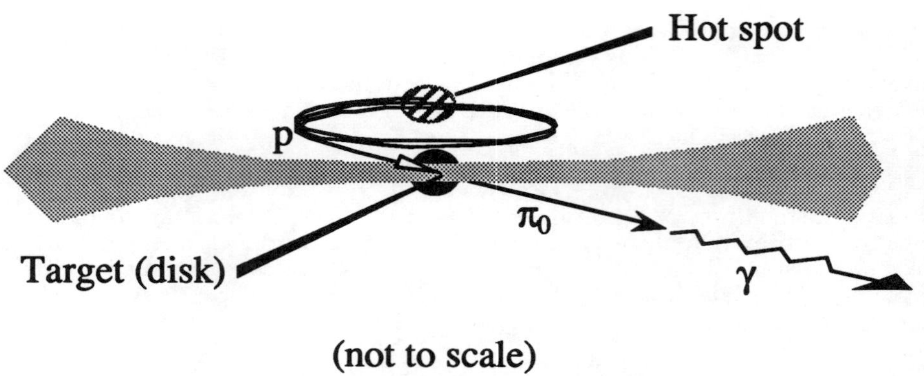

Fig. 6.2.—Perspective schematic view of γ-ray production by a magnetic neutron star accreting from a disk. Protons are accelerated in a resistive region, double layer, or reconnection zone within the magnetosphere. They then enter the disk at an oblique angle where the highly twisted magnetic field from the acceleration region threads the disk. The accelerated protons collide with protons in the disk, producing energetic π_0's. When the π_0's decay, they generate a beam of γ-rays close to the disk plane.

Figure 6.2 shows a possible scenario for producing energetic γ-rays with the right beam direction by accelerating protons in a resistive region, double layer, or reconnection zone within the magnetosphere. The protons enter the disk at an oblique angle where the highly twisted magnetic field from the acceleration region threads the disk. Collisons of the energetic protons with protons in the disk produce π_0's which then decay, generating a beam of γ-rays close to the disk plane. If the acceleration is electrodynamic, only those stars with $\boldsymbol{\mu} \cdot \boldsymbol{\Omega}_{\mathbf{s}} > 0$ can produce γ-rays in this way, since only such stars will accelerate protons *toward the disk*.

Because the potential drop across the acceleration region necessarily leads to differential rotation within the magnetosphere, we do not expect these mechanisms to produce γ-radiation that oscillates in intensity at *exactly* the stellar spin frequency. Indeed, unless the energy release is relatively steady and the magnetosphere is quiescent, the spread in pattern frequencies is likely to be comparable to the spread in the Keplerian frequencies of the disk plasma that is threaded by the magnetic flux from the acceleration region. However, the long-term average of the frequencies of any oscillations in the intensity of the γ-rays should be equal to the stellar spin frequency.

7. Proton Deceleration at the Stellar Surface

Plasma falling onto a neutron star is decelerated by interaction with radiation and plasma near the stellar surface. During the deceleration process, the kinetic energy of the accreting plasma is converted into heat and radiation. The structure of the resulting hot, relatively dense plasma layer near the stellar surface, and the spectrum of the X-rays produced in the layer, are determined in part by the depth at which the infalling protons, which carry almost all the kinetic energy of the accreting plasma, deposit their energy. In this section, we first review the basic physical processes that are thought to be important in decelerating the infalling protons and review previous calculations of stopping by these processes. We then summarize the results of a recent study of proton stopping by Coulomb collisions with electrons in the surface plasma of a strongly magnetic neutron star.

7.1 PREVIOUS WORK

The field-aligned flow of accreting plasma within the neutron star magnetosphere leads to the formation of so-called accretion columns or funnels above the magnetic poles of the star, as shown in Figure 3.1. Which processes are most important in decelerating the infalling plasma in an accretion funnel depends on whether the accretion luminosity is greater or less than the critical luminosity (Arons, Klein, and Lea 1987)

$$L_{\rm crit} \equiv \left(\frac{A_{\rm cap}}{4\pi R^2}\right)\left(\frac{4\pi GMm_pc}{\sigma_{||}}\right) . \tag{7.1}$$

Here $A_{\rm cap}$ is the area of the polar cap onto which the accreting plasma falls, m_p is the mass of the proton, and $\sigma_{||}$ is the angle- and frequency-averaged cross section for scattering of radiation propagating parallel to the magnetic field. $L_{\rm crit}$ is $\sim 10^{36}\,{\rm ergs\,s^{-1}}$ if $A_{\rm cap}/4\pi$ is 0.01 and the polar cap magnetic field strength is 3×10^{12} G. For accretion luminosities $L \gg L_{\rm crit}$, interaction of the infalling plasma with the radiation from the base of the funnel strongly decelerates the plasma above the stellar surface (Davidson 1973; Basko and Sunyaev 1976; Arons, Klein, and Lea 1987; Klein and Arons 1989; Burnard, Arons, and Klein 1991). On the other hand, for luminosities $L \ll L_{\rm crit}$ the radiation stress is relatively unimportant, and the infalling plasma may be slowed by a variety of collisional and collisionless processes. Here we focus on this latter case.

It is possible that the supersonic infall of plasma in the accretion funnels is halted by a collisionless shock wave standing above the stellar surface, as suggested by Shapiro and Salpeter (1975) and discussed further by Langer and Rappaport (1982) and Kirk (1985). It is not clear, however, whether such a shock can actually form under the physical conditions that exist at the base the accretion funnels. Another possibility is that the infalling protons are stopped in a thin, stable, charge-separated "double layer", as proposed by Williams *et al.* (1986).

If the infalling protons are not stopped by these processes, and if they cool rapidly enough that magnetic mirroring does not occur, they will be stopped by collisions with the electrons and protons in the dense "target" plasma near the stellar surface. The beam protons transfer their energy to the electrons in the target plasma through Coulomb collisions, and to the protons in the target plasma primarily through interactions via nuclear forces. When the surface magnetic field of the neutron star is weak ($B_s \ll 10^{12}$ G), the dominant stopping process is small-angle Coulomb scattering of the protons by the electrons in the target plasma. Zeldovich and Shakura (1969) and Lamb, Pethick, and Pines (1973) estimated that the stopping length is $\sim$8–20 $\rm g\,cm^{-2}$ under these conditions. Alme and Wilson (1973) subsequently carried out detailed numerical computations of proton stopping in the surface plasma of a nonmagnetic neutron star and found stopping lengths $\sim$20 $\rm g\,cm^{-2}$.

When the surface magnetic field of the neutron star is strong ($B_s \gtrsim 10^{12}$ G), the situation changes drastically. Quantization of the electron motion perpendicular to the magnetic field strongly suppresses the transfer of energy from the protons to the electrons. Indeed, the protons begin to lose energy only after scattering has increased their pitch angle relative to the magnetic field to values $\gtrsim(m_e/m_p)^{1/4}$. Basko and Sunyaev (1975) first pointed out this effect, and calculated the stopping length of protons in a cold, strongly magnetized plasma using the classical electron-proton Coulomb scattering cross section. They found a very large stopping length, ~200 gpcs, and therefore concluded that the infalling protons are actually stopped by nuclear interaction with protons in the target plasma, which limits the stopping length to no more than ~50 g cm^{-2}.

Pavlov and Yakovlev (1976) improved on the calculations of Basko and Sunyaev by using a quantum-mechanical expression for electron-ion Coulomb scattering cross section. In computing this expression, Pavlov and Yakovlev assumed that the protons occupy plane-wave states rather than the Landau states they actually occupy. They included both "reflection" collisions, in which the electron is scattered backwards, and "fly-by" collisions, in which the electron continues past the proton. They obtained stopping lengths about five times shorter than those found by Basko and Sunyaev. Pavlov and Yakovlev attributed their smaller stopping lengths to their inclusion of reflection collisions and their allowance for the "regular" increase of the proton pitch angle that occurs once it exceeds $\gtrsim(m_e/m_p)^{1/2}$.

Following this work, Kirk (1979) suggested that the proton beam may generate ion-sound turbulence, and attempted to take into account the effect of such turbulence by adding an estimate of the pitch-angle diffusion caused by this turbulence to the pitch angle diffusion caused by Coulomb scattering. However, the degree of turbulence to be expected in the stopping region is not known. Bussard (1980) calculated the cross section for Coulomb scattering relativistically and then made a rough estimate of the stopping length by assuming that the proton pitch angle remains zero throughout the stopping process. As a result, he obtained relatively large stopping lengths, ~60 g cm^{-2}.

Kirk and Galloway (1981, 1982) computed numerically the evolution in time of the proton velocity distribution function during the stopping process, including for the first time several significant effects, such as the thermal spread in electron velocities. They used the Fokker-Planck approximation to the proton kinetic equation. In this approximation, the Boltzmann transport equation, which describes the evolution of the proton distribution in coordinate- and velocity-space, is simplified using the fact that the change in the proton distribution is due almost entirely to small-angle scattering. The collision integral on the right side of the Boltzmann equation is therefore expanded in a power series in the change in the proton velocity and only terms up to order $(\delta V)^2$ are retained. The resulting approximate kinetic equation has the form

$$\left(\frac{\delta f}{\delta t}\right) = -\frac{\partial}{\partial V_\ell}\left(F_\ell f - \tfrac{1}{2}D_\ell\frac{\partial f}{\partial V_\ell} - \tfrac{1}{2}\Delta_x\frac{\partial f}{\partial V_t^2}\right) - \frac{\partial}{\partial V_t^2}\left(\Phi_t f - \tfrac{1}{2}\Delta_t\frac{\partial f}{\partial V_t^2} - \tfrac{1}{2}\Delta_x\frac{\partial f}{\partial V_\ell}\right), \tag{7.2}$$

where the Fokker-Planck coefficients F_ℓ and Φ_t describe advection of protons in velocity space along and transverse to the magnetic field, the coefficients D_ℓ and Δ_t describe diffusion in these directions, and the coefficient Δ_x describes cross-diffusion.

In calculating Fokker-Planck coefficients, Kirk and Galloway treated the accreting protons quantum-mechanically and described plasma collective effects using the quantum-mechanical response function of Kirk (1980). They found proton stopping lengths ~8–16 $\mathrm{g\,cm^{-2}}$ for an initial proton velocity of 0.5 c. These lengths are much shorter than the stopping lengths found previously. Kirk and Galloway attributed their much shorter stopping lengths to their improvements over previous calculations. However, more recent work has shown that they are due largely to inaccuracies in the Fokker-Planck coefficients used by Kirk and Galloway. These include a simple factor-of-two error in calculating the coefficients (see Harding *et al.* 1984), neglect of an exponential cutoff in the sum over the change in the proton quantum number, use of an approximate form for the collision-rate integral, and a simplified treatment of Debye screening (see Pakey 1990 and Pakey, Bussard, and Lamb 1991). All of these errors and approximations tend to increase the Fokker-Planck coefficients, decreasing the calculated stopping distance.

Miller, Salpeter, and Wasserman (1987) used a Monte-Carlo method to determine the stopping distance for protons moving through a uniform target plasma. In calculating the effects of proton-electron Coulomb collisions, they described the motion of the electrons in the target plasma using Landau wave functions, but used plane waves to describe the motion of the infalling protons. As shown by Pakey (1990), the error introduced by the latter approximation is generally small, but can be as large as 30% or more in some cases. Miller *et al.* included proton-electron collisions that excite an electron from its ground state to the next excited state and proton deceleration by interaction with collective modes of the target plasma, as well as proton-electron collisions that leave the electron in its ground state. Excitation of collective modes can be important when the infall velocity of the protons is greater than the electron thermal velocity. Miller *et al.* treated plasma collective effects by separating the long-wavelength ($\lambda \gg \lambda_{\mathrm{Debye}}$), collective part of the Coulomb interaction from the short-wavelength part. Their stopping distances are generally shorter than those reported by Pavlov and Yakovlev, but longer than those found by Kirk and Galloway. Miller *et al.* were unable to determine the shape of the proton distribution function, because they could not perform a sufficient number of trials.

7.2 RECENT RESULTS

Recently Pakey (1990) and Pakey, Bussard, and Lamb (1991) have completed an extensive study of the time evolution of the proton velocity distribution function during the stopping process, using a Fokker-Planck approach. In calculating the Fokker-Planck coefficients, they assumed that the target electrons are in the lowest Landau level before the collision and remain in the lowest Landau level afterwards, that the scattering process can be treated non-relativistically, and that the Born approximation can be used to calculate the collision rate (see also Pavlov and Yakovlev 1976; Canuto and Ventura 1977; Bussard 1980). These approximations are valid for a wide range of proton beam and target plasma conditions, although they are not accurate for all conditions that may be encountered in low-luminosity accreting neutron stars, especially if the magnetic field is relatively weak and the proton beam velocity is relatively high. Pakey *et al.* used Landau wave functions to describe the infalling protons as well as the electrons in the target plasma. In some cases, the resulting Fokker-Planck coefficients are significantly smaller than

TABLE 7.1
Proton Stopping Lengths[a]

Proton Velocity $V_p(c)$	Proton Temperature T_p(keV)	Electron Density $n_e(10^{22}\ \mathrm{cm}^{-3})$	Electron Temperature T_e(keV)	Magnetic Field $B(10^{12}\ \mathrm{G})$	Stopping Length $X(\mathrm{g\,cm}^{-2})$
0.2	50	10	50	5	5.79
0.2	50	10	2	5	0.74
0.25	50	10	10	5	1.36
0.3	50	10	10	1	2.61
0.5	50	10	10	5	38.2
0.5	50	10	50	5	22.4
0.5	50	10	2	5	80.7
0.5	50	10	10	1	43.6
0.5	50	10	10	10	36.4
0.5	50	10^{-2}	50	5	15.9
0.5	50	10^{4}	50	5	39.2
0.5	500	10	10	5	33.6

[a]From Pakey (1990).

the more approximate coefficients used by Kirk and Galloway, particularly when the proton beam velocity is high.

The initial proton velocity distribution was assumed to be a Maxwellian with temperature T_p drifting along the magnetic field with the initial proton infall velocity V_i. The evolution of the proton distribution was then followed in time by integrating the Fokker-Planck equation numerically on a Cray X-MP, using a forward-time, centered-space (FTCS) differencing scheme. The distance that the beam had travelled at any given time was determined by integrating the average proton longitudinal velocity. Pakey *et al.* explored the stopping of protons with initial infall velocities in the range 0.1–0.5 c, a wider range than had been studied previously. They also investigated initial proton temperatures as low as 10 keV, much lower than the quite high proton temperature of 500 keV assumed by Kirk and Galloway, and target electron temperatures as low as 2 keV. Pakey *et al.* were able to follow the proton distribution until it had almost in kinetic equilibrium with the electron distribution in the target plasma.

Typical examples of the stopping lengths found by Pakey *et al.* are shown in Table 7.1. Several trends are worth noting. First, protons with large initial velocities have shorter stopping lengths in high-temperature plasmas than in low-temperature plasmas, whereas the reverse is true for low-velocity protons. Second, increasing the target plasma electron density increases the stopping length only indirectly, by producing increased screening. Third, increasing the magnetic field strength decreases the stopping length in these calculations, due to neglect of collisions that leave the electron in an excited transverse (Landau) state, and the quantization of the proton motion.

As expected, the proton stopping lengths found by Pakey *et al.* are 3–4 times greater than those reported by Kirk and Galloway, when the initial proton beam velocity V_i is high and the electron thermal velocity $v_{\mathrm{th,e}} \approx 0.1(T_e/20\,\mathrm{keV})^{1/2}\,c$ is low. The stopping lengths found by Pakey *et al.* are, however, generally consistent with the Monte Carlo results reported by Miller *et al.*, when the additional stopping proceses considered by the latter authors are unimportant. Pakey *et al.* found that the proton distribution function evolves in two qualitatively different ways, depending on whether the initial proton beam velocity V_i is less than or greater than $v_{\mathrm{th,e}}$.

If the initial proton beam velocity is less than the typical electron thermal velocity, the protons lose their longitudinal kinetic energy without any significant change in their transverse energy distribution, as shown in Figure 7.1. In this case, the protons deposit their kinetic energy fairly uniformly with distance, as shown in the top left panel of Figure 7.2. This behavior is similar to the behavior of protons stopped by interaction with a nonmagnetic plasma.

If instead the initial proton beam velocity is much larger than the electron thermal velocity, the transverse energy of the protons increases without much loss of longitudinal energy at first, as shown in Figure 7.3. Only after the protons have acquired a substantial transverse energy do they begin to slow significantly. Because the width of the transverse energy distribution remains narrow, the proton transverse energy distribution is inverted during much of the slowing process, as reported previously by Kirk and Galloway. This inversion occurs because advection dominates diffusion once the proton pitch angle exceeds $(m_e/m_p)^{1/2}$. Only at the very end of the evolution, when the proton beam velocity has fallen to a value that is comparable to the thermal velocity of the electrons in the target plasma, does diffusion again become dominant. As a result, the protons deposit most of their kinetic energy near the end of their paths, deep inside the target plasma, as shown in Figure 7.2(d). The inversion of the proton transverse energy distribution may decrease or disappear when proton energy loss via cyclotron emission is taken into account.

These results have several implications for models of the X-ray emitting regions of accretion-powered pulsars. First, for target electron temperatures $\lesssim 10\,\mathrm{keV}$, stopping by Coulomb collisions is less effective than stopping by strong interactions with the protons in the target plasma. Second, comparison of the Fokker-Planck and Monte Carlo results suggests that the stopping power of Coulomb collisions that excite electrons and excitation of plasma collective modes taken together is comparable to the stopping power of Coulomb collisions that do not excite the electrons, when the initial proton velocity or the density of the target plasma are large. This conclusion may be altered when the effect on screening of the high proton velocity is taken into account. Third, the error introduced by neglecting proton quantization may be significant. Finally, most of the kinetic energy of the accreting protons is deposited at ~5–20 Thomson scattering optical depths. Thus, most of the X-radiation is produced at depths where the continuum optical depth is small but the optical depth at the cyclotron frequency and its harmonics is very large. Hence, accurate, reliable predictions of the spectrum and angular distribution of radiation from the base of accretion funels will probably require self-consistent modeling of the temperature and density structure of the target plasma.

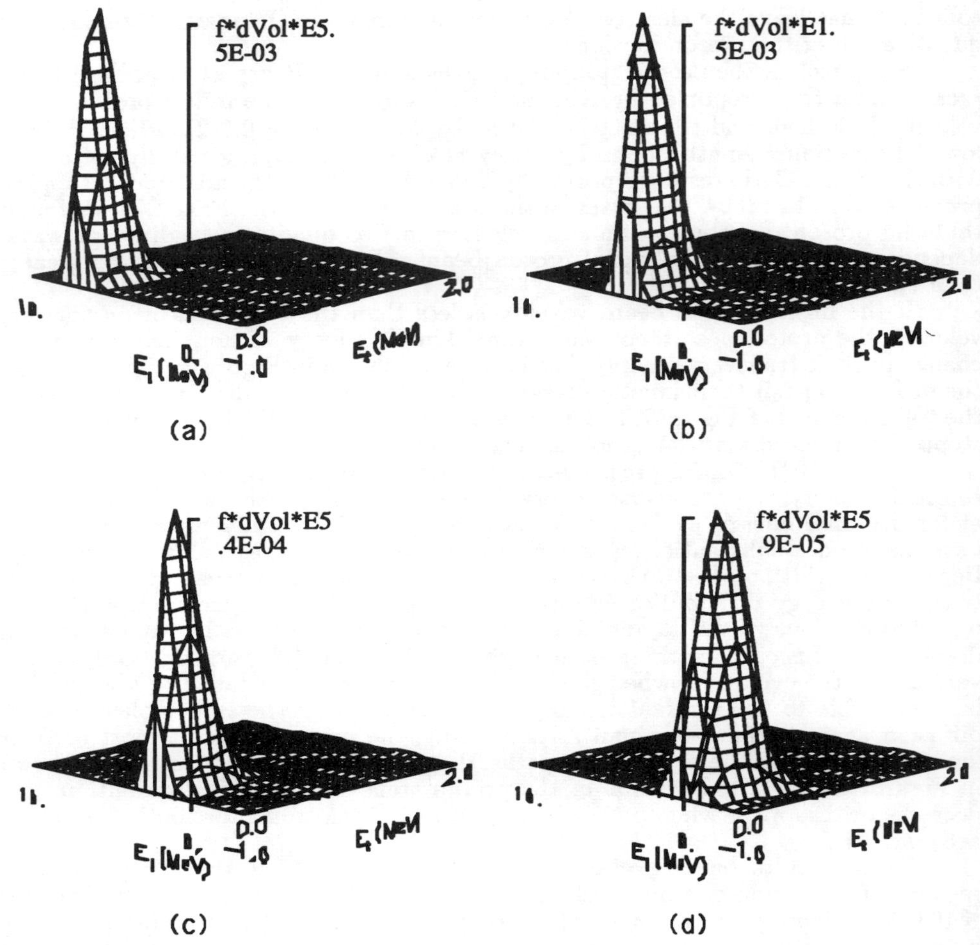

Fig. 7.1.—Snapshots of the distribution of proton kinetic energy in E_ℓ and E_t at four successive times (a)–(d), for $B = 5 \times 10^{12}$ G, $n_e = 10^{23}$ cm^{-3}, $T_p = 50$ keV, $V_i = 0.1$ c, and $T_e = 50$ keV. The vertical axis shows the dimensionless proton energy density in velocity space, which is equal to the distribution function at the center of a given cell in velocity space times the energy of the protons at the center of the cell times the volume of the cell. To distinguish protons moving downward from protons moving upward, the longitudinal kinetic energy E_ℓ of the latter has been assigned a negative value. During the evolution (a)–(d), the average longitudinal kinetic energy of the protons falls from its initial value of 4.7 MeV to 26 keV, while the column density increases from 0 to 2.9 g cm^{-2}. The average transverse kinetic energy of the protons remains almost equal to its initial value of 50 keV throughout the evolution. The proton distribution shown in panel (d) is close to kinetic equilibrium with the electrons in the target plasma. From Pakey (1990).

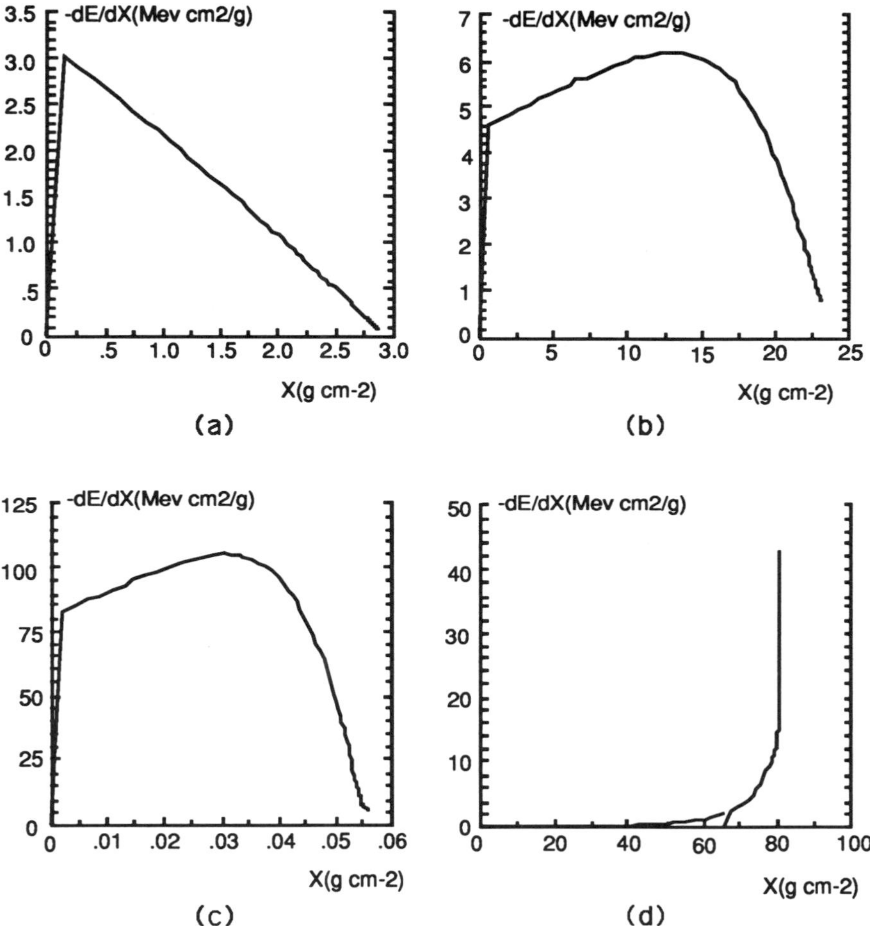

Fig. 7.2.—Proton kinetic energy deposition profiles for four combinations of the initial proton infall velocity V_i and target plasma electron temperature T_e. The combinations are: (a) $V_i = 0.1\,c$ and $T_e = 50\,\mathrm{keV}$; (b) $V_i = 0.5\,c$ and $T_e = 50\,\mathrm{keV}$; (c) $V_i = 0.1\,c$ and $T_e = 2\,\mathrm{keV}$; (d) $V_i = 0.5\,c$ and $T_e = 2\,\mathrm{keV}$. The magnetic field strength, initial proton temperature, and electron density are the same as in Fig. 7.1. In case (a), V_i is less than the thermal velocity of the electrons in the target plasma whereas in case (d) it is much greater. Panels (b) and (c) show two intermediate cases; the shapes of these profiles are similar because the stopping process has an exact scaling (see Pakey 1990) that is approximately satisfied for these two cases. From Pakey (1990).

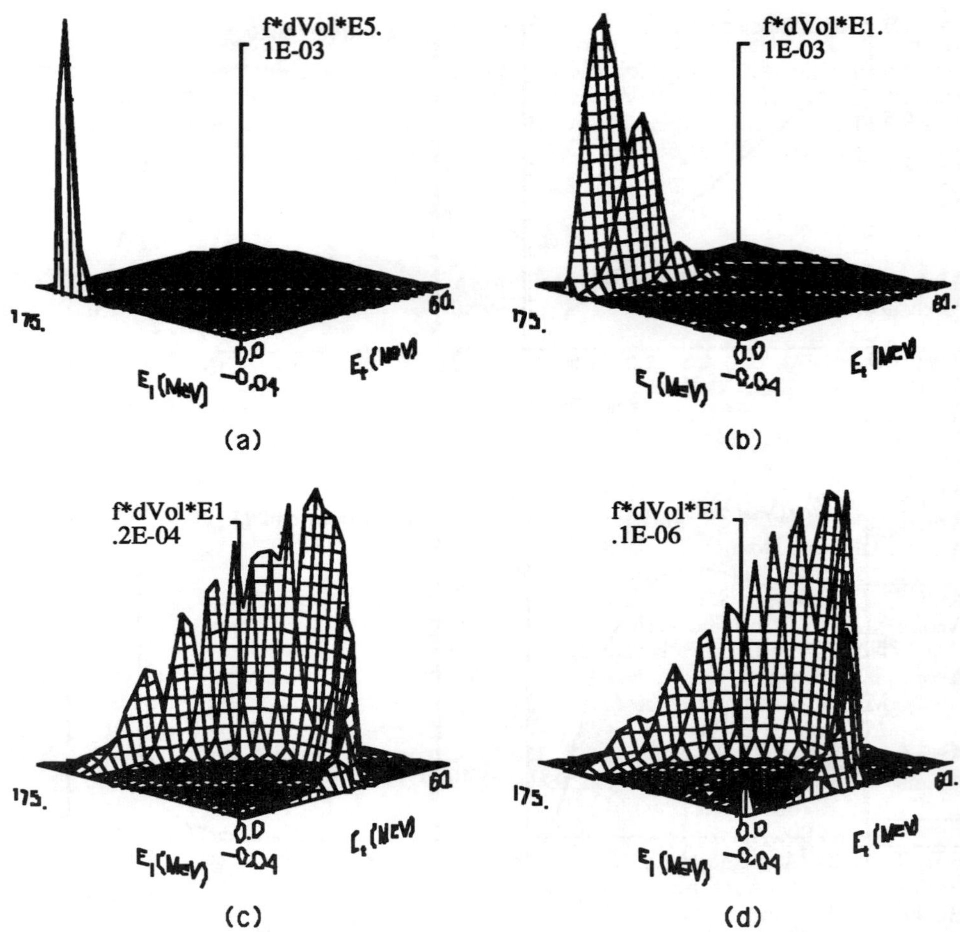

Fig. 7.3.—Snapshots of the distribution of proton kinetic energy in E_l and E_t at four successive times (a)–(d). The apparent raggedness of the distribution is an artifact of the plotting routine, which plots sections through the distribution. The axes have the same meaning as in Figure 7.1, although the scale is much expanded. The magnetic field strength, initial proton temperature, and electron density are also the same as in Fig. 7.1. However, V_i is five times larger and T_e is 25 times smaller. As a result, the proton beam velocity is much larger than the thermal velocity of the electrons in the target plasma. During the evolution (a)–(d), the average longitudinal kinetic energy of the protons falls from its initial value of 117 MeV to 160 keV, while the column density increases from 0 to 81 g cm^{-2}. The average transverse kinetic energy of the protons first increases from its initial value of 50 keV in panel (a) to 3.9 MeV in panel (b) and 14 MeV in panel (c). It then decreases, so that by the time of panel (d), it is only 160 keV. From Pakey (1990).

It is a pleasure to thank B. Fortner, O. de Jager, M. van der Klis, G. S. Miller, C. J. Pethick, M. Ruderman, J. Shaham, N. Shibazaki, and R. Staubert for stimulating discussions of the topics discussed in these lectures. We are also grateful to the directors J. Ventura and D. Pines for their kind hospitality at the Advanced Study Institute in Agia Pelagia. This research was supported in part by NASA grant NAGW 1583 and NSF grant PHY 86-00377 at Illinois.

8. References

Abramowicz, M. A., Calvani, M., and Nobili, L. 1980, *Astrophys. J.*, **242**, 772.
Abramowicz, M. A., Czerny, B., Lasota, J. P., and Szuszkiewicz, E. 1988, *Astrophys. J.*, **332**, 646.
Abramowicz, M. A., Jaroszyński, M, and Sikora, M. 1978, *Astron. Ap.*, **63**, 221.
Adam, J. A. 1978, *J. Plasma Phys.*, **19**, 77.
Alme, M. L., and Wilson, J. R. 1973, *Astrophys. J.*, **186**, 1015.
Alpar, M. A., and Shaham, J. 1985, *Nature*, **316**, 239.
Aly, J.-J. 1980, *Astron. Ap.*, **86**, 192.
Aly, J.-J., and Kuipers, J. 1990, *Astron. Ap.*, **227**, 473.
Angelini, L., Stella, L., and Parmar, A. N. 1989, *Astrophys. J.*, **346**, 906.
Anzer, U., and Börner, G. 1980, *Astron. Ap.*, **83**, 133.
________. 1983, *Astron. Ap.*, **122**, 73.
Anzer, U., Börner, G., and Monaghan, J. J. 1987, *Astron. Ap.*, **176**, 235.
Arons, J. 1987, in *The Origin and Evolution of Neutron Stars, IAU Symposium No. 125*, ed. D. J. Helfand and J.-H. Huang (Dordrecht: Reidel), p. 207.
Arons, J., Burnard, D. J., Klein, R. I., McKee, C. F., Pudritz, R. E., and Lea, S. M. 1984, in *High Energy Transients in Astrophysics, AIP Conf. Proc. No. 115*, ed. S. E. Woosley (New York: American Institute of Physics), p. 15.
Arons, J., Klein, R. I., and Lea, S. 1987, *Astrophys. J.*, **312**, 666.
Arons, J., and Lea, S. M. 1976a, *Astrophys. J.*, **207**, 914.
________. 1976b, *Astrophys. J.*, **210**, 792.
________. 1980, *Astrophys. J.*, **235**, 1016.
Bai, T. 1980, *Astrophys. J.*, **239**, 334.
Bailes, M. 1989, *Astrophys. J.*, **342**, 917.
Basko, M. M., and Sunyaev, R. A. 1975, *Soviet Phys. JETP*, **41**, 52.
________. 1976, *Mon. Not. R. astr. Soc.*, **175**, 395.
Becker, R. H., *et al.* 1977, *Astrophys. J.*, **214**, 879.
Björnsson, G. 1990, Ph.D. Thesis, University of Illinois at Urbana-Champaign.
Björnsson, G., and Svensson, R. 1991, *Astrophys. J.*, in press.
Blondin, J. M., Kallman, T. R., Fryxell, B. A., and Taam, R. E. 1990, *Astrophys. J.*, **356**, 591.
Bondi, H., and Hoyle, F. 1944, *Mon. Not. R. astr. Soc.*, **104**, 273.
Boynton, P. E., Deeter, J. E., Lamb, F. K., Zylstra, G., Pravdo, S. H., White, N. E., Wood, K. S., and Yentis, D. J. 1984, *Astrophys. J.*, **238**, L35.
Bradt, H., and McClintock, J. E. 1983, *Ann. Rev. Astron. Astrophys.*, **21**, 13.
Burnard, D. J., Arons, J., and Klein, R. I. 1990, *Astrophys. J.*, in press.
Burnard, D. J., Lea, S. M., and Arons, J. 1983, *Astrophys. J.*, **266**, 175.
Bussard, R. W. 1980, *Astophys. J.*, **237**, 970.
Campbell, C. G. 1987, *Mon. Not. R. astr. Soc.*, **229**, 405.
Canuto, V., and Ventura, J. 1977, *Fundamentals of Cosmic Physics*, **2**, 203.

Chadwick, P. M., McComb, T. J. L., and Turver, K. E. 1990, *J. Phys. G*, **16**, 1773.
Chanmugam, G., and Brecher, K. 1985, *Nature*, **313**, 767.
Chapman, S., and Ferraro, V.C.A. 1931, *Terr. Mag.*, **36**, 171.
Cheng, K. S., and Ruderman, M. 1989, *Astrophys. J.* (*Letters*), **337**, L77.
Cheng, K., Ho, C., and Ruderman, M. 1986a, *Astrophys. J.*, **300**, 495.
________. 1986b, *Astrophys. J.*, **300**, 522.
Clark, G. W., Woo, J. W., Nagase, F., Makishima, K., and Sakao, T. 1990, *Astrophys. J.*, **353**, 274.
Cowie, L. L. 1977, *Mon. Not. R. astr. Soc.*, **180**, 491.
Cronin, J. 1990, Proc. XIV International Symposium on Lepton and Photon Interactions, in press.
Cunningham, C. T. 1973, Ph.D. Thesis, University of Washington.
Davidson, K. 1973, *Nature Phys. Sci.*, **246**, 1.
Davidson, K., and Ostriker, J. P. 1973, *Astrophys. J.*, **179**, 585.
Davies, R. E., and Pringle, J. E. 1980, *Mon. Not. R. astr. Soc.*, **191**, 599.
de Jager, O. C. 1991, This volume.
Dodd, K. N., and McCrea, W. H. 1952, *Mon. Not. R. astr. Soc.*, **112**, 205.
Eichler, D., and Verstrand, W. T. 1985, *Nature*, **318**, 345.
Elsner, R. F. 1976, Ph.D. thesis, University of Illinois at Urbana-Champaign.
Elsner, R. F., Ghosh, P., and Lamb, F. K. 1978, *Astrophys. J.*, **241**, L155.
Elsner, R. F., and Lamb, F. K. 1976, *Nature*, **262**, 356.
________. 1977, *Astrophys. J.*, **215**, 897.
________. 1984, *Astrophys. J.*, **278**, 326.
Fenimore, E. E., *et al.* 1988, *Astrophys. J.* (*Letters*), **335**, L71.
Fortner, B., Lamb, F. K., and Zylstra, G. 1984, in *Numerical Astrophysics*, ed. J. M. Centrella, J. M. LeBlanc, and R. L. Bowers (Boston: Jones and Bartlett), p. 128.
Frank, J. 1979, *Mon. Not. R. astr. Soc.*, **187**, 833.
Fryxell, B. A., and Taam, R. E. 1988, *Astrophys. J.*, **335**, 862.
Fryxell, B. A., and Taam, R. E., and McMillan, S.L.W. 1987, *Astrophys. J.*, **315**, 536.
Gething, J.P.D. 1951, *Mon. Not. R. astr. Soc.*, **111**, 468.
Ghosh, P., and Lamb, F. K. 1978, *Astrophys. J.*, **223**, L83 (GL).
________. 1979a, *Astrophys. J.*, **232**, 259 (GL).
________. 1979b, *Astrophys. J.*, **234**, 296 (GL).
________. 1991, in preparation.
Ghosh, P., Lamb, F. K., and Pethick, C. J. 1977, *Astrophys. J.*, **217**, 578.
Giacconi, R. 1975, in *Proc. 8th Texas Symp. on Relativistic Astrophysics* (*Ann. NY Acad. Sci.*, **262**, 312).
Giacconi, R., Gursky, H., Kellogg, E., Schreier, E., and Tananabaum, H. 1971, *Astrophys. J.* (*Letters*), **167**, L67.
Hasinger, G., Langmeier, A., Sztajno, M., Trümper, J., Lewin, W.H.G., and White, N. E. 1986, *Nature*, **319**, 469.
Ho, C. 1988, *Mon. Not. R. astr. Soc.*, **232**, 91.
Ho, C., and Arons, J. 1987, *Astrophys. J.*, **321**, 404.
Ho, C., Taam, R. E., Fryxell, B. A., Matsuda, T., Koide, H., and Shima, E. *et al.* 1989, *Mon. Not. R. astr. Soc.*, **238**, 1447.
Hoyle, F., and Lyttleton, R. A. 1939, *Proc. Cam. Phil. Soc.*, **35**, 405.
Hunt, R. 1971, *Mon. Not. R. astr. Soc.*, **154**, 141.
Illarionov, A. F., and Sunyaev, R. A. 1975, *Astron. Astrophys.*, **39**, 185.

Jaroszyński, M., Abramowicz, M. A., and Paczyński, B. 1980, *Acta Astr.*, **30**, 1.
Joss, P. C., and Rappaport, S. A. 1984, *Ann. Rev. Astron. Astrophys.*, **22**, 537.
Katz, J. I., and Smith, I. A. 1988, *Astrophys. J.*, **326**, 733.
Kazanas, D., and Ellison, D. C. 1986, *Nature*, **319**, 380.
Kennel, C. F., and Coroniti, F. V. 1975, in *The Magnetospheres of the Earth and Jupiter*, ed. V. Formisano (Dordrecht: Reidel), p. 451.
________. 1977, *Ann. Rev. Astron. Astrophys.*, **15**, 389.
Kiraly, P., and Mészáros, P. 1988, *Astrophys. J.*, **333**, 719.
Kirk, J. G. 1979, *Plasma Phys.*, **21**, 1021.
________. 1980, *Plasma Phys.*, **22**, 639.
________. 1985, *Astron. Ap.*, **142**, 430.
Kirk, J. G., and Galloway, D. J. 1981, *Mon. Not. R. astr. Soc.*, **195**, 45P.
________. 1982, *Plasma Phys.*, , 339.
Kirk, J. G., and Trümper, J. E. 1983, in *Accretion-Driven Stellar X-Ray Sources*, ed. W.H.G. Lewin and E.P.J. van den Heuvel (Cambridge: Cambridge University Press), p. 261.
Klein, R. I., and Arons, J. 1989, in *Proc. 23rd ESLAB Symp. on X-ray Astronomy*, ed. J. Hunt and B. Battrick (ESA SP-296), p. 89.
Kozlowski, M., Jaroszyński, M., and Abramowicz, M. A. 1978, *Astron. Ap.*, **63**, 209.
Kuipers, J. 1990, in *Active Close Binaries*, ed. C. Ibanoğlu and I. Yavuz (Dordrecht: Kluwer Academic Publ.).
Kulkarni, S. R. 1986, *Ap. J. (Letters)*, **306**, L85.
Kundt, W., and Robnik, M. 1980, *Astron. Ap.*, **91**, 305.
Kusunose, M., and Takahara, F. 1988, *Pub. Astr. Soc. Japan*, **40**, 435.
________. 1989, *Pub. Astr. Soc. Japan*, **41**, 263.
________. 1990, *Pub. Astron. Soc. Japan*, in press.
Kylafis, N. D., and Lamb, D. Q. 1979, *Astrophys. J.*, **228**, L105.
Lamb, D. Q. 1988a, in *Nuclear Spectroscopy of Astrophysical Sources, AIP Conf. Proc.*, ed. N. Gehrels and G. Share (New York: American Institute of Physics), in press.
Lamb, F. K. 1975a, in *X-Rays in Space*, ed. D. Venkatesan (Calgary: University of Calgary), p. 613.
________. 1975b, in *Proc. 7th Texas Symposium on Relativistic Astrophysics* (*Ann. NY Acad. Sci.*, **262**, 331).
________. 1977, in *Proc. 8th Texas Symposium on Relativistic Astrophysics* (*Ann. NY Acad. Sci.*, **302**, 482).
________. 1978, unpublished.
________. 1984, Lecture series presented at the Workshop on Astrophysical Magnetospheres, Taos, New Mexico, August 1984.
________. 1988b, in *Physics of Compact Objects: Theory versus Observations*, ed. L. Filipov and N. White (Oxford: Pergamon), in press.
________. 1989a, in *Timing Neutron Stars*, ed. H. Ögelman and E.P.J. van den Heuvel (Dordrecht: Kluwer Academic Publ.), p. 649.
________. 1989b, in *Proc. 23rd ESLAB Symp. on X-ray Astronomy*, ed. J. Hunt and B. Battrick (ESA SP-296), p. 215.
________. 1991, This volume.
Lamb, F. K., Fabian, A. C., Pringle, J. E, and Lamb, D. Q. 1977, *Astrophys. J.*, **217**, 197.
Lamb, F. K., Ghosh, P., Hamilton, R., and Miller, M. C. 1991, in preparation.

Lamb, F. K., and Pethick, C. J. 1974, in *Astrophysics and Gravitation, Proc. 16th Internat. Solvay Congress* (Brussels: L'Université de Bruxelles), p. 135.
Lamb, F. K., Pethick, C. J., and Pines, D. 1973, *Astrophys. J.*, **184**, 271.
Lamb, F. K., Pines, D., and Shaham, J. 1978, *Astrophys. J.*, **224**, 969.
Lamb, F. K., *et al.* 1985, *Nature*, **317**, 681.
Lewin, W.H.G., and Joss, P. 1983, in *Accretion-Driven Stellar X-Ray Sources*, ed. W.H.G. Lewin and E.P.J. van den Heuvel (Cambridge: Cambridge University Press), p. 41.
Lewin, W.H.G., and van Paradijs, J., and van der Klis, M. 1988, *Space Sci. Rev.*, **46**, 273.
Liang, E.P.T. 1979, *Astrophys. J.*, **234**, 1105.
Lightman, A. P. 1990, personal communication.
Lightman, A. P., and Eardley, D. M. 1974, *Astrophys. J.* (*Letters*), **187**, L1.
Livio, M., 1991, in *Frontier Objects in Astrophysics and Particle Physics, Proc. Vulcano Workshop*, ed. F. Giovannelli, in press.
Livio, M., Soker, N., de Kool, M., and Savonije, G. J. 1986a, *Mon. Not. R. astr. Soc.*, **218**, 593.
________. 1986b, *Mon. Not. R. astr. Soc.*, **222**, 235.
Lovelace, R.V.E 1976, *Nature*, **262**, 649.
Makishima, K., *et al.* 1987, *Astrophys. J.*, **314**, 619.
Mason, K. O. 1977, *Mon. Not. R. astr. Soc.*, **178**, 81P.
Matsuda, T., Inoue, M., and Sawada, K. 1987, *Mon. Not. R. astr. Soc.*, **226**, 785.
Matsuda, T., Sekino, N., Shima, E., and Sawada, K. 1989, *Mon. Not. R. astr. Soc.*, **236**, 817.
Matsuda, T., Sekino, N., Sawada, K., Shima, E., Livio, M., Anzer, U., and Börner, G. 1991, *Astron. Ap.*, submitted.
Mészáros, P. 1982, in *Accreting Neutron Stars*, ed. W. Brinkmann and J. Trümper (Garching, FRG: Max-Planck-Institut fur Physik und Astrophysik), p. 253.
________. 1986, in *Plasma Penetration into Magnetospheres*, ed. N. Kylafis *et al.* (Iraklion, Crete: Crete University Press), p. 143.
Michel, F. C. 1977a, *Astrophys. J.*, **213**, 836.
________. 1977b, *Astrophys. J.*, **214**, 261.
________. 1977c, *Astrophys. J.*, **216**, 838.
Middleditch, J., and Priedhorsky, W. C. 1986, *Astrophys. J.*, **306**, 230.
Mihara, T., *et al.* 1990, *Nature*, **346**, 250.
Miller, G. S., and Lamb, F. K. 1991, in preparation.
Miller, G. S., Wasserman, I., and Salpeter, E. E. 1989, *Astrophys. J.*, **346**, 405.
Muchotrzeb, B., and Paczyński, B. 1982, *Acta Astr.*, **32**, 1.
Murakami, T., Ikegami, T., Inoue, H., and Makishima, K. 1987, *Pub. Astr. Soc. Japan*, **39**, 253.
Murakami, T., *et al.* 1988, *Nature*, **335**, 234.
Nagase, F. 1989, *Pub. Astr. Soc. Japan*, **41**, 1.
Ögelman, H. 1989, in *Timing Neutron Stars*,ed. H. Ögelman and E.P.J. van den Heuvel (Dordrecht: Kluwer Academic Publ.), p. 169.
Paczyński, B. 1980, *Acta Astr.*, **30**, 347.
Paczyński, B., and Bisnovatyi-Kogan, G. 1981, *Acta Astr.*, **31**, 283.
Paczyński, B., and Wiita, P. J. 1980, *Astron. Ap.*, **88**, 23.
Pakey, D. D. 1990, Ph.D. thesis, University of Illinois at Urbana-Champaign.
Pakey, D. D., Bussard, R. W., and Lamb, F. K. 1991, in preparation.

Parmar, A., White, N., Stella, L., Izzo, C., and Ferri, P. 1989, *Astrophys. J.*, **338**, 359.

Pavlov, G. G., and Yakovlev, D. G. 1976, *Sov. Phys. JETP*, **43**, 389.

Piran, T. 1978, *Astrophys. J.*, **221**, 652.

Pringle, J. E. 1976, *Mon. Not. R. astr. Soc.*, **177**, 65.

________. 1981, *Ann. Rev. Astron. Astrophys.*, **19**, 137.

Pringle, J. E., and Rees, M. J. 1972, *Astron. Astrophys.*, **21**, 1.

Pringle, J. E., Rees, M. J., and Pacholczyk, A. G. 1973, *Astron. Ap.*, **29**, 179.

Rappaport, S., and Joss, P. C. 1977, *Nature*, **266**, 683.

Riffert, H. 1980, *Astrophys. Space Sci.*, **71**, 195.

Rose, L. A., *et al.* 1979, *Astrophys. J.*, **231**, 919.

Różyczka, M., and Muchotrzeb, B. 1982, *Acta Astr.*, **32**, 285.

Ruderman, M. 1989, Talk presented at the NASA EGRET Symposium, Greenbelt, Maryland.

________. 1990, Proc. XIV International Symposium on Lepton and Photon Interactions, in press.

Sato, N., Nagase, F., Kawai, N., Kelly, R. L., Rappaport, S., and White, N. E. 1986, *Astrophys. J.*, **304**, 241.

Sawada, K., Matsuda, T., Anzer, U., Börner, G., and Livio, M. 1989, *Astron. Ap.*, **231**, 263.

Schreier, E. J., *et al.* 1972, *Astrophys. J.*, **172**, L79.

Shakura, N. I. 1972, *Astron. Zh.*, **49**, 921 [Engl. transl. *Sov. Astron.–AJ*, **16**, 756].

Shakura, N. I., and Sunyaev, R. A. 1973, *Astron. Ap.*, **24**, 337.

________. 1976, *Mon. Not. R. astr. Soc.*, **175**, 613.

Shapiro, S. L., and Lightman, A. P. 1976, *Astrophys. J.*, **204**, 555.

Shapiro, S. L., Lightman, A. P., and Eardley. D., M. 1976, *Astrophys. J.*, **204**, 187.

Shapiro, S. L., and Salpeter, E. E. 1975, *Astrophys. J.*, **198**, 671.

Shibazaki, N., and Lamb, F. K. 1987, *Astrophys. J.*, **318**, 767.

Shima, E., Matsuda, T., Takeda, H., and Sawada, K. 1985, *Mon. Not. R. astr. Soc.*, **217**, 367.

Slane, P., and Fry, W. F. 1989, *Astrophys. J.*, **342**, 1129.

Soker, N. 1991, preprint.

Soker, N., and Livio, M. 1984, *Mon. Not. R. astr. Soc.*, **211**, 927.

Soker, N., Livio, M., de Kool, M., and Savonije, G. J. 1986, *Mon. Not. R. astr. Soc.*, **221**, 445.

Spicer, D. S. 1982, *Space Sci. Rev.*, **31**, 351.

Stella, L. 1988, *Memorie della Società Astr. Italiana*, **59**, 185.

Stella, L., and Rosner, R. 1984, *Astrophys. J.*, **277**, 312.

Stenger, V. J. 1984, *Astrophys. J.*, **284**, 810.

Szuszkiewicz, E. 1990, *Mon. Not. R. astr. Soc.*, **244**, 377.

Taam, R. E. 1984, in *High Energy Transients in Astrophysics, AIP Conf. Proc. No. 115*, ed. S. E. Woosley (New York: American Institute of Physics), p. 1.

________. 1985, *Ann. Rev. Nucl. Part. Sci.*, **35**, 1.

Taam, R. E., and Fryxell, B. A. 1988, *Astrophys. J.*, **327**, L73.

Taam, R. E., and Lin, D. N. C. 1984, *Astrophys. J.*, **287**, 761.

Tanaka, Y. 1991, in *X-Ray Binaries and the Formation and Evolution of Binary and Millisecond Pulsars*, ed. E.P.J. van den Heuvel (Dordrecht: Kluwer Academic Publ.), in press.

Tananbaum, H., Gursky, H., Kellogg, E. M., Levinson, R., Schreier, E., and Giacconi, R. 1972a, *Astrophys. J. (Letters)*, **174**, L143.

Treves, A., Maraschi, L., and Abramowicz, M. 1988, *Pub. Astron. Soc. Pacific*, **100**, 427.
Tritz, B., and Tsuruta, S. 1989, *Astrophys. J.*, **340**, 203.
Trümper, J., Pietsch, W., Reppin, C., Voges, W., Staubert, R., and Kendziorra, E. 1978, *Astrophys. J. (Letters)*, **219**, L105.
Tuohy, I. R., *et al.* 1981, *Astrophys. J.*, **245**, 183.
van den Heuvel, E.P.J. 1989, in *Timing Neutron Stars*, ed. H. Ögelman and E.P.J. van den Heuvel (Dordrecht: Kluwer Academic Publ.), p. 523.
_______. 1991, This volume.
van den Heuvel, E.P.J., van Paradijs, J., and Taam, R. E. 1986, *Nature*, **322**, 153.
van der Klis, M. 1989, *Ann. Rev. Astr. Ap.*, **27**, 517.
_______. 1991, This volume.
van der Klis, M., Jansen, F., van Paradijs, J., Lewin, W.H.G., van den Heuvel, E.P.J., Trümper, J., and Sztajno, M. 1985, *Nature*, **316**, 225.
Voges, W., Pietsch, W., Reppin, C., Trümper, J., Kendziorra, E., and Staubert, R. 1978, *Astrophys. J.*, **263**, 803.
Wang, Y.-M. 1981, *Astron. Astrophys.*, **102**, 36.
_______. 1987, *Astron. Astrophys.*, **183**, 257.
Wang, Y.-M., and Nepveu, M. 1983, *Astron. Astrophys.*, **118**, 267.
Wang, Y.-M., Nepveu, M., and Robertson, J. A. 1984, *Astron. Astrophys.*, **135**, 66.
Wang, Y.-M., and Robertson, J. A. 1984, *Astron. Astrophys.*, **139**, 93.
Wang, Y.-M., and Welter, G. M. 1982, *Astron. Astrophys.*, **113**, 113.
Waters, L.B.F.M. and van Kerkwijk, M. H. 1989, *Astron. Ap.*, **223**, 196.
Weekes, T. C. 1988, *Phys. Rep.*, **160**, 1.
Wheaton, W. A., *et al.* 1979, *Nature*, **282**, 240.
White, N. E. 1982, in *Accreting Neutron Stars*, ed. W. Brinkmann and J. Trümper (Garching, FRG: Max-Planck-Institut fur Physik und Astrophysik), p. 29.
White, N. E., and Stella, L. 1987, *Mon. Not. R. astr. Soc.*, **231**, 325.
White, T. R., and Lightman, A. P. 1989,*Astophys. J.*, **340**, 1024.
_______. 1990, *Astophys. J.*, **352**, 495.
Wiita, P. J. 1982, *Astrophys. J.*, **256**, 666.
Williams, A. C., Weisskopf, M. C., Elsner, R. F., Darbro, W., and Sutherland, P. G. 1986, *Astrophys. J.*, **305**, 759.
Williams, D. J. 1975, *Mon. Not. R. astr. Soc.*, **171**, 537.
Zel'dovich, Ya. B., and Shakura, N. I. 1969, *Sov. Astron. - AJ*, **13**, 175.
Zylstra, G. 1988, Ph.D. thesis, University of Illinois at Urbana-Champaign.
Zylstra, G., Lamb, F. K., and Aly, J.-J. 1991, in preparation.

UNIFIED MODEL OF X-RAY SPECTRA AND QPOs IN LOW MASS NEUTRON STAR BINARIES

FREDERICK K. LAMB
Departments of Physics and Astronomy
University of Illinois at Urbana-Champaign
1110 W. Green St., Urbana, IL 61801
U.S.A.

ABSTRACT. Recent work on a unified model of accretion flows and X-ray emission in low-mass X-ray binaries is summarized. In this model, a weakly magnetic neutron star accretes gas simultaneously from a Keplerian disk and a corona above the inner part of the disk. Photons are produced primarily in the inner disk and a compact central corona surrounding the small neutron star magnetosphere. The photons then escape through an approximately radial inflow of gas captured from the inner disk corona. Changes in the optical depths of the central corona and the radial flow may explain the Z-shaped hardness-intensity and color–color tracks observed in the most luminous sources. The ~20–55 Hz intensity oscillations observed on the horizontal spectral branch are attributed to interaction of the small, rotating magnetosphere with plasma in the inner disk. The ~6–10 Hz oscillations seen on the normal branch are attributed to overstable oscillation of the radial flow. Numerical simulations show that the radial flow oscillates when the luminosity rises to within ~10 percent of the Eddington critical luminosity L_E, and that the oscillation frequency is ~5–10 Hz if the radial flow develops ~300 km from the neutron star as expected. Numerical calculations give oscillation amplitudes and phases that agree with observation. The 10–20 Hz oscillations observed in Sco X-1 when it is on the flaring branch may be caused by photohydrodynamic modes that grow in amplitude when the luminosity exceeds L_E.

1. Introduction

Although the luminous low-mass X-ray binaries (LMXBs) in the galactic bulge were among the first X-ray sources discovered outside the solar system, until recently they were something of an enigma. The main reason for the slow progress in understanding these objects was that little or no systematic behavior had been uncovered. However, about five years ago this situation began to change. Detailed studies of the X-ray spectra of individual LMXBs with the *OSO 8*, *Hakucho*, *Tenma*, *EXOSAT*, and *Ginga* satellites showed that there are several distinct components (for reviews, see Swank and Serlimitsos 1985; White and Mason 1985; Lamb 1986; Mitsuda and Tanaka 1986; White 1986). More recent studies (Hasinger 1987b, 1988a,b; Hasinger *et al.* 1989; Schulz *et al.* 1990; Hasinger

J. Ventura and D. Pines (eds.), Neutron Stars: Theory and Observation, 445–481.

and van der Klis 1989; Hasinger *et al.* 1990) have shown further that many of the most luminous LMXBs have three distinct spectral states, called the horizontal, normal, and flaring branches. Systematic studies of the X-ray spectra of collections of LMXBs (White *et al.* 1988; Schulz *et al.* 1989) have made possible direct intercomparisons of their spectra.

The unexpected discovery of quasi-periodic intensity oscillations (QPOs) in several luminous LMXBs (van der Klis *et al.* 1985; Middleditch and Priedhorsky 1986; Hasinger *et al.* 1986; Norris and Wood 1987) has provided a wealth of additional information. (The term QPO is used in this context simply to denote the presence of a prominent but relatively broad peak in the power density spectrum of the X-ray intensity time series.) QPOs have now been detected in most of the luminous LMXBs. Further synthesis was achieved when it was realized that the spectral and temporal properties of the luminous LMXBs are closely correlated (Priedhorsky *et al.* 1986; van der Klis *et al.* 1987a; Hasinger 1987a,b, 1988a,b; Lamb 1988a; Hasinger *et al.* 1989; Hasinger *et al.* 1990). For recent reviews of QPO observations, see Lewin, van Paradijs, and van der Klis (1988), Stella (1988), and van der Klis (1989, 1991).

Here I summarize the current status of efforts at the University of Illinois to develop a consistent physical model of the luminous LMXBs that can account for their X-ray spectral states and quasi-periodic intensity oscillations (Lamb 1988b, 1989b,c, 1991; Miller 1990; Park and Miller 1991; Fortner *et al.* 1989, 1991a,b; Ghosh *et al.* 1991a,b; Miller and Lamb 1988, 1991). In this model, the neutron star has a relatively weak magnetic field and accretes plasma simultaneously from a Keplerian disk and from a corona above the inner part of the disk. Photons are produced primarily in the inner disk and a small central corona surrounding the neutron star magnetosphere. The photons then escape through an approximately radial inflow of material captured from the inner disk corona.

When the luminosity is moderate, the optical depth of the inner disk corona and radial inflow is low and has only a small effect on the spectral and temporal properties of the emerging radiation. However, when the luminosity is within $\sim 10\%$ of the Eddington critical luminosity L_E, the outward force of the escaping radiation slows the radial inflow, the scattering optical depth of the inflow becomes $\sim$5–15, and the inflow substantially alters the X-ray spectrum. The Z-shaped patterns observed in hardness-intensity and color–color diagrams of the luminous LMXBs are attributed to changes in the X-ray spectrum caused by changes in the scattering optical depth of the inner disk and the radial flow.

The $\sim$20–50 Hz quasi-periodic oscillations observed primarily in the horizontal branch spectral state are assumed to be luminosity oscillations caused by interaction of the neutron star magnetosphere with the inner accretion disk (the magnetospheric beat-frequency modulated-accretion model). The $\sim$5–10 Hz quasi-periodic intensity oscillations observed in the normal branch spectral state are attributed to optical depth oscillations caused by interaction of escaping radiation with the radial inflow from the corona above the inner disk when the luminosity rises to within $\sim 10\%$ of L_E. The $\sim$10–20 Hz quasi-periodic intensity oscillations observed in the flaring branch spectral state are attributed to photohydrodynamic modes that are excited by the oscillations in the radial flow and grow when the luminosity equals or exceeds the Eddington luminosity.

The remainder of the paper is organized as follows. Some key observational results and the basic elements of the model are summarized in §2 and §3, respectively. I discuss the origin of the horizontal, normal, and flaring branch

X-ray spectral states in §4. The beat-frequency model of horizontal branch oscillation is summarized in §5, while the overstable-radial-flow model of normal branch oscillation is described in §6. In §7 I discuss the possibility that the 10–20 Hz intensity oscillations observed in Sco X-1 when it is on the flaring branch are a manifestation of photohydrodynamic modes that are excited when the luminosity approaches or exceeds the Eddington critical luminosity. §8 contains some concluding remarks.

2. Summary of Observations

2.1 X-RAY LUMINOSITIES

The persistent luminosities of LMXBs range from moderate ($\sim 3 \times 10^{37}$ $\mathrm{ergs\,s^{-1}}$) to high ($\sim 10^{38}$ $\mathrm{ergs\,s^{-1}}$) (for reviews, see Bradt and McClintock 1983; White 1986; White *et al.* 1988; van der Klis 1989). The majority of moderate-luminosity LMXBs produce regular X-ray bursts, and have peak burst fluxes that are at least a factor of two larger than their persistent fluxes. Their X-ray orbital periods, when detected, are usually $\lesssim 10$ hours. Such luminosities and orbital periods are consistent with mass transfer driven by orbital angular momentum loss (Rappaport *et al.* 1983).

High-luminosity LMXBs are generally among the brightest sources in the X-ray sky and rarely produce X-ray bursts. So far, the ~ 6 and ~ 20–55 Hz QPOs discussed below have been observed only in these systems. Orbital periods have rarely been detected in the high-luminosity LMXBs but are thought to be ~ 1 day or longer. Orbital periods of this length and high luminosities are consistent with mass transfer driven by nuclear evolution of the companion (Webbink *et al.* 1983). Examples of high-luminosity LMXBs include Cyg X-2 and LMC X-2. Other LMXBs with similar properties, such as Sco X-1, are also thought to have high luminosities, even though their distances are not well determined (Schulz *et al.* 1989).

2.2 X-RAY SPECTRA

LMXB X-ray spectra do not depend strongly on the inclination of the system (White 1986), which suggests that the central X-ray emission region is relatively spherical. Swank and Serlemitsos (1985) found that the X-ray spectra of many LMXBs cannot be adequately described by a single spectral function and suggested that a spectral function composed of two terms be used instead. By now, there is an extensive literature concerned with the representation of LMXB spectra by multi-term spectral functions (see White 1986; Lamb 1986; White *et al.* 1988; Schulz *et al.* 1989; Lamb 1989b,c, 1991; van der Klis 1989 and references therein). One convenient function that provides a fairly accurate description of the photon number spectra of LMXBs over the energy range 2–30 keV is (White 1986; White *et al.* 1988; Schulz *et al.* 1989)

$$\frac{dN}{dE} \approx \left\{ A\,(E/E_a)^{-\alpha} \exp\left(-E/E_a\right) + \frac{B\,(E/E_b)^2}{[\exp(E/E_b)-1]} \right\} \exp\left[-N_H \sigma(E)\right] . \quad (1)$$

Here N_H is the equivalent hydrogen column density to the source and $\sigma(E)$ is the interstellar absorption cross section. In fits to the spectra of moderate-luminosity LMXBs, the B term is very small or absent; typical values of the parameters in the A term are $\alpha \approx 1$–2 and $E_a \approx 10$–30 keV (Schulz *et al.* 1989; White *et al.* 1988). In fits to the spectra of high-luminosity LMXBs, the B term contributes $\lesssim 10$–30% of the total flux; E_b is typically ~ 1 keV, and typical values of the parameters in the A term are $\alpha \approx 1$–2 and $E_a \approx 4$–7 keV (Schulz *et al.* 1989).

2.3 X-RAY SPECTRAL STATES

Recent surveys of LMXB X-ray spectra (Schulz *et al.* 1989; Hasinger and van der Klis 1989) indicate that many of the high-luminosity sources exhibit two or three distinct spectral branches, as illustrated in Figure 1. The different spectral states are apparent both in X-ray hardness–intensity diagrams (Schulz *et al.* 1989) and in color–color diagrams (Hasinger 1988a; Hasinger *et al.* 1986; Hasinger *et al.* 1989; Schulz *et al.* 1989). However, color–color diagrams of different sources appear more similar than hardness–intensity diagrams.

Hardness–intensity diagrams are affected by variations in the source intensity caused by changes in the projected area of the emitting region or the fraction of the X-ray flux that is scattered out of the line of sight by intervening plasma, as well as by changes in the X-ray hard color. Thus, such diagrams are sensitive to the inclination and geometry of the source. In contrast, color–color diagrams depend only on the shape of the X-ray spectrum and are unaffected by variations in intensity. Since plasma near the neutron star is highly ionized and therefore does not absorb X-rays appreciably, obscuration by such plasma can reduce the intensity of the source without significantly changing its X-ray colors. Obscuration by cooler plasma located far from the neutron star would significantly change the X-ray colors, as well as reducing the intensity, but such changes are only expected on time scales comparable to the orbital period. The fact that the color–color diagrams of different sources appear more similar than their hardness–intensity diagrams is probably caused in large part by differences in our viewing angle (Hasinger 1988a; Lamb 1988a; Hasinger *et al.* 1989).

The existence of three distinct spectral states is most apparent in Cyg X-2 (Hasinger 1988a,b; Schulz *et al.* 1989; Hasinger and van der Klis 1989; Hasinger *et al.* 1990). When plotted in an X-ray color–color diagram, data from Cyg X-2 form a Z-shaped curve (see Fig. 1). The top, diagonal, and bottom legs of the Z have been designated the horizontal branch (HB), the normal branch (NB), and the flaring branch (FB), respectively. Once the effects of interstellar absorption have been removed, data from all the luminous LMXBs fall in a relatively small area of the X-ray color–color diagram, indicating that these sources have similar intrinsic properties (Schulz *et al.* 1989).

Sources apparently do not jump from one spectral branch to another, but move only along the branches (Hasinger 1988a,b; Schulz *et al.* 1989). This motion occurs irregularly, with appreciable displacements on time scales of minutes to hours (Hasinger *et al.* 1990). Movement of a source from left to right across the horizontal branch corresponds, loosely speaking, to an increase in the size of the B-term in equation (1), with the other parameters remaining approximately constant, whereas movement down the normal branch corresponds to a decrease in the cutoff energy E_a, with the other parameters approximately constant (Schulz *et al.* 1989). Movement of a source from the bottom left toward the top right of the flaring

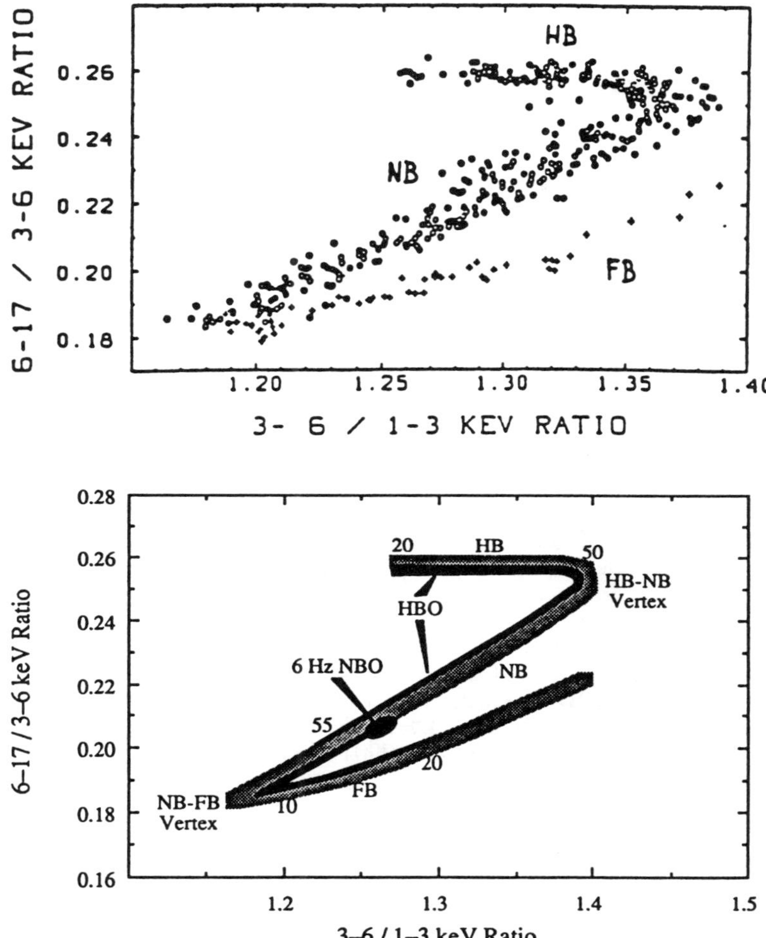

Fig. 1.—X-ray color-color diagrams. *Top*: Cyg X-2 data from Schulz *et al.* (1989), showing the horizontal branch (HB), normal branch (NB), and flaring branch (FB). *Bottom*: Schematic Z-track, showing the connection between X-ray spectral state and QPO. As a source moves from left to right across the horizontal branch, the frequency of the horizontal branch oscillation (HBO) increases from ~ 20 Hz to ~ 50 Hz while its strength decreases; the HBO is weaker but still detectable at ~ 55 Hz near the middle of the normal branch. The normal branch oscillation (NBO) begins near the middle of the normal branch with a frequency ~ 6 Hz, which increases to ~ 10 Hz near the bottom of the normal branch. Sco X-1 and Cyg X-2 power spectra show a peak at ~ 10 Hz near the bottom of the flaring branch. As these sources move up the flaring branch, the centroid of this peak increases to ~ 20 Hz and the peak steadily broadens, until it eventually dissolves into a broad continuum of excess power density.

branch can be reproduced by a monotonic increase in E_a, B, and E_b (Schulz *et al.* 1989). Indications are that the mass flux and bolometric luminosity increase as a source progresses from the horizontal branch to the normal branch and then to the flaring branch (Priedhorsky *et al.* 1986; Hasinger 1988a,b; Lamb 1988a; Hasinger *et al.* 1989; Hasinger *et al.* 1990).

2.4 POWER DENSITY SPECTRA

Figure 2 shows power density spectra like those that led to the discovery of QPOs in the luminous LMXBs GX 5−1 and Sco X-1. The widths of the peaks in these power spectra indicate that the oscillations have Q values ~ 3. The total power in the peaks is equivalent to the power in a sinusoidal oscillation with an rms amplitude $\sim 5\%$. Although easily detected in power spectra, such amplitudes are sufficiently small that the oscillations are not discernable in the X-ray intensity time series. Thus, the waveforms that give rise to these peaks are at present unknown.

Power spectra like those shown in Figure 2 already hinted that there are at least two distinct types of QPOs in the luminous LMXBs. The spectrum of GX 5−1 shows a high-frequency (~ 35 Hz) peak rising above a continuum that decreases steeply with increasing frequency. In contrast, the spectrum of Sco X-1 shows a lower-frequency (~ 6 Hz) peak rising above a continuum that is relatively flat. Further observations (discussed below) have established that these two peaks are indeed produced by different mechanisms.

The higher-frequency QPO is strongest on the horizontal branch, although it has been seen at a lower amplitude on the normal branch. It is therefore referred to as a "horizontal branch oscillation" (HBO). The lower-frequency QPO is seen only on the lower half of the normal branch and hence is called a "normal branch oscillation" (NBO). In addition to these oscillations on the horizontal and normal branches, $\sim$10–20 Hz "flaring branch oscillations" (FBOs) have been seen in Sco X-1 (Middleditch and Priedhorsky 1986; Priedhorsky *et al.* 1986; Hasinger *et al.* 1989) and Cyg X-2 (Hasinger *et al.* 1990) when they are on the lower part of the flaring branch.

The relation between the spectral and temporal properties of the luminous LMXBs is illustrated by the Cyg X-2 horizontal, normal, and flaring branch power spectra shown in Figure 3. All three have a "very low frequency noise" (VLFN) component that dominates the power density at frequencies below ~ 0.1 Hz, but they differ qualitatively in most other respects. The horizontal branch power spectrum shows a distinct "low frequency noise" (LFN) component, which dominates the continuum power density in the range 0.1–20 Hz but falls off approximately exponentially above ~ 20 Hz. The HBO peak is at ~ 55 Hz. The NBO peak at ~ 6 Hz is clearly visible in the power spectrum from the middle of the normal branch. LFN and HBO are also present, but with about half their power on the horizontal branch. The power spectrum from the bottom of the flaring branch shows a broader FBO peak at ~ 10 Hz.

Sco X-1 power spectra also show VLFN, plus a flatter noise component with an e-folding cutoff frequency of ~ 60 Hz (Hasinger *et al.* 1989). The latter has been dubbed "high frequency noise" (HFN). Sco X-1 power spectra from the right end of the horizontal branch (near the HB-NB vertex) show LFN, which weakens as Sco X-1 moves onto the normal branch. Spectra from the bottom half of the normal branch show ~ 6 Hz NBOs, while spectra from the bottom half of the flaring branch show an FBO peak at ~ 10 Hz (Priedhorsky *et al.* 1986; Hasinger *et*

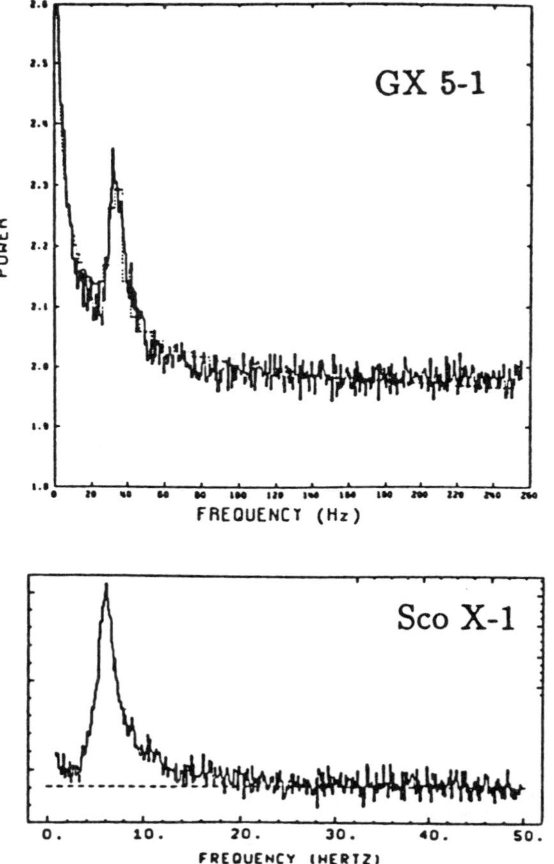

Fig. 2.—Power density spectra of GX 5−1 and Sco X-1 X-ray intensity time series, showing QPO peaks. Note the pronounced decrease of the GX 5−1 continuum power density with increasing frequency and the absence of such a decrease in Sco X-1. From van der Klis (1989).

al. 1989). Thus, power spectra like those of Figure 3 appear to be characteristic of the Z-class LMXBs (Hasinger *et al.* 1989; Hasinger *et al.* 1990).

The HBO frequency is strongly positively correlated with X-ray intensity, increasing from ~20 Hz at the left of the horizontal branch to ~50 Hz at the right. The NBO frequency increases from ~6 Hz at the middle of the normal branch to ~10 Hz at the bottom (near the NB-FB vertex), but is only weakly correlated with X-ray intensity. The recent confirmation (Hasinger *et al.* 1990) of earlier hints (Norris and Wood 1987; Mitsuda and Dotani 1989) that HBO and NBO are both visible in Cyg X-2 power spectra from the upper half of the normal branch

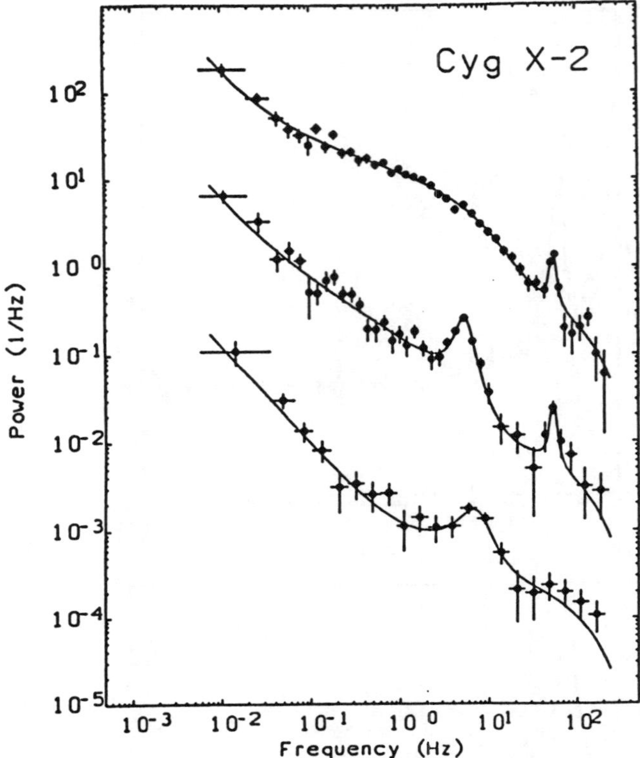

Fig. 3.—Cyg X-2 horizontal branch (top), normal branch (middle), and flaring branch (bottom) power density spectra. Note the VLFN components in all three, the LFN component at 0.1–20 Hz and the HBO at 50 Hz in the horizontal branch spectrum, the weaker LFN and HBO and the strong NBO at 6 Hz in the normal branch spectrum, and the broader FBO peak at $\sim$10 Hz in the flaring branch spectrum. From Hasinger *et al.* (1990).

(see Fig. 3) shows conclusively that these two types of oscillations have physically distinct origins.

In both Sco X-1 (Priedhorsky *et al.* 1986; Hasinger *et al.* 1989) and Cyg X-2 (Hasinger *et al.* 1990), the FBO centroid frequency increases smoothly from $\sim$ 10 Hz at the NB-FB vertex to $\sim$20 Hz halfway up the flaring branch. The width of the FBO peak also appears to increase steadily. The FBO peak in Sco X-1 eventually dissolves into a broad continuum of excess power density near the middle of the flaring branch (Middleditch and Priedhorsky 1986; Priedhorsky *et al.* 1986; Hasinger *et al.* 1989).

Periodic oscillations have not been detected so far in any of the Z-class LMXBs. Current upper limits on the rms amplitudes of such oscillations range from 0.4 to 2.5% for GX 5−1 (van der Klis *et al.* 1985), from 0.3 to 0.5% for Cyg X-2

Table 1. Properties of QPOs in Z-Class Low Mass X-Ray Binaries[a]

Source Property	X-Ray Spectral Branch Horizontal	Normal	Flaring
QPO frequency (Hz)	20–60	5–10	10–20
Low frequency noise (LFN)	Strong	Weak	Weak
Very low frequency noise (VLFN)	Present	Present	Present
QPO frequency–X-ray intensity correlation	Strong, positive	Weak, variable	Highly variable
Energy spectrum of QPO photons	Relatively hard	Concave upward	Relatively hard?
Magnitude and sign of QPO phase shift	$\sim$0.2–4 ms hard lag	$\sim$80 ms hard lag	?

[a]Based on Hasinger (1988a,b), Lewin and van Paradijs (1988), Hasinger *et al.* (1989), van der Klis (1989), and Hasinger *et al.* (1990).

(Norris and Wood 1987), from 0.5 to 0.7% for Sco X-1 (Hasinger *et al.* 1989), from 0.5 to 1.9% for GX 17+2 (Stella *et al.* 1987), and from 0.5 to 0.7% for GX 340+0 (van Paradijs *et al.* 1988), depending on the assumed oscillation frequency.

The properties of the QPOs associated with the horizontal, normal, and flaring branch spectral states of the Z-class LMXBs are summarixed in Table 1. Question marks indicate uncertainty, either because the property is undetermined at present (FBO phase shift), or because it has been determined only for one source (FBO photon energy spectrum).

3. Unified Model

The model currently being explored at Illinois assumes that the X-ray source is a neutron star with a relatively weak ($B \sim 10^8$–10^9 G) magnetic field accreting matter from a disk fed by the companion star. The disk is disrupted by the magnetic field of the neutron star at $\sim$15–20 km (Lamb 1988b, 1989b,c, 1991; see also Ghosh and Lamb 1979a,b; Ghosh *et al.* 1991a). The energy to power the X-ray emission comes from release of the gravitational potential energy of the accreting matter as it approaches the neutron star.

The inner disk probably consists of dense clumps of cooler plasma in Keplerian orbit, surrounded by a hot-proton plasma, as shown schematically in Figure 4. Protons in the hot plasma are heated by viscous energy dissipation and interaction with the magnetosphere and neutron star, and cooled by collisions with electrons.

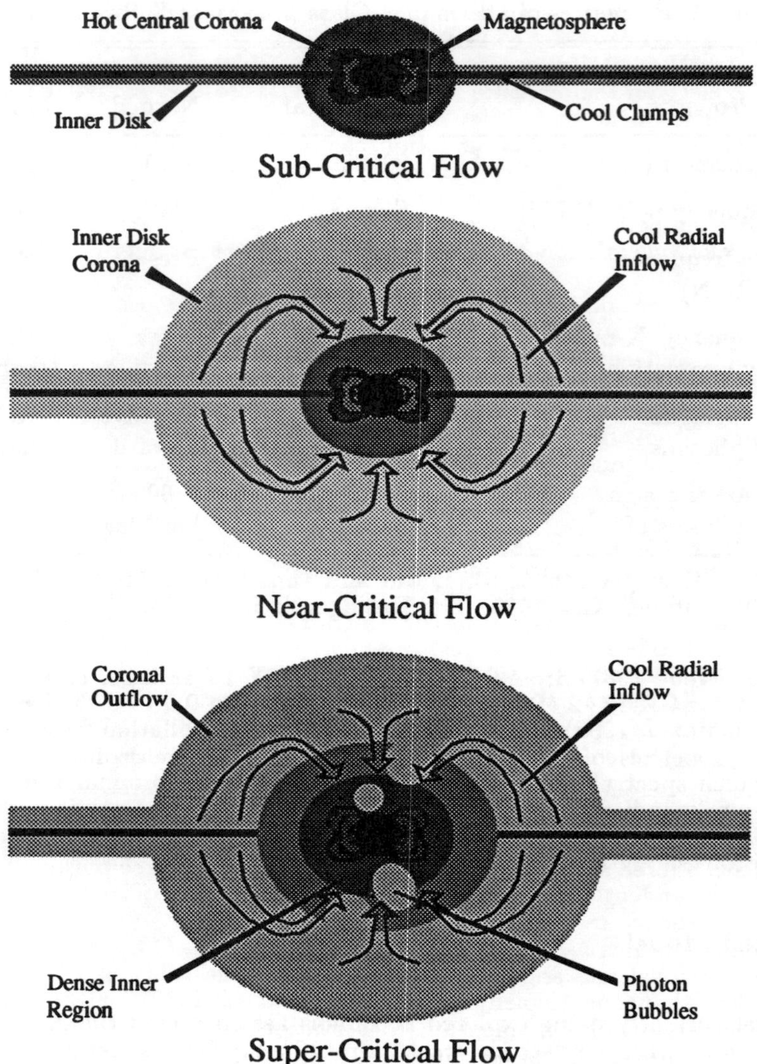

Fig. 4.—Schematic side views of the accretion flows discussed in the text. At moderate luminosities (top), the inner accretion disk consists of clumps of cooler plasma in Keplerian orbit, surrounded by a hot-proton plasma. Radiation forces cause the disk to thicken around the small neutron star magnetosphere, creating a compact central corona. At higher luminosities (middle), the inner disk expands vertically, forming a cool inner disk corona surrounding the hot central corona. The neutron star then accretes plasma from the inner disk corona as well as the disk (arrows). At luminosities above the Eddington limit, the radial flow near the central corona becomes unstable to growth of photohydrodynamic modes, which segregate the accreting plasma and outflowing radiation (bottom), allowing accretion to continue. After Lamb (1989b, c, 1991).

The resulting proton temperature in the hot plasma is ~ 100 MeV. The electrons in this plasma are heated by collisions with the protons and cooled by interaction with the ~ 1 keV radiation coming from the neutron star and magnetosphere. As a result, the electron temperature is ~ 10–20 keV (Ghosh *et al.* 1991b).

The neutron star magnetosphere is filled with inflowing plasma, which is slowed by the outward force of radiation coming from near the star. Because of the high plasma density, the X-ray photosphere is some distance above the neutron star surface. Radiation produced in this region has a thermal spectrum with a characteristic temperature comparable to the effective temperature (~ 1 keV). High-harmonic cyclotron emission in the outer magnetosphere produces a substantial flux of soft photons.

The accretion flow pattern depends on the luminosity. At moderate luminosities (~ 0.5–$0.9\, L_E$), the outward momentum of the radiation escaping from the inner disk and neutron star causes the inner disk to thicken, forming a hot *central corona* around the small (radius ~ 15–20 km) magnetosphere of the neutron star, as shown schematically in the top panel of Figure 4. The electron scattering optical depth τ_{CC} of the central corona is probably ~ 5. The radiation that emerges from the inner disk and central corona has a spectrum that is produced largely by unsaturated comptonization and therefore has a cutoff energy ~ 10–20 keV (Lamb 1989b,c, 1991).

At luminosities $\gtrsim 0.9\, L_E$, the pressure of radiation escaping from the inner disk drives some plasma into a more extensive *inner disk corona* surrounding the central corona, as shown in the middle panel of Figure 4. Radiation drag causes plasma in the inner disk corona near the neutron star to lose its angular and vertical momentum and to fall approximately radially toward the star (Fortner *et al.* 1989, 1991a; Lamb 1989c, 1991).

The size of the radiation drag and the structure of the radial flow depend on the mass flux $\dot{M}_d$ through the disk and the mass flux $\dot{M}_r$ in the radial flow. These are conveniently specified by the dimensionless parameters $\mu_d \equiv \dot{M}_d/\dot{M}_E$ and $\mu_r \equiv \dot{M}_r/\dot{M}_E$, where $\dot{M}_E$ is the mass flux that produces an accretion luminosity L_E. The total mass flux is then specified by the dimensionless parameter $\epsilon \equiv 1-\mu_d-\mu_r$. Since $L_\infty/L_E \approx 1-\epsilon$, the parameter ϵ is also a measure of the relative importance of the radiation force acting on the flow. Only two of the three parameters μ_r, μ_d, and ϵ are independent. When there is an adequate supply of matter, one expects $L_\infty \approx L_E$, and hence $\epsilon \ll 1$.

Inside the radius $r_{\rm radial} \approx 2\pi(1-\epsilon)^2 r_2$, where $r_2 \equiv GM/\epsilon c^2$, radiation drag removes the vertical and angular momentum of plasma orbiting in the inner disk corona in less than one Kepler period (Fortner *et al.* 1989, 1991a; Lamb 1989c, 1991). However, angular momentum conservation limits the radial mass flux to $\mu_r \leq \phi_s/c^2$, where ϕ_s is the gravitational potential at the neutron star surface (Fortner *et al.* 1989, 1991a; Lamb 1989c, 1991). Thus, at most $\sim 30\%$ of the total mass flux onto the neutron star may come from the inner disk corona. If the inner disk corona is sufficiently dense, the actual radial mass flux should be near this maximum. Since the maximum is proportional to the luminosity, the radial mass flux may be approximately constant for luminosities near L_E.

Plasma in the inner disk corona that has lost angular momentum accelerates inward until its radial velocity reaches the value $\epsilon c/2$ at which the comoving luminosity equals L_E and the acceleration vanishes (Miller 1990). Inside the inner critical radius $r_1 \equiv (\mu_r/\epsilon)R$ the flow becomes optically thick, radiation trapping is small but significant, and the flow velocity decreases linearly with radius

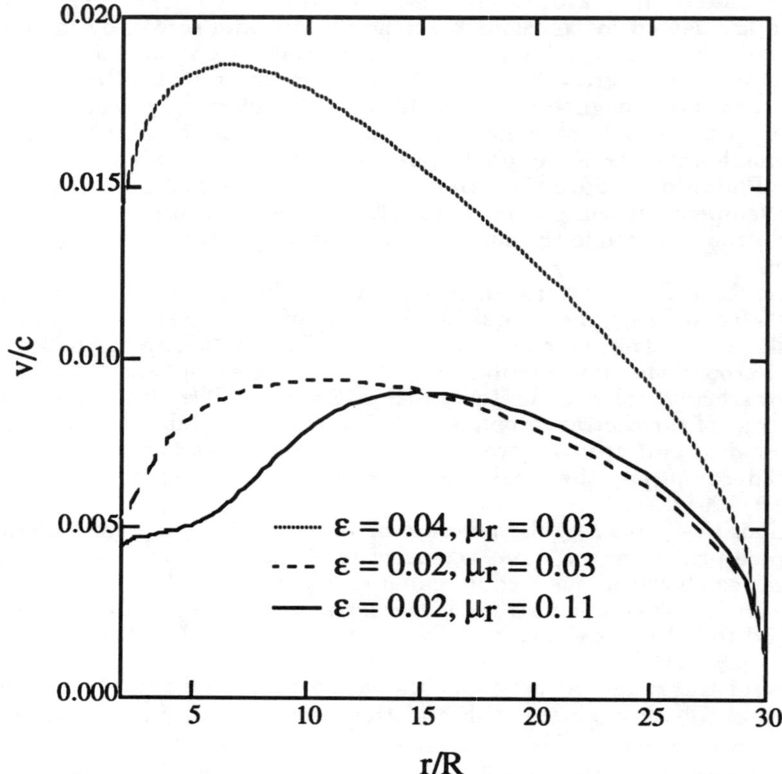

Fig. 5.—Velocity profiles of three selected near-critical radial flows, illustrating the dependence of the velocity structure on the relative difference ϵ between the total mass flux and the critical flux and on the radial mass flux μ_r. The maximum velocity increases with increasing ϵ. From Fortner, Lamb, and Miller (1991a).

(Miller 1990) until gas pressure gradients become important (Fortner *et al.* 1989, 1991a,b). The time t_f required for gas captured from the inner disk corona to reach the neutron star is dominated by the inflow time from the outer part of the radial flow.

Flows with total mass fluxes in the range $0.01 \leq \epsilon \leq 0.25$ and radial mass fluxes in the range $0 \leq \mu_r \leq 0.5$ have been simulated by Fortner *et al.* (1989, 1991a,b). In these simulations, all of the radiation was assumed to come from the surface of a compact central corona that forms the inner boundary of the radial flow. Of course, the interaction between the radiation and the flow is most pronounced for small ϵ and large μ_r. The velocity, stationary-frame luminosity, and comoving-frame luminosity profiles of three such flows are shown in Figures 5–7. These figures illustrate several important features of near-critical radial flows.

Figure 5 displays the radial velocity profiles of three selected flows. All three show the three distinct regions characteristic of a near-critical radial flow: an outer region, where the flow is optically-thin and the velocity increases with decreasing

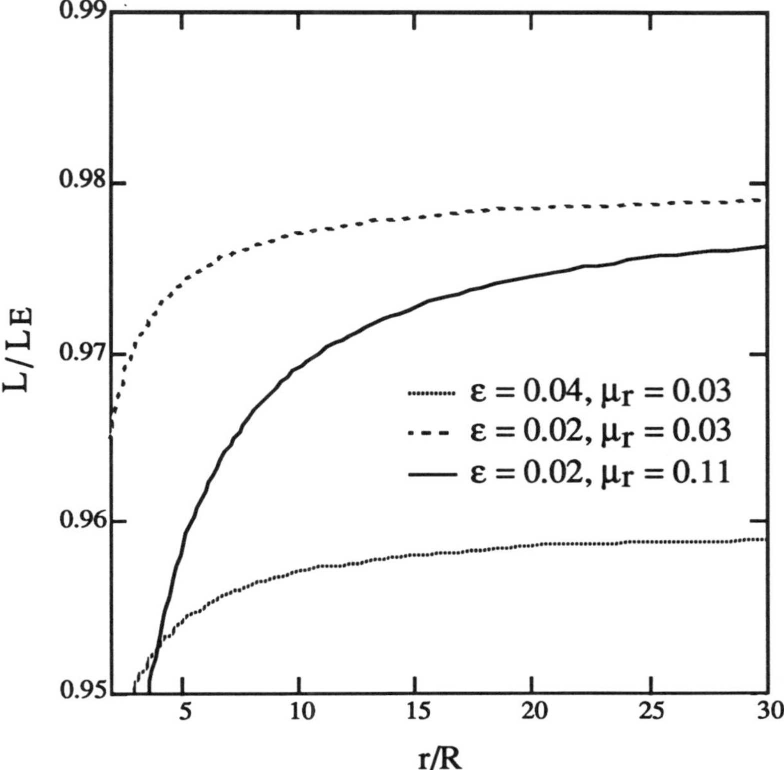

Fig. 6.—Stationary-frame luminosity profiles for the three flows plotted in Fig. 5, illustrating the work done on the radiation field by the converging flow. The radiation field at a larger radius has had more work done on it, and hence the luminosity there is greater. The two curves for $\epsilon = 0.02$ show that flows with larger radial mass fluxes do more work on the radiation, other things being equal. From Fortner, Lamb, and Miller (1991a).

radius; a middle region, where the velocity is approximately constant; and an inner region, where the flow is optically-thick and the velocity decreases with decreasing radius.

The increase in the stationary-frame luminosity with radius, which is evident in Figure 6, can be understood as follows. As the gas in the converging radial flow scatters photons, it does work on the radiation field. The stationary-frame luminosity therefore decreases with decreasing radius, since the radiation field at a larger radius has had more work done on it. Also, flows with larger radial mass fluxes do more work on the radiation, other things being equal.

In contrast to the stationary-frame luminosity, the comoving-frame luminosity *increases* with decreasing radius throughout the outer part of the flow. This is evident in Figure 7, which displays the radial profiles of the comoving-frame luminosity for the three flows plotted in Figure 5. Moreover, the comoving-frame luminosity at the outer boundary falls as the radial mass flux increases, other things

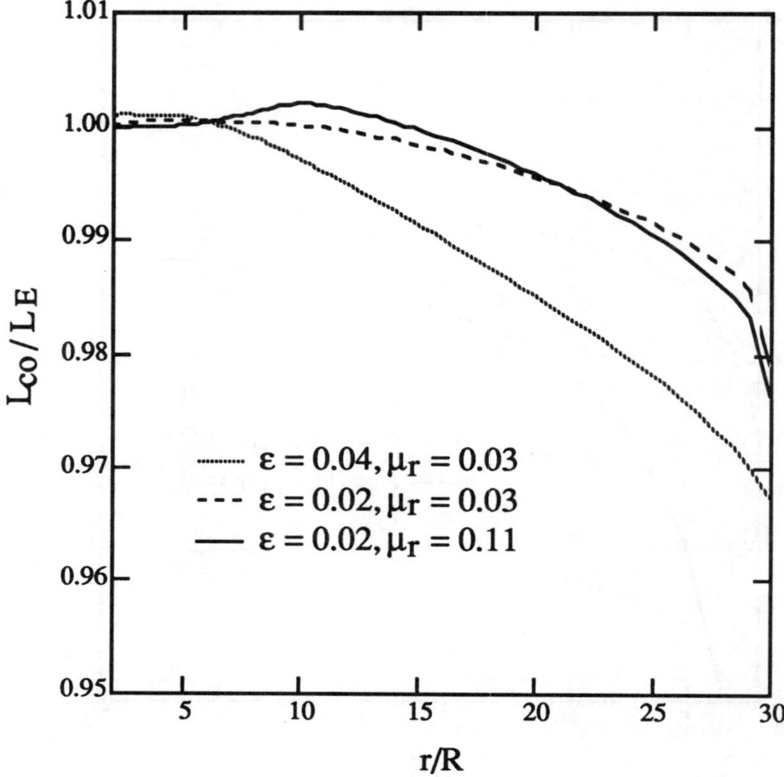

Fig. 7.—Comoving-frame luminosity profiles for the three flows plotted in Fig. 5, showing that the luminosity in the comoving frame increases with decreasing radius throughout the outer part of the flow. In the inner part of the flow, the comoving-frame luminosity tends to overshoot the Eddington luminosity by a small amount and then to approach the Eddington luminosity from above as the radius decreases further. From Fortner, Lamb, and Miller (1991a).

being equal. In the inner part of the flow, the comoving-frame luminosity tends to overshoot the Eddington luminosity by a small amount and then to approach the Eddington luminosity from above as the radius decreases further. It is this overshoot that causes the radial velocity profile to turn over (see Fig. 5). The overshoot is due to the rapid increase of the radiation stress and energy density with decreasing radius; the flow must see a mildly supercritical comoving luminosity in order to decelerate sufficiently to avoid seeing an even more supercritical comoving luminosity at still smaller radii. The three profiles show that the size of the overshoot (and hence the magnitude of the deceleration—again see Fig. 5) is greater for flows with larger radial mass fluxes and occurs at a larger radius for flows with greater total mass fluxes, other things being equal.

A more accurate treatment of the radial flow would include both general relativistic effects and the distributed character of the emission from the accretion disk. These effects will alter the quantitative properties of the near-critical flows shown in Figures 5–7, but will leave the qualitative features largely unchanged, in part because the two effects tend to offset one another (see Park and Miller 1991; Fortner, Lamb, and Miller 1991a; Lamb 1991).

Consider first the effects of general relativity on the radial flow, when all radiation comes from the inner boundary. This is the simplified model that was used in constructing the radial flows shown in Figures 5–7. In newtonian theory, the stationary-frame luminosity always increases with increasing radius, because of the work done on the radiation by the flow. Thus, if the stationary-frame luminosity is subcritical at infinity, it is subcritical everywhere and the accretion flow can always reach the star. However, in general relativity the luminosity near the star is diminished by gravitational redshift and time dilation. Therefore, if the work done on the radiation field by the flow is ignored, the stationary-frame luminosity *decreases* with increasing radius, and the stationary-frame luminosity can become supercritical at small radii even if it is subrcritical at large radii. However, as shown by Park and Miller (1991), the work done on the radiation by the radial flow and the general relativistic correction to the newtonian gravitational force more than compensate for the gravitational redshift and time dilation, so long as a substantial fraction of the total luminosity is powered by the radial mass flux. For such flows, the critical luminosity in general relativity is the same as in newtonian theory.

In the unified model of LMXB X-ray emission, most of the luminosity comes from the mass flux through the accretion disk, since the mass flux in the radial flow is at most $\sim 30\%$ of the total mass flux. As a result, if all the radiation is treated as coming from the inner boundary, the work done on the radiation by the radial flow is marginally or even appreciably too small to compensate fully for the effects on the luminosity of the gravitational redshift and time dilation. The critical luminosity is then determined by force balance at the the inner boundary, and is less than the critical luminosity in newtonian theory.

However, a more accurate treatment of the unified model would take into account the fact that the emission from the disk is distributed in radius. If the disk emission is treated in this more realistic way, the resulting increase in luminosity with increasing radius offsets the gravitational redshift and time dilation, causing a given general relativistic flow to resemble the corresponding newtonian flow (see Park and Miller 1991). Thus, the radial flows in the LMXBs are likely to be qualitatively similar to the flows shown here. These considerations do show, however, that the precise value of the critical luminosity in the LMXBs is sensitive to the structure of the emission region.

The most important properties of the radial flow can be summarized as follows (Lamb 1988b, 1989b,c, 1991):

- The radial flow is optically thin to absorption at X-ray energies, that is, $\tau_* \equiv (3\tau_{es}\tau_{ff})^{1/2} \ll 1$. Hence, the plasma and radiation in the radial flow are *not* in local thermodynamic equilibrium.
- The compton time t_c in the radial flow is $\sim 10^{-6}$ s and the electron-ion energy exchange time is only a little longer. These times are very short compared to the inflow time.
- Energy dissipation in the radial flow is small and the radiation coming from the inner disk and central corona therefore quickly cools the electrons and protons in the radial flow to temperatures very close to the local compton temperature $T_c \equiv \langle E^2 \rangle / 4\langle E \rangle$. (Here the angle brackets denote an average over

the photon number spectrum.) Observed values of T_c are ~ 1 keV, although the spectrum of most LMXBs is rather uncertain below 1 keV.

- The specific gravitational potential energy released near the neutron star is $\sim 10^4$ times the specific thermal energy in the radial flow. Thus, the luminosity required to maintain the plasma in the radial flow at the local compton temperature is a negligible fraction of the total luminosity.
- At luminosities $\lesssim 0.9\,L_E$, the scattering optical depth τ_R of the radial flow is of order unity. Thus, at these luminosities the flow further isotropizes the radiation emerging from the central corona, but has only a modest effect on the X-ray spectrum, which is still formed primarily in the inner disk and central corona.
- At luminosities within a few percent of L_E, the momentum of the outflowing radiation and radiation drag greatly slow the radial flow, causing τ_R to rise sharply. As a result, the electrons in the flow comptonize the radiation from the inner disk and central corona, altering its spectral and temporal properties.
- Photons escape from the radial flow in $\sim 10^{-3}$ s, which is $\lesssim 10^{-2}$ inflow times. Thus, photons are never trapped in the radial flow and the radiation field adjusts quasi-statically to changes in the flow.

Although the number of photons produced in the radial flow is very small, the work done by the flow on the radiation from the inner disk and central corona contributes a significant fraction of the luminosity of the source, when the luminosity is near L_E.

In this model, the horizontal branch intensity oscillation is attributed to modulation of the accretion rate to the neutron star surface by interaction of the clumps in the inner disk with the magnetosphere (Alpar and Shaham 1985; Lamb *et al.* 1985; Shibazaki and Lamb 1987). The normal branch intensity oscillation is due to quasi-periodic oscillation of the optical depth of the radial flow, caused by interaction of the flow with escaping radiation when the luminosity is within a few percent of L_E (Lamb 1988b, 1989b,c, 1991; Fortner *et al.* 1989, 1991b).

When the luminosity is $\gtrsim L_E$, nonaxisymmetric photohydrodynamic modes (Spiegel 1976, 1977; Hameury *et al.* 1980; Wang 1982) develop in the slow radial flow near the neutron star magnetosphere (Lamb 1989c, 1991; Fortner *et al.* 1989). These modes may be metastable (Spiegel 1976, 1977), requiring a finite-amplitude perturbation of the flow in order to grow. The normal branch oscillation of the radial flow provides such a perturbation, exciting photohydrodynamic modes of the same frequency (Lamb 1989c, 1991; Fortner *et al.* 1989). Growth of these modes segregates the accreting plasma and outflowing radiation as shown in the bottom panel of Figure 4, allowing the neutron star to continue to accrete even when its luminosity exceeds L_E(Spiegel 1976, 1977; Hameury *et al.* 1980; Wang 1982). The optical depth oscillation produced by the dominant photohydrodynamic mode creates the X-ray intensity oscillation seen on the lower flaring branch (Lamb 1989c, 1991; Fortner *et al.* 1989).

In the next four sections I discuss the spectral states of the Z-class sources and the horizontal, normal, and flaring branch intensity oscillations in more detail, showing how the model sketched in this section may be able to provide a unified explanation of these diverse phenomena.

4. X-Ray Spectral States

According to the unified model described in §3, the X-ray spectrum of LMXBs is formed in three physically distinct regions: (1) relatively cool ($T_e \sim 1$ keV), dense plasma near the neutron star, where comptonization is partially or fully saturated; (2) hotter ($T_e \sim 10$–30 keV), less dense plasma in the inner disk and central corona, where comptonization is unsaturated; and, when the luminosity approaches and exceeds L_E, (3) a relatively cool ($T_e \sim 1$ keV) inner disk corona and radial flow, which partially degrades the radiation produced near the star. The relative contributions of these three regions depend on the luminosity of the source.

As noted in §2, moderate-luminosity LMXBs typically have a power-law X-ray spectrum with a relatively high cutoff energy $E_a \sim 15$–30 keV. According to the unified model, this spectrum is produced by unsaturated comptonization in region 2, where the electron temperature is $\sim$15–30 keV. This process generates a stable power-law spectrum with an index $\sim$1–2 for a wide range of electron temperatures and luminosities (Katz 1976; Shapiro *et al.* 1976), if there is an adequate source of soft photons.

The sequence of horizontal, normal, and flaring branch spectral states observed in the luminous Z-class sources is produced by the changing contributions of all three regions, as the mass accretion rate increases (Lamb 1988b, 1989b,c, 1991; Miller and Lamb 1988, 1991).

4.1 HORIZONTAL BRANCH

The horizontal branch spectral state corresponds to luminosities $\sim$0.6–0.9 L_E. At these luminosities, the density and optical depth of plasma near the neutron star is high. Thus, the neutron star surface cannot be seen directly. Moreover, comptonization in the cool, dense plasma of region 1 partially or completely saturates, producing a more or less fully developed Wien photon distribution with a characteristic energy $E_b \sim 1$ keV. Such a distribution has a peak in the photon number spectrum near the energy $E \sim 2E_b$. In the unified model, it is this peak, and not blackbody emission from the surface of the neutron star, that gives rise to a significant B term when mathematical functions of the form (1) are fit to X-ray spectra of sources on the horizontal branch.

The power-law term in such fits is produced by unsaturated comptonization of photons from the neutron star magnetosphere and inner disk in the hotter plasma of region 2. The relatively low ($\sim$7–15 keV) value of the cutoff energy E_a that is typically observed in the horizontal branch spectra of the most luminous sources may indicate a relatively low electron temperature in region 2 and/or some compton degradation by cool electrons in an inner disk corona.

Region 1 can be inside region 2, because the observed distortion of the power-law spectrum by the Wien peak is relatively small, and comptonization in region 2 is incomplete. Region 2 *cannot* be inside region 1, because region 1 would then completely comptonize the photons from region 2. The model requires that soft photons be supplied to region 2 in order to produce a power-law component that is relatively stable and dominates the Wien peak. One obvious possibility for this source is high-harmonic cyclotron emission in the neutron star magnetosphere.

The increase in X-ray intensity that occurs as a source moves from left to right across the horizontal branch reflects an increase in the luminosity of the source,

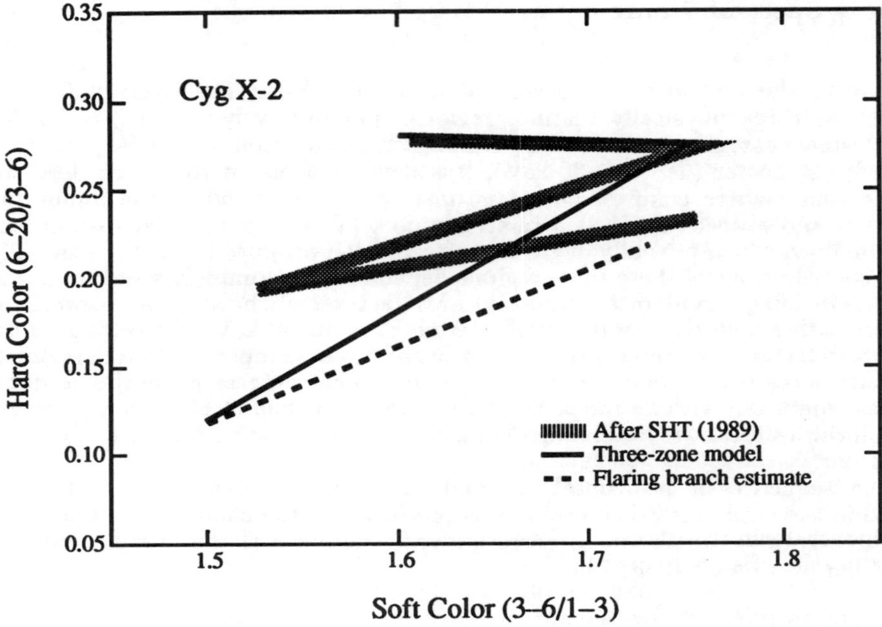

Fig. 8.—X-ray color–color diagram comparing the Z-track given by preliminary numerical calculations based on the unified model discussed in the text (solid and dashed lines; Miller and Lamb 1988, 1991) with observations of the Z-track in Cyg X-2 (wide shaded line; Schulz *et al.* 1989).

produced by a rise in the mass accretion rate. The change in the X-ray spectrum is caused by an increase in the size of the Wien distribution produced in region 1 relative to the unsaturated compton spectrum produced in region 2. This increase could be the result of (a) an increase in the degree of saturation of comptonization in region 1 as the mass flux increases, and/or (b) a small change in the structure of region 2 that causes the electron scattering optical depth to drop slightly as the luminosity increases. In either case, the distortion of the spectrum by the Wien distribution will be greater, and the B term in spectral fits will be larger. Either possibility can probably produce a horizontal branch similar to that observed in Cyg X-2, as indicated in Figure 8.

4.2 NORMAL BRANCH

A source moves from the horizontal to the normal branch when its luminosity rises to within $\sim 10\%$ of L_E, and a substantial inner disk corona forms. At these luminosities, the scattering optical depth τ_R of the radial flow is greater than unity, and scattering by the relatively cool electrons in the flow substantially degrades the hard part of the spectrum. Even if the radial mass flux remains nearly constant, τ_R rises rapidly as the luminosity approaches L_E and can produce a hard vertex in X-ray hardness–intensity and color–color diagrams, as shown schematically in Figure 8.

In Cyg X-2, the change in the X-ray spectrum from the top to the bottom of the normal branch produces a decrease in the cutoff energy E_a of the A term from ~ 7 to ~ 4 keV when equation (1) is fit to the spectrum; the other parameters in such fits remain approximately constant (Schulz *et al.* 1989). As a result, the X-ray spectrum pivots about an energy $E_p \sim 5$ keV, as shown in Figure 9. The stability of the HBO frequency on the normal branch, when interpreted in terms of the beat-frequency model (see below) suggests that the luminosity of Cyg X-2 increases only slightly as Cyg X-2 moves from the top to the bottom of the normal branch (compare Hasinger *et al.* 1990). The normal branch spectral evolution of other sources is qualitatively similar (Schulz *et al.* 1989).

Numerical calculations of comptonization by an extended cloud of cooler plasma (Miller and Lamb 1988, 1991) produce just this kind of rocking, as illustrated by the result shown in Figure 9. In this example, the spectrum of the radiation from the central corona was chosen to be the spectrum actually observed at the HB-NB vertex. This spectrum was then held fixed while the optical depth of the extended cloud was increased, in order to mimic the increase in the optical depth of the radial flow that occurs when the luminosity approaches L_E. The temperature of the cloud was calculated self-consistently.

In reality, it is quite likely that the spectrum of the radiation that enters the radial flow from the inner disk and central corona changes at least somewhat as a source moves down the normal branch. If the spectrum of this radiation is adjusted arbitrarily, the observed spectral evolution can of course be reproduced exactly. However, it is not clear what one has then learned. It has therefore seemed to us better to explore how well one can reproduce the observed normal branch spectral evolution with a fixed input spectrum from the central corona.

Such a simple model can correctly reproduce the observed decrease in soft color from the top to the bottom of the normal branch (Fig. 8) and the correct spectral rocking point (Fig. 9) if τ_R increases by ~ 15. However, such a large increase in τ_R causes the spectrum to rock more at high energies than is observed in Cyg X-2, as indicated by the large fall in hard color (see Fig. 8). If τ_R instead increases by only ~ 10, the fall in hard color is close to that observed, but the decrease in soft color is too small and the rocking energy is a little too high. Despite these quantitative deficiencies, the remarkable qualitative agreement of such a simple model with the observations suggests that the spectral evolution from the top to the bottom of the normal branch is indeed caused by comptonization in a radial flow that increases in optical depth from ~ 1 to ~ 10.

Several relevant effects were not included in the spectrum shown in Figure 9. These include (a) the work done on the radiation by the converging radial flow, (b) the spatial variation of the electron density and temperature in the flow, (c) scattering of photons from the radial flow into the inner disk, (d) injection of photons from the inner disk into the radial flow, (e) changes in the spectrum of the radiation entering the radial flow, (f) induced scattering, and (g) atomic absorption and emission processes and bremsstrahlung cooling.

Other numerical calculations of radiative transfer through a static, spherical scattering envelope have recently been reported (Ponman *et al.* 1990). The envelope assumed in these calculations is not based on a dynamical model, but has properties somewhat similar to the radial flow discussed here. The calculations include atomic absorption and emission processes and bremsstrahlung cooling as well as compton scattering, and adequately describe the X-ray spectra of a variety of LMXBs, when the input spectrum from the central source is adjusted. Further numerical

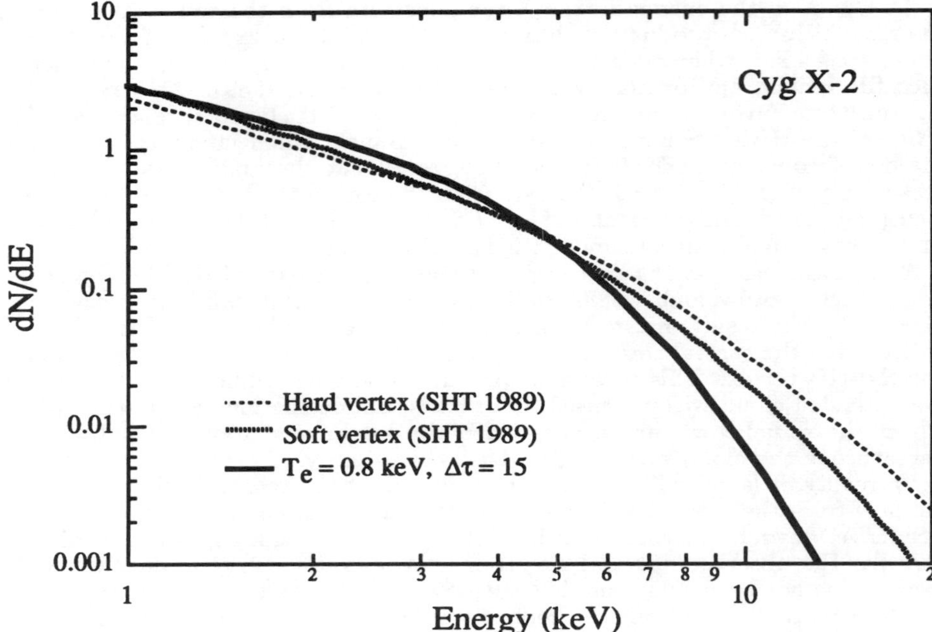

Fig. 9.—Dashed curve: Cyg X-2 photon number spectrum at the hard (HB-NB) vertex, after removal of interstellar absorption (see Schulz *et al.* 1989). Light solid curve: Cyg X-2 spectrum at the soft (NB-FB) vertex, after removal of interstellar absorption (see Schulz *et al.* 1989). Heavy solid curve: hard vertex spectrum, after comptonization by a cloud of plasma with properties like those of the radial inflow that occurs when sources are near the bottom of the normal branch (Miller and Lamb 1988, 1991).

calculations that treat atomic absorption and emission processes, bremsstrahlung cooling, induced scattering, the work done on the radiation by the radial flow, and the resulting variation of temperature and density in the radial flow are currently in progress (Miller and Lamb 1991; Kylafis 1991). Inclusion of these effects is not expected to change the spectrum substantially.

4.3 FLARING BRANCH

The transition from the normal branch to the flaring branch occurs when the photon luminosity exceeds L_E (Hasinger 1987b; Lamb 1988a; Hasinger 1988b; Hasinger *et al.* 1989), causing photohydrodynamic (PHD) modes to grow rapidly in the slow radial inflow near the neutron star magnetosphere (Lamb 1989c, 1991; Fortner *et al.* 1989), as discussed in §3. The growth of these modes increasingly separates the inflowing plasma and the outflowing radiation (Spiegel 1976, 1977). This separation allows the neutron star to continue to accrete plasma even though its luminosity exceeds L_E, and explains why the properties of Sco X-1 and Cyg X-2 change, but not radically, at the soft (NB-FB) vertex. The growth of PHD modes also explains the evolution of the X-ray spectrum that occurs on the flaring branch.

As Sco X-1 and Cyg X-2 move up the flaring branch, their X-ray spectra evolve so that when equation (1) is fit, the cutoff energy E_a of the A term and the size and characteristic energy E_b of the B term all increase (Schulz *et al.* 1989). As a result, these sources move upward and to the right in the color–color plane. Sco X-1 also moves up and to the right in the hardness–intensity plane (Priedhorsky *et al.* 1986; Schulz *et al.* 1989). In contrast, Cyg X-2 moves erratically in the hardness–intensity plane (Schulz *et al.* 1989; Hasinger *et al.* 1990). This difference in behavior is most likely due to partial and erratic obscuration of the central source in Cyg X-2, which is viewed at a higher inclination than is Sco X-1 (Hasinger 1988a; Lamb 1988a; Hasinger *et al.* 1989; Hasinger *et al.* 1990).

This spectral evolution can be understood in terms of the growth in amplitude of the PHD modes (Lamb 1989c, 1991). At luminosities close to L_E, the density contrast between the plasma fingers and the photon bubbles is small. At the moderate densities expected in the radial flow, the bubbles rise slowly, and the radiation percolates between them (Spiegel 1977). As the luminosity increases, the density contrast increases, causing the average scattering optical depth seen by the radiation to decline. Compton degradation of the spectrum therefore diminishes, and the hard color increases.

The increase in luminosity also causes the intensity to rise, increasing the size and characteristic energy of the underlying Wien distribution. As a result, the soft color also increases, and the source moves upward and to the right in the color–color plane, almost retracing the path it followed down the normal branch, but shifted to somewhat greater soft color by the increase in the effective temperature of the spectrum. If the source is otherwise unobscured, it should also move upward and to the right in the hardness–intensity plane, as Sco X-1 does.

Instabilities in the supercritical accretion flow may be responsible for the intensity fluctuations on all time scales observed on the flaring branch.

5. Horizontal Branch Oscillations

A successful model of HBOs must explain the following key properties (see Lewin and van Paradijs 1988; van der Klis 1989):

- The total power in a typical HBO is equivalent to the total power in a sinusoidal oscillation with an rms amplitude of 3–7%.
- The frequency of the HBO in a single source increases from ~ 20 Hz to ~ 55 Hz as the X-ray intensity increases by $\sim 30\%$.
- The FWHM of a typical HBO peak in the power spectrum is ~ 0.15–0.3 times the central frequency.
- Power at overtones of the fundamental HBO frequency is modest (overtones are sometimes seen, but their power is typically <20% of the power at the fundamental).
- The total power in the HBO peak varies approximately in step with the total power in the LFN component of the power spectrum.

At present, the most promising model is the magnetospheric beat-frequency modulated-accretion model (Alpar and Shaham 1985; Lamb *et al.* 1985; Shibazaki and Lamb 1987). The key ideas of this model may be summarized as follows:

- The neutron star rotates in the same sense as the disk flow, with angular velocity Ω_s.
- The relatively weak magnetic field of the neutron star disrupts the disk at an inner radius r_0 only slightly greater than the radius of the neutron star.
- Instabilities and interaction of the neutron star magnetosphere with the inner disk cause plasma density and magnetic field fluctuations in the boundary layer at r_0.
- The fluctuations in the boundary layer are carried around the neutron star at the local Kepler frequency. Thus, on average a given fluctuation reappears at the same *magnetospheric* azimuth with frequency

$$\omega_B \equiv \Omega_0 - \Omega_s \,, \tag{2}$$

  where Ω_0 is the mean orbital frequency of the plasma in the boundary layer at r_0.
- The interaction of the rotating magnetospheric field with the plasma and magnetic field fluctuations in the boundary layer causes the mass flux from the boundary layer to the star to vary at the beat frequency (2).
- The resulting quasi-periodic modulation of the accretion rate creates a quasi-periodic modulation of the *luminosity* of the neutron star. This luminosity modulation causes the X-ray intensity seen by a distant observer to oscillate quasi-periodically.

This model accounts in a natural way for many of the observed properties of the HBOs.

The beat-frequency model gives HBO frequencies in the range 20–55 Hz for neutron star rotation rates ~ 100 Hz and magnetic field strengths $\sim 10^9$ G. These rotation rates and field strengths are consistent with evolutionary arguments, which suggest that the luminous low-mass X-ray binaries are progenitors of the millisecond rotation-powered pulsars (Alpar and Shahm 1985; see van den Heuvel 1988, 1991). Magnetic field strengths of this order are also consistent with recent work which suggests that the magnetic dipole moment of a neutron star does not decay below 10^8–10^9 G on the relevant evolutionary time scales (Kulkarni 1986; van den Heuvel *et al.* 1986).

In the beat-frequency model, the narrowness of the oscillation peaks is explained by the narrowness of the boundary layer at r_0, while the hardness of the HBO photons is accounted for by the fact that they come from near the neutron star. The presence of LFN in horizontal branch power spectra and the fact that the strength of the HBO rises and falls with the strength of the LFN follows from the fact that the fluctuations in the boundary layer which make the beat frequency visible also produce the LFN. The VLFN is attributed to variations in the accretion disk flow much further from the neutron star.

The beat-frequency model also accounts naturally for the steep increase in the HBO frequency with increasing intensity. The reason is that an increase in the mass flux through the boundary layer compresses the magnetosphere. At this smaller radius, the fluctuations in the boundary layer orbit the neutron star at a higher frequency, causing the beat frequency to increase, as shown in Figure 10 (the rotation frequency of the neutron star remains essentially constant on this time scale). But the increase in the mass accretion rate also increases the X-ray luminosity of the neutron star, causing the X-ray intensity seen by a distant observer

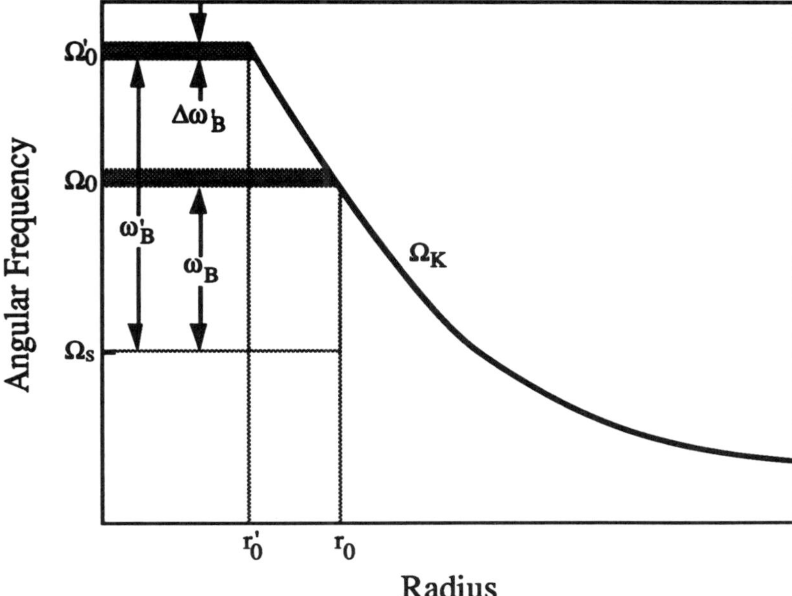

Fig. 10.—Plot of the orbital frequency Ω_K of plasma inhomogeneities in the inner disk and boundary layer as a function of radius, showing how the beat-frequency increases from $\omega_B = \Omega_0 - \Omega_s$ to $\omega_B' = \Omega_0' - \Omega_s$ as the radius of the boundary layer decreases from r_0 to r_0'. Here Ω_0 and Ω_0' are the mean orbital frequencies in the boundary layer, Ω_s is the stellar rotation frequency, and $\Delta\omega_B$ and $\Delta\omega_B'$ are the ranges of the beat frequencies produced by fluctuations within the boundary layer. After Lamb (1989b,c, 1991).

to increase. Thus, a strong positive correlation between the HBO frequency and the X-ray intensity is expected.

Figure 11 shows that the simplest version of the beat frequency model adequately describes the HBO frequency–intensity relation in GX 5−1. If the beat-frequency model proves correct, fits such as this one can be used to constrain the structure of the inner disk and outer magnetosphere of the neutron star.

Initially, the absence of detectable periodic oscillations at the rotation frequency of the neutron star appeared puzzling to some workers. The expectation that magnetic neutron stars in LMXBs would show significant periodic oscillations was based on experience with canonical accretion-powered pulsars. However, the size of the magnetosphere, the accretion flow pattern, and the flow of radiation in the luminous LMXBs (Lamb *et al.* 1985; Lamb 1986, 1988a) differ in crucial ways from those in canonical accretion-powered pulsars (Lamb 1989a).

For example, the density of plasma in and around the magnetosphere is much higher, due to the higher mass accretion rate and the greater outward force of the radiation, which slows the accretion flow (Lamb *et al.* 1985; Lamb 1986). Recent theoretical work has shown that a central corona with a scattering optical depth as small as 5 is likely to reduce the amplitude of periodic intensity oscillations caused by beaming to levels that are currently undetectable, while leaving quasi-periodic intensity oscillations caused by luminosity variations essentially unaffected

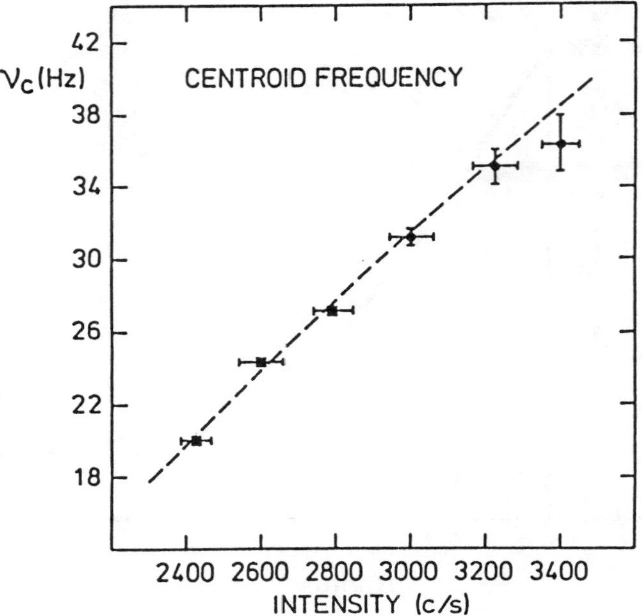

Fig. 11.—Observed horizontal branch oscillation frequency as a function of X-ray intensity in GX 5−1 (data points) and the relation predicted by the simplest version of the beat-frequency model (curve). From van der Klis *et al.* (1985).

(Brainerd and Lamb 1987; Kylafis and Klimis 1987; Kylafis 1988; Kylafis and Phinney 1989). Other effects, such as the Doppler shift caused by orbital motion of the neutron star, are likely to reduce further the amplitude of periodic oscillations, to levels below current detection thresholds (see Lamb *et al.* 1985; Lamb 1986, 1988a; Shibazaki and Lamb 1987).

The magnetospheric beat-frequency modulated-accretion model has gained further support with the recent discovery (Angellini, Stella, and Parmar 1989) of quasi-periodic intensity oscillations in the 45 s accretion-powered pulsar EXO 2030+375 at a frequency and with a strength consistent with the model. Beat-frequency intensity oscillations in accretion-powered pulsars had been predicted in the original papers proposing the model (Alpar and Shaham 1985; Lamb *et al.* 1985).

Recent more detailed studies of horizontal branch power spectra (Mitsuda *et al.* 1988) have measured more accurately the total power in the LFN and in the first overtone of the HBO, confirming earlier evidence (van der Klis *et al.* 1985; Lamb *et al.* 1985; Shibazaki and Lamb 1987) that some coherence is required in order to fit the HBO with an oscillating-shot model (Lamb *et al.* 1985; Lamb 1986; Shibazaki and Lamb 1987). The results of such studies must be interpreted cautiously, because of the difficulty of separating the LFN (which is supposed to be associated with the HBO) from the VLFN (which is not), the likelihood of a wide range of shot lifetimes (Shibazaki and Lamb 1987), and broadening of the HBO peak by frequency modulation (Shibazaki and Lamb 1987).

Efforts are currently underway (Norris *et al.* 1990) to determine whether the amplitudes of the intensity variations that contribute to the LFN are correlated with the HBO amplitude for time differences less than or comparable to the shot lifetime, as expected in the oscillating shot model. The effects of this correlation are difficult to detect with current experiments, because the oscillating shot model predicts that a large number of shots are occurring simultaneously ($\sim$200, when the fractional modulation in the X-ray intensity associated with the LFN is $\sim$5%, and even more, when the fractional modulation associated with the LFN is less; see Lamb *et al.* 1985; Lamb 1986; Shibazaki and Lamb 1987).

6. Normal Branch Oscillations

A successful model of NBOs must explain the following key properties (see Lewin and van Paradijs 1988; van der Klis 1989):

- NBOs have been detected in a half-dozen Z-class sources and may be present in all such LMXBs. Thus, they are a generic feature of the X-ray emission from these objects.
- The total power in a typical NBO is equivalent to the total power in a sinusoidal oscillation with an rms amplitude of 3–7%.
- The total power in HBO and LFN components seen on the normal branch is substantially less than on the horizontal branch.
- The frequency of the NBO at its onset near the middle of the normal branch is $\sim$6 Hz in all the Z-class sources. Thus, the physics that produces the NBO must be relatively insensitive to properties of the luminous LMXBs that vary from source to source.
- The frequency of the NBO in a given source increases from $\sim$6 Hz near the middle of the normal branch to $\sim$10 Hz near the bottom.
- The FWHM of a typical NBO peak in the power spectrum is $\sim$0.2–0.6 times the central frequency.
- The NBO amplitude in Cyg X-2 is modest at X-ray energies below 5 keV, has a minimum at an energy E_m near 5 keV, and increases steeply with X-ray energy for energies above 5 keV (Mitsuda 1988a,b; Mitsuda and Dotani 1989). The NBO amplitude in GX 5$-$1 varies similarly, with a minimum at $\sim$2 keV (Mitsuda 1988a).
- NBOs at X-ray energies well above a phase shift energy E_s appear to lag the NBOs at X-ray energies below E_s by $\sim 180°$ in both Cyg X-2 and GX 5$-$1 (Mitsuda 1988a; Mitsuda and Dotani 1989). E_s is $\sim$5 keV in Cyg X-2 (Mitsuda 1988a,b; Mitsuda and Dotani 1989) and $\sim$2 keV in GX 5$-$1 (Mitsuda and Dotani 1989).

The more precise phase data available for Cyg X-2 are consistent with a constant oscillation phase up to $\sim$5 keV; above 5 keV the phase begins to shift, with the lag increasing steeply with increasing X-ray energy, until it again becomes constant at $\sim 150°$ above 12 keV (Mitsuda 1988a,b; Mitsuda and Dotani 1989).

At first, oscillations or instability of the inner disk appeared to be promising NBO mechanisms (van der Klis *et al.* 1985; van der Klis *et al.* 1987b; Lamb 1988a). However, further investigations have brought to light several difficulties with such models:

- The time scales characteristic of the inner disk are either too slow (radial drift) or too fast (dynamical) to explain the ~ 6 Hz characteristic frequency.
- The strongly sheared flow in the disk makes it difficult to explain the relatively narrow range of observed NBO frequencies and the relatively small width of the NBO peak.
- The small solid angle of geometrically-thin disks makes oscillations difficult to produce in systems viewed at small inclinations.
- The relatively constant energy-integrated X-ray flux and the phase and amplitude variations with X-ray energy indicate that NBOs are primarily spectral, rather than luminosity phenomena, contrary to what is expected for radial flow instabilities in accretion disks.
- The fact that the change in the X-ray spectrum down the normal branch gives a rocking energy that is the same as the minimum amplitude energy and the phase shift energy of the NBOs strongly suggests a close connection between these phenomena. However, in geometrically thin disk models it is difficult to explain these spectral changes, and why they are almost independent of viewing angle.

These difficulties have caused attention to shift from oscillations or instability of the inner disk to other mechanisms.

It has been suggested that the NBOs are produced by interaction of beamed X-rays from the neutron star with an oscillating, geometrically-thick region of the accretion disk $\sim 10^8$ cm from the star. In this scenario, the NBOs are caused by oscillations in the Compton scattering optical depth of the disk at the local dynamical frequency, due to excitation of internal sound waves. This mechanism also has several difficulties:

- Geometrically-thick disks are likely to have a strong velocity shear in the vertical as well as the radial direction (see Ghosh and Lamb 1991). This makes it difficult to propagate sound waves through the flow. Moreover, the wide frequency range of the internal modes in such a disk make it difficult to explain the relatively narrow range of observed NBO frequencies and the relatively small width of the NBO peak.
- Comparison of Compton scattering spectral models with the normal branch X-ray spectrum and with the oscillation in the spectrum that occurs during each cycle of the NBO indicates that the X-radiation comes from a region with an electron-scattering optical depth ~ 10. However, most of the X-rays falling on the geometrically-thick part of the accretion disk would suffer only a few scatterings before leaving the disk.
- The large optical depth that is required to explain the X-ray spectrum means that the plasma in the disk orbits the neutron star at something like the ordinary—rather than the modified—Keplerian frequency. The ~ 6 Hz frequency of the NBO then implies that the geometrically-thick region of the disk must be $\sim 10^8$ cm from the neutron star. This radius is much greater than the radius at which prevously proposed geometrically-thick disk models have there maximum thickness. Thus, either the disk structure is quite different

from that of existing models, or the radiation from the neutron star is blocked from reaching the disk at the appropriate radius.

- In order for beamed X-radiation from the neutron star to reach the geometrically-thick part of the disk, the intervening region must be optically-thin. This makes it difficult to explain the absence of significant periodic X-ray intensity oscillations at the spin frequency of the neutron star.

Because of these difficulties with disk models of the NBOs, we have turned our attention to the interaction of the escaping radiation with the approximately radial accretion flow described in §3 when the luminosity of the neutron star is near the Eddington critical luminosity.

We find that the accretion flow model described in §3, which was developed to explain the X-ray spectral evolution on the normal branch, also provides a natural explanation of the NBOs (Lamb 1988b, 1989b,c, 1991; Fortner *et al.* 1989, 1991a,b). In this model, the radial flow is very sensitive to the luminosity when the luminosity is within $\sim 10\%$ of L_E. This sensitivity, and the unavoidable time lag between changes in the inward mass flux from the corona and the resulting luminosity changes, suggested that the radial flow would oscillate for luminosities sufficiently close to L_E (Lamb 1988b, 1989b,c, 1991). Simple estimates of the expected frequency of the oscillation and the variation with photon energy of its amplitude and phase agreed qualitatively with the observed properties of the NBOs.

Recently, we have carried out extensive simulations of the radial inflow of plasma from the inner disk corona, using a one-dimensional, time-dependent radiation hydrocode (Fortner *et al.* 1989, 1991a,b). First, steady, near-critical radial flows were constructed. The stability of such flows was then studied by disturbing the luminosity of the central corona or the inward mass flux from the outer boundary. The response of steady flows to perturbations of the luminosity or the mass flux was found to be very similar. When the luminosity is within $\sim 10\%$ of L_E and the fraction of the total mass flux in the radial flow is sufficiently high, both types of perturbations cause the radial flow to develop quasi-periodic oscillations.

Figures 12 and 13 show the results of an early simulation in which regular oscillations developed. For this particular run, $\mu_r = 0.05$ and $\epsilon = 0.01$. Similar results were obtained for ϵ values as large as 0.10. The inflow time t_f from the outer edge was ~ 0.3 s and the simulation was followed for ~ 10 inflow times.

From $t = 0$ to 0.6 s, an interval corresponding to two inflow times, mass shells were added to the outer edge of the flow at a constant rate and the flow was allowed to relax to a steady-state solution by adjusting the luminosity at the inner boundary so that the luminosity at infinity remained constant. Beginning at 0.6 s, the luminosity at the inner boundary was calculated self-consistently from the radial inflow.

At 1.0 s, the flow was disturbed by adding 30 extra mass shells (equivalent to 5% of the mass in the radial flow) to the outer edge of the flow; after this time, mass shells were again added to the outer edge at a constant rate. One inflow time later, at 1.3 s, the density perturbation reached the inner boundary and created a brief, small increase in the luminosity of the central corona. This luminosity perturbation caused a small amount of gas to accumulate in the outer part of the flow and near 150 km, where the inflow velocity is most sensitive to the luminosity.

When these density maxima subsequently reached the inner boundary, they created primary and secondary luminosity peaks, causing the cycle to repeat. The

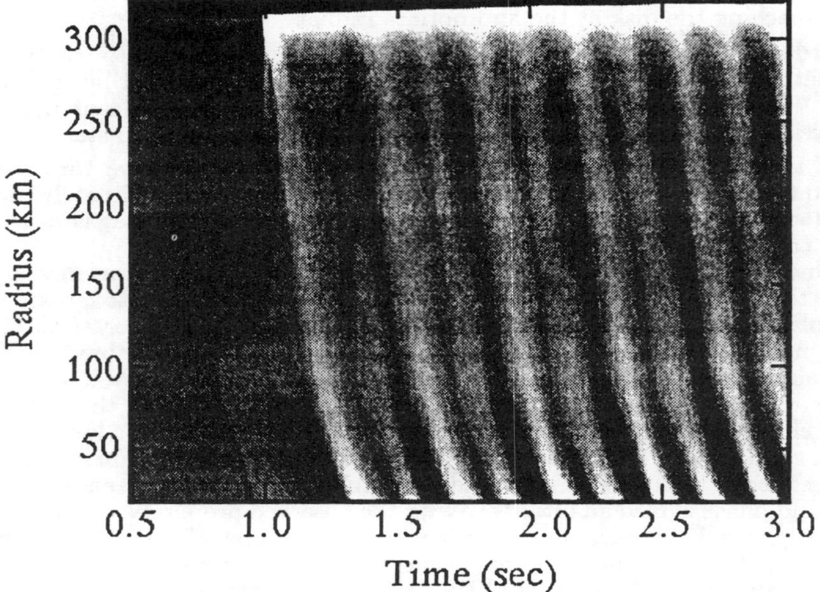

Fig. 12.—Grayscale plot showing the gas density in a typical simulation relative to the gas density in the corresponding unperturbed flow, as a function of radius and time. Higher density regions are lighter; lower density regions, darker. The flow was disturbed at 1.0 s, when a mass equal to 5% of the total mass in the flow was suddenly added at the outer edge of the flow; afterward, mass was again added to the flow at a constant rate. The leftmost light striation shows the progress of the resulting density enhancement as it falls from the outer edge of the flow at $\sim$300 km toward the compact central corona at 20 km. When the enhancement reaches the inner boundary at 1.3 s, it causes a brief luminosity peak that creates two regions of enhanced density, at $\sim$150 and $\sim$300 km. The result is a continuing quasi-periodic oscillation of the flow, which appears as a series of light striations descending to the right. From Fortner *et al.* (1989).

secondary peak grew to about the same size as the primary peak within a few cycles, and the flow subsequently oscillated with frequency

$$f_{\rm osc} \approx 2/t_f \approx 6 \text{ Hz}. \tag{3}$$

All the simulations that displayed regular oscillations had an oscillation frequencies $\approx 2/t_f$.

For the accretion flow model and zoning used in these simulations, perturbations in the mass flux or luminosity are damped for $\mu_r/\epsilon \lesssim 4$ but grow exponentially for $\mu_r/\epsilon \gtrsim 4$. For $\mu_r/\epsilon \approx 4$, disturbances of the mass flux or luminosity produce regular oscillations. These stability properties can be understood simply as follows.

Any density perturbation at the outer edge of the radial flow of relative size $\Delta\rho_i/\rho_i$ at time t_i is advected toward the neutron star, producing a change in the mass flux at the inner boundary of a similar relative size one inflow time later, at time t_{i+1}. The resulting relative change in the luminosity is $\Delta\epsilon_{i+1} \propto -\mu_r \Delta\rho_i/\rho_i$.

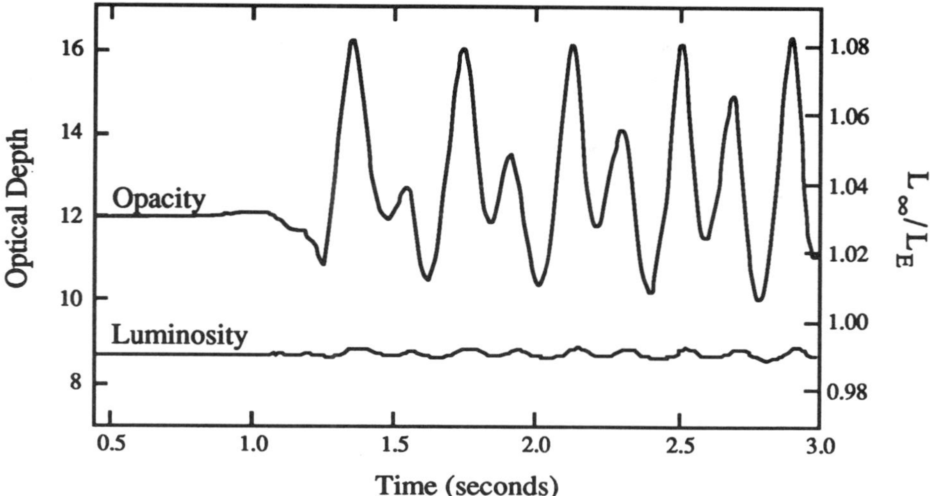

Fig. 13.—Variation of the electron scattering optical depth (left scale) and total luminosity (right scale) during the simulation of Figure 12. Note the stability of the flow before it was disturbed at 1.0 s, and the quasi-periodic oscillation that developed afterwards. The optical depth varies by $\sim 30\%$ whereas the luminosity varies by less than 1%. From Fortner *et al.* (1989).

This change is communicated to the outer boundary on the (very short) radiation diffusion time scale.

Since the inward mass flux at the outer boundary is constant, the density of the flow there depends only on the local velocity, which varies with luminosity as ϵ, as discussed above. Thus, the change in luminosity produces a change in the density at the outer boundary at time t_{i+1} of size $\Delta\rho_{i+1}/\rho_{i+1} \propto (\mu_r/\epsilon)\Delta\rho_i/\rho_i$, and hence the amplitude of the oscillation grows exponentially at a rate $\gamma \sim \ln(\mu_r/\epsilon)/t_f$.

The actual boundary between stability and instability ($\mu_r/\epsilon \approx 4$) is determined by the competition between this growth mechanism and the damping effects of compton cooling, spreading of the density perturbations caused by gas pressure gradients and radiation drag, and numerical viscosity. Due to these sources of damping, even a sharp disturbance in the mass flux or luminosity produces a relatively sinusoidal response. Because the radiation hydrocode omits some physics and includes damping effects, such as numerical viscosity, that are not present in the real system, the boundary between stability and instability cited here should be viewed as indicative, rather than definitive. It is possible that oscillations may develop at smaller luminosities.

For radial flows beginning at $\sim$200–300 km, the simulations that exhibited regular oscillations all had oscillation frequencies $f_{\rm osc} \sim$ 5–10 Hz. These frequencies are comparable to the observed frequencies of NBOs at their onset near the middle of the normal branch. The scattering optical depths of the inflows that exhibited regular oscillations were all ~ 10.

The relative amplitude $\Delta L/L$ of the luminosity oscillations tends to be much smaller than the relative amplitude $\Delta\tau_{es}/\tau_{es}$ of the oscillations in the scattering

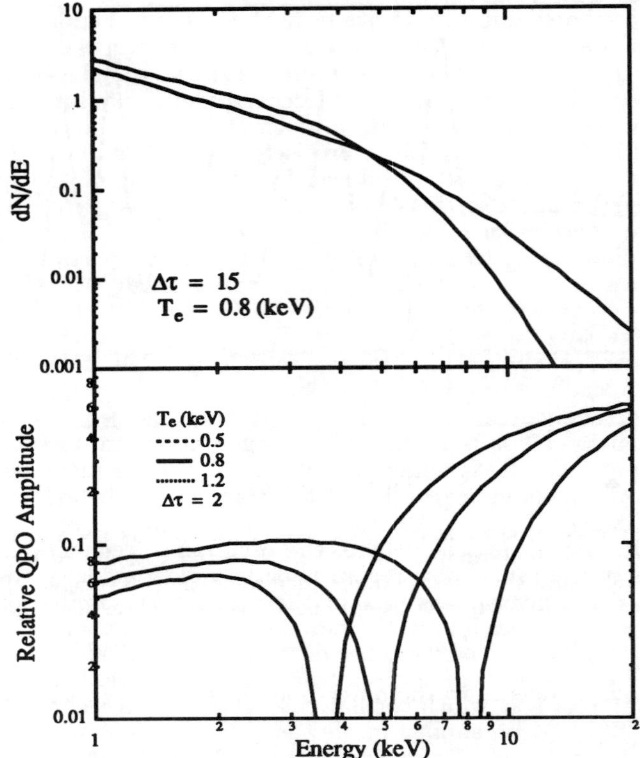

Fig. 14.—Top: X-ray spectral variation produced by a variation of 15 in the optical depth with $T_e = 0.8$ keV, illustrating the change in the X-ray spectrum from the top to the bottom of the normal branch. Note how the spectrum rocks around the pivot energy $E_p \sim 5$ keV. Bottom: oscillation amplitude as a function of X-ray energy, for an optical depth of 10, an optical depth oscillation of 2, and electron temperatures of 0.5, 0.8, and 1.2 keV. The amplitude of the oscillation goes through a minimum at an energy E_m determined by the pivot energy E_p of the spectrum. E_m rises as T_e increases. From Miller and Lamb (1991).

optical depth. This is to be expected, since the luminosity cannot vary by an amount greater than $\sim \epsilon L$ without disrupting the flow, whereas variations in the luminosity of this magnitude produce large variations in the density and optical depth of the flow. Thus, these simulations support the earlier suggestion (Lamb 1988b, 1989b,c, 1991) that NBOs are primarily oscillations in optical depth rather than luminosity.

Numerical calculations of the effect of optical depth oscillations on the X-ray spectrum (Miller and Lamb 1991) show that such oscillations cause the spectrum to rotate about a pivot energy E_p that depends on the spectrum (see top panel of Fig. 14), but is in the range 2–7 keV for observed spectra. The quasi-periodic oscillations in the optical depth of the radial flow therefore cause the X-ray spectrum to rock quasi-periodically about the energy E_p, in effect moving the source a small distance up and down the normal branch. A byproduct of this motion is that the X-ray intensity oscillates quasi-periodically in most energy bands.

This NBO mechanism also predicts that the energy E_m at which the oscillation amplitude goes through a minimum is close to the pivot energy E_p of the X-ray spectrum on the normal branch. Moreover, the X-ray intensity oscillations above and below E_p should be $\sim 180°$ out of phase, and hence the phase shift energy E_s should also be $\sim E_p$. More precisely, the NBOs above E_s should lead the NBOs below E_s by the 180° produced by the rocking of the spectrum, plus the time required for photons that start out with energies greater than E_s to be downscattered to energies less than E_s.

These predictions agree quantitatively with the properties of the NBO in Cyg X-2, as shown in Figure 15. The amplitude minimum at ~ 5 keV indicates that the electron temperature in the radial flow is ~ 0.8 keV. The $\sim 150°$ phase *lag* of the NBOs above E_s relative to the NBOs below E_s reported by Mitsuda (1988a,b) and Mitsuda and Dotani (1989) is, in this model, actually a $\sim 210°$ phase *lead*, composed of the 180° phase lead produced by the rocking of the spectrum, plus an additional $\sim 20°$ phase lead (equivalent to ~ 10 ms) due to the time required for photons to downscatter to energies less than E_s. These predictions also agree quantitatively with observations of GX 5−1, which has a normal branch pivot energy ~ 2 keV (Hasinger, personal communication) and a minimum in the NBO amplitude at about the same energy (Mitsuda 1989).

In the NBO model described here, the oscillation frequency is governed by the outer radius of the radial flow region and the inflow velocity there. Both are functions of ϵ. For a given mass flux in the radial flow, oscillations begin at a definite value of ϵ. The mass flux in the radial flow is determined by the luminosity of the star and the structure of the inner disk corona far inside the outer radius of the disk, but still far outside the neutron star magnetosphere. Therefore, the radial mass flux, the value of ϵ at which oscillations develop, and the frequency of these oscillations are likely to be similar in all the luminous low-mass X-ray binaries. Thus, the mechanism described here accounts in a natural way for the fact that the NBOs in all Z-class sources have about the same frequency where they begin, near the middle of the normal branch. If the extent of the radial flow is ~ 300 km, the model gives an NBO frequency ~ 6 Hz at this point.

The increase in the NBO frequency from ~ 6 to ~ 10 Hz that occurs as sources approach the bottom of the normal branch may be caused by shrinking of the radial flow region as the luminosity approaches L_E. The radial flow region will shrink if mass loss from the inner disk creates an inner disk corona that is so dense that the extent of the radial flow region is determined not by r_2, but by the limitation (see §3) of the radial mass flux to $\sim 20\%$ of the total mass flux (Lamb 1989c, 1991).

To summarize, regular oscillations in the scattering optical depth of the radial flow are expected to occur with a well-defined frequency when the mass flux onto the neutron star is in a narrow interval near but below the critical mass flux $\dot{M}_E$. The frequency of these oscillations is comparable to the observed NBO frequency for plausible radial flow dimensions and is likely to be similar in all the luminous LMXBs.

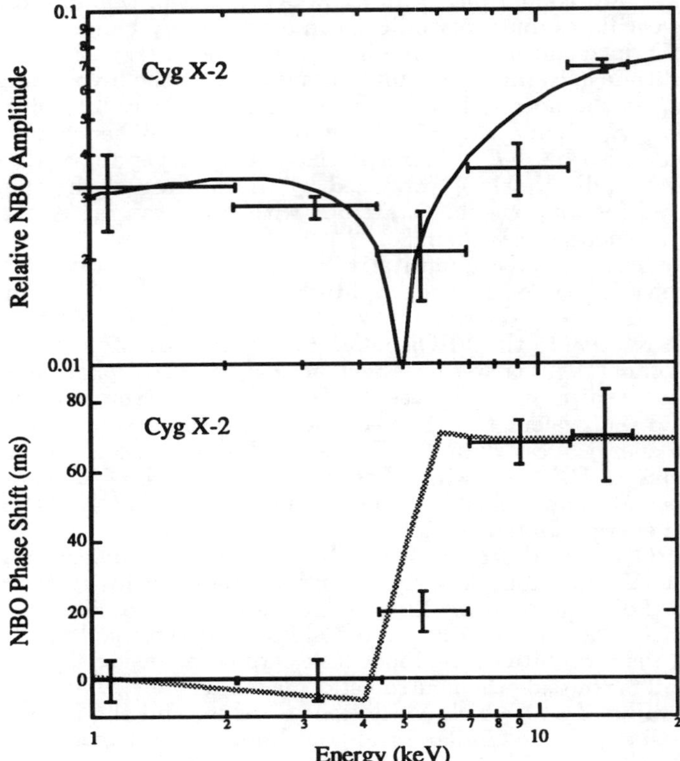

Fig. 15.—Observed (Mitsuda and Dotani 1989) and calculated (Miller and Lamb 1991) Cyg X-2 normal branch oscillation amplitude (top) and phase (bottom), as a function of X-ray energy. The amplitude calculation assumes an optical depth of 10, an optical depth oscillation of 2, and an electron temperature of 0.8 keV. The phase shift at ~ 5 keV is less than 180° because of the time required for photons with energies above 5 keV to be scattered down to energies below 5 keV.

7. Flaring Branch Oscillations

Any model of the FBOs must explain the following key observations (see Lewin and van Paradijs 1988; van der Klis 1989):

- The FBO frequency near the bottom of the flaring branch (the NB-FB vertex) is close to the ~ 10 Hz NBO frequency near the bottom of the normal branch (the same vertex). Stated differently, as a source moves from the bottom of the normal branch to the bottom of the flaring branch, the oscillation frequency appears to change smoothly.
- The oscillation frequency and the width of the oscillation peak both increase as a source moves up the flaring branch.

- As Sco X-1 moves up the flaring branch, the oscillation peak broadens and eventually dissolves into a broad continuum of excess power density in the interval ~10–20 Hz.

These observations point to a connection between the normal branch and flaring branch oscillations, at least when the source is near the NB-FB vertex.

As discussed in §3 and §4, the slow radial flow near the neutron star magnetosphere is expected to become unstable to growth of photohydrodynamic (PHD) modes, when the luminosity is $\gtrsim L_E$ and there are finite amplitude perturbations in the flow. The oscillation in the radial flow that occurs on the normal branch (see §6) is therefore likely to excite PHD modes that have about the same frequency (Lamb 1989c, 1991; Fortner *et al.* 1989). Besides allowing accretion to continue even when the luminosity exceeds L_E, these asymmetric modes cause the average optical depth and hence the X-ray intensity to oscillate. The excitation of PHD modes by the radial oscillation explains why the ~10 Hz oscillation seen at the bottom of the normal branch goes over smoothly to the ~10 Hz oscillation seen at the bottom of the (supercritical) flaring branch.

As the mean density of the radial flow rises, the density contrast between the photon bubbles and the plasma spikes increases (Spiegel 1977). The outflowing radiation therefore flows more and more through the lower density photon bubble regions (Spiegel 1977), increasing the effective gravity in the plasma spikes, and causing the frequency of the largest amplitude PHD modes to rise (Lamb 1989c, 1991). This accounts for the increase in the FBO frequency as a source moves up the flaring branch (Lamb 1989c, 1991).

At sufficiently high luminosities, a broad spectrum of PHD modes will be excited (Spiegel 1976, 1977). This explains why the Sco X-1 FBO peak widens and eventually dissolves into a broad continuum of excess power density as Sco X-1 moves further up the flaring branch (Lamb 1989c, 1991; Fortner *et al.* 1989).

8. Concluding Remarks

Hard work has produced a wealth of new data on the luminous LMXBs and especially the relationship between their spectral and temporal behavior. The accretion flow and X-ray emission model described in the previous sections provides a promising theoretical framework for understanding the shape of the X-ray spectrum and the three spectral branches of the Z-class sources. The beat-frequency model of the horizontal branch intensity oscillations can be incorporated in this framework. The framework also suggests a natural explanation for the ~6–10 Hz normal branch intensity oscillations, including their frequency and coherence, and the observed variations of the oscillation amplitude and phase with photon energy. Finally, the framework suggests an explanation of the FBOs.

Study of the luminous LMXBs is fascinating in part because we are attempting to apply our understanding of physics to conditions many orders of magnitude different from any terrestrial or solar system experience. The results have important implications for much of neutron star physics, including the formation of neutron stars, the creation and evolution of neutron star magnetic fields and spins, the birth of millisecond pulsars, and the structures of neutron star magnetospheres.

Study of the luminous LMXBs is also important because the results can be applied to other astrophysical systems, such as active galactic nuclei and quasars. The luminous LMXBs provide nearby, smaller-scale examples of physical phenomena, such as radiation-pressure dominated accretion flows, near-Eddington radiation fluxes, and collimated jets, that are thought to be important in these extragalactic systems. The presence at the center of this activity of a neutron star with a weak magnetic field provides diagnostics that are absent when the accreting object is a black hole.

Important questions about the Z-class sources remain. One is whether the theoretical framework described here will be consistent with the new data that are now being gathered and the more detailed theoretical calculations that are now underway. Another is the amplitude of periodic oscillations at the rotation frequencies of these neutron stars. Detection of such oscillations would immediately show that these stars do have appreciable magnetic fields and would also fix their spin frequencies, further constraining the parameters of the beat-frequency model. Until a spin frequency is detected in a Z-class source, the beat-frequency model of the horizontal branch oscillations must remain to some extent in doubt.

There are also important unsettled questions concerning those LMXBs that do not show Z-shaped tracks in hardness-intensity or color–color diagrams. Examples include the Rapid Burster, which exhibits very complicated behavior, including intensity oscillations with properties quite different from those discussed here (Tawara *et al.* 1982; Stella *et al.* 1988a,b; Dotani 1989), and Cir X-1, which makes very large excursions in the X-ray color–color plane and has a variety of other exceptional properties (Tennant 1987, 1988). There are also other LMXBs that are similar to the Z-class sources in some respects, but do not show standard QPOs or a Z-track in X-ray color–color plots (Hasinger and van der Klis 1989; van der Klis 1991). Whether the observed differences between these 'atoll' sources and the Z-class sources are due to differences in mass accretion rate and magnetic field strength, or to some other cause(s), is at present unknown (Hasinger and van der Klis 1989; van der Klis 1991). Despite the differences between atoll and Z-class sources, the unified model described here may provide a helpful framework for comparison.

It is a pleasure to thank A. Alpar, R. Epstein, B. Fortner, P. Ghosh, G. Hasinger, N. Kylafis, D. Q. Lamb, G. S. Miller, M.-G. Park, W. C. Priedhorsky, N. Schulz, N. Shibazaki, and M. van der Klis for many fruitful discussions of X-ray binaries and QPO sources. I especially thank my collaborators B. Fortner, P. Ghosh, G. S. Miller, and M.-G. Park for help in understanding the physics of near-critical accretion flows and for permission to use results in advance of publication. This research was supported in part by NASA grant NAGW 1583 and NSF grant PHY 86-00377.

9. References

Alpar, M. A., and Shaham, J. 1985, *Nature*, **316**, 239.
Angelini, L., Stella, L., and Parmar, A. N. 1989, *Ap. J.*, **346**, 906.
Bradt, H.V.D., and McClintock, J. E. 1983, *Ann. Rev. Astr. Ap.*, **21**, 13.
Brainerd, J., and Lamb, F. K. 1987, *Ap. J.* (*Letters*), **317**, L33.

Dotani, T. 1989, in *Proc. 23rd ESLAB Symp. on X-Ray Astronomy*, ed. N. E. White (ESA SP-296), p. 209.

Fortner, B. F., Lamb, F. K., and Miller, G. S. 1989, *Nature*, **342**, 775.

________. 1991a, *Ap. J.*, submitted.

________. 1991b, in preparation.

Ghosh, P., and Lamb, F. K. 1979a, *Ap. J.*, **232**, 259.

________. 1979b, *Ap. J.*, **234**, 296.

________. 1991, This volume.

Ghosh, P., Lamb, F. K., and Miller, G. S. 1991, in preparation.

Ghosh, P., Lamb, F. K., Miller, G. S., and Park, M.-G. 1991, in preparation.

Hameury, J. M., Bonazzola, S., and Heyvaerts, J. 1980, *Astr. Ap.*, **90**, 359.

Hasinger, G. 1987a, *IAU Symposium* **125**, *The Origin and Evolution of Neutron Stars*, ed. D. J. Helfand and J. H. Huang (Dordrecht: Reidel). p. 333.

________. 1987b, *Astr. Ap.*, **186**, 153.

________. 1988a, in *Physics of Compact Objects*, ed. N. E. White and L. G. Filipov (*Adv. Space Res.*, **8**), p. 377.

________. 1988b, *Physics of Neutron Stars and Black Holes*, ed. Y. Tanaka (Tokyo: Universal Academy), p. 97.

Hasinger, G., Langmeier, A., Sztajno, M., Trümper, J., Lewin, W.H.G., and White, N. E. 1986, *Nature*, **319**, 469.

Hasinger, G., Priedhorsky, W. C., and Middleditch, J. 1989, *Ap. J.*, **337**, 843.

Hasinger, G., and van der Klis, M. 1989, *Astr. Ap.*, **225**, 79.

Hasinger, G., van der Klis, M., Ebisawa, K., Dotani, T., and Mitsuda, K. 1990, *Astr. Ap.*, **235**, 131.

Katz, J. I. 1976, *Ap. J.*, **206**, 910.

Kulkarni, S. R. 1986, *Ap. J.* (*Letters*), **306**, L85.

Kylafis, N. D. 1988, in *Physics of Compact Objects*, ed. N. E. White and L. G. Filipov (*Adv. Space Res.*, **8**), p. 455.

________. 1991, in preparation.

Kylafis, N. D., and Klimis, G. S. 1987, *Ap. J.*, **323**, 678.

Kylafis, N. D., and Phinney, E. S. 1989, in *Timing Neutron Stars*, ed. H. Ögelman and E.P.J. van den Heuvel (Dordrecht: Kluwer), p. 731.

Lamb, F. K. 1986, in *The Evolution of Galactic X-Ray Binaries*, ed. J. Trümper, W.H.G. Lewin, and W. Brinkmann (Dordrecht: Reidel), p. 151.

________. 1988a, in *Physics of Compact Objects*, ed. N. E. White and L. G. Filipov (*Adv. Space Res.*, **8**), p. 421.

________. 1988b, Talk presented at the Los Alamos Workshop on QPOs, La Cienega, N.M., September 1988.

________. 1989a, in *Timing Neutron Stars*, ed. H. Ögelman and E.P.J. van den Heuvel (Dordrecht: Kluwer), p. 649.

________. 1989b, in *Proc. 14th Texas Symp. on Relativistic Astrophysics*, ed. E. J. Fenyves (*Ann. NY Acad. Sci.*, **571**), p. 347.

________. 1989c, in *Proc. 23rd ESLAB Symp. on X-Ray Astronomy*, ed. N. E. White (ESA SP-296), p. 215.

________. 1991, *Ap. J.*, in press.

Lamb, F. K., Shibazaki, N., Alpar, M. A., and Shaham, J. 1985, *Nature*, **317**, 681.

Lewin, W.H.G., and van Paradijs, J., and van der Klis, M. 1988, *Space Sci. Rev.*, **46**, 273.

Middleditch, J., and Priedhorsky, W. C. 1986, *Ap. J.*, **306**, 230.

Miller, G. S., and Lamb, F. K. 1988, Talk presented at the Los Alamos Workshop on QPOs, La Cienega, N.M., September 1988.
________. 1990, *Ap. J.*, **356**, 572.
Miller, G. S., and Lamb, F. K. 1991, in preparation.
Mitsuda, K. 1988a, in *Physics of Compact Objects*, ed. N. E. White and L. G. Filipov (*Adv. Space Res.*, **8**), p. 391.
________. 1988b, in *Physics of Neutron Stars and Black Holes*, ed. Y. Tanaka (Tokyo: Universal Academy), p. 117.
________. 1989, in *Proc. 23rd ESLAB Symp. on X-Ray Astronomy*, ed. N. E. White (ESA SP-296), p. 197.
Mitsuda, K., and Dotani, T. 1989, *Pub. Astr. Soc. Japan*, **41**, 557.
Mitsuda, K., Dotani, T., and Yoshida, A. 1988, in *Physics of Neutron Stars and Black Holes*, ed. Y. Tanaka (Tokyo: Universal Academy), p. 133.
Mitsuda, K., and Tanaka, Y. 1986, in *The Evolution of Galactic X-Ray Binaries*, ed. J. Trümper, W.H.G. Lewin, and W. Brinkmann (Dordrecht: Reidel), p. 195.
Norris, J. P., Hertz, P., Wood, K. S., Vaughan, B. A., Michelson, P. F., Mitsuda, K., and Dotani, T. 1990, *Ap. J.*, **361**, 514.
Norris, J. P., and Wood, K. S. 1987, *Ap. J.*, **312**, 732.
Park, M.-G., and Miller, G. S. 1991, *Ap. J.*, in press.
Ponman, T. J., Foster, A. J., and Ross, R. R. 1990, *M.N.R.A.S.*, **246**, 287.
Priedhorsky, W. C., Hasinger, G., Lewin, W.H.G., Middleditch, J., Parmar, A., Stella, L., and White, N. 1986, *Ap. J.* (*Letters*), **306**, L91.
Rappaport, S. A., Verbunt, F., and Joss, P. C. 1983, *Ap. J.*, **275**, 713.
Schulz, N. S., Hasinger, G., and Trümper, J. 1989, *Astr. Ap.*, **225**, 48.
Shapiro, S. L., Lightman, A. P., and Eardley, D. M. 1976, *Ap. J.*, **204**, 187.
Shibazaki, N., and Lamb, F. K. 1987, *Ap. J.*, **318**, 767.
Spiegel, E. A. 1976, in *Physique des Mouvements dans les Atmosphères Stellaires* (Paris: CNRS), p. 19.
________. 1977, in *IAU Colloq.* **38**, *Problems of Stellar Convection*, ed. J. Ehlers, K. Hepp, R. Kippenhahn, H. A. Weidenmüller, and J. Zittartz (Berlin: Springer-Verlag), p. 19.
Stella, L. 1988, *Memorie della Società Astr. Italiana*, **59**, 185.
Stella, L., Haberl, F., Lewin, W.H.G., Parmar, A. N., van der Klis, M., and van Paradijs, J. 1988a, *Ap. J.* (*Letters*), **327**, L13.
Stella, L., Haberl, F., Lewin, W.H.G., Parmar, A. N., van Paradijs, J., and White, N. E. 1988b, *Ap. J.*, **324**, 379.
Stella, L., Parmar, A. N., and White, N. E. 1987, *Ap. J.*, **321**, 418.
Swank, J. H., and Serlemitsos, P. J. 1985, in *Galactic and Extragalactic Compact X-ray Sources*, ed. Y. Tanaka and W.H.G. Lewin (Tokyo: ISAS), p. 175.
Tawara, Y., *et al.* 1982, *Nature*, **299**, 38.
Tennant, A. F. 1987, *M.N.R.A.S.*, **226**, 971.
________. 1988, *M.N.R.A.S.*, **230**, 403.
van den Heuvel, E.P.J. 1988, in *Physics of Compact Objects*, ed. N. E. White and L. G. Filipov (*Adv. Space Res.*, **8**), p. 355.
van den Heuvel, E.P.J. 1991, This volume.
van den Heuvel, E.P.J., van Paradijs, J., and Taam, R. E. 1986, *Nature*, **322**, 153.
van der Klis, M. 1989, *Ann. Rev. Astr. Ap.*, **27**, 517.
________. 1991, This volume.
van der Klis, M., Jansen, F., van Paradijs, J., Lewin, W.H.G., Sztajno, M., and Trümper, J. 1987a, *Ap. J.* (*Letters*), **313**, L19.

van der Klis, M., Jansen, F., van Paradijs, J., Lewin, W.H.G., van den Heuvel, E.P.J., Trümper, J., and Sztajno, M. 1985, *Nature*, **316**, 225.
van der Klis, M., Stella, L., White, N., Jansen, F., and Parmar, A. N. 1987b, *Ap. J.*, **316**, 411.
van Paradijs, J., Hasinger, G., Lewin, W.H.G., van der Klis, M., Sztajno, M., Schulz, N., and Jansen, F. 1988, *M.N.R.A.S.*, **231**, 379.
Wang, Y.-M. 1982, *Astr. Ap.*, **112**, 24.
Webbink, R. F., Rappaport, S. A., and Savonije, G. J. 1983, *Ap. J.*, **270**, 678.
White, N. E. 1986, in *The Physics of Accretion onto Compact Objects*, ed. K. O. Mason, M. G. Watson, and N. E. White (Heidelberg: Springer-Verlag), p. 377.
White, N. E., and Mason, K. O. 1985, *Space Sci. Rev.*, **40**, 167.
White, N. E., Stella, L., and Parmar, A. N. 1988, *Ap. J.*, **324**, 363.

ACCRETING PULSARS, GAMMA-RAY BURSTERS AND LMXB AS HIGH ENERGY POLARIZED SOURCES

P. MÉSZÁROS
Harvard-Smithsonian Center for Astrophysics, and
Pennsylvania State University (permanent address)
Astron. & Astrophysics Dpt., 525 Davey Lab.
University Park, PA 16802, U.S.A.

ABSTRACT. We discuss the significance of polarization producing mechanisms in determining the physical conditions in models of neutron star high energy sources, and the possible impact of polarization measurements at high energies in determining the geometry of the emission regions. Current experiments and their capabilities are reviewed. We discuss calculations of the expected polarization from nonmagnetic neutron stars with accretion disks, and the pulse phase dependence of the degree and angle of polarization in magnetized accreting X-ray pulsars. Calculations of the polarization of gamma-ray burst sources and accreting pulsars are presented, with particular attention to the cyclotron line energy range, using relativistic two-photon processes.

1. Introduction

The X-ray and gamma-ray emission from neutron stars (NS) is expected to be polarized as a consequence of a variety of mechanisms. The most important factor leading to polarization is the strong magnetic field expected in many NS, but even for weak magnetic field NS such as X-ray bursters or other low mass binary NS sources one expects some polarization as a result of scattering in the asymmetric accretion flow around the NS. The fact that the emission is polarized affords a valuable tool for studying the geometry of the emission and/or reprocessing region of the NS. Direct imaging is of course impossible, due to the compactness of these sources, so currently most of the data relating to the physics of these sources comes from spectral or temporal studies, both of which can give only indirect and in some cases ambiguous information about the geometry. Roughly speaking, the spectra give information about the optical depth, which is related to the total integrated mass density along one coordinate, while the time variability can give in some cases information about the absolute physical dimension of the object along this coordinate. While model fitting using only these two variables can be used to determine the shape or orientation of the emission region, the information so obtained is usually not unique. The inclusion of two more parameters, the degree and direction of polarization, provides qualitatively different information that relates to the geometry. The latter is crucial for determining the beaming factor, and thus the total

J. Ventura and D. Pines (eds.), Neutron Stars: Theory and Observation, 483–492.

luminosity and energetics of the source. This in turn affects our understanding of the statistics of NS, which is one of the major observational checks on our understanding of the late stages of stellar evolution, with repercussions for the problems of the galactic chemical evolution, the energy balance of the interstellar medium, the X- and γ-ray luminosity of distant galaxies, and the contributions made by these to the corresponding diffuse radiation backgrounds.

2. Basics of Polarization Phenomena

2.1 POLARIZATION MECHANISMS

2.1.1. *Radiation in a Magnetic Field.* In the classical limit, the radiation from charged particles moving in a magnetic field is polarized. The reason is that the particles are constrained in their motion by the field, so that the second derivative of their electric dipole moment (determining the electric field vector of the radiated intensity) has a well defined set of directions. Examples of this are synchrotron and curvature radiation. In synchrotron radiation, where the emission is caused by the circular motion component in the plane perpendicular to the field, the projection of the electric dipole seen by an observer oscillates back and forth in a direction which on average is perpendicular to the field, and so is the electric vector of the radiation. When the field is not perpendicular to the line of sight, the polarization is in general elliptical, but the major axis of the ellipse is still perpendicular to the projected direction of the field. Curvature radiation, which is due to the curved motion of the charged particles along the electric field, also produces polarized radiation, but in this case the dipole moment projection is along the field, and so is the polarization direction.

2.1.2 *Scattering in a field-free region.* Even in the absence of a field, or for low fields (which for X-rays can mean $B \lesssim 10^{11}$ G), one expects some polarization to arise in the presence of scattering, provided the scattering region is significantly non-spherical. The reason for this is that in a scattering event, the reradiation occurs more efficiently for radiation with electric vector perpendicular to the plane of scattering. As a consequence, the scattered radiation is predominantly polarized in a direction perpendicular to the plane of scattering (the ratio being $1 : \cos^2\theta$). Thus radiation escaping from a scattering dominated accretion disk which is Thomson optically thick will be predominantly polarized parallel to the disk plane, because the radiation tends to escape perpendicular to the disk surface, and the photons observed have a last scattering plane which is perpendicular to the disk plane. On the other hand, a Thomson optically thin disk will be polarized perpendicular to the disk, because the radiation observed had to travel a long distance along the disk before being scattered toward the observer, and the scattering plane is almost coincident with the disk plane.

2.1.3. *Scattering in strong magnetic field.* In the strong magnetic field, the motion of the electron perpendicular to the field is constrained. Classically one has circular orbits, while in the quantum regime the transverse momentum is quantized and one has discrete Landau orbitals of energy $n\hbar\omega_c = \hbar eB/m_e c$, with $n = 1, 2, ..$ corresponding to successive cyclotron harmonics. An incoming photon with energy $\hbar\omega$

much less than the ground cyclotron energy $\hbar\omega_c$ and electric vector perpendicular to the field (extraordinary polarization) tries to jiggle the electron perpendicular to the field, but is not able to do so effectively because it does not have enough energy to lift he electron up to the next excited level $2\hbar\omega_c$. The scattering cross section for this polarization is therefore suppressed, $\sigma_X \simeq \sigma_T(\omega/\omega_c)^2 \ll \sigma_T$. On the other hand, a photon of the same frequency but electric vector parallel to the field (ordinary polarization) would have no trouble jiggling the electron along the field, and the scattering cross section is $\sigma_O \simeq \sigma_T$. As a result of this difference in the scattering cross section for the two polarizations, the flux at $\omega \ll \omega_c$ escaping from a magnetized scattering atmosphere is strongly polarized.

2.2 X-RAY POLARIMETERS

2.2.1. *Scattering Polarimeter.* This type of polarimeter is possibly the simplest, and relies on the polarization dependence of the Compton scattering cross section in an unmagnetized medium (in this case the detector), $d\sigma/d\Omega = \sigma_T(1+\cos^2\theta)$. The photons scattered into various angles are measured by a position sensitive detector encircling the scattering medium, and this translates into a measure of the polarization. The scattering medium is usually a low Z material to minimize absorption effects, and the whole system is slowly rotated to obtain a modulation and eliminate systematic errors. This is the principle of the polarimeter successfully flown on the OSO-8 mission, and is also the basis for the COMPTEL low energy gamma-ray telescope to be flown on GRO. The advantage of this type of polarimeter is that it operates over a broad band of energies, limited below by the photoabsorption energy (few keV) and above by the telescope reflectivity, if any (e.g. 20 keV), the photon detector sensitivity, or the source spectrum.

2.2.2. *Bragg crystal polarimeter.* This uses the well known property of regular crystals that for light incident at 45° coherent reflection occurs only for the polarization perpendicular to the plane of scattering, the Bragg reflection phenomenon. The corresponding Bragg condition is $\hbar\omega = n\pi\hbar c/d\sin(\pi/4)$, where n is the scattering order and d is the lattice spacing. The Bragg condition limits the photon energy for which this occurs, for a pure crystal, to a relatively narrow band of energies, which for graphite is $\Delta E \sim$ few eV centered around energies of a few keV. On the other hand, the crystal does not take up much space and can be positioned at the focus of a collecting mirror or telescope, thus increasing the photon statistics.

2.2.3. *Photoelectric polarimeter.* Some solid state X-ray detectors, which measure the current from electrons dislodged by photons via the photoelectric effect, produce a current which is polarization sensitive (Fraser, *et al.*, 1988). The photocurrent is especially polarization sensitive at shallow angles of incidence, $\theta \lesssim 10^o$. The photocathodes used so far have been mostly CsI, combined with a microchannel plate detector. The advantage of this type of polarimeter, which is under active development, is that the quantum efficiency is very high, and large modulation factors can be obtained.

2.2.4. *Observing Prospects.* The Institute of Cosmic Research (Moscow) and a NASA consortium with Columbia and Marshall as lead institutions, are jointly developing a multiple X-ray polarimeter to be flown on the soviet Spektrum X-Gamma mission to measure distant high energy sources, including neutron stars.

This polarimeter (e.g. Kaaret, *et al.*, 1990), whose acronym is SXRP for spectral X-ray polarimeter, consists of three modules with a) a lithium scattering device, b) a mosaic graphite Bragg polarimeter, and c) a photoelectric polarimeter (PEP). These are positioned on the optical axis of the SODART X-ray telescope, and the whole is rotated at 0.1 RPM. The scattering polarimeter is sensitive over the band 2 keV $\lesssim$ E $\lesssim$ 18 keV, operating in the single scattering regime. The modulation factor is 18%, going up to 81% in conjunction with the PSPC around it. The mosaic graphite Bragg reflector achieves a much broader energy band by the fact that it is an imperfect crystal, with many crystalets at random orientations to satisfy the Bragg condition over a wider range of angles and energies. The modulation factors expected here are close to 99%. The PEP detector consists of a CsI block with a square-pore multichannel plate, and two different filters to achieve two-color photometry, 0.3 − 1.3 keV and 1.3 − 3 keV. The modulation factors here are about 30%. It is expected that polarization degrees of less than a few percent can be measured on bright galactic sources with observation times of 10^5 s. The flight schedule envisages a 1993 launch.

3. Neutron Star Spectral and Polarization Calculations

3.1.- LOW MAGNETIC FIELD NEUTRON STARS

In this category, which includes mostly low mass X-ray binaries, a variety of unpolarized spectral calculations have been made by many groups. Among these sources are steady and transient galactic bulge X-ray sources, X-ray bursters and quasi-periodic oscillators (QPO). The mass transfer in these binaries is Roche lobe overflow type, leading to an accretion disk around the NS.

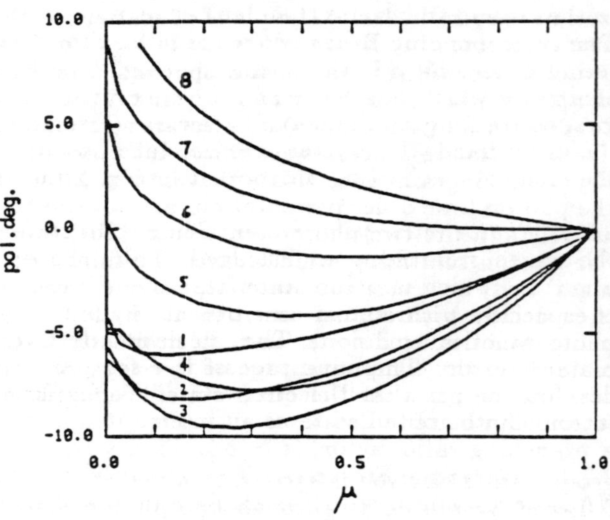

Fig. 1.- *Newtonian accretion disk model, showing the degree of linear polarization (%) as a function of the cosine μ of the angle respect to the disk normal, for disk optical depths τ_T = 0.1, 0.2, 0.5, 1., 2., 4., 10., 100., numbered 1 through 8. Positive (negative) polarization values have electric vector parallel (perpendicular) to the surface. From Phillips and Mészáros, 1986.*

The X-ray emission comes partly from the accretion disk (which due to the low field extends down to the stellar surface), partly from the boundary layer and partly from the stellar surface. If the contribution from the star is dominant, in principle one might expect no polarization because of the spherical symmetry. The presence of a disk, however, breaks this symmetry, since it blocks or reflects part of the light. The degree of polarization expected from this star-disk interaction is model dependent. However, in some cases an upper limit to the degree of polarization may be estimated to be that which would be obtained from a pure disk system. Calculations of the disk polarization for the case of a rotating or nonrotating central black hole have been considered in the GR case by Connors, *et al.*, 1980, while detailed calculations of the degree and direction of polarization as a function of angle for NS in the Newtonian limit have been calculated by Lapidus and Sunyaev, 1985, and Phillips and Mészáros, 1986. Figure 1 shows the degree of linear polarization as a function of the cosine μ of the angle between the observer and the disk normal, for various disk optical Thomson thickness values τ_T. Positive values (for high τ_T) are polarized parallel to the disk surface, while negative values (low τ_T) are polarized perpendicular to the disk surface. These results indicate that linear polarization degrees of up to 8-10% may be expected, for disks observed not too far from edge-on. More realistically, and taking into account the other components, values of 3-5% are probably reasonable.

3.2 . MAGNETIZED NEUTRON STARS - RADIATION PHYSICS

The early work on magnetized neutron stars used nonrelativistic magnetic cross sections with first order relativistic and quantum corrections (e.g. Mészáros and Nagel, 1985a). Subsequent studies have dealt with deriving the magnetic QED cross sections (Bussard, Alexander and Mészáros, 1986, Daugherty and Harding, 1986). Their application for accreting pulsars (AXP) and gamma-ray bursters (GRB) was emphasized by Alexander, Mészáros and Bussard (1989), Harding and Preece (1989) and Alexander and Mészáros (1989). One process is the regular one-photon Compton scattering (Fig. 2a), where both the initial and final electron are in the ground transverse (Landau) state, and there is one photon in and one photon out; since this process has two propagator-photon vertices, the cross section is $\propto \alpha_f^2$. Two other processes are the two-photon scattering (Fig. 2b) discussed by Bussard, *et al.*(1986) and Daugherty and Harding (1986), and the two-photon emission (Fig. 2c) discussed by Alexander and Mészáros (1990a), c.f. also Melrose and Kirk (1986). These two latter processes are *compound* processes, i.e. Figs. 2b and 2c contain *two* Feynman diagrams each (Note that in Fig. 2 the additional diagrams consisting of the photon time order permutations are not shown, but are used in the calculations as well). In the two-photon scattering, (2b), the ground state electron undergoes one-photon scattering which leaves the final electron in an excited (real, on the mass shell) final electron state (this is one cross section); the real excited electron then decays with almost unit probability with emission of a second photon (this being cyclotron emission). The calculation of the cross section for the two processes making up the compound process is essentially decoupled from each other by the real, on the mass shell electron state. To emphasize this, we have drawn the real intermediate excited state as an interrupted line. The compound rate, aside from a branching ratio factor, is $\propto \alpha_f^2$. Similarly, in the two-photon emission, (2c), a ground state electron is excited to a real excited final state (cyclotron excitation), followed by the decay through two photon emission of the real excited electron (a separate cross section calculation). The compound rate, aside from a different branching ratio, is again $\propto \alpha_f^2$. At resonance, all three processes approximate

the cyclotron absorption $\propto \alpha_f$, modified by energy-momentum restrictions and branching ratio factors.

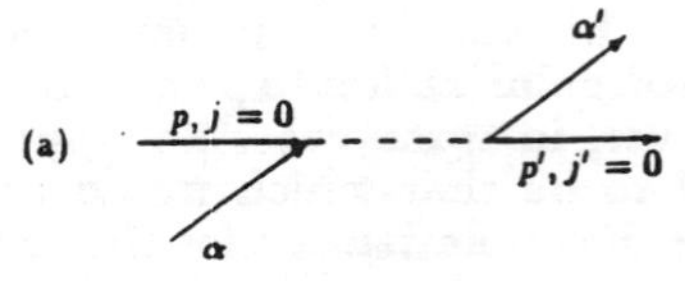

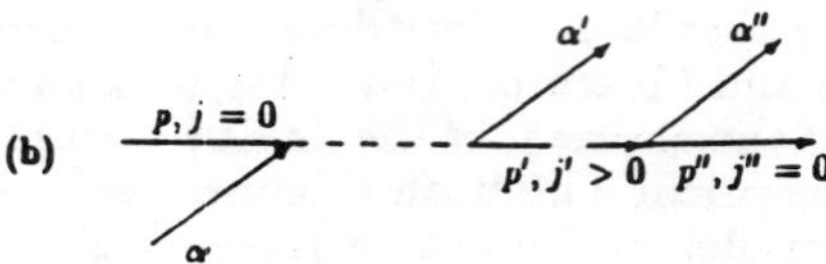

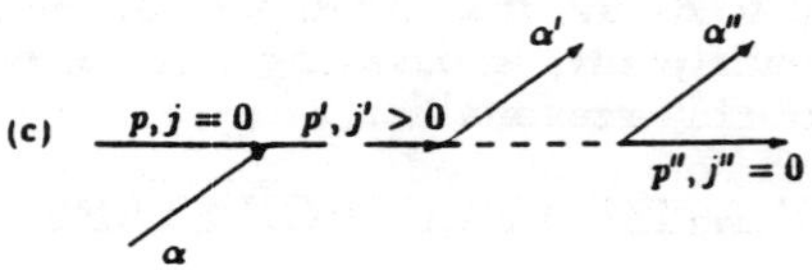

Fig. 2.- ***a) One-photon Compton scattering. b) Two-photon scattering, i.e. scattering leaving final real electron in excited state, followed by cyclotron emission. c) Two-photon emission, i.e. cyclotron excitation, followed by two-photon deexcitation. From Alexander and Mészáros (1991a).***

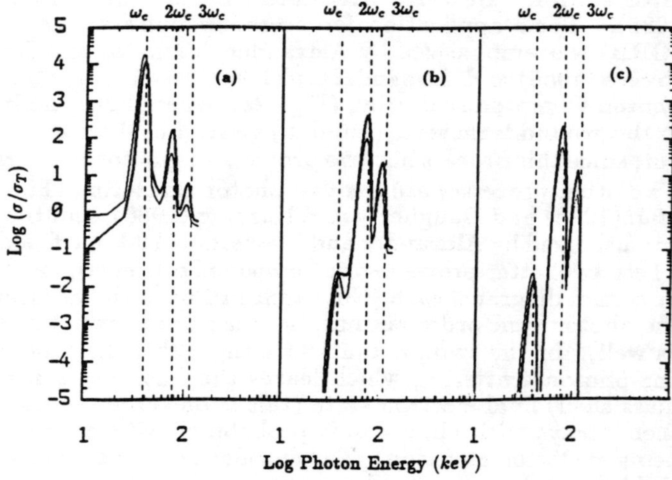

Fig. 3.- ***Opacities for a) one-photon scattering, b) two-photon scattering, c) two-photon emission. Notice the different value of the opacity for the two normal polarization modes (vacuum and plasma effects are included). From Alexander and Mészáros (1991a)***

The opacities for these three processes are shown in Figure 3, where one sees that at the first harmonic ($\omega \sim \omega_c$), the one-photon (ground to ground) scattering (3a) is more important than either the two-photon scattering (3b) or the two-photon emission (3c), but at the higher harmonics $\omega \sim n\omega_c$ ($n \geq 2$) the two-photon processes (3b) and (3c) dominate the one-photon scattering (3a). At the resonances, (3c) is weaker than (3b), and because of strict energy-momentum conservation in

the initial cyclotron absorption, the compound two-photon emission opacity (3c) does not have a continuum contribution as (3b) does.

3.3 . GAMMA-RAY BURSTERS

Gamma-ray bursters (GRB) are currently believed to be magnetized NS, based on their shortest burst rise times close to NS crossing times, average estimated luminosities close to a one solar mass Eddington luminosity, and several observations of first and second cyclotron harmonics in their spectra (Murakami, *et al.*, 1988, Yoshida, *et al.*, 1990) which indicate magnetic fields of magnitude $1-2\times 10^{12}$ G, typical of NS. Spectral calculations of the cyclotron line spectrum in these GRB have been carried out by Wang, *et al.*(1989) and Lamb, *et al.*(1990), using polarization-averaged non-relativistic cross sections with relativistic kinematics and a Monte Carlo code. Using relativistic polarization dependent QED cross sections, Alexander and Mészáros (1989) carried out polarized transfer calculations using a Feautrier method including one-photon and two-photon scattering. The importance of the two-photon processes, which arise following scattering that leaves the electron in an excited state, is that they lead to a photon multipication (one photon in, two photons out). Alexander and Mészáros (1991a, 1991b) did a more detailed calculation of the higher harmonics, including also two-photon emission which has a frequency behavior differing from that of two-photon scattering, both being non-linear in the radiation field. With these changes, the total optical depth required to fit the line ratios is lower than that of the earlier calculation (Alexander and Mészáros, 1989). The results for a model of the burster GB880205 are shown in Fig. 4.

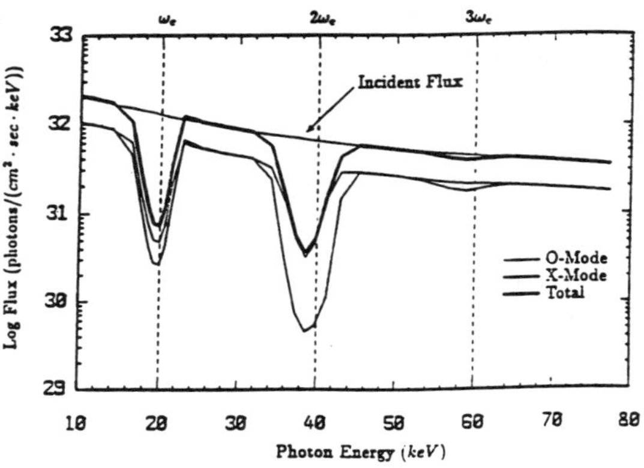

Fig. 4.- *Theoretical gamma-ray burster cyclotron line spectrum with* $\hbar\omega_c = 20\ keV$, $\rho = 6\times 10^{-4} g\ cm^{-3}$, $kT = 5.2\ keV$, *and* $\tau_T = 1.1\times 10^{-2}$. *This uses a scattering angle of* 60°, *including two-photon scattering and two-photon emission, and shows the polarizations separately. The flux normalization is arbitrary. From Alexander and Mészáros(1991b).*

In this figure one sees the similar depth of the first and second harmonic, in agreement with the observations of Murakami, *et al.*(1988), a feature which is only obtained when the photon spawning two-photon processes are used, and one sees also the characteristic asymmetric line profiles introduced by the relativistic effects. The polarization degree depends sensitively on the total optical depth, which for this polarized calculation is $\tau_T = 1.1 \times 10^{-2}$. (As a control, we have also done a polarization averaged calculation, and find that in this case, in order to get similar line ratios, the optical depth needed would be reduced from the previous value by a factor ~ 3). The difference in the cross sections of the two polarization modes and the interplay between polarizations in the transfer (e.g. scattering with mode switching) both play a significant role in determining the appropriate optical depth that will reproduce the observed line ratios. Of course another important factor affecting the optical depth is the angle of incoming and outgoing photons, here taken to be $\pm 60^o$. In fig. 2 the two polarizations are plotted separately, and for an incident continuum assumed to be unpolarized, the resulting cyclotron harmonic lines are strongly polarized. The degree of polarization here is about 30%, 80% and 10% for the first, second and third harmonics.

3.4. ACCRETING X-RAY PULSARS

Accreting X-ray pulsar (AXP) model spectra have been calculated in a polarization dependent manner, using nonrelativistic one-photon scattering, by Mészáros and Nagel (1985a, 1985b), including vacuum and plasma polarizability effects. More approximate polarized calculations, as well as extensive polarization averaged calculations including higher order relativistic corrections were done by Wang, *et al.*(1988). The pulse shapes, degree of polarization and angle of polarization as a function of pulse phase were calculated for pencil and fan beam AXP models by Mészáros, *et al.*(1988). An example for a pencil beam with a particular set of observation angles is shown in Fig. 5.

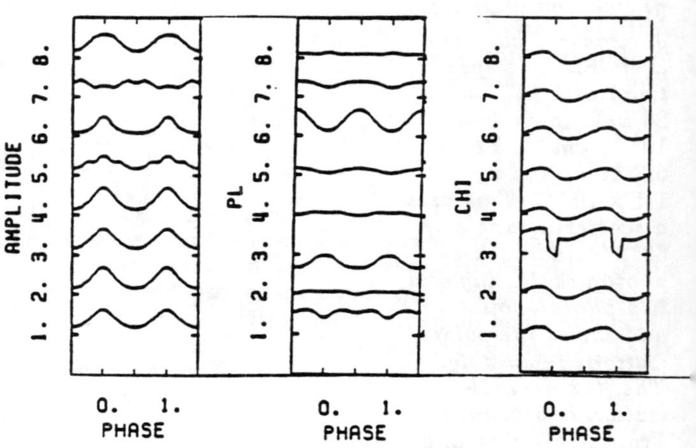

Fig. 5.- *Accreting X-ray pulsar model, showing the pulse phase dependence of the flux (normalized), the degree of linear polarization (+100% to -100%) and electric vector position angle χ ($+2\pi$ to -2π) for 8 upward increasing energies (cyclotron energy is sixth up). The inclination angles are ($50^o, 20^o$). From Mészáros, et al., 1988.*

When at a particular energy there is a change in the dominance of one polarization over the other, the degree of polarization goes through zero and the electric vector direction jumps by π. In pencil beams, the angle χ goes from positive to negative at pulse maxima (phase 0 and 1) and from negative to positive at pulse minima (phase 0.5). In fan beam models, the direction is reversed, χ going from negative to positive at pulse maxima and from positive to negative at pulse minima. For both pencils and fans, χ goes through zero twice in a full phase cycle, the rate of change being fast in one crossing and slow in the other. However for pencils, the fast change of χ occurs at pulse maxima, while for fans it occurs at pulse minima. These characteristics can be used to distinguish pencil beams from fan beams. More recently, Alexander and Mészáros (1991b) performed calculations with the fully relativistic QED cross sections (Alexander and Mészáros (1990a), including two-photon scattering and two-photon emission as well as stimulated scattering, using a polarized Feautrier transfer scheme. These calculations, unlike the previous ones, include also higher harmonics. The shape of the ground harmonic, as well as the spectral region above it are affected by these effects. This region is important in view of the recent observation or reobservation of cyclotron lines in several AXP (Clark, *et al.*, 1990, Mihara, *et al.*, 1990, Makishima, *et al.*, 1990, Yoshida, *et al.*, 1990, Staubert, *et al.*, 1990). A model calculation is shown in Fig. 6.

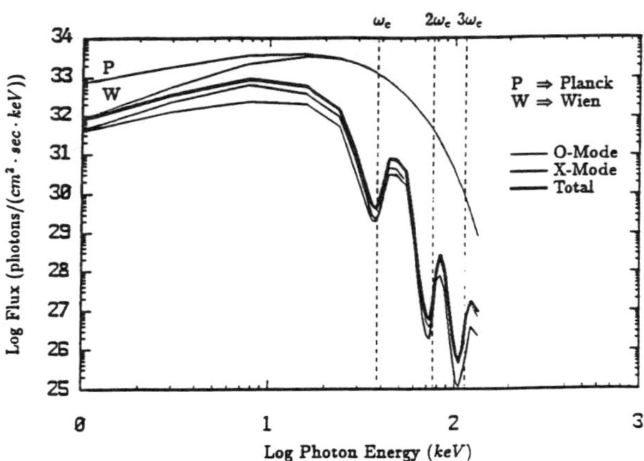

Fig. 6.- *Relativistic calculation of an accreting X-ray pulsar model, showing the polarized spectrum for an optical depth* $\tau_T = 2 \times 10^4$. *Processes included are spontaneous and stimulated one-photon scattering, two-photon scattering and two-photon emission. From Alexander and Mészáros, 1991b.*

Compared to earlier calculations with one harmonic only, these here show a much steeper flux drop above the first harmonic blue shoulder, which is also narrower since it can not go up much in energy before dipping again towards the second harmonic. The degree of polarization, as can be seen, is larger for the second and third harmonic than for the first, and the continuum is also strongly polarized.

4. Discussion

A consideration of the polarization dependence of the cross sections and its influence on the formation of spectral features is important for a detailed understanding of the physics of magnetized NS, since these phenomena play a significant role in determining the optical depth at which thermalization occurs, affecting in particular the cyclotron line depths and harmonic ratios. Polarization averaged calculations can give similar spectra, but the optical depths needed are different. The second and more far reaching use of polarization dependent model calculations is that, combined with observations which are expected to be forthcoming within the next few years, they should provide information about the geometrical structure of the emission regions, with implications for the disk orientation in nonmagnetic NS or the physical mechanisms operating in magnetic NS (e.g. radiation or binary collision deceleration, radiation mechanisms), and the beaming properties, duty cycle and total energetics.

Acknowledgements: This research has been supported in part through grants NASA NAGW-1522, NSF AST 8815266, and the CfA Smithsonian Visiting Scientist program. Local support through NATO ASI and the University of Crete is appreciated.

5. References

Alexander, S. and Mészáros, P., 1989, Ap.J.(Lett.), 344, L1
Alexander, S. and Mészáros, P., 1991a, Ap.J., in press
Alexander, S. and Mészáros, P., 1991b, Ap.J., in press
Clark, G., *et al.*, 1990, Ap.J., Ap.J., 353, 274
Connors, P.A., Piran, T. and Stark, R.F., 1980, Ap.J., 235, 224
Bussard, R., Alexander, S. and Mészáros, P., 1986, Phys.Rev., D34, 440
Daugherty, J. and Harding, A.K., 1986, Ap.J., 309, 362
Fraser, G.W., *et al.*, 1988, Proc. SPIEE, 982, 98-107.
Kaaret, P., *et al.*, 1990, to appear in *Proc. IAU Symp. No. 123: "Observatories in Earth Orbit and Beyond".*
Harding, A.K. and Preece, R., 1989, Ap.J.(Lett.), 338, L21
Lamb, D.Q., *et al.*, 1990, Ap.J., 363, 370
Lapidus, I.I. and Sunyaev, R.A., 1985, M.N.R.A.S., 217, 291
Makishima, K., *et al.*, 1990, subm. to Ap.J.(Lett.)
Melrose, D. and Kirk, J.G., 1986, Astron.Ap., 156, 268
Mészáros, P. and Nagel, W., 1985a, Ap.J., 298, 147
Mészáros, P. and Nagel, W., 1985b, Ap.J., 299, 138
Mészáros, P., *et al.*, 1988, Ap.J., 324, 1056
Mihara, T., 1990, Nature, 346, 250
Murakami, T., *et al.*, 1988, Nature, 335, 234
Phillips, K.C. and Mészáros, P., 1986, Ap.J., 310, 284.
Staubert, R., Kendziorra, E. and Mony, B., 1991, in *Proc. Crete Symp. on Neutron Stars*, ed. J. Ventura (Kluwer, Dordrecht).
Wang, J.C., *et al.*, 1988, Ap.J. (Supp.), 68, 735
Wang, J.C., *et al.*, 1989, Phys.Rev.(Lett.), 63, 1550
Yoshida, A., *et al.*, 1990, to appear in *Proc. Los Alamos Workshop on Gamma Ray Bursters*, eds. R. Epstein, *et al.*, (Cambridge)

ACCRETION FLOWS ON TO NEUTRON STARS

A. R. KING
Astronomy Group,
University of Leicester,
Leicester LE1 7RH, U.K.

ABSTRACT. This article briefly reviews our understanding of accretion in both high– and low–mass X–ray binaries. In the first class the angular momentum of the captured material is rather uncertain, and the accretion flow may be very complex: the spin rates of the neutron stars in both supergiant and Be–star binaries suggest that the X–ray stage is very short. The nature of the accretion discs in the second group is a matter of debate, as observations imply vertical structure subtending large angles at the central neutron star.

1. Introduction

Close binaries containing neutron stars are readily found as X–ray sources, and fall naturally into two groups distinguished by the mass of the companion star. In high–mass systems, this is an early–type, massive star; either an O or B supergiant, or a Be star. In the first case the binary has period $P \sim 10$ d; in the second, $P \sim 30-200$ d, and the orbit can be significantly eccentric. In low–mass systems the period may be short ($P \sim$ a few hours down to 10s of minutes) with the companion presumably fairly close to the main sequence, or degenerate for $P \lesssim 80$ min; a second group of low–mass systems have $P \sim 2-10$ d and must have evolved low–mass subgiant companions. One property of the neutron star appears to correlate well with the nature of the companion, presumably for evolutionary reasons: in the high–mass systems the neutron star is usually strongly magnetic (polar fields $\sim 10^{11}$ - 10^{12} G), whereas in low–mass systems it is generally not. The strong fields are revealed by regular pulsing of the X–ray emission at the spin period of the neutron star, and in some cases can be directly measured because of the presence of cyclotron features at X–ray energies. In one picture, the quasi–periodic oscillations (QPOs) observed in the X–ray emission of some low–mass systems are manifestations of rather weaker ($\sim 10^8$ G) fields, but there is as yet no direct evidence supporting this.

The nature of the accretion flows is rather different in the high– and low–mass X–ray binaries. In the former case, accretion via Roche lobe overflow would be dynamically unstable since the companion is more massive than the neutron star. Hence in observed systems accretion is usually from the wind of the companion: either the usual spherical outflow from an O or B supergiant, or a slow dense equatorial wind from a Be star. The X–ray phase is therefore a rather brief interlude

J. Ventura and D. Pines (eds.), Neutron Stars: Theory and Observation, 493–500.

between the epoch when the companion first develops a sufficiently strong wind, and that when this wind disappears or the star overflows the Roche lobe and very rapid mass transfer swamps the neutron star. Accordingly, in high–mass X–ray binaries the major problem is that we do not have a good understanding of the amount of specific angular momentum which the neutron star captures from the wind; thus it is not straightforward to predict whether accretion should occur via a disc for example. In the low–mass case this must be true, as the companions fill their Roche lobes. These systems ought then to be the neutron–star (and black–hole) analogues of the cataclysmic variables (CVs), in which the accreting star is a white dwarf: here accretion discs conforming quite well to a simple theoretical picture are inferred from observation. However the observed X–ray light curves of neutron–star systems exhibit orbital modulations which are curiously difficult to square with this standard picture. In this short review I shall discuss the present status of these two problems, and the ways in which progress may come about.

2. Accretion in High–Mass X–ray Binaries.

As described above, accretion here is from the stellar wind of the companion. This process (see Livio, 1990, for a recent review) is usually discussed in terms of the simple picture first suggested by Hoyle and Lyttelton (1939) in which matter is captured from within a cylinder about the neutron star, with size given by the "accretion radius"

$$r_{\rm acc} = \frac{2GM}{v^2}. \tag{1}$$

Here M is the neutron star mass, and v the relative velocity of the stellar wind at the neutron star. Thus

$$v^2 = v_{\rm w}^2 + v_{\rm n}^2, \tag{2}$$

where $v_{\rm w}$ is the wind's velocity with respect to the companion, and $v_{\rm n}$ the neutron star's orbital velocity about this star. Numerical work (e.g. Shima *et al.* , 1985) gives reasonable agreement with the accretion rate predicted by this picture. However the amount of angular momentum captured by the neutron star is far less certain. The simplest possible assumption is that the accretion cylinder is symmetrical about the neutron star, which thus captures all the wind angular momentum entering the cylinder. It is usual to parametrize in terms of this, and write the captured specific angular momentum as

$$j_{\rm capt} = \frac{\eta \pi r_{\rm acc}^2}{2P}. \tag{3}$$

Here η is an efficiency factor which would be unity if the symmetry assumption were valid. However Davies and Pringle (1980) pointed out that velocity or density gradients across the cylinder can have a marked effect on this estimate. In particular it might be that the captured matter is focussed essentially into a line, which would therefore deposit very little angular momentum. Clearly this is a problem which can only be handled numerically, and equally clearly, it is very difficult. However, there appears now to be rather general agreement (e.g. Livio *et al.* , 1986a, b; Anzer *et al.* , 1987) that η is considerably less than unity, perhaps of order 5 - 10%.

The reduction in η means that in a steady state, it will be very difficult for an accretion disc to form around the neutron star. This requires the captured matter to have sufficient angular momentum to orbit this star rather than crash directly into its magnetosphere, i.e.

$$j_{\rm capt} > (GMr_\mu)^{1/2} \qquad (4)$$

where r_μ is the magnetospheric radius. For typical parameters, (4) now fails by a wide margin for $\eta \lesssim 0.1$. It appears however that the assumption of a steady state also fails in some cases. Higher–resolution hydrodynamical calculations (Matsuda *et al.* , 1987; Taam & Fryxell, 1988; Fryxell & Taam, 1988) suggest that the accretion flow sometimes results in a wake behind the neutron star which oscillates from side to side. The captured specific angular momentum changes sign at each oscillation, fluctuating with very large amplitude ($\sim$ 40 times the average $j_{\rm capt}$) about the mean value (3), on a typical timescale $\sim 4r_{\rm acc}/v$. In consequence, a short–lived accretion disc forms on each oscillation, changing its sense of rotation as $j_{\rm capt}$ reverses. This behaviour explains, at least qualitatively, the observed fact that the neutron stars sometimes fractionally change their spin periods on timescales $(\Delta\Omega/\Omega)t_{\rm spin}$ far shorter than implied by the *long–term average* spinup timescale

$$t_{\rm spin} = \frac{I\Omega}{\dot{M} j_{\rm capt}}, \qquad (5)$$

where I and Ω are the moment of inertia and spin angular velocity of the neutron star, and $\dot{M}$ is the average accretion rate. Another consequence of the angular momentum reversals is that as the disc begins to accrete the opposite sign of angular momentum, it rapidly collapses on to the neutron star, resulting in flares.

Although Ω can fluctuate fractionally, its observed mean value must of course be determined by the long–term behaviour of the accretion torque. For supergiant binaries it is well known that $t_{\rm spin}$ is longer than the lifetime as an X–ray binary (e.g. Waters & van Kerkwijk, 1989), so that the observed values of Ω must reflect an earlier evolutionary phase. Interestingly, it appears that for Be/X–ray binaries $j_{\rm capt}$ is probably too low for the neutron stars to be presently spinning in equilibrium, so that here too Ω may be determined by an earlier evolutionary phase (King, 1991, in preparation). Further, the X–ray stage must be short compared with $t_{\rm spin}$, so that Be – neutron star binaries must spend most of their lifetimes not emitting X–rays strongly.

3. Accretion in Low–Mass X–ray Binaries.

In low–mass systems, the companion star must fill the Roche lobe to provide an accretion rate on to the neutron star sufficient to produce the observed X–ray luminosities. Mass transfer is driven in one of two main ways (see e.g. King, 1988 for a recent review).

(a) For orbital periods $\gtrsim$1 d the companion star is evolved (e.g. a subgiant). Mass transfer proceeds on a nuclear timescale as the stellar radius expands, with the binary evolving to longer orbital periods. Once the companion's envelope has been transferred we are left with a wide white dwarf – neutron star binary. These systems are good candidates for the progenitors of millisecond radio pulsars with long–period circular orbits.

(b) For orbital periods $\lesssim 1$ d (typically a few hours) the binary evolution must be driven by angular momentum losses from the orbit, via e.g. magnetic braking of the companion star or gravitational radiation directly from the orbit.

Apart from the substitution of a neutron star for a white dwarf as the accreting object, the short–period systems should be identical to cataclysmic variables. The latter are well–studied observationally, as there are ~ 150 with known orbital periods. All the available evidence supports a picture conforming to simple theoretical expectations: accretion takes place via a disc, which is geometrically thin and consists of material moving in Kepler orbits. A basic property of such discs is the dependence of the effective temperature $T_{\rm eff}$ on radial distance R in a steady state:

$$T_{\rm eff} \propto R^{-3/4}. \qquad (6)$$

This can be confirmed in eclipsing systems by using the occultation by the companion as a probe of the disc's surface brightness (Horne & Marsh, 1986).

One should not be too surprised to find deviations from the radial dependence (6) in low–mass X–ray binaries, as the central object is enormously brighter than in cataclysmics. Irradiation by even a small fraction of the total X–ray emission can affect the surface brightness, and may also lead to very substantial mass loss, breaking the assumption of constant mass inflow as a function of R which underlies (6) (Czerny & King, 1989). However to cause substantial deviations from Kepler flow in the orbital plane, irradiation would have to produce disc temperatures which are implausibly large (see below). Hence a geometrically thin Keplerian disc ought to provide a good model for accretion also in low–mass X–ray binaries.

X–ray observations of the light curves of low–mass X–ray binaries produce results which are rather surprising, given the conclusion above. (see Mason, 1986 for a review). If the X–rays came as expected from a point source with a Keplerian disc around it, we should see either no orbital modulation at all, or a total eclipse with near–instantaneous ingress and egress, depending on the system's inclination. Yet eclipses of this type are very rare amongst observed orbital modulations, and indeed *partial* X–ray eclipses are sometimes observed. This is very suggestive that the central X–ray source is extended, presumably in the form of some kind of scattering corona about the neutron star, the dimensions of this region being comparable to the radius of the companion. If the corona subtends a larger angle than the companion at the neutron star, then at inclinations too low for an eclipse we may observe continuous period modulation of the X–rays caused by any persistent structure in the corona. This behaviour is seen, as well as "dipping": irregular decreases in X–ray intensity accompanied by increased photoelectric absorption, occurring in a rather stable interval of orbital phase. A second effect of the corona is to scatter X–rays preferentially away from the disc plane, so that edge–on systems are relatively faint in X–rays. This explains why eclipses by the companion are relatively rare. This general scheme was first proposed by White & Holt (1982), and has proved quite successful in explaining the light curve morphologies. Further support for it comes from the fact that the ratio of X–ray to optical brightness is significantly lower in systems where only the corona is seen (i.e those showing continuous modulations) than in face–on systems where we presumably see the neutron star directly. This is expected, since the optical emission comes mainly from the accretion disc and should not be very strongly dependent on the inclination, in contrast to the marked difference in X–ray fluxes between the two classes.

The White & Holt picture as it stands is purely empirical: there is no physical model showing that the existence of the posited structures is reasonable. Theoretical attempts to do this have been only partially successful. There seems good reason to expect a corona to exist, provided some of the central X–rays irradiate the central parts of the disc (Fabian *et al.* , 1982; Czerny & King, 1989), and it can quite plausibly have the required scattering optical depth $\tau \sim 0.1$. The main difficulty is in modelling the vertical structure required to produce the observed periodic modulations such as dips. White & Holt (1982) suggested that this structure was at the edge of the disc. This proposition is open to several objections.

1. The vertical angular extent required of this structure implies

$$\frac{H}{R} \gtrsim 0.15, \tag{7}$$

where H is its scale height and R its radial position. If H is assumed to be the thermal scale height of a disc, the standard relation $H/R \sim c_S/v_\phi$ (e.g. Frank *et al.* , 1985), where c_S is the sound speed and v_ϕ the Kepler velocity, implies a required disc temperature

$$T \sim \frac{0.014 GMm_H}{kR} \gtrsim 4.5 \times 10^5 \text{ K} \tag{8}$$

at a typical disc outer radius of 3×10^{10} cm. By contrast, the radiation temperature from even an Eddington–limited object of 1 $M_\odot$ at this radius is only

$$T_{\rm rad} = \left(\frac{L_{\rm edd}}{4\pi R^2 \sigma}\right) = 1.6 \times 10^5 \text{ K}. \tag{9}$$

White & Holt (1982) and others suggested that turbulence might be responsible for the large scale height. Clearly, this only shifts the problem elsewhere, because (8) shows that the turbulence would have to be highly supersonic. Since irradiation is the one way in which the fact that the central star is a neutron star rather than a white dwarf can be transmitted out to the edge of the disc, any way of providing the vertical structure at the disc edge which does not invoke it is open to the objection that we should see similar phenomena in cataclysmics, for which there is little evidence.

2. Even leaving this problem aside, the only plausible way of getting material to these heights is to have it follow ballistic orbits. The phase dependence suggests that this might occur as the result of the impact of the mass transfer stream from the companion at the edge of the disc. But ballistic orbits can only result in structure which is rather smooth in the azimuthal direction, as shown by the fact that a circular orbit out of the disc plane will intersect it at two points separated by 0.5 in orbital phase. In particular the highly structured "Manhattan skyline" disc edges inferred for this picture would require material to orbit in circles parallel to, but above or below, the orbital plane. These are not permissible test particle orbits, and it is difficult to see how any gaseous material could be made to form itself stably into such shapes.

3. We know from observation that the X–ray dips are accompanied by photoelectric absorption. Since this operates only at lower X–ray energies, most of the loss of X–ray flux is caused by electron scattering; equivalent hydrogen column densities $\gtrsim 10^{25}$ cm^{-2} are needed to give the observed depths. But these imply that the photoelectric absorption at low energies should be considerably more severe than observed. There are two ways to escape from this difficulty: either the

elements (CNO, He) giving rise to the absorption are underabundant relative to H, or their K–shells, which do the absorbing, are ionized. But this latter possibility is denied if the material causing the dips is located at the disc edge, for the ionization parameter

$$\xi = \frac{L_X}{NR^2} \lesssim \frac{L_X}{10^{25}R} \lesssim \frac{10^{12}}{R} \sim 30 \tag{10}$$

falls well below the values $\sim$ a few $\times 10^2$ required to ensure that photoionization by X–rays will do this (in (10) I have taken $NR > 10^{25}$ and an X–ray luminosity $L_X \sim 10^{37}$ erg s^{-1}). Putting the dip material at the disc edge thus forces one into assuming rather large underabundances (factors up to $\sim 10^3$). While not impossible, in the absence of compelling independent evidence this seems a rather contrived expedient.

In view of these difficulties it is reasonable to look for ways of providing the required vertical structure at positions closer to the neutron star. The main problem here is to reproduce the phase dependence of the structure. A knowledge of the orbital phase can be transmitted to inner disc regions either via tidal effects from the secondary star, or by part of the gas stream from this star penetrating or skimming over the faces of the disc. Tides are expected to work much more effectively in the outer regions of the disc, so most attention has been paid to the second possibility. Frank *et al.* (1987) showed that if enough of the gas stream follows a ballistic trajectory, the self intersections near the circularization radius ($R \sim 10^{10}$ cm) will lead to matter being thrown to large heights ($\sim R$) above the disc. Most importantly, for accretion on to neutron stars, this material is subject to a two–phase instability, independently of the accretion rate. This means that much of it will condense into cool blobs, which will absorb low–energy X–rays photoelectrically. The model thus gives a physical reason for the prime observational feature of dip sources, namely cool material at large scale heights, at a fixed phase location. The model also predicts the rough number, size, and column density of the blobs, in quite good agreement with the fairly sparse observational evidence. Lubow (1989) suggests slightly different ballistics for the gas stream, which will alter the orbital phase dependence somewhat, but a two–phase instability is likely here too. In cataclysmics the lower ionizing luminosity of the central object for a given accretion rate means that there will be no two–phase instability. But the regions of the disc which are obscured by this structure would be observable mainly in the EUV and soft X–rays. The faintness and short orbital periods of these objects have so far severely restricted phase–resolved observations at these wavelengths.

The main objection to this type of model is that while it provides physical reasons for the vertical structure and dip behaviour, it does not straightforwardly reproduce all of the orbital phase dependence observed in some low–mass X–ray binary light curves. It should be pointed out however that the disc–edge picture only achieves this by *fiat*, since it is physically unconstrained.

4. The Future.

It is apparent that our understanding of accretion in neutron–star binaries is still rather rudimentary. For instance, we still have little idea of how disc accretion is broken up by a central magnetic field, or the possibly related question of the accretion behaviour responsible for the quasi–periodic oscillations discussed elsewhere in this volume. This is perhaps unsurprising, as we know the orbital periods of only some ~ 20 low–mass systems and the orbital and spin periods of a similar number of high–mass ones. For the latter systems, the increasing sensitivity of radio pulsar searches might provide a much clearer picture, for example if some of the "missing" Be – neutron star binaries were to turn up. In both cases however we can expect considerable progress from the much larger sample which the ROSAT X–ray sky survey will provide.

Acknowledgments

I thank the organisers for support, and them and the other participants for many fruitful discussions.

References

Anzer, U., Boerner, G. and Monaghan, J. J. 1987, *Astron. Astrophys.*, **176**, 235.

Czerny, M. and King, A. R. 1989, *Mon. Not. R. astr. Soc.*, **241**, 839.

Davies, R. E. and Pringle, J. E. 1980, *Mon. Not. R. astr. Soc.*, **191**, 599.

Fabian, A. C., Guilbert, P. and Ross, R. 1982, *Mon. Not. R. astr. Soc.*, **199**, 1041.

Frank, J., King, A. R. and Lasota, J. P., 1987, *Astron. Astrophys.*, **178**, 137.

Frank, J., King, A.R. and Raine, D. J. 1985, *Accretion Power in Astrophysics*, 1st Edition, Cambridge University Press, Ch. 5

Fryxell, B. A. and Taam, R. E. 1988, *Astrophys. J.*, **335**, 862.

Horne, K. and Marsh, T. 1986, In: *The Physics of Accretion onto Compact Objects*, p. 1, eds Mason, K.O., Watson, M.G. & White, N.E., Springer-Verlag, Berlin.

Hoyle, F. and Lyttelton, R. A. 1939, *Proc. Camb. Phil Soc.* **39**, 405

King, A. R. 1988, *Q. Jl. R. astr. Soc.*, **29**, 1.

Livio, M., 1991, In *Frontier Objects in Astrophysics and Particle Physics*, Vulcano Workshop 1990, ed F. Giovannelli, in press

Livio, M., Soker, N., de Kool, M. and Savonije, G. J. 1986a, *Mon. Not. R. astr. Soc.*, **248**, 593.

Livio, M., Soker, N., de Kool, M. and Savonije, G. J. 1986b, *Mon. Not. R. astr. Soc.*, **222**, 235.

Lubow, S. H., 1989, *Astrophys. J.*, **340**, 1064.

Mason, K. O., 1986, In: *The Physics of Accretion onto Compact Objects*, p. 29, eds Mason, K.O., Watson, M.G. & White, N.E., Springer-Verlag, Berlin.

Matsuda, T., Inoue, M. and Sawada, K. 1987, *Mon. Not. R. astr. Soc.*, **226**, 785.

Shima, E., Matsuda, T., Takeda, H. and Sawada, K. 1985, *Mon. Not. R. astr. Soc.*, **217**, 367.

Tamm, R. E. and Fryxell, B. A. 1988, *Astrophys. J. Lett.*, **327**, L73.

Waters, L. B. F. M. and van Kerkwijk, M. H. 1989, *Astron. Astrophys.*, **223**, 196.
White, N. E. and Holt, S. S. 1982, *Astrophys. J.*, **257**, 318.

SOFT X–RAY TRANSIENTS

J. M. HAMEURY
DAEC
Observatoire de Paris
Section de Meudon
F–92195 Meudon Cedex
France

ABSTRACT. I present the observational characteristics of soft X–ray transients and discuss the two classes of models that have been put forward to explain them, namely the irradiation instability of the secondary and the disc instability model.

1. Introduction

X–ray transients are systems in which emit most their energy during outbursts. Just as for X–ray binaries, X–ray transient fall into two groups: hard X–ray transients and soft X–ray transients. Hard transients occur in massive X–ray binaries, in which the orbit is very eccentric. The compagnon is usually a Be star, and accretion occurs when the compact object is at periastron. The hardness of the spectrum is explained, as in massive X–ray binaries, by the fact that the primary is a strongly magnetic neutron star.

On the other hand, soft X–ray transients (SXTs) occur in low–mass X–ray binaries. They can be divided into two subgroups: soft and ultrasoft X–ray transients, that are most probably due to the same mechanism, the basic difference being the different nature of the compact object, a neutron star for soft transients and a black–hole in the case of ultrasoft transients. Up to now, about a dozen of such sources are known, but it is quite difficult to determine their total number in the galaxy.

2. Observations

The light curve of a typical ultrasoft transient source, 4U1630–47 is shown in Fig. 1. Four outbursts were found during a period of five years, and, very probably, none was missed. In general, recurrence times vary from half a year to 50 years in the case of the source A0620–00, but of course this upper value might not be real, but simply due to a selection effect.

2.1 OUTBURST

The total energy emitted during an outburst is about 10^{44} erg, with a maximum

J. Ventura and D. Pines (eds.), Neutron Stars: Theory and Observation, 501–514.

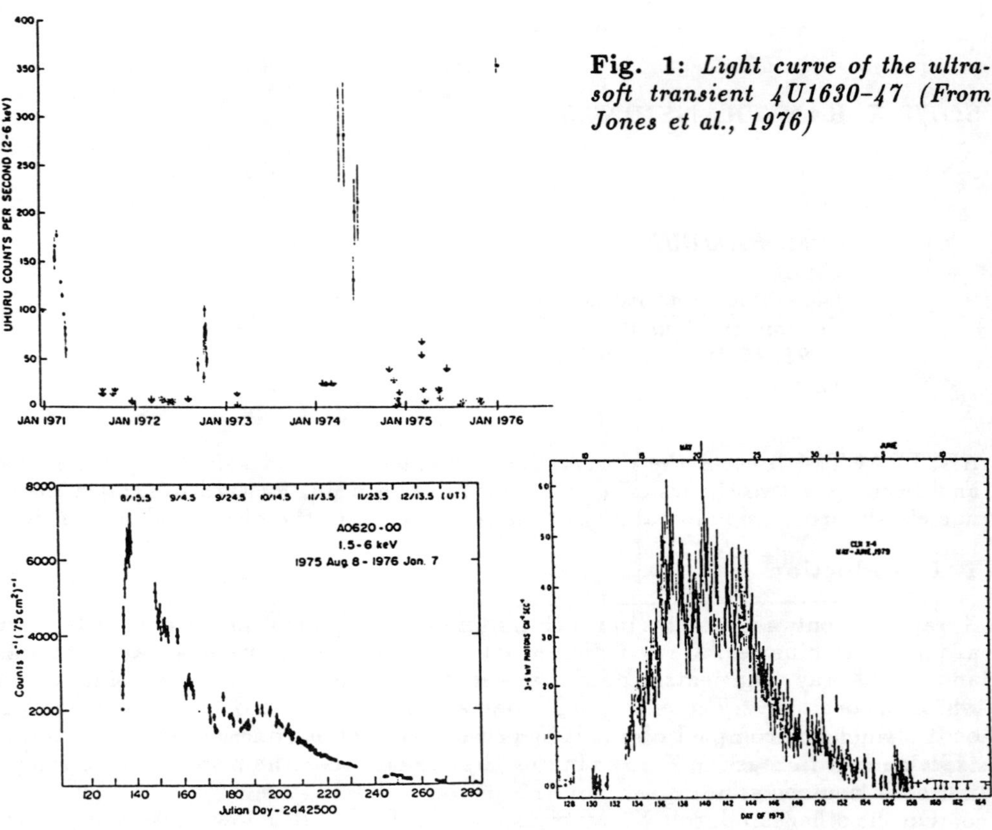

Fig. 1: *Light curve of the ultra-soft transient 4U1630-47 (From Jones et al., 1976)*

Fig. 2: *Light curves of two soft X-ray transients during outburst: A0620-00 (left, from Matilsky et al., 1976) and Cen X-4 (right, from Kaluznienski et al., 1975). A0620-00 is an ultrasoft transient, which has a long recurrence time, while Cen X-4 is a soft transient, with a short recurrence time.*

luminosity of $10^{37} - 10^{38}$ erg s^{-1}. Figure 2 shows two examples of light curves of SXT's during outburst, and one can see that the rise time, in the range 2 – 10 days, is much shorter than the decay time, which is in the range 10 – 50 days.

The spectra are different for the two subclasses of SXT's: ultrasoft X-ray transients spectra can be fitted by a blackbody with a temperature less than 1 keV, plus a hard tail, that becomes more and more prominent as the flux decreases (Fig. 3). This is quite reminiscent of black holes candidates such as Cyg X-1, and is one of the reasons for assuming that the primary in ultrasoft transients might be a black hole, although the origin of this hard component is still uncertain (see e.g. Guilbert and Fabian, 1982; Kazanas, 1986). The spectra of soft transients are the sum of a blackbody with $kT \sim 2$ keV at maximum, and a harder component that could originate from a corona around the accreting neutron star (King and Lasota,

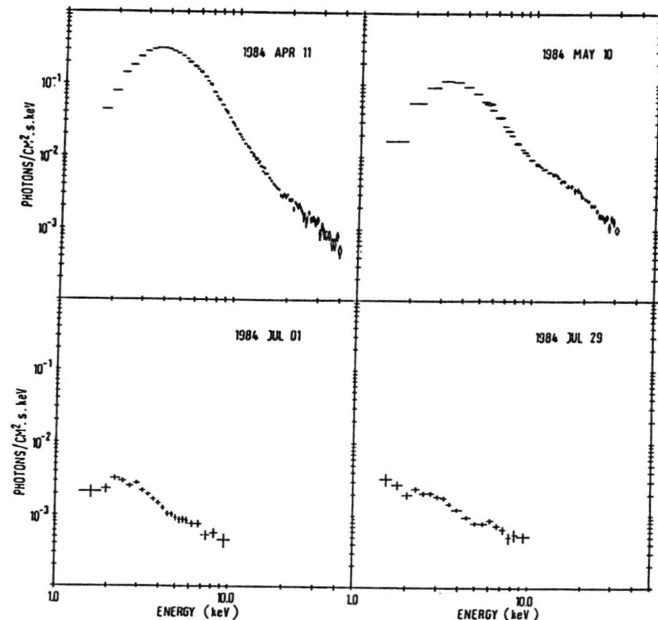

Fig. 3: *Spectral evolution of the ultrasoft X-ray transient 4U 1630-47 (from Parmar, Stella and White, 1986)*

1987).

It should finally be noted that X-ray bursts are usually observed in soft transients during decline, implying that the accreting object is a neutron star; the fact that they have never been seen in ultrasoft sources strengthens the idea that these contain black holes.

2.2 QUIESCENCE

In quiescence, these objects are very faint, both in X-ray and optical. The X-ray emission can be undetectable, with upper luminosity limits of 10^{32} erg s^{-1} in the case of A0620-00 (Long, Helfand and Grabelsky, 1981) and 10^{33} erg s^{1} for Aql X-1 (van Paradijs et al., 1987), or very weak, as in Cen X-4, with a luminosity of (1 - 3) $\times 10^{33}$ erg s^{-1} (van Paradijs et al., 1987).

In contrast with the outburst, during which the optical luminosity comes from the accretion disc, the optical light during quiescence essentially reveal a companion, which is almost always a late-type main-sequence star. Orbital periods can be deduced from the light curves; they can be as long as 15.1 hr, in the case of Cen X-4 (Chevalier et al., 1989), implying that the secondary does not fill its Roche lobe, or , more probably, that it is very slightly evolved. Shorter periods (7.8 hr for A0620-00, Mc Clintock and Remillard, 1986; 3.8 hr from the X-ray eclipses of EXO 0748-67, Parmar et al., 1986) are also found. There is however a selection effect against finding short periods, since short period systems have very low mass companions that are faint.

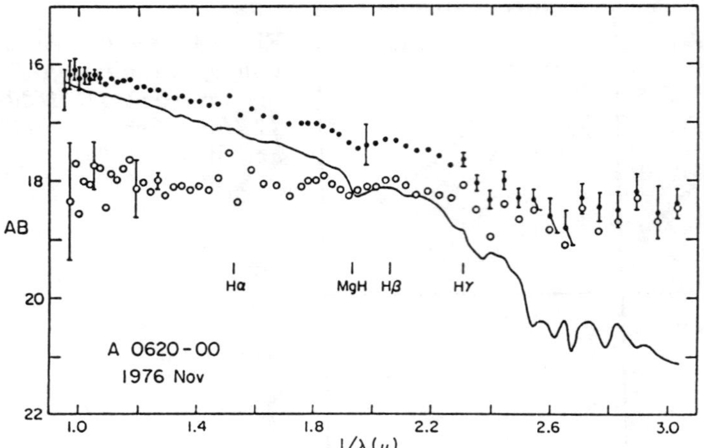

Fig. 4: *Optical spectrum of the SXT A0620-00 in quiescence (from Oke, 1977)*

Optical observations of relatively strong sources can provide radial velocity curves, from which one can obtain the mass function:

$$f(M) = \frac{M_1^3 \sin^3 i}{(M_1 + M_2)^2} \leq M_1 \tag{1}$$

which sets a lower limit on the primary mass; in the case of A0620–00, Mc Clintock and Remillard (1986) found $f(M) = 3.18 \pm 0.16$ M$_\odot$, implying that this source is a black hole.

Finally, a blue excess has been found in A0620–00 (see Fig. 4, Oke, 1977), which is attributed to the presence of an accretion disc; the accretion rate onto the compact object, is about 10^{-11} M$_\odot$ yr^{-1}, if the disc is steady. This does not contradict the X–ray upper limit of 10^{32} erg s^{-1} (de Kool, 1988), since, for those low accretion rates, the temperature at the inner edge of the disc is one tenth of a keV, and the bulk of the luminosity is absorbed by the interstellar medium.

3. Models

Soft X–ray transients have similar properties as dwarf novae, which are a subgroup of cataclysmic variables (CVs) (see van Paradijs and Verbunt, 1984 for a detailed comparison of these two classes of objects). CVs are semi–detached binary systems in which the primary is a white dwarf and the secondary a low–mass star. Dwarf novae outbursts last a few days and are recurrent on time scales of weeks to a few months, and have been much more thouroughly studied than SXTs, because they are much closer, and hence much brighter despite the fact that the compact object is a white dwarf.

Two classes of models have been proposed to explain the dwarf novae outbursts: a mass transfer instability, in which the instability arises in the ionisation zones of the secondary (Bath, 1975), and an accretion disc instability, in which the mass loss from the secondary is constant, but the accretion disc itself is thermally unstable (Osaki, 1974; Meyer and Meyer–Hofmeister, 1981; see Smak, 1984, and Verbunt,

1986 for reviews). Similarly, two types of models have been proposed for SXTs: a mass transfer instability model (Osaki, 1985; Hameury, King and Lasota, 1986) in which the mass transfer outburst is caused by the illumination of the secondary by X–rays from the primary, and an accretion disc instability (Canizzo, Wheeler and Ghosh, 1982, 1985; Lin and Taam, 1984).

4. Illumination instability model

During quiescence, the X–rays emitted by the compact object are hard, and can penetrate below the photosphere of the secondary, heating it up to a temperature T given by:

$$\sigma T^4 = \Phi_* + (1 - a_X)\Phi_X \tag{2}$$

where Φ_* and Φ_X are the intrinsic stellar flux and X–ray flux at the Lagrangian point L_1 respectively, and $a_X \sim 0.3$ the X–ray albedo. The X–ray flux is proportional to the accretion rate $\dot{M}$ according to:

$$\Phi_X = \frac{\eta \dot{M} c^2}{4\pi b^2} \tag{3}$$

$\eta \leq 0.1$ being the efficiency for hard X–ray production and b the distance of the compact star to the Lagrangian point L_1. This causes the atmosphere of the secondary to expand, and changes the mass overflow rate, as (Lubow and Shu, 1975):

$$\dot{M} = Q\rho_0 \exp\left[-(\Delta r/H)^2\right] \tag{4}$$

where $Q = 2 \times 10^{17}(T/10^4\mathrm{K})(P/1\ \mathrm{hr})^2$ is the cross section of the mass transfer throat at L_1, P the orbital period, H the pressure scale height, ρ_0 the maximum density at which X–ray can penetrate, and Δr the distance from that point to L_1.

4.1. THE LIMIT CYCLE

Combining Eqs. (2) to (4) yields a relation between $\dot{M}$ and Δr (Fig. 5). The curve has an Z–shape, and it can easily be shown that the central part is thermally unstable. However, the average mass transfer rate is imposed by the mechanisms driving the binary evolution (angular momentum losses dure to gravitational radiation or magnetic braking, nuclear evolution of the secondary, ... This situation leads to limit cycle behaviour if the imposed average mass transfer rate is in the unstable range $[\dot{M}_A, \dot{M}_B]$. The lower limit corresponds to $\Phi_* = \Phi_X$ at L_1; because of gravity effects, Φ_* at L_1 is about one third of the average stellar flux $< \Phi_* >$, and the average illumination effect is small:

$$\frac{L_X}{L_*} = \frac{1}{4}\frac{b^2}{D^2}\frac{\Phi_X}{< \Phi_* >} \sim (1-5) \times 10^{-2} \tag{5}$$

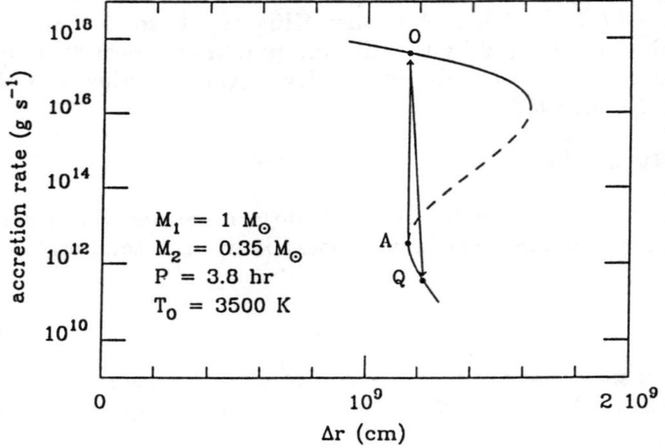

Fig. 5: *Accretion rate versus Δr assuming an efficiency η of 0.1. The dashed portion is thermally unstable (from Hameury, King and Lasota, 1990).*

where D is the orbital separation; the model does not contradict the observations that do not show any significant illumination effect in the ellipsoidal light curves.

The upper limit corresponds to the point at which the sensitivity of the mass transfer rate versus Δr is not large enough; this occurs for $(\Delta r/H)^2 = 3$, corresponding to $\dot{M} \sim 10^{16}$ g s^{-1}.

4.2. THE OUTBURST

If one is in a situation where $\dot{M}$ is smaller than what it should be on average, the secondary expansion or the system contraction will dominate and force Δr to decrease, until point A is reached. Mass transfer becomes then unstable, and would tend to reach the corresponding point O on the upper branch, in the thermal time of the atmosphere, which is exceedingly small. However, because of the presence of an accretion disc, the response is not instantaneous, and the rise time will be of the order of the viscous time of the disc. This disc will build up and thicken as the mass transfer rate increases, and would ultimately reach a steady state corresponding to point O, where its thickness h/r is:

$$\frac{h}{r} = 2.6 \times 10^{-2} \alpha^{-0.1} M_1^{-3/8} \left(\frac{\dot{M}}{10^{18}\text{g s}^{-1}}\right)^{3/20} \left(\frac{r}{10^{11}\text{cm}}\right)^{1/8} \tag{6}$$

where α is the standard viscous disc parameter. This would be enough to shield entirely the L_1 region, which has a radius of $(Q/\pi)^{1/2}$ and stop the instability. This occurs when h/r reaches ~ 0.03. By that time, the total amount of transferred matter is of the order of the disc mass, $\sim 10^{24}$ g.

These rough estimates are quite well confirmed by more accurate numerical calculations of the light curves produced in this model (Hameury, King and Lasota, 1988). These calculations, that take into account the structure of the illuminated mass loosing envelope during the outburst and the presence of an accretion disc, treated as described in Bath and Pringle (1981), produce light curves with rise times

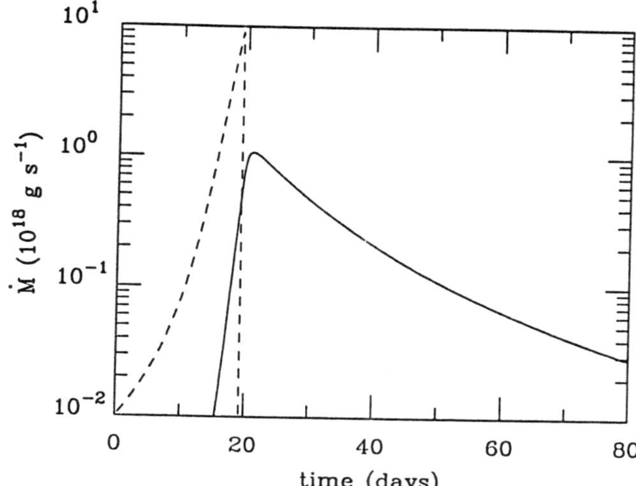

Fig. 6: *Evolution of the mass overflow rate from the secondary (dashed curve) and of the accretion rate onto the compact object (solid curve), for a K5 main sequence secondary, an orbital period of 7.8 hr, and* $\alpha = 1$. *(from Hameury, King and Lasota, 1988)*

of the order of days and decay times of weeks, for values of the viscosity parameter α of 0.1 – 1 (see Fig. 1). Because of the presence of the accretion disc, that delays accretion onto the compact object by a few days, the mass overflow outburst is in fact finished before the bolometric luminosity reaches its maximum. As the disc is very far from steady state during the rise, there are strong surface density gradients that cause the viscous time to be short; on the other hand, the disc empties during decline, and is much closer to steady state, so that the viscous time is much longer than during the rise.

4.3 SXTs IN QUIESCENCE

After the outburst, matter in the L_1 region is replaced by matter from below and away from L_1. Calculations of stellar models in Roche potential show that this fresh matter has less specific entropy than the accreted matter by an amount $\Delta S \sim 0.1\ k/\mu m_p$, so that Δr changes by 20% of a scale height, bringing the system to point Q in Fig. 5. Therefore, during quiescence, the mass overflow rate is close to the critical rate $\dot{M}_A$.

The X–ray illumination during quiescence causes the superadiabatic layers of the secondary to expand on a short time scale (one to ten years), so that the system will eventually reach point A where a new outburst will occur. The recurrence time is determined solely by the amount of mass transferred during one outburst and the average mass transfer rate, imposed by the binary evolution:

$$t_{rec} = \frac{\Delta M}{< \dot{M} >} \sim 1 - 10 \text{ yr} \tag{7}$$

again in agreement with observations.

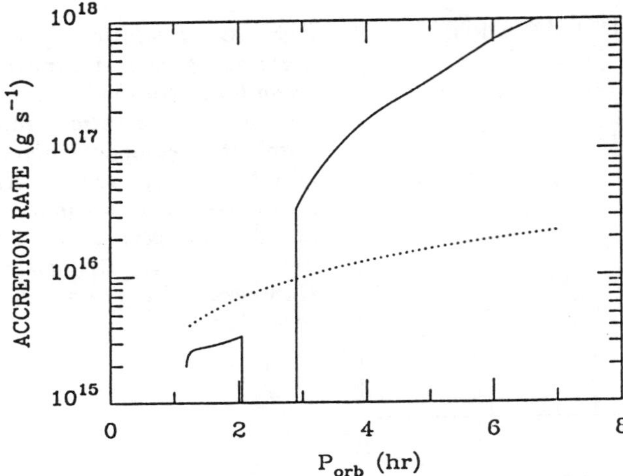

Fig. 7: *Average mass transfer rate (solid line) deduced from evolutionary calculations and critical accretion rate $\dot{M}_B$ (dotted line) versus period for low-mass X-ray binaries with unevolved companions (from Hameury, King and Lasota, 1987).*

This average mass transfer rate, that can be computed using binary evolution models, can be compared with $\dot{M}_B$, the critical rate above which the mass transfer is stable (Hameury, King and Lasota, 1987). Figure 6 shows that only systems with periods less than 3 hr are expected to be transients; this seems to contradict the fact that all measured orbital periods in SXTs are longer than 3 hr. However, in those systems, the observed average mass transfer rate is $< \dot{M} >\simeq 10^{16}$ g s^{-1}, far below the value expected from evolutionary calculations; the observed $< \dot{M} >$ is in fact below $\dot{M}_B$. Similar fluctuations of the same order are observed in cataclysmic variables, which are very similar to LMXBs, the only difference being that the compact object is a white dwarf; nevertheless, these systems must establish secular mean values of $\dot{M}$ which are the same for all systems (see Ritter, 1985). It is also worth noting that *all* systems below 2 hr should be transients; in these conditions, their orbital period would be very difficult to determine if they are not eclipsing systems; this would explain the lack of observed LMXBs with periods in the range 1 – 3 hr.

Finally, SXTs with long orbital periods ($P > 10$ hr) require evolved companions. The same criteria as for systems containing unevolved secondaries can be applied to determine if they are transients. The condition that $\dot{M}$ must be in the range $[\dot{M}_A, \dot{M}_B]$ is quite easy to fulfill; however, the photosphere of subgiant stars is more extended than that of main sequence stars, and the condition that X–rays must penetrate below the photosphere is much more stringent. It is found (Hameury, King and Lasota, 1987) that only systems with companions K0IV – K2IV and orbital periods 9 – 70 hr can be transients, thus accounting for Cen X–4.

4.4 COMPARISON WITH OBSERVATIONS

The illumination instability model accounts for the main observed characteristics of SXTs: the light curves (rise-time, duration), the recurrence time, the mean accretion rate, the X–ray observations in quiescence. However, the optical observation of A0620–00 during quiescence imply a mass transfer rate of the order of 10^{15} g s^{-1}, far exceeding $\dot{M}_A$. As mentioned above, this system most probably contains a black hole of about 7 $M_\odot$ or more. A standard (one temperature) disc would have a spectrum too soft for the model to work, and no boundary layer is present around a black hole. It is possible however that for $\dot{M} > \dot{M}_{TT} \simeq 10^{16}\alpha^2 M/M_\odot$ g s^{-1} (White and Lightman, 1989) a two temperature inner disc is formed. This inner disc emits hard X–rays, as observed during decline (see fig. 3) where the hard component dominates. For $\dot{M} < \dot{M}_{TT}$, the system behaves as if it were unilluminated; when $\dot{M}$ reaches $\dot{M}_{TT}$, the system is unstable, and goes into outburst. Hameury, King and Lasota (1990) have shown that this modification to the neutron star SXT model is sufficient to account for the observations of A0620–00.

This model also predicts that there should be a cutoff in the distribution of LMXBs at an X–ray luminosity of about 10^{36} erg s^{-1}, sources with lower luminosity being transients (King, Hameury, and Lasota, 1987); this agrees with the results of the *EINSTEIN* survey of globular clusters, which found no source in the range $10^{34.5} < L_X < 10^{36}$ erg s^{-1} (Hertz and Grindlay, 1983); the low luminosity sources in clusters are probably transients in quiescence (Verbunt, van Paradijs, and Elson, 1984), in which the secondary is evolved.

There are two critical tests of this model: first, the X–ray flux in quiescence should be of the order of 10^{32} erg s^{-1} for short period systems (main sequence companions), and higher (up to 10^{34} erg s^{-1} if the secondary is evolved; the positive detection of Cen X–4 in quiescence is very encouraging from this point of view. This model also predicts that the X–ray emission should lag behind the optical light curve by a few days, as the disc first builds up before accretion onto the compact object occurs.

5. Disc instability model

5.1. DWARF NOVAE MODELS

This model is essentially based on the model for dwarf novae. The partial ionization regions of the accretion disc are found to be thermally unstable, since, in these regions, the effective temperature versus surface density curve ($T_{eff}(\Sigma)$) has a negative slope (Fig. 8). This instability is related to the presence of the ion H^-, which strongly increases the opacity in the regions where it is present, causing a steep temperature dependence of the opacity. This results in a limit cycle: on the cool branch, the local $\dot{M}$ is lower than the steady value; matter accumulates until Σ reaches the turning point A; it then jumps to the upper branch in a thermal time; $\dot{M}$ is now higher than the steady value, so that it moves back to point B where it returns to the cool branch. This instability is found for a given range of effective temperatures, corresponding to a range in mass accretion rates $[\dot{M}_A, \dot{M}_B]$ given by:

$$(2-10) \times \dot{M}_A \sim \dot{M}_B \sim 10^{16} \left(\frac{r}{10^{10}\ \mathrm{cm}}\right)^{2.6} \left(\frac{M_1}{M_\odot}\right)^{0.87} \ \mathrm{g\ s^{1}} \quad (8)$$

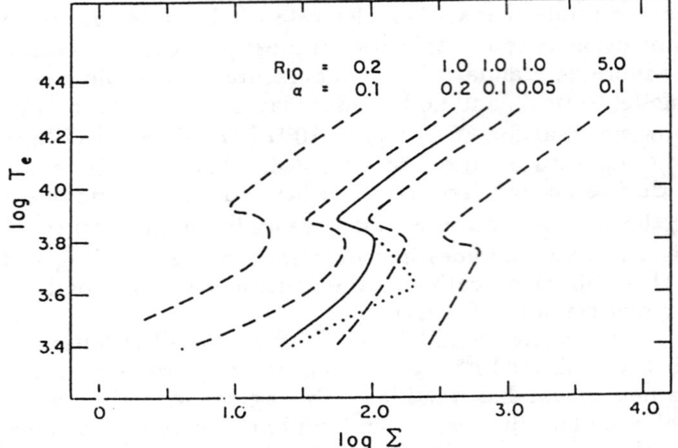

Fig. 8: *effective temperature as a function of surface density in a disc, for various of the radial distance R_{10} (in 10^{10} cm) and of the viscosity parameter α (from Smak, 1984)*

The disc is stable everywhere if either $\dot{M} < \dot{M}_A$ or $\dot{M} > \dot{M}_B$ everywhere. As the radius of the inner edge of the disc is essentially that of the compact object, the first condition can never be fulfilled; therefore, the stability condition is:

$$\dot{M} > 10^{16} \left(\frac{r_{max}}{10^{10}\,\mathrm{cm}}\right)^{2.6} \left(\frac{M_1}{M_\odot}\right)^{0.87} \mathrm{g\ s}^{1} \qquad (9)$$

where r_{max} is the radius of the outer edge of the disc. This agrees with the fact that only CVs with low mass transfer rates show dwarf novae outbursts. However, the existence of local instabilities does not nececessarily imply large amplitude global instabilities. In fact, time–dependent disc calculations show that these local instabilities result only in flickering if α is kept constant (Smak, 1984); the main reason is that matter returns to the cool branch on time scales shorter than the time on which the instability would propagate away. One requires that α has a larger value α_h on the hot branch than the value α_c on the cool branch for the limit cycles to be synchronous throughout the disc. Values of α_h and α_c of 0.1 and 0.02 respectively can reproduce the observed time scales.

5.2. APPLICATION TO SXTs

The application of this model to soft X–ray transients is straightforward. The model should however account for the fact that the recurrence time is much longer in SXTs than in dwarf novae. A first important difference between these two classes of objects is the mass of the compact object: it is ~ 1.4 $M_\odot$ in SXTs, while it is only ~ 0.7 $_\odot$ in dwarf novae. As a consequence, the accretion disc is larger in SXTs, and hence the amount of mass transferred during one outburst; the recurrence time is thus longer in SXTs. In addition, the radius of the inner edge of the disc is of the order of 10^6 cm in LMXBs, much smaller than in dwarf novae; this also leads to an increase of the recurrence time. Those two facts together are however not sufficient to account for the very long recurrence time scales observed in some transient

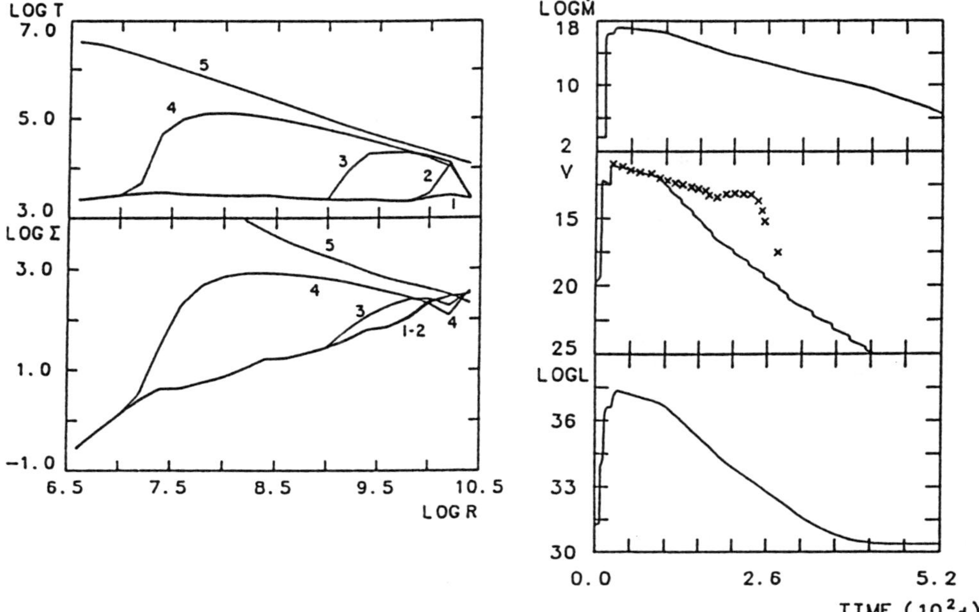

Fig. 9: *left: evolution of the surface density and effective temperature as a function of radius. The elapsed times since the begening of the outburst are – 100, 0, 2, 5 and 18 days.* $\alpha = 100(h/r)^{1.5}$, *where h is the scale height of the disc and r the radius; this corresponds to* $\alpha_h = 0.003$ *and 0.4 at the inner and outer edge of the disc, and to* $\alpha_c = 0.0002$ *and 0.016 at those same positions.*
Right panel: evolution of the mass accretion rate, visual magnitude and bolometric luminosity of the outburst (From Mineshige and Wheeler, 1990).

sources, such as A0620–00; one must then assume that the viscosity parameter α is very depressed during the low states. Numerical calculations (Mineshige and Wheeler, 1989, see Fig. 9) show that long recurrence times can be obtained this way. For the case shown in Fig. 9, the recurrence time is in the range 4 – 16 yr, i.e. corresponding to Cen X–4. Longer recurrence times can be obtained by increasing the amplitude of α_h/α_c.

Another important difference between SXTs and dwarf novae is the presence of a strong X–ray flux, in particular during outburst, that significantly alters the thermal structure of the disc; it has been argued thet the dwarf–nova type instability might be suppressed in low–mass X–ray binaries (Meyer and Meyer–Hofmeister, 1984). Tuchman, Mineshige and Wheeler (1990) have shown that only the convective portion of the disc will be significantly altered; X–ray illumination does not inhibit the local instability provided that

$$T_{irr} = (L_{irr}/\sigma)^{1/4} < 10^4 \; K \tag{10}$$

The amplitude of the instability is however quite significantly changed (see Fig.

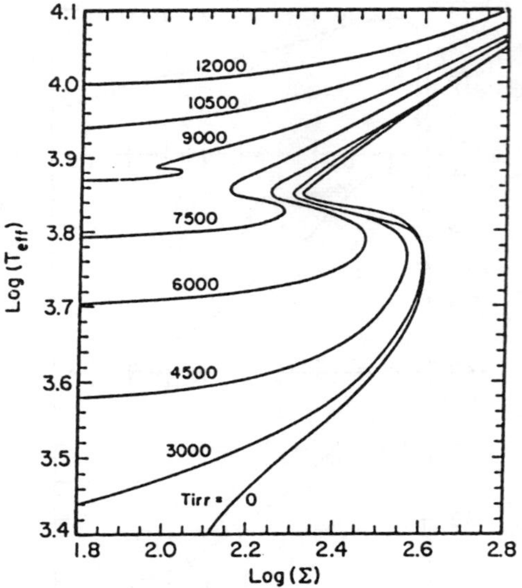

Fig. 10: *Thermal equilibrium curves in the* $(\log T_{eff}, \log \Sigma)$ *plane for various irradiation temperatures. The radius is* 3×10^{10} *cm, and* α *= 0.1 (from Tuchman, Mineshige and Wheeler, 1990).*

10), but, because X–ray illumination is negligible in quiescence, the disc is still locally unstable. Mineshige, Tuchman, and Wheeler (1990) have shown that the non–linear evolution of the instability leads to light curves with a plateau in the decay, due to the delay between $\dot{M}$ at a given distance and the X–ray illumination. Their calculation was however restricted to an annulus instead of a full disc; recent disc calculations by Meyer and Meyer–Hofmeister (1990) showed that, if a constant value of α is used, one finds a diffusive instability instead of dwarf novae outbursts. There is clearly a need for more detailed calculations that allow α to vary.

5.3. COMPARISON WITH OBSERVATIONS

The characteristic timescales (rise–time, decay, recurrence), as well as the shape of the light curves produced by the disc–instability model are in good agreement with observations. There is however a difficulty in explaining the observed luminosities in the low state, both in optical and in X–ray. In the case of Cen X–4, the X–ray luminosity is of the order of 10^{33} erg s^{-1} in quiescence, while the disc instability model in its current form would predict a bolometric luminosity less than 10^{32} erg s^{-1}; in addition, most of the bolometric luminosity comes from the outer parts of the disc, and the accretion rate onto the compact object, that is responsible for the observed X–ray emission, is less than 10^{-21} $M_\odot$ yr^{-1}. There is a similar problem in A0620–00, which is much less severe, as this source is not detected in X–rays, and it is possible to argue that the observed light comes from the hot spot. In order to solve this difficulty, one would like to either increase the value of α_c, or to assume that the whole disc is not on the cool branch; both possibilities would lead to a decrease of the recurrence time. It seems that it would be difficult, although not necessarily impossible, to reconcile long recurrence times and significant X–ray and optical luminosity in quiescence.

6. Conclusion

As can be seen, none of the two proposed models can be definitevely ruled out for the time being; the defenders of the disk instability model argue that the mass transfer instability model does not take into account the fact that in their low state, the disc would be unstable, while the defenders of the mass transfer instability argue that the disc instability model cannot account both for the observed fluxes in the low state and the recurrence times.

Observations of soft X–ray transients in the low state are clearly required, both in X–rays and in optical, as they can provide decisive tests of the models.

References

Bath, G. T. 1975, *Monthly Notices Roy. Astron. Soc.* **194**, 967

Bath, G. T., Pringle, J. E. 1981, *Monthly Notices Roy. Astron. Soc.* **194**, 967

Bradt, H. D. V., McClintock, J. E. 1983, *Ann. Rev. Astron. Astrophys.* **21**, 13

Canizzo, J. K., Wheeler, J. C., Ghosh, P. 1982, in *Pulsations in classical and cataclysmic variables stars*, ed. J. P. Cox and C. J. Hansen, University of Colorado Press, p. 13

Canizzo, J. K., Wheeler, J. C., Ghosh, P. 1985, in *Cataclysmic Variables and Low Mass X–ray Binaries*, eds. J. Patterson and D. Q. Lamb, Reidel, p. 307

Chevalier, C., Ilovaisky, S. A., van Paradijs, J., Pedersen, H., van der Klis, M. 1988, *Astron. Astrophys.* **210**, 114

Cominsky, L.R., London, R.A., Klein, R.I.: 1987, *Astrophys. J.* **315**, 162

Guilbert, P. W., Fabian, A. G. 1982, *Nature* **296**, 226

Hameury, J. M., King, A. R., Lasota, J. P. 1986, *Astron. Astrophys.* **162**, 71

Hameury, J. M., King, A. R., Lasota, J. P. 1987, *Astron. Astrophys.* **171**, 140

Hameury, J. M., King, A. R., Lasota, J. P. 1988, *Astron. Astrophys.* **192**, 187

Hameury, J. M., King, A. R., Lasota, J. P. 1990, *Astrophys. J.* **353**, 585

Hertz, P., Grindlay, J. E. 1983, *Astrophys. J.* **275**, 105

Jones, C., Forman, W., Tananbaum, H., Turner, M. J. L. 1976, *Astrophys. J.* **210**, L9

Kaluznienski, L. J., Holt, S. S., Boldt, E. A., Serlemitsos, P. J., Eadie, G., Pounds, K. A., Ricketts, M. J., Watson, M. 1975, *Astrophys. J.* **201**, L121

Kazanas, D. 1986, *Astron. Astrophys.* **166**, L19

King, A. R., Hameury, J. M., Lasota, J. P. 1987, in *variability of galactic and extragalactic X–ray sources*, ed. A. Treves, p. 193

King, A. R., Lasota, J. P. 1987, *Astron. Astrophys.* **185**, 155

de Kool, M. 1988, *Astrophys. J.* **334**, 336

Lin, D. N. C., Taam, R. E. 1984, in *High Energy Transients in Astrophysics*, ed. S. E. Woosley, AIP Conf. Proc. No. 115, p. 83

Long, K. S., Helfand, D. S., Grabelsky, D. A. 1981, *Astrophys. J.* **248**, 925

Lubow, S. H., Shu, F. H. 1975, *Astrophys. J.* **198**, 383

Matilsky, T., Bradt, H. V., Buff, J., Clark, G. W., Jernigan, J. G., Joss, P. C., Laufer, B., McClintock, J., Zubrod, D. 1976, *Astrophys. J. Letters* **210**, L127

McClintock, J. E., Remillard, R. A. *Astrophys. J.* **308**, 110
Melia, F., Rappaport, S., Joss, P.C. 1986, *Astrophys. J. Letters* **305**, L51
Meyer, F, Meyer–Hofmeister, E. 1981, *Astron. Astrophys.* **104**, L10
Meyer, F, Meyer–Hofmeister, E. 1984, *Astron. Astrophys.* **140**, L35
Meyer, F, Meyer–Hofmeister, E. 1990, *Astron. Astrophys.*, submitted
Mineshige, S., Wheeler, C. J. 1989, *Astrophys. J.* **343**, 241
Mineshige, S., Tuchman, Y., Wheeler, J. C.. 1990, *Astrophys. J.* **359**, 176
Oke, J. B. 1977, *Astrophys. J.* **217**, 187
Osaki, Y. 1974, *Publ. Astron. Soc. Japan* **26**, 429
Osaki, Y. 1985, *Astron. Astrophys.* **144**, 369
Paczynski, B. 1969, *Acta Astron.* **19**, 1
Parmar, A. N., Stella, L., White, N. E. 1986, *Astrophys. J.* **304**, 664
Parmar, A. N., White, N. E., Giommi, P., Gottwald, M. 1986, *Astrophys. J.* **308**, 199
van Paradijs, J., Verbunt, F. 1984, in *High Energy Transients in Astrophysics*, ed. S. E. Woosley, AIP Conf. Proc. No. 115, p. 49
van Paradijs, J., Verbunt, F., Shafer, R. A., Arnaud, K. A. 1987, *Astron. Astrophys.* **182**, 47
Ritter, H. 1985, *Astron. Astrophys.* **145**, 227
Smak, J., 1984, *Publ. Astron. Soc. Pacific* **96**, 5
Tuchman, Y., Mineshige, S., Wheeler, J. C. 1990, *Astrophys. J.* **359**, 164
Verbunt, F. 1986, in *The physics of accretion onto compact objects*, ed. K. O. Mason, M. G. Watson and N. E. White, Springer, p. 59
Verbunt, F., van Paradijs, J., Elson, R. 1984, *Monthly Notices Roy. Astron. Soc.* **210**, 899
White, N.E., Kaluzienski, J.L., Swank, J.H. 1984, in *High Energy Transients in Astrophysics*, ed. S.E. Woosley, AIP Conf. Proc. No. 115, p. 31
White, T. R., Lightman, A. P. 1989, *Astrophys. J.* **340**, 1024

SLOWLY ACCRETING NEUTRON STARS ("SANS")

J.P. Lasota
UPR–176, C.N.R.S.; D.A.R.C.
Observatoire de Paris, Section de Meudon
92195 Meudon Cedex, France

...sans teeth, sans eyes, sans taste, sans everything,
W. Shakespeare ("As You Like It", II, vii)

ABSTRACT. We discuss some physical processes occuring at the surface of a magnetized, slowly accreting neutron star (hereafter called SANS). We show that thermonuclear runaway in the accreted matter may produce events very similar to γ–ray bursts

1. Introduction

I will discuss physical processes which may occur at the surface of an old (age $> 10^9$ yr, strongly magnetized ($B \gtrsim 10^{11}$ G), slowly accreting neutron star (SANS). There should be $\sim 10^9$ old neutron stars in the Galaxy with a density within ~ 0.5 kpc of the Sun equal to $n_{NS} \approx 0.0014$ (pc^{-3}) (Paczyński 1990; see however Hartmann, Epstein and Woosley 1990, which obtain lower estimates). According to Paczyński (1990) the half–density half–thickness of the old neutron stars distribution is $z_{1/2} \approx 193$ pc. Until recently the assumption that old neutron stars would maintain large magnetic fields was considered to be extravagant, since it was thought that the magnetic field of radio–pulsars undergoes an exponential decay on a time–scale $\sim 10^6$ yr. One of the main conclusions of the present conference was that the evidence of the exponential decay is in the end rather weak and that the discovery of cyclotron features in spectra of some γ–ray bursts (Murakami *et al.* 1988) is widely considered to be the evidence for the presence of strong magnetic fields in old neutron stars. Although in the present article I will argue that at least some of the γ–ray bursts may originate at the surface of old, strongly magnetized SANS, I would like to point out that the conclusive evidence connecting γ–ray burtst to *old* neutron stars is still to be found (see e.g. van Paradijs 1989). One should also note that some recent models of evolution of magnetic moments

J. Ventura and D. Pines (eds.), Neutron Stars: Theory and Observation, 515–527.

of neutron stars predict that strong fields will still be present even after 10^{10} years (Chanmugam and Sang 1990; Romani 1990).

2. Accretion from the interstellar medium

A neutron star moving through the insterstellar medium (ISM) will accrete matter at the rate (Alcock and Illarionov 1980):

$$\dot{M} = 7.25 \times 10^{-16} \left(\frac{20\,\mathrm{km/s}}{v} \right)^3 \left(\frac{M_*}{\mathrm{M}_\odot} \right)^2 n_1 \ \mathrm{M}_\odot \mathrm{yr}^{-1}, \tag{1}$$

where $n_1 = \left(\frac{n}{1\ \mathrm{cm}^{-3}} \right)$, v is the neutron star velocity with respect to the ISM and M_* is the mass of the star. It is assumed in eq. (1) that the ISM is ionized by the accreting neutron star. Eq. (1) does not result from the classical Strömgren analysis of an H II region but takes into account the fact that the neutron star leaves the classical Strömgren zone before it can be ionized.

For typical values of $n_1 \lesssim 1$ only very slowly moving neutron stars are of interest (see below). Let us note however that slow neutron stars could be an important fraction of the total population of old neutron stars (Paczyński 1990).

In the case of a rotating magnetized neutron star accretion may be prevented by the "propeller mechanism" if the spinning rate is too high. A simple method of estimating the required spin periods can be obtained by requiring the corotation radius $R_{co} = (GM_* P_{\mathrm{spin}}/4\pi)^{1/3}$ to be larger than then the magnetospheric radius at which the ram pressure of the accreting matter is equal to the magnetic pressure. For a dipolar magnetic field one obtains:

$$P_{\mathrm{spin}} \gtrsim 800\ \mu_{30}^{6/7} \dot{M}_{15}^{-3/7} \left(\frac{M_*}{\mathrm{M}_\odot} \right)^{-9/14} \quad \mathrm{s}, \tag{2}$$

where $\dot{M}_{15} = \left(\frac{\dot{M}}{10^{-15} \mathrm{M}_\odot\ \mathrm{yr}^{-1}} \right)$. It is not clear however that an isolated neutron star can be spun–down in a reasonable time to spin periods as long as those required by eq. (2) if its initial spin period was as fast as that found in radio pulsars.

A straightforward application of the propeller mechanism formula of Illarionov and Sunyaev (1975) gives spin–down time $\sim 10^8 - 10^9$ yr (Blaes *et al.* 1989). A more detailed treatment of pulsar spin–down history (Davies and Pringle 1981) suggests that in reality the spin–down time of an isolated neutron star acreting from the ISM could be longer than the Hubble time. Clearly this problem requires a more detailed and carefull treatment.

3. Do SANS exist?

The accretion luminosity ($L_{acc} = GM\dot{M}/R$) of a slowly accreting neutron star is

$$L_{acc} = 8.47 \times 10^{30}\ \dot{M}_{15} \frac{M_*}{M_\odot} R_6^{-1}\ \ \text{erg s}^{-1}. \tag{3}$$

The effective temperature $T_{eff} = (L_{acc}/\sigma A)^{1/4}$, where A is the accreting surface, is equal to

$$T_{eff} = 1.97 \times 10^6 \dot{M}_{15}^{1/4} \left(\frac{M_*}{M_\odot}\right)^{1/4} R_6^{-1/4} A_1^{1/4}\ \ \text{K}, \tag{4}$$

with $A_1 = (A/1\,\text{km}^2)$. If most of the accretion energy is thermalised the emitted flux should appear in soft X-rays. Fluxes of $\sim 10^{-12} - 10^{-13}$ erg cm^{-2} s^{-1} corresponding to distances of a few hundred parsecs could have been observed during deep X-ray surveys by EXOSAT (see §§10 and 11).

Recently the X-ray source H0253+193 has become a serious canditate for an example of a SANS (Bhatt 1990). The position of this HEAO-A-2 hard X-ray source coincides with the peak in the CO map of the molecular cloud MBM 12, which is identical to the dark cloud L 1457. The recent discovery by GINGA (Takano *et al.* 1989) of X-ray pulsations (P_{spin} = 206 s) from H0253+193 suggested that this object could be an isolated magnetized neutron star accreting from ISM. Its average 2–20 keV X-ray flux is 6×10^{-11} erg cm^{-2} s^{-1}, which at the distance of 65 pc to the molecular cloud L 1457 gives an X-ray luminosity of $\sim 3 \times 10^{31}$ erg s^{-1}, which could correspond to an accretion rate $\sim 10^{-15} M_\odot$ yr^{-1} (compare with eq. (1)).

Bhatt (1990) shows that H0253+193 can be neither a T Tauri star nor a Be/X-ray binary imbedded in the molecular cloud. He concludes that the X-ray source is an isolated neutron star spun down by a propeller type mechanism in the interstellar medium. This attractive suggestion should however be subject to caution: for the moment observations do not preclude the possibility that H0253+193 is an intermediate polar (IP) a close binary in which a spinning magnetized white dwarf accretes matter from a Roche-lobe filling low-mass companion. IPs are hard X-ray sources (see e.g. Norton and Watson 1989) with spectra showing low-energy cutoffs corresponding to photoelectric absorption by an equivalent hydrogen column density $N_H \sim 10^{23}$ cm^{-2}. For comparison the column density to H0253+193 is estimated to be $2 \times 10^{22} < N_H < 1.5 \times 10^{23}$ cm^{-2}. The 2–20 keV fluxes observed in IPs are $\gtrsim 10^{-11}$ erg cm^{-2} s^{-1}. Finally a range of spin periods from $\sim$ 300 to 4000 s is observed (or even from 33 s if one includes the weak X-ray sources: the so-called DQ Her systems). At first sight therefore H0253+193 has almost all the typical characteristics of an IP. However no evidence of binarity has yet been found. Roche filling companions in IPs should be late type (K or M) dwarfs and in the case of the system GK Per ($P_{orb} \approx$ 48h, $P_{spin} \approx$ 351 s) the companion is a K2 subgiant.

In the case of H0253+193 there is no IRAS (12 μm band) point source at the X–ray source position but the 12 μm flux of the presumed low–mass companion would be $< 10^{-2}$ mJy. On the other hand at a distance of few hundred parsecs its 2.2 μm magnitude should be K= 13 – 19, which could be detected by infrared imaging (Lasota, Boisson and Hameury 1990).

4. The structure of the accreted layer

I shall consider accretion rates $\gtrsim 10^{-15}\,\mathrm{M}_\odot\,\mathrm{yr}^{-1}$ which, as we shall see below, lead to a thermonuclear flash in the matter accreted on the surface of the SANS. Lower accretion rates have been considered by Blaes *et al.* (1990) (see also Blaes 1990). Slightly higher have been discussed by Stan Woosley and collaborators (for a review see e.g. Hartmann and Woosley 1988).

The accreting matter is channelled by the strong magnetic fields and accumulates on a small part of the neutron star surface (~ 1 km^2). The thermal structure of the accreted layer is determined by the balance between heating due to reactions of the *pp* cycle, the not β–limited CNO cycle and electron capture and conductive cooling towards the neutron star interior.

The observed accretion flux should be close to $F_X \approx 5 \times 10^{-12} d_{100}^{-2}\,\mathrm{erg\,cm^{-2}\,s^{-1}}$, where the distance $d_{100} = (d/100\,\mathrm{pc})$. The effective temperature is given by eq. (4) but the exact form of the emitted spectrum corresponding to slow accretion in a strong magnetic field is not known.

The magnetic field acts as a stiff container and is not distorted on a large scale during the accumulation of matter prior to the thermonuclear runaway. Indeed, the deformation of the magnetic field to material stresses is

$$\theta \sim \frac{\delta B_r}{B_z} \approx \frac{h}{R_{\mathrm{cap}}} \times \frac{P_{\mathrm{gas}}}{P_{\mathrm{mag}}} \qquad (5).$$

where B_z is the vertical component of the magnetic field, δB_r is its radial distortion, h is the height of the accreted layer, R_{cap} its radius, P_{gas} and P_{mag} are the gas and the magnetic pressures respectively. Numerical calculations (Hameury *et al.* 1983a) show that for parameters of interest the confinement is effective as long as the magnetic field is greater than 3×10^{11} G. On the other hand the diffusion time accross the magnetic field is much longer than the characteristic accretion time.

The gravitational settling of helium and other heavy elements is very efficient (Hameury, Bonazzola and Heyvaerts 1983) as the characteristic time for element separation is much shorter than the accretion time. As a result an almost pure hydrogen layer (with a metalicity $Z \sim 10^{-4}$) is formed above a layer in which helium and metals do not have time to separate. The two layers are separated by a very thin transition zone, which is few centimeters thick and which contains a high metal content ($Z \sim 0.1$).

5. The hydrogen flash

When the mass of the hydrogen layer reaches the critical value $M_{\rm crit} \sim 2.5 \times 10^{20}$ g km^{-2}, hydrogen burning becomes unstable, if the accretion rate is larger than a critical value $\dot{M}_{\rm crit} \sim 10^{-15} {\rm M}_\odot\,{\rm km}^{-2}\,{\rm yr}^{-1}$. A more precise value of $\dot{M}_{\rm crit}$ was given by Hameury, Bonazzola and Heyvaerts (1983):

$$\dot{M}_{crit} = 0.1(T_6 - T_{\rm i})\frac{M_*}{R_6^2}\ {\rm g\ cm^{-2}\ s^{-1}} \tag{6}$$

where the temperature in units of 10^6 K is:

$$T_6 = 7.311\ \log\left\{55.28\ \frac{M_*^2}{R_6^2}\left[1+\frac{0.049(T_6^2 - T_{\rm i}^2)T_6^{-4/3}}{1-0.02057T_6}\beta\frac{M_*}{R_6^2}\right]^{-1}\right\}. \tag{7}$$

$T_{\rm i}$ is the temperature of the neutron star interior in units of 10^6 K, M_* is the mass of the neutron star in solar units, $R_6 = (R/10^6\ {\rm cm})$ is the neutron star radius and $\beta \sim 1$. Two typical $\dot{M}_{\rm crit}(T_{\rm i})$ curves are shown on Fig. 1.

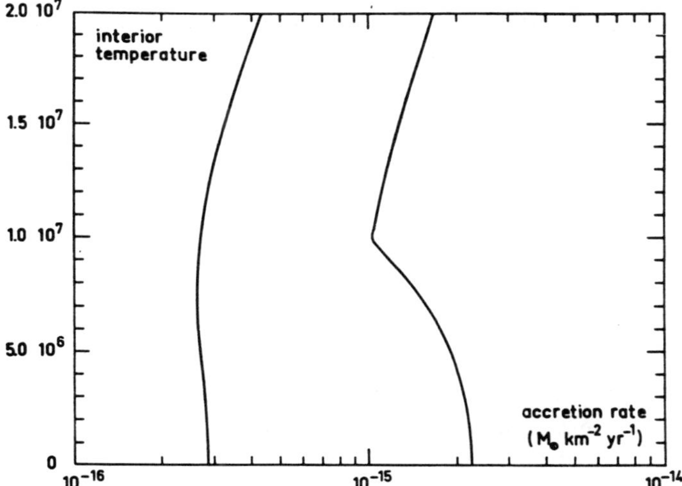

Fig. 1–Stability regions in the plane $\dot{M}, T_{\rm i}$. The region to the right of a given curve represent acrretion regimes leading to a hydrogen flash. The curve to the right is for $M_* = 1, R_6 = 1$; the left curve is for $M_* = 0.7, R_6 = 1.5$. Eqs. (6) and (7) are the result of numerical fits.

The characteristic accretion time between thermonuclear runaways is therefore $\sim$ 100 yr.

The hydrogen flash corresponds to a transition to the β–limited CNO cycle. In the pure hydrogen layer the metallicity is very low which makes long both the heating time $t_{\mathbf{heat}} \sim 2 \times 10^4 T_8$ s and the burning time $t_{\mathbf{burn}} \sim 10^6$ s, where T_8 is the layer's temperature in units of 10^8 K. In the thin transition layer the metallicity is high. In this conditions the burning time is about 10^3 s, assuming a steady state β–limited CNO cycle. The local heating time is $t_{\mathbf{heat}} = 20\ T_8$ s. The actual heating time will be longer by a factor $\sim$ 2 due to cooling by heat conduction. The energy flux released in this layer is $\Phi_{\mathbf{trans}} \gtrsim 10^{34}$ erg km^{-2} s^{-1} and the corresponding temperature $T_{\mathrm{eff}} \gtrsim 1$ keV. The timescale, energy flux and effective temperature are in a very good agreement with the properties of the X–ray precursor observed by *GINGA* in one γ–ray burst (Murakami *et al.* 1991). However this observation cannot be treated as real confirmation of the thermonuclear model before calculations of the burning front propagation are performed.

6. The helium flash

When the temperature in the layer reaches about 1.5×10^8 K helium ignites. The nature of the explosion which follows is difficult to describe. In principle, it can be either a detonation (propagation by a shock wave) or a deflagration (subsonic burning). In the case of detonation the burning front moves at about the sound speed, the rise time can be as short as $\sim 10^{-7}$ s and whole layer can be ignited in about 10^{-4} s. The case of deflagration is complicated by the presence of a strong magnetic field which influences both the convection and electron conduction. The rise time can be estimated as $\sim 10^{-2}$ s (Fryxell and Woosley 1982), no reliable estimate exists for the ignition time of the whole layer (note that estimates in Bonazzola *et al.* 1984 deal with the ignition time in the vertical direction).

Both modes of helium ignition give the same final result: a 9 m layer of ^{56}Ni at a temperature of 5.5×10^9 K. The energy released is $\sim 10^{38}$ erg km^{-2}. The hot nickel layer containing the energy released by explosion is buried deep ($\sim$ 60 m) below the surface of the cool hydrogen layer. At first sight one could think that this energy released in a thermal form will have to be emitted by a purely thermal mechanism. However as we shall see in the next paragraph, the hot *magnetized* nickel layer covered by cold hydrogen is a powerful electromechanical machine which can produce non–thermal energy transport and emission.

7. Energy transport

Cooling by heat conduction is in this conditions very inefficient, the characteristic timescale being $\tau \sim 10^3 - 10^4$ s (Bonazzola *et al.* 1984). The temperature gradient in the nickel layer is superadiabatic but steady convection is inhibited by the strong magnetic field. One can show however that oscillatory ("overstable") convection will set in (Bonazzola *et al.* 1984). This process both shakes the magnetic field lines and produces acoustic waves. The oscillations frequency is $\nu_c \sim \frac{v_A}{2d} \sim 10^6\ s^{-1}$, where v_A is the Alfvén speed and d is the thickness of the layer ($d \sim 10^2$ cm). The lateral dimension of the convective cells is $\lambda \sim 10^{-2} d \sim 1$ cm. The magnetic field perturbations due to oscillatory convective motions is then $\frac{\delta B}{B} \approx \frac{\lambda}{d} \sim 10^{-2}$. This magnetic field perturbations will propagate along the the fieldlines (as "near zone" Alfvén waves) carrying away an energy flux equal to:

$$F_A = \frac{(\delta B)^2}{8\pi} v_A \approx 10^{37} B_{12}^3 \ \mathrm{erg\,km^{-2}\ s^{-1}} \tag{8}$$

where B_{12} is the magnetic field measured in units of 10^{12} G. This is remarkably close (for a field of few 10^{12} G) to fluxes that a γ–ray bursts would emit at a distance of few hundred parsecs (see e.g. Higdon and Lingenfelter 1990). The energy contained in Alfvén waves has to be transformed into high energy photons (see §8).

The efficiency of conversion of heat released by nuclear reactions into mechanical waves is at most:

$$\eta < 1 - \frac{T_2}{T_1} = 0.95 \tag{9}$$

where T_1 and T_2 are the temperatures of nickel and hydrogen layers respectively. In principle, there should be no problems in converting most of the released thermal energy into waves. On the other hand some numerical simulations (see e.g. Proctor and Weiss 1982) showed that in the non–linear regime energy transport by oscillatory convection is not very efficient. The so–called Nusselt number, equal to the ratio of the total (convective plus conductive, but without the contribution of the Alfvén waves) flux to the conductive flux is close to unity and often oscillatory convection develops into steady convection. These calculations were performed in a different physical regime from the one considered here. First they assume steady energy production so there is time for the steady convection to develop. Second, and more important, the strongest magnetic fields considered were many orders of magnitude lower than 10^{12} G, so no "interpolation" to present case is possible.

One should note that the energy flux given by eq. (8) is two to three orders of magnitude larger than the conductive flux. One can therefore expect that most of the cooling is due to Alfvén and acoustic waves.

8. Coronal heating

Acoustic waves and Alfvén waves release their energy in different regions of the neutron star surface:

- Acoustic waves steepen into shocks ans dissipate at large optical depths. This produces a thermal component of the radiated spectrum.

- Alfvén waves propagate to optically thin regions will dissipate their energy via magnetic field reconnection.

The magnetic field perturbations will be able to reconnect if the characteristic time for reconnection is shorter than the characteristic time of Alfvén waves oscilattions:

$$\tau_{rec} < \nu_{\mathrm{A}}^{-1}. \tag{10}$$

It was shown by Hameury *et al.* (1985) that inequality (10) is satisfied in regions where the Thomson optical depth is low:

$$\tau_{\mathrm{Th}} \lesssim 2 \times 10^{-2} \left(\frac{F_A}{10^{37}} \right)^{1/2} B_{12}^{-3/4}. \tag{11}$$

The reconnection of magnetic field perturbations creates an electric field parallel to the unperturbed fieldlines. This electric field accelerates electrons to relativistic energies:

$$E_{\parallel} \approx 100 \; T_8^{3/4} B_{12}^{5/4} \tau_{\mathrm{Th}}^{-1} \;\mathrm{keV} \tag{12}$$

along the field lines. The initial electron distribution thus formed is unstable since the electron drift velocity is much larger than the ion thermal velocity. As a result a broad "thermal–like" distribution is formed. Note that this distribution is one–dimensional, along the magnetic field. The charateristic time of those instabilities is much shorter than reconnection time:

$$t_{\mathrm{ins}} \sim \omega_{\mathrm{p}}^{-1} \sim 10^{-16} \left(\frac{\rho}{1 \mathrm{gcm}^{-3}} \right)^{-1/2} \mathrm{s} \; \ll t_{\mathrm{rec}}, \tag{13}$$

so that a steady state can be reached. As a result a hot corona is formed.

9. Coronal cooling

Hameury *et al.* 1985 (see also Hameury and Lasota 1989) were the first to point out that resonant scattering of photospheric photons in the hot strongly magnetized corona is an efficient cooling mechanism which could explain energy spectra observed in γ–ray bursts. In particular, the so–called "X–ray paucity" problem invoked in the context of γ–ray bursts emission (see Epstein, 1986), ceases to be a problem at all: emitted spectra contain a very low fraction of soft photons. Indeed due to this resonant scattering (in some parts of New Mexico and California known under the name of "CUSP" – see e.g. Fenimore 1991) a photon with an initial energy $h\nu$ is upscattered to an energy of $\frac{(h\nu_B)^2}{h\nu}$ for a transition to the first Landau level. One has $h\nu_B = 11.6 B_{12}$ keV so a 5 keV photon gets upscattered in one collision to ~ 500 keV. Transitions involving the second Landau level may produce two photons of twice that energy. The scattered photons are highly collimated: they are contained in a narrow cone of a solid angle $d\Omega \approx (E_B/E)^2$. This collimated primary photon emission is obviously difficult to observe. Half of it is directed upwards and escapes freely; the other half is directed downwards and is intercepted by the photosphere. As a result an electron population with high transverse momenta is created and synchronous radiation is emitted in addition to the thermal photospheric emission.

The electrons in the corona can be continously energized and they are able to reach an equilibrium distribution because the following inequalities are satisfied:

$$t_{\rm inst} \ll t_{\rm cool} \ll t_{\rm rec}, \tag{14}$$

where $t_{\rm cool}$ is the cooling time in the corona. The schematic view of the emitting region is shown on Fig. 2.

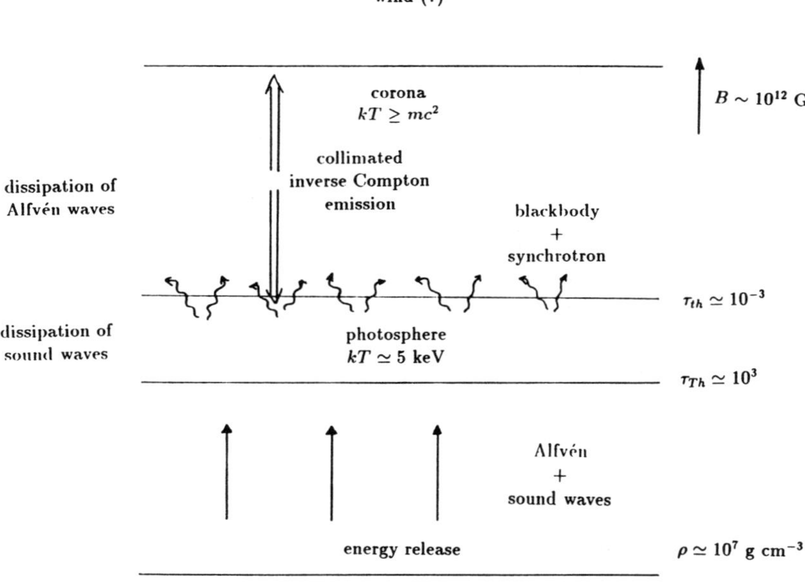

Fig. 2. A schematic view of the emitting region at the surface of a strongly magnetized neutron star (not to scale). The vertical extent of the emitting region is in reality much smaller than its diameter.

10. Emitted spectra

Hameury *et al.* (1985) and Hameury and Lasota (1989) used Monte–Carlo simulations to treat the transfer in the corona (see Dermer 1989 for a semi–analytical treatment). These calculations take into account both the resonant and Compton scattering. The injected spectrum is calculated consistently with the backscattered flux. There are four free parameters:

- the magnetic field strength: B,
- the effective temperature of the photosphere: T_{BB},
- the ratio of the emitted flux to the Alfvén flux: F/F_A
- the Thomson optical depth of the corona as compared to the maximum value given by eq.(11): $\tau/\tau_{\rm max}$.

The last two parameters are not really free, they are rather unknown since we are unable to obtain them by reliable calculations. The assumed range of parameters was the following: $B \sim 1 - 5 \times 10^{12}$ G, $T_{BB} \sim 2 - 5$ keV, $F/F_A \sim 0.1 - 1$, $\tau/\tau_{\rm max} \sim 0.1 - 1$. An example of calculated energy spectrum is given on Fig. 3. It is a spectrum similar to those observed in some γ–ray bursts.

One sees therefore that a SANS can be a source of γ-ray bursts: the thermonuclear bursts occuring their surface have the right energy, the right recurrence rate, the correct timescales and the emitted spectra are similar to those observed.

Observations of γ–ray burst error boxes found no point sources but with one exception the upper limits on the accretion rates are consistent with the thermonuclear model (Boer *et al.* 1988, 1990). As we shall see observations of error boxes of recent γ–ray bursts are probably not a serious test of the thermonuclear model since accretion could be stopped for some 10 yr by the burst itself.

11. The influence of bursting SANS on the surrounding medium

It is possible that observations of γ–ray error boxes could fail to detect an X–ray emission because the accretion process on to the neutron star had been stopped by the burst itself. In order to estimate the possibility of this process and its characteristic timescales let us consider a simple model in which $\sim 0.1\%$ of the energy released in the burst is contained in a supersonic wind. Although there is no direct evidence for the existence of winds connected with γ–ray bursts, they are a natural ingredient of the thermonuclear model by analogy with winds from thermonuclear explosions associated with X–ray bursts (for a model of X–ray burst wind see Paczyński and Anderson 1986; see also Hartmann and Woosley 1988).

The wind energy would then be $E_{\rm wind} \approx 10^{35}$ erg. Supersonic expansion of the wind into the ISM leads to the formation of a strong shock in a sort of miniature version of a supernova shell expansion. By analogy the shock propagation can be represented by the Sedov solution (see e.g. Landau and Lifshitz 1959) describing the selfsimilar motion of the shock–front into the surrounding medium.

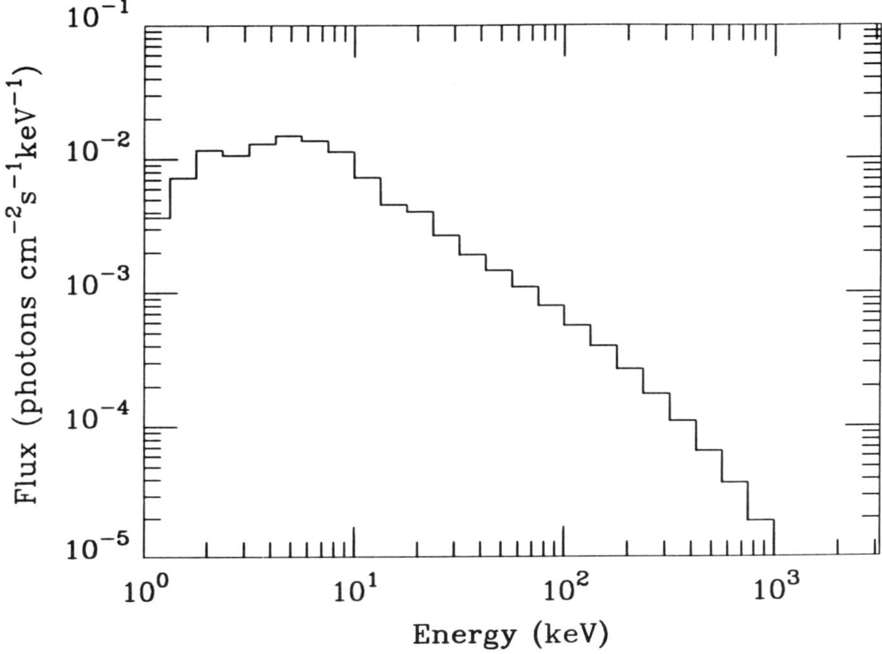

Fig. 3. A typical energy spectrum from a burst at the surface of a highly magnetized neutron star according to the model by Hameury *et al.* (1985).

The evolution of the shock radius $R_{\rm sh}$ is given by:

$$R_{\rm sh} = 7.6 \times 10^{14} E_{35}^{1/5} n_1^{-1/5} t_{\rm y}^{2/5} \text{ cm}, \tag{15}$$

where $t_{\rm y}$ is time measured in years, and $E_{35} = \dfrac{E_{\rm wind}}{10^{35}\ \text{erg}}$. The shock speed is described by:

$$\dot{R}_{\rm sh} = 97 E_{35}^{1/5} n_1^{-1/5} t_{\rm y}^{-3/5} \quad \text{km s}^{-1}. \tag{16}$$

The post-shock temperature is

$$T_{\rm sh} = 2.2 \times 10^5 \ E_{35}^{2/5} n_1^{-2/5} t_{\rm y}^{-6/5} \quad \text{K} \tag{17}$$

The cooling time at $T \approx 10^5$ K is rather long compared with the timescale of interest ($t_{\rm cool} \approx 600 n_1^{-1}$ yr) so that post-shock cooling can be neglected in the description of shock propagation. Finally the post-shock velocity of the compressed ISM is equal to $v \approx 70\ t_{\rm y}^{-3/5}$ km s^{-1}.

Before the burst the flux due to accretion on to the neutron star ionizes the ISM to a distance (Alcock and Illarionov 1980):

$$R_{\rm i} \approx 3.7 \times 10^{14} \left(\frac{N_{\rm i}}{10^{39}} \right) \left(\frac{20 \ {\rm kms}^{-1}}{v} \right) \ {\rm cm} \tag{1}$$

where $N_{\rm i}$ is the number of ionizing photons. For $\dot{M} \gtrsim 10^{-15}\,{\rm M}_{\odot}\ {\rm yr}^{-1}$ the ionization radius is bigger than the accretion radius. The shock will reach $R_{\rm i}$ in about one year and die, since the non–ionized medium is collisionless. The neutron star itself will move through a distance $\sim R_{\rm i}$ in about 6 years. According to this simple model accretion would cease for at least a few years after the γ–ray burst. It is not therefore surprising that EINSTEIN and EXOSAT were unable to detect X–ray emission from γ–ray burst error boxes.

The study of γ–ray burst error boxes determined by SIGMA or GRO in the ROSAT X–ray survey could, in principle, impose constraints on accretion rates *prior* to burst and provide a compelling test of models of γ–ray bursts from accreting neutron stars (Boer 1990).

12. Conclusions

Slowly accreting, highly magnetized neutron stars moving through the interstellar medium, may from time to time produce high energy events which are similar to γ–ray bursts.

Acknowledgments I thank Jean–Marie Hameury for useful discussions and comments. Andrew King contributed the quotation from Shakespeare.

References

Alcock, C., and Illarionov, A. 1980, *Astrophys. J.*, **235**, 541.

Bhatt, H. C. 1990, *Astron. Astrophys.*, **232**, 367.

Blaes, O. 1991, Taos Meeting on *Gamma–ray Burts*, in press.

Blaes, O., Blanford, R. .D., Goldreich, P., and Madau, P. 1989, *Astrophys. J.*, **343**, 839.

Blaes, O., Blanford, R. .D., Madau, P., and Koonin, S. 1990, *Astrophys. J.*, **363**, 612.

Boer, M. 1990, private communication.

Boer, M. *et al.* 1988, *Astron. Astrophys.*, **202**, 117.

Boer, M., Gottardi, M., Hurley, K., and Pizzichini G. 1988, *Astron. Astrophys.*, , submitted.

Bonazzola, S., Hameury, J. .M., Heyvaerts, J. and Lasota, J .P. 1984, *Astron. Astrophys.*, **136**, 89.

Chanmugam, G. and Sang, Y. 1990, *Astrophys. J.*, , in press.

Davies, R. E., and Pringle, J. E. 1981, *Mon. Not. R. astr. Soc.*, **196**, 209.

Fenimore,E.E. 1991, Taos Meeting on *Gamma-ray Bursts*, in press.

Fryxell, B.A., and Woosley, S.E. 1982, *Astrophys. J.*, **261**, 332.

Hameury, J. M., and Lasota, J. P. 1989, *Astron. Astrophys.*, **211**, L18.

Hameury, J. M., and Lasota, J. P. 1986, in *Gamma-Ray Bursts, AIP Conference Proceedings 141*, ed. E. P, Liang and V. Petrosian (New York: American Institute of Physics.

Hameury, J. M., Bonazzola, S., and Heyvaerts, J. 1983, *Astron. Astrophys.*, **121**, 259.

Hameury, J. M., Bonazzola, S., Heyvaerts, J. and Lasota, J.P. 1983b, *Astron. Astrophys.*, **128**, 369.

Hameury, J. M., Lasota, J. P., Bonazzola, S. and Heyvaerts, J., 1985, *Astrophys. J.*, **293**, 56.

Hartmann, D., and Woosley, S. E. 1988, in *Multiwavelength Astrophysics*, ed. F. Cordova (Cambridge: Cambridge University Press), p. 189

Hartmann, D., Epstein, R. I. and Wossley, S. E., 1990, *Astrophys. J.*, **348**, 625.

Higdon, J. C., and Lingenfelter, R. E. 1990, *Ann. Rev. Astr. Astrophys.*,**28**,401.

Illarionov, A. and Sunyaev, R.A. 1975, *Astron. Astrophys.*, **39**, 18.

Landau, L. D., and Lifshitz, E. M. 1959, *Fluid Mechanics* (London: Pergamon Press)

Lasota, J. P., Boisson, C., and Hameury, J. M. 1990, in preparation.

Murakami, T. *et al.* 1991, Taos Meeting on *Gamma-ray Bursts*, in press.

Murakami, T. *et al.* 1988, *Nature*, **335**, 234.

Norton, A. J., and Watson, M. G. 1989, *Astron. Astrophys.*, **237**, 853.

Paczyński B. 1990, *Astrophys. J.*, **348**, 485.

Paczyński B., and Anderson N. 1986, *Astrophys. J.*, **302**, 1.

Proctor, M.R.E. and Weiss, N.O. 1982, *Rep. Prog. Phys.*, **45**, 1317.

Romani, R.W. 1990, *Nature*, **347**, 741.

Takano, S. *et al.* 1989, *IAU Circular* 4745

van Paradijs, J. 1989, *Mon. Not. R. astr. Soc.*, **238**, 45p.

5. BURSTS AND TeV EMISSION

X-RAY BURSTS

D. Q. LAMB
Department of Astronomy and Astrophysics
and Enrico Fermi Institute
University of Chicago
5640 S. Ellis Avenue
Chicago, IL 60637, U.S.A.

ABSTRACT. We summarize the observed features of X-ray bursts and the key elements of the thermonuclear flash model of this phenomenon. We then elucidate the full range of possible burst behavior using a global analysis of the stability of nuclear burning on accreting neutron stars.

1. Introduction

Compelling evidence exists that X-ray bursts are due to thermonuclear flashes on the surface of accreting neutron stars. Much of this evidence comes from comparison of observations with theoretical calculations of the properties of individual bursts (see Lewin and Joss 1981, 1983; Taam 1985 for references). Global analysis of the stability of nuclear burning on accreting neutron stars [Fujimoto, Hanawa, and Miyaji 1981 (hereafter FHM); Hanawa and Fujimoto 1982 (hereafter HF); Paczyński 1983; Fushiki and Lamb 1987b (hereafter FL)] is an alternative theoretical approach which can map the full range of possible X-ray burst behavior.

Following the latter approach, FL showed that, given a composition for the accreting matter and a model of the neutron star, the nuclear burning behavior depends on the mass accretion rate per unit area $\dot{\sigma}$, the temperature T at the bottom of the accreted matter, and the column density of accreted matter σ, and is thus intrinsically a *three-dimensional* problem. They therefore introduced the concepts of a nuclear fuel surface S and an ignition surface $S^{\rm ign}$ in $(\dot{\sigma}, T, \sigma)$-space. The former is the column density σ at which the nuclear fuel is exhausted, or at which the sum of the nuclear burning luminosity $L_{\rm nuc}$ and compressional heating luminosity $L_{\rm g}$ equals the radiated luminosity L of the neutron star, whichever comes first. The latter is the column density σ at which the nuclear burning is thermally unstable; it is of crucial significance in determining the properties of X-ray bursts.

Certain cuts taken through the ignition surface ($\dot{\sigma}$, T, σ)-space correspond to particular physical situations, such as the "steady-state" case discussed by FHM, HF, and FL, and the thermal equilibrium and compressional heating cases discussed by FL (see below). Observed bursting sources need not lie on any of these cuts, but they do lie on the ignition surfaces $S^{\rm ign}$ in $(\dot{\sigma}, T, \sigma)$-space. As FL emphasized, such a global analysis thus provides a complete framework for interpreting the observations.

In this lecture, we summarize the observed features of X-ray bursts and the key elements of the thermonuclear flash model of this phenomenon. We then elucidate the full range of possible burst behavior using FL's global analysis of the stability of nuclear burning on accreting neutron stars.

J. Ventura and D. Pines (eds.), Neutron Stars: Theory and Observation, 531–544.

2. Summary of Observations

X-ray bursts were discovered in 1975 independently by Grindlay *et al.* (1976) and Belian, Conner, and Evans (1976). Lewin, and co-workers subsequently studied the X-ray burst phenomenon intensively using the SAS-3 satellite (for comprehensive reviews of X-ray burst observations, see Lewin and Joss 1981, 1983).

X-ray bursts have rise times $\lesssim 1$ sec, durations $\sim 3 - 1000$ sec, and recur on a time scale $t_r \sim 10^3 - 10^6$ sec. Burst sources have persistent X-ray luminosities $L_o \sim 10^{36} - 10^{38}$ erg s^{-1}. The bursts themselves have luminosities $L_b \sim 10^{39}$ erg s^{-1} and total energies $E_b \sim 10^{39} - 10^{40}$ erg s^{-1}. The ratio α of the time-averaged burst luminosity E_b/t_r to the persistent luminosity L_o is $\sim 20 - 300$.

Approximately 50 burst sources are now known; Figure 1 shows the locations in galactic coordinates of 32 of these. The burst sources are concentrated toward the galactic center, and there are at least 10 in globular clusters.

Figure 2 shows the time histories in different X-ray energy ranges of bursts from five different sources. These time histories are typical, and show the distinct spectral softening evident in the tail of the burst. Swank *et al.* (1977) showed that the spectral evolution of a particular long ($\sim$ 600 sec duration) burst was consistent with the cooling of a blackbody of ≈ 10 km radius, assuming a distance of ≈ 10 kpc (see Figure 3). This provided an important piece of evidence that X-ray bursts come from accreting neutron stars.

A number of X-ray bursts are double-peaked (see Figure 4). Such a double-peaked profile is now understood as being due to photospheric expansion. The radius of the neutron star never changes during the burst, but the luminosity during the most luminous phase of the burst may be sufficiently close to the Eddington luminosity to drive a strong wind off the surface of the neutron star. The wind is optically thick to scattering of X-rays. Consequently, the X-ray photospheric radius temporarily increases to a value of $\sim$ 100 km or more, while the burst luminosity remains near the Eddington value. The temperature of the photosphere therefore briefly falls. This moves flux out of the X-ray and into the extreme ultraviolet energy band, producing the observed double-peaked X-ray time history.

3. Thermonuclear Flash Model

In the thermonuclear flash model, matter accretes onto a neutron star from a binary companion. The accreted matter is usually taken to consist of H, He, and very small amounts of heavier elements, since these binary systems are thought to be old and to belong to Population II. As the accreted matter builds up, the density and temperature at the bottom of the accreted layer increases until eventually nuclear burning of the accreted H occurs. If the temperature in the neutron star envelope is sufficiently low, the H burns steadily into He in a thin shell via electron captures or the pycnonuclear p-p chain. At higher densities, the He burns steadily into C in a thin shell via the pycnonuclear triple-α reaction. At still higher densities, a sequence of electron capture reactions occurs, transforming C into heavy elements (see, e.g. Lamb and Lamb 1977, 1978).

At higher temperatures in the neutron star envelope, H burning occurs via the thermonuclear CNO cycle, and He burning occurs via the triple-α reaction. Such burning is thermally unstable (Hansen and Van Horn 1975), and leads to thermonuclear H and He flashes, which appear as X-ray bursts (Woosley and Taam 1976; Maraschi and Cavaliere 1977; Joss 1977, 1978; Lamb and Lamb 1977, 1978).

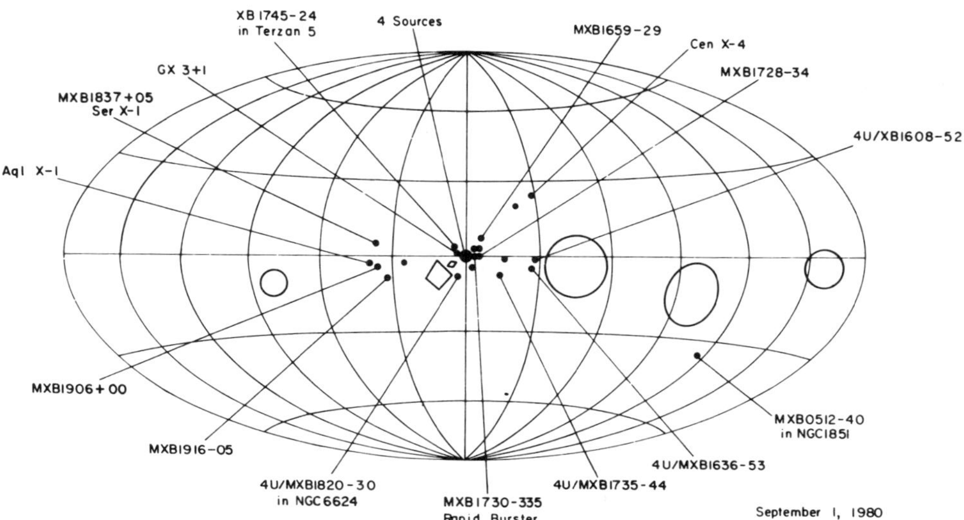

Figure 1. Sky map (galactic coordinates) of 32 X-ray burst sources. (From Lewin and Joss 1981.)

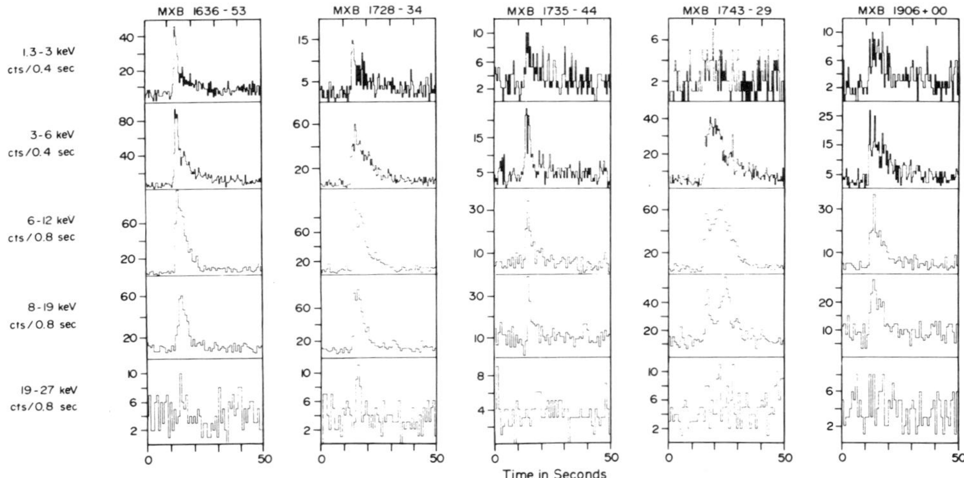

Figure 2. Profiles of X-ray bursts from five different sources. Note that the gradual decay (the burst "tail") persists longer at low energies than at high energies, indicating cooling of the burst emission region. (From Lewin and Joss 1981.)

Figure 3. Average spectra, in three time intervals, of a very long (~ 600 s) X-ray burst from 1724-30. Time zero is near the burst onset. The solid curves show the best fits to blackbody spectra. The values for kT are ~ 0.9 keV (0-20s), ~2.3 keV (40-70 s) and ~ 1.2 keV (150-440 s). Assuming a spherical emitting surface and a source distance of 10 kpc, the best-fit blackbody radii are ~ 100 km during th first 20 s of the burst and ~ 15 km thereafter. (From Swank *et al.* 1977.)

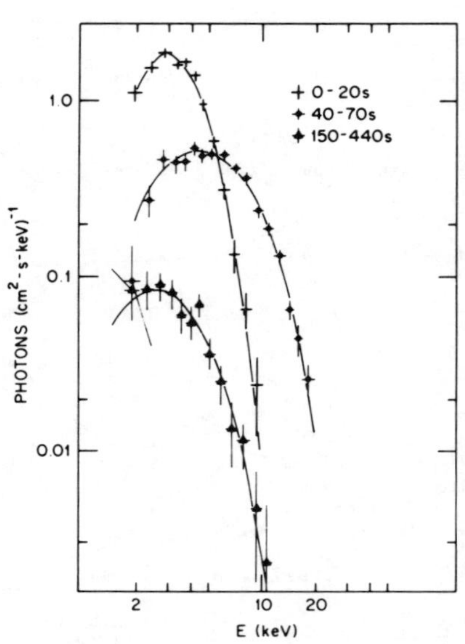

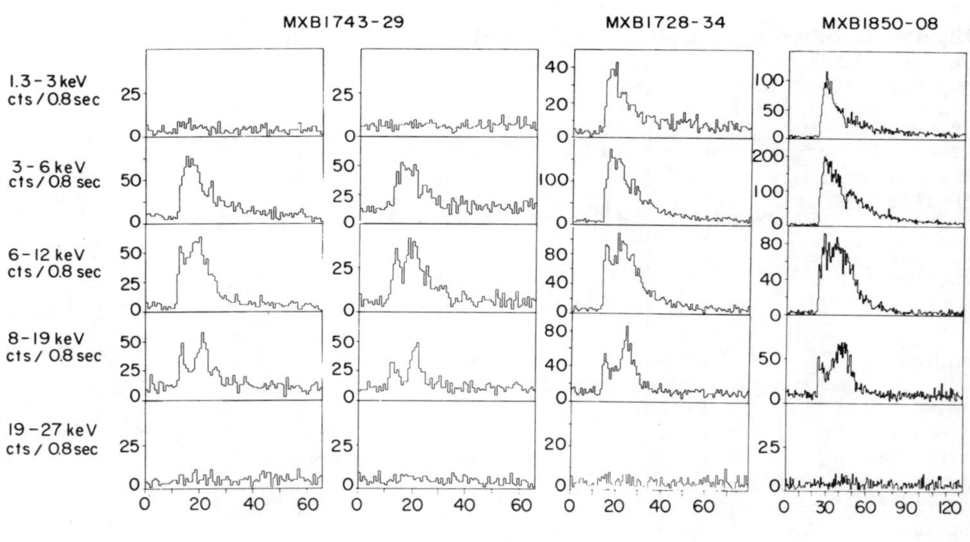

Figure 4. Profiles of X-ray bursts showing double peaks from three sources. The spectrum from 1743-29 (located near the galactic center) is cut off below 3 keV due to interstellar absorption. (From Lewin and Joss 1981.)

The energy released in an X-ray burst is

$$E_b \sim Q \frac{4\pi R^2 H_P \rho}{m_H} \sim 2 - 8 \times 10^{39} \text{erg} , \tag{3.1}$$

where $R = 10$ km is the radius of the neutron star, and $H_P = 10^2$ cm and $\rho = 10^6$ g cm^{-3} are the pressure scale height and density at the bottom of the accreted layer. The quantity Q is the energy per nucleon released by the nuclear burning ($Q \approx 7$ MeV nucleon^{-1} and $Q \approx 1.5$ MeV nucleon^{-1} for H and He burning, respectively). The above range of values for E_b agrees with observation.

The persistent luminosity due to accretion is

$$L_{\rm acc} = \frac{GM\dot{M}}{R} , \tag{3.2}$$

where M is the mass of the neutron star and $\dot{M}$ is the mass accretion rate. The time-averaged burst luminosity is approximately

$$L_b \approx Q \frac{\dot{M}}{m_H} . \tag{3.3}$$

Taking the ratio of equations (3.2) and (3.3) gives

$$\alpha \approx 30 - 120 (\frac{M}{M_\odot}) (\frac{R}{10\text{km}})^{-1} . \tag{3.4}$$

This range of values for α agrees with observation.

Figure 5 shows the behavior of the temperature, density, and composition in the neutron star envelope through a burst. Figure 6 shows an example burst time history found from detailed numerical calculations for the case of a combined hydrogen-helium thermonuclear flash. The burst luminosities, temperatures, and rise and decay time scales found from calculation agree with observation. For comprehensive reviews of X-ray burst theory, see Lewin and Joss (1981, 1983) and Taam (1985).

4. Global Analysis of Nuclear Burning

We now describe a global analysis of nuclear burning on accreting neutron stars, following FL.

Because matter accreting onto a neutron star forms an envelope whose thickness is less than 1% of the stellar radius, FL approximate it as plane parallel. General relativity then enters the Tolman-Oppenheimer-Volkoff equations of stellar structure only as a constant numerical factor (Thorne 1977). With this approximation, integrating the equation of hydrostatic equilibrium gives the relation $P = g_s \sigma$, where P is the pressure and g_s is the local gravitational accceleration at the stellar surface. FL neglect the inward heat flux due to the accretion luminosity, since the surface temperature $T(0)$ resulting from it is in all cases much smaller than the envelope temperatures of interest. Introducing the photon flux $F = L/4\pi R^2$, where L is the photon luminosity and R is the stellar radius, FL write the energy conservation and heat transport equations in integral form as

$$F(\sigma) = F(0) - \int_0^\sigma (\epsilon_{\rm nuc} + \epsilon_{\rm g} - \epsilon_\nu)\, d\sigma' , \tag{4.1}$$

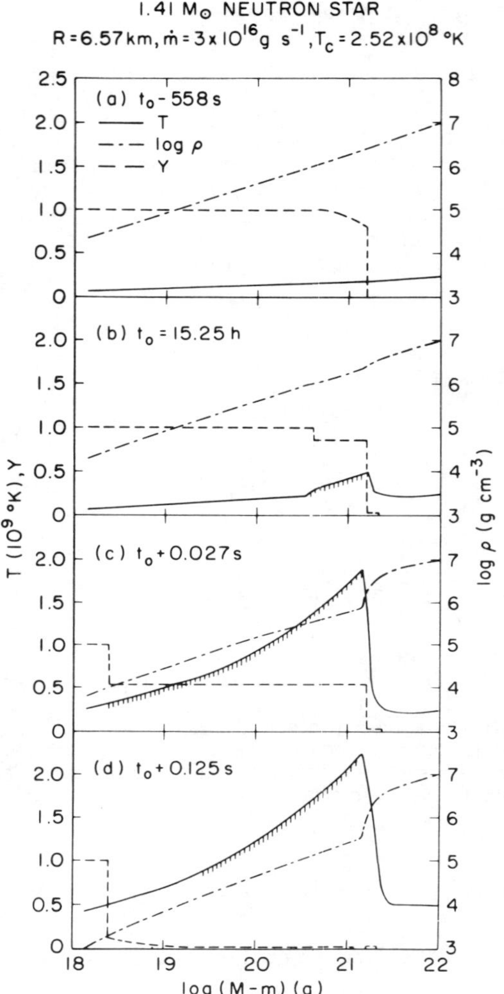

Figure 5. Structure of the envelope of an accreting neutron star prior to and during a pure He-burning flash, which begins near time t_o. M is the total mass of the neutron star and $m(r)$ is the mass enclosed within a sphere of radius r; thus $(M-m)$ is the mass of the envelope above radius r. T is the temperature (left-hand scale), ρ is the density (right-hand scale), and Y is the fractional abundance of He by mass (left-hand scale). The hatched regions indicate the extent of the convection zone generated by the flash. (a) Just prior to the flash; (b) near the start of the flash; (c) when 50% of the helium fuel has been consumed; near the time of peak shell-burning temperature and peak surface luminosity. (From Joss 1978.)

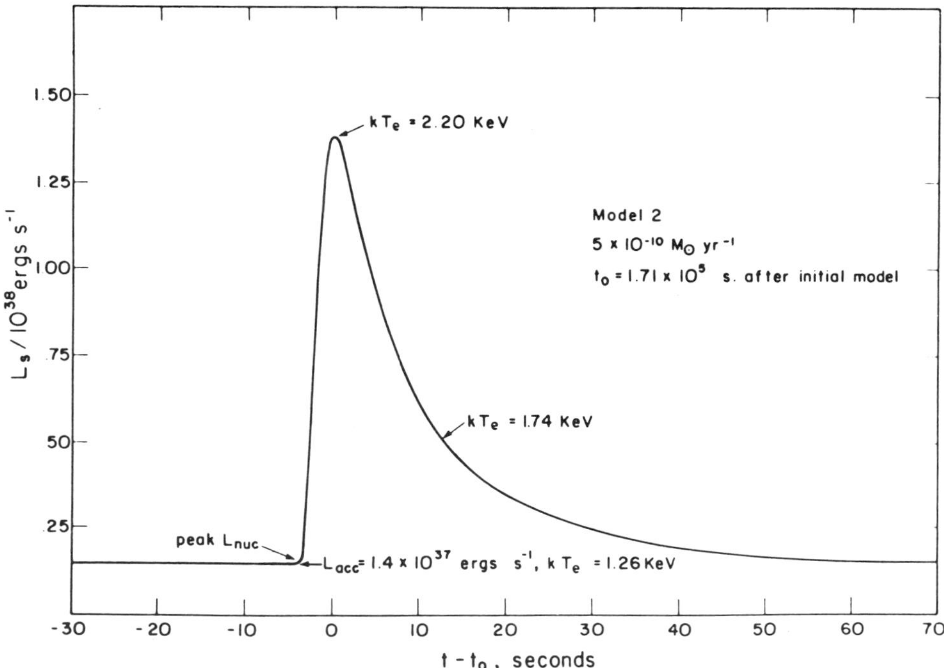

Figure 6 The behavior of the surface luminosity L_s following a thermonuclear flash. The accretion rate is as indicated. The accreted matter is assumed to have an initial heavy-element abundance $Z = 0.004$ by mass. Time $t = 0$ is at the start of accretion onto the neutron star surface, $L_{\rm acc}$ is the level of the persistent accretion-driven lumnosity, and the effective temperature T_e is indicated at a few points in time. The flash is driven by He burning, but the entrainment of H into the flash has a strong effect upon its properties. (From Taam 1980.)

$$T(\sigma) = T(0) + \frac{3}{4ac}\int_0^{\sigma} \frac{\kappa F(\sigma')}{T^3} d\sigma' \; . \tag{4.2}$$

where $\epsilon_{\rm nuc}$, $\epsilon_{\rm g}$ and ϵ_ν are the energy generation rates per unit mass due to nuclear reactions, to compressional heating (i.e., the increase in the entropy of the accreted matter due to gravitational contraction), and to neutrino energy losses, respectively, and κ is the opacity per unit mass.

At the surface of the neutron star, the accretion rate per unit area $\dot{\sigma}$ is given by $\dot{\sigma} = \dot{M}/4\pi R^2$, where $\dot{M}$ is the local mass accretion rate. The integral equations for the hydrogen abundance X and the helium abundance Y are

$$X(\sigma) = X(0) - \frac{1}{\dot{\sigma} q_{\rm H}}\int_0^{\sigma} \epsilon_{\rm H} d\sigma' \; , \tag{4.3}$$

$$Y(\sigma) = Y(0) - \frac{1}{\dot{\sigma} q_{\rm He}}\int_0^{\sigma} \epsilon_{\rm He} d\sigma' \; , \tag{4.4}$$

where $\epsilon_{\rm H}$ and $\epsilon_{\rm He}$ are the nuclear energy generation rates per unit mass due to hydrogen burning (the pp-chain and the CNO-cycle) and helium burning (the triple-α

reaction), respectively. The quantities $q_{\rm H}$ and $q_{\rm He}$ are the Q-values per unit mass for hydrogen and helium burning, respectively. Although FL wrote equations (4.1) – (4.4) in integral form for simplicity of presentation, they calculated the full density, temperature, and composition structure of the envelope in order to solve them (cf. Paczyński 1983).

As discussed by HF, the equation $P = g_s \sigma$ and equations (4.1) – (4.4) depend on only the two parameters g_s and $\dot{\sigma}$, instead of on the stellar mass M, the stellar radius R, and the mass accretion rate $\dot{M}$. This formulation is also convenient for treating situations in which accretion occurs only over a small fraction of the stellar surface, as in nuclear flash models of γ-ray bursts. FL fix g_s and leave $\dot{\sigma}$ as a free parameter.

4.1 NUCLEAR FUEL SURFACES

Often the bottom of the accreted hydrogen-rich layer has been taken to be $X(\sigma) \approx 0$, corresponding to the column density σ at which hydrogen is consumed and the flux $L^{\rm H}_{\rm nuc}/4\pi R^2$ is produced. However, equation (4.1) shows that outward flux is also produced by compressional heating. Therefore, given a surface flux $F(0)$, the flux $F(\sigma)$ in the envelope may reach zero before hydrogen is consumed if the accretion rate is sufficiently high. Since almost all ($\gtrsim 90$ %) of the flux produced in the envelope goes out (Ayasli and Joss 1982, Fujimoto *et al.* 1984), the bottom of the hydrogen-rich layer is given by the condition $F(\sigma) \approx 0$ in this situation. Therefore, the column density σ at the bottom of the hydrogen-rich layer is given by $X(\sigma) \approx 0$ or $F(\sigma) \approx 0$, whichever comes first (FL).

FL calculate the column density σ as a function of mass accretion rate per unit area $\dot{\sigma}$ and temperature T as follows. Equations (4.1) and (4.3) determine the column density σ at the bottom of the hydrogen layer as a function of $\dot{\sigma}$ and $F(0)$. FL use equation (4.2) to determine T as a function of $\dot{\sigma}$ and $F(0)$ [the dependence on $\dot{\sigma}$ comes from the compressional heating term in the integral expression for $F(\sigma)$], and invert to find σ as a function of $\dot{\sigma}$ and T. This defines the ***hydrogen fuel surface*** $\sigma = S_{\rm H}(\dot{\sigma}, T)$ in $(\dot{\sigma}, T, \sigma)$-space.

Applying similar conditions to Y and $F(\sigma)$, FL find the column density σ at the bottom of the helium layer, which defines the ***helium fuel surface*** $\sigma = S_{\rm He}(\dot{\sigma}, T)$ in $(\dot{\sigma}, T, \sigma)$-space.

4.2 IGNITION SURFACES

Nuclear burning in the envelope of accreting neutron stars is often unstable (Hansen and Van Horn 1975). FL determine the column density σ at which the burning becomes thermally unstable, if it does, as follows. Writing equation (4.1) in differential form, the left hand side $dF(\sigma)/d\sigma$ can be regarded as the cooling rate. The nuclear energy generation rate $\epsilon_{\rm nuc}$ is the dominant term on the right hand side of the ***first variation*** of this differential equation, since it is very temperature sensitive, whereas $\epsilon_{\rm g}$ is not and ϵ_ν is very small. Any difference in the variation of the cooling rate and the nuclear energy generation rate generates a variation in the entropy of the accreting matter. If the difference is positive, the internal temperature of the matter increases and a thermonuclear runaway ensues. If the difference is negative, the internal temperature of the matter decreases and nuclear burning is quenched. The boundary of thermal stability of the nuclear burning therefore corresponds to the condition,

$$\frac{\delta \epsilon_{\rm nuc}(\sigma, T)}{\delta T} = \frac{\delta \epsilon_{\rm cool}(\sigma, T)}{\delta T} \ . \tag{4.5}$$

Since equation (4.5) is not local (the cooling rate $dF(\sigma)/d\sigma$ depends on the first and second derivatives of the temperature), thermal stability depends on the overall temperature and density structure of the envelope. The most unstable mode is the global temperature variation $d\delta T/d\sigma = 0$.

FL calculate the column density σ at which thermonuclear runaway occurs for hydrogen and helium burning using equation (4.5). This defines the *ignition surfaces* $\sigma = S_{\rm H}^{\rm ign}(\dot{\sigma}, T)$ and $\sigma = S_{\rm He}^{\rm ign}(\dot{\sigma}, T)$ in $(\dot{\sigma}, T, \sigma)$-space.

5. Results

As an illustrative example, FL consider matter with one tenth the solar abundance of metals ($X = 0.749, Y = 0.25$, and $Z = 0.001$) accreting onto a neutron star with $g_s = 2.45 \times 10^{14}$ cm s^{-2}, appropriate for, e.g., $M = 1.4$ M$_\odot$ and $R = 10$ km. Specializing to these particular values of M and R, the relationship between the observed radiated luminosity L^∞ and the local interior temperature T is $L^\infty/L_\odot \approx 3.4\ (T/10^8{\rm K})^{2.98}$. The observed accretion luminosity is $L_{\rm acc}^\infty/L_\odot = 0.40\ \dot{\sigma}$, and the observed Eddington luminosity is $L_{\rm E}^\infty/L_\odot = 3.4 \times 10^4$. For $\dot{\sigma} > 8.6 \times 10^4$ g cm^{-2}s^{-1}, the observed accretion luminosity exceeds the Eddington limit. The conversion between the local accretion rate per unit area $\dot{\sigma}$ (g cm^{-2}s^{-1}) and the observed accretion rate $(dM/dt)^\infty$ (M$_\odot$ yr^{-1}) is $(dM/dt)^\infty = 1.5 \times 10^{-13}\dot{\sigma}$.

5.1 NUCLEAR FUEL SURFACES

Figure 7(a) shows the nuclear fuel surfaces and ignition surfaces for hydrogen and helium in ($\dot{\sigma}$, T, σ)-space. In order to make all four surfaces visible, FL take the $\dot{\sigma}$ and σ axes in non-conventional directions. Their choice of the range for the accretion rate per unit area cuts off the hydrogen and helium fuel surfaces at $\dot{\sigma} = 10^{-1}$ and 10^6 gm cm^{-2} s^{-1}, while those for the temperature and column density cut off the hydrogen fuel surface at $T = 10^6$ K and $\sigma = 10^4$ gm cm^{-2}. At $T \sim 10^9$ K, the luminosity $L = 4\pi R^2 F(0)$ of the envelope solution reaches the Eddington luminosity $L_{\rm E}$ and the hydrogen and helium fuel surfaces terminate.

The shape of the hydrogen fuel surface $S_{\rm H}$ arises as follows. At very low $\dot{\sigma}$ and T, compressional heating is small and $S_{\rm H}$ is determined by the exhaustion of hydrogen ($X \approx 0$) due to (stable) pycnonuclear and p-p burning at large σ. At moderate T, thermonuclear burning due to the CNO-cycle becomes important in exhausting hydrogen ($X \approx 0$); its dependence on a high power of the temperature drives $S_{\rm H}$ to lower σ as T increases. At still higher T, the CNO-cycle saturates due to inverse β-decay (Joss 1977, 1978; Lamb and Lamb 1977, 1978) and becomes independent of the temperature; $S_{\rm H}$ therefore becomes flat ($\sigma \approx$ constant).

As $\dot{\sigma}$ increases, compressional heating becomes important at low and moderate T, and the location of $S_{\rm H}$ is determined by the zero flux condition $F(\sigma) \approx 0$. We term this the compressional heating region. At high T, exhaustion of hydrogen ($X \approx 0$) due to the β-limited (hot) CNO-cycle determines $S_{\rm H}$; it is driven deeper in σ as $\dot{\sigma}$ increases. FL term this the stable hydrogen-burning regime.

At low and moderate T ($\lesssim 3 \times 10^8$ K), the helium fuel surface $S_{\rm He}$ lies at or below $S_{\rm H}$ because the triple-α reaction is less efficient than the hydrogen-burning CNO-cycle, and requires a higher column density to exhaust the helium ($Y \approx 0$). Even at high T, $S_{\rm He}$ lies at or below $S_{\rm H}$ because the carbon produced by the triple-α reaction enhances the (hot) CNO-cycle. Because of the latter effect, $S_{\rm H}$ and $S_{\rm He}$ are essentially coincident for $T \gtrsim 2 \times 10^8$ K and $\dot{\sigma} \gtrsim 10^3$ gm cm^{-2} s^{-1}.

The shape of the helium fuel surface $S_{\rm He}$ arises as follows. At very low $\dot{\sigma}$ and T, compressional heating is small. Equation (4.1) shows that in order for $S_{\rm He}$ to lie below $S_{\rm H}$, the surface flux $F(0)$ must exceed that produced by compressional

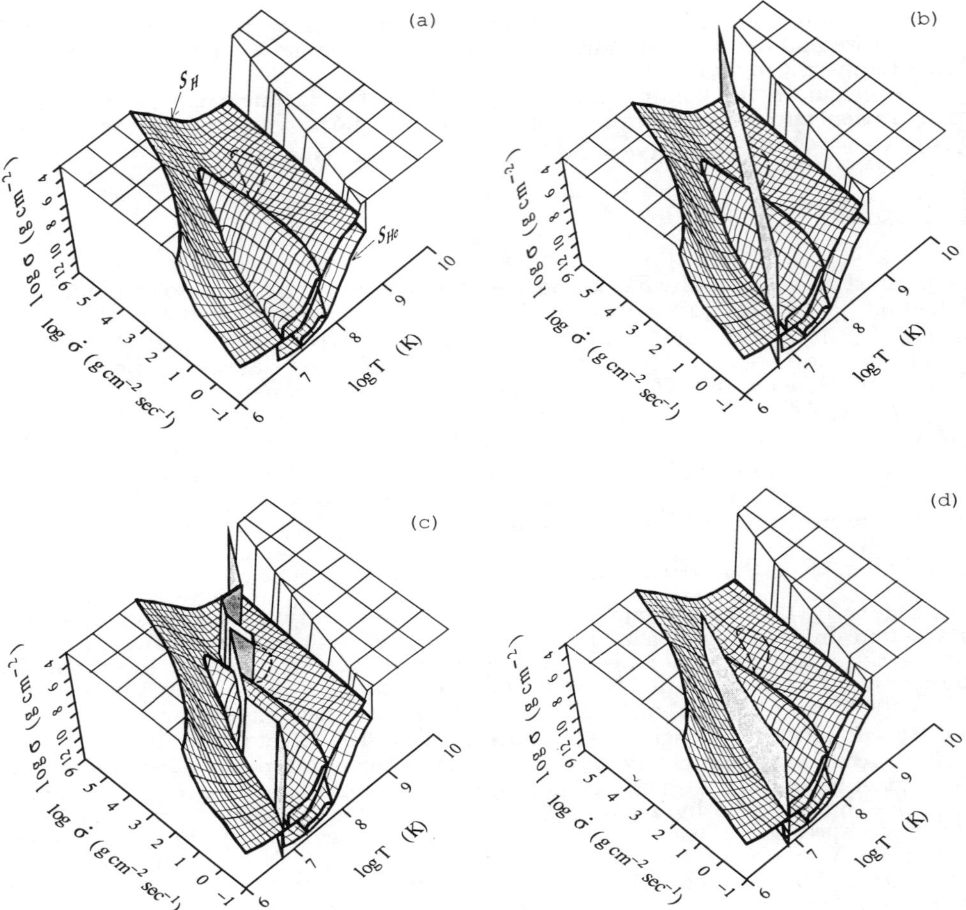

Figure 7. (a) Nuclear fuel and ignition surfaces for hydrogen and helium in $(\dot{\sigma}, T, \sigma)$-space. The hydrogen fuel surface is labelled S_{H}, and the hydrogen ignition surface $S_{\mathrm{H}}^{\mathrm{ign}}$ is the inverted v-shaped surface above it. The helium fuel surface is labelled S_{He}, and the helium ignition surface $S_{\mathrm{He}}^{\mathrm{ign}}$ is the surface barely visible above it. The broken line shows the region where $S_{\mathrm{He}}^{\mathrm{ign}}$ lies very close to S_{H}. (b) Same as (a), except showing the cut corresponding to the steady-state case. (c) Same as (a), except showing the cut corresponding to the thermal equilibrium case. (d) Same as (a), except showing the cut corresponding to the compressional heating case.

heating *and* hydrogen burning, i.e., it must exceed $q_{\mathrm{H}} X \dot{\sigma}$. At low T [and hence low $F(0)$], this does not occur and S_{He} is coincident with S_{H}. At $T \approx 5 \times 10^6$ K, this does occur and S_{He} quickly moves inward to higher σ. The (stable) pycnonuclear triple-α reaction (Cameron 1959; Nomoto, Thielemann, and Miyaji 1985; Fushiki and Lamb 1987a; Schramm and Koonin 1990) is important in exhausting helium ($Y \approx 0$) in a narrow band lying just above this transition in T. At higher T, thermonuclear burning due to the triple-α reaction becomes important; its dependence on a high power of the temperature drives S_{He} to lower σ as T increases, as in the case of the thermonuclear CNO-cycle. At very high T, the kinetic energy of the three reacting α-particles approaches the resonant energies of ^{8}Be and ^{12}C, the triple-α reaction becomes insensitive to the temperature (Fushiki and Lamb 1987a), and S_{He} becomes flat in σ.

As $\dot{\sigma}$ increases, compressional heating becomes important at low and moderate T, and the zero flux condition $F(\sigma) \approx 0$ determines the location of S_{He}. FL again term this the compressional heating region. At high T, exhaustion of helium ($Y \approx 0$) due to the temperature-insensitive triple-α reaction determines S_{He}; it is driven deeper in σ with increasing $\dot{\sigma}$, as in the case of the β-limited CNO-cycle. FL term this the stable helium-burning regime.

5.2 IGNITION SURFACES

The shapes of the hydrogen and helium ignition surfaces S_H^{ign} and S_{He}^{ign} arise as follows. At very low $\dot{\sigma}$, the low temperature boundaries of the two ignition surfaces correspond to the onset of the thermonuclear regimes for the pp-chain and triple-α reaction, respectively. At higher $\dot{\sigma}$, compressional heating is important, and both low temperature boundaries are driven to higher T by the flux condition $F(\sigma) \approx 0$. The high temperature boundaries of the two ignition surfaces are due to the onset of the β-limited regime of the (hot) CNO-cycle and to the temperature insensitivity of the triple-α reaction at $T \gtrsim 5 \times 10^8$ K. The band at moderate temperature in which the CNO-cycle and the triple-α reaction depend on temperature is squeezed between the compressional heating and the temperature insensitive regimes at high temperature as $\dot{\sigma}$ increases. Both ignition surfaces therefore narrow with increasing $\dot{\sigma}$, terminating at $\dot{\sigma} \approx 2 \times 10^4$ gm cm^{-2} s^{-1}.

At low $\dot{\sigma}$, the hydrogen ignition surface S_H^{ign} hugs the hydrogen fuel surface S_H, but at higher $\dot{\sigma}$, S_H^{ign} becomes sharply peaked [the steepness of its high-T side can be inferred from Figure 7(a), although it is largely hidden]. The helium ignition surface S_{He}^{ign} hugs the helium fuel surface S_{He} for all $\dot{\sigma}$. A broken line on S_H in Figure 7(a) shows the region, centered at $\dot{\sigma} \approx 3 \times 10^3$ gm cm^{-2} s^{-1} and $T \approx 3 \times 10^8$ K, where S_{He}^{ign} lies very close to S_H. In this region, the enhanced carbon produced by helium burning pushes S_H outward to lower σ, an effect which can be discerned in Figure 7(a). By connecting this region with the low and high temperature boundaries of S_{He}^{ign} at $\dot{\sigma} = 10^{-1}$ gm cm^{-2} s^{-1}, one can get a sense of the shape of S_{He}^{ign}. At low $\dot{\sigma}$, it is roughly coincident in T with S_H^{ign}, while at higher $\dot{\sigma}$, it lies at higher T.

6. Discussion

Since the hydrogen and helium fuel and ignition surfaces lie at column densities σ which are deep in the region of electron degeneracy, except along the Eddington luminosity boundary at very high T, the trajectory of the radiative-zero envelope solution for a given T approaches the surfaces nearly vertically from above. Thus accreting matter, which follows the radiative-zero trajectory, moves nearly vertically downward. If the accreting matter intercepts S_H or S_{He}, stable burning comences and the matter continues nearly vertically downward. If the accreting matter intercepts S_H^{ign} or S_{He}^{ign} instead, a thermonuclear runaway ensues and the temperature of the matter increases suddenly while $\dot{\sigma}$ and σ remain constant. We see that whether an X-ray burst occurs depends on whether the accreting matter intercepts S_H^{ign} or

S_{He}^{ign}, or not. Whether the burst is a pure He flash or a combined H-He flash depends on where the two ignition surfaces lie in ($\dot{\sigma}$, T, σ)-space in relation to each other. Hence the location and the shape of the ignition surfaces are of crucial significance in determining the properties of X-ray bursts.

The hydrogen and helium fuel and ignition surfaces also determine the observable burst properties, including the recurrence time between bursts,

$$\tau^{\infty} = (1+z)S^{ign}/\dot{\sigma} \ , \tag{6.1}$$

the burst energy,

$$E_b^{\infty} = 4\pi R^2 \int_0^{S_{He}^{ign}} (Xq'_H + Yq'_{He})d\sigma'/(1+z_s) \ , \tag{6.2}$$

and the "α-value,"

$$\alpha = E_b^{\infty}/(L_{acc}^{\infty}\tau^{\infty}) \ . \tag{6.3}$$

Here z_s is the redshift factor at the neutron star surface and the Q-values per unit mass of hydrogen and helium, q'_H and q'_{He}, are larger than those in equations (4.3) and (4.4) because the bursts burn hydrogen and helium all the way to iron-peak nuclei. All vary continuously (sometimes dramatically) with changes in $\dot{\sigma}$ and T.

Using the three-dimensional perspective provided by Figure 7(a), FL identified three regions of unstable burning, corresponding to three distinct types of X-ray bursts. (i) Inside the broken line, centered at $\dot{\sigma} \approx 3 \times 10^3$ and $T \approx 3 \times 10^8$ K, helium flashes take place in a bath of *partially depleted* hydrogen. (ii) In a region extending diagonally downward toward lower $\dot{\sigma}$ and T, and corresponding roughly to the projection of S_{He}^{ign} upward on S_H, and in a region covering the upper (in T) surface of S_H^{ign}, stable hydrogen burning eventually leads to pure helium flashes. Over most of the remainder of S_H^{ign}, *a succession of hydrogen flashes leads eventually to a pure helium flash.* (iii) In a region on the lower (in T) surface centered at $\dot{\sigma} \approx 1 \times 10^2$ gm cm^{-2} s^{-1} and $T \approx 2 \times 10^7$ K and extending over the range $\dot{\sigma} \approx 30 - 1 \times 10^3$ gm cm^{-2} s^{-1}, hydrogen flashes immediately trigger helium flashes. Elsewhere hydrogen and helium burning are both stable; in particular, they are stable *for all* T when $\dot{\sigma} \gtrsim 2 \times 10^4$ gm cm^{-2} s^{-1}.

Changes in $\dot{\sigma}$ on time scales less than the thermal time scale of the envelope, correspond to slices at constant T in Figure 7(a). Changes in T reflect the thermal evolution of the neutron star core and/or envelope. If they occur on time scales less than that of $\dot{\sigma}$, they correspond to slices at constant $\dot{\sigma}$ in Figure 7(a). More generally, changes in $\dot{\sigma}$ lead to changes in T, so that the two variables are correlated. The "steady-state" case $L = L_{nuc}$ discussed by FHM and HF corresponds to the diagonal cut $\dot{\sigma} \approx 404(T/10^8\ \text{K})^{2.98}$ shown in Figure 7(b). Their shell curves and ignition curves correspond to the intersection of the nuclear fuel and ignition surfaces with this cut. According to FL, the types of burst behavior expected in the steady-state case are qualitatively similar to those outlined in FHM and HM and encompass the three distinct types identified above. However, the burst properties differ greatly because FL correctly determine the boundaries by increasing T *at constant* $\dot{\sigma}$ *and* σ, whereas FHM and HF do so by (implicitly) increasing T along the steady-state cut.

The thermal equilibrium case, in which the surface luminosity L equals the nuclear burning luminosity L_{nuc} *averaged over a burst cycle*, corresponds to the complicated cut shown in Figure 7(c). It represents the highest temperature to which the envelope and core of an accreting neutron star can be raised by nuclear burning. At very high $\dot{\sigma}$ (or T), hydrogen and helium burning are stable, and the thermal equilibrium and steady-state cuts coincide. In the range of $\dot{\sigma}$ (or T) where steady hydrogen burning leads to pure helium flashes, the two cuts lie close together, since the helium flashes contribute little to the average nuclear luminosity. Finally, at very low $\dot{\sigma}$ (or T), hydrogen and helium burn stably via pycnonuclear reactions and the two cuts again coincide. In the other ranges of $\dot{\sigma}$, luminosity drains out of the envelope and core of the star between bursts, unchecked by any steady nuclear luminosity. Since only a fraction ≈ 0.1 of the burst luminosity flows inward to heat the envelope and core (Ayasli and Joss 1982, Fujimoto *et al.* 1984), the thermal equilibrium case corresponds to $L \approx 0.1 L_{\rm nuc}$ and the envelope temperature is a factor ~ 2 less than that assumed in the steady-state case $L = L_{\rm nuc}$. Thus when bursts occur, the *steady-state case is physically realizeable only in the small range of $\dot{\sigma}$ where steady hydrogen burning leads to pure helium flashes, and then only approximately.* The types of burst behavior expected in the thermal equilibrium case are the same as those for the steady-state case.

The compressional heating case, in which the surface luminosity L equals the rate $L_{\rm g}$ at which energy is liberated by compressional heating of the accreted matter at the moment the nuclear fuel reaches the ignition surface, corresponds to the cut shown in Figure 7(d). The types of bursts expected in this case are (i) hydrogen flashes immediately triggering helium flashes for $\dot{\sigma} \approx 30 - 10^3$ gm cm^{-2} s^{-1}, (ii) a succession of hydrogen flashes leading eventually to a helium flash at $\dot{\sigma} \approx 1 \times 10^3 - 2 \times 10^4$ gm cm^{-2} s^{-1}, and, possibly, (iii) helium flashes in a bath of hydrogen in a small region at $\dot{\sigma} \approx 2 \times 10^4$ gm cm^{-2} s^{-1}.

The entire nuclear burning luminosity $L_{\rm nuc}$ heats the envelope in the stable-burning region, while only $\approx 0.1 L_{\rm nuc}$ does so in the regions corresponding to the ignition surfaces $S_{\rm H}^{\rm ign}$ and $S_{\rm He}^{\rm ign}$, as discussed above. If a source lies near the high-T boundary of $S_{\rm H}^{\rm ign}$ or $S_{\rm He}^{\rm ign}$, a small change in $\dot{\sigma}$ produces a large change in T, and hence in burst properties. If a source lies near the low-T boundary a large change in $\dot{\sigma}$ produces no change in T, and hence in the burst properties. Thus, e.g., in the thermal equilibrium case, we expect highly erratic behavior near each of the corners in the cut [see Figure 7(c)] (FL). Such behavior is relevant to the erratic behavior observed in sources such as Ser X-1 and 1735-44 (Lewin and Joss 1981, 1983).

Earlier global analyses (FHM; HF) of nuclear burning on accreting neutron stars dealt only with a single cut (the steady-state case) in the two-dimensional $(\dot{\sigma}, T)$-plane, and missed most of the possible parameter space. Of particular interest are the regions which lie away from the steady-state cut, and which were revealed for the first time in FL's analysis. The region at low $\dot{\sigma}$ and high T corresponds to the X-ray burst behavior exhibited by young neutron stars, and includes a large region where a succession of hydrogen flashes leads eventually to a helium flash. This region is inaccessible to old neutron stars, since it lies above the thermal equilibrium curve, which represents the highest temperature to which the envelope and core of an accreting neutron star can be raised by nuclear burning. The region at high $\dot{\sigma}$ and low T corresponds to the X-ray burst behavior exhibited by cold neutron stars and is relevant to transient X-ray sources which exhibit X-ray bursts, such as 0748-67, 1608-52, 1659-29, and Aql X-1, while the low $\dot{\sigma}$ part of this region is relevant to X-ray transients, such as 28 June 1976 and 7 February 1977 (Lewin and Joss 1981, 1983). Compressional heating, which is negligible along the whole steady-state cut (FHM,HF), dominates this entire region (FL). Accreting matter, which moves nearly vertically downward along the trajectory of a radiative-zero solution, is deflected upward in T along $S_{\rm H}$ until it reaches the cut shown in Figure 7(d). All sources in this region ignite along this "compressional heating" cut. Their burst behavior is therefore *a direct function* of the observed mass accretion rate just prior to the burst, independent of the thermal state of the neutron star interior.

We gratefully acknowledge the contributions of Ikko Fushiki to the work described in this lecture. We thank Walter Lewin, Jan van Paradijs, and Nick White for numerous informative discussions about observations of X-ray bursts, and Fred Lamb, Paul Joss, and Ron Taam for valuable discussions of the physics of nuclear burning on accreting neutron stars. This work was supported in part by NASA grants NAGW-830 and NAGW-1284.

7. References

Ayasli, S. and Joss, P. C. 1982, *Ap. J.*, **256**, 637.

Belian, R. D., Conner, J. P., and Evans, W. D. 1976, *Ap. J.* (*Letters*), **206**, L135.

Cameron, A. G. W. 1959, *Ap. J.*, **130**, 916.

Fujimoto, M. Y., Hanawa, T., Iben, I. and Richardson, M. B. 1984, *Ap. J.*, **278**, 813.

Fujimoto, M. Y., Hanawa, T. and Miyaji, S. 1981, *Ap. J.*, **246**, 267 (FHM).

Fushiki, I., and Lamb, D. Q. 1987a, *Ap. J.*, **317**, 368.

Fushiki, I., and Lamb, D. Q. 1987b, *Ap. J.* (*Letters*), **323**, L55.

Grindlay, J.E., *et al.* 1976, *Ap. J.* (*Letters*), **205**, L127.

Hanawa, T. and Fujimoto, M. Y. 1982, *Publ. Astron. Soc. Japan*, **34**, 495.

Hansen, C. G. and Van Horn, H. M. 1975, *Ap. J.*, **195**, 735.

Joss, P. C. 1977, *Nature*, **270**, 310.

Joss, P. C. 1978, *Ap. J.* (*Letters*), **225**, L123.

Lamb, D. Q. and Lamb, F. K. 1977, *Ann. N. Y. Acad. Sci.* **302**, 261.

Lamb, D. Q. and Lamb, F. K. 1978, *Ap. J.*, **220**, 291.

Lewin, W. H. G., and Joss, P. C. 1983, in *Accretion Driven Stellar X-Ray Sources*, ed. W. H. G. Lewin and E. P. J. van den Heuvel (Cambridge: Cambridge U. Press), p. 41.

Lewin, W. H. G., and Joss, P. C. 1981, *Space Sci. Rev.*, **28**, 3.

Maraschi, L., and Cavaliere, A. 1977, *in Highlights in Astronomy* **4**, Part I, ed. E. A. Müller (Dordrecht: Reidel), p. 127.

Nomoto, K., Thielemann, F.-K., and Miyaji, S. 1985, *Astr. Ap.*, **149**, 239.

Paczyński, B. 1983, *Ap. J.*, **264**, 282.

Schramm, S., and Koonin, S. E. 1990, *Ap. J.*, **365**, 296.

Swank, J. H., *et al.* 1977, *Ap. J.* (*Letters*), **212**, L73.

Taam, R. E. 1980, *Ap. J.*, **241**, 358.

Taam, R. E. 1985, *Ann. Rev. Nucl. Part. Sci.*, **35**, 1.

Thorne, K. 1977, *Ap. J.*, **212**, 825.

Woosley, S. E., and Taam, R. E. 1976, *Nature*, **263**, 101.

GAMMA-RAY BURSTS

D. Q. LAMB
Department of Astronomy and Astrophysics
and Enrico Fermi Institute
University of Chicago
5640 S. Ellis Avenue
Chicago, IL 60637, U.S.A.

ABSTRACT. We summarize the observed features of γ-ray bursts, with particular emphasis on the cyclotron lines seen in their spectra. We then discuss the theory of cyclotron resonant scattering, and compare the theoretical and observed line spectra. Finally, we mention some implications.

1. Introduction

Recent *Ginga* observations extending to lower energies and having higher spectral resolution (Murakami *et al.* 1988, Fenimore *et al.* 1988; Yoshida *et al.* 1990), and radiation transfer calculations treating resonant cyclotron scattering in superstrong magnetic fields (Lamb *et al.* 1989; Wang *et al.* 1989; Lamb, Wang, and Wasserman 1990; see also Alexander and Mészáros 1989) provide compelling evidence that many of the bursts come from strongly magnetic neutron stars.

Three bursts seen by *Ginga* exhibit harmonically spaced dips at $\approx 20-25$ keV and $\approx 40-50$ keV: GB870303, GB880205, and GB890929 (Yoshida *et al.* 1990). Wang *et al.* (1989) carried out explicit fits of theoretical spectra from radiation transfer calculations to the *Ginga* data for GB880205. They showed that cyclotron resonant scattering can explain the observed positions, strengths, and widths of the dips in the spectrum. The success of the model convincingly demonstrates the existence of a superstrong magnetic field, $B \approx 2 \times 10^{12}$ G, in this source. This result, taken together with the *Ginga* (Yoshida *et al.* 1990) and Konus data (Mazets *et al.* 1981), which show low-energy dips in approximately 20% of all classical γ-ray bursts, provides powerful evidence that many γ-ray bursts come from strongly magnetic neutron stars. These neutron stars must belong to the Galaxy, resolving a long-standing controversy about the distances to γ-ray burst sources.

In this lecture, we summarize the observed features of γ-ray bursts, with particular emphasis on the cyclotron lines seen in their spectra. We then discuss the theory of cyclotron resonant scattering, and compare the theoretical and observed line spectra. Finally, we mention some implications.

2. Summary of Observations

Gamma-ray bursts were discovered in 1969 and reported in 1973 by Klebesadel, Strong, and Olson (1973). More than 500 have now been seen. [For comprehensive discussions of γ-ray burst observations and theory, see the conference proceedings edited by Lingenfelter, Hudson, and Worrall (1982); Woosley (1984); and Liang

J. Ventura and D. Pines (eds.), Neutron Stars: Theory and Observation, 545–560.

and Petrosian (1986); and the recent reviews by Higdon and Lingenfelter (1990), Harding (1991), and Hurley and Lamb (1991).]

Gamma-ray bursts have rise times as short as ~ 0.1 msec and durations that range from ~ 0.01 sec to at least 1000 sec. No "classical" γ-ray bursts have been seen to repeat. The bursts have luminosities $L_b \sim 10^{37} - 10^{38}$ erg s^{-1} and total energies $E_b \sim 10^{37} - 10^{40}$ erg, assuming distances $d \sim 100 - 300$ pc.

Figure 1 shows the locations in galactic coordinates of 175 γ-ray burst sources for which the direction to the source is known. The sources are distributed uniformly on the sky. The time histories of many, but not all, bursts exhibit rapid time variability (see Figure 2).

The continuum energy spectrum of γ-ray bursts is roughly a broken power law, $F_E = EdN/dE \propto E^{-\alpha}$, with an X-ray spectral index $\alpha_X \sim 0$ and a γ-ray spectra index $\alpha_\gamma \sim 0.5 - 2$ (Epstein 1986, Zdziarski 1987, Lamb 1988). The transition between the X-ray and γ-ray regimes occurs between 100 keV and 1 MeV. This spectral form implies that most of the γ-ray burst luminosity is emitted in γ-rays (Epstein 1986). The typical observed ratio of X-ray to γ-ray luminosities is $L_X(3 - 10 \text{ kev})/L_\gamma(> 100 \text{ keV}) \sim 0.02$. Many bursts detected by the Solar Maximum Mission extend to energies of $\sim 5 - 10$ MeV or more, and show no high energy cutoffs (Matz *et al.* 1985). Figure 3 compares the power spectrum ($P = E^2dN/dE$) of four γ-ray bursts with the spectra of other high energy sources. It shows that γ-ray burst spectra are harder than those of any other astrophysical source.

The challenge of deciphering the nature of γ-ray bursts is exacerbated by the fact that one cannot predict when or from where the bursts will occur, and the fact that it has been impossible to date to find quiescent counterparts of the bursts at radio, infrared, optical, ultraviolet, X-ray, or γ-ray energies. The latter puts a premium on garnering knowledge from the bursts themselves.

The shape of the continuum spectrum can provide important constraints on theory, but inverting it uniquely to determine the radiation mechanism, let alone physical parameters like the density and temperature, is exceedingly difficult. In contrast, the power of lines is well known: analyses of atomic lines transformed astronomy into astrophysics. Because atomic lines are unavailable above ~ 7 keV, studies of γ-bursts must rely on cyclotron and, possibly, pair annihilation lines. Analyses of these lines can yield not only the optical depth (or column density) and temperature (or velocity) but also the strength and orientation of the magnetic field, and the geometry of the emission region.

3. Observations of Cyclotron Lines

Some time ago Mazets *et al.* (1981, 1982) reported single low-energy ($E \approx 50$ keV) dips in the spectra of approximately 20% of the bursts observed with the Konus experiment on Venera 11-14. Figure 4 shows the spectra of two such bursts, GB790412 and GB790612. Hueter (1988) reported single low-energy dips in several bursts observed with HEAO-1 A4. Figure 5 shows the spectrum of the strongest of these bursts, GB780325. These features were interpreted as due to absorption at the cyclotron first harmonic (fundamental), $\hbar\omega_B \approx 11.6B_{12}$ keV, where B_{12} is the magnetic field strength in units of 10^{12} G. They constituted the strongest evidence in favor of a magnetic neutron star origin for γ-ray bursts [see, *e.g.*, the reviews by Mazets and Golenetskii (1981, 1982); Cline (1981); Lamb (1982, 1983); and Hurley (1982, 1983)].

However, the statistical significance of the dips is low. Moreover, their properties and significance depend sensitively on the assumed form of the continuum spectrum (Fenimore *et al.* 1983; Teegarden 1984). Finally, the horizontal and ver-

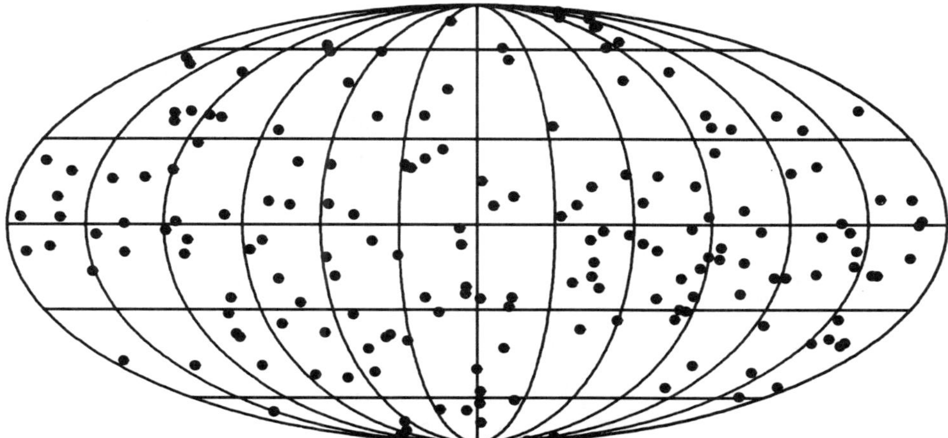

Figure 1. Sky map (galactic coordinates) of 175 γ-ray burst sources. (From Hurley and Lamb 1991.)

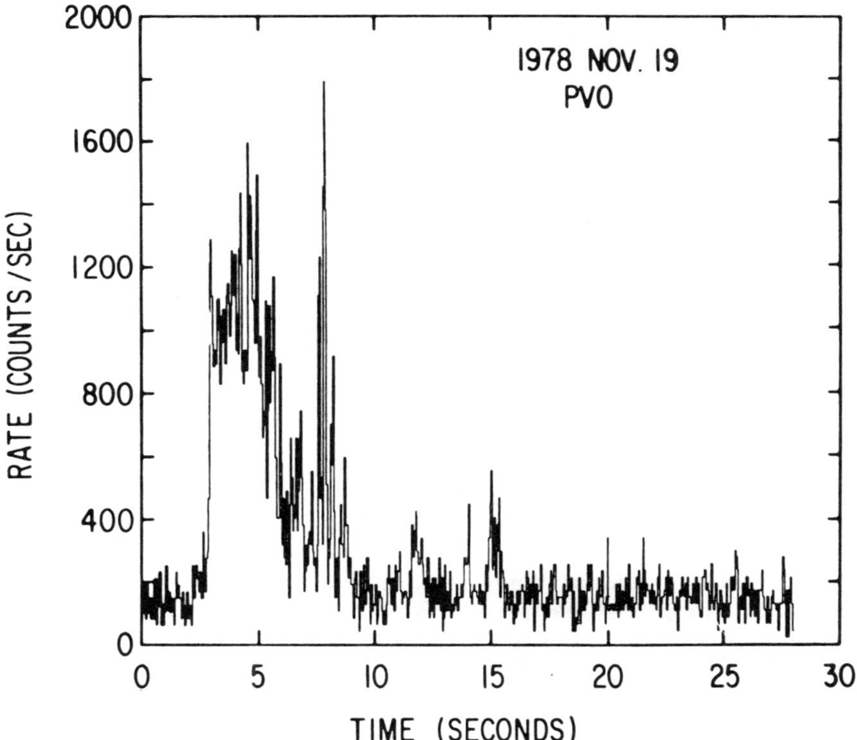

Figure 2. Profile of the 19 November 1978 burst, showing the temporal structure of this rapidly varying and intense event. (From Evans *et al.* 1980.)

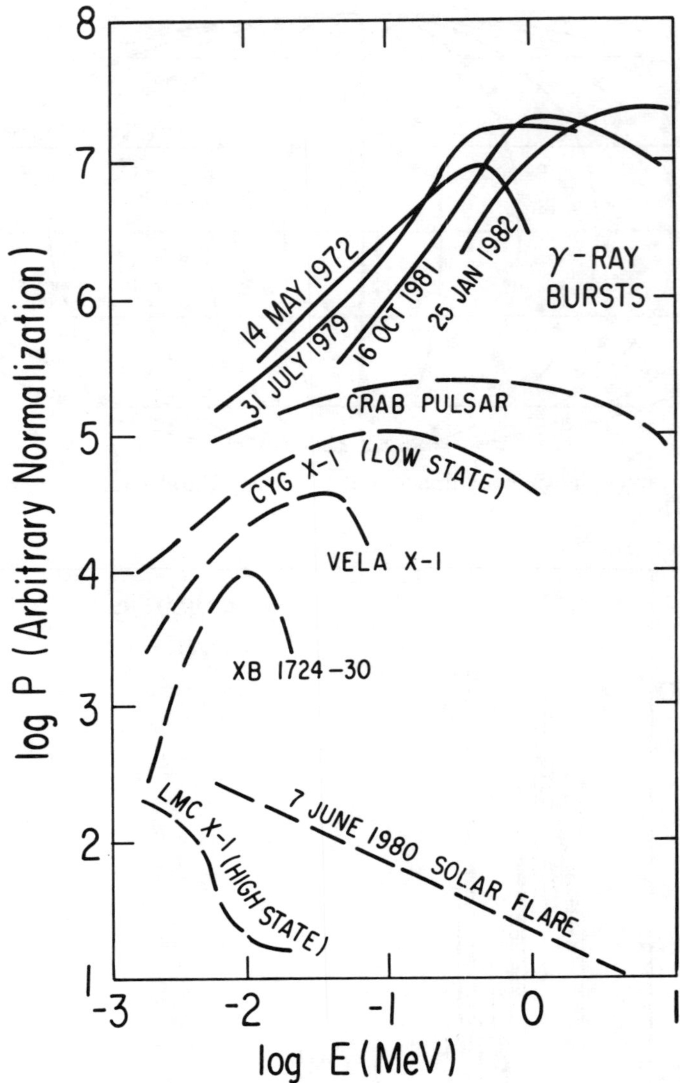

Figure 3. Comparison of the power spectra ($P = E^2 dN/dE$) of four γ-ray bursts with that of the rotation-powered pulsar in the Crab Nebula, the black hole candidate Cyg X-1 when its spectrum is hardest, the rotation-powered pulsar Vela X-1, a burst from the X-ray source XB 1724-30, one of the hardest solar flares observed, and the X-ray source LMC X-1 in the Large Magellanic Cloud. (After Epstein 1986.)

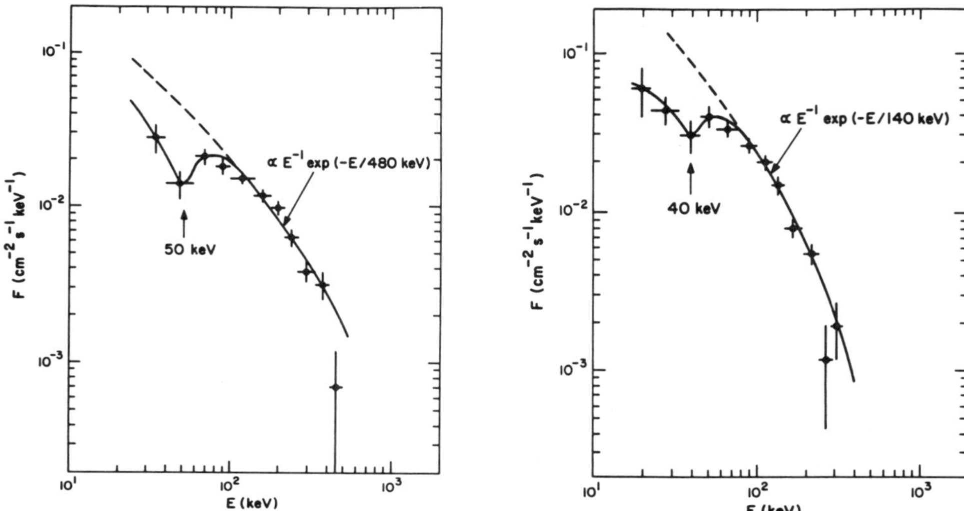

Figure 4. Konus spectra of γ-ray bursts GB790412 (left) and GB790612 (right). The vertical positions and error bars of the data points correspond to the theoretical spectrum multiplied by the ratio of the observed and theoretical count rate spectrum at that energy, and are therefore model-dependent. The horizontal positions of the data points correspond to the detector energy channel centers and the horizontal error bars correspond to the expected photon energy loss in the corresponding channel, and do not measure the spectral resolution of the detector; the actual spectral resolution of the detector is significantly worse. (After Mazets *et al.* 1981.)

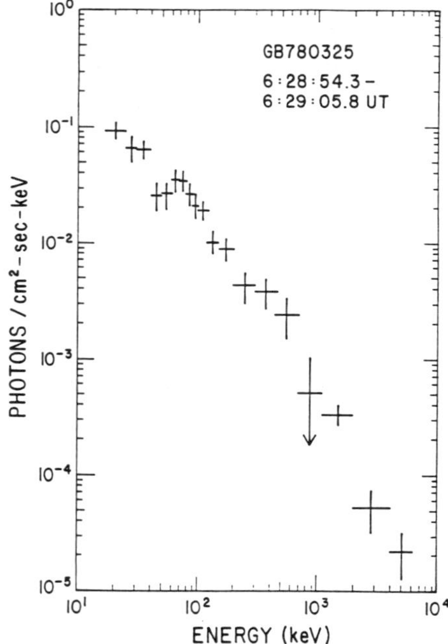

Figure 5. HEAO-1 A4 spectrum of the second intensity peak of γ-ray burst GB780325, showing a low-energy dip at ≈ 50 keV. The horizontal positions and horizontal error bars of the data points have the same meaning as in Figure 4. (From Hueter 1988.)

tical error bars of the data points in the photon number spectrum that have been derived using the model-fitting method employed in analyzing these bursts have no straight-forward statistical meaning. In particular, the vertical positions and error bars of the data points are model dependent, and the actual spectral resolution can be much worse than the horizontal error bars indicate (Loredo and Epstein 1989). Consequently, the existence and interpretation of these dips, and therefore the magnetic neutron star model of γ-ray bursts, remained controversial.

Low-energy dips have been seen in the spectra of three bursts detected recently by the *Ginga* satellite (Yoshida *et al.* 1990). The raw count rate spectrum of one of these bursts, GB880205, is shown in Figure 6. These dips are better resolved than previously: They span six or eight detector energy channels, not just one as in the Konus data. The evidence that the features are real is thus much stronger. Furthermore, the spectra of the three bursts show two strong dips – not just one, as in the Konus data and the HEAO-1 A-4 data – and the features are harmonically spaced at $\approx 20-25$ and $40-50$ keV, as is expected for cyclotron features.

The positions of the dips at ≈ 20 and 40 keV indicate a magnetic field strength $B \approx 2 \times 10^{12}$ G. The presence of strong dips at the cyclotron first and second harmonics, but not at the third, suggests that the resonant scattering optical depth of the line-forming region is modest.

Two aspects of the dips might seem puzzling. First, the dips are narrow, indicating that the line-forming region is much cooler than the typical photon energy $E_\gamma \gtrsim 1$ MeV in the continuum spectrum. Second, the strengths of the dips at the first and second harmonics are similar, despite the fact that the cyclotron resonant scattering cross section decreases rapidly with increasing harmonic number when $B \ll B_c \approx 4.4 \times 10^{13}$ G.

Fenimore *et al.* (1988) fit a variety of analytic models to the spectrum for the 5-s interval of GB880205 labeled b in Figure 6. This allowed them to assess the statistical significance of the two dips and to provide quantitative estimates of the physical parameters in the scattering region. Using an F test, they found that the (corrected) probability that the six additional parameters needed to describe the dips as Gaussian line features are unnecessary is 9×10^{-6}. Further, they obtained an excellent fit to the spectrum by (1) approximating cyclotron resonant scattering as cyclotron absorption, which is not valid for the first harmonic but is a good approximation for higher harmonics; (2) allowing different column densities for each harmonic; and (3) assuming nonrelativistic kinematics. Their analysis strengthened the cyclotron interpretation of the dips. They found best fit values and 68% confidence intervals $B_{12} = 1.69 \pm 0.04$ G, $T\mu^2 = 6.6 \pm 2.4$ keV, and (correcting for a factor of 2π) $N_e^{\rm los}(1-\mu^4) \approx 2.4 \times 10^{21}$ electrons cm^{-2}, where $N_e^{\rm los}$ is the column density along the line of sight. The quantity μ, which is the cosine of the viewing angle θ relative to the magnetic field, was not determined by the fit and T and $N_e^{\rm los}$ were therefore poorly constrained.

4. Theory of Cyclotron Resonant Scattering

In a superstrong magnetic field, the motion of electrons along the magnetic field remains classical. For cool electrons ($kT \ll \hbar\omega_B \approx 11.6\, B_{12}$ keV),

$$\left(\frac{v_\parallel}{c}\right)^2 \sim \left(\frac{v_{th}}{c}\right)^2 \sim 0.01 \ll 1, \tag{4.1}$$

where $v_{th} = (2kT_\parallel/m_e)^{1/2}$ is the electron thermal velocity parallel to the magnetic field. Despite this inequality, a fully relativistic treatment of the photon-electron scattering kinematics is necessary (see below).

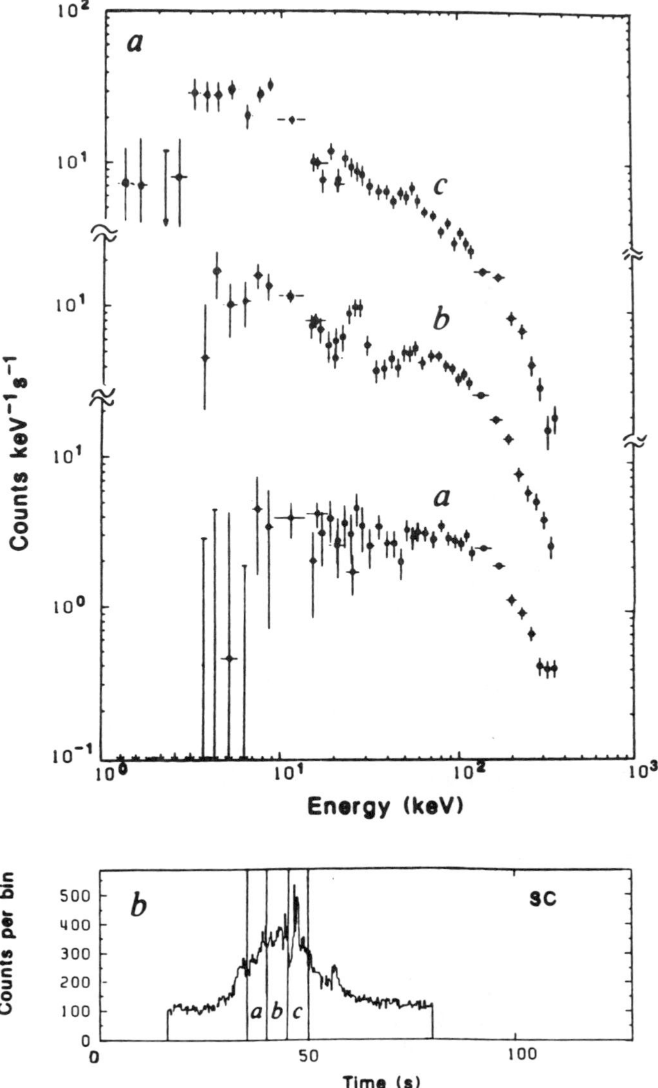

Figure 6. *Ginga* observation of γ-ray burst GB880205. (a) Raw count rate spectra after background subtraction, normalized to the width of the detector energy loss channels (not corrected for detector response). (b) Raw count rate history. The spectra in (a) correspond to the intervals labelled a, b, and c in (b). Note the spectral features at $\approx$ 20 and 40 keV, each spanning 6-8 energy loss bins. (From Murakami *et al.* 1988.)

The motion of electrons perpendicular to the magnetic field is quantized in Landau levels with characteristic gyration velocity,

$$\left(\frac{v_\perp}{c}\right)^2 \sim N\left(\frac{B}{B_c}\right) \ll 1, \tag{4.2}$$

where N is the number of the Landau level and $B_c = m_e^2c^3/e\hbar = 4.414 \times 10^{13}$ G. A weak-field quantum-mechanical treatment of cyclotron scattering is therefore adequate.

For cool electrons, $kT \ll \hbar\omega_B$, the electrons are largely confined to the lowest $N = 0$ Landau level. In fact, the radiative decay rate $\Gamma_{\rm rad}$ of the $N \neq 0$ excited states is much faster than the collisional decay rate $\Gamma_{\rm coll}$ at the high magnetic field strengths and low densities relevant here. Therefore the populations of the upper levels are determined by radiative, not thermal, equilibrium and are much smaller than thermal. Thus, the electrons form a one-dimensional plasma, which significantly alters the scattering kinematics, and therefore both the resonant scattering line profiles and the photon frequency redistribution in each scattering. Furthermore, the inequality $\Gamma_{\rm rad}/\Gamma_{\rm coll} \gg 1$ means that absorption (in which radiative excitation of an electron is followed by collisional decay) is entirely negligible and photons and electrons interact purely through scattering.

The narrow widths of the low-energy features in γ-ray burst spectra are a long-standing puzzle, given the high color temperature $T_{\rm color} \gtrsim 1$ MeV of the continuum spectra. Lamb, Wang, and Wasserman (1990) explain the narrow widths as follows. A low density atmosphere is naturally expected at the outer edge of the X-ray emission region. Heating by Compton scattering and cooling by inverse Compton scattering will bring the material in this layer to the Compton temperature T_C at which these processes balance on a time scale $t_C \sim 10^{-8}$ s, assuming characteristic γ-ray burst source parameters. At small enough column depths $\tau_{\rm con} \ll 1$ and $\tau_{\rm line} \gg \tau_{\rm con}$, and cyclotron resonant scattering dominates the heating and cooling. Hence X-ray photons near the cyclotron first harmonic, *not* the more plentiful high energy X and γ rays in the continuum spectrum, determine the temperature of the scattering medium.

The Compton cooling experienced by a one-dimensional thermal electron distribution is $\propto (T/m_e)(dn_\gamma/d\ln E)$, where $n_\gamma \propto N_\gamma/E^2$ is the photon occupation number and N_γ is the photon number flux. The Compton heating experienced by the electrons due to recoil is $\propto n_\gamma E/m_e$. Balancing Compton cooling and heating gives the Compton equilibrium temperature,

$$T_C \sim 1/(d\ln n_\gamma/dE) \sim T_{\rm color}\ , \tag{4.3}$$

where $T_{\rm color}$ is the color temperature of the radiation field *at the resonance*. Resonant scattering at the cyclotron fundamental is the dominate process at small enough column depths. Evaluating the derivative in equation (4.3) at $\sim E_B$ gives the *resonant* Compton temperature $T_C \sim E_B$, for the line-dominated region. For $B \ll B_c = 4.414 \times 10^{13}$ G, $T_C \sim E_B \ll m_e$, and the line-dominated region has a much smaller temperature than the continuum color temperature ($\gtrsim m_e$) characteristic of γ-ray bursts.

At larger column depths, continuum Compton heating dominates line Compton heating, and the temperature increases ($T \propto N_e$). At sufficiently large column depths, the matter temperature approaches the continuum color temperature ($\gtrsim m_e$). At such high temperatures, Doppler broadening washes out the cyclotron scattering lines. Lamb, Wang, and Wasserman (1990) therefore conjecture that the *line-forming region* is the low column depth $\tau_{\rm con} \ll 1$ *line-dominated region*. They show that this conjecture is in agreement with the column depth of the line-forming region derived from detailed fits to the spectrum of GB880205 (see below).

Three dimensionless parameters govern resonant line transfer at the cyclotron first harmonic: the line optical depth τ_1, the natural line width parameter

$$a \equiv \Gamma_{\rm rad}/2\omega_B(v_{th}/c), \tag{4.4}$$

which is the ratio of the radiative line width to the Doppler width, and the recoil parameter

$$\Delta \equiv \frac{(\hbar\omega_B)^2/2m_e c^2}{\hbar\omega_B(v_{th}/c)} = \frac{\frac{1}{2}B/B_c}{v_{th}/c}, \tag{4.5}$$

which is the ratio of the recoil energy shift to the Doppler energy shift. For the physical conditions relevant to GB880205, $\tau_1 \sim 10 - 100 \gg 1$, $\tau_2 \sim (B/B_c)\tau_1 \sim 0.1 - 10 \sim 1$, $\tau_3 \sim (B/B_c)^2\tau_1 \sim 0.01 - 0.1 \ll 1$. By comparison, the optical depth in the continuum $\tau_{\rm con} \sim 10^{-3} \ll 1$. The natural line width parameter $a \sim 10^{-3}$, and the recoil parameter $\Delta \sim 0.1 - 1$.

Resonant scattering may be treated as resonant absorption followed immediately by radiative reemission, which redistributes the frequency and direction of the incident photon but conserves energy and momentum. The scattering profile gives the probability of resonant absorption and the redistribution function gives the frequency and direction of the scattered photon.

A fully trelativistic treatment of the photon-electron scattering kinematics is essential. Relativistic kinematics in one dimension produces a sharp cutoff in the scattering profile at an energy $\hbar\omega = \hbar\omega_N/\sqrt{1-\mu^2}$, where $\hbar\omega_N = [(1 + 2NB/B_c)^{1/2} - 1]m_e c^2$, leading to a strongly asymmetric line profile.

Since $a\tau_1 \ll 1$ in GB880205, the line wings are optically thin and natural line width may be ignored in treating the scattering profiles (Wang, Wasserman, and Salpeter 1988). However, natural line width cannot be ignored for the photon frequency redistribution, despite its small magnitude. When $a = 0$, one-dimensional kinematics fixes the velocity of the electrons that can undergo resonant scattering. This in turn constrains the photons' frequency after many resonant scatters to lie within $\sim \Delta \lesssim 1$ Doppler widths of its initial frequency (Wasserman and Salpeter 1980). The photon is therefore trapped in frequency space and escapes only by spatial diffusion after $\sim \tau^2$ scatters. When $a \neq 0$, trapping is moderated because there is now a probability per scatter $\sim a$ that a photon can scatter off an electron with a velocity $\gtrsim v_{th}$ away from the resonant velocity. Thus in $\sim 1/a$ scatters a photon trapped in the line core can scatter into the line wings in a single scatter and escape. If $a\tau^2 \gtrsim 1$, escape by such rare core-wing transitions is more efficient than spatial diffusion and natural line width must be included in the frequency redistribution (Wasserman and Salpeter 1980; Wang, Wasserman, and Salpeter 1988). This is the case in the present problem, where $a\tau_1^2 \sim 0.1 - 10$.

The dip observed at $\approx 20 - 25$ keV in the spectra of γ-ray bursts GB870303, GB880205, and GB890929 is interpreted as cyclotron *resonant scattering*, in which electrons undergo radiative $0 \to 1 \to 0$ Landau transitions triggered by photons near the first harmonic frequency $\omega \approx \omega_B$ (Fenimore *et al.* 1988, Wang *et al.* 1989, Alexander and Mészáros 1989). No simple description can be used to explain the first harmonic line, whose appearance depends critically on the outcome of the multiple resonant scatters required in order for individual photons to escape (Wasserman and Salpeter 1980; Wang, Wasserman, and Salpeter 1988), as well as on the introduction of new photons at energies near that of the first harmonic which are "spawned" by Raman scattering at the higher harmonics (Bussard and Lamb 1982), as described below.

The dip at $\approx 40 - 50$ keV is attributed to *Raman scattering*, in which electrons undergo $0 \to 2 \to 1 \to 0$ radiative transitions triggered by photons with energies near that of the second harmonic ($\omega \approx 2\omega_B$) (Fenimore *et al.* 1988; Wang *et al.* 1989;

Alexander and Mészáros 1989). Resonant scattering of second harmonic photons, in which electrons undergo radiative $0 \to 2 \to 0$ Landau transitions, occurs with a branching ratio $\sim \hbar\omega_B/m_e c^2 \sim 0.1$ (Daugherty and Ventura 1977). It happens rarely because $B/B_c \ll 1$; when it does occur, the spawned photons must be treated in the same way as photons at the first harmonic. Because most of the photons which undergo scattering at the second harmonic are destroyed, the resulting line feature is approximately that for absorption (Fenimore *et al.* 1988, Wang *et al.* 1989, Harding and Preece 1989). This is a general property of higher harmonic cyclotron lines when the magnetic field strength $B \ll B_c$.

The observed strengths of the features at the first and second harmonics might seem puzzlingly similar, given the rapid decrease in the cyclotron resonant scattering cross section with increasing harmonic number (see *e.g.*, Harding and Preece 1989). If they were due to ***absorption*** in an optically thin medium, the strength of the second harmonic relative to the first would indeed be $\approx 2\tau_2/\tau_1 \approx \sigma_2/\sigma_1 \sim 2(B/B_c) \ll 1$, where σ_1 and σ_2 are the resonant cross sections at the first and second harmonics. However, the first harmonic is always optically thick at line center. If the features were due to ***absorption*** at angles $\mu \approx 1$, where the first harmonic is optically thick but the second is optically thin, the relative strength would be $\approx \sqrt{\pi}\tau_2/(\ln \tau_1)^{1/2} \ll 1$. If they were due to ***absorption*** at angles $\mu \approx 0$, where the first and second harmonics are both optically thick, the relative strength would be $\approx 2(\ln \tau_2/\ln \tau_1)^{1/2} \lesssim 1$. However, the observed ratio of equivalent widths $EW_2/EW_1 \approx 9.1$ keV/3.7 keV ≈ 2.5 (Fenimore *et al.* 1988). Thus, even were the harmonics due to absorption, the finite optical depth makes the strengths of the first and second harmonics similar but cannot account for the observed ratio of equivalent widths. Wang *et al.* (1989) demonstrate that resonant scattering at the first harmonic, which does not destroy the original photon, and photon spawning due to Raman scattering at higher harmonics, which fills in the first harmonic, successfully do so (see Figure 7).

Figure 7 shows what the relative strengths of the first and second harmonics would be, were they due to absorption (bottom line) or to resonant scattering and Raman scattering without photon spawning (middle line). The (heavy) top line gives the actual profiles, which are due to resonant scattering and Raman scattering with photon spawning included. Figure 7 also shows that the profiles of the second and third harmonics closely resemble those for absorption; in particular, they show the strong asymmetry due to relativistic kinematics in one dimension.

Resonant scattering of photons near the first harmonic is only moderately peaked parallel and antiparallel to the magnetic field because the scattering cross section $\propto (1 + \cos^2\theta)$. By contrast, scattering of photons at higher harmonics is highly anisotropic because the scattering cross section $\propto \sin^{2N-2}\theta$ for the Nth harmonic. However, the photons spawned by Raman scattering at higher harmonics are created with the angular distribution $1 + \cos^2\theta$ characteristic of the first harmonic. Therefore Raman scattering primarily removes photons travelling perpendicular to the magnetic field and spawns photons with an angular distribution moderately peaked along the magnetic field. Figure 7 shows that these effects lead to pronounced variations in the line profiles of the various harmonics as a function of viewing angle θ, the angle between the line of sight to the viewer and the direction of the magnetic field.

5. Comparison Between Theory and Observation

Wang *et al.* (1989) compared the spectrum expected from cyclotron resonant scattering and the spectrum observed in interval b of GB880205 as follows. They first fit the X-ray continuum outside the first- and second-harmonic line features

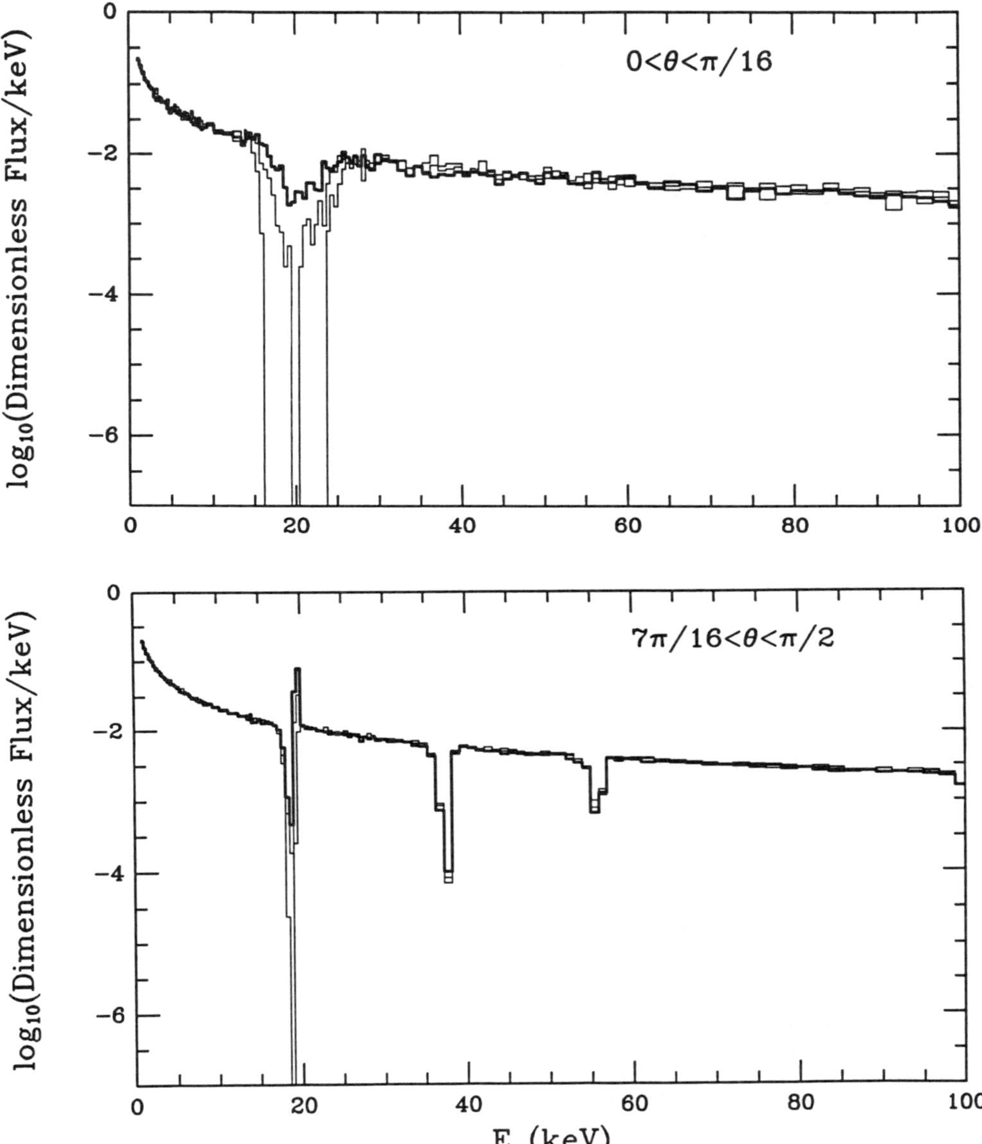

Figure 7. Theoretical photon number spectra for $B = 1.71 \times 10^{12}$ G, $N_e = 1.2 \times 10^{21}$ electrons cm^{-2}, $T = 5.3$ keV, and two different viewing angles θ relative to the magnetic field. The bottom and middle lines show the line profiles that would result from absorption or resonant scattering and Raman scattering without photon spawning, respectively. The (heavy) top line gives the actual profiles, which are due to resonant scattering and Raman scattering with photon spawning included. (From Lamb *et al.* 1989.)

using an analytical broken power-law form, $N(E) = A(E/E_b)^{-\alpha}$, for the photon number spectrum. They took the detector response function to be that for an angle of 37° between the direction of the burst source and the normal to the detectors, the same as assumed in Fenimore *et al.* (1988). They found best-fit parameter values $A = 0.08$ cm^{-2} s^{-1} keV^{-1}, $E_b = 101.3$ keV, and $\alpha = 0.846$ (1 keV $\leq E \leq$ 101.3 keV) and 1.174 (101.3 keV $\leq E \leq$ 1000 keV), with $\chi^2 = 14.2$ for 23 degrees of freedom. They then injected photons with this spectrum isotropically into a slab with constant B, T and $N_e(= \mu N_e^{\rm los})$, with the slab normal parallel to B. They followed the resulting cyclotron resonant line transfer with their Monte Carlo code.

Using more than 100 Monte Carlo spectra, each with $(1-3) \times 10^6$ photons, and folding these spectra through the *Ginga* detector response functions, Wang *et al.* (1989) fit the observed photon count rate spectrum. They found best-fit values $B = (1.71 \pm 0.07) \times 10^{12}$ G, $N_e = (1.2 \pm 0.6) \times 10^{21}$ electrons cm^{-2}, and $\mu = 0.31 \pm 0.05$, with $\chi^2 = 43.9$ for 36 degrees of freedom [P($> \chi^2$) = 0.17]. The resonant cyclotron Compton temperataure is not a free parameter, but is fixed by their model to be $T_C = 5.3^{+0.3}_{-0.2}$ keV. Figure 8 shows the predicted and observed photon count-rate spectra, the residuals, and the incident photon-number spectrum for the best-fit parameters. Figure 9 shows the 68.3%-, 95.4%-, and 99.7%-confidence regions in (B, N_e, μ)-space.

6. Discussion

The above work has profound implications for the nature of γ-ray burst sources. It offers a physically self-consistent model for the formation of the narrow cyclotron scattering features in γ-ray burst spectra, and demonstrates that cyclotron resonant and Raman scattering can explain the observed positions, strengths, and widths of the dips seen in the spectrum of GB880205. The success of the model convincingly demonstrates the existence of a superstrong magnetic field ($B \approx 2 \times 10^{12}$ G) in this source. This result, taken together with the *Ginga* (Yoshida *et al.* 1990) and Konus data (Mazets *et al.* 1981), which show low-energy dips in approximately 20% of all classical γ-ray bursts, provides powerful evidence that many γ-ray bursts come from strongly magnetic neutron stars. These neutron stars must belong to the Galaxy, resolving a long-standing controversy about the distance to γ-ray burst sources.

The magnetic fields of many neutron stars have been measured indirectly through the spin-down rate of rotation-powered pulsars and through the spin behavior of accretion-powered pulsars (see *e.g.*, Michel 1991). Magnetic field values have been measured directly for nine accretion-powered pulsars which show cyclotron scattering features in their X-ray spectra (Makishima 1991). The above work suggests that X-ray observations of γ-ray bursts may provide direct, and therefore more accurate, measurements of the magnetic fields of many more neutron stars than any other means.

If γ-ray bursts originate from magnetic neutron stars which lie at distances of $\lesssim 300$ pc, as several physical arguments suggest (see, *e.g.*, Zdziarski 1984; Epstein 1985; Lamb, Wang, and Wasserman 1990), some of these neutron stars must be much older than 2×10^7 years. This follows from comparing the number of distinct γ-ray sources now known (Higdon and Lingenfelter 1990, Harding 1991, Hurley and Lamb 1991) and the birth rate of neutron stars (Paczyński 1990; Hartmann, Epstein, and Woosley 1990). Such a conclusion contradicts the conventional wisdom that the surface magnetic fields of neutron stars decay on a time scale $\tau_{\rm decay} \lesssim 10^7$ years (see also Michel 1991, Lamb 1991).

We gratefully acknowledge the contributions of our collaborators, particularly

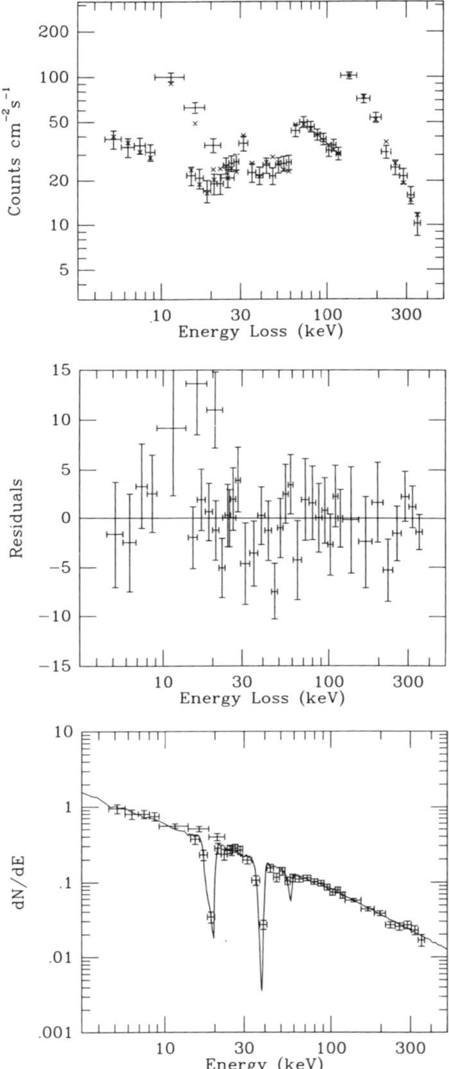

Figure 8. (top) Observed count rate spectrum (crosses) and best-fit theoretical count rate spectrum (x's) for the 5.0 s interval of GB880205 labelled b in Figure 6 for the proportional counter (PC) and scintillation counter (SC) on *Ginga*. (middle) Residuals for the PC and SC counters. (bottom) Best-fit theoretical photon number spectrum (solid curve) and *Ginga* PC and SC data (crosses). The vertical and horizontal positions and error bars of the data points have the same meaning as in Figure 4. (From Wang *et al.* 1989.)

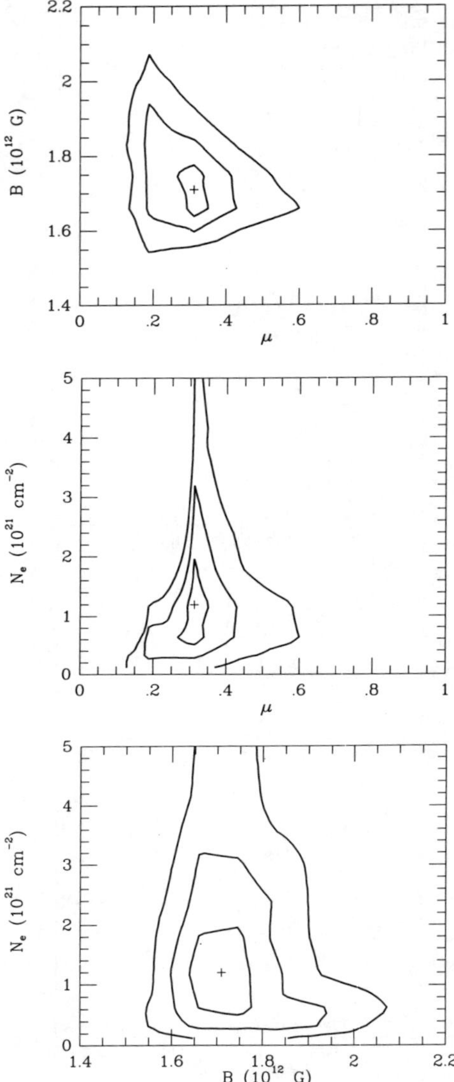

Figure 9. The 68.3%, 95.4%, and 99.7% confidence regions in (B, N_e, μ)-space for the 5.0-s interval of GB880205 labeled b in Figure 6, as determined by χ^2 fits of theoretical angle-dependent photon number spectra to the *Ginga* data: (top) projected in the (μ, B)-plane; (middle) projected in the (μ, N_e)-plane; (bottom) projected in the (B, N_e)-plane. (From Wang *et al.* 1989.)

Ed Fenimore, Tom Loredo, Toshio Murakami, John Wang, Ira Wasserman, and Atsumasa Yoshida, to the work described in this lecture. We thank Rich Epstein for wide-ranging discussions about models of γ-ray bursters, and Fred Lamb for valuable discussions about cyclotron resonant and Raman scattering. This work was supported in part by NASA grants NAGW-830 and NAGW-1284.

7. References

Alexander, S. G., and Mészáros, P. 1989, *Ap. J. (Letters)*, **344**, L1.

Bussard, R. W., and Lamb, F. K. 1982, in *Gamma-Ray Transients and Related Astrophysical Phenomena*, AIP Conference Proceedings No. 77, ed. R. E. Lingenfelter, H. S. Hudson, and D. M. Worrall (New York: American Institute of Physics), p. 189.

Cline, T. L. 1981, in *Proceedings Tenth Texas Symposium on Relativistic Astrophysics*, ed. R. Ramaty and F. C. Jones, *Ann. N.Y. Acad. Sci.*, **375**, 314.

Daugherty, J. K., and Ventura, J. 1977, *Astr. Ap.*, **61**, 723.

Epstein, R. 1985, *Ap. J.*, **297**, 555.

Epstein, R. 1986, in *Radiation Hydrodynamics in Stars and Compact Objects*, ed. K.-H. Winkler and D. Mihalas (Heidelberg: Springer-Verlag), p. 305.

Evans, W. D., Glore, J. P., Klebesadel, R. W., Laros, J. G., Tech, E. R., and Spalding, R. E. 1980, *Science*, **205**, 119.

Fenimore, E. E., Klebesadel, R. W., and Laros , J. G. 1983, Adv. Space Res. **3**, 207.

Fenimore, E. E., *et al.* 1988, *Ap. J. (Letters)*, **328**, L51.

Harding, A. K. 1991, *Phys. Repts.*, in press.

Harding, A. K., and Preece, R. 1989, *Ap. J. (Letters)*, **338**, L21.

Hartmann, D. H., Epstein, R. I., and Woosley, S. E. 1990, *Ap. J.*, **348**, 625.

Higdon, J. C., and Lingenfelter, R. E. 1990, *Ann. Rev. Astr. Ap.*, **28**, 401.

Hueter, G. J. 1988. Ph.D. Thesis. University of California, San Diego.

Hurley, K. 1982, in *Accreting Neutron Stars*, MPE Report No. 177, ed. W. Brinkmann and J. Trümper (Garching: Max-Planck Institut für Extraterrestrische Physik), p. 161.

Hurley, K. 1983, in *Positron-Electron Pairs in Astrophysics*, AIP Conference Proceedings No. 101., ed. M. L. Burns, A. K. Harding, and R. Ramaty (New York: American Institute of Physics), p. 21.

Hurley, K., and Lamb, D. Q. 1991, *Phys. Repts.*, in press.

Klebesadel, R. W., Strong, I. B., and Olson, R. A. 1973, *Ap. J.*, **182**, L85.

Lamb, D. Q. 1982, in *Gamma-Ray Transients and Related Astrophysical Phenomena*, AIP Conference Proceedings No. 77, ed. R. E. Lingenfelter, H. S. Hudson, and D. M. Worrall (New York: American Institute of Physics), p. 249.

Lamb, D. Q. 1983, in *Proceedings Eleventh Texas Symposium on Relativistic Astrophysics*, ed. D. Evans, *Ann. N.Y. Acad. Sci.*, **422**, 237.

Lamb, D. Q. 1988, in *Nuclear Spectroscopy of Astrophysical Sources*, AIP Conference Proceedings No. 170, ed. N. Gehrels and G. H. Share (New York: American Institute of Physics), p. 265.

Lamb, D. Q. 1991, in *Frontiers in X-Ray Astronomy*, Proceedings of the 28th Yamada Conference, ed. Y. Tanaka and K. Koyama (Tokyo: Universal Academy Press), in press.

Lamb, D. Q., Wang, J. C. L., Loredo, T. J., Wasserman, I., Fenimore, E. E. 1989, in *Proceedings of the Fourteenth Texas Symposium on Relativistic Astrophysics*, ed. E. Fenyves, Ann. NY Acad. Sci., **571**, 460.

Lamb, D. Q., Wang, J. C. L., and Wasserman, I. 1989, *Ap. J.*, **363**, 670.

Liang, E. P., and Petrosian, V., ed. 1986, *Gamma-Ray Bursts*, AIP Conference Proceedings No. 141 (New York: American Institute of Physics).

Lingenfelter, R. E., Hudson, H. S., and Worrall, D. M., ed. 1982, *Gamma-Ray Transients and Related Astrophysical Phenomena*, AIP Conference Proceedings No. 77 (New York: American Institute of Physics).

Loredo, T. J., and Epstein, R. 1989, *Ap. J.*, **336**, 896.

Makishima, K. 1991, in *Frontiers in X-Ray Astronomy*, Proceedings of the 28th Yamada Conference, ed. Y. Tanaka and K. Koyama (Tokyo: Universal Academy Press), in press.

Matz, S. M., *et al.* 1985, *Ap. J. (Letters)*, **288**, L37.

Mazets, E. P., Golenetskii, S. V., Aptekar', R. L., Gur'yan, Yu. A., and Il'inskii, V. N. 1981, *Nature*, **290**, 378.

Mazets, E. P., Golenetskii, S. V., Il'inskii, V. N., Gur'yan, Yu. A., Aptekar', R. L., Panov, V. N., Sokolov, I. A., Sokolova, Z. Ya., and Kharitonova, T. V. 1982, *Ap. Space Sci.*, **82**, 261.

Michel, F. C. 1991, *Theory of Neutron Star Magnetospheres* (Chicago: University of Chicago Press).

Murakami, T., *et al.* 1988, *Nature*, **335**, 234.

Paczyński, B. 1990, *Ap. J.*, **348**, 485.

Teegarden, B. J. 1984, in *High-Energy Transients in Astrophysics*, AIP Conference Proceedings No. 115, ed. S. E. Woosley, (New York: American Institute of Phys.), 352.

Wang, J. C. L., Wasserman, I., and Salpeter, E. E. 1988, *Ap. J. Suppl.*, **68**, 735.

Wang, J. C. L., *et al.* 1989, *Phys. Rev. Letters*, **63**, 1550.

Wang, J. C. L., Wasserman, I., and Salpeter, E. E. 1989, *Ap. J.*, **338**, 343.

Wasserman, I., and Salpeter, E. E. 1980, *Ap. J.*, **241**, 1107.

Woosley, S. E., ed. 1984, *High Energy Transients in Astrophysics*, AIP Conference Proceedings No. 115 (New York: American Institute of Physics).

Yoshida, A., Murakami, T., Nishimura, J., Kondo, I., and Fenimore, E. E. 1991, in *Proceedings of the Taos Workshop on Gamma-Ray Bursts*, ed. R. I. Epstein, E. E. Fenimore, and C. Ho (New York: Cambridge University Press), in press.

Zdziarski, A. A. 1984, *Astr. Ap.*, **134**, 301.

Zdziarski, A. A. 1987, in *Proceedings Thirteenth Texas Symposium on Relativistic Astrophysics*, ed. M. Ulmer (Singapore: World Scientific.), p. 563.

ANNIHILATION RADIATION IN STRONG MAGNETIC FIELDS AND GAMMA-RAY BURST SPECTRA

G.G. Pavlov [1], A.D. Kaminker [1], and P.G. Mamradze [2],
[1] *A.F. Ioffe Institute of Physics and Technology 194021, Leningrad, USSR*
[2] *Abastumani Astrophysical Observatory 380008, Tbilisi, USSR*

Abstract

We calculate the spectra of intensity and polarization of the annihilation radiation produced by one-dimensional thermal or power-law distributions of $e^{\mp}$ on the ground Landau level in strong magnetic fields $B \sim 10^{12}$ G. The spectra depend strongly on the angle θ between the magnetic field and line-of-sight. The results are used for interpreting the annihilation features observed in some gamma-ray burst spectra. Narrowest features can be interpreted in terms of the one- dimensional thermal $e^{\mp}$ distributions. Broad features with power-law tails can be produced by power-law-distributed $e^{\mp}$ if the detected annihilation radiation is strongly beamed along the magnetic field or if it comes from an extended region with sufficiently inhomogeneous magnetic field.

1 Introduction

It is widely accepted now that gamma-ray bursts (GRB) originate from galactic neutron stars with rather strong magnetic fields, $B \gtrsim 10^{12}$ G. The main evidence for such fields is provided by the cyclotron absorption lines which appear in $\sim$ 20% of the GRB spectra at energies $\sim$(20 - 60) keV (Mazets *et al.* 1981, Murakami *et al.* 1988) and correspond to $B \sim (2-6) \times 10^{12}$ G. An extensive review of theoretical interpretations of the cyclotron lines was presented at this Workshop by Don Lamb and Peter Mészáros.

In this talk we shall deal with the spectra at higher photon energies, $E \gtrsim 100$ keV, which also are very important for understanding the GRB phenomena. Firstly, the emission spectral features centered at $\sim$ 350-500 keV were discovered in $\sim$5 % of the spectra detected by the KONUS experiment (Mazets *et al.* 1981, Golenetskii *et al.* 1986). The features were interpreted as electron-positron annihilation lines redshifted by gravitational field of the neutron stars. Secondly, hard power-law-like tails ($dN/dE \propto E^{-\delta}, \delta \approx 2-4$) at 1MeV $\lesssim E \lesssim$ 10MeV were observed in some spectra by the Solar Maximum Mission (Nolan *et al.* 1984, Matz *et al.* 1985). The hard-energy tails probably correlate with the annihilation features (Golenetskii *et al.* 1986). This means that there should be relativistic particles (probably e^+e^--pairs) in GRB radiating regions. Thus, the problems arise

J. Ventura and D. Pines (eds.), Neutron Stars: Theory and Observation, 561–569.

(i) ***in which way the annihilation features can be affected by strong magnetic fields***, and

(ii) ***which information can one obtain from the annihilation features and hard-energy tails.***

To solve the problems, one should take into account that, owing to a strong cyclotron cooling, e^- and e^+ occupy mainly the ground Landau level, i.e., their motion is one-dimentional, with spins oriented in opposite directions. Then the spectrum, angular distribution and polarization of the annihilation radiation can be obtained from the differential annihilation rate

$$A_\lambda(\omega,\theta) = I_0 \int_{-\infty}^{+\infty} \int_{-\infty}^{+\infty} dp_-\, dp_+\, \eta_-(p_-)\, \eta_+(p_+)\ |\, v_- - v_+ \,|\ (d\sigma_\lambda/d\Omega\, d\omega), \qquad (1)$$

where ω is photon energy (in units of mc^2), θ is the angle between a wave-vector and the magnetic field, $I_0 = N_- N_+ (e^2/mc^2)^2 (mc)^{-1}$; N_-, p_-, v_- and $\eta_-(p_-)$ are the number density, longitudinal momentum (in mc), longitudinal velosity (in c) and distribution function of electrons , respectively; + labels the same quantities for positrons; $d\sigma_\lambda/d\Omega\, d\omega$ is the differential cross-section, and λ is the polarization index.

Up to recently only case of the "annihilation at rest" has been thoroughly investigated (Wunner, 1979; Daugherty and Bussard, 1980; Kaminker *et al.*, 1987). It corresponds to $\eta_\pm(p_\pm) = \delta(p_\pm)$ in eq. (1). It has been shown that the one-photon annihilation is negligible unless $B > 10^{13}$ G. Moreover, if the cyclotron energy $\hbar\omega_B$ is much lower than the typical $e^\pm$ kinetic and rest energies, one can neglect the influence of the magnetic field on the cross-section of the two-photon annihilation (Kaminker *et al.* 1990). Then the magnetic field manifests itself only through one-dimensionality of the $e^\pm$ motion and polarization of their spins.

In this approximation we have considered the annihilation rates for the following $e^\pm$ distributions

$$\eta_\pm = [2\, K_1(T^{-1})]^{-1}\ \exp(-\epsilon_\pm/T) \qquad (2)$$

and

$$\eta_\pm = \pi^{-1/2}\, \Gamma\left(\frac{k}{2}\right) \left[\Gamma\left(\frac{k-1}{2}\right)\right]^{-1} \epsilon_\pm^{-k}, \qquad (3)$$

where $\epsilon_\pm = \sqrt{1+p_\pm^2}$ is the dimensionless energy, T is the (longitudinal) temperature (in mc^2), $k > 1$ is the power-law index, and K_1 and Γ are McDonald and gamma functions, respectively. Substituting these distributions into eq.(1), one can find the spectra and angular distributions of the radiation intensity $I = A_1 + A_2$, Stokes parameter $Q = A_2 - A_1$ and linear polarization $P = Q/I$ ($P > 0$ if the polarization vector is in the 'magnetic field - wave vector' plane). Also we can determine the mean energy ϖ and mean-square width $\Delta\omega$ of the spectra,

$$\varpi(\theta) = \int_0^\infty I\,\omega\, d\omega \Big/ \int_0^\infty I\, d\omega, \qquad \Delta\omega(\theta) = \sqrt{\overline{\omega^2} - \varpi^2}, \qquad (4)$$

and other energy- and/or direction-integrated characteristics (see Kaminker *et al.* 1990). We shall present some computed annihilatuon spectra in Section 2, and compare them with GRB observations in Section 3.

2 Annihilation Spectra

Figures 1 and 2 show the intensity spectra for different angles θ and different temperatures (or power-law indices). The widths and blue-shifts of the spectra grow with increasing T (decreasing k) as well as with decreasing θ. If $\theta \neq 0$, the spectra show logarithmic singularities at $\omega = (\sin\theta)^{-1}$ ($I \propto \ln(T/\mid \omega\ \sin\theta - 1\mid)$) associated with square-root singularities of the $e^{\pm}$ phase-space densities in the case of one-dimentional motion (Kaminker *et al.* 1990). One-dimensional thermal e^+ and e^- generate exponential wings of the annihilation spectra, e.g.,

$$I = \frac{I_0\,\omega^{3/2}}{\sqrt{\pi T}\cos^2\theta}\exp\left[-\frac{(\omega-1)^2}{\omega T\cos^2\theta}\right] \quad \text{for} \quad T \ll 1, \quad \cos\theta \gg \sqrt{T}; \tag{5}$$

$$I \propto \exp[-\omega\,(1+\sin\theta)/T] \quad \text{for} \quad \omega \gg T \gg 1. \tag{6}$$

On the other hand, the power-law distributions lead to the power-law wings,

$$I \propto \omega^{-(k-1)} \quad \text{and} \quad I \propto \omega^{-(2k+3)} \quad \text{for} \quad \omega \gg 1 \tag{7}$$

at $\theta = 0$ and $\theta = \pi/2$, respectively.

Figures 3 and 4 show the direction-integrated spectra corresponding to extended radiating regons with randomly oriented magnetic field,

$$S(\omega) = \frac{1}{2}\oint I\,\mathrm{d}\Omega \tag{8},$$

where 1/2 is due to the photon identity. We see that (finite) maxima of these spectra lie at $\omega = 1$ for any T and k. The blue wings are softer than at $\theta = 0$ but harder than at $\theta = \pi/2$: $S(\omega \gg 1) \propto \omega^{-2}\ I(\omega \gg 1,\ \theta = 0)$. Figure 3 demonstrates also how the annihilation intensity grows in comparison with that for the isotropic $e^{\pm}$ distribution at the same temperature.

Spectra of the polarization degree P for the power-law distributions are shown in Figure 5. The strongest polarization occurs in blue wings of the spectra at large θ: $P_{max} \sim 60-70\%$ at $\omega \gg 1$ and $\theta = \pi/2$. The energy-integrated polarization also is rather high (up to 20-30 %) in a wide cone around $\theta = \pi/2$, especially for relativistic $e^{\pm}$ distributions.

3 Application to Gamma-Ray Bursts

While interpreting the annihilation features (AF) observed, one should bear in mind that strong gravitational field of a neutron star redshifts all the frequences by a factor of $g = \sqrt{1 - R_g/R}$, where $R_g = 2.95(M/M_{\odot})$ km is the Schwarzschild radius, and R is the radius of the radiating region. Kaminker *et al.* (1990) has suggested a simple method for evaluating the parameters of the $e^{\mp}$ distributions, the angle θ between the magnetic field and direction of radiation, and the redshift factor g. The first step consists in determining the mean frequency $\varpi_a = g\,\varpi$ and mean-square width $\Delta\omega_a = g\ \Delta\omega$ of the observed AF, where the subscript 'a' denotes the 'apparent' quantities as measured by a distant observer. The ratio $\Delta\omega_a/\varpi_a = \Delta\omega/\varpi$ is independent of the gravity effect. Hence one can use the theoretical dependences of this ratio on T or k (shown in Figure 6 for the power-law $e^{\mp}$ distributions)

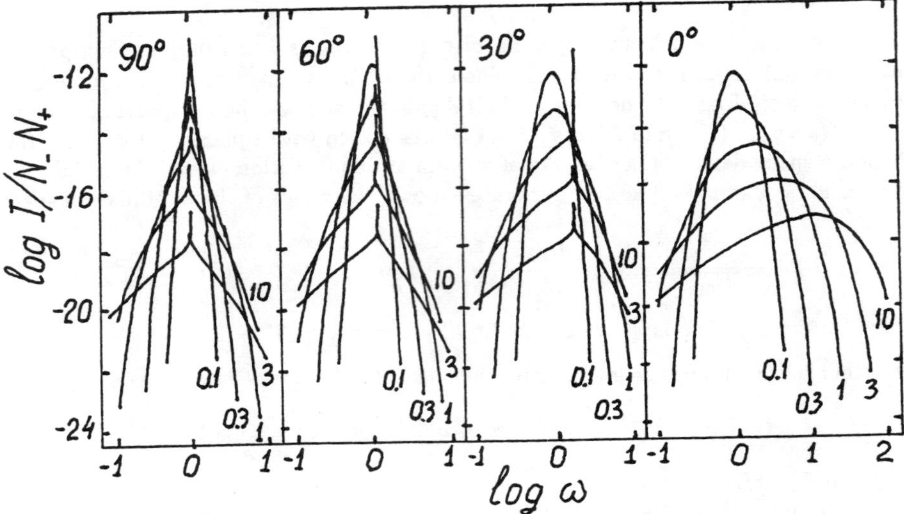

Figure 1: *Intensity spectra (in units of* $\mathrm{cm^3\ s^{-1}\ MeV^{-1}\ sr^{-1}}$*) for different angles* θ *and longitudinal temperatures* T *(figures near the curves). The values of* I *for* $T=$ *0.1, 0.3, 1, 3, and 10 are multiplied by* $10^{-2}, 10^{-1}, 10^{0}, 10^{1}$*, and* 10^{2}*, respectively.*

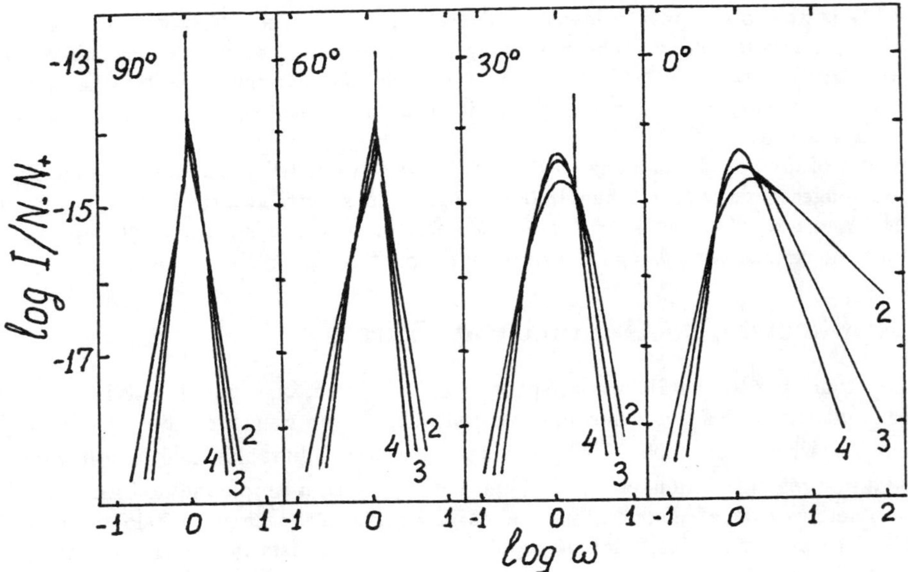

Figure 2: *Same as in Figure 1 for different* $e^{\mp}$ *power-law indices (figures near the curves).*

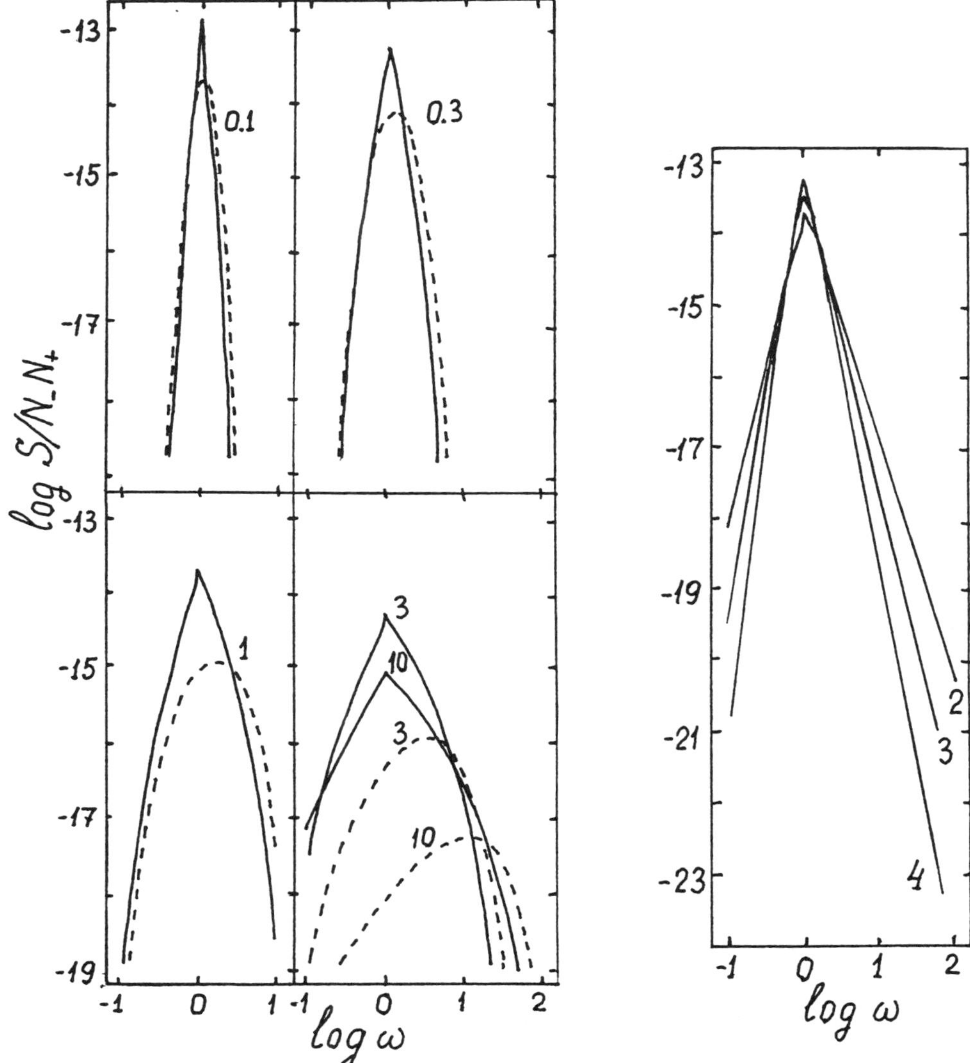

Figure 3: *Spectra of the direction-integrated intensity (in* cm^3 s^{-1} MeV^{-1}*) for different* T *(figures near the curves). Dashed curves show the spectra for isotropic* $e^{\mp}$ *distributions.*

Figure 4: *Same as in Fig. 3 for power-law* $e^{\mp}$ *distributions.*

Table 1: *Temperatures, true mean energies, true mean-square widths and redshift factors determined for two sets of* $\overline{\omega}_a$ *and* $\Delta\omega_a$ *(about the same as for GB780918 and GB781104b) for different assumptions on* θ. *The lower line corresponds to the case of randomly oriented magnetic fields*

θ	$\overline{\omega}_a = 0.8,\ \Delta\omega_a = 0.2$				$\overline{\omega}_a = 0.8,\ \Delta\omega_a = 0.4$			
(°)	T	$\overline{\omega}$	$\Delta\omega$	g	T	$\overline{\omega}$	$\Delta\omega$	g
0	0.13	1.21	0.30	0.66	0.71	2.31	1.16	0.35
5	0.13	1.21	0.30	0.66	0.85	2.47	1.27	0.33
15	0.15	1.21	0.30	0.66	5.42	3.64	1.82	0.22
30	0.24	1.24	0.31	0.64	5.45	2.18	1.09	0.37
60	0.88	1.17	0.29	0.68	5.16	1.36	0.68	0.59
90	1.02	1.07	0.27	0.75	5.06	1.20	0.60	0.67
⋆	0.30	1.15	0.28	0.70	1.08	1.48	0.74	0.54

to find these parameters for some assumed value of θ. Then T or k can be used to determine 'true' $\Delta\omega$ and $\overline{\omega}$ from the $\Delta\omega(T,k)$ or $\overline{\omega}(T,k)$ curves for corresponding θ (see Figure 7). This allows one to evaluate $g = \overline{\omega}_a/\overline{\omega} = \Delta\omega_a/\Delta\omega$.

Observational data on AFs were mostly provided by the KONUS experiments aboard Venera 11-14 (Mazets *et al.* 1981, Golenetskii *et al.* 1986). Mazets *et al.* (1981) fitted AFs in 7 spectra by Gaussians. The fits yield $\overline{\omega}_a = \omega_{ma} = 0.8 - 0.9$, $< \overline{\omega}_a >= 0.84$, $\Delta\omega_a = 0.2 - 1.0$, $< \Delta\omega_a >= 0.4$, where $\omega_{ma} = g\omega_m$ is the apparent energy of the AF maximum, $< x >$ is the average value of x. It is natural to interprete these data in terms of the 1D thermal $e^{\mp}$ distribution. Two examples are presented in Table 1. We see that the narrower AF (GB 790919) is consistent with wide ranges of T and θ, $0.1 \lesssim T \lesssim 1$ and $0° < \theta < 90°$; the redshift parameter remains almost constant with increasing θ (and T) until θ approaches 90°. Note that this narrow AF can be produced by distributions with relativistic temperatures if θ is large.

On the other hand, the broader AF (GB 781104) corresponds to unrealistically small g if small angles are assumed (a neutron star is unstable if $g < 0.53$ – see Lindblom 1984). This indicates that either the temperatures and angles are large, $T \approx 5$ and $\theta > 60°$ (which is hardly likely) or the adopted hypothesis on thermal distribution of annihilating particles is invalid. Larger values of $\Delta\omega_a$ would correspond to higher temperatures and larger θ, whereas the broadest AF (GB781119 with $\Delta\omega_a \approx 1$) is inconsistent with thermal $e^{\mp}$ at any T and θ (*cf.* Klužniak 1989).

Another assumption on the shapes of AFs detected in the KONUS experiments was adopted by Golenetskii *et al.* (1986). Basing on the fact that at least some (if not all) spectra with AFs had hard spectral tails, they suggested that the tails are also caused by the annihilation, and they fitted AFs with the following law

$$\frac{dN}{d\omega} \propto \omega^{-\delta} \left[1 + \frac{\delta}{\alpha}\left(\frac{\omega_{ma}}{\omega}\right)^{\alpha+\delta}\right]^{-1}, \tag{9}$$

where α, δ and ω_{ma} are the fitting parameters. The fits of AFs in 38 spectra from 20 GRBs yield $\delta = 1 - 5$, $< \delta > \approx 3$, $\omega_{ma} = 0.7 - 1.0$, $< \omega_{ma} >= 0.84$. These results can naturally be interpreted in terms of 1D power-law $e^{\mp}$ distribution. This immediately

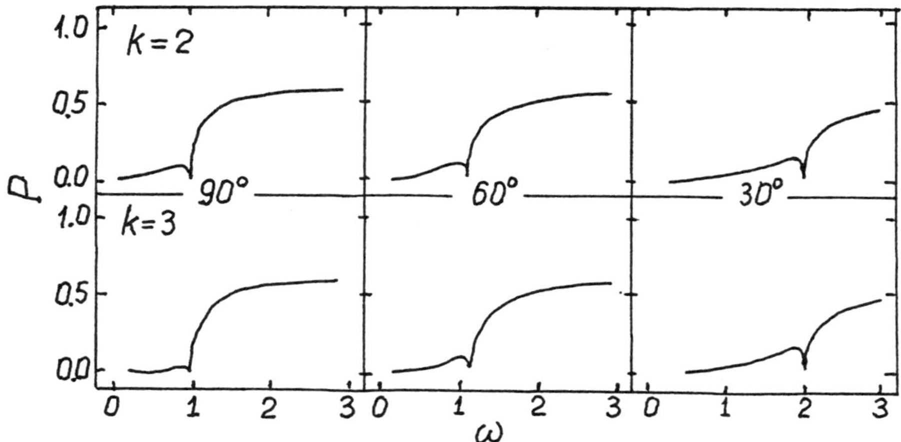

Figure 5: *Spectra of the linear polarization degree for different angles θ and* $e^{\mp}$ *power-law indices k.*

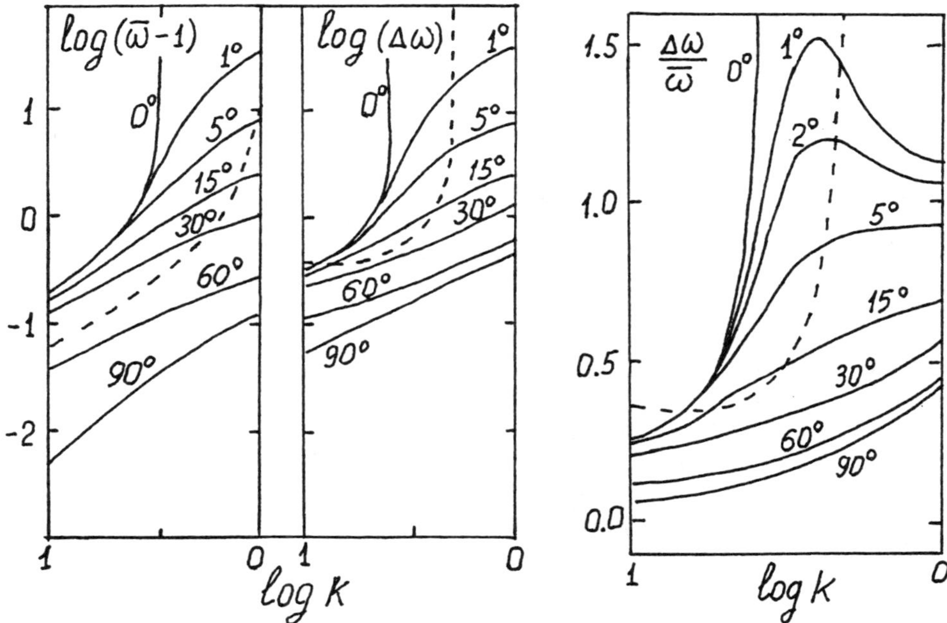

Figure 6: *Mean energy and mean-square width for different θ and k.*
Figure 7: $\Delta\omega/\bar{\omega}$*-ratio for different θ and k.*

Table 2: *Power-law indices of the $e^{\mp}$ distributions, redshift factors and true maximum energies determined for $\theta = 0$ and for randomly oriented magnetic fields*

GB	δ	α	ϖ_a	$\Delta\omega_a$	$\theta = 0$			⋆(extended region)		
					k	g	ω_m	k	g	ω_m
790113	3.5	3.5	1.27	1.23	4.2	0.68	1.30	2.3	0.89	1.00
790418	3.2	3.2	1.09	1.74	4.1	0.54	1.26	2.1	0.71	0.96
820827	3.7	2.5	1.38	1.14	4.5	0.77	1.26	2.5	0.97	1.00
820908	3.2	3.2	1.21	1.94	4.1	0.61	1.26	2.1	0.79	0.96

excludes the angles $\theta \gtrsim {\omega_{max}}^{-1}$ (ω_{max} is a maximal energy of the power-law tail) since at larger angles the typical power-law tails, $\delta \sim 2 - 4$, would be produced by improbably hard $e^{\mp}$ distributions with $k = (\delta - 3)/2$ in the range $-0.5 \leq k \leq 0.5$. This means that either the annihilation emission is beamed along the magnetic field ($\theta \lesssim 3° - 10°$), when $k \approx \delta + 1 = 3 - 5$, or it is generated in an extended region around the neutron star with essentially inhomogeneous magnetic field, when $k \approx \delta - 1 = 1 - 3$. The above method of estimating k and g can be used for AF with $\delta > 3$ only (otherwise the integral in eq.(4) diverges), which holds for 14 spectra. Four representative examples are given in Table 2. We see that in the case of small θ the values of k obtained from ϖ_a and $\Delta\omega_a$ are fairly close to δ-1 (with average value $< k >$ =4.4). In the case of randomly distributed θ one has $k \approx \delta + 1$ ($< k >$=2.5). In both cases the 'true' maximum energies ω_m almost coincide with the corresponding theoretical values (see Figures 2 and 4). This means that both hypotheses look at least selfconsistent. For small θ, the derived values of the redshift parameter, $0.53 \leq g \leq 0.77$ ($< g >= 0.64$), do not contradict to the current models of neutron stars for all the spectra examined. For randomly oriented field (extended radiating region), we obtain essentially higher g, $0.69 \leq g \leq 0.98$ ($< g >= 0.83$), which might mean that the radiating region is situated above the neutron star surface (e.g., for $g = 0.64$ at $R = R_S$, one has $g = 0.83$ at $R = 1.9\ R_S$). Thus, both hypotheses do not contradict to the observational data on the broad AFs, although they lead to essentially differnent power-law indices and redshift factors.

In conclusion it is worthwhile to stress that the observational data on annihilation radiation in GRBs as well as their interpretation are far from being unambiguous. In particular, it seems necessary to answer the following questions :

Is the correlation between the AFs and power-law tails real?
Are the tails really caused by the annihilation radiation?
What are the real shapes of the annihilation features?
Where are the emitting regions located?
How are e^- e^+-pairs produced?
How can the one-dimensional power-law and/or thermal $e^{\mp}$ distributions be produced?

Note that polarization measurements of the GRB radiation might be very useful to answer the second and third questions since the polarization of the annihilation radiation is orthogonal to that of, e.g., cyclotron radiation.

REFERENCES

Daugherty, J.K. and Bussard, R.W. (1980) 'Pair annihilation in superstrong magnetic fields', *Astrophys. J.* **238**, 296.

Golenetskii, S.V., Mazets, E.P., Aptekar, R.L., Gurian, Yu.A.,and Ilyinskii, V.N. (1986) 'Annihilation radiation in cosmic gamma-ray bursts', *Astrophys. Space Sci.* **124**, 243.

Kaminker, A.D., Pavlov, G.G.,and Mamradze, P.G. (1987) 'Two-photon annihilation radiation in strong magnetic fields: The case of small longitudinal velocities of electrons and positrons', *Astrophys. Space Sci.*, **38**, 1.

Kaminker, A.D., Pavlov, G.G., and Mamradze, P.G. (1990) 'Annihilation radiation from thermal electron-positron plasma on the ground Landau level; The case of low magnetic fields', *Astrophys. Space Sci., to be published.*

Klužniak, V. (1989) 'Redshifts derived from gamma-ray bursts spectra are inconclusive', *Astrophys.J.*,**336**, 367.

Lindblom, L. (1984) 'Limits on the gravitational redshifts from neutron stars', *Astrophys. J.*, **278**, 364.

Matz, S.M., Forrest, D.J., Vestrund, W.T.,Chupp, E.L.,Share, G.H., and Rieger, E. (1985) 'High-energy emission in gamma-ray bursts', *Astrophys. J.*, **288**, L37.

Mazets, E.P., Golenetskii, S.V., Aptekar, R.L., Gurian, Yu.A.,and Ilyinskii, V.N. (1981) 'Cyclotron and annihilation lines in γ -ray bursts', *Nature*, 290, 378.

Murakami,T., Fujii, M., Hayashida, K., and Itoh, M. (1988) 'Evidence for cyclotron absorption from spectral features in gamma-ray bursts seen by Ginga', *Nature* ,**335**, 234.

Nolan, P.L., Share, G.N., Matz, S.M., Chupp, E.L., Forrest, D.L., and Rieger, E. (1984) 'High-energy emission from gamma-ray bursts', in S.E.Woosley (ed.) *High-Energy Transients in Astrophysics*, AIP Conf. Proc. No. 115, New York, p.399.

Wunner,G. (1979) 'Comparison of 1γ and 2γ pair annihilation in strong magnetic field', *Phys. Rev. Letters*, **42**, 79.

THE SEARCH FOR TeV EMISSION FROM NEUTRON STARS

O.C. De Jager,
Dept. of Physics, PU for CHE,
Potchefstroom 2520,
South Africa

ABSTRACT. We present a review of the observation of very high energy γ-rays (above 100 GeV) from radio and X-ray pulsars. The best candidates for such emission are Crab, Her X-1, Vela X-1 and a few others discussed in this paper. However, the pulsed radiation is not always seen and none of the detections are of an exceedingly high statistical significance, except for the Crab nebula's unpulsed emission. The low signal to noise ratio usually encountered requires a careful approach to periodicity searches from X-ray pulsars where the spin periods change with accretion rate. We therefore give constraints and general hints for period searches from such objects.

1. Introduction

The detection of TeV gamma-rays above 100 GeV from neutron stars is proof for the existence of particle accelerators associated with the neutron star's magnetosphere. A TeV gamma-ray entering the atmosphere of Earth produces a Cerenkov shower covering a large area ($\approx 100m^2$) at ground level which can be detected with ordinary light collectors (mirrors) having a field of view of $< 2^\circ$ (see ref. [1] for a detailed review of this field). The main problem however is the presence of the isotropic cosmic ray background which also initiate such Cerenkov showers and this results in a signal to noise ratio which is usually less than unity. This is the main reason for slow growth of this field, despite the fact that it is now in operation for nearly three decades.

2. Observations of radio pulsars

A periodic signal may be reliably extracted if the pulsar period P is known with an error $\delta P < P^2/T$, where T is the duration of the observation. However, efforts are usually made to obtain near contemporary radio measurements to minimize δP and exclude any search. If the period is unkown or has an error $\gg P^2/T$, then one has to search in period. Anyone ignorant of statistics may overestimate the significance of a result and give a wrong detection. About 46 radio pulsars have been observed via the Atmospheric Cerenkov Technique since the late sixties, and in the following we give a review of some important reports.

2.1 The Crab pulsar

Most TeV observations of radio pulsars have been made of the Crab pulsar since 1972. The picture is confusing in the sense that all TeV observations did not reveal consistent pulse profiles and the signal is not always detectable. Four groups [1] have reported results on Crab observations before 1980: A narrow pulse was mostly detected, but only on two occasions did the pulse(s) coincide with the radio main and/or interpulse positions. On other occasions the peaks did not align as

J. Ventura and D. Pines (eds.), Neutron Stars: Theory and Observation, 571–577.

expected. However, the era of better electronics and more stable timekeeping systems started after 1980 [1] and since then (up to February 1987) four reports have been made by the Durham [2], Tata ([3],[4]) and California [5] groups of a pulse in phase with the radio main pulse position, but the pulse widths varied from 1% up to 20% over a timescale of a year. From the published COS-B pulse profiles [6] above 50 MeV the pulse width appears to be variable, but not as dramatic as seen at TeV energies. Reports have also been made of bursts from the Crab on timescales of a few minutes [1], but such phenomena have not been seen at lower energies. Since 1969 seven reports [7] by five independent experiments have been made of unpulsed TeV emission from the Crab nebula at significance levels between 3 and 15σ. The most remarkable feature thereof was the improvement in sensitivity when the Whipple group obtained a significance level as high as 15σ after applying the so-called 'imaging technique' which rejects at least 98% of the cosmic ray proton showers while retaining at least 30% of the gamma-ray signal from Crab [8]. This technique was implemented since August 1986, but no pulsed signal from Crab could be identified. The pulsed flux limit in this case is less than any other previously detected pulsed flux from Crab.

2.2 The Vela pulsar

The Vela pulsar is the brightest gamma-ray source in the sky above 50 MeV, showing a pulsed light curve similar to that of Crab, except that the region between the two pulses is filled at the positions where the two optical pulses is found. The high energy gamma-ray spectral index was found to be variable at the optical position [9]. The first discovery of pulsed TeV gamma-rays was made by the SAO-Sydney group in 1972. A narrow pulse (4σ level) above 0.3 TeV was seen 3 ms before the radio pulse position [10]. This position does not fall into the phase interval where the 100 MeV gamma-ray emission is found. Subsequent observations in 1973 and 1974 by the same group have revealed a single pulse (3σ) above 5 TeV coinciding with one of the optical pulses [1]. Confirmation for the latter feature around 5 TeV was given by the Tata group from observations taken between February 1981 and 1985 when good timing information was avaiable [11]. Observations by the Durham [12] and Potchefstroom groups [13] have failed to provide evidence for pulsed emission from Vela, but the threshold energy in both cases was much lower than 5 TeV so that a large noise component from lower energy cosmic ray primaries may hide the signal if it is there. Observations at larger threshold energies may clarify the situation.

2.3 Millisecond pulsars

An interesting scenario is developing around the millisecond pulsars with reports of TeV gamma-rays from PSR1937+214, PSR1953+29, PSR1855+09, PSR1821-24 and the eclipsing binary containing PSR1957+20. From the latter [14] an unpulsed excess was seen above 50 MeV (3σ) by COS-B and above 1 TeV (3σ) by the Potchefstroom group at the fourth Lagrange point in this system. It was shown that this position may be a stable point for residing matter [15], but it remains to be shown how such matter can be sustained against radiation pressure and replenished after evaporation. In the case of PSR1953 a single pulse was both seen at TeV [16] and radio wavelengths, whereas a double pulse is seen at both wavelenghts for PSR1937 [17], PSR1855 ([18],[12]) and PSR1821 [13]. The pulsars PSR1953 and PSR1855 have comparable spindown powers, but the TeV luminosity of PSR1953 is $\approx$ 15 times larger (and comparable to its spindown power [18]) than that of PSR1855. The radio luminosity of PSR1953 is also at least 100 times that of PSR1855. However, all these results need furhter confirmation, but the apparent TeV variability makes this task difficult.

2.4 Other radio pulsars

More than one positive detection have been made of PSR1509-58 by the Potchefstroom group [19] with a peak flux during 1985 and the pulse profile has a dominating third harmonic component

- a feature which is not seen at radio and X-ray wavelengths. The White Cliffs [20] and Durham groups [12] however failed to detect pulsed emission after 1986. The latest report of TeV gamma-rays from a radio pulsar was given for PSR0355+54 [21]: The pulse is narrow but not in phase with the radio pulse.

A summary of all the reported TeV observations (see also [1],[12],[13],[22],[23]) of radio pulsars is given in Figure 1. In most cases no searches in period have been made. It is encouraging to see that the reported positive detections are clustered at those pulsars having the shortest periods and/or largest available flux values (i.e. spindown power / distance squared). The latter is also a measure of the available spindown power at Earth. If they were statistical fluctuations, one would have expected a random scattering of positive detections on the Figure. One could argue against this saying that observers would spend more time (and hence greater probability of making an error) on the most energetic pulsars. However the Potchefstroom group have conducted a search on 22 radio pulsars and each pulsar was treated similarly. The reassuring point is that positive indications were found from only those few pulsars having the largest fluxes as discussed above, but why don't any of them show a continuously detectable TeV signal?

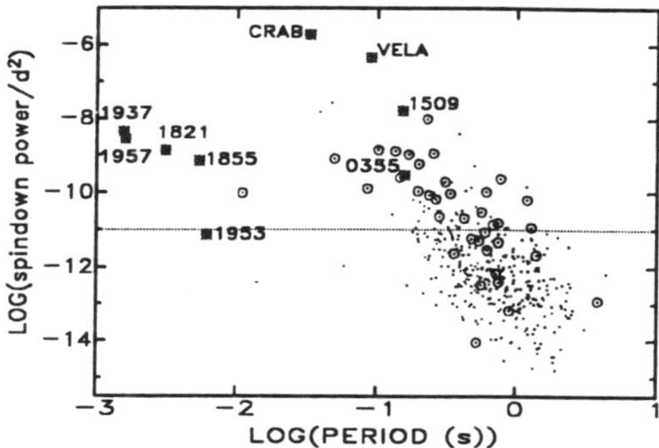

Figure 1. The distribution of spindown power/distance squared (in $erg.cm^{-2}.s^{-1}$) versus period for 450 radio pulsars (dots). The open circles represent those which have been observed above 0.1 TeV and indications of signals have been seen from the crossed circles. The horisontal dotted line represents the approximate sensitivity level for most TeV telescopes.

3. Observations of X-ray pulsars above 0.1 TeV

X-ray pulsars, in contrast to radio pulsars, are known to emit thermal radiation due to the accretion of matter from the secondary onto the neutron star but a TeV detection from such a system would hint at the coexistence of a particle accelerator. Most TeV observations of X-ray sources have concentrated on pulsars with large magnetic fields (such as Her X-1, 4U0115+63, Vela X-1, Cen X-3 and others [24]) in which case the accretion flow is channeled along the magnetic field lines towards the polar caps to give the pulsed X-rays.

3.1 Her X-1

This well known system with a relative low mass companion contains a neutron star which is spinning at a period of 1.23 s. The system has an orbital period of 1.7 days and a 35 day period (probably due to precession or obscuration) showing two phases of activity. TeV emission from Her X-1 was discovered by the Durham group and nine groups have reported detections of Her X-1 afterwards [24]. In one case there was a simultaneous detection [1] of the spin period by the Whipple and Durham groups on April, 4 1984 with a combined statistical significance of 4×10^{-5}. The outstanding features of the TeV reports on Her X-1 are the following: (1) The TeV pulse profile is similar to the X-ray profile, (2) the TeV activity is sporadic on timescales less than an hour and such activity is detectable in less than 10% of the observations, (3) the positive detections are clustered at those times when Her X-1 is in an ON-state within the 35-day cycle, (4) the TeV detection of Her X-1 during source eclipse indicates that the periodic TeV source is not associated with the neutron star and (5) some reports have been made of blueshifted periodicities with $\Delta\nu/\nu \approx 0.16\%$. Attempts have been made to explain the latter ([25],[26]) but there are no clear guidelines to facilitate a search for such frequency shifts given the low signal to noise ratios and low duty fraction of activity.

3.2 Vela X-1

This massive X-ray binary has a wind driven X-ray pulsar spinning at a period of ≈ 283 s which experiences a random walk in time. Pulsed TeV emission was discovered by the Potchefstroom group [27] and confirmed independently by the Durham group [28], but the JANZOS group has reported an upper limit [29]. The double pulse profile found at hard X-rays implies that the neutron star is close to being an orthogonal rotator [30]. However, a near-sinusoidal pulse profile was found by the Potchefstroom and Durham groups which is in conflict with this interpretation. It was also shown [31] that some of the period measurements by the Potchefstroom group are incompatible with the X-ray period history. It should be noted that a whole year's data base has been treated as a single set of arrival times (i.e. a 'coherent analysis') covering more than one month and a single representative period has been searched. In Section 3.4 we shall show that this approach may give wrong results given the random walk in spin period and alternative suggestions for data analyses will be given.

3.3 Other X-ray pulsars

Reports of TeV emission from other X-ray sources [24] such as 4U0115+63, Cen X-3, SMC X-1, 1E2259+59, 4U1626-67, LMC X-4, X0021.8-7221 and 4U1145-61 were made and upper limits were given for A0538-66, V0332+53, 1E1048.1, 2S1417-62, GX1+4, 1E1145.1, 4U0352+30 and A0535+26.

Pulsed emission from 4U0115+63 was reported by four groups [24] and based on the positional coincidence between this source and Cas Gamma-1 (detected as a DC source by the Crimean group in 1971 and 1972) it has been suggested that the Crimean group could have identified the X-ray source instead [32].

The Durham group have identified pulsed emission from Cen X-3 which is centered in a 5% band around orbital phase 0.73 (ascending node) [24]. The Potchefstroom group found that most of the emission is confined to the orbital phase interval 0.6 to 0.8 which may be associated with the accretion wake found between phases 0.55 and 0.75 [33]. However, the signal found by the Potchefstroom group was weak and not always at the X-ray period, but one strong frequency blueshift of 0.16% (i.e. similar to those of Her X-1) was found [7] during an observation covering the orbital phases 0.7 to 0.8.

Simultaneous observations of 1E2259+59 have been made by the Whipple and Durham group. The latter group got an indication of TeV emission from this source, while no evidence for emission had been found by the Whipple group when the imaging technique was applied [34]. This same technique also failed to provide evidence for gamma-radiation from 4U0115+63 and Her X-1 whereas evidence was found in the raw uncut data.

3.4 Periodicity searches for X-ray pulsars

Data from more than one night are sometimes linked together to improve the sensitivity as well as the period resolution given that the expected period range is less than the '1-cycle' range of $\Delta P_{1cyc} = P^2/T_0$ (where T_0 is the average time span of the data gaps which are typically one day). It is important to satisfy the latter constraint since the number of statistical penalties increases rapidly with increasing period range, until the search range equals ΔP_{1cyc}, after which they accumulates slowly (provided that the search range is still less than P^2/T_s, where T_s is the duration of a single observation which is a few hours) [35].

Two quantities should be known before attempting a coherent analysis in the absence of contemporary X-ray measurements: (1) the timescale for the pulsar to loose spin coherency. This also gives the upper limit on the timespan for coherent analyses, and, (2) the search range in period provided that one assumes that the TeV period has to be the same as the spin period. X-ray observations of Vela X-1 (as an example) resulted in n=101 pulse period measurements up to the last Ginga observations on JD2447717 (see e.g. refs. [31],[36] and the newest Ginga measurements). By considering the n(n-1)/2 independent pair combinations one may identify the range of period changes $\Delta P = P_j - P_i$ versus the time interval $\Delta T = t_j - t_i$ (for $i = 1, ..., n-1$ and $j = i+1, ...n$). The results are shown in Figure 2. Each time interval ΔT is accompanied by a distribution of ΔP. If one adds the maximum scatter in ΔP (positive and negative) to the applicable X-ray period measurement, the range of possible periods for the time of TeV observation is obtained. One can see from Figure 2 that the maximum search range is 0.4 s for Vela X-1, but this interval can be considerably reduced if an X-ray measurement close to, or, before and after the TeV observations are available, giving a search range $\ll \Delta P_{1cyc} = 0.9s$ and hence a minimum number of penalties.

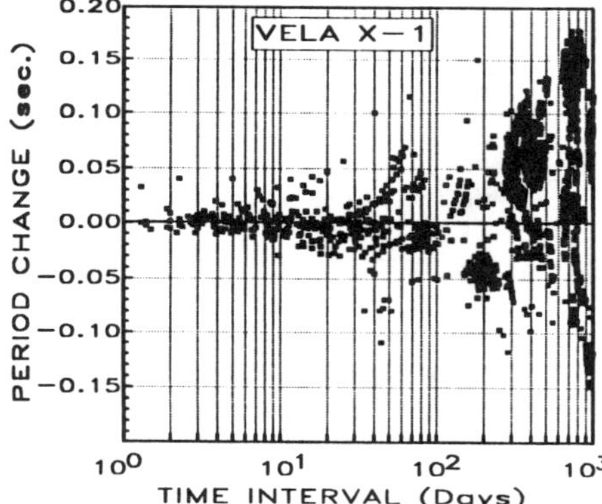

Figure 2. The period change versus time between all independent pair combinations of X-ray pulse period measurements.

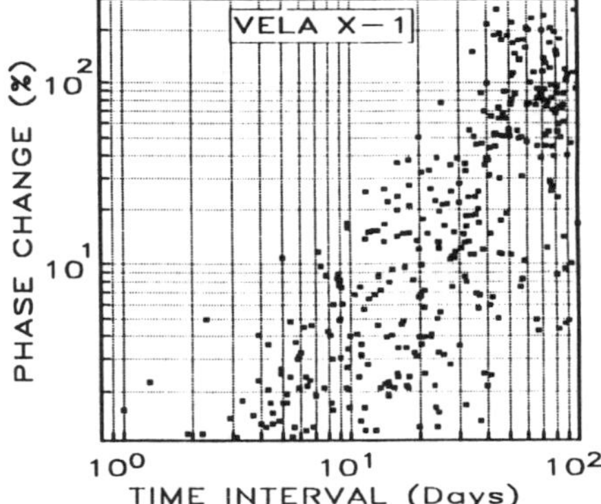

Figure 3. The % phase dispersion versus time between all independent pair combinations of X-ray pulse period measurements,

Secondly, a conservative estimate of the allowable timespan per coherent analysis is given by that ΔT such that the phase dispersion $\Delta P \Delta T/2P^2$ due to a change in period is much less than unity. Figure 3 gives the distribution of phase dispersion as a function of time and one can see that the maximum dispersion is less than 20% for $\Delta T < 10$ days but as large as 50% for 20 days. We therefore suggest that one restricts the Vela X-1 'coherent analyses' to 10 day groups, or, to dark moon periods only. If one uses the Rayleigh test (which tests for power in the fundamental harmonic) it would be least sensitive for phase dispersion in comparison to other tests which add power from higher harmonics. One may improvise on this method for disk fed pulsars by including an average $\dot{P}$ term, but still including the additional period noise component.

To quantify the significance of a result from a period search one should avoid relying completely on a theoretical distribution for a statistical test, especially if the pulsar period is in the red noise region and/or for gapped data. Detailed Monte Carlo simulations may be used to establish the significance of a result, but such computer intensive simulations do not take all the data structures, atmospheric and zenith effects into account. We suggest that one obtain an estimate of the significance by obtaining the maximum of the test statistic in each search interval placed at frequencies smaller and larger than the spin frequency (excluding the region of spectral leakage around the spin frequency), but spaced independently from each other by amounts of $1/T_s$. The significance of the maximum within the X-ray search range can then be obtained from the distribution of these maxima.

It is well known that a better sensitivity is obtained from a periodic analysis using data from ON-source tracking, than would be the case for ON/OFF observations to obtain a DC signal. This interpretation is correct for radio pulsars where a narrow pulse is usually seen in the pulse profile. However, all TeV observations of X-ray pulsars have revealed sinusoidal light curves in the fundamental or second harmonic. Is one then still better off with periodic analyses only? Consider the case where N_{on} and N_{off} events have been collected ON- and OFF-source respectively with equal exposure times. The signal strength is then $p = (N_{on} - N_{off})/N_{on}$ and the significance of the DC signal is $\sigma = (N_{on} - N_{off})/\sqrt{N_{on} + N_{off}}$. If this excess is pulsed with a sinusoidal modulation, the Rayleigh significance of the periodicity (given that the period is known) may be derived from ref. [37] giving

$$-log_{10}(Probability) = 0.434[(2-p)\sigma^2/4+1] \pm 0.434\sqrt{(2-p)\sigma^2/2+1}$$

What would one gain from a periodic analysis in terms of significance if a DC signal of say 3σ has been obtained? Since the signal to noise ratio is usually small ($p \ll 2$), a significance of $-log_{10}(Probability) = 2.4 \pm 1.4$ would result which is on average less significant than the DC measurement! Thus, contrary to the radio pulsar case, the significance of a DC signal would be comparable to that of a periodic signal (or better given a search range in period), but the scatter (i.e. "Rayleigh fluctuations") in the significance of the periodicity is expected to be much larger.

4. Conclusions and prospects for the future

Most detections in this field are of transient nature, low statistical significance and some may even be false. This is a characteristic of all new Astronomies, but once the sensitivity is at an acceptable level the false sources will be eliminated in due time. It is quite possible that the 'once only and never again' detections may mostly be false, but those which were seen on more than one occasion by more than one group may be real. However, the signal from the Crab nebula is steady and is seen consistently by the Whipple group at a significance rate of 2.5σ per hour when the imaging technique is applied. However, this technique did not confirm any of the other reports of pulsed emission from radio and/or X-ray pulsars. It is our feeling that this technique would be generally established if it can identify at least one periodic cosmic source.

In this paper we have also attempted to provide guidelines for the analysis of TeV data from X-ray pulsars w.r.t. coherency timescales, the choice of a search range in period and significance estimates for complicated searches in gapped data. Unfortunately no clear prescription exists for period searches if the TeV period is significantly different from the X-ray period and such a result would be believable if the corresponding TeV signal has a high statistical significance at the shifted period. Since the pulse profiles of X-ray pulsars are broad, one may expect the DC significance (from ON/OFF measurements) to exceed (on average) the Rayleigh significance (even for a 'no-search') from the ON-source data. This is especially true when one has to search in period since the significance of the periodicity is reduced accordingly. We therefore suggest that OFF-source channels removed by $\approx 5^{\circ}$ from the ON-source channels (to eliminate all source contamination) will *mostly* establish the existence of any TeV signal from an X-ray pulsar, but not necessarily for a radio pulsar having a small pulse duty cycle.

High quality TeV work may be done if simultaneous TeV and X-ray observations may be scheduled. In this way we eliminate the effect of searching and the pulse phases may be directly compared. However, the time schedules of X-ray sattelites are tight, but simultaneous ground based optical and TeV observations are easier to do and the appropriate candidate sources in this case are some types of Cataclysmic Variables: The sensitivity of an ordinary TeV telescope is significantly improved if one selects only those times when optical observations reveal the transient states of flaring activity when the source may be accelerating paricles to give gamma-rays.

Acknowledgements: I thank B.C Raubenheimer for his critical comments and useful suggestions concerning this work, H.I. Nel for gathering the material on radio pulsars and providing new unpublished results, K.E. Turver, G. Vacanti and T.C. Weekes for some useful discussions.

References

[1] Weekes, T.C. (1988) Physics Reports 160, 1.
[2] Dowthwaite, J.C. et al. (1984) Astrophys. J. Lett. 286, L35.
[3] Bhat, P.N. et al. (1986) Nature 319, 127.
[4] Protheroe, R.J. (1987) Proc. 20th ICRC (Moscow) 8, 21.
[5] Tumer, O.T. et al. (1985) Proc. 19th ICRC (La Jolla) 1, 139.
[6] Clear, J. et al. (1987) Astron. Astrophys. 174, 85.
[7] Fegan, D.J. (1990) Proc. 21st ICRC (Adelaide) 11, 23.
[8] Weekes, T.C. et al. (1989) Astrophys. J. 342, 379.
[9] Grenier, I.A., Hermsen, W. and Clear, J. (1988) Astron. Astropys. 204, 117.
[10] Grindlay, J. et al. (1975) Astrophys. J. 201, 82.
[11] Bhat, P.N. et al. (1987) Astron. Astrophys. 178, 242.
[12] Brazier, K.T.S. et al. (1990) Proc. 21st ICRC (Adelaide) 2, 304.
[13] Nel, H.I. et al. (1990) Poster session, this conference.
[14] Brink, C. et al. (1990) to appear in Astrophys. J.
[15] Shapiro, S.L. and Teukolsky, S.A. (1988) Nature 333, 213.
[16] Chadwick, P.M. et al. (1985) Nature 317, 236.
[17] Chadwick, P.M. et al. (1986) Proc. NATO Workshop on V.H.E. Gamma-Ray Astronomy, ed. K.E. Turver, Reidel, Dordrecht, p.151.
[18] De Jager, O.C. et al. (1990) Nucl Phys. B (Proc. Suppl.) 14A, 169.
[19] Nel, H.I. et al. (1990) Astrophys. J.
[20] Clay et al. (1987) Proc. 20th ICRC (Moscow) 1, 271.
[21] Bhat, P.N. et al. (1990) Astron. Astrophys. 236, L1.
[22] Porter, N.A. and Weekes, T.C. (1978) Smithsonian Astr. Obs. Spec. Rep. No. 381.
[23] Chadwick, P.M. et al. (1985) Proc. 19th ICRC (La Jolla) 1, 155.
[24] Turver, K.E. (1990) to appear in 'Particle Acceleration in Accreting Compact Objects' (Amsterdam).
[25] Cheng, K.S. and Ruderman, M. (1989) Astrophys. J. 337, L77.
[26] Slane, P.O. and Fry, W.F. (1989) Astrophys. J. 342, 1129.
[27] North, A.R. et al. (1987) Nature 326, 567.
[28] Carraminana, A. et al. (1989) Astrophys. J. 346, 967.
[29] Bond, I.A. et al. (1990) Proc. 21st ICRC (Adelaide) 2, 271.
[30] Riffert, H. and Meszaros, P. (1988) Astrophys. J. 325, 207.
[31] Raubenheimer, B.C. and Ögelman, H. (1990) Astron. Astrophys. 230, 73.
[32] Lamb, R.C. and Weekes, T.C. (1986) Astrophys. Lett. 25, 73.
[33] Raubenheimer, B.C. et al. (1991) to appear in "High energy Gamma-ray Astrophysics" (Ann Arbor).
[34] Cawley, M.F. et al. (1991) to appear in Astrophys. J.
[35] Lewis, D.A., Lamb, R.C. and Biller, S.D. (1991) to appear in Astrophys. J.
[36] Nagase, F. (1989) Publ. Astron. Soc. Japan 41, 1.
[37] De Jager, O.C., Raubenheimer, B.C. and Swanepoel, J.W.H (1988) Data Analysis in Astronomy III, eds. Di Gesu et al. Plenum Publ. Corp. p.21.

SUBJECT AND OBJECT INDEX